HANDBOOK OF
NUCLEIC ACID
PURIFICATION

HANDBOOK OF
NUCLEIC ACID
PURIFICATION

Edited by
Dongyou Liu

CRC Press
Taylor & Francis Group
Boca Raton   London   New York

CRC Press is an imprint of the
Taylor & Francis Group, an **informa** business

CRC Press
Taylor & Francis Group
6000 Broken Sound Parkway NW, Suite 300
Boca Raton, FL 33487-2742

First issued in paperback 2017

© 2009 by Taylor & Francis Group, LLC
CRC Press is an imprint of Taylor & Francis Group, an Informa business

No claim to original U.S. Government works

ISBN 13: 978-1-138-11387-9 (pbk)
ISBN 13: 978-1-4200-7096-5 (hbk)

Library of Congress Cataloging-in-Publication Data

Handbook of nucleic acid purification / editor, Dongyou Liu.
    p. ; cm.
  Includes bibliographical references and index.
  ISBN 978-1-4200-7096-5 (hardcover : alk. paper)
  1. Nucleic acids--Purification--Handbooks, manuals, etc. I. Liu, Dongyou.
  [DNLM: 1. Nucleic Acids--isolation & purification. QU 58 H236 2009]

QP620.H35 2009
572'.33--dc22   2008029325

Visit the Taylor & Francis Web site at
http://www.taylorandfrancis.com

and the CRC Press Web site at
http://www.crcpress.com

Contents

PART III   Purification of Nucleic Acids from Fungi

PART IV   Purification of Nucleic Acids from Parasites and Insects

PART V   Purification of Nucleic Acids from Mammals

PART VI   Purification of Nucleic Acids from Plants

PART VII   Purification of Nucleic Acids from Miscellaneous Sources

Preface

Deoxyribonucleic acid (DNA) and ribonucleic acid (RNA) are macromolecules composed of single or double strands of nucleotides, with each nucleotide consisting of a nitrogenous base (a derivative of purine [adenine and guanine] or pyrimidine [thymine, uracil, and cytosine]) and a sugar (a pentose deoxyribose or ribose, together referred to as a nucleoside) as well as a phosphate group. Although the first isolation of DNA was reported by Friedrich Miescher in 1869, it was only in 1953 that the molecular structure of the DNA duplex was elucidated by James Watson and Francis Crick. This landmark discovery uncovered the role of DNA as the chemical bearer and transmitter of hereditary features, and expedited the development of contemporary molecular biology and biotechnology, with its full impact yet to be felt in the years to come.

Because of the structural and functional complexities within prokaryotic and eukaryotic cells, purification of nucleic acids often forms a vital first step in the study of molecular biology of living organisms as well as in the evolutionary/phylogenetic analysis of ancient specimens. To this end, many innovative nucleic acid isolation methods have been designed, which are treated in a variety of professional journals and books. This book aims to be all encompassing on nucleic acid purification strategies for viruses, bacteria, fungi, parasites, insects, mammals, and plants as well as ancient samples, with an additional emphasis on sample preparation methods for direct molecular applications.

Each chapter begins with informative coverage of the biological background followed by an expert review of basic principles and current techniques for isolation of nucleic acids from specific sample types along with an insightful discussion of future development trends. Besides providing a comprehensive, reliable reference on nucleic acid purification for anyone with an interest in molecular biology, this book is a practical guide for clinical, forensic, and research scientists involved in molecular analysis of biological specimens; a convenient textbook for prospective undergraduate and graduate students intending to pursue a career in molecular biology, microbiology, and forensic science; and an indispensable roadmap for upcoming and experienced researchers wishing to acquire or sharpen their skills in nucleic acid preparation.

The scope and depth of the topics covered in this book would clearly have been impossible without a concerted team effort. I am fortunate and honored to have a panel of international scientists as contributors with expertise in their respective fields of molecular biology, whose knowledge and technical insights have enriched this book tremendously. In addition, the professionalism and dedication of editorial staff at CRC Press have further enhanced presentation. I hope the readers through the perusal of this book will find it rewarding as it develops an understanding of the theory and practice of nucleic acid purification from virtually all sample types in a few weeks that may otherwise take many years.

Editor

Dongyou Liu is currently a member of the research faculty in the Department of Basic Sciences, College of Veterinary Medicine at Mississippi State University in Starkville, Mississippi. He obtained a veterinary science degree from Hunan Agricultural University in Changsha, China in 1982. After a one-year postgraduate training at Beijing Agricultural University in Beijing, China, he completed his PhD on the immunological diagnosis of human infectious disease due to the parasitic tapeworm *Echinococcus granulosus* at the University of Melbourne School of Veterinary Science in Parkville, Victoria, Australia in 1989.

During the past two decades, Dr. Liu has worked in several research and clinical laboratories in Australia and the United States, with a particular emphasis on molecular microbiology, especially in the development of nucleic acid–based assays for species- and virulence-specific determination of microbial pathogens such as the ovine foot rot bacterium (*Dichelobacter nodosus*), and the dermatophyte fungi (*Trichophyton, Microsporum,* and *Epidermophyton*), and listeriae (*Listeria* species). He is the editor of *Handbook of* Listeria monocytogenes and a forthcoming CRC book entitled *Molecular Detection of Foodborne Pathogens.*

Contributors

Jack Ballantyne
Department of Chemistry
National Center for Forensic Science
University of Central Florida
Orlando, Florida

Nicolas Berthet
Génie et Microbiologie des
  Procédés Alimentaires
INRA
AgroParisTech
Thiverval Grignon, France

Armelle Bigot
Faculté de Médecine René Descartes
Université Paris Descartes
Paris, France

and

Unité de Pathogénie des Infections
  Systémiques
Paris, France

Marc Buée
Genomique, Ecophysiologie
  et Ecologie Fonctionnelles
INRA de Nancy
Champenoux, France

Jennifer A. Byrne
Molecular Oncology Laboratory
Oncology Research Unit
The Children's Hospital at Westmead
Westmead, New South Wales, Australia

and

Discipline of Paediatrics and Child Health
The University of Sydney
Westmead, New South Wales, Australia

Vitaliano A. Cama
Division of Parasitic Diseases
Centers for Disease Control
  and Prevention
Atlanta, Georgia

Paula F. Campos
Ancient DNA and Evolution Group
Centre for Ancient Genetics
Biological Institute
University of Copenhagen
Copenhagen, Denmark

Trinad Chakraborty
Institute for Medical Microbiology
Justus Liebig University
Hesse, Germany

Alain Charbit
Faculté de Médecine René Descartes
Université Paris Descartes
Paris, France

and

Unité de Pathogénie des Infections
  Systémiques
Paris, France

Som Subhra Chatterjee
Institute for Medical Microbiology
Justus Liebig University
Hesse, Germany

Brian F. Cheetham
Department of Molecular
  and Cellular Biology
University of New England
Armidale, New South Wales, Australia

Zhen F. Fu
Department of Pathology
College of Veterinary Medicine
University of Georgia
Athens, Georgia

M. Thomas P. Gilbert
Ancient DNA and Evolution Group
Centre for Ancient Genetics
Biological Institute
University of Copenhagen
Copenhagen, Denmark

Grant S. Hansman
Department of Virology II
National Institute of Infectious Diseases
Tokyo, Japan

Erin K. Hanson
Department of Chemistry
National Center for Forensic Science
University of Central Florida
Orlando, Florida

Larry A. Hanson
Department of Basic Sciences
College of Veterinary Medicine
Mississippi State University
Starkville, Mississippi

Renaud Ioos
Laboratoire National de la Protection des
  Végétaux
Unité de Mycologie agricole et Forestière
Génomique, Ecophysiologie et Ecologie
  Fonctionnelles
Malzéville, France

Françoise Irlinger
UMR782 Génie et Microbiologie des
  Procédés Alimentaires
INRA
AgroParisTech
Thiverval Grignon, France

Akira Ito
Department of Parasitology
Asahikawa Medical College
Hokkaido, Japan

Kjetill S. Jakobsen
Department of Biology
Centre for Ecology and
  Evolutionary Synthesis
University of Oslo
Oslo, Norway

Margaret E. Katz
Department of Molecular and
  Cellular Biology
University of New England
Armidale, New South Wales, Australia

Sergey Kovalenko
Department of Molecular Pathology
Peter MacCallum Cancer Centre
East Melbourne, Victoria, Australia

Yaning Li
Department of Plant Pathology
Biological Control Center of Plant Diseases
  and Plant Pests of Hebei Province
Agricultural University of Hebei
Hebei, China

Dongyou Liu
Department of Basic Sciences
College of Veterinary Medicine
Mississippi State University
Starkville, Mississippi

Jérôme Mounier
Laboratoire Universitaire de Biodiversité
  et Ecologie Microbienne
ESMISAB
Technopôle de Brest Iroise
Université Européenne de Bretagne/
  Université de Brest
Plouzané, France

Rosely Angela Bergamin Nichols
Scottish Parasite Diagnostic
  Laboratory
Stobhill Hospital
Glasgow, Scotland, United Kingdom

Munehiro Okamoto
Department of Parasitology
School of Veterinary Medicine
Tottori University
Tottori, Japan

Karen Page
Department of Cancer Studies and
  Molecular Medicine
University of Leicester
Leicester, United Kingdom

Theo Papakonstantinou
Centre for Green Chemistry
Monash University
Clayton, Victoria, Australia

Nipuna B. Parahitiyawa
Faculty of Dental Sciences
University of Peradeniya
Peradeniya, Sri Lanka

Dmitrii V. Pyshnyi
Laboratory of Bionanotechnology
Institute of Chemical Biology and
  Fundamental Medicine
Siberian Division of the Russian
  Academy of Sciences
Novosibirsk, Russia

Knut Rudi
MATFORSK Norwegian Food
  Research Institute
Ås, Norway

and

Department of Natural Science
  and Technology
Hedmark University College
Hamar, Norway

Lakshman P. Samaranayake
Division of Oral Biosciences
Faculty of Dentistry
The University of Hong Kong
Hong Kong, China

C. Jayampath Seneviratne
Division of Oral Biosciences
Faculty of Dentistry
The University of Hong Kong
Hong Kong, China

Jacqueline Amanda Shaw
Department of Cancer Studies and
  Molecular Medicine
University of Leicester
Leicester, United Kingdom

Huw Vaughan Smith
Scottish Parasite Diagnostic
  Laboratory
Stobhill Hospital
Glasgow, Scotland, United Kingdom

Tatiana Vallaeys
Génie et Microbiologie des
  Procédés Alimentaires
INRA
AgroParisTech
Thiverval Grignon, France

Valérie Vasseur
Laboratoire Universitaire
  de Biodiversité et Ecologie
  Microbienne
ESMISAB
Technopôle de Brest Iroise
Université Européenne de Bretagne/
  Université de Brest
Plouzané, France

Valentin V. Vlassov
Institute of Chemical Biology and
  Fundamental Medicine
Siberian Division of the Russian
  Academy of Sciences
Novosibirsk, Russia

Pavel E. Vorobjev
Laboratory of Bionanotechnology
Institute of Chemical Biology and
  Fundamental Medicine
Siberian Division of the Russian
  Academy of Sciences
Novosibirsk, Russia

Judith Weidenhofer
School of Biomedical Sciences
University of Newcastle
Callaghan, New South Wales,
  Australia

Song Weining
College of Agronomy
Northwestern A&F University
Shaanxi, China

Eske Willerslev
Ancient DNA and Evolution Group
Centre for Ancient Genetics
Biological Institute
University of Copenhagen
Copenhagen, Denmark

Robert C. Wilson
Department of Natural Science
  and Technology
Hedmark University College
Hamar, Norway

Lihua Xiao
Division of Parasitic Diseases
Centers for Disease Control and Prevention
Atlanta, Georgia

Chunfang Zhang
Centre for Green Chemistry
Monash University
Clayton, Victoria, Australia

Hong-Bin Zhang
Department of Soil and Crop Sciences
Texas A&M University
College Station, Texas

Meiping Zhang
College of Life Sciences
Jilin Agricultural University
Jilin, China

and

Department of Soil
  and Crop Sciences
Texas A&M University
College Station, Texas

Ling Zhao
Department of Pathology
University of Georgia
Athens, Georgia

Kun Yan Zhu
Department of Entomology
Kansas State University
Manhattan, Kansas

1 Nucleic Acids: Structures, Functions, and Applications

Valentin V. Vlassov, Dmitrii V. Pyshnyi, and Pavel E. Vorobjev

CONTENTS

1.1   INTRODUCTION

Nucleic acids are wonder molecules, the most important molecules of life. Beginning with the landmark discovery by Watson and Crick of the principle of complementary interaction of nucleic acids, and continuing with the technological breakthrough of the following decades, research of nucleic acid structures and functions revolutionized all aspects of biology [1]. Elucidation of the mechanism of gene expression and discovery of enzymes manipulating deoxyribonucleic acid (DNA) laid foundation to contemporary molecular biology and biotechnology. Genome sequencing data provided a basis for detailed research into metabolism, gene regulation, evolution, and pathology of biological organisms. Advances in nucleic acid syntheses, elaboration of polymerase chain reaction (PCR), and molecular selection techniques have resulted in the development of a number of nucleic acid-based technologies. The enormous specificity of the complementary interactions of ribonucleic acid (RNA) and DNA fragments and oligonucleotides has provided the possibility of designing new materials, molecular machines and devices for the detection, isolation and sequencing analysis of nucleic acids, and manipulation of DNA and RNA and proteins. In this chapter, we present a brief outline concerning properties of nucleic acids, their roles in biological systems, and growing number of applications.

1.2 STRUCTURES

1.2.1 SIZES AND CLASSES

Nucleic acids are polymers consisting of nucleotides. Natural specimens of nucleic acids vary in length from tens of nucleotides in some RNAs to tens of millions in prokaryotic genomes and hundreds of millions in eukaryotic chromosomes. The number of nucleotides in complete genomes of plants and animals approaches to the value of 10 billions [2–6]. Figure 1.1 presents the information about the size of cellular nucleic acids. Nucleic acids are present in cells in single-stranded or double-stranded (duplex) state. MicroRNA (miRNA), transfer RNA (tRNA), messenger RNA (mRNA), ribosomal RNA (rRNA), as well as a number of viral RNA- and DNA-containing genomes are single stranded. Small interfering RNA (siRNA), the rest of viral genomes, and genomic DNA of prokaryotes and eukaryotes, including genomes of mitochondria and chloroplasts, are double stranded.

1.2.2 COMPOSITIONS

A nucleotide unit consists of a monosaccharide residue (ribose or 2'-deoxyribose), a nitrogen base (purine or pyrimidine), and a phosphate residue. One of the nitrogen atoms (N1 in pyrimidines and N9 in purines) forms *N*-glycosidic bond with anomeric carbon of the sugar residue. Phosphoric acid forms esters with 3'-hydroxyl group of one nucleotide and with 5'-hydroxyl group of another. This phosphodiester is generally referred to as internucleotide phosphodiester bond. The third acidic function remains free and is ionized under normal conditions (pK ~1.2). Thus, each polynucleotide strand is a polyanion. Two polynucleotide strands can form a continuous helical complex due to complementarity of nitrogen bases. Strands are oriented in antiparallelly; the bases are located inside the helix, sugar-phosphate backbone on the outer surface (Figure 1.2) of the helix. Two distinct grooves are present on the outer surface of the helix due to asymmetry in arrangement of glycosidic bonds within the Watson–Crick pairs. At first glance, the double-stranded nucleic acid helix represents a regular structure. The main parameters including helical twist, rise per base pair, and helix pitch are invariant within the whole helix. Base pairs A-T and G-C have similar shape and dimensions, which allow them to fill up the inner space of the helix without distorting outer helical contour [7].

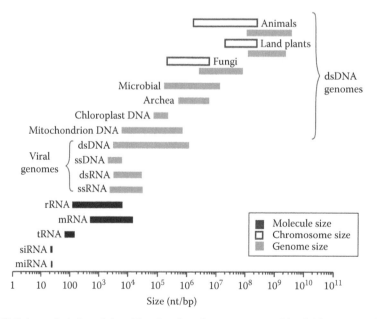

FIGURE 1.1   Cellular and viral nucleic acid molecules, chromosomes, and haploid genomes size.

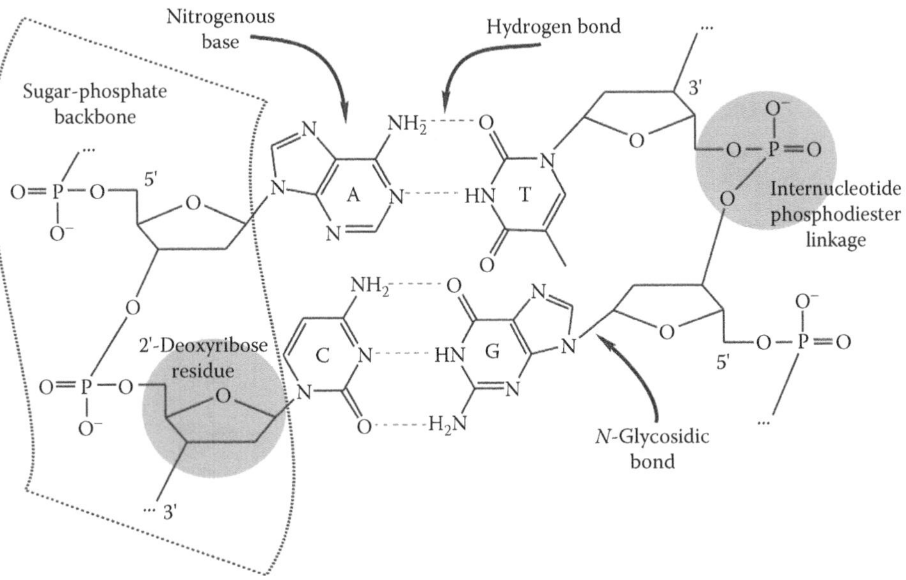

FIGURE 1.2 Chemical structure of dsDNA.

The double helix is formed due to relatively weak noncovalent forces, predominantly short-range dispersion forces sensitive to temperature [8]. Watson–Crick hydrogen bonds between bases maintain the alignment of two strands. The major sources of stabilization for the native double helix are vertical stacking interactions between neighboring base pairs. The energy of stacking depends on the type of nearest neighbor. These differences influence local stability of the helix. They also determine the structural state of each pair as function of nearest and following neighbors' state.

The main forces destabilizing the nucleic acid duplex are electrostatic repulsions between anionic phosphate groups. Counterions (e.g., Na^+, K^+, Mg^{2+}) can significantly screen repulsing groups and compensate the destabilization. *In vivo*, these counterions are often represented by proteins as can be illustrated by the example of eukaryotic chromatin, a complex of DNA with a number of highly basic proteins—histones. When destabilizing forces take over stabilizing ones, the nucleic acid is denatured, and each strand assumes the conformation of random coil [8]. However, such a deep denaturation is rarely if ever accessible. More often nucleic acids turn into complex disordered structure with remaining random noncovalent interactions. Other altered structures with incomplete or imperfect helical structure unrecognizable by specific proteins or enzymes are also considered as denatured. Noncovalent stabilizing forces can be broken by rising temperature, extreme pH values, low ionic strength, as well as in the presence of substances competing with nitrogen bases for hydrogen bonding (e.g., dimethyl sulfoxide, urea, or guanidine). When isolated, nucleic acids are often denatured.

More severe process, called degradation, can also take place. In degraded nucleic acids, covalent bonds can be cleaved leading to formation of apurinic/apyrimidinic sites or strand breaks. Extreme pH values are the most dangerous factors when handling nucleic acids. At pH < 5.5, DNA is apurinized; at pH > 8.5, phosphodiester bonds in RNA are hydrolyzed. It should be noted that random damage of nucleic acids takes place under ambient conditions (although with low frequency). Longer exposures at higher temperatures can significantly increase the frequency of random breaks thus leading to significant damage [9].

1.2.3 Conformations

Native double helix can adopt various conformations (forms). Most crystallographic forms can be divided into three families: A, B, and Z. Generally, nucleic acid structures are described in terms

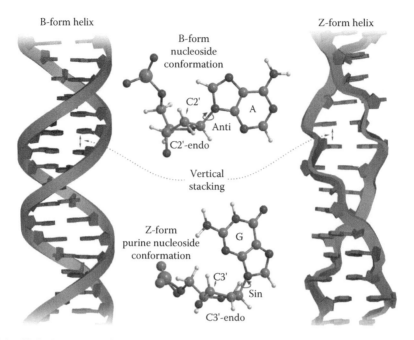

FIGURE 1.3 Helical structure of B and Z DNAs.

of torsion angles. They can also be described in terms of particular conformations of nucleosides and position of helix elements relative to the helix axis. For example, B-form of DNA can be described as follows: right-handed helix, single turn contains 10 base pairs, and planes corresponding to the base pairs are perpendicular to the helix axis (tilt value—0°). All nucleosides are in anti-conformation, sugar puckering—C2'-endo. (Figure 1.3, Table 1.1) [7,10].

Z-form is a left-handed helix. This conformation can be adopted only by DNAs with specific sequences [e.g., $(GC)_n$]. Depending on the conditions (salt concentration, methylation, supertension degree), DNA with PuPy repeats sequences can assume in B- or Z-form. B-form is more typical for double-stranded DNA (dsDNA) under normal conditions. However, DNA duplexes can also adopt A- and Z-form under specific salt and pH conditions. For RNA duplexes A-family structures are typical, RNA duplexes cannot assume B-form because of steric hindrance caused by 2'-hydroxyl group.

TABLE 1.1
Typical Parameters Associated with Three DNA Families

Parameters	A-DNA	B-DNA	Z-DNA
Helix sense	Right handed	Right handed	Left handed
Base pairs per turn	11	10	12
Helical twist	33°	36°	10°, −50°
Rise per base pair	2.9 Å	3.4 Å	3.7 Å
Helix pitch	32 Å	34 Å	35 Å
Propeller twist	15.4°±6.2°	11.7°±4.8°	4.4°±2.8°
Glycosidic conformation	anti-	anti-	anti-, sin-
Tilt	13°	0°	−7°
Roll	6.0°±4.7°	−1°±5.5°	3.4°±2.1°
Sugar puckering	C3'-endo	C2'-endo	C3'-endo, C2'-endo

For the most part of double-stranded molecules, DNA and RNA are found in genomes. They can be circular (closed) or linear (open-ended). Genomic nucleic acids are very long, within the cell or viral capsid these molecules are compacted. Prokaryotic genomes are compacted due to supercoiling and formation of complexes with proteins. A key topological property of DNA is its linking number (Lk), which is equal to the number of times a strand of DNA winds around the helix axis when the axis is constrained to lie in a plane. The linking number can be changed, for example, by specific enzyme—topoisomerase. Provided that strand ends are fixed (by proteins in linear molecules or by themselves in circular molecules), the arising tension should be compensated by formation of Z-form regions, cruciforms, open loops and by supercoiling. The number of supercoils, formed by the helix axis, also referred to as writhe (Wr) depends also on the twist (Tw), which is the measure of helical winding of the DNA strand around each other. Actually, twist and writhe are related as follows: $Lk = Tw + Wr$. Supercoiled DNA is more compact than a relaxed DNA molecule of the same length [11].

Eukaryotic genomes are generally present as a chromatin—nucleoprotein of complex architecture. Packaging of nucleic acids into chromatin allows to achieve the highest level of compaction. Summary contour length of DNA from human metaphase chromosomes amounts to 2 m, whereas the sum of linear dimensions in compacted state does not exceed 200 μm. Moreover, these chromosomes are packed up into nucleus with a diameter of 0.5 μm [11].

When discussing nucleic acids structure, attention should also be paid to motifs consisting of more than two strands. Nucleic acid helices formed by three strands are called triplexes. Stable triplex structures can be formed on polypurine tracts, because only purine bases have additional donors and acceptors for hydrogen bonding. The third strand can contain both purines and pyrimidines. Triplets are formed by addition of the third base to canonical Watson–Crick pair. Only triplets with two hydrogen bonds between the purine and the third base are stable. Cytidine forms a triplet with G-C pair only when protonated, thus giving rise to pH-dependent triplexes, which are formed at pH < 6.0. All other triplets are pH-independent. The third strand can be fully pyrimidine, fully purine, or mixed. Depending on the composition of the third strand, it can bind to the polypurine tract in parallel or antiparallel orientation. Stability of the triplex is also determined by its composition. Only Py-Pu/Py triplets are isomorphous. Alteration of nonisomorphous triplets leads to distortion of backbone and significantly decreases the stability of the triplex [12].

Structures containing four strands are referred to as quadruplexes. Four guanines can form a quartet, where each base is bound by two hydrogen bonds with two other guanines. Piles of three or four quartets are stable enough to compete with complementary duplexes. Additional stabilization arises when potassium ion is bound between two quartets. Both triplexes and quadruplexes can be inter- or intramolecular. dsDNA molecules with specific sequences undergo transition from duplex to triplex or from duplex to quadruplex depending on the environment or structure tensions [13].

Single-stranded molecules possess complex secondary and tertiary structures, which allows them to implement various functions. Most cellular RNAs are present in single-stranded state.

Secondary structures of nucleic acids are formed as a result of complementations between various regions of one strand. These complementations turn a single-stranded molecule into folded structure, consisting of a number of stems (helical structures) and other nonhelical structures, including bulges, inner loops, hairpin loops, and multibranched loops or junctions (Figure 1.4). Pseudoknots also can be considered as a secondary structure element. Inner loops are formed when unpaired bases are located on both sides of the helix. Depending on the number of bases on each side of the loop, inner loops can be symmetrical or asymmetrical. Bulges are structures, where only one side of the helix contains unpaired bases. Hairpin loop is located at the end of the helix, when the sugar-phosphate backbone folds back on itself, forming an open loop [14].

The tertiary structure of single-stranded nucleic acids is also formed due to noncovalent forces: hydrogen bonds and stacking interactions. It is also necessary to mention the role of environment (e.g., proteins, water molecules, counterions). The efficiency of stacking is determined by the type of bases and geometry of interaction. Hydrogen bonds can be divided into three groups: base–base,

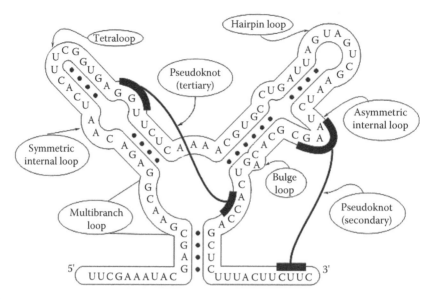

FIGURE 1.4 Typical RNA structure motifs.

base–backbone, and backbone–backbone. Within helical regions only base–base bonds are found. Within nonregular regions, the backbone conformation allows various functional groups to be closed to each other. As a consequence, all three types of hydrogen bonds are realized. Another variant of backbone–backbone interaction is closing of two phosphate groups so that a site of strong binding for metal ions is formed. An important role in formation of the RNA spatial structure is played by a number of rare or minor nucleosides incorporated into their sequence (e.g., dihydrouridine, pseudo-uridine, and inosine). These minor nucleosides provide the tertiary RNA structure with additional diversity [15]. Probably, the most impressive example of minor nucleosides incorporation represents tRNAs serving as an adaptor at the translation (Figure 1.5).

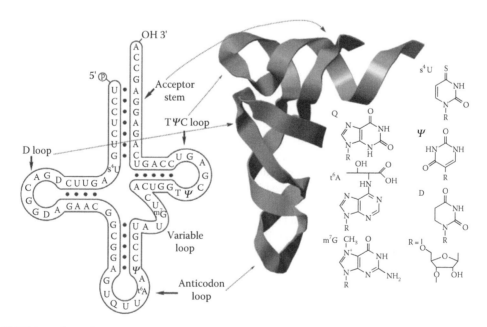

FIGURE 1.5 Secondary and tertiary structures of *Escherichia coli* ^AsntRNA, minor nucleosides.

Tertiary structure of single-stranded nucleic acids is extremely important for all their interactions. It is a precise recognition of tertiary structure of tRNA that determines its aminoacylation. The accuracy of this process presets the level of correctness at translation. The precision of the tertiary structure is also very important in the case of rRNAs. rRNAs serve as a skeleton arranging ribosomal proteins and also catalyze transpeptidation. The structures of rRNAs are very conservative, indicating the importance of the RNA tertiary structure for proper functioning [16].

1.3 FUNCTIONS

1.3.1 GENETIC STORAGE AND TRANSMISSION

With the exception of some viruses, DNA constitutes the molecule of heredity in all living organisms. DNA keeps the information of organisms in the form of its nucleotide sequence, using a four-letter nucleotide language. In prokaryotes, genome consists of one chromosome that is attached to the inner cell membrane. In addition to this major genetic structure, prokaryotic cells contain plasmids that are small (1–200 kb), autonomously replicating extrachromosomal carriers of genetic information. Plasmids can be composed of DNA or RNA, they can be double stranded or single stranded, linear or circular. They can be transmitted from one cell to other cells and play important roles in distribution of genetic programs among bacterial cells. In eukaryotes, majority of DNA is in the nucleus in the form of nucleoprotein complex and is referred to as nuclear genomic DNA. Plant and animal cells also contain DNA outside the nucleus, in organelles like mitochondria, and chloroplasts that have their own genomes. Viruses exhibit a variety of genetic strategies. Genomes of viruses can be composed of RNA or DNA, viral genomes can be linear or circular, single or double stranded; RNA genomes can be segmented nonsegmented.

DNA provides storing genetic information and transmission of the information between generations of a species. This molecule is replicated before the cell division provides offspring with the same genetic information. The possibility of duplication is provided by the double-stranded nature of DNA, one strand serving as template for the making of its copy. The genetic information stored in DNA is not directly usable for making proteins. It must be extracted by transcription, by converting genetic information stored in the form of base pairs in DNA into the sequence of bases in mRNA [17]. Some transcripts function as such (e.g., tRNA, rRNAs, and snRNAs) or as mRNAs that after processing are used as programs for synthesis of proteins. In prokaryotes, modifications or processing of the primary transcripts are not required and they can be used directly as mRNAs. In eukaryotes, mRNAs are usually synthesized as precursors, containing coding sequences, exons, and sequences that are not represented in the mature mRNA, introns (intervening sequences). The number of introns and their size vary considerably between genes. Processing of mRNA precursors in the nucleus by RNA splicing removes introns and joins exons to yield the mature mRNA that is transported to cytoplasm. Sequence of mRNAs is identical to the sequence of the sense, or positive strand of the DNA template. The complementary strand of DNA is called antisense, or negative strand.

During translation, the sequence of bases in the mRNA is converted (translated) into the sequence of amino acids in the protein product according to the rules of genetic code [18]. Reading of the genetic message (translation) occurs on molecular machines known as ribosomes. A number of RNA molecules (tRNA, rRNAs, small nuclear RNAs) are the key participants of the translation process. tRNAs function as adaptors by mediating the incorporation of amino acids into proteins according to the sequence of trinucleotide codons in mRNA. How rRNAs are important for the cell can be seen from the amount in which they are synthesized. Prokaryotic ribosomes can constitute up to one fourth of the mass of an entire cell. rRNAs constitute two thirds of the mass of each ribosome.

1.3.2 GENETIC REGULATION

For a long time, it was believed that genetic information always flowed from DNA to RNA to protein, according to the so-called central dogma of molecular biology, and RNA was considered as the genetic intermediary between DNA and protein. However, later it was found that in retroviruses, the viral RNA genome is copied to produce DNA in the course of reverse transcription. Then it was discovered that mobile elements of genome called retrotransposons move from one site of the genome to other site via synthesis of DNA copies of the RNA transcribed from the element. Discovery of catalytic RNAs and RNAs folded into complex structures endowed with ability of binding to specific proteins and small molecules demonstrated that besides their role in information transport, RNA molecules can participate in the cellular metabolism as proteins. Further studies led to recognition of RNA as a unique class of molecules because it serves several fundamentally distinct functions. The double capacity of the RNA molecule, both to carry information and to be catalytically active, is the foundation for its many functions in the cell.

RNA molecules called ribozymes can catalyze specific chemical reactions within cells, and represent one of the key classes of molecules in the biochemistry of life [19–21]. RNA-cleaving ribozymes are essential for replication of viral RNA molecules. Catalytic RNA motifs containing introns catalyze the splicing steps that remove introns from mRNAs in the nucleus of eukaryotic cells. Peptidyl transferase activity of ribosomes appears to be provided by the compactly folded domain of 23S rRNA, that is, the RNA is a natural ribozyme catalyzing chemical reaction joining amino acids to form a new protein [22].

Some cellular RNAs are bound to specific proteins, forming complexes that are endowed with catalytic activity. Telomerase enzyme responsible for maintenance of telomeres, the structures that cap the ends of eukaryotic chromosomes, is a ribonucleoprotein comprising a reverse transcriptase protein and a telomeric RNA containing 11-base template CUAACCCUAAC. The RNA template binds to the 3'-end of the chromosomal DNA and directs synthesis of the proper telomeric DNA consisting of repeats of the sequence TTAGGG [23].

RNA is considered to have been critical for the evolution of life because it appears to be the most self-sufficient substance of the living matter, capable of serving as genetic material and capable of performing almost all functions of contemporary proteins. It is believed that at the very first steps of evolution, before the genetic code had evolved, when neither DNA nor proteins existed, there was the ancient RNA world. In this world, there were just ensembles of replicating RNA molecules that could catalyze their own formation and replication from simple precursors [24,25]. The RNA world in some form has been preserved as relics and evolved as molecules in the contemporary cells (Figure 1.6) [26,27]. One example of the RNAs that seem to be remnants

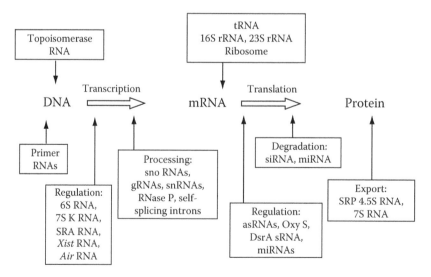

FIGURE 1.6   Contemporary RNA world.

of the RNA world are plant pathogens like viroids [28]. Viroids are composed of small (246–401) nucleotide single-stranded circular RNAs that do not encode proteins and are able to replicate autonomously in susceptible hosts. Some viroids behave as catalytic RNAs and they can elicit RNA silencing, which mediates their pathogenic attack.

Contemporary cells contain a great variety of functional RNAs [26]. rRNAs, mRNAs, and tRNAs are directly involved in synthesis of proteins. Different RNAs are involved in processes of DNA replication, mRNA processing, regulation of translation, transport of proteins, cell differentiation, embryogenesis, etc. [29]. Each year, more and more new species of RNAs are being discovered. Analysis of genomes of higher eukaryotic organisms resulted in finding that only a small part of DNA (about 1.2%) is coding for proteins. However, it is evident that most of the genomes are transcribed to yield complex patterns of transcripts encoding a great number of short and long noncoding RNAs (ncRNAs).

In the last years, various families of ncRNAs have been identified in most eukaryotic genomes [30]. ncRNAs control a remarkable range of biological processes. They are involved in regulation of translation, transposon jumping, development of muscles and brain, oncogenesis, resistance to viral infections, and chromosome architecture.

Antisense RNAs, which are polynucleotides with base sequences complementary to mRNAs, have been found in both prokaryotes and eukaryotes [31]. They bind to complementary mRNAs and repress their translation. A considerable proportion of the mammalian transcriptome comprises small regulatory RNAs, these are siRNAs and miRNAs. The discovery of siRNAs, which silence gene expression at the posttranscriptional level and at the transcriptional level in a sequence-specific way, has revolutionized the biological sciences. siRNAs are 21–28 nucleotide long RNA duplexes [32]. One strand of the siRNA is incorporated in multiprotein complex (RNA-induced silencing complex) guiding it to target enzymatic degradation of RNA with perfect or near-perfect complementarity, resulting in cleavage of the RNA. siRNAs are formed through cleavage of double-stranded RNAs (dsRNA). siRNAs can be derived from viruses (exogenous siRNAs) or derived from double-stranded cellular RNA transcripts with hairpin structures (endogenous siRNAs). In some systems, siRNAs can function as primers for an RNA-dependent RNA polymerase that synthesizes additional siRNAs, which result in great enhancement of the effect. Suppression of expression of a target gene by dsRNA is known as RNA interference (RNAi). RNAi is considered to be an ancient and ubiquitous mechanism for sequence-specific posttranscriptional gene silencing among species from various kingdoms that protects cells against viruses and regulates gene expression.

MicroRNAs are expressed as highly structured hairpin transcripts [33]. Enzymatic processing of these transcripts in the nucleus and further in cytoplasm yields short stem loop miRNA that can bind to complementary sequences in mRNAs and modulate their translation. miRNAs can be processed by the same enzymatic machinery used for generation of siRNA from long dsRNA, to yield siRNAs triggering degradation of the complementary RNAs. It is estimated that there are about 500 different miRNA genes in humans, and each miRNA can affect expression of hundreds of different genes. miRNAs regulate different processes ranging from apoptosis and immune response to cell differentiation and cell metabolism. They have important roles in the pathogenesis of several human diseases, including metabolic disorders and cancer.

Some small RNAs were shown to induce methylation in sequence-homologous DNA and chromatin modification converting it to the heterochromatin form that is not transcribed [34]. In some lower eukaryotes, small RNAs direct DNA rearrangements [35]. Small nucleolar RNAs (snoRNAs) are almost all derived from introns. Members of this family are involved in processing of mRNAs, epigenetic imprinting, posttranscriptional cleavage, and modification of rRNA precursor [30]. Little is known about the functions of large ncRNAs. The few long ncRNAs that have been characterized to date exhibit a diverse range of functions and are expressed in specific cell types [36].

Many of the rapidly growing number of genome sequences analyzed today encode for active RNAs, which still have to be discovered. Apparently, the protein network functions in concert with highly organized regulatory RNA network of the cell.

1.4   APPLICATIONS

Nucleic acids are one of the most interesting and important objects for studying. They also provide one of the most prospective functional materials for use in many fields of science and technology. The area of nucleic acids application is extremely wide and continues to grow (Figure 1.7).

1.4.1   RAW MATERIALS FOR MOLECULAR MANIPULATIONS

First of all, natural nucleic acids provide a source of nucleosides and nucleotides. These building blocks serve as a raw material for synthesizing countless analogs and derivatives of nucleic acids for bioorganic and medicinal chemistry.

The information about the nucleic acid sequences from various species is a subject of investigation in such specialties as bioinformatics and computational biology. Development of biophysics for over 50 years in many respects was determined by growing interest to spatial organization of nucleic acids and their supramolecular complexes. This interest also stimulates development of methods for computer modeling of biomolecules and instrumental basis for structural investigations.

Impressive progress in molecular and cell biology propelled development of a wide spectrum of methods for manipulation with nucleic acids and thus facilitated formation of biotechnology, genetic engineering, and nucleic-acid-based therapy.

The possibility to chemically synthesize DNA, extract specific DNA sequences from genomes and produce DNA in bacteria provided DNA material for construction of novel genes and genetically modified organisms. These DNA molecules can be manipulated with the help of restriction endonucleases and ligases to produce recombinant DNA molecules. The DNA of interest can be inserted into cloning vectors, plasmid, or virus, and amplified in appropriate host cell. The isolated DNA constructs can be directly used for the purposes of genetic therapy and genetic immunization. Recombinant DNA molecules are used for production of genetically modified prokaryotic or eukaryotic cells. These genetically modified organisms are the backbones of the contemporary bioprocessing industry and represent a source of important therapeutics and new materials. Recent successes in chemical synthesis of large DNA molecules opened up possibilities for designing completely artificial genomes of the future artificial organisms for biotechnology application [37].

1.4.2   BIOLOGICAL MARKERS

One of the most important applications for nucleic acids from any source is a role of biological marker [38,39]. First of all, nucleic acids carry genetic information that allows to discover characteristic features distinguishing one genome from another [40,41]. On the other hand, nucleic acids participate in expression of genetic information, thus making possible to analyze the level of single gene expression or profiling expression of a number of genes in response to internal or external factors [42].

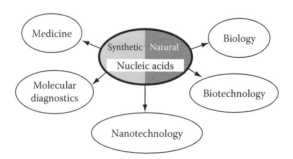

FIGURE 1.7   Areas of nucleic acids application.

Bioanalytical techniques designed for these purposes have become an integral part of such research fields as paleobiology, archaeology, forensics, molecular diagnostics, epidemiology, and pharmacology [43]. They also offer a prospect of personalized medicine [44]. The availability of human genome sequence and sequences of genomes of various infectious agents presents researchers with a new important tool for discovering genetically based diseases and for development of new therapeutic approaches. Analysis of nucleic acids isolated from biological samples opens a possibility to identify pathogens and disease-associated mutations. DNA typing is used for the purpose of distinguishing between individuals of the same species. DNA released from cancer cells can be isolated from blood and analyzed to detect the presence of specific tumor markers—mutated genes and epigenetic markers, aberrant methylation patterns within specific genes. Every parameter listed can serve as a marker: the length of a complete nucleic acid sequence or its specific fragment, the number of restriction sites and their arrangement, unique sequence, the presence of point mutations (microdeletions and microinsertions as well), single nucleotide polymorphism within the known sequences, and specific modification of nucleotide residues. In other words, bioanalytical applications of DNA reveal markers characteristic of a biological object due to unique structure of its nucleic acids.

Since nucleic acids have become molecules that can be synthesized and manipulated easily and reliably, they have found a variety of applications in different molecular devices and techniques. Synthetic fragments of nucleic acids are widely used as molecular tools in every field of life sciences. Development of efficient synthetic techniques allowed to reproduce nucleic acid fragments of desirable length and sequence in easy and cheap computer-controlled process of chemical synthesis [45]. Phosphoramidite approach gives perfect results when synthesizing both DNA and RNA fragments up to 100 nt. The availability of oligonucleotides stimulated the development of algorithms for high-precision forecasting of duplex structure and stability [46,47]. This makes possible designing nucleic acids with predictable properties for a number of experimental applications. They make up a basis for therapeutics, capable of directed intervention into intracellular genetic programs, which results in expression of pathogens.

1.4.3 NANOTECHNOLOGY

Predictability of intra- and intermolecular folding in nucleic acids reveals new aspects for their application in nanotechnology [48]. DNA as well as RNA is considered as a promising building material for various nanostructures and nanodevices [49]. Since spatial structure of nucleic acids depends strongly on their nucleotide sequence, various two- and three-dimensional nano- and even microstructures (consisting of nanoblocks) can be designed due to preset complementarity of certain sequence regions. Experimentation with RNA nanostructures resulted in development of RNA units capable of assembling to form two-dimensional arrays of oriented filaments [50]. Great number of artificial DNA-based nanostructures of different geometry can be created either using only synthetic nucleic acid fragments or with participation of longer nucleic acid molecules with known sequence, which act as a framework. High-molecular assemblies of various shapes have been obtained, thus giving rise to DNA origami as a part of nucleic-acid-based nanoarchitectonics [51]. Nanoboxes containing various molecules (e.g., proteins) in the internal space can also be created from DNA [52]. Nanoconstructions based on nucleic acids can implement various functions. They can be used for targeted delivery of biologically active compounds into certain tissue, cell, or specific compartment. Attempts are made to create nucleic acid-based biosensors, catalysts, and machines of computational and mechanical nature. Oligonucleotides and large DNA programmed into high-order structures can be used to encode mathematical computation [53].

Nucleic acids are exceptionally well suited for the design of recognition elements. Synthetic oligonucleotides are unique molecules for the design of molecular sensors, capable of detecting molecular targets with virtually any chemical structure. To be used as nanosensor, oligonucleotide can be modified in several ways, depending on the type of signal that has to be generated. Formation of specific intermolecular complex consisting of two or more components, specific chemical, or enzymatic reaction (formation or breakage of chemical bond) or just rearrangements of nanosensor spatial structure can result in a sensor's signal.

1.4.4 Diagnostic and Therapeutic Targets

DNA and RNA fragments and oligonucleotides have found a variety of applications in molecular biology. Materials with covalently attached single- and double-stranded oligonucleotides provide an efficient and simple means for isolation of nucleic acids containing specific nucleotide sequences and enzymes capable of recognizing and binding to the immobilized oligonucleotide structure. They are used as primers in the course of nucleic acid sequencing and in multiplication of RNA and DNA using nucleic acid sequence-based amplification or PCR methodology and for targeted mutagenesis [54,55].

PCR is a powerful technique that is used for amplification of DNA sequences [55]. PCR makes use of two primers that sequence-specifically bind to opposite strands of dsDNA, which are elongated by thermostable DNA polymerase in opposite directions such that the elongated products form two complementary strands. Thermal dissociation of the DNA and repeated synthesis under the conditions of excess of the primers results in exponential growth of the number of the duplexes of desired structure in the system. PCR is an enzymatic process that is carried out in discrete cycles of amplification, each cycle doubling the amount of DNA. PCR is used in a number of applications concerned with detection of trace amounts of specific nucleic acids, for preparation of highly specific DNA probes for different molecular hybridization techniques, and for amplification of DNA fragments bearing required genetic information for needs of biotechnology. Before the PCR was invented, genes of interest were isolated by laborious and costly procedures from complementary DNA (cDNA) and genomic libraries. PCR opened up a possibility to rapidly amplify and isolate the needed genes without cloning and other time-consuming operations.

Real-time PCR is one of the most suitable approaches to quantify the amount of specific DNA or RNA with the detection limit of only a few copies of the molecule. There are a number of strategies for performing the readable and comprehensive PCR. The strategy based on the use of Taqman probes seems to be the most powerful [56]. Numerous probes of different designs have also been successfully developed as molecular diagnostics tools (e.g., molecular beacons, Scorpion, and MagiProbe) [57–59]. PCR also allows to find the differences in genomic organization of related biological species. The use of arbitrary or rationally designed partially randomized primers makes possible to identify unknown differences in the primary sequences of compared samples [60]. It is the invention of PCR that made possible the development of the molecular selection technique for production of novel ribozymes and aptamers.

Even relatively short single-stranded nucleic acid molecules (15–60 nt) are capable of forming manifold spatial structures due to complex intramolecular interactions. These structures are characterized by a set of well-ordered functional groups or by the presence of internal cavities of a certain topology, which can be complementary to the structure of target molecule as described by key–lock model. Structural diversity of tertiary structures allows to identify specific ligands to various chemical or biological targets. The pool of oligonucleotides of certain length (e.g., 25 nt) containing all possible sequences can be a source of tightly binding traps or inhibitors for a wide spectra of targets starting from molecules with low molecular weight up to biopolymers including proteins, polysaccharides, and even nucleic acids. Such molecules are referred to as aptamers. The stability of the complex formed by aptamer with the target depends on the number of specific contacts between the two molecules.

Aptamers are obtained by a procedure called *in vitro* selection or molecular selection [61,62]. This technique also referred to as SELEX (selective evolution of ligands by exponential enrichment) allows to select aptamers consisting of DNA, RNA, and modified nucleic acids. This iterative PCR-based Darwinian-type process provides a possibility to select from large libraries of RNA or

DNA sequence-specific oligonucleotides, which are folded into unique structures and display specific properties. One of the important applications of SELEX is generation of unique oligonucleotide structures capable of specific binding to different proteins and other molecules, aptamers. Aptamers are similar to antibodies in terms of specificity and possess affinity in the low nanomolar range. Compared with antibodies, aptamers have some advantages: besides being nonimmunogenic, they are much smaller, cheaper, and more rapid to synthesize and introduce chemical modification. Aptamers find applications as ligands for design of aptasensors in analytical systems [63]. Using aptamers as sensor elements is a successfully developing area of research. Aptamers also can be modified to produce different signals such as fluorescent and electrochemical signals. Aptamers can be designed not only on the basis of natural oligonucleotides, but also spiegelmers are biostable aptamers built of L-nucleotides, the mirror images of natural nucleotides [64]. Unnatural configuration of sugar-phosphate backbone makes this type of oligonucleotides highly resistant to abundant nucleases that results in possibility to be bioactive even at nanomolar concentration.

Oligonucleotides are used as specific tags for labeling micro- and nanoparticles. Oligonucleotides or DNA fragments tagged with various labels (radiolabeled groups, fluorescent groups, haptens recognized by specific antibodies, or even complete proteins) have found applications as diagnostic probes and for isolation and identification of functionally significant RNA- and DNA-based elements [65]. Fluorescently labeled oligonucleotides are used for localization of specific nucleic acids within cells and tissues [66].

Synthetic oligonucleotides and analogs as well as enzymatically prepared DNA fragments are widely used as capture probes in a variety of techniques based on heterogeneous hybridization [67]. The main analytical tool in this case is the biochip or nucleic acid microarray. Biochip is the planar solid support carrier bearing in the preset places of surface various specific capture probes. Glass slides, silicon wafer, and polymeric membranes (e.g., nylon filters) can serve as a carrier. There are a number of strategies for producing biochips. Some of them are based on immobilization of preliminary synthesized and purified oligonucleotides or DNA fragments onto solid support with or without chemical activation. This strategy is convenient for preparation of custom biochip with low- or medium-density probe array with the number of capturers varying from tens to few thousands. High-density arrays are available due to development of methods allowing to synthesize the probes in parallel mode directly on the surface of support [68]. Application of photolithographic technique drastically enhanced the available number of spots per microarray up to hundreds of thousands and even millions [69]. Such devices are difficult to customize, but a wide range of accurately designed high-density arrays for standard applications are commercially available.

The main advantage of both types of biochips consists of the possibility to carry out the multiplex parallel analysis of a sample in terms of presence of specific markers from defined set and their relative amount [70]. Another important problem, that is much more easily solvable with biochips, is determination of unknown characteristics of both nucleic acid–nucleic acid hybridization and affinity binding of various ligands with nucleic acids [71,72]. Nowadays biochips have become powerful tools for a number of research fields.

Therapeutic group of nucleic acids is represented by immunostimulating CpG oligonucleotides, aptamers, antisense and antigene oligonucleotides, ribozymes, dsRNAs, and functional DNA constructs.

Some poly- and oligonucleotides can serve as antiviral therapeutics by mimicking nucleic acids of viral and bacterial pathogens that are recognized by specific cellular Toll-like receptors and trigger activation of innate immune defense [73]. dsRNAs are known inducers of interferon synthesis in mammalian cells. Formation of dsRNA during viral infection is recognized by the cell as a signal for viral genes activity and triggers a cascade of defensive reactions leading to production of type I interferons. It was shown that double-stranded polyribonucleotide poly(rI):poly(rC) is an efficient interferon inducer; however, toxicity of this complex prevented its use in therapy.

Less toxic mismatched complex poly(rI):poly(rC12U) displayed high activity and reduced toxicity and is considered as an antiviral therapeutics [74].

1.4.5 REGULATORS OF HOST IMMUNE RESPONSE AND GENE EXPRESSION

In contrast to bacterial genomes, vertebrate genomes contain less CpG sequences and they are usually more heavily methylated. Bacterial and viral DNAs with unmethylated CpG motifs interact with Toll-like receptor 9 and trigger reactions that lead to production of type I interferon and stimulation of B-cells and dendritic cells. Short oligodeoxynucleotides harboring CpG motif and some other motifs were shown to produce the same effects as bacterial DNA and they already find applications as adjuvants and nonspecific immune modulators [75].

Oligonucleotides targeted to specific nucleic acids represent a highly promising class of potential therapeutic agents that could be designed to affect a variety of host and infectious disease targets. Attention to oligonucleotides as potential gene-targeted therapeutics was attracted by the seminal works of Grineva on the synthesis of reactive conjugates of oligonucleotides for targeting nucleic acids and by experiments of Zamecnik on inhibition of viral proliferation by oligonucleotides [76,77].

Oligonucleotides can be used to control gene expression according to several mechanisms [78,79]. The antisense strategy makes use of oligonucleotides binding to mRNA, pre-mRNA, and viral RNAs. Binding of oligonucleotides can arrest mRNA translation, pre-mRNA processing, and expression of viral RNAs. The hybridization of antisense molecules to mRNA can result in inhibition of translation by physically interfering with essential proteins binding to the target RNAs. Oligonucleotides can cause translation arrest by binding to the 5'-end of mRNAs and inhibiting binding of proteins required to cap the mRNA and translation initiation factors. Binding of deoxyoligonucleotides and some analogs to RNA induces degradation of the target RNA by endogenous enzyme ribonuclease H that cleaves the RNA strand in DNA–RNA hybrids. Antisense oligonucleotides are well suited to suppress the expression of specific genes giving rise to knockdown phenotypes of the cells. This provides an important technology for studying biological functions of unknown genes and can also have an impact on the treatment of diseases caused by aberrant gene expression. Antisense therapy against different viruses and cancer cells is currently being attempted by targeting oligonucleotide analogs against viral nucleic acids and oncogenes.

The interest and activity in the oligonucleotide therapeutics have grown amazingly rapidly during the past years and have directed much attention to synthesis of oligonucleotide analogs [79]. Development of such analogs was motivated to find the reagents that exhibit increased biostability and supersede the hybridization properties of parent oligonucleotides in terms of stability and specificity. Oligonucleotides were conjugated to different ligands in order to improve their cellular uptake and target to specific cells. Oligonucleotide conjugates equipped with different reactive groups were synthesized to produce the targeted agents capable of cleaving or chemically damaging specific viral nucleic acids and oncogenes [78,79].

Ribozymes, catalytically active oligoribonucleotides, can be considered as an improved version of antisense oligonucleotides capable of cleaving target RNA molecules [32]. Synthetically designed ribozymes can act as molecular scissors destroying unwanted RNA of specific sequence such as viral RNA or mutant RNA associated with a disease. Targeted ribozymes have been used scientifically and therapeutically to cleave viral RNAs and mRNAs associated with disease.

Oligonucleotides capable of triple helix formation with homopurine–homopyrimidine stretches in dsDNA can be used as gene-targeted molecules and represent another class of potential oligonucleotide-based therapeutics targeted to specific DNA sequences [80]. Oligonucleotides that bind to specific sequences of dsDNA by forming triple helices are called antigene oligonucleotides or triplex forming oligonucleotides (TFOs). A number of oligonucleotide analogs and conjugates demonstrating enhanced stability of triple-stranded complexes with DNA have been developed. TFOs were shown to affect transcription of the target genes and to induce targeted mutations, although with low frequency [81].

Strong binding oligonucleotide analogs displaying high affinity to complementary DNA can locally unwind the double helix and invade into the DNA structure [82]. Examples of such analogs are polyamide nucleic acids (PNA) and locked nucleic acids (LNA). In the PNA, the entire ribose phosphate backbone is replaced by an electroneutral polyamide backbone with the side groups, pyrimidine and purine bases. In the LNA, nucleotides contain a methylene bridge that connects the 2-oxygen of the ribose with the 4-carbon. This bridge results in a reduced conformational flexibility of the ribose, which yields a remarkable increase in hybridization affinity of LNA. It was shown that PNA and LNA oligonucleotides targeted to transcription start sites within promoters DNA can inhibit expression of specific genes [83,84].

Efficient technologies for aptamers production provided a possibility to develop protein-targeted and small molecules-targeted nucleic acid-based therapeutics. Aptamers can bind target proteins with very high affinity and interfere with their functions [85]. RNA and DNA aptamers have been isolated against several viral proteins and were demonstrated to interfere with the functions of the proteins *in vitro*. Aptamer inhibitors of specific proteins can function as therapeutics modulating physiological processes in the organism, for example, aptamer inhibitors of von Willebrandt factor can inhibit process of arterial thrombosis [86].

Currently, siRNAs represent the most promising class of potential oligonucleotide drugs capable to efficiently knockdown a specific gene in somatic cells and inhibit virus replication [32,87]. siRNAs can be synthesized chemically or can be induced endogenously by intracellular expression of small hairpin RNAs (shRNAs) and miRNA precursors from plasmid or viral vectors delivered to the cells. Introduction of siRNAs to mammalian cells leads to sequence-specific destruction of endogenous RNA molecules complementary to siRNA. Any disease-causing gene and any cell type can potentially be targeted. Harnessing oligonucleotide analogs and siRNA holds great promise for therapy. The most important obstacle in this field is the problem of efficient delivery of oligonucleotides into cells.

DNA constructs containing functional genes can serve as informational therapeutics in gene therapy. The concept of gene therapy is to introduce into target cell a piece of genetic material that will cure the disease. Molecular studies have revealed a genetic basis for many human diseases caused by single gene mutations and multiple mutations in different genes. The diseases can be treated only by gene therapy, by correcting the genetic defects in the cells by the addition or replacement of the mutant gene with its normal type [88].

One approach to gene therapy is the expression of a recombinant gene in a patient. This gene can be delivered to the cells of a patient directly or it can be delivered into cells *in vitro*, followed by transfer of the modified cells to the patient. DNA construct can be a plasmid with inserted gene of interest. In this case, the construct is delivered into cells in the form of a complex with synthetic carriers. More efficient, although associated with some side effects, are the constructs based on viruses that can enter the cells spontaneously. Viral vectors are derived from viruses with either RNA or DNA genomes and are presented as both nonintegrating and integrating vectors, capable of inserting their genes into the genome of the target cells. Retroviral and adenoviral vectors serve as the most convenient vectors for transfection for the purpose of gene targeting. In the vectors, the interesting protein-encoding gene is inserted under the control of appropriate eukaryotic promoter that brings about the expression of the gene in the cell.

Transfer of normal genes can provide organism with required functional product, but it does not permit correction of the original defect, and proper regulation of the introduced gene is a problem. The ideal approach to gene therapy could be correction of the mutation in the defective gene. In theory, transfer of nucleic acid into cells can be aimed at correction of individual mutations. The correction can be achieved by homologous recombination with oligonucleotides that can interact with homologous sequence due to their flanking homologous sequences and replace the defective sequence by new correct sequence, contained in the center of the oligonucleotide [89]. Development of methods for genetic therapy met a number of problems such as the problem of efficient delivery of DNA into cells and problem of biological safety.

Genetic immunization appeared to be an effective and practical procedure based on therapeutic gene expression [90]. Nucleic acid vaccines are based on the finding that injection into skeletal muscle of mRNA or a plasmid DNA encoding a gene leads to significant expression of heterologous genes within the muscle cells. This elicits immune responses against the expressed protein. An advantage of genetic immunization is the persistence *in vivo* of antigen-producing cells that could provide continuous immunization for extended time intervals. It provides a possibility to express an exogenous antigen in the authentic tissue environment, to present antigen in a native conformation and can induce both humoral and cellular immune responses that is not achievable with inoculated protein immunogens. Because preparation and isolation of expression plasmids for DNA vaccination are simple and inexpensive, genetic immunization can be a rapid and cheap approach compared with the preparation of protein vaccines.

One more example of using nucleic acids as informational drugs is the development of cell vaccines, genetically modified dendritic cells. Dendritic cells are the most potent antigen-presenting cells for the initiation of antigen-specific immune response. Transfection of dendritic cells with mRNA for specific antigene, for example, tumor antigene or corresponding DNA construct or virus vector, results in loading of the cells with the antigene. The modified dendritic cells in the organism of patients initiate cytotoxic attack of the immune cells on the tumor [91].

1.5 CONCLUSION

The last years demonstrated that nucleic acid research remains one of the hot spots of the contemporary science and technology. Genome studies and sequencing continue to grow at an amazing rate. siRNAs are at the forefront of biomedical research, new areas where RNA has been found to play important roles indicate that the rapid expansion of this exciting field will last on.

Significant efforts are already being deployed toward the development of efficient platforms and instrument systems for high-throughput analysis of various biological objects. Nucleic acid technology is exponentially growing in the area of biosensors, bio- and nanotechnology, and therapeutics. The future use of these specific nucleic acid-based techniques will facilitate control of bioprocesses and thus result in high productivity of all aspects of bioprocessing. Automation and design of lab-on-a-chip analytical devices capable of performing rapid PCR amplification, electrophoresis, DNA sequencing, and real-time data capture will provide the researchers with powerful tools for nucleic acid analysis virtually in all areas of application.

Innovations in microarray design are anticipated to enhance our ability to find out distinguishing features of individual genomes that is crucial for the development of personalized medicine. Oligonucleotides and siRNA designed to affect specific genetic programs represent a revolutionary advance in pharmacotherapy. Successes in this field are raising hopes in bringing nucleic acid therapeutics from the bench to the bedside.

ACKNOWLEDGMENT

We apologize to colleagues for referring to recent reviews instead of original papers due to space limitation.

Valentin V. Vlassov received his chemical education in Novosibirsk State University (NSU), Novosibirsk, Russia. He did his PhD training in the biochemical department of the Novosibirsk Institute of Organic Chemistry, Novosibirsk, Russia. With the expansion of this department a new institute was developed—Institute of Bioorganic Chemistry (since 2003—Institute of Chemical Biology and Fundamental Medicine, Siberian Division of the Russian Academy of Sciences), in which he became the vice director and head of Laboratory of Nucleic Acids Biochemistry. Currently, he is the director of this institute and head of the molecular biology department at the NSU. His earlier research interests were mostly in biochemistry of nucleic acids, design of chemical methods for investigation of nucleic acid–protein interactions, and development of antisense oligonucleotide-based therapeutics. More recently, he has focused on the development of methods for gene therapy, genetic analysis, and approaches to molecular diagnostics. He has published over 300 scientific papers, and invited reviews and book chapters.

Dmitrii V. Pyshnyi received his chemical education at the Novosibirsk State University in Novosibirsk, Russia, and began his scientific work at the Institute of Chemical Biology and Fundamental Medicine (former Novosibirsk Institute of Bioorganic Chemistry, Novosibirsk, Russia) in 1991, with a focus on nucleic acid chemistry and complementary addressed modification of nucleic acids. After completing his PhD training, he concentrated on the development of systems for hybridization analysis of nucleic acids and physicochemical description of nucleic acid complexation. Since 2006, he has been working as a deputy director of the institute and a head of the laboratory of bionanotechnology. He has published over 90 scientific papers.

Pavel E. Vorobjev received his chemical education at the Novosibirsk State University (NSU), Novosibirsk, Russia, and started his work at the Institute of Chemical Biology and Fundamental Medicine (ICBFM SD RAS, former Novosibirsk Institute of Bioorganic Chemistry, Novosibirsk, Russia) in 1993, with an emphasis on the chemistry of oligonucleotides derivatives, including their reactive conjugates. Upon finishing his PhD training, he worked as a researcher in the laboratory of bionanotechnology, ICBFM, focusing on nucleic acid aptamers and their applications. Since 1996, he has been working as a teacher/lecturer in biochemistry and molecular biology at NSU.

REFERENCES

1. Watson, J.D. and Crick, F.H.C., Molecular structure of nucleic acids; a structure for deoxyribose nucleic acid, *Nature*, 171: 737, 1953.
2. Wuyts, J., Perrière, G., and Van De Peer, Y., The European ribosomal RNA database, *Nucleic Acids Res.*, 32: D101, 2004. Available at http://bioinformatics.psb.ugent.be/webtools/rRNA/index.html.
3. Lowe, T.M. and Eddy, S.R., tRNAscan-SE: A program for improved detection of transfer RNA genes in genomic sequence, *Nucleic Acids Res.*, 25: 955, 1997. Available at http://lowelab.ucsc.edu/GtRNAdb/.
4. HUGE Protein Database. Available at http://www.kazusa.or.jp/huge/.
5. Organelle Genome Resources. Available at http://www.ncbi.nlm.nih.gov/genomes/ORGANELLES/organelles.html.
6. The NCBI Entrez Genome Project Database. Available at http://www.ncbi.nlm.nih.gov/sites/entrez?db=genomeprj.
7. Sanger, W., *Principles of Nucleic Acids Structure*, Springer, New York, 1984, Chapter 2.
8. Cantor, C.R. and Schimmel, P.R., *Biophysical Chemistry*, W.H. Freeman, San Francisco, CA, 1980, Chapter 22.
9. Malacinski, G.M., *Essentials of Molecular Biology*, Jones & Bartlett Publishers, Sadbury, MA, 2003, p. 112.
10. Dickerson, R.E. et al., The geometry of A, B, and Z DNA, in *Nucleic Acid Research. Future Development*, Mizobuchi, K., Watanabe, I., and Watson, J.D. (Eds.), Academic Press, Tokyo, Japan, 1983, pp. 35–59.
11. Stryer, L., *Biochemistry*, 4th ed., W.H. Freeman, New York, 1995, Chapter 37.
12. Giovannangeli, C. et al., Triple-helix formation by oligonucleotides containing the three bases thymine, cytosine, and guanine, *Proc. Natl. Acad. Sci. USA*, 89: 8631, 1992.
13. Burge, S. et al., Quadruplex DNA: Sequence, topology and structure, *Nucleic Acids Res.*, 34: 5402, 2006.
14. Tinoco, I., Jr. and Bustamante, C., How RNA folds, *J. Mol. Biol.*, 193: 271, 1999.
15. Normanly, J. and Abelson, J., Transfer RNA identity, *Ann. Rev. Biochem.*, 58: 1029, 1989.
16. Noller, H.F., RNA structure: Reading the ribosome, *Science*, 309: 279, 2005.
17. White, R.J., *Gene Transcription: Mechanism and Control*, Blackwell Scientific, Oxford, United Kingdom, 2000, p. 273.
18. Wilson, D.N. et al., Protein synthesis at atomic resolution: Mechanistics of translation in the light of highly resolved structures for the ribosome, *Curr. Protein Peptide Sci.*, 3: 1, 2002.

19. Kruger, K. et al., Self-splicing RNA: Autoexcision and autocyclization of the ribosomal RNA intervening sequence of *Tetrahymena*, *Cell*, 31: 147, 1982.
20. Guerrier-Takada, C. et al., The RNA moiety of ribonuclease P is the catalytic subunit of the enzyme, *Cell*, 35: 849, 1983.
21. Krupp, G. and Gaur, R.K. (Eds.), *Ribozyme, Biochemistry and Biotechnology*, Earon Publishing, BioTechniques Books Division, Natick, MI, 2000.
22. Nissen, P. et al., The structural basis of ribosome activity in peptide bond synthesis, *Science*, 289: 920, 2000.
23. White, L.R., Wright, W.E., and Snay, J.W., Telomerase inhibitors, *Trends Biotechnol.*, 19: 114, 2001.
24. Gilbert, W., Origin of life: The RNA world, *Nature*, 319: 618, 1986.
25. Puglisi, J.D. and Williamson, J.R., *The RNA World*, 2nd ed., Cold Spring Harbor Laboratory Press, Cold Spring Harbor, NY, 1999, p. 403.
26. Spirin, A.S., Omnipotent RNA, *FEBS Lett.*, 530: 4, 2002.
27. Brosius, J., Echoes from the past—are we still in an RNP world? *Cytogenet. Genome. Res.*, 110: 8, 2005.
28. Daròs, J.A., Elena, S.F., and Flores, R., Viroids: An Ariadne's thread into the RNA labyrinth, *EMBO Rep.*, 7: 593, 2006.
29. Grosshans, H. and Filipowicz, W., Molecular biology: The expanding world of small RNAs, *Nature*, 451: 414, 2008.
30. Michalak, P., RNA world—the dark matter of evolutionary genomics, *J. Evol. Biol.*, 9: 1768, 2006.
31. Crooke, S.T. and Lebleu, B. (Eds.), *Antisense Research and Applications*, CRC Press, Boca Raton, FL, 2000.
32. Sioud, M. (Ed.), *Ribozymes and siRNA Protocols, Methods in Molecular Biology*, 2nd ed., Humana Press, Totowa, NJ, 2004, p. 252.
33. Marquez, R.T. and McCaffrey, A.P., Advances in microRNAs: Implications for gene therapists, *Hum. Gene Ther.*, 19: 27, 2008.
34. Rassoulzadegan, M. et al., RNA-mediated non-mendelian inheritance of an epigenetic change in the mouse, *Nature*, 441: 469, 2006.
35. Nowacki, M. et al., RNA-mediated epigenetic programming of a genome-rearrangement pathway, *Nature*, 451: 153, 2008.
36. Mercer, T.R. et al., Specific expression of long noncoding RNAs in the mouse brain, *Proc. Natl. Acad. Sci. USA*, 105: 716, 2008.
37. Gibson, D.G. et al., Complete chemical synthesis, assembly, and cloning of a *Mycoplasma genitalium* genome, *Sci. Express*, January 24, 2008.
38. Gillespie, D. and Spiegelman, S., A quantitative assay for DNA–RNA hybrids with DNA immobilized on a membrane, *J. Mol. Biol.*, 12: 829, 1965.
39. Klee, E.W., Data mining for biomarker development: A review of tissue specificity analysis, *Clin. Lab. Med.*, 28: 127, 2008.
40. Pardue, M.L. and Gall, J.G., Molecular hybridization of radioactive DNA to the DNA of cytological preparations, *Proc. Natl. Acad. Sci. USA*, 64: 600, 1969.
41. Kearney, L. and Horsley, S.W., Molecular cytogenetics in haematological malignancy: Current technology and future prospects, *Chromosoma*, 114: 286, 2005.
42. Schena, M. et al., Quantitative monitoring of gene expression patterns with a complementary DNA microarray, *Science*, 270: 467, 1995.
43. Brettell, T.A., Butler, J.M., and Almirall, J.R., Forensic science, *Anal. Chem.*, 79: 4365, 2007.
44. Agrawal, S. and Khan, F., Human genetic variation and personalized medicine, *Indian J. Physiol. Pharmacol.*, 51: 7, 2007.
45. Beaucage, S.L. and Caruthers, M.H., Deoxynucleoside phosphoramidites—a new class of key intermediates for deoxypolynucleotide synthesis, *Tetrahedron Lett.*, 22: 1859, 1981.
46. SantaLucia, J., Jr. and Hicks, D., The thermodynamics of DNA structural motifs, *Annu. Rev. Biophys. Biomol. Struct.*, 33: 415, 2004.
47. Mathews, D.H. et al., Expanded sequence dependence of thermodynamic parameters improves prediction of RNA secondary structure, *J. Mol. Biol.*, 288: 911, 1999.
48. Mathews, D.H., Revolutions in RNA secondary structure prediction, *J. Mol. Biol.*, 359: 526, 2006.
49. Yan, H., Materials science: Nucleic acid nanotechnology, *Science*, 306: 2048, 2004.
50. Chworos, A. et al., Building programmable jigsaw puzzles with RNA, *Science*, 306: 2068, 2004.

51. Rothemund, P.W.K., Folding DNA to create nanoscale shapes and patterns, *Nature*, 440: 297, 2006.
52. Erben, C.M., Goodman, R.P., and Turberfield, A.J., Single-molecule protein encapsulation in a rigid DNA cage, *Angew. Chem. Int. Ed.*, 45: 7414, 2006.
53. Liedl, T., Sobey, T.L., and Simmel, F.C., DNA-based nanodevices, *Nanotoday*, 2: 36, 2007.
54. Kievits, T. et al., NASBA isothermal enzymatic in vitro nucleic acid amplification optimized for the diagnosis of HIV-1 infection, *J. Virol. Method*, 35: 273, 1991.
55. Saiki, R. et al., Enzymatic amplification of beta-globin genomic sequences and restriction site analysis for diagnosis of sickle cell anemia, *Science*, 230: 1350, 1985.
56. Lee, L.G., Connell, C.R., and Bloch, W., Allelic discrimination by nick-translation PCR with fluorogenic probes, *Nucleic Acids Res.*, 21: 3761, 1993.
57. Tyagi, S. and Kramer, F.R., Molecular beacons: Probes that fluoresce upon hybridization, *Nat. Biotechnol.*, 14: 303, 1996.
58. Whitcombe, D. et al., Detection of PCR products using self-probing amplicons and fluorescence, *Nat. Biotechnol.*, 17: 804, 1999.
59. Yamane, A., MagiProbe: A novel fluorescence quenching-based oligonucleotide probe carrying a fluorophore and an intercalator, *Nucleic Acids Res.*, 30: e97, 2002.
60. Williams, J.G. et al., DNA polymorphisms amplified by arbitrary primers are useful as genetic markers, *Nucleic Acids Res.*, 8: 6531, 1990.
61. Ellington, A.D. and Szostak, J.W., In vitro selection of RNA molecules that bind specific ligands, *Nature*, 346: 818, 1990.
62. Tuerk, C. and Gold, L., Systematic evolution of ligands by exponential enrichment: RNA ligands to bacteriophage T4 DNA polymerase, *Science*, 249: 505, 1990.
63. Willner, I. and Zayats, M., Electronic aptamer-based sensors, *Angew. Chem. Int. Ed. Engl.*, 46: 6408, 2007.
64. Wlotzka, B. et al., In vivo properties of an anti-GnRH Spiegelmer: An example of an oligonucleotide-based therapeutic substance class, *Proc. Natl. Acad. Sci. USA*, 99: 8898, 2002.
65. Protocols for nucleic acid analysis by nonradioactive probes: Methods and protocols, 2nd ed., in *Methods in Molecular Biology*, Hilario, E. and Mackay, J. (Eds.), Humana Press, Totowa, NJ, 2007, p. 353.
66. Sokol, D.L. et al., Real time detection of DNA RNA hybridization in living cells, *Proc. Natl. Acad. Sci. USA*, 95: 11538, 1998.
67. Microarrays, in *Synthesis Methods, Methods in Molecular Biology*, Rampal, J.B. (Ed.), 2nd ed., Vol. 1, Humana Press, Totowa, NJ, 2007, p. 381.
68. Fodor, S.P. et al., Light-directed, spatially addressable parallel chemical synthesis, *Science*, 251: 767, 1991.
69. Lipshutz, R.J. et al., High density synthetic oligonucleotide arrays, *Nat. Genetics*, 21: 20, 1999.
70. Schena, M., Genome analysis with gene expression microarrays, *Bioessays*, 18: 427, 1996.
71. Kolchinsky, A. and Mirzabekov, A., Analysis of SNPs and other genomic variations using gel-based chips, *Hum. Mutat.*, 19: 343, 2002.
72. Zasedateleva, O.A. et al., Specificity of mammalian Y-box binding protein p50 in interaction with ss and ds DNA analyzed with generic oligonucleotide microchip, *J. Mol. Biol.*, 324: 73, 2002.
73. Cheng, G. et al., Double-stranded DNA and double-stranded RNA induce a common antiviral signaling pathway in human cells, *Proc. Natl. Acad. Sci. USA*, 104: 9035, 2007.
74. Essey, R.J., McDougall, B.R., and Robinson, W.E. Jr., Mismatched double-stranded RNA (polyI-polyC(12)U) is synergistic with multiple anti-HIV drugs and is active against drug-sensitive and drug-resistant HIV-1 in vitro, *Antiviral Res.*, 51: 189, 2001.
75. Klinman, D.M., Use of CpG oligodeoxynucleotides as immunoprotective agents, *Expert. Opin. Biol. Ther.*, 4: 937, 2004.
76. Belikova, A.M., Zarytova, V.F., and Grineva, N.I., Synthesis of ribonucleosides and diribonucleoside phosphates containing 2-chloroethylamine and nitrogen mustard residues, *Tetrahedron Lett.*, 37: 3557, 1967.
77. Zamecnik, P.C. and Stephenson, M.L., Inhibition of Rous sarcoma virus replication and cell transformation by a specific oligodeoxynucleotide, *Proc. Natl. Acad. Sci. USA*, 75: 280, 1978.
78. Knorre, D.G. et al., *Design and Targeted Reactions of Oligonucleotide Derivatives*, CRC Press, Boca Raton, FL, 2000.
79. Crooke, S.T. (Ed.), *Antisense Drug Technology, Principles, Strategies, and Applications*, Marcel Dekker, New York, 2001.

80. Malvy, C., Harel-Bellan, A., and Pritchard, L.L. (Eds.), *Triple Helix Forming Oligonucleotides*, Kluwer Academic, Boston, MA, 1999.
81. Macris, M.A. and Glazer, P.V., Targeted genome modification via triple helix formation, in *Antisense Drug Technology, Principles, Strategies, and Applications*, Crooke, S.T., Eds., Marcel Dekker, New York, 2001, Chapter 32.
82. Demidov, V.V. et al., Kinetics and mechanism of the DNA double helix invasion by pseudocomplementary peptide nucleic acids, *Proc. Natl. Acad. Sci. USA*, 99: 5953, 2002.
83. Beane, R.L. et al., Inhibiting gene expression with locked nucleic acids (LNAs) that target chromosomal DNA, *Biochemistry*, 46: 7572, 2007.
84. Hu, J. and Corey, D.R., Inhibiting gene expression with peptide nucleic acid (PNA)—peptide conjugates that target chromosomal DNA, *Biochemistry*, 46: 7581, 2007.
85. Gold, L. et al., Diversity of oligonucleotide functions, *Annu. Rev. Biochem.*, 64: 763, 1995.
86. Oney, S. et al., Antidote-controlled platelet inhibition targeting von Willebrandt factor with aptamers, *Oligonucleotides*, 17: 265, 2007.
87. Dykxhoorn, D.M. and Lieberman, J., Running interference: Prospects and obstacles to using small interfering RNAs as small molecule drugs, *Annu. Rev. Biomed. Eng.*, 8: 377, 2006.
88. Verma, I.M. and Weitzman, M.D., Gene therapy: Twenty-first century medicine, *Annu. Rev. Biochem.*, 74: 711, 2005.
89. Sorrell, D.A. and Kolb, A.F., Targeted modification of mammalian genomes, *Biotechnol. Adv.*, 23: 431, 2005.
90. Liu, M.A., Wahren, B., and Karlsson Hedestam, G.B., DNA vaccines: Recent developments and future possibilities, *Hum. Gene. Ther.*, 17: 1051, 2006.
91. Ballestrero, A. et al., Immunotherapy with dendritic cells for cancer, *Adv. Drug. Deliv. Rev.*, 60: 173, 2008.

Part I

*Purification of Nucleic Acids
from Viruses*

2 Isolation of Viral DNA from Cultures

Larry A. Hanson

CONTENTS

2.1 INTRODUCTION

Viruses are submicroscopic infectious particles that are incapable of growing or reproducing outside host cells. Each viral particle (or virion) is composed of a genetic core in the form of DNA or RNA and a protective protein coat called capsid, which often measures between 10 and 300 nm in diameter. The viral capsid may be of simple helical and icosahedral (polyhedral or near-spherical) forms, or of more complex structures with tails (or envelope). The viral envelope often protects a virion from enzymes and certain chemicals, and also functions as receptor molecules to allow recognition and uptake by host cells. This puts enveloped virus in an advantaged position over other capsid-only virions. Being noncellular organisms, viruses possess genes that enable them to reproduce in host cells by creating multiple copies of themselves through self-assembly and to evolve in infected cells by natural selection. While many more viruses contain RNA and belong to RNA virus category (see Chapter 3), some viruses possess a double-stranded DNA (dsDNA) or single-stranded DNA (ssDNA), and are thus called DNA viruses. DNA viruses often have larger genomes due to the high fidelity of their replication enzymes—DNA polymerases.

Viruses have important impacts on human health (as wells as companion animal and food animal health) invertebrates, and plants. Viruses are also important pathogens of microbes, fungi, and protozoans. The influence of microbial pathogens is one of the primary driving forces in evolution and biological diversity. The obligatory pathogenic nature and the intricate interaction of the viruses with their host genomes make the viruses a source of selective pressure and as well as a mechanism for providing genetic diversity to the host organism. The marine environment viruses represent 94% of the nucleic acid containing particles and are thought to be a major driving force in microbial biodiversity. The human genome contains at least 50,000 retroviral elements, the most common of the integrated viral sequences [1].

Viruses serve as useful functional genome cassettes for studying basic mechanisms in molecular biology and they are often used as gene expression systems. DNA viruses vary greatly in replication strategy, genome structure, and virion structure. Different DNA isolation techniques are used for different types of virus, different applications, and the structural form of DNA desired. In this chapter, we focus on viruses of vertebrates and insects (hosts from which there are well established cell lines) and review procedures that are applied for the purification of virus genomic DNA from eukaryotic cell cultures.

2.1.1 CLASSIFICATION OF DNA VIRUSES

Viruses are classified based on the structure of the infectious particle (virion), the composition and structure of the genome, and the virus's replication strategy. The genomes of viruses can be fragmented (consist of more than one genomic fragments) or consist of a single molecule. The replication can involve host and viral-encoded enzymes. Furthermore, the genomic replication and virion assembly can occur in the cytoplasm and the nucleus. The virions may be variable in shape (pleomorphic), fusiform or rodlike, or they may be icosahedral. These particles may consist of just a protein capsid surrounding the genome (naked) or they may be encased in a phospholipid bilayer membrane (envelope). The genomic structure of DNA viruses ranges from small ssDNA as in parvoviruses, partial dsDNA as in hepadnaviruses, small dsDNA viruses such as papillomaviruses and large dsDNA as herpesviruses, poxviruses, and iridoviruses (Table 2.1).

Based on the Baltimore classification system, DNA viruses are divided into two groups: Group I (consisting of dsDNA viruses) and Group II (consisting of ssDNA viruses). The two important ssDNA virus families include Circoviridae and Parvoviridae, which are responsible for causing a number of diseases in mammals including humans (Table 2.1). Other two notable ssDNA virus families are Inoviridae and Microviridae, which cover some diverse bacterial viruses called bacteriophages (or phages). Phages usually measure between 5 and 500 kb, and can be composed of single stranded RNA, double stranded DNA, ssDNA, and ddDNA with either circular or linear arrangement. On the other hand, many more economically significant DNA virus families are double-stranded (Table 2.1).

2.1.2 BIOLOGY AND LIFE CYCLE OF DNA VIRUSES

The physical and biological properties of DNA virus families from animals are summarized in Table 2.1. These families can be broadly grouped into ssDNA viruses, hepadnaviruses, small dsDNA viruses, large nuclear replicating dsDNA viruses, and large cytoplasmic replicating dsDNA viruses.

Small ssDNA viruses such as circoviruses and parvoviruses primarily use the host's enzymes for DNA replication and thus may be very specific for cell type or replicative stage for productive virus infection. Neither type of virus appears to have strong mechanisms to drive the host cell to a more permissive cell cycle phase. Some of the parvoviruses require coinfection by adenoviruses (AdVs) or herpesviruses to replicate. The members of the parvovirus group may establish latency by site-specific integration into the genome. This characteristic has made the use of adeno-associated viruses (AAVs) more attractive for gene therapy applications and thus much work has been done with generating recombinant viruses. The circoviruses have a circular dsDNA replicative form and replication apparently

TABLE 2.1
Families of DNA Viruses of Animals

DNA Virus Family with Subfamilies	Genome Size and Structure	Replication Site	Virion Structure and Density[a]	Host and Important Pathogens
Circoviridae	2 kb ss circular, infectious	Nucleus	Icosahedral naked Density 1.33–1.37	Vertebrates—humans (transfusion transmitted virus), swine (porcine circovirus), and birds (chicken anemia virus)
Parvoviridae Parvovirinae Densovirinae	4–6 kb ss linear, infectious	Nucleus, required S-phase cells or helper virus	Icosahedral naked Density 1.39–1.42	Vertebrates—humans (parvovirus B19—fifth disease), other mammals (canine parvovirus, swine parvovirus, feline parvovirus), and birds Invertebrates—arthropods
Hepadnaviridae	3 kb partially ds circular not covalently closed, infectious	Nucleus and cytoplasm, DNA produce by rt of RNA	Icosahedral enveloped Density 1.25	Vertebrates—humans (HBV), other mammals, and birds (duck HBV)
Polyomaviridae	5 kb ds circular, infectious	Nucleus	Icosahedral naked Density 1.34–1.35	Vertebrates—humans (JCV, BKV), other mammals (SV-40, bovine polyomavirus), and birds (avian polyomavirus)
Papillomaviridae	7–8 kb ds circular	Nucleus	Icosahedral naked Density 1.34–1.35	Vertebrates—humans (human papilloma viruses 1–100—cervical cancer) and other mammals (bovine papilloma virus 1)
Adenoviridae	28–45 kb ds linear, infectious	Nucleus	Icosahedral naked Density 1.30–1.37	Vertebrates—humans (HAdVs A–F), other mammals (canine AdV 1 and 2, equine AdVs 1), birds (fowl AdVs A–E), reptiles, and amphibians
Baculoviridae Nucleopolyhedro-virus Granulovirus	80–180 kb ds circular, supercoiled, infectious	Nucleus	Rod, enveloped; two forms: occluded 1.18–1.25 and budded 1.47	Invertebrates—insects (autographa californica multicapsid nucleopolyhedrovirus) and crustaceans (Baculovirus penaei, Penaeus monodon baculovirus)
Ascoviridae	100–180 kb ds circular	Nucleus and cytoplasm	Rod, enveloped	Invertebrates—insects
Herpesviridae	125–300 kb ds linear, infectious	Nucleus	Icosahedral enveloped Density 1.22–1.28	Vertebrates—humans (human herpesvirus 1–8—oral and genital herpes, chicken pox, shingles, mononucleosis, roseola, Kaposi's sarcoma), and all other vertebrate types Invertebrates—mollusks

(continued)

TABLE 2.1 (continued)
Families of DNA Viruses of Animals

DNA Virus Family with Subfamilies	Genome Size and Structure	Replication Site	Virion Structure and Density[a]	Host and Important Pathogens
Iridoviridae Ranavirus[b] Lymphocystivirus[b] Megalocytivirus Chloriridovirus Iridovirus	140–383 kb ds linear, circularly permuted, noninfectious	Nucleus and cytoplasm	Icosahedral enveloped or naked Density 1.35–1.6	Vertebrates, fish, amphibians, and reptiles (epizootic hematopoietic necrosis virus, frog virus 3, red sea bream iridovirus) Invertebrates—insects
Polydnaviridae	150–250 kb multiple segments 2–31 kb each	Nucleus	Rod, enveloped	Invertebrates—insects
Poxviridae Cordopoxvirinae Entomopoxvirinae	130–375 kb ds linear, covalently closed, terminal inverted repeats, noninfectious	Cytoplasm	Rod, enveloped Density 1.30	Vertebrates—humans (variola—small pox, vaccinia, molluscum contagiosum virus), other mammals, birds, and reptiles Invertebrates—insects
Asfarviridae	170–190 kb ds linear, covalently closed, terminal inverted repeats	Cytoplasm	Icosahedral, enveloped Density 1.19–1.24	Mammals (African swine fever virus) Invertebrates[c]—ticks

Source: Van Regenmortel, M.H.V. et al. (Eds.), *Seventh Report of the International Committee on Taxonomy of Viruses,* Academic Press, San Diego, CA, 2000.
Notes: kb, kilobase; ss, single-stranded; ds, double-stranded; rt, reverse transcription.
[a] Boyant density in CsCl.
[b] Genome heavily methylated.
[c] Invertebrate host is vector for vertebrate host.

involves a rolling circle mechanism. The parvoviruses remain linear during replication and have terminal repeats that are self-complementary, and generate hairpin structures that allow self-priming. The genomes of both virus types are infectious but require the production of dsDNA before they are transcribed, and the Rep genes needed for helicase and genome replication are expressed. Both virus families generate nonenveloped virions that remain infectious when exposed to low pH (pH 3.0), alcohols, lipid solvents, detergents, and high temperatures (56°C for 1 h).

Hepadnaviruses are a cross between RNA viruses and DNA viruses and are not considered true DNA viruses by Baltimore classification because they do not replicate using a DNA-dependant DNA polymerase. The genomic DNA is transcribed into pregenomic RNA in the nucleus and this is packaged into capsids in the cytoplasm where the pregenomic RNA is reverse transcribed into genomic DNA. An envelope is acquired as virus egresses by budding through the cytoplasmic membrane. Purification involves isolation of cytoplasmic nucleocapsid protected DNA. The extract is then digested with DNase 1 [3] or micrococcal nuclease [4], and then the DNA is extracted by proteinase digestion and phenol extraction.

Small dsDNA viruses include polyomaviruses and papillomaviruses. These viruses replicate in nucleus and utilize the host's DNA replicative machinery. These viruses can influence the replicative stage of the host cell and thus may cause cancers. Papillomaviruses are especially important to human health, but difficulty in culturing these viruses in established cell lines has hampered research. These viruses are not enveloped. The genome consists of a single circular 5–8 kb dsDNA molecule that is amenable to many DNA purification methods developed for plasmids.

AdVs are medium-sized dsDNA viruses with linear genomes that replicate in the nucleus. The viruses are nonenveloped and develop crystalline arrays in the nucleus before being released by lysing the host cell. These viruses replicate using a viral-encoded DNA polymerase that uses a unique protein primer. The replicative form of the genome is linear as is the mature form. The relative ease in purifying the virus, its stability, and the infective nature of its genome make the AdVs useful as virus expression vectors.

Large dsDNA viruses that occur in insects but not in vertebrates include ascoviruses, baculoviruses, and polydnaviruses. All of these replicate in the genome and are enveloped. The polydnaviruses are unique viruses that exist as proviruses integrated into the genomes of parasitic wasps. They release viral particles containing DNA into the lepidopteran host that the wasp larvae parasitize but do not release infectious particles and are not culturable in cell cultures [5]. Ascoviruses also involve a parasitic wasp and its host but, unlike the polydnaviruses, ascoviruses replicate in both hosts and can be expressed in cell culture [6]. Baculoviruses are the most intensely studied of the three. Baculoviruses are important insect pathogens and are important pathogens of crustaceans. The insect baculoviruses have been used as biological pesticides to control economically devastating insect pests. These viruses are easily cultured and manipulated to improve their effectiveness for biological control. The most recognized application of baculoviruses is their widespread use for protein expression applications in a variety of research areas. The baculoviruses are rod-shaped viruses that undergo DNA replication in the nucleus. The mature form of the genome is circular and it is infectious when transfected into susceptible cells. Some of the assembled virions are released from the cell by budding and others are retained in the nucleus and are incorporated into a proteinaceous occlusion body. The occlusion body embedded viruses are resistant to environmental factors and provide protection for the virus to allow transmission to a new host insect by ingestion. The inclusion body gene is expressed at a high level and is therefore a gene that is often modified for gene expression work. Purifying the occlusion bodies is a useful method for purifying nonrecombinant baculoviruses especially in field applications [7].

Herpesviruses are the only known group of large DNA viruses of vertebrates that replicate solely in the nucleus. This group of viruses is widespread among vertebrates and is responsible for many important diseases. Herpesviruses have also been found in mollusks. These viruses have linear genomes that circularize in the infected cell. During replication, branching concatamers are formed and they are then packaged as linear molecules. After entering a cell, herpesviruses can establish latency or initiate lytic replication. Latent herpesviruses are maintained in the infected cell as a low copy-number circular plasmid. Herpesviruses can immortalize the cells that they are latent in. The virus can be induced to undergo replicative stage in these cells using chemical or physiological stimuli. Herpesviruses are the largest vertebrate viruses with infectious genomes. The ability to produce live progeny from purified virus has greatly facilitated recombinant herpesvirus production and marker rescue studies. In order to accomplish this, the virus genome must be intact and special precautions must be taken to prevent damaging the DNA during DNA purification. Additionally, the use of cloned genomes in bacterial artificial chromosomes (BACs) has proven very useful. The isolation of a herpesvirus BAC for transforming *Escherichia coli* requires obtaining circular DNA from the nucleus and not the linear form found in the virion.

The cytoplasmic replicating DNA viruses, Iridoviridae, Poxviridae, and Asfarviridae, are all large DNA viruses that encode many genes. Iridoviruses are pathogens in insects, fish, amphibians, and reptiles. The genomes of Iridoviridae are carried to the nucleus where they are initially transcribed, then they travel to the cytoplasm for replication. Replication involves a homologous recombination process that generates large concatamers and the DNA is packaged using a headfull mechanism that results in circular permutation of the genome sequences (each isolate has different sequence at the ends). The genomes of some iridoviruses are heavily methylated making them resistant to many restriction enzymes. The capsid is icosahedral and may bud from the infected cell and obtain an envelope. Nonenveloped viruses released by cell lysis are also infectious.

Poxviruses are pathogens of birds, mammals, reptiles, and insects. Asfarviridae are pathogens of domestic and wild pig species and soft ticks that are vectors of these viruses. Both families of viruses have large linear dsDNA genomes that replicate solely in the cytoplasm.

2.1.3 ECONOMIC IMPACT OF DNA VIRUSES

DNA viruses represent some of the most important pathogens for human and veterinary medicine. Examples of important diseases caused by these viruses are listed in Table 2.1. The ssDNA parvovirus B19 targets the human erythroid progenitor cell, and causes a variety of clinical illnesses in man including several hematological diseases, with symptoms ranging from transient aplastic crisis in a host with underlying hemolysis, hydrops fetalis in the midtrimester fetus to pure red cell aplasia in an immunocompromised patient. Due to the expression of the viral nonstructural protein, infection with B19 is cytotoxic. Although host humoral immunity often leads to the termination of the infection, commercially available immunoglobulin and recombinant capsids produced in a baculovirus system can be used to either treat persistent infection or prevent infection.

Hepatitis B virus (HBV) in the virus family Hepadnaviridae is one of the smallest enveloped animal viruses with a virion diameter of 42 nm. The genome of HBV is made of circular DNA, comprising a full-length strand of 3020–3320 nucleotides and a short-length strand of 1700–2800 nucleotides. HBV infects the liver and causes an inflammation called hepatitis in humans, which results in loss of appetite, nausea, vomiting, jaundice, and rarely, death in acute infection, and liver cirrhosis and liver cancer in chronic infection. With about 350 million people being chronic carriers, this virus poses a significant health problem worldwide [8].

Polyomavirus BK virus (BKV) in the Polyomaviridae family is a dsDNA virus with a 5 kb genome. The virus is transmitted via transplacental, kidney transplantation, and possibly fecal-oral routes, and can be detected in urine, blood, semen, genital tissues, and normal skin biopsies, as well as urban sewage. While the most common symptom associated with BKV infection is an upper respiratory infection, acute cystitis with or without hematuria is also reported. With the resolution of primary infection, the virus enters a latent phase and is found in urogenital sites (e.g., kidneys, urinary bladder, prostate, cervix, and vulva, as well as testes, prostate, seminiferous tubules, and semen) and hematolymphoid tissues (e.g., peripheral blood mononuclear cells, tonsils). BK infection in kidney transplant recipients does not show fever, malaise, myalgias, leukopenia, anemia, thrombocytopenia, or other symptoms or signs typical of viral infection. The human neurotropic polyomavirus, JC virus (JCV), is another nonenveloped icosahedral dsDNA in the family Polyomaviridae that causes a fatal demyelinating disease called progressive multifocal leukoencephalopathy in the central nervous system of mainly immunosuppressed patients. In addition, JCV has also been associated with human tumors of the brain and other organs [9].

Papillomaviruses in the family Papillomaviridae are nonenveloped dsDNA viruses with icosahedral symmetry and 72 capsomeres that surround the genome of approximately 8000 bp. Being infective to many vertebrate species, papillomaviruses mainly cause papillomas (benign epithelial growths), such as skin and genital warts. The human papillomaviruses (HPVs) consist of three categories: genital-mucosal types, nongenital cutaneous types, and types specific for epidermodysplasia verruciformis (which is a rare genetic skin condition characterized by widespread chronic nongenital HPV lesions). While most infections are asymptomatic, the lesions usually appear approximately 3 months after infection of genital tract cells. HPV is responsible for nearly 500,000 new cases of cervical cancer and 250,000 cervical cancer deaths worldwide each year. Indeed, oncogenic HPV types 16 and 18 cause up to 70% of HPV-related cancers, with invasive cervical carcinoma being the second commonest malignancy in women worldwide. Nononcogenic HPV types 6 and 11 cause around 90% of cases of genital warts as well as cause cutaneous lesions and respiratory papillomatosis [10].

AdVs in the family Adenoviridae are nonenveloped, lytic DNA viruses with a linear double-stranded genome and icosahedral symmetry. There are at least 51 human adenovirus (HAdVs) serotypes that have been described to date, which are grouped into six species (A–F) based on genome size, composition and organization, DNA homology, hemagglutinating properties, and oncogenicity in rodents. Whereas AdVs C, E, and some B species typically infect the respiratory tract, other B species the urinary tract; species A and F target the gastrointestinal tract and species D the eyes. Common symptoms of HAdVs infections include mild respiratory, gastrointestinal, urogenital, and ocular diseases. In infants, the HAdVs often cause viral gastroenteritis, and in immunocompetent individuals, they produce diverse clinical syndromes such as upper and lower respiratory tract disease, (kerato)conjunctivitis, gastroenteritis, and hemorrhagic cystitis, with hepatitis, myocarditis, meningoencephalitis, or nephritis being encountered occasionally. In immunocompetent individuals, AdV disease is mostly mild and self-limiting with few long-term consequences. One HAdV, Ad-36, has been associated with obesity in humans. It appears that Ad-36 is capable of turning on the enzymes of fat accumulation and recruitment of new adipocytes [11].

Human herpesviruses in the family Herpesviridae are large dsDNA viruses consisting of eight distinct members, namely, herpes simplex virus type 1 (HSV1), herpes simplex virus type 2 (HSV2), varicella-zoster virus (VZV), Epstein–Barr virus (EBV), cytomegalovirus (CMV), human herpesvirus 6 (HHV6), human herpesvirus 7 (HHV7), and human herpesvirus 8 (HHV8). These viruses cause a variety of disorders affecting skin, eyes, urogenital organs, central and peripheral neural systems, lymphatic system, and others. Besides being a major source of oral and sexually transmitted genital ulcers worldwide, HSV1 and HSV2 are also responsible for causing encephalitis, corneal blindness, and several disorders of the peripheral nervous system [12]. VZV, the pathogen of varicella (chickenpox), is associated with herpes zoster (shingles) and central nervous system complications such as myelitis and focal vasculopathies. EBV is associated with a large number of benign and malignant diseases (e.g., Burkitt lymphoma, Hodgkin lymphoma, nasopharyngeal carcinoma). CMV causes a common congenital infection worldwide that can lead to permanent disabilities. HHV6 and HHV7 infections occur in early childhood and cause short-febrile diseases, sometimes associated with cutaneous rash (exanthem subitum). HHV-8 causes kaposis sarcoma.

2.1.4 APPLICATION OF DNA VIRUSES IN MOLECULAR RESEARCH AND THERAPIES

Viruses provide simple systems that can be exploited to manipulate and investigate the functions of cells, contributing to our understanding of the basic mechanisms of DNA replication, transcription, RNA processing, translation, and protein transport. Viruses are naturally evolved vehicles that efficiently transfer their genes into host cells. Thus, they can be used as vectors for the delivery of therapeutic genes. In particular, members of the DNA virus family such as AdV, AAV, or herpesvirus are attractive candidates for efficient gene delivery. Among the currently developed virus vectors, the AAV vectors are regarded as ideal due to their lack of pathogenicity and toxicity, their ability to infect dividing and nondividing cells of various tissue origins, their remarkable genome stability, low levels of vector genome integration, and well-characterized virus biology. The first-generation AdV vectors contain deletion in the *E1* gene region (required for replication in permissive cells) and optionally in the *E3* gene region, with a transgene capacity of up to 8 kb. These vectors provide strong short-term transgene expression in immunocompetent animals with potential for use in genetic vaccination. The second-generation AdV vectors have deletions or inactivations in the *E2* and/or *E4* genes in addition to the *E1* deletion leading to reduced expression of viral genes. The third-generation AdV vectors harbor deletions in all viral coding sequences leading to the development of helper-dependent, high-capacity, or gutless AdV vectors, with a cargo capacity of up to 36 kb for nonviral DNA. As no viral gene expression exists in these vectors, their *in vivo* toxicity and immunogenicity are substantially reduced, permitting long-term transgene expression in liver, eye, and brain [13–15].

Baculoviruses contain covalently closed circles of dsDNA of between 80 and 180 kb and are pathogenic for insects predominantly of the orders Lepidoptera, Hymenoptera, and Diptera. Owing to their ease of cloning and propagation combined with the eukaryotic posttranslational modification machinery of the insect cell, recombinant baculoviruses have been utilized as vectors for expression of a large variety of foreign proteins in insect cell cultures. Furthermore, baculovirus surface display system has been developed using different strategies for presentation of foreign peptides and proteins on the surface of budded virions. These surface display strategies enhance the efficiency and specificity of viral binding and entry to mammalian cells. Additionally, baculovirus surface display vectors have been engineered to contain mammalian promoter elements designed for gene delivery both *in vitro* and *in vivo*. Together, these attributes make baculoviruses important tools in functional genomics [16].

Vaccinia virus is a large dsDNA virus in the Poxvirus family. It has also been engineered to express foreign genes for production of recombinant proteins and to serve as vaccine delivery system for heterologous antigens. The modified vaccinia virus Ankara (MVA) represents a vaccinia virus strain of choice for clinical investigation because of its avirulence and its deficiency to productively grow in human cells. In comparison to replication competent vaccinia viruses, nonreplicating MVA engenders similar levels of recombinant gene expression even in nonpermissive cells. MVA-based vaccines have been shown to protect against immunodeficiency viruses, influenza, parainfluenza, measles virus, flaviviruses, or plasmodium parasites in animal models. In recent years, the Western Reserve strain vaccinia has been explored as a replicating oncolytic virus for cancer virotherapy due to its exceptional ability to replicate in tumor cells. This strain contains deletion in the viral genes for thymidine kinase and vaccinia growth factor resulting in a vaccinia mutant with enhanced tumor targeting activity and fully retaining its efficiency of replication in cancer cells [17].

2.1.5 CURRENT TECHNIQUES FOR PURIFICATION OF VIRAL DNA FROM CULTURES (IN-HOUSE AND KITS)

Genomic purification schemes take advantage of the physical or spatial difference between virus genome and the host DNA. Many times the method of choice depends on the end use of the genomic DNA and the quantity needed (Table 2.2). Simple methods are generally adequate for small genomes and when fragmented DNA is acceptable. More involved protocols are generally needed if whole genomes from large DNA viruses are needed such as for the production of infectious herpesvirus DNA. Furthermore physical state of the genomic DNA may influence the extraction protocol. The mature

TABLE 2.2
Types of Virus DNA Extraction Protocol Based on Application

Use of Genome	Product Needed	Extraction Methods	DNA Source
Polymerase chain reaction (PCR), gene cloning, RFLP (4–6 bp cutters), small viruses	~50 kb	Rapid, general protocols	Mature form—cytoplasmic or extracellular
Infectious genome—marker rescue, recombinant virus production Whole genome size evaluation, PFGE	>50 kb (whole genome—linear or circular)	Gentle methods. Avoid mechanical shearing (vortexing excessive pipetting)	Mature form—cytoplasmic or extracellular
Virus BAC transfer into *E. coli*	>50 kb (whole genome—circular)	Gentle specialized methods. Avoid mechanical shearing (vortexing excessive pipetting) and nicking	Circular form required—for herpesviruses, this is an early replicative form found in the nucleus

form of the packaged genome of many viruses is linear, whereas the replicative form may be circular, concatameric, or branching. The isolation of a circular form is needed for the production of viable plasmids or BACs that can be replicated in bacteria.

2.2 METHODS

This chapter focuses on methods for isolating genomic DNA from viruses with DNA genomes produced in cell cultures. The purification schemes in general take advantage of the unique structural, packaging, or spatial characteristics of the viral genome when compared to the host DNA. However, specialized purification schemes may be necessary to address specific applications.

Reagents and solutions

20x Tris–EDTA (TE): 200 mM Tris-HCl pH 8.0, 20 mM ethylenediaminetetraacetic acid (EDTA) pH 8.0 (store at room temperature).
1x TE: 10 mM Tris-HCl pH 8.0, 1 mM EDTA pH 8.0 (store at room temperature).
Tris–Nacl–EDTA (TNE) buffer: 100 mM NaCl, 10 mM Tris-HCl pH 8.0, 1 mM EDTA pH 8.0 (store at room temperature).
Tris buffered saline (TBS): 100 mM Tris-HCl pH 8.0, 150 mM NaCl (store at room temperature).
10x DNase buffer: 400 mM Tris-HCl pH 8.0 (at 25°C), 100 mM $MgSO_4$, 10 mM $CaCl_2$ (store at room temperature).
Lysis buffer: 10 mM Tris-HCl pH 7.5, 1 mM EDTA pH 8.0, 1% Nonidet P-40 (store at room temperature).
Hirt solution: 10 mM Tris-HCl pH 7.5, 10 mM EDTA pH 8.0, 0.6% sodium dodecyl sulfate (SDS) (store at room temperature).
Proteinase K stock solution: 20 mg/mL in water (store at −20°C).
0.5 M EDTA pH 8.0 (store at room temperature).
5 M NaCl (store at room temperature).
Phenol:chloroform:isoamyl alcohol (25:24:1 ratio): This is made with phenol that has been equilibrated to a pH of 7.8–8.0 with 100 mM Tris buffer pH 8.0 and contains 0.1% hydroxyquinoline. The solution can be stored up to 1 month at 4°C under a layer of 100 mM Tris buffer.
Water saturated ether: Store in a small glass bottle over a layer of water in an explosion proof refrigerator.
TE saturated butanol: Store butanol in a small glass bottle over a layer of TE buffer.
Ethidium bromide: 10 mg/mL in water.
Cesium trifluoroacetate (CsTFA) solution: Dissolve 100 g cesium trifluoroacetate in 30 mL of TE buffer add buffer until the specific gravity is 2 g/mL (store at room temperature).
Chloroform.
Isopropanol.
100% ethanol.
70% ethanol.
Dialysis tubing or cassettes: Tubing should be boiled in 500 mL of 2% sodium bicarbonate and 1 mM EDTA, rinsed in distilled water, then boiled for 10 min in 1 mM EDTA, and then stored submerged in this solution at 4°C.

Equipment

High-speed centrifuge and rotor
Ultracentrifuge and swinging bucket rotor for concentrating virus
Vertical rotor if gradient purification used
Microfuge

Different DNA isolation techniques are used for different types of virus, different applications, and the structural form of DNA desired. Most common virus DNA isolation methods make use of

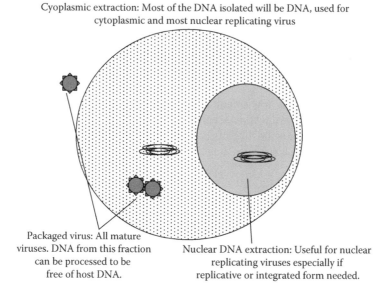

Cyoplasmic extraction: Most of the DNA isolated will be DNA, used for
cytoplasmic and most nuclear replicating virus

Packaged virus: All mature
viruses. DNA from this fraction
can be processed to be
free of host DNA.

Nuclear DNA extraction: Useful for nuclear
replicating viruses especially if
replicative or integrated form needed.

FIGURE 2.1 Virus DNA can be isolated from mature virions by the cytoplasm of the cell or the nucleus. The
choice of extract to use for DNA isolation depends on the end use, the virus type, and the purity of the product
(free from host DNA).

a characteristic of the virus DNA that is unique from that of the host genomic DNA (Figure 2.1).
The most simplistic assays make use of the unique packaging and transport of virus DNA out of
the infected cell in the virion. In these assays, the virion is concentrated or purified, then the DNA
is extracted. This general method is especially useful for assays that seek to correlate genome
copies with infectious units, in assays where exclusively mature form of viral genomic DNA is
desired or where extremely pure virus DNA is desired (exclusive of host genome or mitochondrial
DNA). However, higher yields are often obtained if protocols that purify viral genomes from
infected cells are used. The most common of these protocols extract viral DNA from the host cell
cytoplasm. These protocols are especially appropriate for viruses that undergo genome replication
in the cytoplasm. However, many viruses that undergo nuclear replication are highly cell associated and very high virus concentration can be found in the host cell cytoplasm. Small amounts of
host genomic and mitochondrial DNA may contaminate these preps. Protocols that separate virus
DNA from genomic DNA based on size are also relatively common. These protocols are often used
when a replicative (nonmature form) of a virus that replicates in the nucleus is desired. The actual
protocols used also depend on the size of the genome and end use of the virus DNA. PCR and gene
cloning can make use of relatively small fragments of DNA. Whereas the production of infectious
genomic DNA for marker rescue experiments and for generating viral recombinants require intact
whole genomes and are very sensitive to protocols that shear the genomic DNA. Furthermore,
protocols that involve transfer of BAC virus clones into bacteria require a circular form of the
genomic DNA.

2.2.1 Processing of the Infected Cell Culture

2.2.1.1 Virion Isolation

In general, if the objective is isolating viral DNA, virion samples can withstand substantial physical
abuse without fear of damaging the DNA. However, the infectivity of the virus may be compromised
especially in enveloped viruses.

1. The cells are infected with virus (generally a multiplicity of infection of about 0.1 plaque forming unit per cell) and allowed to reach maximum cytopathic effect (CPE).
2. If CPE causes loss of adherence,* dislodge remaining adhered cells using a scraper or by pipetting pellet cells at $300\,g$ for $10\,min$.
3. Centrifuge the cell culture supernatant at $10,000\,g$ for $10\,min$ to remove any debris. Then centrifuging the medium at $100,000\,g$ for $1\,h$ can be used to concentrate the virus.† Disperse pellet in DNase 1 buffer by sonication or vigorous pipetting.
4. Resuspend the cell pellet in a small amount of TBS ($1\,mL/150\,cm^2$ flask). Completely freeze and thaw this cell suspension three times by cycling the tube from ultralow freezer or dry ice to a $37°C$ water bath. Then centrifuge the medium at $10,000\,g$ for $10\,min$ to remove the nuclei and cellular debris. Supernatant can then be pooled and virus concentrated by centrifugation at $100,000\,g$.† Disperse pellet with extracellular virus concentrate in 1xDNase 1 buffer (Step 3).
5. The virion DNA can be easily separated from any contaminating DNA by pretreating the concentrated virion suspension with DNase 1. Add 250 units of DNase 1 per mL, incubate at $37°C$ for $1\,h$, and then add $50\,\mu L$ of $0.5\,M$ EDTA. The product can now be used for DNA extraction.

2.2.1.2 Cytoplasmic and Nuclear DNA Isolation

Many viruses undergo DNA replication in the cytoplasm. The extraction of DNA directly from the cytoplasmic fraction of infected cell will be predominantly virus DNA with some contaminating mitochondrial DNA, but often the yield will be higher than virion-based methods. The following method uses mild detergent to preferentially lyse the cytoplasmic membrane while leaving the nucleus intact. Alternatives include freeze thaw protocols similar to Step 4 above and a glycerol shock protocol [18]. Also this mild detergent lysis method can be used for isolating virions from the cytoplasm [3]. When working with released or nonpackaged DNA, you must avoid sonication, excessive pipetting, and excessive vortexing, especially if you are trying to obtain a product that is over $50\,kb$. In order to isolate the cytoplasmic fraction of infected cells, the infected cell flask should be washed with TBS, drained and lysed with lysis buffer ($4\,mL/150\,cm^2$ flask), incubated at $4°C$ for $10\,min$, and then centrifuged for $15\,min$ at $1500\,g$. Save the supernatant for cytoplasmic virus DNA extraction. The pellet will contain nuclei and can be used for DNA extraction for special applications (such as circular forms of herpesvirus genomes).

2.2.2 Small-Scale Extraction of Viral DNA

These methods can be used to extract small quantities of DNA for viruses with medium or small genomes (less than $50\,kb$) or for applications on large DNA viruses where whole genomes are not needed. There are several references on the use of rapid extraction methods [19,20].

---

* Alternatively, if cells remain adherent, they may be rinsed with TBS, overlayed with a small amount of TBS ($4\,mL/150\,cm^2$ flask), and frozen. The cells can then be dislodged by rocking the flask while it is thawing. Then this cell suspension can be collected in a tube for the two subsequent freeze thaw cycles.

† Lower centrifugation speeds ($30,000\,g$) can be used for large viruses such as herpesviruses and poxviruses. Virus can also be concentrated using polyethylene glycol (PEG) precipitations. This eliminates the need for ultracentrifuge. To precipitate virion particles, adding 1/10 volume of 10% NaCl, mix, add dry PEG (MW 6000) 10% w/v, and dissolve by gentle vortexing. The sample is then incubated on ice for $2\,h$ followed by centrifugation at $10,000\,g$ for $20\,min$. Discard the supernatant (remove as much as possible) and resuspend the pellet in DNase buffer.

2.2.2.1   Traditional Extraction

1. Transfer 400 μL aliquots of virus or cytoplasmic extract to each microfuge tube, add 40 μL (1/10 volume) of 10% SDS and 4 μL (1/100 volume) of proteinase K. Incubate at 56°C for 2 h.
2. Extract with phenol:chloroform. Add 400 μL (1 volume) of phenol:chloroform vortex at medium speed to mix. Centrifuge (maximum speed in a microfuge 15 s, 1600 g in low-speed centrifuge for 3 min) and transfer aqueous (clear) phase to a new tube.
3. Repeat phenol:chloroform extraction two more times.
4. Extract with 400 μL (1 volume) of chloroform.
5. Adjust the NaCl concentration to 0.2 M (if the starting solution is low salt, add 16 μL [1/25 volume] of 5 M NaCl). Add 800 μL (2 volumes) of cold ethanol. Incubate on ice for 1 h. Then centrifuge at 13,000 g at 4°C for 20 min.
6. Pour off the supernatant and add 500 μL of cold 70% ethanol. Centrifuge at 4°C for 2 min. Pipette off supernatant, and allow remaining fluid to evaporate at room temperature.
7. Add the desired volume of TE buffer, warm to 37°C to hydrate pellet for 1 h, and gently resuspend.

Modifications to obtain intact genomes of large viruses:

1. Use large bore pipette tips when transferring DNA containing solutions.
2. Draw solutions into the pipettes slowly.
3. Avoid high-speed vortexing, mix solutions by stirring or inverting the tube, or low-speed vortexing.
4. If possible, do not precipitate DNA—instead of Step 5 above, trace phenol and chloroform can be removed by extracting twice with 400 μL (1 volume) of water saturated ether (ether will be on top). Then the residual ether can be removed by heating the tube to 70°C for 10 min.

2.2.2.2   Commercial Kits

Commercial DNA extraction kits that are designed for PCR template preparations work well for small-scale extractions (1–10 μg). Because of relatively harsh handling, the use of these methods is restricted to downstream applications that can use DNA of less than 30 kb. One such kit that is widely used is the QIAamp Kit (Qiagen). Also rapid plasmid prep systems (QIAPrep, Qiagen) actually work very well for small viruses [19]. Both of these kits are limited by size of DNA that is extracted because they utilize silica columns. Rapid DNA extraction protocols that do not use columns are required to produce whole genomes from large DNA viruses. One such kit that provides larger product is the Gentra Puregene Kit (Qiagen). It uses a modified salting-out precipitation method to purify the DNA.

QIAPrep can be used on DNA viruses of less than 10 kb* and involves four sequential processes: (1) alkaline lysis of cells, (2) addition of a high salt buffer to neutralize the pH and precipitate the genomic DNA, (3) the low molecular weight virus DNA binds to the silica membrane of the column under high salt conditions, and (4) after washes with high salt buffer the DNA is eluted in Tris buffer. The method used by Ziegler et al [20] on 60 mm diameter plates of mammalian cell cultures infected with polyoma virus is outlined below. Their yield was 5–6 μg of DNA per plate of infected cells.

---

* The manufacturer indicates that DNA from 10 to 50 kb can be effectively purified using this method if the elution buffer is preheated to 70°C before use.

Supplies and Solutions for Qiagen Protocols (Sections 2.2.2.2 and 2.2.3.3):

Buffer C1: 1.28 M sucrose; 40 mM Tris·Cl pH 7.5; 20 mM MgCl$_2$; 4% Triton X-100 (store at 4°C)

Buffer G2: 800 mM guanidine HCl; 30 mM Tris·Cl pH 8.0; 30 mM EDTA, pH 8.0; 5% Tween-20; 0.5% Triton X-100 (store at room temperature)

Buffer QIAGEN equilibration (QBT): 750 mM NaCl; 50 mM 3-(N-morpholino)-propane sulfonic acid (MOPS), pH 7.0; 15% isopropanol; 0.15% Triton X-100 (store at room temperature)

Buffer QC: 1.0 M NaCl; 50 mM MOPS, pH 7.0; 15% isopropanol (store at room temperature)

Buffer QF: 1.25 M NaCl; 50 mM Tris·Cl, pH 8.5; 15% isopropanol (store at room temperature)

Buffer P1: 50 mM Tris·Cl, pH 8.0; 10 mM EDTA; 100 μg/mL RNase A (store at 4°C)

Buffer P2: 200 mM NaOH; 1% SDS (store at room temperature)

Buffer P3: 3.0 M CH$_3$COOK, pH 5.5 (store at room temperature)

Proteinase K solution: Cat. # 19131

QiaPrep Kit: Cat. # 27104

QIAGEN mini Genomic-tip 20/G: Cat. # 10223

QIAGEN midi Genomic-tip 100/G: Cat. # 10243

QIAGEN Genomic-tip 500/G: Cat. # 10262

1. Wash the plates with phosphate buffered saline (PBS).
2. Add 250 μL each of buffers P1 and P2 directly to the plates, incubate at room temperature for 5 min. Scrape the plates using a rubber policeman and gently transfer to a microcentrifuge tube.
3. Add 20 μL of proteinase K stock solution, mix gently, and incubate at 55°C for 1 h.
4. Add 350 μL of buffer N3 and immediately mix by inverting the tube five times.
5. Incubate on ice for 5 min to precipitate cellular DNA and cell debris.
6. Centrifuge at ~18,000 g for 10 min.
7. Transfer the supernatant onto Qiaprep spin column and centrifuge at ~18,000 g for 1 min.
8. Wash spin column by adding 0.5 mL buffer PB. Centrifuge for 1 min and discard the flow through.
9. Wash spin column with 0.75 mL buffer PB. Centrifuge for 1 min and discard the flow through.
10. Centrifuge for 1 min to remove residual wash buffer.
11. Transfer column to a clean microfuge tube, add 80 μL elution buffer (10 mM Tris-HCl pH 8.5) to the center of the column, incubate at 37°C for 5 min and centrifuge at ~18,000 g for 1 min. The flow through is the purified viral DNA.

2.2.3 MEDIUM- AND LARGE-SCALE PURIFICATION OF VIRAL DNA

2.2.3.1 Hirt Extraction

The chromosomal DNA can be separated from the lower molecular weight DNA by differential precipitation in the presence of 1 M NaCl [21]. This method is widely used for a variety of DNA viruses and is especially useful for viruses that are present in the highest concentrations in the nucleus. The basic protocol is as follows:

1. Rinse an infected monolayer of cells in a 75 cm^2 flask with 5 mL of cold PBS.
2. Lyse the cells by adding 700 μL of Hirt solution and incubate at room temperature for 10 min.
3. Add 350 μL of 5 M NaCl and rock the flask for 2 min.
4. Scrape the cell lysate from the plate with a rubber policeman, and transfer to a microcentrifuge tube and incubate for 18 h at 4°C.
5. Centrifuge at 18,000 g at 4°C for 40 min.
6. Transfer the supernatant to a new tube and add 20 μL of proteinase K and incubate at 56°C for 1 h.

7. Extract twice with phenol:chloroform:isoamyl alcohol and once with chloroform. Precipitate the viral DNA by adding two volumes of ethanol incubating for 1 h on ice, then centrifuge at 18,000 g at 4°C for 1 h. Wash the pellet with 70% ethanol and centrifuge for 5 min. Resuspend the DNA in TE buffer.

2.2.3.2 Cesium Trifluoroacetate Gradient Centrifugation

Isopycnic gradient centrifugation using high concentrations of cesium chloride or sodium iodide salts is relatively common and has been used for years for isolating DNA and separating the DNA species according to structure and density [22]. Recently, the use of cesium trifluoroacetate gradients has proven even better because this chemical solubilizes the viral proteins [23,24].

1. Concentrate virus from medium and cytoplasmic extract as described in Section 2.2.1.2.
2. Resuspend the pellet in 1 mL of TNE buffer and add 50 μL of 10% NLS and 5 μL of proteinase K solution.
3. Digest at 56°C for 1 h.
4. Mix with 2.8 mL of CsTFA (adjusted to a specific gravity of 2.0 g/cm^3).
5. Transfer to a quick seal tube for a Beckman NVT 65.2 rotor.
6. Fill with 20x TE buffer (1.2 mL) and add 10 μL of ethidium bromide.
7. Centrifuge at 15°C for 6 h at 316,000 g.
8. Illuminate with long-wave UV light and slowly draw off the predominant DNA band (if two bands are visible, this should be the lower band) with an 18 gauge needle.
9. Extract with 1 volume of TE saturated butanol and repeat until no color remains from the ethidium bromide.
10. Aliquot 400 μL per microfuge tube, add 20 μL (0.05 volume) of 10% lithium chloride and 0.8 mL (2 volumes) of ethanol, and centrifuge at ~18,000 g for 10 min at 4°C.*

2.2.3.3 Commercial Kits

The choice of commercial kits depends on the downstream application. For larger product sizes such as production of whole genomes from large DNA viruses for transfections, products designed for gentle extraction should be used, such as QIAGEN Genomic-tips. QIAGEN Genomic-tips are gravity-flow, anion-exchange tips that enable purification of DNA of up to 150 kb. They are available in three sizes according to DNA output: mini, midi, and maxi prep for 20, 90, and 400 μg maximum yields, respectively. For productive virus systems, expected yields would justify the use of mini columns for 75 cm^2 flask, midi columns for two 150 cm^2 flasks, and maxi columns for nine 150 cm^2 flasks.

Isolating DNA from Concentrated Virus (for a Mini Column)

1. Add 1 mL of buffer G2, and completely resuspend the viral pellet by vortexing for 10–30 s at maximum speed.
2. Add 25 μL of proteinase K stock solution, and incubate at 56°C for 60 min.
3. Pellet the particulate matter by centrifugation at 4°C for 10 min at 5000 g. Use the supernatant in the column (Step 3 below).

---

* Alternatively, for reduced physical damage to large products, dialyze against three changes of 500 mL TE buffer at 4°C, with at least 4 h before each buffer change.

Isolating DNA from Infected Cells (for a Mini Column)

1. Obtain cells from an infected culture (75 cm² flask) by washing the monolayer two times with 4 mL of cold PBS.
2. Scrape off the flask into 2 mL of PBS and rinse flask with 2 mL and add to the centrifuge tube.
3. Pellet cells at 300 g for 5 min, discard supernatant.
4. Resuspend cells in 2 mL of PBS.
5. Add 2 mL of ice-cold buffer C1 and 6 mL of ice-cold distilled water. Mix by inverting the tube several times. Incubate for 10 min on ice.
6. Centrifuge the lysed cells at 4°C for 15 min at 1300 g. Save the supernatant as the cytoplasmic fraction and the pellet as the nuclear fraction.
7. Completely resuspend the nuclear pellet in 4 mL of buffer P1. Add 4 mL buffer P2, mix gently but thoroughly by inverting four to six times, and incubate at room temperature for 5 min.
8. Add 4 mL chilled buffer P3, mix immediately but gently by inverting four to six times, and incubate on ice for 10 min.
9. Centrifuge at ≥20,000 g for 30 min at 4°C. The pellet will contain host cell DNA and integrated DNA, and the supernatant will contain nonintegrated virus DNA. Remove the supernatant quickly and use it on the column (Step 3 below).
10. To the cytoplasmic fraction, add RNase A to a final concentration of 10–20 μg/mL. Add 25 μL proteinase K solution. Incubate for 1 h at 50°C. Add NaCl to a final concentration of 750 mM and adjust the pH to 7.0. Load onto an equilibrated column (Step 3 below).

Isolate DNA from Virion Pellet (1), Nuclear Extract (2 g), or Cytoplasmic Extract (2 h) by Anion-Exchange Chromatography

1. Equilibrate a QIAGEN Genomic-tip 20/G with 1 mL, of buffer QBT, and allow the tip to empty by gravity flow. (Allow the tip to drain completely. The tip will not run dry. Do not force out the remaining buffer.)
2. Vortex the sample (from the last step of the specific sample preparation) for 10 s at maximum speed and apply it to the equilibrated tip. Allow it to enter the resin by gravity flow. Vortexing for 10 s causes a small amount of size reduction, but it helps with column flow rate. If the object is to get whole genomic DNA from large viruses, the sample should be centrifuged briefly to remove any particulates and immediately loaded on to the column.
3. Wash the tip three times each with 1 mL of buffer QC using gravity flow.
4. Elute the genomic DNA two times each with 1 mL of buffer QF prewarmed to 50°C.
5. Precipitate the DNA by adding 1.4 mL (0.7 volume) room temperature isopropanol to the eluted DNA, mix and centrifuge immediately at >5000 g for 15 min at 4°C. Carefully remove the supernatant.
6. Wash the DNA pellet with 1 mL of cold 70% ethanol. Vortex briefly and centrifuge at >5000 g for 10 min at 4°C. Carefully remove the supernatant without disturbing the pellet. Air-dry for 5–10 min, and resuspend the DNA in 0.1 mL of TE buffer. Dissolve the DNA overnight or at 55°C for 1–2 h.

2.3   FUTURE DEVELOPMENT TRENDS

The ever-increasing sophistication of molecular biology assays in combination with relatively simple, robust amplification methods is reducing the need for large-scale DNA purification protocols. The use of methods such as traditional cloning, Southern blots, and restriction fragment polymorphism analysis has largely given way to PCR, amplified fragment length polymorphism (AFLP), microarrays, tiling arrays, and direct sequencing from PCR products. The new generation high-throughput sequencers will allow whole genomes to be sequenced from less than 5 µg of input DNA. Furthermore, many viral genomes are conducive to PCR and *in-vitro* rolling circle amplification allowing the use of very small quantities of starting material. Often we use cell cultured virus as a way to amplify the genome to generate the amount of DNA that we need. As DNA requirements decrease, the scale of the cell cultures will likewise decrease and many of the applications will be replaced using DNA directly from field isolations. Current trends are on the use of high throughput, rapid, and automated DNA extraction methods to augment high-throughput data acquisition platforms and high-throughput and rapid diagnostic platforms. The DNA extraction systems will be increasingly geared toward automated simultaneous processing of large numbers of samples each representing small numbers of cells in small volumes from wells of multichambered plates. This shift in scale along with the requirement of reproducibility and efficiency provides new opportunities for the integration of new, less traditional separation technology into DNA purification [25,26].

Larry A. Hanson received his PhD in veterinary medical sciences from the Louisiana State University, Baton Rouge, Louisiana. His dissertation research involved developing DNA libraries and whole genome extraction to identify the thymidine kinase gene in Ictalurid herpesvirus 1 using marker rescue analysis. He has been on faculty at the College of Veterinary Medicine, Mississippi State University, Starkville, Mississippi for 18 years where he has been active in virus research and diagnostics. His focus has been on recombinant herpes virology and broad spectrum molecular diagnostic assays of fish pathogens. He has worked with several fish herpesviruses, ranaviruses, lymphocystis virus, adenoviruses, and poxviruses.

REFERENCES

1. Smit, A.F., The origin of interspersed repeats in the human genome, *Curr. Opin. Genet. Dev.*, 6: 743, 1996.
2. Van Regenmortel, M.H.V. et al. (Eds.), Virus taxonomy. *Seventh Report of the International Committee on Taxonomy of Viruses*, Academic Press, San Diego, CA, 2000.
3. Calvert, J. and Summers, J., Two regions of an avian hepadnavirus RNA pregenome are required in cis for encapsidation, *J. Virol.*, 68: 2083, 1994.
4. Habig, J.W. and Loeb, D.D., Small DNA hairpin negatively regulates in situ priming during duck hepatitis B virus reverse transcription, *J. Virol.*, 76: 980, 2002.
5. Rodriguez-Perez, M.A. and Beckage, N.E., Comparison of three methods of parasitoid polydnavirus genomic DNA isolation to facilitate polydnavirus genomic sequencing, *Arch. Insect Biochem. Physiol.*, 67: 202, 2008.
6. Asgari, S., Replication of *Heliothis virescens* ascovirus in insect cell lines, *Arch. Virol.*, 151: 1689, 2006.
7. Ebling, P.M. and Holmes, S.B., A refined method for the detection of baculovirus occlusion bodies in forest terrestrial and aquatic habitats, *Pest Manag. Sci.*, 58: 1216, 2002.
8. Lin, C.L. and Kao, J.H., Hepatitis B viral factors and clinical outcomes of chronic hepatitis B, *J. Biomed. Sci.*, 15: 137, 2008.
9. Kuber, W. et al., Progressive multifocal leukoencephalopathy: Value of diffusion-weighted and contrast-enhanced magnetic resonance imaging for diagnosis and treatment control, *Eur. J. Neurol.*, 13: 819, 2006.
10. Cutts, F.T. et al., Human papillomavirus and HPV vaccines: A review, *Bull. World Health Organ.*, 85: 719, 2007.
11. Atkinson, R.L., Viruses as an etiology of obesity, *Mayo Clin. Proc.*, 82: 1192, 2007.
12. Gupta, R., Warren, T., and Wald, A., Genital herpes, *Lancet*, 370: 2127, 2007.
13. Sakurai, F., Kawabata, K., and Mizuguchi, H., Adenovirus vectors composed of subgroup B adenoviruses, *Curr. Gene Ther.*, 7: 229, 2007.
14. Hartman, Z.C., Appledorn, D.M., and Amalfitano, A., Adenovirus vector induced innate immune responses: Impact upon efficacy and toxicity in gene therapy and vaccine applications, *Virus Res.*, 132: 1, 2008.
15. Kreppel, F. and Kochanek, S., Modification of adenovirus gene transfer vectors with synthetic polymers: A scientific review and technical guide, *Mol. Ther.*, 16: 16, 2008.

16. Hu, Y.C., Baculoviral vectors for gene delivery: A review, *Curr. Gene Ther.*, 8: 56, 2008.
17. Guo, Z.S. and Bartlett, D.L., Vaccinia as a vector for gene delivery, *Expert Opin. Biol. Ther.*, 4: 901, 2004.
18. Sarmiento, M. and Batterson, W.W., Glycerol shock treatment facilitates purification of herpes simplex virus, *J. Virol. Methods*, 36: 151, 1992.
19. Orlando, S.J., Nabavi, M., and Gharakhanian, E., Rapid small-scale isolation of SV40 virions and SV40 DNA, *J. Virol. Methods*, 90: 109, 2000.
20. Ziegler, K. et al., A rapid *in vitro* polyomavirus DNA replication assay, *J. Virol. Methods*, 122: 123, 2004.
21. Hirt, B., Selective extraction of polyoma DNA from infected mouse cell cultures, *J. Mol. Biol.*, 26: 365, 1967.
22. Walboomers, J.M.M. and Schegget, J.T., A new method for the isolation of herpes simplex virus type 2 DNA, *Virology*, 74: 256, 1976.
23. Christensen, L.S. and Normann, P., A rapid method for purification of herpesvirus DNA, *J. Virol. Methods*, 37: 99, 1992.
24. Ling, J.Y., Kienzle T.E., and Stroop, W.G., An improved rapid method for purification of herpes simplex virus DNA using cesium trifluoroacetate, *J. Virol. Methods*, 58: 193, 1996.
25. Chen, X. et al., Continuous flow microfluidic device for cell separation, cell lysis and DNA purification, *Analyt. Chim. Acta*, 584: 237, 2007.
26. Wolfe, K.A. et al., Toward a microchip-based solid-phase extraction method for isolation of nucleic acids, *Electrophoresis*, 23: 727, 2002.

3 Isolation of Viral RNA from Cultures

Ling Zhao and Zhen F. Fu

CONTENTS

3.1 INTRODUCTION

3.1.1 CLASSIFICATION OF RNA VIRUSES

Viruses are submicroscopic infectious agents that are usually made up of a nucleic acid core (DNA or RNA) with a protective protein coat called capsid. The capsid may appear helical or icosahedral (polyhedral or near spherical). Some viruses have more complex structures with tails or an envelope that is derived from the host cellular membrane. Based on the type of nucleic acids they contain, viruses are classified into two major groups: DNA viruses that possess deoxyribonucleic acid (Chapter 2), and RNA viruses that have ribonucleic acid as their genetic materials.

RNA viruses can further be classified in accordance with the strandedness of the nucleic acid (i.e., single-stranded RNA [ssRNA] or double-stranded RNA [dsRNA]); the sense or polarity of the RNA (i.e., negative-sense, positive-sense, or ambisense); and the mode of replication. The genomic RNA of positive-sense RNA viruses is identical to viral mRNA, which can be directly translated into proteins by the host ribosomes. The resultant proteins then direct the replication of the genomic RNA. The genomic RNA of the negative-sense RNA viruses is complementary to mRNA and it thus must be transcribed to positive-sense mRNA by an RNA-dependent RNA polymerase (RdRp). Therefore,

TABLE 3.1

Major RNA Virus Families

Family	Form (Copy)	Strand	Polarity	Size (kb)	Notable Pathogens
Picornaviridae	Linear (1)	Single	Positive	7	Polio virus; human enterovirus; hepatitis A virus; foot-and-mouth disease virus
Astroviridae	Linear (1)	Single	Positive	7–8	Astrovirus
Caliciviridae	Linear (1)	Single	Positive	8	Norovirus; sappovirus
Retroviridae	Linear (1)	Single	Positive	7–11	Human T-lymphotropic virus; HIV
Flaviviridae	Linear (1)	Single	Positive	10–12	Dengue virus; yellow fever virus; WNV; hepatitis C virus
Togaviridae	Linear (1)	Single	Positive	10–12	Rubella virus
Arteriviridae	Linear (1)	Single	Positive	13–16	Equine arteritis virus; porcine reproductive and respiratory syndrome virus
Coronaviridae	Linear (1)	Single	Positive	20–33	SARS virus
Arenaviridae	Linear (2)	Single	Negative	5–7	Lassa fever virus
Bornaviridae	Linear (1)	Single	Negative	9	Borna disease virus
Rhabdoviridae	Linear (1)	Single	Negative	11–15	Rabies virus
Orthomyxoviridae	Linear (8)	Single	Negative	12–15	Influenza virus
Paramyxoviridae	Linear (1)	Single	Negative	15–16	Mumps virus; measles virus; Newcastle disease virus; human parainfluenza virus
Filoviridae	Linear (1)	Single	Negative	19	Marburg virus; Ebola virus
Bunyaviridae	Linear (3)	Single	Negative or ambisense	10–23	California encephalitis virus
Birnaviridae	Linear (1)	Double	Both	6	Infectious bursal disease virus
Reoviridae	Linear (10–12)	Double	Both	18–30	BTV; rotavirus; Colorado tick fever virus

purified RNA of a positive-sense virus may be infectious when transfected into cells. On the other hand, purified RNA of a negative-sense virus is not infectious.

The ssRNA viruses consist of both positive-sense ssRNA and negative-sense ssRNA viruses. The ssRNA viruses encompass a large number of viral families, the most significant of which are listed in Table 3.1 while there are only two major families of the dsRNA viruses (*Birnaviridae* and *Reoviridae*) that are important human and animal pathogens. For example, rotavirus is the common cause of gastroenteritis in young children and animals and bluetongue virus (BTV) is an economically important pathogen for cattle and sheep (Table 3.1) [1].

In comparison with DNA viruses that have considerably lower mutation rates due to the proofreading ability of DNA polymerases within the host cell, RNA viruses generally have very high mutation rates, as they lack DNA polymerases that are vital for repairing damaged genetic material. Mutation rates of RNA viruses have been shown to be in the range of 10^{-3} to 10^{-5} substitutions per nucleotide copy. Indeed, there is evidence that RNA viruses replicate near at the error threshold, a minimal fidelity that is compatible with their genetic maintenance. At this level of mutation frequencies, most individual genomes of RNA viruses in a virus population will differ in one or more nucleotides from the average or consensus sequence of the population. Being acellular, viral populations do not grow through cell division, and they rely on the machinery and metabolism of a host cell to multiply.

3.1.2 REPLICATION OF RNA VIRUSES

In general, the infectious cycle of RNA viruses involves the following six basic stages:

1. *Attachment.* This occurs as a specific binding between viral capsid or surface proteins and specific receptors on the host cellular membrane. Indeed, the specificity of this binding

determines the host range and tissue tropism of a virus, as exemplified by the human immunodeficiency virus (HIV), which infects only human T cells or macrophages due to the interaction of its surface protein, gp120, with CD4 and coreceptors on the T cells and macrophages. Attachment of virus to the host receptor can lead to changes in the viral-surface protein resulting in fusion of viral and cellular membranes.

2. *Penetration.* Following attachment, viruses gain entry into the host cell through receptor-mediated endocytosis or membrane fusion.

3. *Uncoating.* After entry, the viral capsid is removed and degraded by viral or host enzymes to release the viral genomic nucleic acid.

4. *Replication.* This process involves transcription of viral mRNA, synthesis of viral proteins, and replication of viral genomic RNA.

5. *Assembly.* Viral proteins and viral genome are assembled into virion particles.

6. *Release.* Viruses are released from the host cell by lysis or budding (most of the enveloped viruses) [2].

3.1.3 ECONOMICAL IMPACT OF MAJOR RNA VIRUSES

RNA viruses are responsible for causing many severe diseases in humans and animals worldwide. In fact, over 70% of the known viral pathogens either have RNA as genetic material or replicate via an RNA intermediate. Clearly, the genetic flexibility of RNA viruses including their ability to undergo mutation, homologous and nonhomologous recombination, and segment reassortment may have contributed to the genetic variation and evolution of RNA viruses and their ubiquity. Thus, it is no surprise that a majority of recent emerging infections are caused by RNA viruses such as many of the hemorrhagic fever viruses, HIV, the SARS coronavirus, and avian influenza viruses.

Among the RNA viruses, HIV is a member of the genus *Lentivirus* in the family *Retroviridae* that can lead to acquired immunodeficiency syndrome (AIDS), a life-threatening opportunistic infection. It is estimated that AIDS has killed more than 25 million people since it was first recognized in 1981, which makes it one of the most destructive pandemics in recorded history [3]. HIV is transmitted via body fluids such as blood, semen, vaginal fluid, and breast milk. Following entry into the target cells (e.g., helper T cells, macrophages, and dendritic cells), HIV RNA genome converts to double-stranded DNA with the help of a virally encoded reverse transcriptase. This viral DNA then integrates into the cellular DNA using a virally encoded integrase and host cellular cofactors, leading to the transcription of the genome. As helper T cells, macrophages and dendritic cells are vital components in the human immune system, and their decimation by HIV renders the body susceptible to opportunistic infections and ultimately to death.

Coronaviruses are positive-sense ssRNA in the family *Coronaviridae* that can cause enteric or respiratory tract infection in a variety of animals including humans and birds. In 2003, a new coronavirus was identified as the possible cause of severe acute respiratory syndrome (SARS) in humans, which resulted in 8096 known infected cases and 774 deaths (with a mortality rate of 9.6%) between November 2002 and July 2003 [4]. Initial symptoms are flu-like (e.g., fever, myalgia, lethargy, gastrointestinal symptoms, cough, sore throat). Yellow fever results from infection of a positive ssRNA virus in the family *Flaviviridae*. This disease produces jaundice in some patients and is an important cause of hemorrhagic illness in many African and South American countries. The virus is transmitted to humans via mosquito saliva. After replicating locally, the virus is transported to the rest of the body via the lymphatic system and subsequently establishes itself in heart, kidneys, adrenal glands, liver, and in the blood [5]. The acute phase of yellow fever is normally characterized by fever, muscle pain (with prominent backache), headache, shivers, loss of appetite, and nausea or vomiting.

West Nile virus (WNV) is another viral member in the family *Flaviviridae* that is found in both tropical and temperate regions. Although the virus mainly infects birds, it is known to cause diseases in humans and other mammals through the bite of an infected mosquito. In the initial stage, WNV infection is asymptomatic, which is followed by a mild febrile syndrome (which is also called West Nile fever, showing fever, headache, chills, diaphoresis, weakness, lymphadenopathy, and drowsiness) and finally a neuroinvasive disease (known as West Nile meningitis or encephalitis). The encephalitis caused by WNV is characterized by similar early symptoms together with a decreased level of consciousness (sometimes approaching near-coma) and extrapyramidal disorders. Although WNV was first reported in Africa in the 1930s, the first epidemic occurred in the United States in 1999. From 1999 through 2001, 149 WNV infections were confirmed by CDC, including 18 deaths. In 2002, a total of 4156 cases were reported, including 284 fatalities and the cost of West Nile-related health care was estimated at $200 million [6].

Hepatitis C virus (HCV) of the family *Flaviviridae* causes a blood-borne infectious disease in the liver, leading to liver inflammation (hepatitis), cirrhosis (i.e., fibrotic scarring of the liver), and liver cancer. With approximately 3–4 million new cases of HCV infection each year, it is estimated that 150–200 million people worldwide are chronically infected and are at risk of developing liver cirrhosis and/or hepatocellular carcinoma, which is a significant health care burden globally [7]. In spite of the application of interferon-α as an antiviral control strategy, the morbidity and mortality rates associated with HCV are poised to go up in the coming years.

Dengue viruses of the family *Flaviviridae* cause acute febrile diseases in tropical and subtropical countries, with about 50–100 million individuals being infected, and as many as 500,000 people being admitted to hospital every year. Transmitted by mosquito, human infections with the viruses often display two clearly defined syndromes (dengue fever and dengue hemorrhagic fever/dengue shock syndrome). The disease is manifested by a sudden appearance of fever, severe headache, muscle and joint pains, as well as rashes. The dengue rash usually occurs on the lower limbs and the chest. There may also be gastritis, abdominal pain, nausea, and vomiting. The dengue hemorrhagic fever/dengue shock syndrome is characterized by rapid onset of capillary leakage accompanied by thrombocytopenia, altered hemostasis, and liver damage, which lead to increased aspartate aminotransferase and alanine aminotransferase. Fluids lost into tissue spaces when not replaced promptly can result in shock and gastrointestinal bleeding [8].

Among the negative-sense, ssRNA viral pathogens, Ebola virus belonging to the family *Filoviridae* causes Ebola hemorrhagic fever, which encompasses a range of symptoms including fever, vomiting, diarrhea, generalized pain or malaise, and sometimes internal and external bleeding. Deaths usually results from hypovolemic shock or organ failure. With mortality rates of 50%–89%, Ebola virus represents one of the most deadly epidemic viral pathogens [9]. Marburg virus is another virus in the family *Filoviridae* that is structurally similar to, but antigenically distinct from Ebola virus. The Marburg virus is spread through body fluids such as blood, excrement, saliva, and vomit. Following nonspecific early symptoms including fever, headache, and myalgia, a maculopapular rash often develops on the trunk. The later-stage disease displays jaundice, pancreatitis, weight loss, delirium and neuropsychiatric symptoms, hemorrhagic hypovolemic shock, and multiorgan dysfunction (in particular, liver failure), with fatality rates ranging from 23% to 90% [10].

Measles (also known as rubeola) is a disease caused by a negative-sense, ssRNA virus in the family *Paramyxoviridae*. Being transmitted via respiratory route, measles is highly contagious. The disease is characterized by fever and rash accompanied by secondary complications such as MV-induced immune suppression and the neurological disease postinfectious encephalomyelitis. Measles may have been responsible for killing about 200 million people worldwide during the past 150 years [11]. The isolation of measles virus in 1954 facilitated the development of the vaccines that have played a pivotal role in the reduction of the morbidity and mortality due to measles. Hendra virus and Nipah virus are also negative-sense ssRNA viruses in the family *Paramyxoviridae*, which have recently emerged as zoonotic pathogens capable of causing neurological and respiratory disease and death in domestic animals and humans. Fruit bats may act as the likely virus reservoir.

Rabies is a viral zoonotic neuroinvasive disease causing acute encephalitis (with a variety of neurological disorders) in mammals. This negative-sense ssRNA virus belongs to the genus *Lyssavirus* in the family *Rhabdoviridae*, which also includes Australian bat lyssavirus, European bat lyssaviruses 1 and 2 among others. In contrast to other genera of family *Rhabdoviridae*, all lyssaviruses are transmitted by direct contact, and not associated with transmission by insects in addition to being adapted to replicate in the mammalian central nervous system. Being one of the oldest viral diseases known to man, rabies is almost invariably fatal in nonvaccinated humans after neurological symptoms have developed, but prompt postexposure vaccination may prevent the disease from developing.

Influenza (or flu) is a viral infection of birds and mammals caused by negative-sense ssRNA viruses in the family *Orthomyxoviridae* (the influenza viruses). The viruses are transmitted through respiratory route, feces, and blood, as well as through droppings (birds). Besides common symptoms such as fever, sore throat, muscle pain, severe headache, coughing, and weakness, influenza in humans can also lead to pneumonia, which can be fatal, particularly in young children and the elderly.

Among the dsRNA viruses, rotavirus in the family *Reoviridae* is the leading cause of severe diarrhea in infants and young children. Being transmitted by the fecal-oral route, rotavirus infects cells lining the small intestine and produces an enterotoxin, which induces gastroenteritis with severe diarrhea and sometimes death through dehydration as common outcome. Rotavirus infection continues to have a major global impact on childhood morbidity and mortality. In addition, rotavirus also infects animals and is an important pathogen of livestock.

The BTV is another dsRNA virus in the family *Reoviridae* which causes severe disease in livestock (sheep, goat, cattle, and deer). Being transmitted by the species of *Culicoides*, BTV induces an acute disease in sheep with high morbidity and mortality. Major signs of bluetongue disease include high fever, excessive salivation, swelling of the face and tongue, and cyanosis of the tongue, which gives the tongue its typical blue appearance in a minority of the animals. BTV has been isolated in many tropical, subtropical, and temperate zones.

3.1.4 BASIC PRINCIPLES OF VIRAL RNA ISOLATION

The molecular study of viruses requires the purification of viral nucleic acids such as genomic DNA or genomic RNA or messenger RNA (mRNA) [12]. Below, we will describe various RNA extraction techniques for the isolation of total viral RNA, viral mRNA from infected cells, and viral genomic RNA from purified virus.

All RNA isolation methods include three steps: (1) inactivation of nucleases, (2) dissociation of RNA from proteins, and (3) separation of the RNA from other macromolecules. The first two steps should be immediate and concurrent so that degradation of RNA is minimal [13].

One of the most important concerns in the isolation of RNA is to prevent RNA degradation during the isolation process. Given the wide distribution of nucleases and the potential damage that they can inflict on RNA, RNA breakdown is a universal feature of all cells during RNA processing and metabolism.

Ribonucleases (RNases) are stable and functional even at 95°C and cause the rapid degradation of RNA molecules. It is thus important to ensure that nuclease activity is totally inhibited during the isolation process. A few golden rules can be given and it is important that they are followed if the isolation procedure is to be successful. First, it is important to ensure that all the reagents used are free of nucleases. Solutions used in the initial extraction are often supplemented with nuclease inhibitors. Glassware should be oven baked at 200°C for 4–12 h, whereas plastic is usually autoclaved. Skin is a well-known source of nucleases and so it is essential to wear disposable plastic or latex gloves during all manipulations. Second, a pH range of 6.8–7.2 for solutions is normally a necessity to ensure that no RNA degradation occurs, since RNA is degraded at an alkaline pH as low as 9 [14].

The initial stage for RNA extraction involves the lysis of the cells or virions. The lysis buffers typically fall into one of two categories. The first category consists of harsh chaotropic agents, including guanidinium salts, sodium dodecyl sulfate (SDS), phenol, and chloroform, which disrupt the plasma membrane and subcellular organelles and simultaneously inactivate RNase. Another category of lysis buffer gently solubilizes the plasma membrane while maintaining nuclear integrity, such as hypotonic nonidet P-40 (NP-40) lysis buffers. Nuclei, other organelles, and debris are then moved from the lysate by differential centrifugation, thus making this procedure useful for isolation of RNA from the cytoplasm. The success of this approach is almost entirely dependent on the presence of nuclease inhibitors in the lysis buffer [14]. It is important to avoid procedures that may allow or cause degradation of the RNA.

Treatment of viral RNA with bentonite was found to protect, stabilize, and potentiate the infectivity of TMV-RNA. A modified procedure for the isolation of the RNA by splitting the virus in the presence of bentonite has therefore been developed. This method relies on the use of bentonite as an adsorbent for nucleases during the phenol degradation of the virus [15]. RNA lysis buffers that contain guanidinium thiocyanate or guanidinium-HCl reproducibly yield the highest RNA quality because of the extremely chaotropic nature these chemicals exhibit; they are among the most effective protein denaturants [16]. Since then, guanidinium extraction has become the method of choice for RNA purification, replacing phenol extraction. Although RNA can be precipitated directly from a guanidinium-containing lysis buffer, ultracentrifugation through CsCl is recommended when the concentration of RNA is low, which makes the whole procedure labor intensive. Chomczynski and Sacchi [17] described isolation and purification of RNA with guanidinium thiocyanate containing lysis buffers, but without the need for subsequent CsCl ultracentrifugation of the samples, allowing simultaneous processing of a large number of samples within 4h. This method provides the basis for numerous commercial reagents used to isolate total RNA from various tissues and cells [18].

There are many protocols available for extraction of RNA. However, many of them are labor-intensive or time-consuming. The optimal protocol should provide high sensitivity for extraction of RNA combined with low time consumption, low price, and the potential for automation. By taking advantage of new techniques developed recently, many commercial kits are now available. Many procedures use a precipitation step to improve purity of the nucleic acids. Other protocols take advantage of the nucleic acid binding potential of matrix materials supplied in a column. A recent development is the coverage of magnetic beads with nucleic acid binding matrices, which provides a high potential for automation [19].

3.2 ISOLATION OF VIRAL TOTAL RNA

Some viruses strongly inhibit cellular transcription, allowing synthesis of large quantities of viral mRNAs. In order to isolate viral mRNA, the first step is to extract the total RNA from cells infected by the viruses. Then viral mRNA can be separated by specific affinity methods, such as oligo(dT) cellulose. There are many methods for the extraction of total RNAs from cultured cells. Also many commercial kits are available, making this process easier. Some of the methods will be described in following text.

3.2.1 PHENOL EXTRACTION METHOD

Principle. This procedure was one of the first published methods to prepare nucleic acids free of contaminating proteins. The basis of the method is to suspend the cells in a solution containing a detergent that lyses the cells and dissociates nucleoprotein complexes. It is based on the ability of organic molecules to denature and precipitate proteins in solution while not affecting the solubility of the nucleic acids. Phenol chloroform extraction is carried out by the addition of an equal volume of buffer-saturated phenol–chloroform–isoamyl alcohol (25:24:1) to the sample, vigorous mixing to form an emulsion, and separation of the denser phenol from the aqueous layer by centrifugation.

Reagents.

1. Phenol–chloroform–isoamyl alcohol (25:24:1): It is prepared by equilibrating the pH of the acidic phenol–chloroform–isoamyl alcohol solution with an autoclaved buffer of the

required pH (e.g., 0.1 M Tris buffer pH 7.0). An equal volume of buffer is mixed into the phenol–chloroform–isoamyl alcohol by vigorous shaking. The mixture is allowed to settle and the aqueous buffer is removed. Equilibration with buffer is repeated until the required pH is reached.

2. Sucrose, SDS, 3 M sodium acetate pH 5.2, protein K, ethanol, isopropanol, and diethylpyrocarbonate (DEPC)-treated water.

Procedure. The following steps describe RNA isolation from cells infected with virus using phenol–chloroform [13]:

1. Cells infected with viruses are prepared in various ways according to the types of culture:
   a. Adherent cells: Wash the cells twice with ice-cold PBS and trypsinize the cells with trypsin/EDTA. Pellet the cells by centrifugation at $300\,g$ for 5 min. Aspirate and discard the supernatant and remove all traces of liquid. The cell pellet can be stored at $-70°C$ for later use.
   b. Suspension cells: Cells grown in suspension are pelleted by centrifugation at $300\,g$ for 5 min. Aspirate and discard the supernatant. The cell pellet can be stored at $-70°C$ for later use.
2. Cells are lysed with lysis buffer (0.15 M sucrose, 10 mM sodium acetate, 1% (w/v) SDS). RNase is inactivated by SDS.
3. Proteins are digested over a period of 15 min at room temperature with 0.1 mg/mL proteinase K to remove any potential RNase activity.
4. Proteins are then removed by phenol–chloroform extraction. An equal volume of buffer-saturated phenol–chloroform–isoamyl alcohol (25:24:1) is added and mixed well. The mixtures are shaken either by hand or in orbital incubators for 5–10 min to ensure the formation of a fine emulsion. When the emulsion is centrifuged ($10,000\,g$ for 10 min, 4°C), the phenol–chloroform separates out as a dense bottom layer with the proteins either dissolved in the phenol–chloroform layer or precipitated at the interface, leaving the RNA in the top aqueous layer. Isoamyl alcohol can help to produce a well-defined interface, making recovery of the supernatant easier.
5. Repeat above procedure with equal volumes of buffer-saturated phenol–chloroform–isoamyl alcohol (25:24:1) until there is no protein at the interface after centrifugation. Usually three extractions are sufficient to yield a clear interface.
6. RNA in the aqueous phase is precipitated by addition of sodium acetate pH 5.2 to a final concentration of 0.3 M, followed by either 2 volumes of ethanol or 1 volume of isopropanol. Precipitation is carried out at $-20°C$ for at least 1 h.
7. RNA is pelleted by centrifugation at $10,000\,g$ for 20 min at 4°C. A higher centrifugal force (e.g., $60,000\,g$) is recommended to ensure a high recovery of RNA if only submicrogram amount of RNA is being precipitated.
8. After centrifugation, the aqueous fraction is removed by aspiration and the pellet is washed with ice-cold 70% (v/v) ethanol to remove excess salt. The RNA pellet is then redissolved in proper volume DEPC-treated ddH$_2$O and transferred to a sterile 0.5 mL microcentrifuge tube.
9. The redissolved RNA is then mixed with an equal volume of phenol–chloroform–isoamyl alcohol to remove any residual protein. Centrifugation for 3–5 min at 10,000–14,000 g at room temperature separates the phenol and aqueous layers.
10. To remove traces of phenol, the aqueous layer is recovered and extracted in an equal volume of chloroform–isoamyl alcohol at room temperature.
11. The RNA is precipitated with 2 volume ethanol and 1/10 volume 3 M sodium acetate pH 5.2 for at least 2 h at $-20°C$ before centrifugation at 4°C at 10,000–14,000 g for 10–20 min.
12. Dissolve the RNA pellet in DEPC-treated water or in DEPC-treated 0.5% SDS, depending on the subsequent use of RNA. Store the samples at $-70°C$ or as an ethanol suspension at $-20°C$.

Commentary. This basic phenol extraction procedure has been modified by the inclusion of additives such as 8-hydroxyquinoline to phenol–chloroform–isoamyl alcohol to enhance RNA extraction. A serious problem with phenol extraction is that the RNA can be contaminated with both DNA and polysaccharides as both contaminants can be present in the aqueous layer after phenol extraction. This can be a serious problem because DNA and polysaccharides can also be precipitated along with the RNA by ethanol or isopropanol. Depending on the subsequent use of the RNA, this may affect future analyses. DNA and polysaccharides can be removed from the RNA preparation by enzymatic degradation either by including a digestion step before or after the last phenol–chloroform–isoamyl alcohol extraction. DNA can be removed by addition of DNase I to a final concentration of $2\,\mu g/mL$ and incubated at $37°C$ for 1 h. In addition, phenol is a suspected carcinogen and is hazardous and additional safety measures are required for handling and disposing phenol. It is also a labor-intensive method and has been largely superseded by the acidic guanidinium thiocyanate method.

3.2.2 ACID GUANIDINIUM–PHENOL–CHLOROFORM METHOD

Principle. Guanidinium thiocyanate and chloride are among the most effective protein denaturants. As a strong inhibitor of RNases, guanidinium chloride was first introduced for isolation of RNA by Cox [16]. Since then guanidinium extraction has become the method of choice for RNA purification. Initially, ultracentrifugation of a guanidinium thiocyanate lysate through a CsCl cushion was required for isolation of RNA [20]. In this method, the infected cells are homogenized in a solution of 4 M guanidinium thiocyanate, 30 mM sodium acetate, and 0.14 M 2-mercaptoethanol. The cell homogenate is layered over ½ volume of 5.7 M CsCl and centrifuged in a swing-out rotor at $250,000\,g$ for 3 h at $22°C$. After centrifugation, the DNA, polysaccharides, and proteins remain above the density barrier while the RNA is pelleted at the bottom. This step was eliminated by using an acidic guanidinium thiocyanate–phenol–chloroform mixture for RNA extraction [17], allowing simultaneous processing of a large number of samples. It provides a pure preparation of RNA in high yield and can be completed within 4 h. This single-step method has become widely used for isolating total RNA from biological samples of different sources. The principle basis of the method is that RNA is separated from DNA after extraction with an acidic solution containing guanidinium thiocyanate, sodium acetate, phenol, and chloroform, followed by centrifugation. Under acidic conditions, total RNA remains in the upper aqueous phase, while most of DNA and proteins remain either in the interphase or in the lower organic phase. Total RNA is then recovered by precipitation with isopropanol. It has become the basis for numerous commercial reagents claiming to isolate total RNA from various tissues and cells using a single solution consisting of phenol, guanidinium thiocyanate, buffer, and detergents [21,22].

Reagents.

1. The denaturing solution (solution D: 4 M guanidinium thiocyanate, 25 mM sodium citrate pH 7; 0.5% sarcosyl, 0.1 M 2-mercaptoethanol) is prepared as following: Dissolve 250 g of guanidinium thiocyanate in 293 mL distilled water; 17.6 mL 0.75 M sodium citrate pH 7.0; and 26.4 mL 10% sarcosyl at $60°C–65°C$ by stirring. Then add 4.1 mL 2-mercaptoethanol. This solution can be stored at room temperature for up to 1 month.
2. Phenol, isopropanol, 75% ethanol, 2 M sodium acetate, chloroform:isoamyl alcohol (49:1) and DEPC-treated water.

Procedure. The following steps are needed for RNA isolation using the acid guanidinium–phenol–chloroform (AGPC) method [17]:

1. Cells (either adherent or suspension) infected with viruses are prepared as described above.
2. The cell pellet is lysed with 1 mL of solution D per $5–10 \times 10^6$ cells and then transferred to a 4-mL polypropylene tube.
3. To this solution, 0.1 mL 2 M sodium acetate pH 4, 1 mL phenol (water saturated), and 0.1 mL chloroform–isoamyl alcohol mixture (49:1) is added. The solution is mixed thoroughly by inversion after the addition of each reagent.
4. The final suspension is shaken vigorously for 10 s and cooled on ice for 15 min before centrifugation at 10,000 g for 20 min at 4°C. After centrifugation, RNA is present in the aqueous phase whereas DNA and protein are present in the interphase and phenol phase.
5. The aqueous phase is transferred to a fresh tube, mixed with 1 mL isopropanol, and placed at −20°C for at least 2 h to precipitate RNA.
6. After centrifugation at 10,000 g for 20 min, the resulting RNA pellet is dissolved in 0.3 mL of solution D, transferred into a 1.5 mL Eppendorf tube, and precipitated with 1 volume of isopropanol at −20°C for 1 h.
7. The mixture is centrifuged at 10,000 g for 10 min at 4°C. After discarding the supernatant, the RNA pellet is resuspended in 75% ethanol, votex-mixed, and kept at room temperature for 10–15 min to dissolve residual guanidinium salt.
8. The RNA is pelleted again by centrifugation at 10,000 g for 5 min and the supernatant is discarded.
9. After briefly drying the RNA, it is dissolved in DEPC-treated water or in DEPC-treated 0.5% SDS, depending on the subsequent use of RNA. The samples are stored at −70°C or as an ethanol suspension at −20°C.

Notes

1. Cells should not be washed with PBS.
2. Cells can be stored in denaturing solution at −70 °C for later use.
3. The RNA precipitate forms a white pellet at the bottom of the tube. In some cases, RNA might precipitate along the sides of the tube. Thus, care should be taken not to lose the samples while washing the precipitate.
4. Do not let the pellet dry completely as it will decrease its solubility; however, as much ethanol as possible should be removed without completely drying the pellet.
5. Incubation of 10–15 min at 55°C–60°C may be required to dissolve the RNA pellet.
6. RNA concentration can be determined by measuring the absorbance at 260 nm (1 O.D. = 40 μg of RNA/mL). The quality of the RNA can be checked by measuring absorbance ratio at A260 nm/A280 nm, which should be 1.8–2.0.

Commentary. This single-step method yields the entire spectrum of RNA molecules from cell cultures. The yield of total RNA from cultured cells is in the range of 50–80 μg (fibroblasts, lymphocytes) or 100–120 μg (epithelia cells) per 10^7 cells. The use of phenol is the single disadvantage of this method since phenol is a toxic chemical. A number of manufacturers have improved the original single-step guanidinium–acid–phenol method and marketed products that are simple and that have several procedural advantages. These products are available as a single reagent consisting of phenol, guanidinium salts, buffers, detergents, and stabilizers with stability of 9–12 months at 4°C. TRIzol Reagent from Invitrogen is such a product and is described below.

3.2.3 TRIzol Method

Principle. TRIzol Reagent (Invitrogen), a mono-phasic solution of phenol and guanidine isothiocyanate, is a ready-to-use reagent for the isolation of total RNA from cells and tissues. This reagent is an improvement to the single-step RNA isolation method developed by Chomczynski and Sacchi [17].

During sample homogenization or lysis, TRIzol Reagent maintains RNA integrity, while disrupting cells and dissolving cell components. Addition of chloroform followed by centrifugation separates the solution into an aqueous phase and an organic phase. RNA remains exclusively in the aqueous phase. After transfer of the aqueous phase, RNA is recovered by precipitation with isopropanol or ethanol.

Reagents.   Chloroform, isopropanol, 75% ethanol, and DEPC-treated water.

Procedure.   The following steps describe RNA isolation using TRIzol Reagent [23].

1. Cells (either adherent or suspension) infected with viruses are prepared as described above.
2. The cell pellet is lysed in 1 mL TRIzol Reagent per $5–10 \times 10^6$ cells by pipetting several times. The cell lysate is transferred immediately to a sterile polypropylene tube and incubated for 5 min at room temperature to allow complete dissociation of nucleoprotein complexes.
3. Chloroform (0.2 mL) is added per 1 mL of RNA extraction reagent. After vigorous shaking (or votex-mixing) for 15 s and incubation on ice (or at 4°C) for 5 min, the homogenate is centrifuged at 12,000 g (4°C) for 15 min.
4. The aqueous phase containing RNA is carefully transferred to a sterile microfuge tube. An equal volume of isopropanol is added to the RNA and mixed completely. After incubation at 4°C for 10 min, the sample is centrifuged at 12,000 g (4°C) for 10 min.
5. After the supernatant is removed, the RNA pellet is washed twice with 75% ethanol and centrifuged again for 5 min at 7500 g (4°C).
6. After the pellet is air dried for 5–10 min, RNA is dissolved in DEPC-treated water or in DEPC-treated 0.5% SDS, depending on the subsequent use of RNA and stored at −70°C or as an ethanol suspension at −20°C.

Notes

1. Cells should not be washed with PBS.
2. Cells can be stored in TRIzol Reagent at −70°C for later use.
3. It is preferable to wash the aqueous phase twice with 0.2 volume of chloroform to completely eliminate proteins.
4. The volume of the aqueous phase should be 40%–50% of the total volume of the homogenate plus chloroform. Leave 20% of the aqueous phase over the interphase and avoid disturbing or touching the interphase.

Commentary.   The simplicity of this RNA isolation method makes it possible to process a large number of samples simultaneously, and the excellent recovery of RNA using this method permits the isolation of RNA from very small volume of biological samples. This method is also applicable and is widely used for RNA isolation from serum or other biological fluid samples.

3.2.4   Isolation of Viral RNA from Small Numbers of Cells

Isolation of viral RNA from a small population of cells has become possible by integration of standard techniques with affinity procedures. Many commercial kits are now available and the methodologies of these kits vary. Some of them take advantage of the nucleic acid binding potential of matrix materials supplied in a column. Others make use of magnetic beads with nucleic acid binding matrices. After extraction of cellular and viral RNAs, mRNA can be separated with mRNA selection kits. With some kits, such as Poly(A) Purist mRNA Isolation Kit from Ambion, cellular and viral mRNA can even be separated directly from tissue or cell lysates without the extraction of total RNA. Some of the kits are listed in Table 3.2 together with the suppliers, advantages, and disadvantages.

TABLE 3.2
Microscale RNA Isolation Kits

Kit	Supplier	Advantages	Disadvantages
RNeasy Mini Kit	Qiagen	Entire process takes less than 30 min This kit has an upgraded counterpart in 96-well format for high-throughput analysis	High potential for filter clogging
ChargeSwitch Total RNA Cell Kit	Invitrogen	Rapid and efficient purification of total RNA in less than 15 min Minimal genomic DNA contamination	
High Pure RNA Isolation Kit	Roche	Protocol can be carried out at the bench and at room temperature	Require complicated buffers
		Free from genomic DNA contamination	Require careful pipetting to remove the RNA solution from the silica beads
Absolutely RNA Miniprep Kit	Stratagene	Ideal for postenzymatic RNA cleanup	Potential for filter clogging
Total RNA Purification Kit	Norgen Biotek Corporation	Efficient for total RNA isolation Rapid spin-column or 96-well high-throughput procedure	Potential for filter clogging
Poly(A) Quick mRNA Isolation Kit	Stratagene	High-quality mRNA can be obtained in 30 min Columns can accommodate up to 500 mg of total RNA Decreased RNase contamination	Low yield of mRNA if the total RNA sample is extremely small
Poly(A) Purist mRNA Isolation Kit	Ambion	Isolates mRNA from total RNA with the highest possible yield	
Poly(A) Pure Kits	Ambion	Isolates mRNA directly from tissue or cell lysates Saves time and reduces the potential for sample loss and contamination	Yield and level of purification is somewhat lower

3.2.5 Isolation of Viral Genomic RNA from Cell Culture

Principle.   In addition to isolation and purification of viral total RNA or viral mRNA, viral genomic RNA can be purified for those viruses whose genome is RNA. The first step for isolation of viral genomic RNA is to purify the virions. Ultracentrifugation is the standard method for virus purification. Two methods of centrifugation are available: differential centrifugation and density gradient centrifugation. Differential centrifugation is the simpler form of centrifugation. According to the choice of centrifugation parameter (gravitation force and time), it is possible to separate virus particles from contaminating cellular debris by pelleting the contaminants and leaving the virus in suspension (20,000 g, 15 min) or directly pelleting the virus by centrifugation at higher gravitation force for a longer period of time (100,000 g, 60 min). However, in some cases centrifugation at elevated centrifugal forces is deleterious to viruses, especially enveloped viruses. Instead of pelleting, the virions can be sedimented on a cushion of higher density material such as cesium chloride, sucrose, or potassium tartrate. If a viral suspension is layered carefully over a preformed discontinuous gradient and then centrifuged, the virus particles will migrate according to their sizes until they reach an equilibrium corresponding to their buoyant density [12]. According to the physical characteristics of the viruses, such as size, coefficient of sedimentation, and density, there are different centrifugation

TABLE 3.3
Sucrose Solutions in SB20

Percentage Sucrose (w/v)	Density	Refractive Index
20	1.08	1.3639
55	1.26	1.4307
65	1.32	1.4532

parameters for purification of different viruses. Described below is an example procedure used for the purification of Rous sarcoma virus (RSV). The purified virions can be used for extraction of viral genomic RNA by one of the methods described above.

Reagents.

1. Standard buffer (SB): 0.001 M EDTA, 0.01 M Tris, adjust pH to 7.4, store at 4°C.
2. Saturated ammonium sulfate solution: This solution is made by dissolving 531 g of $(NH_4)_2SO_4$, 2 mL 0.5% (w/v) phenol red in up to 1 L with ddH_2O followed by neutralizing the pH with ammonia, and stored at room temperature.
3. Sucrose solutions in SB20 (Table 3.3).
4. Beckman centrifuge (the data come from Beckman product manual) (Table 3.4).

Procedure.   The following steps describe RSV purification with centrifugation [24].

1. Spin down virus-infected cells and debris for 10 min at 8000 rpm (8000 g) in an SS-34 rotor at 4°C and collect the supernatant.
2. Concentrate the supernatant if necessary. Slowly add an equal volume of saturated ammonium sulfate to the supernatant with stirring. Incubate the sample on ice for 15 min. Then centrifuge for 10 min in an SS-34 rotor at 16,000 rpm at 4°C. Discard the supernatant and remove any remaining drops.
3. Resuspend the virus pellet in SB to 5% of the starting volume. If the suspension is not clear, add more SB or reclarify by repeating the above steps.
4. Prepare a sucrose gradient that consists of a cushion of 65% sucrose in SB, a layer of 20% sucrose in SB, and the virus suspension. The solutions can be overlayed successively, starting with 65% cushion, or underlying successively, starting with the virus suspension.
5. Sediment the virus using the appropriate speeds and times according to the rotor. However, the longer the centrifugation time, the greater the contamination by slower-sedimenting particles in the virus suspension.
6. Collect the interface between 20% and 65% sucrose solutions.

TABLE 3.4
Beckman Centrifuge

Rotor	Tube	Volume (mL)	65% Sucrose (mL)	20% Sucrose (mL)	Virus Suspension (mL)	rpm	Time (min)
SW50.1	0.5" × 2"	5	0.5	2	2	50,000	30
SW41	0.56" × 3.5"	13	0.7	6	6	41,000	60
SW27	1" × 3.5"	38.5	1	18	18	27,000	60
SW25.1	1" × 3"	34	1	16	16	25,000	60

7. Dilute the collected sample 1:4 with SB so that the density is less than that of the 20% sucrose. In order to improve virus purity, the virus can be sedimented through 20% sucrose after the material at the 20%–65% interface has been diluted with SB.
8. Load the virus onto a 20%–55% sucrose gradient.
9. Centrifuge the viral solution for twice the time centrifuged in step 4 using the same rotor. The virus will band at a density of approximately 1.16 (37% sucrose).
10. Collect the gradient and determine the density of each fraction, by weighing 100 μL in a micropipet or by determining the refractive index.
11. Collect the purified virus.

Viral RNA Extraction. After pelleting the viruses, the viral RNA can be extracted by one of the methods as described before, such as APGC or TRIzol method.

3.2.6 ISOLATION OF VIRAL GENOMIC RNA FROM SMALL AMOUNT OF CELL SUPERNATANT

Viral genomic RNA can also be isolated from small amounts of cell culture supernatant by integration of standard techniques with affinity procedures. Many commercial kits are now available. Some of them take advantage of the nucleic acid binding potential of matrix material supplied in a column. A recent development is the coverage of magnetic beads with nucleic acid binding matrices, which promises a high potential for automation. Some of the commercial kits are listed in Table 3.5.

TABLE 3.5
Microscale Isolation Kits for Viral RNA from Culture Supernatant

Kit	Supplier	Advantages	Disadvantages
QIAamp Viral RNA Mini Kit	QIAgen	Easy-to-handle solutions and straightforward protocol	Kit is not designed to separate viral RNA from cellular DNA (DNase is not provided)
		Elution of pure, stable, and intact viral RNA	Problematic for high-throughput applications because of the potential for filter clogging and the risk of cross-contamination
PureLink Viral RNA Kits	Invitrogen	Rapid and efficient purification within 45 min	The kit is not designed to separate viral RNA from cellular DNA (DNase is not provided)
		Ability to elute viral RNA in small elution volumes of 10–50 μL	
		High sensitivity for samples with very low viral titer	Problematic for high-throughput applications because of the potential for filter clogging and the risk of cross-contamination
		Purified RNA free of contaminants such as proteins and nucleases	
High Pure Viral RNA Kit	Roche	Saves time	This method requires that complicated buffers be made consistently and also requires careful pipetting to remove the RNA solution from the silica beads
		Minimizes RNA loss	
		Increases lab safety	
MagMAX Viral RNA Isolation Kit	Ambion	More effective recovery of viral RNA	Ideal for small sample sizes
		Ideal for low viral concentration	
		Improved yields and reproducibility	
		Potential for automation	

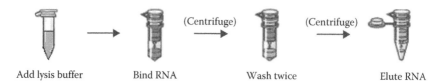

FIGURE 3.1   Viral RNA isolation by spin column.

The methodologies of the kits vary and two of them (one from QIAGEN and another from Ambion) are described below.

3.2.6.1   QIAamp Spin Columns

Principle.   The sample is first lysed under highly denaturing conditions to inactivate RNases and to ensure isolation of intact viral RNA. Buffering conditions are then adjusted to provide optimum binding of the RNA to the QIAamp membrane. When the samples are loaded onto the QIAamp mini spin column, the RNA binds to the membrane and contaminants are efficiently washed away using two different wash buffers with high salts. Then low salt elution removes the RNA from the membrane (Figure 3.1).

Procedure.   The following steps describe viral RNA isolation according to the product manual [25] and the method developed by Shafer et al. [26]:

1. Cell medium aliquots (200–1000 μL) are added into 1.5 mL microcentrifuge tubes and centrifuged for 10 min at 125,000 g at 10°C to concentrate the virus.
2. The supernatant is discarded carefully and the nonvisible virus pellet is resuspended in 140 μL of DEPC-treated H_2O.
3. A total of 560 μL of virus lysis buffer-carrier RNA is added to 140 μL concentrated virus and the mixture is vortexed and kept at room temperature for 10 min.
4. A total of 560 μL of 100% ethanol is added and the mixture is vortexed.
5. The new mixture is added to the spin column and is centrifuged at 6000 g for 1 min in a microcentrifuge.
6. The spin column is washed twice with 500 μL of wash buffer at 6000 g for 1 min.
7. The bound nucleic acid is eluted by adding 50 μL of 80°C DEPC-treated H_2O to the column and centrifuging at 6000 g for 1 min. The eluted RNA is ready to use or stored at −20°C for future use.

Commentary.   The whole process does not require RNA precipitation, organic solvent extractions, or extensive handling of the RNA. However, there are some problems associated with filter-based methods, such as clogging, large elution volumes, and inconsistent RNA yield. It may be problematic for high-throughput applications because of the potential for filter clogging and the risk of cross-contamination.

3.2.6.2   Magnetic Beads

Principle.   Ambion's magnetic bead-based MagMAX Viral RNA Isolation Kit is optimized for isolating viral RNA from biological fluids and cell-free samples. The viral RNA is released in a guanidinium thiocyanate-based solution, which simultaneously inactivates nucleases in the sample matrix. Paramagnetic beads with a nucleic acid binding surface are then added to the solution to bind nucleic acids. The complex of beads and nucleic acids are captured on magnets and proteins and

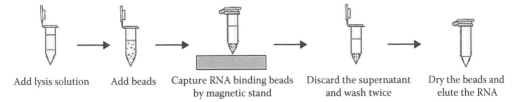

Add lysis solution   Add beads   Capture RNA binding beads   Discard the supernatant   Dry the beads and
by magnetic stand   and wash twice   elute the RNA

FIGURE 3.2 Viral RNA isolation by magnetic beads.

other contaminants are washed away. Then nucleic acids are eluted in a small volume of elution buffer with low salt (Figure 3.2).

Procedure. The following steps describe viral RNA isolation according to the product instruction manual [27]:

1. A 400 µL aliquot of cell medium is added into 802 µL lysis/binding solution and mixed by gently vortexing for 30 s or by flicking the tube several times. After brief centrifugation, the contents are collected at the bottom of the tube.
2. 20 µL bead mix is added into each sample and gently shaken for 4 min on a vortex to completely lyse viruses and allow the RNA bind to the RNA binding beads. After brief centrifugation, the contents in the tube are collected again.
3. The processing tube is moved to a magnetic stand for at least 3 min. After the RNA binding beads form a pellet against the magnet in the magnetic stand, the supernatant will be carefully removed without disturbing the beads.
4. The RNA binding beads are washed twice with 300 µL wash solution 1.
5. Then the RNA binding beads will be washed twice with 450 µL wash solution 2. Wash solution 2 from the sample should be removed as much as possible since it may inhibit downstream applications.
6. The beads are dried in air for 2 min to allow any remaining alcohol from the wash solution 2 to evaporate.
7. 50 µL elution buffer is added into each sample and shaken vigorously on a vortex for 4 min. After the RNA binding beads form a pellet on a magnetic stand, the supernatant is carefully transferred to a nuclease-free container, so that it is appropriate for future use.

Commentary. Magnetic bead-based technology is extremely effective for viral RNA isolation, resulting in better RNA capture and higher RNA yield than glass filter-based techniques. RNA yields are more consistent even from high-volume or low-viral titer samples. Another important benefit of magnetic bead-based RNA capture over glass filter-based methods is that there is no problem with filter clogging with samples. Furthermore, since only a small volume of magnetic beads is needed, bound RNA can even be eluted in as little as 20 µL of nuclease-free water. Because this procedure is not time consuming and does not have difficult-to-automate protease digestion, organic extraction, or centrifugation steps, it is thus adaptable to 96-well plate for high-throughput needs. Since this procedure recovers total nucleic acids, cellular DNA/RNA will be recovered along with the viral RNA if cells are present in the sample.

3.3 PURIFICATION OF VIRAL mRNA

Viral mRNA can be isolated and concentrated from total RNA. Because mRNA is polyadenylated, it can be purified by affinity chromatography using oligo(dT)-cellulose columns or using magnetic beads coated with oligo(dT) (Figure 3.3). This step yields mRNA with higher purity, which ensures better results in subsequent experiments.

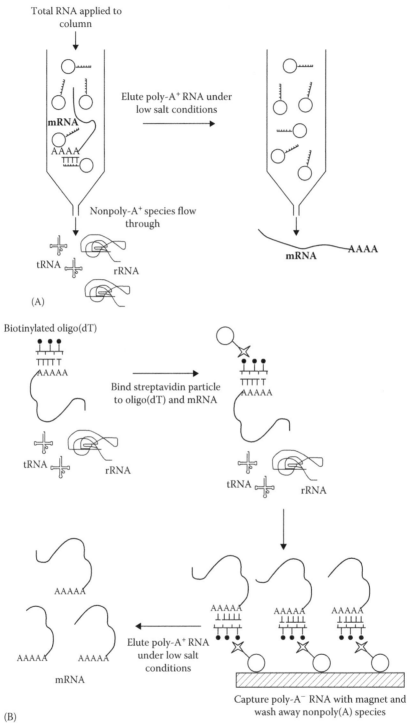

FIGURE 3.3 Diagram of poly-A+ RNA selection using oligo(dT) cellulose column (A) and streptavidin-linked magnetic particles (B). (From Krieg, P.A., *A Laboratory Guide to RNA Isolation, Analysis, and Synthesis*, Wiley-Lies, New York, 1996. With permission.)

3.3.1 Oligo(dT) Cellulose Columns

Principle. Total RNA isolated using one of the above methods can be further fractionated to obtain viral mRNA. The most frequent type of affinity isolation procedure is the purification of eukaryotic mRNA by virtue of the poly(A) tail which is a distinct feature of most mature mRNA. When total RNA is incubated with oligo(dT) in buffers with high salt, the poly(A) at the end of mRNA stably base-pairs with the thymidylate sequence. The binding extent is determined by the length of the poly(A) tail and the capacity of the matrix. Most of the cellular RNAs without poly(A) tail remain unbound. Under low salt conditions, base pairing between the poly(A) and oligo(dT) is unstable and the mRNA is released. Oligo(dT) cellulose is available commercially from several suppliers [28].

Reagents.

Column wash solution (0.1 N NaOH, 5 mM EDTA)
Column loading buffer (20 mM Tris-Cl, pH 7.6; 0.5 M NaCl; 1 mM EDTA, pH 8.0; 0.1% SDS)
Column elution buffer (10 mM Tris-Cl, pH 7.6; 1 mM EDTA, pH 8.0; 0.1% SDS)
Oligo(dT) cellulose and small chromatography columns

Procedure. Oligo(dT) cellulose columns may be used to load high-purity total RNA prepared by using one of the isolation methods as described above. The amount of RNA to be loaded depends on its availability and experimental needs. The following steps describe mRNA isolation using oligo(dT) cellulose columns [28].

Suspend the oligo(dT) cellulose in several milliliters of column wash solution. After 5 min sedimentation, remove the liquid and any cellulose that is still in suspension. Repeat this procedure three times.

1. Suspend the oligo(dT) cellulose in column wash solution once again and quickly load it into a small column.
2. Wash the column with ddH$_2$O until the effluent is less than pH 8.0.
3. Equilibrate the column with at least 5 column volumes of column loading buffer.
4. Suspend the RNA sample to be fractionated in 1 mL column loading buffer. Heat the RNA solution to 65°C for 5 min and cool on ice for 2 min.
5. Carefully load the RNA sample onto equilibrated column and collect the elute. Then reload the elute onto the column and collect the elute again.
6. Wash the column with 5 column volumes of column loading buffer.
7. Elute poly-A$^+$ RNA with 5 column volumes of column elution buffer.
8. Precipitate the RNA by addition of 1 volume isopropanol and 1/10 volume of 3 M NaAc. Then wash with 500 μL aliquots of 70% ethanol to remove residual salt.
9. Further purification can be performed to remove contaminating rRNA by passing the elution back through the column, washing, and eluting as above.

Regeneration of oligo(dT) cellulose column

1. The oligo(dT) cellulose is washed with column wash solution and then resuspended in 10 bed volume column wash solution.
2. The oligo(dT) cellulose is transferred to a centrifuge tube and pelleted by centrifugation at 75 g for 1 min.
3. The oligo(dT) cellulose is washed three more times with 10 volume regeneration solution to remove contaminants.
4. The oligo(dT) cellulose is washed two times with 3 volume loading buffer to remove NaOH.
5. The oligo(dT) cellulose is resuspended in 1 volume loading buffer and stored at 4°C for future use.

Commentary. Oligo(dT) cellulose has a very high binding capacity, which makes it effective for use in small columns and therefore for small volumes. Variations of the traditional oligo(dT) column approach offer different advantages. In one variation, syringes are used to force total RNA solution through the oligo(dT) matrix. This is much quicker than gravity flow chromatography and purification can be finished within 15 min. Another variation combines oligo(dT) sequences to small latex particles or cellulose beads suspended in solutions, which bind to poly-A$^+$ RNA when incubated with total RNA. After centrifugation, these particles are pelleted together with poly-A$^+$ RNA, thus achieving an effective purification and concentration.

3.3.2 MAGNETIC BEADS

Principle. Columns are prone to losses of RNA on the supporting matrix. They tend to become clogged and may flow very slowly. Instead of using a column, methods have been developed using magnetic beads. In this method, mRNA binds to the oligo(dT) attached to magnetic beads.

After incubation with total RNA or cell lysates, the beads associated with poly-A$^+$ RNA are sequestered using a magnet applied to the side of the tube. The supernatant containing the cellular debris and other RNA species is removed. After being washed several times, the mRNA can be eluted from the beads.

Reagents.

Bead prehybridization buffer: 0.1 M Tris-HCl pH 7.5, 0.01 M EDTA, 4% (w/v) fraction V bovine serum albumin (BSA), 0.5% (w/v) sodium lauroyl sarcosine, 0.05% (w/v) bronopol.
Magnetic stand.

Procedure. The following steps describe mRNA isolation using magnetic beads [13]:

1. A guanidinium thiocyanate buffer is adjusted by dilution (62.5 mL water/100 mL 4 M guanidinium thiocyanate extract) to 2.5 M guanidinium thiocyanate, 0.2 M Tris-HCl pH 7.5, 0.04 M EDTA, 0.5% (w/v) sarkosyl, and 10% (w/v) dextran sulfate-5000. Then the cells are homogenized with this adjusted buffer.
2. A 0.5 mL aliquot of oligo(dT)-coated beads (1 mg of beads carrying 62.5 pmol oligo(dT)) kept in bead prehybridization buffer is mixed with 0.25 mL cell lysates, which lowers the guanidinium thiocyanate concentration to 0.83 M.
3. The mixture is incubated for 5 min at 37°C to allow hybridization of mRNA to the oligo(dT).
4. Beads are harvested by using a magnet. The supernatant is discarded and the hybrid–bead complexes are resuspended in a solution suitable for further procedures.

Commentary. There are a number of variations of the method that involve binding of beads coated with biotinylated oligo(dT). In one variation, the poly(A) tails of mRNA bind to the biotinylated oligo(dT) and the hybrid in turn bind to beads coated with streptavidin. Streptavidin can bind more than one biotin per molecule giving a better isolation of mRNA. The magnetic beads can be collected using a magnet, washed, and the RNA can be eluted using a low salt buffer. The yield of mRNA using this technique tends to be higher and the length of time required is less because the hybridization takes place in liquid phase rather than solid phase required by the oligo(dT) column technology.

3.4 FUTURE DEVELOPMENT TRENDS

With the rapid development of biotechnology, more and more efficient ways for the extraction of viral RNA from cultures will be available. The new techniques should provide high sensitivity for extraction of viral RNA combined with short time consumption, reduced hands-on time, low price, and high potential for automation. The magnetic bead technology developed recently is a typical representative of this trend, which is easily adapted for either manual or robotic high-throughput processing [29]. This technology yields RNA of high purity with consistent recovery and it does not require vacuum filtration or centrifugation, making walkaway automation easier to implement. On the basis of this technology, some automated platforms are developed, such as the MagMAX Express platform (Ambion). On this platform, many commercial, high-throughput RNA extraction kits can be efficiently performed by liquid handling robots. It incorporates rapid, reliable, and cost-efficient extraction of RNA.

Considering the high risk of disease outbreak in remote areas and in underdeveloped regions, simple viral RNA extraction procedure that can be performed in the fields has been developed [30]. It is performed with a syringe and a denature buffer even by a layperson with minimal basic training in a very short time. It is suitable for distribution to underdeveloped areas where electricity and sophisticated laboratory facilities are not easily available. The wide variety of available options means that researchers can surely find a system to meet their needs, whether they prefer traditional methods of denaturation and extraction, or fully automated methods that integrate RNA isolation with downstream applications.

ACKNOWLEDGMENT

This work is supported partially by Public Health Service grant AI-051560 from the National Institute of Allergy and Infectious Diseases.

Ling Zhao received his MS and DVM from Central China (Huazhong) Agricultural University, Wuhan, China. During his master dissertation, he focused on the molecular biology of *Mycoplasma*. Since 2004, he has been in the Department of Pathology at the University of Georgia, Athens, Georgia pursuing his PhD study under the direction of Dr. Zhen F. Fu. His PhD study centers on the mechanism of rabies pathogenesis, particularly on the role of chemokines in the development of rabies as well as protection. He applies the state-of-the-art reverse genetics technology to develop vaccines that can activate the innate immunity as well as enhance the adaptive immunity. Currently, he is a member of the American Society for Microbiology and the American Society of Virology.

Zhen F. Fu is a professor in the Department of Pathology, College of Veterinary Medicine, University of Georgia, Athens, Georgia; he obtained his DVM from the Huazhong Agricultural University in Wuhan, China, and his MS and PhD from Massey University in Palmerston, New Zealand. He completed his postdoctoral training with the famous Dr. Hilary Koprowski at the Wistar Institute in Philadelphia, Pennsylvania. He was trained as a virologist and has been engaged in rabies research since 1989. He was an assistant professor at Thomas Jefferson University in Philadelphia, Pennsylvania; an associate professor at Kansas State University, Manhattan, Kansas; and has been a professor in the University of Georgia since 2000. In addition, he holds adjunct appointments in other universities (Huazhong Agricultural University and China Agricultural University), serves as an adviser for other universities and institutions (Virginia Technology Institute, Chinese Academy of Preventive Medicine, and National Veterinary Research & Quarantine Service of Republic of Korea), and works as a reviewer for various funding agencies (National Institutes of Health [NIH], USDA, and the Chinese National Science Foundation) and journals (*Journal of Virology*, *Journal of Neuro-Virology*, and *Virus Research*). His research interests include the neuropathogenesis of rabies, functional genomics and proteomics, development of antiviral vaccines, monoclonal antibodies and agents, and the regulation of rabies virus transcription and replication. His research has been funded by NIH since 1993. He has published more than 90 scientific papers in various journals. He has trained many postdoctoral fellows, doctoral and master students, as well as undergraduate students and professional students for research.

REFERENCES

1. Patton, J.T., *Segmented Double-Stranded RNA Viruses: Structure and Molecular Biology*, Caister Academic Press, Norfolk, pp. 1–12, 2008.
2. Barman, S. et al., Transport of viral proteins to the apical membranes and interaction of matrix protein with glycoproteins in the assembly of influenza viruses, *Virus Res.*, 77: 61, 2001.
3. Joint United Nations Program on HIV/AIDS, Chapter 2: Overview of the global AIDS epidemic update, 2006 Report on the Global AIDS Epidemic, 2006.
4. WHO, Summary of probable SARS cases with onset of illness from 1 November 2002 to 31 July 2003, 2003.
5. Schmaljohn, A.L., et al., Alphaviruses (*Togaviridae*) and flaviviruses (*Flaviviridae*). In: *Baron's Medical Microbiology* (Baron, S. et al., eds.), 4th ed., University of Texas Medical Branch, Galveston, TX, 1996.
6. CDC, West Nile Virus: Statistics, surveillance, and control, 2002.
7. Armstrong, G.L. et al., The past incidence of hepatitis C virus infection: Implications for the future burden of chronic liver disease in the United States, *Hepatology*, 31: 777, 2003.
8. McBride, W.J.H. et al., Dengue viral infections; pathogenesis and epidemiology, *Microbes Infect.*, 2: 1041, 2000.
9. Rouquet, P., et al., Wild animal mortality monitoring and human Ebola outbreaks, Gabon and Republic of Congo, 2001–2003, *Emerg. Infect. Dis.*, 11: 283, 2005.
10. WHO, Epidemic and pandemic alert and response (EPR): Report after final death 2004–2005 outbreak, 2005.
11. Torrey, E.F. and Yolken, R.H., Their bugs are worse than their bite, *Washington Post*, April 3, 2005.
12. Payment, P. and Trudel, M., *Methods and Techniques in Virology*, Marcel Dekker, New York, pp. 242, 1993.
13. Jones, P., Qiu, J., and Rickwood, D., *RNA Isolation and Analysis*, BIOS Scientific Publishers, Oxford, 1994.
14. Farrell, R.E., *RNA Methodologies: A Laboratory Guide for Isolation and Characterization*, Academic Press, San Diego, CA, pp. 15–46, 1993.
15. Fraenkel-Conrat, H., Singer, B., and Tsugita, A., Purification of viral RNA by means of bentonite, *Virology*, 14: 54, 1961.
16. Cox, R.A., *Methods in Enzymology* (Grossman, L. and Moldave, K., eds.), Vol. 12, Part B, pp. 120–129, Academic Press, Orlando, FL, 1968.
17. Chomczynski, P. and Sacchi, N., Single-step method of RNA isolation by acid guanidinium thiocyanate-phenol-chloroform extraction, *Anal. Biochem.*, 162: 156, 1987.
18. Chomczynski, P. and Sacchi, N., The single-step method of RNA isolation by acid guanidinium thio-cyanate-phenol-chloroform extraction: Twenty-something years on, *Nat. Prot.*, 1: 581, 2006.
19. Kleins, M., et al., Efficient extraction of viral DNA and viral RNA by the Chemagic viral DNA/RNA kit allows sensitive detection of cytomegalovirus, hepatitis B virus, and hepatitis G virus by PCR, *J. Clin. Microbiol.*, 41: 5273, 2003.
20. Chirgwin, J.M. et al., Isolation of biologically active ribonucleic acid from sources enriched in ribo-nuclease, *Biochemistry*, 18: 5294, 1979.
21. Chomczynski, P., Product and process for isolating RNA, US Patent 4,843,155, 1989.
22. Chomczynski, P., A reagent for the single-step simultaneous isolation of RNA, DNA and proteins from cell and tissue samples, *BioTechniques*, 15: 532, 1993.
23. Invitrogen, *TRIzol Reagent Manuals*, 2007.
24. Martin, G.S., Purification of Rous sarcoma virus (RSV); available at http://www.bio.com/protocol-stools/protocol.jhtml?id=p301.
25. Qiagen, QIAamp *Viral RNA Mini Handbook*, 2nd ed., 2005.
26. Shafer, R.W. et al., Comparison of QIAamp HCV kit spin columns, silica beads, and phenol-chloroform for recovering human immunodeficiency virus type 1 RNA from plasma, *J. Clin. Microbiol.*, 35: 520, 1997.
27. Ambion, MagMAX Viral RNA Isolation Kit. *Instruction Manual*, 2007.
28. Krieg, P.A., *A Laboratory Guide to RNA Isolation, Analysis, and Synthesis*, Wiley-Liss, New York, pp. 1–12, 1996.
29. Pichl, L. et al., Magnetic bead technology in viral RNA and DNA extraction from plasma minipools, *Transfusion*, 45: 1106, 2005.
30. Zhong, J.F. et al., Viral RNA extraction for in-the-field analysis, *J. Virol. Methods*, 144: 98, 2007.

4 Preparation of Viral Samples for Direct Molecular Applications

Grant S. Hansman

CONTENTS

4.1 INTRODUCTION

Viruses are submicroscopic infectious agents that range in size from about 20 to 400 nm in diameter ($1\,\text{nm} = 10^{-9}$ m), the majority of which are small enough to pass through conventional sterilizing filters ($0.2\,\mu\text{m}$). Each viral particle, or virion, consists of a single- or double-stranded nucleic acid (either deoxyribonucleic acid [DNA] or ribonucleic acid [RNA]) and at least one protein surrounded by a protein shell (known as capsid), whose shape varies from simple helical and icosahedral (poly-hedral or near-spherical) forms to more complex structures with tails or an envelope. In addition, some viruses may possess an outer envelope composed of lipids and proteins. The protein capsid provides protection for the nucleic acid while other proteins (enzymes) enable the virus to enter its appropriate host cell. Viruses are classified on the basis of their nucleic acid content, their size, the shape of the capsid, and the presence of a lipoprotein envelope, and they are divided into two primary classes: RNA viruses and DNA viruses.

As viruses are unable to grow and reproduce outside living host cells, they rely on animal, plant, and bacterial cells exclusively for their lifecycle, and may cause biological, biochemical, and physiological imbalances and diseases of differing severity as a consequence. Common human diseases caused by viruses include influenza (flu), chickenpox, and cold sores; and AIDS, avian influenza, and SARS are examples of more serious human viral infections. Traditionally, viruses are detected by *in vitro* cell culture, biochemical and serological methods, which can be slow and produce variable outcomes. With the advent of nucleic acid amplification technology such as poly-merase chain reaction (PCR), it is possible to detect viruses much more rapidly, sensitively, and precisely. In this chapter, we utilize sapovirus and norovirus within the virus family *Caliciviridae* as a model to illustrate the ways to prepare viral samples for direct molecular detection and identification.

The virus family *Caliciviridae* contains four genera, *Sapovirus, Norovirus, Lagovirus*, and *Vesivirus*, which include Sapporo virus, Norwalk virus, Rabbit hemorrhagic disease virus, and Feline calicivirus, respectively. Noroviruses (formally termed Norwalk-like viruses) and sapovi-ruses (formally termed Sapporo-like viruses) were originally classified as small round-structured viruses based on their morphological appearance using electron microscopy (EM). Noroviruses and sapoviruses are important etiological agents of human gastroenteritis, but some strains can cause disease in animals such as pigs and cows. Viruses belonging to the other two genera (*Lagovirus* and *Vesivirus*) mostly infect animal species. Human gastroenteritis is one of the leading causes of human death by an infectious disease, with more than 700 million cases of acute diarrheal disease occurring annually [1].

Transmission of noroviruses predominately occurs through ingestion of contaminated foods, airborne transmission, and person-to-person contact, whereas little is known about the transmission of sapoviruses. Over the past 15 years or so, numerous molecular biology techniques for detecting noroviruses and sapoviruses have been developed and refined. Molecular epidemiological studies have provided valuable information on their global distribution and contamination in the natural environment. However, different molecular biology techniques, such as RNA extraction and reverse transcription-PCR, are used in different laboratories around the world, and these different tech-niques likely affect the final result (although some progress has been made toward international standardization of the methods) [2–6]. This chapter describes the various norovirus and sapovirus techniques we have used, including sample preparation, RNA extraction, RT-PCR, real-time RT-PCR, and enzyme-linked immunosorbent assay (ELISA). Several other molecular biology techniques are also discussed.

4.1.1 Human Norovirus

Human noroviruses are the leading cause of outbreaks of gastroenteritis in the world and cause outbreaks in various settings, including hospitals, cruise ships, schools, and restaurants [7–13]. The prototype strain of human norovirus is the Norwalk virus (Hu/NV/Norwalk virus/1968/US), which was first discovered in an outbreak of gastroenteritis in an elementary school in Norwalk, Ohio, in 1968 [13]. Numerous molecular epidemiological studies have revealed a global distribution of these viruses [14,15]. In many countries, norovirus infection is prevalent during the winter months [16–18], though several studies showed no seasonal distribution [14–20]. Noroviruses have been detected in environmental samples (e.g., treated and untreated sewage) as well as in contaminated oysters, shellfish, sandwiches, salads, raspberries, and even ice [21–24]. Noroviruses are approximately 38 nm in diameter and possess a single-stranded, positive-sense RNA genome of 7.5–7.7 kb. The norovirus genome is predicted to contain three main open reading frames (ORF1–3), where ORF1 encodes the nonstructural proteins, including the RNA-dependent RNA polymerase (RdRp); ORF2 encodes the major capsid protein (VP1); and ORF3 encodes a small structural protein (VP2). Human and animal noroviruses were recently divided into five genetically distinct genogroups (GI–GV) based on their complete capsid amino acid sequences. The majority of human noroviruses belong to GI and GII, which can be further subdivided into at least 14 and 17 genotypes, respectively (Figure 4.1) [25].

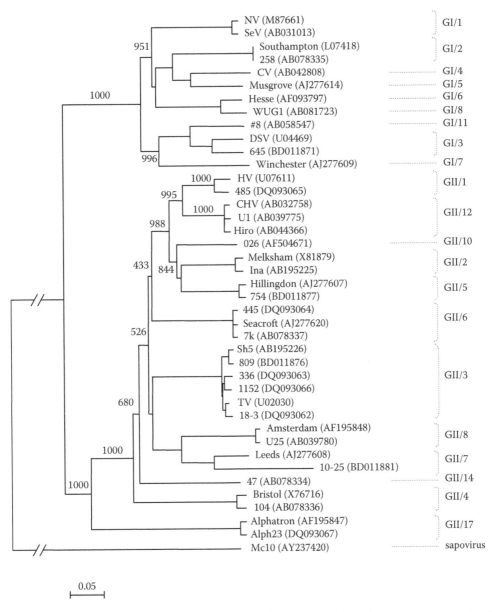

0.05

FIGURE 4.1 Phylogenic tree of the norovirus sequences. Norovirus amino acid sequences were constructed using the entire VP1. The numbers on the branches indicate the bootstrap values for the clusters. Bootstrap values of 950 or higher were considered statistically significant for the grouping.

The nucleotide sequences of norovirus genogroups show up to 60% variation, the genotypes vary between 14% and 44%, and the strains within a genotype vary by up to 14% [26]. Norovirus genotype identities are generally maintained across the ORFs. However, a number of norovirus strains failed to maintain their sequence identities for RdRp and VP1, and they were shown to be recombinant [27–29]. Evidence suggested that the recombination site occurred at the conserved polymerase and capsid junction between ORF1 and ORF2 (described in detail below for sapovirus). At present, norovirus strains belonging to GII/genotype 4 (GII/4) are dominant worldwide. However, in the last 10 years, four global pandemic norovirus GII/4 strains were identified—the US95-96 virus in 2000, Farmington Hills virus in 2002, Hunter virus in 2004, and 2006b virus in 2006. These four GII/4 strains have been termed variant GII/4 strains, since their complete capsid amino acid sequences differ only by approximately 5%, and they form subclusters within the GII/4 genotype. These minor amino acid changes in the norovirus capsid may be likened to influenza A virus evolution, where a single amino acid change in the antigenic region of the protruding glycoprotein can lead to a new antigenic variant capable of causing pandemics, a process known as antigenic drift [30,31]. Additionally, the sequence variation may also be a result of the high mutation rate caused by the lack of proofreading of the RdRp. The reason the variant GII/4 strains and not other genotypes are predominant is unknown, but it has been speculated that the variant GII/4 capsids are stronger in the natural environment; have a mechanism that allows the virus to evade the immune system; or are more virulent.

4.1.2 HUMAN SAPOVIRUS

Sapovirus is an important cause of sporadic gastroenteritis in young children, although recent reports have also found it to be an important cause of outbreaks of gastroenteritis in the general community [32–35]. In recent years, the number of sapovirus-associated outbreaks of gastroenteritis, especially involving adults, appears to be steadily increasing, suggesting that sapovirus virulence and/or prevalence may be increasing [33–36]. The prototype strain of human sapovirus, the Sapporo virus, was originally discovered from an outbreak in an orphanage in Sapporo, Japan, in 1977 [37]. In that study, Chiba et al. identified viruses with the typical animal calicivirus morphology, the Star-of-David structure, by EM. Besides having this classical structure, sapovirus particles are typically 41–46 nm in diameter and have a cup-shaped depression and/or 10 spikes on the outline. Sapovirus was recently detected in water samples, which included untreated wastewater samples, treated wastewater samples, and river samples [38]. Sapovirus was also detected in shellfish samples destined for human consumption, and the sequences detected in the clam samples closely matched sapovirus sequences detected in clinical stool samples [39]. These findings suggested that contaminations in the natural environment could lead to food-borne infections in humans. The sapovirus genome is a single-stranded, positive-sense RNA molecule of approximately 7.5 kb that is polyadenylated at the 3' end. The sapovirus genomes either contain two or three ORFs. The sapovirus ORF1 encodes the nonstructural proteins and the major capsid protein (VP1), while ORF2 and ORF3 encode proteins of unknown function [40–42]. Based on the complete capsid sequence (nucleotide or amino acid), sapovirus can be divided into five genogroups, which can be further subdivided into numerous genotypes (Figure 4.2). Recently, naturally occurring intragenogroup recombinant sapovirus strains were identified (i.e., strains Mc10 and C12) [43]. The overall genomic nucleotide similarity between Mc10 and C12 was 84.3%, while ORF1 and ORF2 shared 85.5% and 73.3% nucleotide identity, respectively, indicating that they were genetically distinct. However, by comparing the sequence similarity across the length of the genomes using SimPlot software [44], a potential recombination site was discovered, at a point where the similarity analysis showed a sudden drop in nucleotide identity after the RdRp region (Figure 4.3). Nucleotide sequence analysis of the ORF1 nonstructural proteins and the VP1 sequences revealed 90.1% and 71.3% nucleotide identity, respectively. These results suggested that a single point recombination event occurred

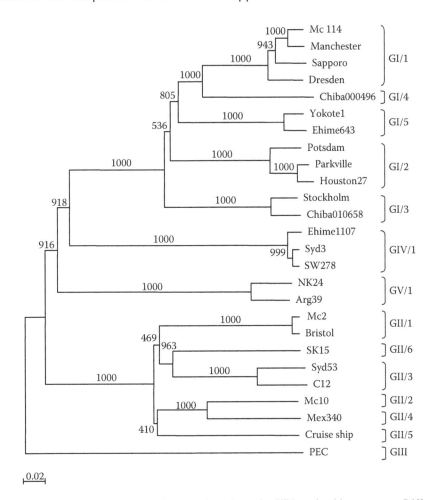

FIGURE 4.2 Phylogenetic tree of sapovirus based on the entire VP1 nucleotide sequences. Different geno-groups and genotypes are indicated. The numbers on each branch indicate the bootstrap values for the genotype. Bootstrap values of 950 or higher were considered statistically significant for the grouping. The scale represents nucleotide substitutions per site.

at the RdRp–VP1 junction. More recently, intergenogroup recombinant sapovirus strains were identified (i.e., strains SW278 and Ehime1107) [45]. Based on the classification scheme of either the partial or complete VP1 sequences, Manchester and Dresden were clustered into GI; Bristol, Mc2, Mc10, C12, and SK15 were clustered into GII; PEC was into clustered into GIII; SW278 and Ehime1107 were clustered into GIV; and NK24 was clustered into GV [6,46,47]. These genogroups were not maintained when the nonstructural region was analyzed, i.e., SW278 and Ehime1107 were clustered into GII for the nonstructural region-based grouping, but were clustered into GIV for the structural region-based grouping. Comparisons of the complete genome sequences showed that SW278 and Ehime1107 shared over 97% nucleotide identity and likely represented the same strain, although isolated from different countries (Sweden and Japan, respectively). By comparing the sequence similarity across the length of the genomes, a sudden drop in nucleotide similarity after the polymerase region for SW278 and Ehime1107 was observed, indicating that a recombination event occurred at the RdRp–capsid junction.

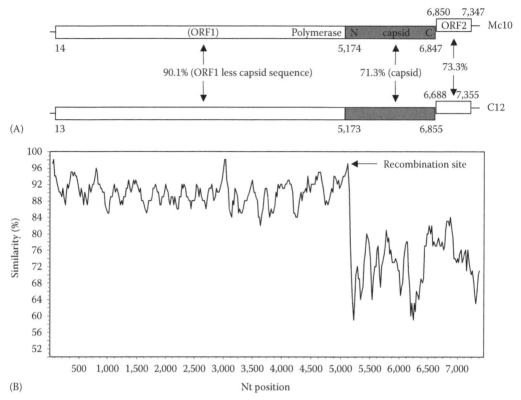

FIGURE 4.3 SimPlot analysis of the sapovirus Mc10 and C12 genomes. (A) Genome organization of the Mc10 and C12 strains, and the nucleotide sequence similarity of different genomic regions. (B) The Mc10 genome sequence was compared to that of C12 by using a window size of 100 bp with an increment of 20 bp. All gaps were removed. The recombination site is suspected to be located between the RdRp and VP1 genes (arrow), as the arrow shows.

4.1.3 SEROLOGICAL STUDIES

Human norovirus and sapovirus cannot be grown in standard cell cultures, as such the cross-reactivity is not completely understood and serotyping based on neutralization is not possible. Earlier cross-challenge studies in volunteers suggested that protective immunity was likely to be short lived and primarily genogroup- or genotype-specific [48,49]. However, recent studies have identified broadly reactive epitopes on the norovirus capsid [50,51], which suggests some level of cross-protection. In an early study of sapovirus antibody prevalence in the general community [52], Sakuma et al. indicated that sapovirus infections were acquired more readily after 2 years of age than before 2 years of age, and especially in infants attending nurseries and children attending kindergarten or primary schools. Despite these viruses remaining noncultivable, expression of the recombinant VP1 in a baculovirus expression system or mammalian expression system results in the self-assembly of virus-like particles (VLPs) that are morphologically and antigenically similar to the native virus particles [46,53–56]. Hyperimmune sera against these purified VLPs can be prepared in rabbits and guinea pigs, which can be used to examine the cross-reactivity among the different genogroups and genotypes. In the case of sapovirus, evidence suggested that there was a likely correspondence between the sapovirus antigenicity and VP1 genogrouping and genotyping [57]. However, in the case of norovirus, the antigenicity showed a more complicated result. The cross-reactivities among 26 different norovirus VLPs (6 GI and 12 GII genotypes) indicated that the antigenicity is affected not only by conserved amino acid residues but also, in part, by VP1 secondary structures [56].

4.1.4 IDENTIFICATION AND DIAGNOSIS

Methods for identification and diagnosis of norovirus and sapovirus include EM, antigen and antibody ELISA, RT-PCR, and real-time RT-PCR. Noroviruses and sapoviruses are usually detected using different reaction mixtures, because of the large genetic and antigenic difference between the two genera. Moreover, the detection method of choice usually depends on a number of factors, including running costs, need for rapid results, basic research, clinical samples or environmental samples, etc.

The first full-length norovirus genome (Norwalk virus strain) was determined in 1990 [58]. A few years later, an RT-PCR detection method with primers designed against the exact Norwalk virus sequence was developed [59]. In that basic research study, stool samples were collected from 10 adult volunteers who were experimentally infected with the purified Norwalk virus. Viral RNA was extracted from 10% to 50% (wt/vol) stool samples using 1,1,2, trichloro-1,2,2-trifluoroethane, and then the virus was precipitated using poly(ethylene glycol) (PEG) 6000. The pellet was added to 10% cetyltrimethylammonium bromide (CTAB) with 4 M NaCl and then the RNA was purified by a phenol-chloroform method. A total of 55 stool samples were collected (three or more from each volunteer) over several days (up to 68 h). All 10 volunteers were positive by RT-PCR, but only 37 of 55 (67%) samples were positive. Several important findings were discovered in this first study. First, it was found that only 9 of 10 volunteers developed symptoms. Second, the sensitivity increased when the authors used samples collected between 25 and 48 h postinfection. Third, they found that the use of separate tubes for RT and PCR slightly increased (10–100 times) the sensitivity, and the number of PCR cycles should be at least 40 in order to give a clear band on the gel. The authors concluded that the quality of the RNA was important for obtaining a positive result, i.e., they found that two steps (PEG precipitation and CTAB extraction) were critical for their RT-PCR.

One of the earliest molecular epidemiological studies of norovirus in the United Kingdom had a low detection rate of EM-positive (i.e., small round-structured viruses) samples [60]. In that study, the authors examined 50 outbreaks of viral gastroenteritis using RT-PCR with specific primers designed against a specific Norwalk virus sequence. The RNA extraction method used guanidinium thiocyanate-Triton (GTC) X-100 and silica beads to which the nucleic acid was bound. Only 3 of 50 outbreaks were positive using this RT-PCR method. They concluded that the Norwalk virus strain against which they designed their primers was infrequent in the United Kingdom, and the low sensitivity was unlikely to have been a result of the extraction method or degraded nucleic acid. Several years later, new primers were developed and these had a higher detection rate and identified local strains circulating in the United Kingdom [61].

Shortly after the United States and United Kingdom results were published, a study was conducted to evaluate four different RNA extraction methods for the detection of small round-structured viruses [62]. One method used GTC with absorption of viral RNA onto silica as in the U.K. study [60] but with slight modifications; one method used the CTAB as in the U.S. study [59], one method purified the virus by exclusion chromatography; and one method used the metal chelating agent Chelex-100. The same set of primers and stool samples was used for each extraction method. The authors found that the GTC and CTAB methods were the most effective at removing inhibitors and were the most appropriate for detection using RT-PCR, but the recoveries of the RNA yield by these methods were lower than those by the exclusion chromatography and Chelex methods.

A study conducted in Japan examined several sets of primers for their efficacy in detecting noroviruses prevailing in Japan [63]. Multiple alignments of published norovirus sequences previously detected in Japan (over a 10 year period) were performed in order to design new sets of consensus primers. Three different primers sets were designed and together with a panel of coded stool samples were sent to 10 different laboratories in Japan. The results reported back by the laboratories indicated that the different primers had different sensitivities. However, they also revealed that some of the laboratories had different results with the same set of primers. The authors concluded that the different results among the laboratories were likely attributable to the different RNA extraction and reverse transcription methods.

A recent large international collaborative study was performed in order to harmonize the methods of norovirus detection [2]. A panel of coded stool samples was distributed to different laboratories in different countries. The laboratories were asked to extract the RNA using different extraction methods and perform RT-PCR with a variety of published primers. The study found that no one single method could detect all the positive samples, although one method was termed "satisfactory." The sensitivity ranged from 52% to 73% overall, and the sensitivities were different for GI and GII viruses. The authors concluded that the different sensitivities might have been attributable to the use of different extraction assays. They also suggested that some samples may have deteriorated during storage.

As the sequence data on norovirus accumulated, countless other primers were designed and then redesigned against the great genetic diversity of norovirus sequences [3,64–68]. Norovirus RT-PCR primers were targeted against different regions of the genome and many were degenerate because of the great genetic variation. Most of the norovirus RT-PCR primers were directed near the RdRp and capsid junction, which was found to be the most conserved region in the norovirus genome [69]. This region has remained popular for designing new primers. Moreover, phylogenetic analysis and genetic classification could be performed using this short region in the genome [69]. Currently, two sets of reaction tubes are generally used for the detection of noroviruses, one tube for GI and one for GII sequences. However, since RT-PCR with degenerate primers and/or low annealing temperatures tends to produce nonspecific bands of the estimated size, the RT-PCR products need to be sequenced before a sample is considered to be true positive [2].

Following the advent of the RT-PCR detection methods, numerous other methods were developed. These included a number of different real-time RT-PCR assays that had high sensitivities, were broadly reactive, and could be used for quantitative analysis [70–74]. These methods were rapid and practical for processing a large number of samples. Different research groups have used different real-time RT-PCR assays, including different extraction methods and different primers. However, many of the real-time RT-PCR primers were directed against the conserved RdRp and capsid junction. Kageyama et al. [74] found that real-time RT-PCR had a higher detection rate and lower detection limit than conventional RT-PCR. The real-time RT-PCR could detect down to 10 RNA copies per tube, which corresponded approximately to 2×10^4 RNA copies per gram of stool [74]. A number of other real-time RT-PCR assays were also capable of detecting low copy numbers. Höhne et al. found that their real-time RT-PCR assay could detect between 10 and 100 copies per reaction mixture, while Pang et al. and Trujillo et al. found that their assays could detect below 10 copies per reaction mixture [73,75,76]. However, these estimated low detection limits are far from definitive, as plasmids harboring the target sequence and not clinical stool samples were used for the evaluations. Nevertheless, the sensitivities and specificities of these real-time RT-PCR assays were equal to or greater than those of conventional RT-PCR assays. The real-time RT-PCR assays also allowed quantitative analysis of each sample, which was useful for understanding norovirus infections. For example, in a recent molecular epidemiological study, we found that asymptomatic individuals had mean viral loads similar to those of symptomatic individuals and thus asymptomatic transmission may account for the increased number of infections and predominance in Japan via an asymptomatic transmission route [4].

4.2 SPECIMEN PREPARATION

4.2.1 Sample Storage

A number of researchers have found that storing clinical samples for extended periods of time may result in RNA degradation and viral instability [2,67,75]. In this regard, different research laboratories have adopted a range of different storage conditions. For example, raw stool samples, 10% stool suspensions, 10% clarified stool suspensions, and extracted RNA have all been stored at such widely divergent temperatures as $-20°C$, $-60°C$, $-70°C$, or $-80°C$ before molecular applications. In general, we store raw stool samples at $4°C$ before diagnostic work.

4.2.2 Electron Microscopy

To obtain a positive result by EM, approximately 10^6 particles per gram of stool are needed as well as a skilled expert who can differentiate between a VLP and an actual virion. The sample preparation is straightforward although time consuming. A 10% stool suspension is usually prepared in phosphate-buffered solution (PBS) or distilled water and then centrifuged. It is important to thoroughly vortex the suspension before centrifuging in order to separate the virus particles from the organic and inorganic material in the stool. The stool suspension is centrifuged at low speed ($3,000 \times g$ for 10 min) and then the supernatant collected and centrifuged at medium speed ($10,000 \times g$ for 30 min). The supernatant is then collected and centrifuged at high speed in a 10%–20% sucrose cushion ($50,000 \times g$ for 2 h) and then the pellet is resuspended in a small volume (20–50 μL) of PBS or distilled water and stored below −30°C. The method of negative staining can vary and can include (2%–4%) uranyl acetate or (2%–4%) phosphotungstic acid. Various kinds of EM grids can also be used, though carbon-coated grids usually provide the best resolution for norovirus and sapovirus (author's opinion). The virus integrity and morphology are usually stable for a certain amount of time; however, freshly prepared samples should be examined and freeze thawing may disrupt the virus morphology. The methodology for applying the sample to the grid can also vary between laboratories. In our laboratory, 10–20 μL of 1:10 sample is placed on Parafilm, 20 μL of water is placed on the Parafilm next to the sample, and then 20 μL of stain is placed on the Parafilm next to the water. A grid is placed in lockable forceps. The grid is touched to the sample (on the carbon side) and left to dry for 15 s, and then the excess sample is removed from the grid with filter paper. This is repeated for the water and then the stain. Finally, the grid is left to dry for 20 min and then examined by EM at a magnification of 30,000–40,000. The resuspended pellet can also be used for molecular biology techniques, including RT-PCR and ELISA as described below.

4.2.3 Sample Processing

Sapovirus and norovirus were recently detected in untreated wastewater, treated wastewater, and a river in Japan [38,77]. The methods used to concentrate the virus were different for each sample site (see below).

4.2.3.1 Primary-Treated Domestic Wastewater

Viruses in 400 mL of primary-treated domestic wastewater were recovered by using the enzymatic virus elution (EVE) method [78]. Primary-treated domestic wastewater was centrifuged at $9000 \times g$ for 15 min and the supernatant was decanted. The pellet was resuspended in the EVE buffer (10 g/L of each of the following enzymes: mucopeptide N-acetylmuramoylhydrolase, carboxylesterase, chrymotrypsin, and papain) and stirred for 30 min. The suspension was centrifuged at $9,000 \times g$ for 30 min and then the supernatant was collected and stored at −20°C until further analysis. The viruses in the supernatant were concentrated using a PEG precipitation method [79]. PEG 6000 was added to the supernatant at a concentration of 8% and stirred at 4°C overnight. The sample was centrifuged at $9000 \times g$ for 90 min, and the pellet was resuspended in 4 mL of 20 mM phosphate buffer (pH 8.0) and briefly agitated using a vortex. The resuspended sample was centrifuged at $9000 \times g$ for 10 min, and the supernatant was collected and stored at −20°C until further analysis.

4.2.3.2 Secondary-Treated Effluent and River Water

Viruses in 1 L of secondary-treated effluent and river water were concentrated by PEG precipitation as follows. One hundred grams of PEG 6000 and 23.4 g of NaCl were added to 1 L of sample and stirred at 4°C overnight. The sample was then centrifuged at $10,000 \times g$ for 60 min and the pellet was resuspended in 4 mL of distilled water. The resuspended sample was centrifuged at $10,000 \times g$ for 10 min and then the supernatant was collected and stored at $-20°C$ until further analysis. The viruses in the supernatant were concentrated using the PEG precipitation method described above.

4.2.3.3 Seawater

Viruses in 20 L of seawater were concentrated according to the method of Katayama et al. [80]. Seawater was filtered with an HA negatively charged membrane (Nihon Millipore, Tokyo, Japan) with a 0.45 µm pore size. Then, 200 mL of 0.5 mM H_2SO_4 was passed through the membrane to rinse out cations. Next, 10 mL of 1 mM NaOH was poured onto the membrane and the filtrate was recovered in a tube containing 0.1 mL of 50 mM H_2SO_4 and 0.1 mL of 100 x TE buffer. Viruses in the filtrate were further concentrated with a Centriprep Concentrator 50 system (Nihon Millipore) according to the manufacturer's instructions to obtain a final volume of 1 mL.

4.2.3.4 Oysters

Norovirus has been detected in oysters from many different countries, and an equal number of methods have been used to extract the viruses. These methods are typically evaluated with a spiked virus in order to determine the efficiency of the RNA recovery. Recently, a Japanese group developed a more efficient method [24]. Fresh oysters were shucked and their stomachs and digestive tracts were removed by dissection. The samples were weighed and homogenized in 10 mM PBS (pH 7.4; without magnesium or calcium) and made into a 10% suspension. The suspension was added to 0.1 mL antifoam B (Sigma, St. Louis, Missouri) and homogenized twice at 30 s intervals at maximum speed using an Omni-mixer (OCI Instruments, Waterbury, Connecticut). Next, 6 mL of choloroform:butanol (1:1 vol) was added, the suspension was homogenized for 30 s, and 170 µL Cat-Floc T (Calgon, Elwood, Pennsylvania) was added. The sample was centrifuged at $3000 \times g$ for 30 min at 4°C and the supernatant was layered onto 1 mL of 30% sucrose solution and ultracentrifuged at $154,000 \times g$ for 3 h at 4°C. Finally, the pellet was resuspended in 140 µL of distilled water DDW and stored at $-80°C$ until further analysis.

4.2.3.5 Shellfish

Recently, we detected sapovirus in clam samples designated for human consumption [39]. Clams were shucked and the digestive diverticulum removed by dissection. Samples were weighed and homogenized in nine times their weight of PBS (pH 7.4; without magnesium or calcium). One gram of digestive diverticulum (approximately 10–15 clams per package were pooled) was homogenized with an Omni-mixer in 10 mL of PBS (pH 7.2). After centrifugation at $10,000 \times g$ for 30 min at 4°C, the supernatant was layered onto 1 mL of 30% sucrose solution and ultracentrifuged at $154,000 \times g$ for 3 h at 4°C. The supernatant was then carefully removed, and the pellet was resuspended in approximately 140 µL of distilled water and stored at $-80°C$ until further analysis.

4.2.3.6 Stool

Stool samples are usually diluted and clarified before RNA can be extracted using the commercial and in-house RNA extraction methods. In our laboratory, a 10% (wt/vol) stool suspension was prepared with sterilized water and centrifuged at $10,000 \times g$ for 10 min at 4°C. However, other research laboratories have prepared the clarified supernatant in different ways. Yuen et al. prepared 10% (w/v) stool suspensions in Hanks' complete balanced salt solution, clarified the solution by centrifuging the sample at $3500 \times g$ for 15 min at room temperature, followed by $7000 \times g$ for 30 min at $4°C$ [66]. Drier et al. prepared 10% (wt/vol) stool suspension in PBS and clarified the solution by centrifuging the sample at $3500 \times g$ for 15 min [81]. Houde et al. prepared 20% (wt/vol) stool suspension in PBS and clarified the solution by centrifuging the sample $2000 \times g$ for 3 min, followed by $16,000 \times g$ for 5 min [82].

4.3 DETECTION

4.3.1 RNA EXTRACTION

RNA for RT-PCR should be free from DNA contamination, as this can generate false positives during the PCR step. Moreover, many substances found in stool samples can inhibit PCR, because the *Taq* polymerases require optimal conditions, including optimal pH and reagent concentrations [62]. Several groups have suggested using a competitive internal control RNA to monitor inhibition of RT-PCR [83–85]. Today, commercial extraction methods have replaced the earlier extraction methods and effectively remove RT-PCR inhibitors.

4.3.1.1 Manual Methods

One of the most popular methods for extracting RNA from stool samples, environmental samples, or food samples is the QIAamp Viral RNA Mini Protocol (Qiagen, Hilden, Germany). The QIAamp Viral RNA Protocol is actually designed to purify viral RNA from plasma, serum, cell-free body fluids, and cell-culture supernatants. However, the method can be modified to purify viral RNA from other samples as well, including stool samples and environmental samples. For a stool sample, a 10% stool suspension was prepared with sterilized water and centrifuged at $10,000 \times g$ for 10 min at $4°C$. For water samples, we used 140 μL of the concentrated water sample. For the shellfish samples, we prepared a 10% (vol/vol) of the resuspended pellet (described above). In our laboratory, this extraction method has been used to successfully detect norovirus, sapovirus, astrovirus, HEV, and Aichi virus in stool samples. It can be carried out in a standard centrifuge or a vacuum manifold, with the vacuum manifold being more useful for 24–48 samples, and the centrifuge more appropriate for a handful of samples. The entire procedure takes approximately 60–90 min (for 24 samples using a vacuum manifold). Briefly, 560 μL of 100% ethanol was added to the sample and vortexed for 15 s. Then, 600 μL of sample was added to the QIAamp spin column and the vacuum turned on. The remaining sample (approximately 660 μL) was added to the column. After the entire sample was passed through the QIAamp spin column, the column was washed with 750 μL of buffer AW1, and then washed with 750 μL of buffer AW2. The spin column was then centrifuged at $10,000 \times g$ for 2 min at $4°C$ in order to remove the residual reagents. Then, 60 μL of buffer AVE was added, the QIAamp spin column was centrifuged at $10,000 \times g$ for 2 min at $4°C$, and the RNA was collected in a clean tube. This method extracts both RNA and cellular DNA. In order to remove the DNA from the sample, it was recommended that the RNA preparation be digested with DNase, followed by heat treatment to inactivate the DNase. The RNA can be stored at $-20°C$ or $-70°C$ and may remain stable for up to 1 year, although we have found it to be stable for up to 2 years when stored at $-80°C$.

4.3.1.2 Automated Systems

A number of companies have designed automated nucleic acid extraction equipment, including Roche and Qiagen. The Roche system (MagNA Pure LC) is user friendly, with easy-to-use controls. The instrument is fully automated and can isolate nucleic acids (DNA, total RNA, and mRNA) from different samples, including whole blood, serum, tissue, and stool. Up to 32 different samples can be processed in 50–180 min. This method uses a 10% (wt/vol) stool suspension. The method is straightforward, but may require 1 day of training in order to perform it successfully. It has been tested with clinical stool samples and real-time RT-PCR and found to provide rapid and reliable results [86]. We found that the MagNA method had sensitivities equal to the Qiagen RNA extraction method (unpublished). However, we also found that the RNA extracted using the Qiagen RNA method could be used to amplify longer (3–5 kb) fragments, while the MagNA method was unable to amplify these long fragments (unpublished). Nevertheless, the MagNA method is useful for epidemiological studies.

4.3.2   Reverse Transcription

Reverse transcription can be performed using random primers, poly(T) primers, or gene-specific primers. A number of different reverse transcriptases are available, each with a different sensitivity. Our reverse transcriptase of choice is Invitrogen Superscript III. Reverse transcription was carried out in a final volume of 20 μL with 10 μL of RNA in 50 pmol of random hexamer (Takara, Tokyo, Japan), 1 × Superscript III RT buffer (Invitrogen, Carlsbad, California), 10 mM DTT (Invitrogen), 0.4 mM of each dNTP (Roche, Mannheim, Germany), 1 U RNase inhibitor (Toyobo, Osaka, Japan), and 10 U Superscript RT III (Invitrogen). Reverse transcription was performed at 50°C for 1 h, followed by deactivation of RT enzyme at 72°C for 15 min. This standard RT method proved useful for epidemiological studies, but a modified RT method was used for preparing high-quantity cDNA for other molecular methods, i.e., amplification of long PCR fragments. Briefly, an RT mix was prepared in a separate tube (1 × Superscript III RT buffer, 10 mM DTT, 0.4 mM of each dNTP, 1 U RNase inhibitor, and 10 U Superscript RT III) and put on ice. Then two heating blocks were prepared, one at 94°C and another at 55°C. Poly(T) reverse primer was added to a new PCR tube with 10 μL of RNA and then two drops of PCR oil were added. The solution was briefly centrifuged and put in the 94°C block for 2 min. Then, this tube was taken out of the block, placed between the thumb and index finger, and allowed to cool briefly (15 s). The RT master mix (9 μL) was then added quickly to the bottom of the tube. The tube was briefly centrifuged and put in the 55°C block for 2–3 h. The RT was deactivated at 94°C for 15 min and then placed on ice. The cDNA was stored at −20°C. This cDNA was useful for producing long DNA fragments (over 3 kb) and was used in many studies, including full-length genome analyses, expression of the capsid protein, and replication studies. However, the ability to determine the full-length norovirus genome sequence relied on the knowledge of partial sequences and degenerate primers, and on knowledge of the fact that the 5' untranslated region (UTR) of the norovirus genome is usually conserved at the ORF2 start.

4.3.3   PCR Detection

4.3.3.1   Sapovirus Nested PCR

Numerous PCR primers that detect different regions (polymerase or capsid) have been designed and improved. Okada et al. designed a set of primers (for nested PCR) that could detect strains from all human genogroups as well as many of the different genotypes [87]. This modified primer set has also detected novel genotypes. Briefly, for the first PCR, F13, F14, R13, and R14 primers were used, while for the nested PCR, F22 and R2 primers were used (Figure 4.4 and Table 4.1) [87]. The first PCR was carried out with 5 μL of cDNA in a PCR mixture containing 20 pmol of each primer, 1 × *Taq* DNA polymerase buffer B, 0.2 mM of each dNTP, 2.5 U *Taq* polymerase, and up to 50 μL of distilled water. PCR was performed at 94°C for 3 min followed by 35 cycles at 94°C for 30 s, 48°C for 30 s, 74°C for 45 s, and a final extension of 5 min at 74°C. The nested PCR used 5 μL of the first PCR in a second reaction mixture (identical to the first except for the primers) and the same PCR conditions. For molecular methods that require high fidelity, we used KOD DNA polymerase, which has a unique proofreading ability and is faster and more accurate than conventional DNA polymerases. The proofreading ability results in a lower PCR mutation frequency and considerably higher elongation rates than those achieved by conventional DNA polymerases.

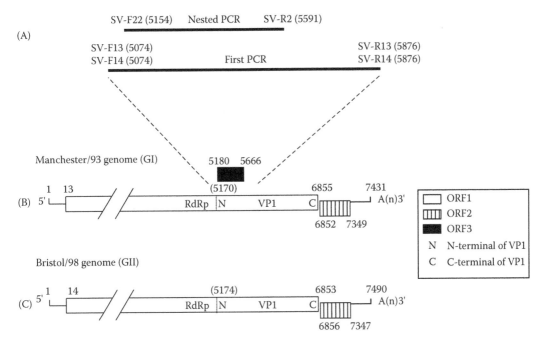

FIGURE 4.4 Detection of sapovirus by nested RT-PCR. The location of the primers in the RdRp/capsid junction.

TABLE 4.1
Primers for Sapovirus Nested RT-PCR and Primers and Probes for Sapovirus Real-Time PCR

Application	Target	Primer	Sequence (5' to 3')	Polarity	Location
		SV-F13	GAYYWGGCYCTCGCYACCTAC	+	5074[a]
		SV-F14	GAACAAGCTGTGGCATGCTAC	+	5074[a]
PCR	GI, GII, GIV, GV	SV-R13	GGTGANAYNCCATTKTCCAT	–	5861[a]
		SV-R14	GGTGAGMMYCCATTCTCCAT	–	5861[a]
		SV-F22	SMWAWTAGTGTTTGARATG	+	5154[a]
		SV-R2	GWGGGRTCAACMCCWGGTGG	–	5572[a]
	GI, GII, GIV	SaV124F	GAYCASGCTCTCGCYACCTAC	+	5078[b]
	GI	SaV1F	TTGGCCCTCGCCACCTAC	+	700[c]
Real-time PCR	GV	SaV5F	TTTGAACAAGCTGTGGCATGCTAC	+	5112[d]
	GI, GII, GIV, GV	SaV1245R	CCCTCCATYTCAAACACTA	–	5163[b]
	GI, GII, GIV	SaV124TP	FAM-CCRCCTATRAACCA-MGB-NQF	–	5105[b]
	GV	SaV5TP	FAM-TGCCACCAATGTACCA-MGB-NQF	–	5142[d]

Note: FAM, 6-carboxyfluorescein (reporter dye); MGB, minor groove binder; NQF, nonfluorescent quencher.

[a] Manchester virus (X86560).
[b] Mc10 virus (AY237420).
[c] Parkville virus (U73124).
[d] NK24 virus (AY646856).

4.3.3.2   Norovirus Nested PCR

For norovirus PCR, primers were designed to amplify the 5′ prime end of the capsid gene [3,25]. For norovirus GI, COG1F and G1SKR primers were used for the first PCR and then G1SKF and G1SKR primers were used for the nested PCR. For norovirus GII, COG2F and G2SKR were used for the first PCR and then G2SKF and G2SKR primers were used for the nested PCR (Figure 4.5 and Table 4.2). Other groups have detection primers that amplify within the RdRp [88–90]. However, we found it difficult to use the RdRp sequences for phylogenetic analysis and genogrouping [69].

4.3.3.3   Sapovirus Real-Time RT-PCR

Recently, we developed a real-time RT-PCR method that could detect all human sapovirus genogroups [91]. In the first instance, a nucleotide alignment of full-length sapovirus genome sequences was subjected to similarity plot analysis in order to identify the most conserved site. In the case of sapovirus this was the polymerase capsid junction. Following this result, multiple alignments of partial sequences in this site were performed and primer sequences were designed. A number of sets of primers were developed and evaluated using control plasmids and then clinical samples. The sapovirus real-time RT-PCR had 100% specificity and sensitivity equal to that of nested RT-PCR. The advantage of real-time RT-PCR is that it can give a rapid result and can be used to determine the number of copies of cDNA per gram of stool sample (equal to the number of RNA copies per gram of stool) [32,33]. The detection limit of the real-time RT-PCR was found to be approximately 1.3×10^5 copies of sapovirus RNA per gram of stool sample.

   The RNA extraction method was identical to the Qiagen method, except to prevent nonspecific amplification the extracted viral RNA was treated with DNase I before RT. Then, viral RNA ($10\,\mu L$) was added to a reaction mixture ($5\,\mu L$) containing DNase I buffer ($150\,mM$ Tris-HCl, pH 8.3; $225\,mM$ KCl; $9\,mM$ MgCl$_2$) and 1 unit of RQ1 DNase. The reaction mixture was incubated first at $37°C$ for $30\,min$ to digest DNA and then at $75°C$ for $5\,min$ to inactivate the enzyme. DNase I-treated RNA ($15\,\mu L$) was added to $15\,\mu L$ of another mixture containing RT buffer, $1\,mM$ of each dNTPs,

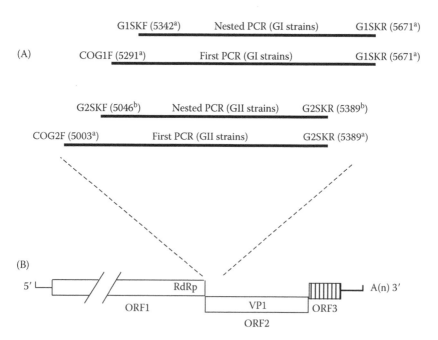

FIGURE 4.5   Detection of norovirus by nested RT-PCR. The location of the primers in the RdRp/capsid junction.

TABLE 4.2

Primers for Norovirus RT-PCR and Primers and Probes for Norovirus Real-Time PCR

Application	Target	Primer	Sequence (5' to 3')	Polarity	Location
		COG1F	CGYTGGATGCGNTTYCATGA	+	5291[a]
	GI	G1SKF	CTGCCCGAATTYGTAAATGA	+	5342[a]
		G1SKR	CCAACCCARCCATTRTACA	−	5671[a]
PCR					
		COG2F	CARGARBCNATGTTYAGRTGGATGAG	+	5003[b]
	GII	G2SKF	CNTGGGAGGGCGATCGCAA	+	5046[b]
		G2SKR	CCRCCNGCATRHCCRTTRTACAT	−	5389[b]
		COG1F	CGYTGGATGCGNTTYCATGA	+	5291[a]
		COG1R	CTTAGACGCCATCATCATTYAC	−	5375[a]
	GI	Probe RING1(A)-TP	FAM-AGATYGCGATCYCCTGTCCA-TAMRA	−	5340[a]
Real-time PCR		Probe RING1(B)-TP	FAM-AGATCGCGGTCTCCTGTCCA-TAMRA	−	5340[a]
		COG2F	CARGARBCNATGTTYAGRTGGATGAG	+	5003[b]
	GII	COG2R	TCGACGCCATCTTCATTCACA	−	5100[b]
		Probe RING2-TP	FAM-TGGGAGGGCGATCGCAATCT-TAMRA	+	5048[b]

Note: FAM, 6-carboxyfluorescein (reporter dye); TAMRA, 6-carboxy-tetramethyrhodamione (quencher dye).

[a] Norwalk virus (M87661).

[b] Camberwell virus (AF145896).

10 mM dithiothreitol, 50 pmol of random hexamers, 30 units of RNase OUT, and 200 units of Super-Script III RNase H-RT. RT was performed at 37°C for 15 min followed by 50°C for 1 h, and then the solution was stored at −20°C. Quantitative real-time RT-PCR was carried out in a 25 µL of a reaction volume using a QuantiTect Probe PCR Kit containing 2.5 µL of cDNA, 12.5 µL of Quanti-Tect Probe PCR Master Mix, 400 nM of each primer (SaV124F, SaV1F, SaV5F and SaV1245R), and 5 pmol of TaqMan MGB probes (SaV124TP and SaV5TP). Several primer sets and probes were designed (using multiple alignment analysis of 27 SaV sequences) to hybridize against the highly conserved nucleotides between 5078 and 5181 with respect to Mc10 virus. For amplification, we designed four primers (SaV124F, SaV1F, SaV5F, and SaV1245R), and for detection, we designed two TaqMan MGB probes (SaV124TP and SaV5TP). These primers and probes were designed and mixed to detect sapovirus GI, GII, GIV, and GV sequences in a single reaction tube. PCR amplification was performed with a 7500 Fast Real-Time PCR System (Applied Biosystems, Foster City, CA) under the following conditions: initial denaturation at 95°C for 15 min to activate DNA polymerase, followed by 40 cycles of amplification with denaturation at 94°C for 15 s, and annealing and extension at 62°C for 1 min. Amplification data were collected and analyzed with Sequence Detector software version 1.3 (Applied Biosystems). A 10-fold serial dilution of standard cDNA plasmid (2.5×10^7 to 2.5×10^1 copies) was used to quantify the viral copy numbers in reaction tubes.

4.3.4 ELISA DETECTION

4.3.4.1 In-House Methods

ELISAs can screen for specific antibodies or virus particles [92–96] and are useful for screening a large number of samples. One of the first norovirus ELISA detection methods was developed using the Norwalk virus VLPs expressed in insect cells. Antiserum (rabbit and guinea pig) was raised against these purified VLPs. However, the developers found that the antigen ELISA had a low detection rate and suggested that the great genetic variability in the capsid gene of the test samples (taken from outbreaks and sporadic infections) was the reason for the low sensitivity. Recently, a sapovirus ELISA based on hyperimmune rabbit and guinea pig antisera raised against sapovirus VLPs was developed [97,98]. After a number of optimization steps, the ELISA was prepared for clinical testing. Briefly, the wells of 96-well microtiter plates were coated overnight with hyperimmune rabbit (capture) antiserum. The wells were washed and blocked, and then a clarified 10% stool suspension was added. The wells were washed again, and then hyperimmune guinea pig (detector) antiserum was added. The wells were washed a third time, and then diluted horseradish peroxidase-conjugated goat anti-guinea pig immunoglobulin G was added. The wells were washed a final time, o-phenylenediamine and H_2O_2 were added, and then the reaction was stopped with H_2SO_4. The absorbance was measured at 492 nm (A_{492}), the cutoff value was defined as the mean plus three standard deviations, and P/N ratios over the cutoff value were considered significantly positive [97,98]. Our results also showed a low sensitivity, although we found that the sapovirus ELISA was useful in detecting sapovirus GI antigens in clinical stool samples collected 2 days after the onset of illness. A number of commercial ELISA methods are now available for norovirus (based on a method similar to that described above). However, the sensitivity in these methods (lower than that of the RT-PCR methods) still remains the biggest issue [99–102].

4.3.4.2   Commercial Methods

A number of commercial norovirus ELISA kits have been developed and tested [101,103]. In general, the commercial ELISAs have a lower sensitivity and detection limit than the RT-PCR methods. However, the ELISAs are useful for preliminary screening and can be used to screen a large numbers of samples [101]. Denka Seiken developed a norovirus antigen ELISA kit (NV-AD version II, Denka Seken, Tokyo, Japan) that can detect both norovirus GI and GII antigens. Murine monoclonal antibodies (which were raised against GI and GII VLPs) capture norovirus antigens in clinical stool samples and then a mixture of peroxidase-labeled rabbit polyclonal (raised against GI and GII VLPs) and peroxidase-labeled murine monoclonal antibodies (raised against GI and GII VLPs) bind to the norovirus antigens. The ELISA kit was found to have a sensitivity of 72.4% and specificity of 98.9% (compared to RT-PCR).

4.4   FUTURE DEVELOPMENT TRENDS

New methods and technologies are continuously being developed and tested. More rapid, reliable, and sensitive detection systems will benefit diagnostic laboratories and may improve outbreak management. Methods that can reduce the time needed for complete analysis will allow for more samples to be processed, which in turn will provide better information on these viruses. Reliable and highly sensitive methods that can be used worldwide, will improve surveillance and reporting.

A number of companies have developed kits that can extract RNA from up to 96 samples at one time. However, slight modifications may be necessary to extract RNA from stool samples since some of the kits were designed to extract RNA from tissue or cells. Ambion developed a rapid high-throughput (96 samples) viral RNA isolation kit (MagMax-96 Viral RNA Isolation Kit; Ambion, Austin, TX) that captures both viral RNA and DNA onto microspherical paramagnetic beads. The Ambion kit can be used with either multichannel pipettors or an automated KingFisher 96 magnetic particle processor (Thermo Scientific, Waltham, MA). Promega developed an automated system that can extract total RNA from 96 samples (SV 96 Total RNA Isolation System; Promega, Madison, WI) and capture the extracted RNA onto glass fibers. This kit can also be used with automated platforms such as the Biomek FX automated platform. Beckman Coulter Qiagen developed a kit that can extract DNA and RNA (96 samples) from plasma, serum, and other cell-free body fluids (QIAamp Virus BioRobot 9604 Kit; Qiagen) and capture the nucleic acid onto a silica-gel membrane. This method can be used with an automated BioRobot Universal System (Qiagen). Roche has developed a system that can completely process up to 96 samples at one time (COBAS AmpliPrep/COBAS TaqMan System; Roche). This system has no manual transfer of samples between the RNA purification, amplification, and detection steps, allowing continuous operation and rapid results. However, currently this system can be used only for HIV-1, hepatitis C virus, and hepatitis B virus using the built-in real-time RT-PCR.

In recent years, other detection methods have been developed, including isothermal nucleic acid sequence-based amplification (NASBA), real-time reverse transcription-loop mediated isothermal amplification (LAMP), and real-time LightCycler RT-PCR [76,104,105]. The NASBA method uses three enzymes (reverse transcriptase, T7 RNA polymerase, and RNaseH) and two target-specific oligonucleotide primers. The antisense primer contains a bacteriophage T7 promoter sequence, while the sense primer contains an electrochemiluminescent (ECL) tail that is used in the detection of the amplified product. A NASBA method for the detection of norovirus was recently designed and was found to have a detection limit similar to RT-PCR [105]. This method could be completed in 4–6 h; however, further refinements will be necessary because the method is specific for the Norwalk virus sequence. A LAMP for the detection of norovirus was also recently developed as a one-step, single-tube method that could be completed within 90 min [104]. This method was designed to detect both GI and GII viruses in one tube. The LAMP detection limit was between 10^2 and 10^3 copies per tube for GI and GII, respectively. Interestingly, the sensitivity of the LAMP method was found to be different for each genotype. An evaluation using clinical stool samples showed a good concordance between the LAMP and conventional RT-PCR (95%). A LightCycler RT-PCR method was also recently developed for the detection of noroviruses [76]. The developers similarly found that the LightCycler RT-PCR had 100% concordance with conventional RT-PCR, but had a 4-log higher sensitivity.

Immunochromatography (IC) is another rapid detection method, and a method to detect norovirus by IC was recently developed by a company in Japan (Immuno-Probe Company, Saitama, Japan). This method uses polyclonal rabbit antisera raised against VLPs (either GII/3 or GII/4) expressed in a baculovirus expression system [106]. For a positive result, the system captures norovirus antigens with a polyclonal antibody (either GII/3 or GII/4) on a test line, which then turns pink. The current method requires two different strips, is specific for either GII/3 or GII/4 strains, and has a low sensitivity against GI strains or other GII genotypes (except GII/3 or GII/4). Briefly, a 10% stool sample is clarified by centrifugation at $10,000 \times g$ for 10 min. Then, 50 µL of clarified stool sample is mixed with 50 µL of reaction buffer (0.2 M NH_4Cl buffer containing 0.15 M NaCl and 0.5% Tween 20). The mixture is added to a sample pad on the IC strip and left for approximately 15 min. The sensitivity and specificity of this test have been reported to be approximately 70% and 94%, respectively, although only a handful of clinical stool samples have been tested. Several improvements to the method have been made, although further refinements are necessary [107].

Recently, a microarray assay was developed for the detection and typing of noroviruses [108]. Briefly, RNA was extracted from stool samples using a commercial method (Tripure Isolation Reagent; Roche) and then amplified using a monoplex RT-PCR (designed to amplify both norovirus and astrovirus). The RT-PCR product was transcribed into single-stranded RNA (ssRNA) and then the ssRNA was hybridized to short detection primers on the microarray. Reverse transcriptase was used to add fluorescent nucleotides to the hybridized RNA, which was then measured by a microarray scanner. The microarray assay was able to differentiate between 12 of 13 different norovirus genotypes, and it has the potential to rapidly detect and genotype large numbers of samples, reducing the time needed for conventional sequencing and analysis.

Grant S. Hansman graduated from Macquarie University, Sydney, New South Wales, Australia with a BSc; he then started his medical research career at the Prince of Wales Hospital in New South Wales, Australia, where he investigated norovirus infections in hospitalized patients. He then moved to Japan for pursuing his PhD at the University of Tokyo, Tokyo, Japan, where he performed molecular epidemiological studies in developing countries with unreported norovirus and sapovirus infections. After completing his PhD, he began working at the National Institute of Infectious Diseases, Tokyo. His current research interests include viral epidemiology, expression, and cross-reactivity.

REFERENCES

1. Murray, C.J. and Lopez, A.D., Evidence-based health policy—lessons from the Global Burden of Disease Study, *Science*, 274, 740, 1996.
2. Vinje, J. et al., International collaborative study to compare reverse transcriptase PCR assays for detection and genotyping of noroviruses, *J. Clin. Microbiol.*, 41, 1423, 2003.
3. Kojima, S. et al., Genogroup-specific PCR primers for detection of Norwalk-like viruses, *J. Virol. Methods*, 100, 107, 2002.
4. Ozawa, K. et al., Norovirus infections in symptomatic and asymptomatic food-handlers in Japan, *J. Clin. Microbiol.*, 45, 3996, 2007.
5. Hansman, G.S. et al., Viral gastroenteritis in Mongolian infants, *Emerg. Infect. Dis.*, 11, 180, 2005.
6. Hansman, G.S. et al., Genetic diversity of norovirus and sapovirus in hospitalized infants with sporadic cases of acute gastroenteritis in Chiang Mai, Thailand, *J. Clin. Microbiol.*, 42, 1305, 2004.
7. Inouye, S. et al., Surveillance of viral gastroenteritis in Japan: Pediatric cases and outbreak incidents, *J. Infect. Dis.*, 181(Suppl 2), S270, 2000.
8. McEvoy, M. et al., An outbreak of viral gastroenteritis on a cruise ship, *Commun. Dis. Rep. CDR Rev.*, 6, R188, 1996.
9. Russo, P.L. et al., Hospital outbreak of Norwalk-like virus, *Infect. Control Hosp. Epidemiol.*, 18, 576, 1997.
10. McIntyre, L. et al., Gastrointestinal outbreaks associated with Norwalk virus in restaurants in Vancouver, British Columbia, *Can. Commun. Dis. Rep.*, 28, 197, 2002.
11. Beuret, C. et al., Virus-contaminated oysters: A three-month monitoring of oysters imported to Switzerland, *Appl. Environ. Microbiol.*, 69, 2292, 2003.
12. Johansson, P.J. et al., Food-borne outbreak of gastroenteritis associated with genogroup I calicivirus, *J. Clin. Microbiol.*, 40, 794, 2002.
13. Kapikian, A.Z. et al., Visualization by immune electron microscopy of a 27-nm particle associated with acute infectious nonbacterial gastroenteritis, *J. Virol.*, 10, 1075, 1972.
14. Nakata, S. et al., Prevalence of human calicivirus infections in Kenya as determined by enzyme immunoassays for three genogroups of the virus, *J. Clin. Microbiol.*, 36, 3160, 1998.
15. Noel, J.S. et al., Identification of a distinct common strain of "Norwalk-like viruses" having a global distribution, *J. Infect. Dis.*, 179, 1334, 1999.
16. Mounts, A.W. et al., Cold weather seasonality of gastroenteritis associated with Norwalk-like viruses, *J. Infect. Dis*, 181(Suppl 2), S284, 2000.
17. Hedlund, K.O. et al., Epidemiology of calicivirus infections in Sweden, 1994–1998, *J. Infect. Dis.*, 181(Suppl 2), S275, 2000.
18. Koopmans, M. et al., Molecular epidemiology of human enteric caliciviruses in the Netherlands, *J. Infect. Dis.*, 181(Suppl 2), S262, 2000.
19. White, P.A. et al., Norwalk-like virus 95/96-US strain is a major cause of gastroenteritis outbreaks in Australia, *J. Med. Virol.*, 68, 113, 2002.
20. O'Ryan, M.L. et al., Human caliciviruses are a significant pathogen of acute sporadic diarrhea in children of Santiago, Chile, *J. Infect. Dis.*, 182, 1519, 2000.
21. Koopmans, M. et al., Molecular epidemiology of human enteric caliciviruses in the Netherlands, *J. Infect. Dis.*, 181(Suppl 2), S262, 2000.
22. Lawson, H.W. et al., Waterborne outbreak of Norwalk virus gastroenteritis at a southwest US resort: Role of geological formations in contamination of well water, *Lancet*, 337, 1200, 1991.
23. Gallimore, C.I. et al., Detection of multiple enteric virus strains within a foodborne outbreak of gastroenteritis: An indication of the source of contamination, *Epidemiol. Infect.*, 133, 41, 2005.
24. Nishida, T. et al., Genotyping and quantitation of noroviruses in oysters from two distinct sea areas in Japan, *Microbiol. Immunol.*, 51, 177, 2007.

25. Kageyama, T. et al., Coexistence of multiple genotypes, including newly identified genotypes, in zoutbreaks of gastroenteritis due to Norovirus in Japan, *J. Clin. Microbiol.*, 42, 2988, 2004.
26. Zheng, D.P. et al., Norovirus classification and proposed strain nomenclature, *Virology*, 346, 312, 2006.
27. Katayama, K. et al., Phylogenetic analysis of the complete genome of 18 Norwalk-like viruses, *Virology*, 299, 225, 2002.
28. Lochridge, V.P. and Hardy, M.E., Snow Mountain virus genome sequence and virus-like particle assembly, *Virus Genes*, 26, 71, 2003.
29. Vinje, J. et al., Genetic polymorphism across regions of the three open reading frames of "Norwalk-like viruses," *Arch. Virol*, 145, 223, 2000.
30. Levin, S.A. et al., Evolution and persistence of influenza A and other diseases, *Math. Biosci.*, 188, 17, 2004.
31. De Jong, J.C. et al., Influenza virus: A master of metamorphosis, *J. Infect.*, 40, 218, 2000.
32. Hansman, G.S. et al., Recombinant sapovirus gastroenteritis, Japan, *Emerg. Infect. Dis.*, 13, 786, 2007.
33. Hansman, G.S. et al., An outbreak of gastroenteritis due to Sapovirus, *J. Clin. Microbiol.*, 45, 1347, 2007.
34. Noel, J.S. et al., Parkville virus: A novel genetic variant of human calicivirus in the Sapporo virus clade, associated with an outbreak of gastroenteritis in adults, *J. Med. Virol.*, 52, 173, 1997.
35. Johansson, P.J. et al., A nosocomial sapovirus-associated outbreak of gastroenteritis in adults, *Scand. J. Infect. Dis.*, 37, 200, 2005.
36. Hansman, G.S. et al., Recombinant Sapovirus gastroenteritis, Japan, *Emerg. Infect. Dis.*, 13, 786, 2007.
37. Chiba, S. et al., An outbreak of gastroenteritis associated with calicivirus in an infant home, *J. Med. Virol.*, 4, 249, 1979.
38. Hansman, G.S. et al., Sapovirus in water, Japan, *Emerg. Infect. Dis.*, 13, 133, 2007.
39. Hansman, G.S. et al., Human sapovirus in clams, Japan, *Emerg. Infect. Dis.*, 13, 620, 2007.
40. Oka, T. et al., Identification of the cleavage sites of sapovirus open reading frame 1 polyprotein, *J. Gen. Virol.*, 87, 3329, 2006.
41. Oka, T. et al., Proteolytic processing of sapovirus ORF1 polyprotein, *J. Virol.*, 79, 7283, 2005.
42. Oka, T. et al., Cleavage activity of the sapovirus 3C-like protease in Escherichia coli, *Arch. Virol.*, 150, 2539, 2005.
43. Katayama, K. et al., Novel recombinant sapovirus, *Emerg. Infect. Dis.*, 10, 1874, 2004.
44. Lole, K.S. et al., Full-length human immunodeficiency virus type 1 genomes from subtype C-infected seroconverters in India, with evidence of intersubtype recombination, *J. Virol.*, 73, 152, 1999.
45. Hansman, G.S. et al., Intergenogroup recombination in sapoviruses, *Emerg. Infect. Dis.*, 11, 1916, 2005.
46. Hansman, G.S. et al., Cross-reactivity among sapovirus recombinant capsid proteins, *Arch. Virol.*, 150, 21, 2005.
47. Guntapong, R. et al., Norovirus and sapovirus infections in Thailand, *Jpn. J. Infect. Dis.*, 57, 276, 2004.
48. Gray, J.J. et al., Detection of immunoglobulin M (IgM), IgA, and IgG Norwalk virus-specific antibodies by indirect enzyme-linked immunosorbent assay with baculovirus-expressed Norwalk virus capsid antigen in adult volunteers challenged with Norwalk virus, *J. Clin. Microbiol.*, 32, 3059, 1994.
49. Wyatt, R.G. et al., Comparison of three agents of acute infectious nonbacterial gastroenteritis by cross-challenge in volunteers, *J. Infect. Dis.*, 129, 709, 1974.
50. Parker, T.D. et al., Identification of Genogroup I and Genogroup II broadly reactive epitopes on the norovirus capsid, *J. Virol.*, 79, 7402, 2005.
51. Kitamoto, N. et al., Cross-reactivity among several recombinant calicivirus virus-like particles (VLPs) with monoclonal antibodies obtained from mice immunized orally with one type of VLP, *J. Clin. Microbiol.*, 40, 2459, 2002.
52. Sakuma, Y. et al., Prevalence of antibody to human calicivirus in general population of northern Japan, *J. Med. Virol.*, 7, 221, 1981.
53. Jiang, X. et al., Expression and characterization of Sapporo-like human calicivirus capsid proteins in baculovirus, *J. Virol. Methods*, 78, 81, 1999.
54. Numata, K. et al., Molecular characterization of morphologically typical human calicivirus Sapporo, *Arch. Virol.*, 142, 1537, 1997.
55. Oka, T. et al., Expression of sapovirus virus-like particles in mammalian cells, *Arch. Virol.*, 151, 399, 2006.
56. Hansman, G.S. et al., Genetic and antigenic diversity among noroviruses, *J. Gen. Virol.*, 87, 909, 2006.
57. Hansman, G.S. et al., Antigenic diversity of human Sapoviruses, *Emerg. Infect. Dis*, 13, 1519, 2007.

58. Jiang, X. et al., Norwalk virus genome cloning and characterization, *Science*, 250, 1580, 1990.
59. Jiang, X. et al., Detection of Norwalk virus in stool by polymerase chain reaction, *J. Clin. Microbiol.*, 30, 2529, 1992.
60. Green, J. et al., Norwalk-like viruses: Demonstration of genomic diversity by polymerase chain reaction, *J. Clin. Microbiol.*, 31, 3007, 1993.
61. Green, S.M. et al., Capsid diversity in small round-structured viruses: Molecular characterization of an antigenically distinct human enteric calicivirus, *Virus Res.*, 37, 271, 1995.
62. Hale, A.D. et al., Comparison of four RNA extraction methods for the detection of small round structured viruses in faecal specimens, *J. Virol. Methods*, 57, 195, 1996.
63. Kawamoto, H. et al., Nucleotide sequence analysis and development of consensus primers of RT- PCR for detection of Norwalk-like viruses prevailing in Japan, *J. Med. Virol.*, 64, 569, 2001.
64. Honma, S. et al., Evaluation of nine sets of PCR primers in the RNA dependent RNA polymerase region for detection and differentiation of members of the family Caliciviridae, Norwalk virus and Sapporo virus, *Microbiol. Immunol.*, 44, 411, 2000.
65. Honma, S. et al., Sensitive detection and differentiation of Sapporo virus, a member of the family Caliciviridae, by standard and booster nested polymerase chain reaction, *J. Med. Virol.*, 65, 413, 2001.
66. Yuen, L.K. et al., Heminested multiplex reverse transcription-PCR for detection and differentiation of Norwalk-like virus genogroups 1 and 2 in fecal samples, *J. Clin. Microbiol.*, 39, 2690, 2001.
67. Jiang, X. et al., Design and evaluation of a primer pair that detects both Norwalk- and Sapporo-like caliciviruses by RT-PCR, *J. Virol. Methods*, 83, 145, 1999.
68. Saito, H. et al., Application of RT-PCR designed from the sequence of the local SRSV strain to the screening in viral gastroenteritis outbreaks, *Microbiol. Immunol.*, 42, 439, 1998.
69. Katayama, K. et al., Phylogenetic analysis of the complete genome of 18 Norwalk-like viruses, *Virology*, 299, 225, 2002.
70. Jothikumar, N. et al., Rapid and sensitive detection of noroviruses by using TaqMan-based one-step reverse transcription-PCR assays and application to naturally contaminated shellfish samples, *Appl. Environ. Microbiol.*, 71, 1870, 2005.
71. Richards, G.P. et al., A SYBR green, real-time RT-PCR method to detect and quantitate Norwalk virus in stools, *J. Virol. Methods*, 116, 63, 2004.
72. Richards, G.P. et al., Genogroup I and II noroviruses detected in stool samples by real-time reverse transcription-PCR using highly degenerate universal primers, *Appl. Environ. Microbiol.*, 70, 7179, 2004.
73. Hohne, M. and Schreier, E., Detection and characterization of norovirus outbreaks in Germany: Application of a one-tube RT-PCR using a fluorogenic real-time detection system, *J. Med. Virol.*, 72, 312, 2004.
74. Kageyama, T. et al., Broadly reactive and highly sensitive assay for Norwalk-like viruses based on real-time quantitative reverse transcription-PCR, *J. Clin. Microbiol.*, 41, 1548, 2003.
75. Trujillo, A.A. et al., Use of TaqMan real-time reverse transcription-PCR for rapid detection, quantification, and typing of norovirus, *J. Clin. Microbiol.*, 44, 1405, 2006.
76. Pang, X. et al., Evaluation and validation of real-time reverse transcription-pcr assay using the LightCycler system for detection and quantitation of norovirus, *J. Clin. Microbiol.*, 42, 4679, 2004.
77. Ueki, Y. et al., Norovirus pathway in water environment estimated by genetic analysis of strains from patients of gastroenteritis, sewage, treated wastewater, river water and oysters, *Water Res.*, 39, 4271, 2005.
78. Sano, D. et al., Detection of enteric viruses in municipal sewage sludge by a combination of the enzymatic virus elution method and RT-PCR, *Water Res.*, 37, 3490, 2003.
79. Lewis, G.D. and Metcalf, T.G., Polyethylene glycol precipitation for recovery of pathogenic viruses, including hepatitis A virus and human rotavirus, from oyster, water, and sediment samples, *Appl. Environ. Microbiol.*, 54, 1983, 1988.
80. Katayama, H. et al., Development of a virus concentration method and its application to detection of enterovirus and norwalk virus from coastal seawater, *Appl. Environ. Microbiol.*, 68, 1033, 2002.
81. Dreier, J. et al., Enhanced reverse transcription-PCR assay for detection of norovirus genogroup I, *J. Clin. Microbiol.*, 44, 2714, 2006.
82. Houde, A. et al., Comparative evaluation of RT-PCR, nucleic acid sequence-based amplification (NASBA) and real-time RT-PCR for detection of noroviruses in faecal material, *J. Virol. Methods*, 135, 163, 2006.
83. Atmar, R.L. and Estes, M.K., Diagnosis of noncultivatable gastroenteritis viruses, the human caliciviruses, *Clin. Microbiol. Rev.*, 14, 15, 2001.

84. Wang, Q.H. et al., Development of a new microwell hybridization assay and an internal control RNA for the detection of porcine noroviruses and sapoviruses by reverse transcription-PCR, *J. Virol. Methods*, 132, 135, 2006.

85. Schwab, K.J. et al., Use of heat release and an internal RNA standard control in reverse transcription-PCR detection of Norwalk virus from stool samples, *J. Clin. Microbiol.*, 35, 511, 1997.

86. Schmid, M. et al., Fast detection of Noroviruses using a real-time PCR assay and automated sample preparation, *BMC Infect. Dis*, 4, 15, 2004.

87. Hansman, G.S. et al., Genetic diversity of Sapovirus in children, Australia, *Emerg. Infect. Dis.*, 12, 141, 2006.

88. Vinje, J. and Koopmans, M.P., Simultaneous detection and genotyping of "Norwalk-like viruses" by oligonucleotide array in a reverse line blot hybridization format, *J. Clin. Microbiol.*, 38, 2595, 2000.

89. Green, J. et al., Broadly reactive reverse transcriptase polymerase chain reaction for the diagnosis of SRSV-associated gastroenteritis, *J. Med. Virol.*, 47, 392, 1995.

90. Fankhauser, R.L. et al., Epidemiologic and molecular trends of "Norwalk-like viruses" associated with outbreaks of gastroenteritis in the United States, *J. Infect. Dis.*, 186, 1, 2002.

91. Oka, T. et al., Detection of human sapovirus by real-time reverse transcription-polymerase chain reaction, *J. Med. Virol.*, 78, 1347, 2006.

92. Nakata, S. et al., Microtiter solid-phase radioimmunoassay for detection of human calicivirus in stools, *J. Clin. Microbiol.*, 17, 198, 1983.

93. Farkas, T. et al., Prevalence and genetic diversity of human caliciviruses (HuCVs) in Mexican children, *J. Med. Virol.*, 62, 217, 2000.

94. Wolfaardt, M. et al., Incidence of human calicivirus and rotavirus infection in patients with gastroenteritis in South Africa, *J. Med. Virol.*, 51, 290, 1997.

95. Chiba, S. et al., Sapporo virus: History and recent findings, *J. Infect. Dis.*, 181(Suppl. 2), S303, 2000.

96. Nakata, S. et al., Detection of human calicivirus antigen and antibody by enzyme-linked immunosorbent assays, *J. Clin. Microbiol.*, 26, 2001, 1988.

97. Hansman, G.S. et al., Development of an antigen ELISA to detect sapovirus in clinical stool specimens, *Arch. Virol.*, 151, 551, 2006.

98. Hansman, G.S. et al., Characterization of polyclonal antibodies raised against sapovirus genogroup five virus-like particles, *Arch. Virol.*, 150, 1433, 2005.

99. Okitsu-Negishi, S. et al., Detection of norovirus antigens from recombinant virus-like particles and stool samples by a commercial norovirus enzyme-linked immunosorbent assay kit, *J. Clin. Microbiol.*, 44, 3784, 2006.

100. Dimitriadis, A. and Marshall, J.A., Evaluation of a commercial enzyme immunoassay for detection of norovirus in outbreak specimens, *Eur. J. Clin. Microbiol. Infect. Dis.*, 24, 615, 2005.

101. de Bruin, E. et al., Diagnosis of Norovirus outbreaks by commercial ELISA or RT-PCR, *J. Virol. Methods*, 137, 259, 2006.

102. Richards, A.F. et al., Evaluation of a commercial ELISA for detecting Norwalk-like virus antigen in faeces, *J. Clin. Virol.*, 26, 109, 2003.

103. Burton-MacLeod, J.A. et al., Evaluation and comparison of two commercial enzyme-linked immunosorbent assay kits for detection of antigenically diverse human noroviruses in stool samples, *J. Clin. Microbiol.*, 42, 2587, 2004.

104. Fukuda, S. et al., Rapid detection of norovirus from fecal specimens by real-time reverse transcription-loop-mediated isothermal amplification assay, *J. Clin. Microbiol.*, 44, 1376, 2006.

105. Greene, S.R. et al., Evaluation of the NucliSens Basic Kit assay for detection of Norwalk virus RNA in stool specimens, *J. Virol. Methods*, 108, 123, 2003.

106. Takanashi, S. et al., Development of a rapid immunochromatographic test for noroviruses genogroups I and II, *J. Virol. Methods*, 148, 1, 2007.

107. Khamrin, P. et al., Evaluation of immunochromatography and commercial enzyme-linked immunosorbent assay for rapid detection of norovirus antigen in stool samples, *J. Virol. Methods*, 147, 360, 2008.

108. Jaaskelainen, A.J. and Maunula, L., Applicability of microarray technique for the detection of noro- and astroviruses, *J. Virol. Methods*, 136, 210, 2006.

Part II

*Purification of Nucleic Acids
from Bacteria*

5 Isolation of Bacterial DNA from Cultures

Dongyou Liu

CONTENTS

5.1 INTRODUCTION

5.1.1 CLASSIFICATION

Bacteria are small unicellular organisms that belong taxonomically to the domain Bacteria (or Eubacteria) in the kingdom Prokaryotae (also known as Prokaryota or Monera, which includes a second domain Archaea, or Archaebacteria for ancient bacteria). Both the domains Bacteria and Archaea appear to have evolved independently from an ancient common ancestor. With sizes ranging from 10^{-7} to 10^{-4} mm, prokaryotes are bigger than viruses (10^{-8} to 10^{-6} mm), but smaller than eukaryotes (10^{-5} to 10^{-3} mm). Indeed, eukaryotes may have arisen from ancient bacteria entering into endosymbiotic associations with the ancestors of eukaryotic cells (possibly related to the Archaea) to form either mitochondria or hydrogenosomes. A subsequent independent engulfment of cyanobacterial-like organisms by some mitochondria-containing eukaryotes may have led to the formation of chloroplasts in algae and plants.

In contrast to the organisms in the eukaryotic kingdoms Protista, Fungi, Plantae, and Animalia, those in the kingdom Prokaryotae lack nuclear membrane (with their DNA usually in a loop or coil), contain few independent membrane-bounded cytoplasmic organelles (e.g., vacuole, endoplasmic reticulum, Golgi apparatus, and mitochondria) apart from two discernable organelles (chromosome and ribosome), have no unique structures in their plasma membrane and cell wall, and do not undergo endocytosis and exocytosis. More specifically, whereas eukaryotic chromosome resides within a membrane-delineated nucleus, bacterial chromosome is located inside the bacterial cytoplasm. This entails that all cellular events (e.g., translational and transcriptional processes, and interaction of chromosome with other cytoplasmic structures) in prokaryotes occur in the same compartment. Furthermore, while eukaryotic chromosome is packed with histones to form linear chromatin, bacterial chromosome assumes a highly compact supercoiled structure in circular form (and rarely in linear form).

Although Archaea are similar to Bacteria in most aspects of cell structure and metabolism, they differ from Bacteria in that being extremophiles, they can live in extreme environments where no other life forms exist. This may be due to the unique structure in archaeal lipids in which the stereochemistry of the glycerol is the reverse of that found in bacteria and eukaryotes, resulting possibly from an adaptation on the part of Archaea to hyperthermophily. In addition, archaeal cell wall does not contain muramic acid, which is commonly present in Bacteria. The archaeal RNA polymerase core is composed of ten subunits in comparison with four subunits in Bacteria. Besides possessing distinct tRNA and rRNA genes, Archaea uses eukaryotic-like initiation and elongation factors in protein translation, and their transcription involves TATA-binding proteins and TFIIB as in eukaryotes.

Bacteria have been classified on the basis of their differences in morphology (e.g., rod, cocci, spirilla, and filament), cell wall structure (e.g., Gram-negative and Gram-positive), growth characteristics (e.g., aerobic and anaerobic), and genetic features (e.g., 16S and 23S rRNA). Whereas the domain Bacteria (Eubacteria) consists of 26 phyla (i.e., Acidobacteria, Actinobacteria, Aquificae, Bacteroidetes, Chlamydiae, Chlorobi, Chloroflexi, Chrysiogenetes, Cyanobacteria, Deferribacteres, Deinococcus-Thermus, Dictyoglomi, Fibrobacteres, Firmicutes, Fusobacteria, Gemmatimonadetes, Lentisphaerae, Nitrospirae, Planctomycetes, Proteobacteria, Spirochaetes, Tenericutes, Thermodesulfobacteria, Thermomicrobia, Thermotogae, and Verrcomicrobia), the domain Archaea (Archaeobacteria) comprises 2 phyla (i.e., Crenarchaeota and Euryarchaeota). Interestingly, among the 26 phyla in the domain Bacteria, Proteobacteria and Firmicutes contain the largest numbers of genera and species followed by Cyanobacteria, Bacteroidetes, Spirochaetes, and Flavobacteria. Bacteria from other phyla are comparatively rare, from which fewer genera and species have been described.

5.1.2 Morphology and Structure

Being usually 0.2–2.0 μm in width and 2–8 μm in length, bacteria are about 10 times smaller than eukaryotic cells. On the one extreme, there are a few bacterial species (e.g., *Thiomargarita namibiensis* and *Epulopiscium fishelsoni*) that measure up to half a millimeter long and are visible to the naked eye. On the other extreme, the smallest bacteria in the genus *Mycoplasma* are only 0.3 μm in size, which are as small as the largest viruses.

Bacteria typically come in four shapes: rod-like bacilli, spherical cocci, spiral bacteria (also called spirilla), and filamentous bacteria. Occasionally, a small number of bacterial species may assume tetrahedral or cuboidal shapes. While many bacterial species exist as single cells, others form characteristic patterns such as diploids (pairs) by *Neisseria*, chains by *Streptococcus*, and clusters (bunch of grapes) by *Staphylococcus*. Additionally, some bacteria may be elongated to form filaments (e.g., *Actinobacteria*), which are often surrounded by a sheath containing many individual cells. The elaborated, branched filaments formed by *Nocardia* may even resemble fungal mycelia in appearance. Frequently, bacteria attach to solid surfaces to form dense aggregations called biofilms (or bacterial mats), which may measure a few micrometers in thickness up to half a meter in depth, and which comprise multiple species of bacteria, archaea, and protists. The formation of biofilms by pathogenic bacteria plays a critical role in chronic bacterial infections and infections relating to implanted medical devices because bacteria are well protected within these structures from antibiotic therapy. Furthermore, with limited access to amino acids, Myxobacteria use quorum sensing to detect surrounding cells, migrate toward each other, and aggregate to form fruiting bodies of up to 500 mm in length with approximately 100,000 bacterial cells. Some of the cells in the fruiting bodies differentiate into a specialized dormant state called myxospores, which are highly resistant to desiccation and other adverse environmental conditions.

Cell wall is a rigid layer located external to the cell membrane, providing the cell with structural support and protection, and acting as a filtering mechanism. It is found in the kingdoms Prokaryotae, Fungi, and Plantae, but absent in the kingdoms Animalia and most Protista. While bacterial cell wall is made up of peptidoglycan (also called murein, which in turn is composed of polysaccharide chain cross-linked by peptides containing D-amino acids), archaeal cell wall consists of surface-layer proteins (also known as S-layer), pseudopeptidoglycan (pseudomurein) and polysaccharides. By contrast, fungal cell wall includes chitin, algal cell wall has glycoprotein and polysaccharides, and plant cell wall often incorporates cellulose and proteins such as extensins.

According to their reaction with the Gram stain (with crystal violet as primary stain and Gram's iodine and basic fuchsin as subsequent stains), bacteria are divided into Gram-positive and Gram-negative categories. Gram-positive bacterial cell wall is composed of several layers of peptidoglycan (which is responsible for retaining the crystal violet dyes during the Gram staining procedure, leading to its purple color) surrounded by a second lipid membrane containing lipopolysaccharides and lipoproteins. Located outside the cytoplasmic membrane, peptidoglycan is a large polymer (formed by poly-N-acetylglucosamine and N-acetylmuramic acid) that contributes to the structural integrity of bacterial cell wall, in addition to countering the osmotic pressure of the cytoplasm. Peptidoglycan is predominant in the cell walls of high G+C% and low G+C% Gram-positive organisms (e.g., Actinobacteria and Firmicutes). Also imbedded in the Gram-positive cell wall are teichoic acids, some of which are lipid-linked to form lipoteichoic acids. On the other hand, Gram-negative cell wall has a thin peptidoglycan layer adjacent to cytoplasmic membrane, which is attributable to its inability to retain the crystal violet stain upon decolorization with ethanol during the gram-staining procedure (leading to its red or pink color after restaining with basic fuchsin). Apart from the thin peptidoglycan layer, Gram-negative cell wall also has an outer membrane that is formed by phospholipids and lipopolysaccharides.

Within the Gram-positive bacterial category, there is another distinct group of bacteria (i.e., acid-fast bacteria such as *Mycobacterium* and *Nocardia*) that can resist decolorization with an acid–alcohol mixture during the acid-fast (or Ziehl–Neelsen) staining procedure, and retain the initial dye carbol fuchsin and appear red. The acid-fast cell wall of *Mycobacterium* includes a large amount of glycolipids, especially mycolic acids that make up approximately 60% of the acid-fast cell wall in addition to a thin, inner layer peptidoglycan. The presence of the mycolic acids and other glycolipids impede the entry of chemicals causing the organisms to grow slowly and be more resistant to chemical agents and lysosomal components of phagocytes than most other bacteria.

Whereas a vast majority of bacteria possess the Gram-negative cell wall, the Firmicutes and Actinobacteria (previously known as the low G+C% and high G+C% Gram-positive bacteria, respectively) have the Gram-positive structure, and the genus *Mycoplasma* is devoid of a cell wall. The differences in the cell wall often determine the susceptibility and resistance of bacteria to antibiotics and other therapeutic reagents. Given that *Mycoplasma* species lack a cell wall, they are unaffected by some commonly used antibiotics such as penicillin and streptomycin that target cell wall synthesis. Not surprisingly, with their small size (0.3 μm), *Mycoplasma* species are often identified as a source of contaminating infection in cell culture (where penicillin and streptomycin are incorporated in the culture media) causing retarded growth of cultured cell lines. The cell wall of bacteria forms part of pathogen-associated molecular patterns, which are recognized by pattern-recognition receptors in mammalian hosts to initiate and promote innate and adaptive immune defenses against invading bacteria.

Besides possessing a distinct cell wall, bacteria have several other recognizable extracellular structures (i.e., flagella, pili, and fimbriae), which protrude from bacterial cell wall and are involved in bacterial twitching movement as well as interaction with one another and other organisms.

Bacterial flagellum (measuring 20 nm in diameter and up to 20 μm in length) is a long, whip-like, and helical projection made up of repeating flagellin protein. The numbers and arrangements of flagella vary among bacterial genera and species. Monotrichous bacteria (e.g., *Vibrio cholerae*) have a single flagellum, amphitrichous bacteria contain a single flagellum on each of cell poles, lophotrichous bacteria include multiple flagella that are located at one cell pole, and peritrichous bacteria have multiple flagella that are situated at several locations. Flagella in bacteria are powered by a flow of H⁺ ions (sometimes Na⁺ ions), those in archaea are powered by adenosine 5'-triphoshate. Despite showing similar appearance, eukaryotic flagella (called cilia or undulipodia) differ from prokaryotic flagella in both structure and evolutionary origin. A eukaryotic flagellum is a bundle of nine fused pairs of microtubule doublets surrounding two single microtubules. Eukaryotic flagella are often arranged en masse at the surface of a stationary cell anchored within an organ, and by lashing back and forth, serve to move fluids along mucous membranes such as trachea. Additionally, some eukaryotic cells (e.g., rod photoreceptor cells of eye, olfactory receptor cells of nose, and kinocilium in cochlea of ear) have immotile flagella that function as sensation and signal transduction devices.

Pilus and fimbria are proteinaceous, hair- or thread-like appendages in bacteria (particularly of Gram-negative category) that are much shorter and thinner than flagellum. Bacteria have up to 10 pili (typically 6–7 nm in diameter), whose main function is to connect the bacterium to another of the same species, or of a different species, to enable transfer of plasmids between the bacteria (i.e., conjugation). A fimbria (measuring 2–10 nm in diameter and up to several micrometers in length) is shorter than pilus. It is deployed by the bacterium (which possesses as many as 1000 fimbriae) to attach itself to surface of another bacterium (to form a biofilm) or host cell (to facilitate invasion). Many pilin proteins are characteristic among bacterial species and subgroups, which have been exploited as targets for serological typing of bacteria (serotypes or serovars).

Many bacteria produce capsules or slime layers around their cells, which can protect cells from engulfment by eukaryotic cells (e.g., macrophages), act as antigens for cell recognition, and aid attachment to surfaces and the formation of biofilms. In addition, some Gram-positive bacteria (e.g., *Bacillus, Clostridium*, and *Anaerobacter*) can form highly resistant, dormant structures called endospores, which contain a central core of cytoplasm with DNA and ribosomes surrounded by a cortex layer and protected by an impermeable and rigid coat. Endospores can survive extreme physical and chemical stresses (e.g., UV lights, γ-radiation, detergents and disinfectants, heat, pressure, and desiccation), and may remain viable for millions of years. Endospore-forming bacteria (e.g., *Bacillus anthrax* and *Clostridium tetanus*) are also capable of causing disease.

5.1.3 BIOLOGY

Having many metabolic pathways (e.g., glycolysis, electron transport chains, chemiosmosis, cellular respiration, and photosynthesis), bacteria can use virtually all carbon or energy supplies for their maintenance and growth. In the laboratory, bacteria are easily grown using either solid or liquid media (e.g., Luria Bertani broth) containing high levels of nutrients to produce large amounts of cells rapidly. Solid growth media (e.g., agar plates) are useful for isolation of pure cultures of a bacterial strain, and liquid growth media are employed to generate bulk quantities of bacterial cells. Additionally, selective media (containing specific nutrients and antibiotics) are utilized to assist the identification of specific bacterial organisms.

As single-celled organisms, prokaryotes reproduce by asexual binary fission, which begins with DNA replication within the cell until the entire prokaryotic DNA is duplicated. The two chromosomes then separate as the cell grows and the cell membrane invaginates, splitting the cell into two daughter cells. This reproductive process is highly efficient and leads to exponential growth of bacteria. In fact, under optimal growth conditions, *Escherichia coli* cells can double every 20 min. Because bacteria are able to multiply rapidly with minimal nutritional requirements, they are abundant in every habitat on Earth. In soil, bacteria live by degrading organic compounds and assist soil formation. In aquatic environments such as ponds, streams, lakes, rivers, seas, and oceans, bacteria such as

Cyanobacteria (sometimes called blue–green algae because of their color) utilize their chlorophylls to capture energy from the sunlight. In the depths of the sea, bacteria obtain energy from naturally occuring oxidizing or reducing sulfur compounds. In humans and animals, bacteria are found in large numbers on the skin, respiratory and digestive tracts, and other parts of the body, constituting a normal microbiota in an essentially symbiotic relationship with mutual benefits. Although the vast majority of bacteria are harmless and sometimes even beneficial to its hosts, a few have the capacity to take advantage of temporary weakness in the host (e.g., injury or impaired immune system) to cause diseases of varying severity.

In a high-nutrient environment, the growth cycle of bacteria usually undergoes three phases. The first phase (the lag phase) is a period of slow growth with the bacterial cells adapting to the high-nutrient environment and preparing for fast growth. In the lag phase, the cell replicates its DNA and makes all the other molecules (e.g., ribosomes, membrane transport proteins) needed for the new cell. The second phase (the logarithmic phase, or log phase, also known as the exponential phase) occurs when DNA replication stops, and is characterized by rapid cell division and exponential growth. The rate at which cells grow during this phase is known as the growth rate, and the time it takes for the cells to double is known as the generation time. During the log phase, nutrients are metabolized at maximum speed until one of the nutrients is depleted that poses a negative impact on growth. The final phase (the stationary phase) results from the depletion of nutrients. During the stationary phase, the cells decrease their metabolic activity and consume nonessential cellular proteins. As there is a transition from rapid growth to a stress response state, there is heightened expression of genes involved in DNA repair, antioxidant metabolism, and nutrient transport. Although the entire cycle of bacterial growth takes about an hour, a rapidly growing bacterial cell carries out multiple rounds of replication simultaneously, which helps to shorten the doubling time for most bacteria to about 20 min.

5.1.4 GENETICS

Bacteria often have a single circular chromosome that ranges in size from only 160,000 (e.g., *Candidatus Carsonella ruddii*) to 12,200,000 bp (base pairs) (e.g., *Sorangium cellulosum*). However, *Borrelia burgdorferi*, the causal agent for Lyme disease, contains a single linear chromosome. Additionally, bacteria may possess small extra-chromosomal DNAs called plasmids, which range from 1 to 400 kb in size and comprise genes or gene cassettes for antibiotic resistance or virulence factors (Chapter 7). As plasmids have at least an origin of replication (or *ori*)—a starting point for DNA replication, they are capable of autonomous replication independent of the chromosomal DNA. A plasmid that integrates into the chromosomal DNA is called episome, which permits its duplication with every cell division of the host. Some viruses (bacteriophages or phages) may also exist in bacteria, with some merely infecting and lysing their host bacteria, while others inserting into the bacterial chromosome. Phages are usually made up of a nucleic acid core (e.g., ssRNA, dsRNA, ssDNA, or dsDNA measuring 5–500 bp in length) with an outer protein hull. A phage containing particular genes may contribute to its host's phenotype, as illustrated by the evolution of *E. coli* O157:H7 and *Clostridium botulinum*, which are converted from harmless ancestral bacteria into lethal pathogens through the integration of phages harboring toxin genes.

Without a membrane-bound nucleus, bacterial chromosome is typically located in the cytoplasm in an irregularly shaped body called the nucleoid. Also in the cytoplasm exists an essential cellular organelle called ribosome, which is responsible for protein synthesis in all living organisms. Being the key component of the ribosome, ribosomal RNA (rRNA) molecules consists of two complex folded subunits of differing sizes (small and large), whose main functions are to provide a mechanism for decoding messenger RNA (mRNA) into amino acids (at center of small ribosomal subunit) and to interact with transfer RNA (tRNA) during translation by providing peptidyltransferase activity (large subunit). Whereas the two rRNA subunits in eukaryotes have sedimentation coefficient values of 40S (Svedberg units) and 60S, those in bacteria measure 30S and 50S, respectively. In virtually all organisms, the small rRNA subunit (40S in eukaryotes and 30S in bacteria) contains a single RNA species (i.e., 18S rRNA in eukaryotes and 16S rRNA in bacteria); the large rRNA subunit (60S) in eukaryotes comprises three RNA species (5S, 5.8S, and 25/28S rRNAs) while that (50S) in bacteria contains two RNA species (5S and 23S rRNAs).

Although bacteria do not undergo meiosis nor mitosis, and do not require cellular fusion to initiate reproduction (as bacteria are not diploid), many bacteria do involve in cell to cell transfer of genomic DNA by various mechanisms. These mechanisms may range from the uptake of exogenous DNA from their environment (a process called transformation), the integration of a bacteriophage DNA introduces foreign DNA into the chromosome (a process called transduction), to the acquisition of DNA through direct cell contact (a process called conjugation). The incorporation of genes and DNA from other bacteria or the environment into the recipient cell's DNA is also called horizontal gene transfer. While DNA transfer occurs less frequently per individual bacterium than that among eukaryotes involving obligate sexual reproduction, the much shorter generation times and high numbers associated with bacteria can make the DNA transfer a significant contributor to the evolution of bacterial populations. Gene transfer is vital to the development of antibiotic resistance in bacteria as it allows the rapid transfer of resistance genes between different pathogens.

Regardless of genome size, most organisms show a mutation rate on the order of one mutation per genome per generation. Given their very short generation times (less than 1 h in culture media and a few hours in the wild) and small genomes (which are a 1000 times smaller than most eukaryotes), prokaryotes generally display 1000 times more mutations per gene, per unit time, per individual than eukaryotes. Furthermore, with greater population sizes that result in the absolute amount of mutational variation entering the population, prokaryotes have enormous capacity to adapt and invade new niches, which are the key factors contributing to the evolutionary success of prokaryotes.

Genomic diversity in bacteria comes in two forms: (1) genetic heterogeneity wherein different strains have different alleles of the same gene and (2) genomic plasticity wherein different strains have different genes. Recent studies indicate that each strain (serovar) within a bacterial species receives a unique distribution of genes from a population-based supragenome that is many times larger than the genome of any given strain. Through the autocompetence and autotransformation mechanisms, bacterial strains (or serovars) within the species may evolve and generate diversity *in vivo* to enable them to persist in the face of myriad host defense mechanisms and environmental stresses. In other words, the strain (serovar)-specific genes (as contingency genes) may provide for an increased number of genetic characters that facilitate the population as a whole to adapt rapidly to environmental factors, such as those experienced in the host during chronic infectious processes [1–4]. There is evidence that under arduous external conditions, many bacteria form biofilms that often exchange DNA at rates several orders of magnitude greater than planktonic bacteria, and that are responsible for many chronic bacterial infections in human patients. For example, biofilm-associated growth of *Pseudomonas aeruginosa* has been implicated in several chronic suppurative otitis media [1].

For instance, Pulse field gel electrophoresis (PFGE) analysis of large restriction fragments of different *P. aeruginosa* strains revealed that the genome size of this bacterium could be more than 15% larger with respect to the sequenced *P. aeruginosa* PA01 genome. Further, other members of the *Pseudomonas* genus have genomes that display 30%–40% plasticity in size. The differences measured for *P. aeruginosa* strains equate to a coding capacity for more than 500 proteins with an average size of 50 kDa. Through restriction endonuclease and sequence analyses of 686 random clones from a *P. aeruginosa* library constructed with 12 clinical strains, Erdos et al. [1] demonstrated that 13% of these clones are not represented in the genome of the reference *P. aeruginosa* strain PA01. These findings suggest that reliance on a single laboratory strain, such as PA01, as being representative of a pathogenic bacterial species will fail to identify many important genes, and that to obtain a complete picture of complex phenomena, including bacterial pathogenesis and the genetics of biofilm development, will require characterization of the *P. aeruginosa* population-based supragenome [2].

Similarly, PFGE data of large restriction fragments of different *Haemophilus influenzae* strains suggest that the genome size of this bacterial species could vary by more than 5% of the total genome. This difference in the amount of DNA could represent coding sequence for 200–500 proteins. Using *H. influenzae* as a model, Hogg et al. [4] identified 53 novel genes from 10 nontypeable *H. influenzae* strains that do not exist in the sequenced *H. influenzae* Rd KW20 genome. Amino acid homology searches using hypothetical translations of the open reading frames disclosed amino acid identities to a variety of proteins, including bacterial virulence factors not previously identified in *H. influenzae* isolates. Nine (17%) of the 53 novel genes were identified in all 10 nontypeable *H. influenzae* strains, with each of the remaining 44 being present in only a subset of the strains. Thus, these genic distribution analyses offer a more effective strain discrimination tool than either multilocus sequence typing or 23S ribosomal gene typing methods.

Moreover, after examination of the genome sequences of 17 *Streptococcus pneumoniae* strains, Hiller et al. [3] showed that 1716 (54%) of the 3170 orthologous gene clusters were not found in all strains. Genic differences per strain pair ranged from 35 to 629 orthologous clusters, with each strain's genome containing between 21% and 32% noncore genes. Using the finite-supragenome model, the authors predicted that (1) the *S. pneumoniae* supragenome contains more than 5000 orthologous clusters and (2) 99% of the orthologous clusters (about 3000) that are represented in the *S. pneumoniae* population can be identified if 33 representative genomes are sequenced. These extensive genic diversity data support the supragenome concept and provide a basis for understanding the great differences in clinical phenotype associated with various pneumococcal strains.

Based on the complete genome sequences of four *Listeria monocytogenes* strains (of serovars 1/2a and 4b), it is evident that the supragenome of *L. monocytogenes* is clearly larger than the genome of individual *L. monocytogenes* serovars and strains. Comparison of the four *L. monocytogenes*, one *Listeria innocua* and one *Listeria welshimeri* genomes at the nucleotide and predicted protein levels showed that in addition to many shared genetic components, a total of 51, 97, 69, and 61 strain-specific genes are identifiable from *L. monocytogenes* serovar 1/2a strains EGD-e and F6854, and serovar 4b strains F2365 and H7858, respectively. Further analysis indicated that 83 of these genes are limited to the serovar 1/2a strains (of lineage II), and 51 genes are limited to the serovar 4b strains (of lineage I). In addition, 149 and 311 species-specific genes are recognized in *L. innocua* CLIP 11262 and *L. welshimeri* SLCC 5334, respectively [5–7]. Taken together with previous studies that demonstrate the presence of a supragenome for *P. aeruginosa* and *H. influenzae*, it appears that the possession of a distributed genome is a common host interaction strategy.

5.1.5 ECONOMIC IMPACT

Being ubiquitous in every habitat on Earth, bacteria exist in huge numbers and diversity. A gram of soil typically contains 40 million bacterial cells, and a millimeter of fresh water has a million bacteria. With an estimation of approximately 5 nonillion (5×10^{30}) bacteria in the world, these organisms play critical roles in chemical cycles, environmental maintenance, food production, and human well-being.

Apart from recycling carbon dioxide (CO_2) via photosynthesis, which releases oxygen into Earth's atmosphere, bacteria are involved in the decomposition of dead plant and animal matter, improving soil fertility. Some bacteria (e.g., the genus *Rhizobium*, which forms nodules on the roots of beans and other plants in the legume family) can fix nitrogen gas using the enzyme nitrogenase, and convert nitrogen in Earth's atmosphere into the nitrogen compound ammonia (a process known as nitrogen fixation) for plant growth. Bacteria can help clean up oil spills, pesticides, and other toxic materials by converting the toxic materials to harmless or useful products such as CO_2 and methane gas. Some bacteria (e.g., the genera *Thiobacillus* and *Sulfolobus*) are able to cause a chemical reaction of sulfides with oxygen—yielding sulfuric acid, which removes (leaches) the copper from the ores (copper sulfides). Other bacteria are useful for food production such as yogurt, cheese, cider, and vinegar. Some bacteria (e.g., *Bacillus thuringiensis*, a soil dwelling Gram-positive bacterium) can be used as pesticides (trade names Dipel and Thuricide) in the biological control of Lepidopteran pest.

As fast growers with relatively low demands for nutrients, bacteria represent ideal hosts for mass production of certain plastics, enzymes used in laundry detergents, and antibiotics such as streptomycin and tetracycline. Using recombinant DNA techniques, a variety of specific products with pharmaceutical potentials (e.g., insulin and somostatin—a human growth hormone) can be conveniently generated in large quantity and exact quality. Bacteria can be also engineered to produce fine chemicals such as acetone, ethanol, and gases (methane). In addition, many bacteria (e.g., *Listeria monocytogenes, Mycobacterium bovis* Bacille Calmette–Guerin, *Salmonella*, *Shigella*, and *Escherichia coli*) have been shown to be useful carriers for delivering vaccine molecules against microbial diseases and cancers. In particular, as a robust bacterium with a high safety threshold and remarkable capacity to stimulate all facets of cell-mediated immunity in the absence of adjuvants, *L. monocytogenes* has been recognized as a valuable vehicle for a range of protective bacterial, viral, and parasitic molecules, with encouraging vaccination results based on *L. monocytogenes* delivery vehicle being described frequently. Given *L. monocytogenes*' unsurpassed ability to initiate both CD4 and CD8 T-cell responses, to promote interferon γ (IFN-γ) and interleukin 12 (IL-12) production, and to deliver vaccine molecules into the cytoplasm, it has become an increasingly popular vector for delivering antiinfective and cancer vaccine molecules [8]. Moreover, bacteria such as *L. monocytogenes* have been shown to promote a Th1-cell response in the host through increased production of IFN-γ. This can be exploited to convert an existing inflammatory Th2 response that is characteristic of allergic diseases into a Th1 response [9,10].

Although many bacteria are harmless or beneficial, a few bacteria can have detrimental effects on food and plant production, human and animal health. Some bacteria can cause food spoilage (e.g., *Lactobacillus*) and foodborne diseases (e.g., *Shigella, Campylobacter, Salmonella*, and *Listeria*), and others can harm agriculture by causing major diseases of plants and farm animals. For instance, *Brachyspira hyodysenteria* causes a type of severe diarrhea in pigs with disastrous consequence for pig farmers. Some bacteria are involved in metal corrosion (wearing away) through the formation of rust, especially on metals containing iron.

There are approximately 10 times as many bacterial cells as human cells in the human body, with large numbers of bacteria on the skin and in the digestive tract. The communities of bacteria and other organisms that inhabit the body are sometimes referred to as the normal microflora or microbiota. Some bacteria in human body produce essential nutrients (e.g., vitamin K), which the body itself cannot make. The most common fatal bacterial diseases are respiratory infections, with tuberculosis (caused by *Mycobacterium tuberculosis*) alone killing about 2 million people a year. One of the world's deadliest bacterial diseases today is cholera, which is caused by foodborne *Vibrio cholerae*. Other globally important bacterial diseases include pneumonia caused by *Streptococcus* and *Pseudomonas*, tetanus, typhoid fever, diphtheria, syphilis, and leprosy. Another common bacterial disease is tooth decay, which results from the acids bacteria produce from sugar via fermentation, which dissolves the enamel of the teeth and create cavities (holes) in the teeth.

Considering the frequency and severity of many diseases caused by bacteria, it is important that pathogenic bacteria concerned are promptly identified and appropriate antibacteria regimens initiated. As the conventional phenotypic tests rely on time-consuming *in vitro* culture procedures, they are not only slow, but also potentially variable. The application of new generation genotype-based methods targeting nucleic acids (DNA or RNA) has provided unprecedented levels of sensitivity, specificity, and speed for laboratory detection and identification of pathogenic bacteria. Because of their critical roles in cellular function and maintenance, rRNA molecules are not only the most conserved (i.e., the least variable) gene in all cells, but is also the most abundant (with each living cell containing 10^4 to 10^5 copies of the 5S, 16S, and 23S rRNA molecules). For this reason, the rRNAs (and their genomic coding sequences rDNAs) are often subjected to sequencing analysis to identify and confirm an organism's taxonomic status and species identity, and to estimate rates of species divergence. In addition, many species-, group- and virulence-specific genes have been identified from a large number of bacteria and used as targets for improved characterization, and detection of pathogens concerned. Some of these genes may code for proteins that are essential for bacterial maintenance and survival (e.g., house-keeping genes) and invasion (e.g., internalins), while others are involved in the regulation of proteins of unknown functions.

5.2 FUNDAMENTALS AND CURRENT TECHNIQUES FOR BACTERIAL DNA PURIFICATION

Besides its use as template for molecular identification and diagnostic application, bacterial DNA of sufficient quantity and quality is also required for genetic manipulation and *in vitro* expression of specific protein products with pharmaceutical potential. While it will be ideal to detect the bacteria and amplify the genes of interest directly from uncultured specimens, this is not feasible at this stage considering the low number of target organisms and presence of multiple substances in these samples that are inhibitory to DNA polymerase and other enzymes employed in molecular analysis. *In vitro* culture techniques are designed to promote the growth of bacteria concerned while eliminating the most inhibitory materials present in the original specimens. Isolation of genomic DNA from bacterial isolates can then proceed. This often involves lysis of bacterial cell wall to release nucleic acids, removal of unwanted RNA, disruption of cellular proteins, separation of DNA from degraded proteins and cell debris, and precipitation and concentration of resulting DNA.

5.2.1 CELL LYSIS

Lysis of bacterial cells is often achieved through the use of lytic enzymes such as lysozyme. Lysozyme (also known as mucopeptide *N*-acetylmuramoylhydrolase, mucopeptide glucohydrolas, or murmidase) from egg white is a 14.7 kDa enzyme (with 129 amino acid residues) that recognizes peptidoglycan in the bacterial cell wall. It catalyzes hydrolysis of 1,4-β-linkages between *N*-acetylmuramic acid and *N*-acetyl-D-glucosamine residues in peptidoglycan, and between *N*-acetyl-D-glucosamine residues in chitodextrins. Having multiple layers of peptidoglycan in their cell wall, Gram-positive bacteria are generally susceptible to lysozyme. On the other hand, due to the presence of only a thin layer of peptidoglycan in their cell wall, Gram-negative bacteria are somewhat less susceptible to lysozyme than Gram-positive bacteria. Nonetheless, improved lysis of Gram-negative bacterial cell wall by lysozyme can be accomplished with the addition of ethylene diamine tetraacetic acid (EDTA) that chelates metal ions in the outer membrane and mediates aggregation of nucleic acids to each other and to proteins. Two other common reducing agents found in extraction buffers are β-mercaptoethanol or dithiothreitol. Lysozyme is active over a broad pH range (6.0–9.0). At pH 6.2, maximal activity is observed over a wider range of ionic strengths (0.02–0.10 M) than at pH 9.0 (0.01–0.06 M).

As lysozyme is produced by mammalian innate immune system as a general and nonspecific means of combating invading bacterial pathogens, various ingenious mechanisms have been evolved by pathogenic bacteria to counter the harmful effects of host lysozyme. One such mechanism involves modification of peptidoglycan in the bacterial cell wall. The modified peptidoglycan thus becomes refractory to destruction by lysozyme. For instance, Gram-positive bacterial pathogen *Staphylococcus aureus* exhibits increased resistance to the muramidase activity of lysozyme via *O*-acetylation of its peptidoglycan [11]. Similarly, zoonotic bacterial pathogen *L. monocytogenes* deacetylates *N*-acetyl-glucosamine residues in its peptidoglycan to increase its tolerance of lysozyme [12]. Given that deacetylase gene exists in many other pathogenic bacteria, peptidoglycan *N*-acetyl-glucosamine may constitute a universal mechanism employed by bacteria to avoid detection and prompt elimination by the host's innate immune network. Therefore, there is no surprise that bacteria sensitive to lysozyme treatment are often nonpathogenic while pathogenic bacteria demonstrate enhanced tolerance to lysozyme. As a consequence, some Gram-positive bacterial pathogens such as *Staphylococcus*, *Streptococcus*, and *Listeria* have been found to resist lysozyme treatment, which may require alternative strategies (e.g., use of other lytic enzymes) for cell lysis to maximize subsequent DNA recovery.

Fortunately, there are a number of additional lytic enzymes available for lysis of bacteria that are resistant to lysozyme. These include mutanolysin, lysostaphin, achromopeptidase, labiase, and protease. Mutanolysin from *Streptomyces globisporus* is an *N*-acetylmuramidase. Like lysozyme, it is muralytic enzyme that cleaves the β-*N*-acetylmuramyl-(1,4)-*N*-acetylglucosamine linkage for the bacterial cell wall polymer peptidoglycan-polysaccharide. Its carboxyl terminal moieties are involved in the recognition and binding of unique cell wall polymers. Mutanolysin is effective in lysing *Listeria* and other Gram-positive bacteria such as *Lactobacillus*, *Lactococcus*, and *Streptococcus*. In a recent study, mutanolysin was utilized to digest cell wall peptidoglycan of *L. monocytogenes* in the presence of amidosulfobetaine-14(ASB-14)/urea/thiourea. Cell lysis with mutanolysin followed by solubilization with ASB-14/urea/thiourea generated a high overall protein yield as analyzed by two-dimension electrophoresis. The increase in surface proteome coverage obtained by mutanolysin and ASB-14/urea/thiourea solubilization suggests the utility of this method for further analytical and comparative studies of surface proteins from *Listeria* and possibly other Gram-positive bacteria [13].

Lysostaphin from *Staphylococcus staphylolyticus* is a zinc endopeptidase with a molecular weight of 25 kDa. Because lysostaphin cleaves the polyglycine cross-links in the peptidoglycan layer of the cell wall of *Staphylococcus* species, it is useful as a bacterial cell lysing reagent and also as a potential antimicrobial therapeutic. The optimal pH for lysostaphin activity is pH 7.5. Achromopeptidase from *Achromobacter lyticus* is a lysyl endopeptidase with a molecular weight of 27 kDa. It can be used for lysis of recalcitrant Gram-positive bacteria that are tolerant to lysozyme. The optimal pH for achromopeptidase activity is between pH 8.5 and 9.0. Labiase from *Streptomyces fulvissimus* is an enzyme preparation useful for the lysis of many Gram-positive bacteria such as *Lactobacillus, Aerococccus*, and *Streptococcus*. Labiase possesses β-*N*-acetyl-D-glucosamine and lysozyme activity, with optimal pH for activity at pH 4.0 and for stability between pH 4.0 and 8.0. Apart from being applied individually, lytic enzymes can be used in combination (e.g., lysozyme and mutanolysin) to enhance the breakup of the cell wall of those difficult-to-lyse bacteria.

Protease (e.g., Sigma Cat. # P6911) can be also applied as a low-cost substitute to proteinase K for cell lysis with a similar outcome. In fact, as protease contains a mixture of proteolytic activities, it may provide a higher efficiency for those hard-to-lyse bacteria, such as Gram-positive organisms.

Besides lytic enzymes, other techniques have also been utilized for bacterial lysis. For instance, glass beads, sonication, and French press can be employed for disruption of the bacterial cell wall mechanically. Indeed, glass beads have been used frequently to break up the mycobacterial cell wall. However, there is evidence that bead beating and sonication may break up DNA into <20 kb fragments. A recombinant bacteriophage A118 lysin (PLY118) from *Listeria* has been described for specific lysis of *Listeria* cell wall [14]. Interestingly, bacteriophages generate lysozyme to destroy the peptidoglycan in the bacterial cell wall and inject its DNA. Inside the bacterial host, bacteriophages secrete lysozyme molecules to lyse the bacterial cell wall and release new phages.

5.2.2 DIGESTION OF RNA AND CELLULAR PROTEINS

Following the lysis of bacterial cells, unwanted RNA can be eliminated by treating the lysate with RNase A (free of DNase). RNase A is often prepared at 10 mg/mL and boiled for 10 min to get rid of any potential contaminating DNase. Alternatively, ready-to-use RNase A can be purchased directly from commercial suppliers (e.g., Promega). After lysozyme and RNase A treatment, it is necessary to destroy cellular proteins before purification of DNA. Traditionally, proteinase K has been used for this purpose. Being a stable and highly reactive serine proteinase, proteinase K belongs to the subtilisin family with an active-site catalytic triad (Asp^{39}-His^{69}-Ser^{224}), which is stable in a broad range of pH, buffer salts, detergents, and temperature. Detergents inhibit nucleases and remove endotoxins bound to cationic proteins such as lysozyme and ribonuclease A during separation of the proteins from the nucleic acids. Widely applied detergents include sodium dodecyl sulfate (SDS) and Triton X-100. One other frequently utilized detergent for nucleic acid extraction is cetyltrimethylammonium bromide (CTAB) (hexadecyltrimethyl-ammonium bromide or cetrimonium bromide), which is a cationic surfactant providing a buffer solution for separation of protein from DNA and RNA. CTAB is most often used in a 2% (w/v) solution and it has been applied for preparation of DNA from a variety of biological organisms. In the presence of 0.1%–0.5% SDS or other detergents, proteinase K will digest a variety of proteins and nucleases in DNA preparations without compromising the integrity of the isolated DNA. In fact, Triton X-100 may be preferable to SDS in the extraction of nucleic acids from certain bacteria such as *Listeria* (Chapter 6). However, use of potassium salts or temperature below 10°C with SDS may cause precipitation of the detergent.

Alternatively, a protein denaturing agent guanidine thiocyanate or guanidine hydrochloride can substitute proteinase K for disrupting cell proteins. DNAzol reagent (Life Technologie) containing guanidine isothiocyanate has been used successfully for isolation of DNA from a range of bacteria including *Listeria, Campylobacter,* and *E. coli.* DNA can be also extracted by using a chelating resin Chelex 100 (which is made up of iminodiacetate-ions-containing styrene divinylbenzene copolymers that can chelate polyvalent metal cations) without proteinase K digestion and phenol–chloroform after cell lysis, and nucleic acids (DNA and RNA) remain in the solution, which can be readily separated from other cell components bound to the resin by centrifugation.

5.2.3 Removal of Degraded Proteins and Cell Debris

Organic solvents such as phenol have been long used to remove degraded proteins and other components (e.g., cell debris) from the bacterial lysates. Phenol (also known as carbolic acid, hydroxyl benzene) is an aromatic compound with antioxidant activity and a partial inhibitor of nucleases. Being a strong denaturing agent for proteins, phenol causes proteins to partition into the organic phases (and interface) and nucleic acids into the aqueous phase, and thus it is useful for separating nucleic acids from proteins. Antioxidants such as 8-hydroxyl quinoline or β-mercaptoethanol can be added to phenol. During phenol extraction, the pH of the buffer is important in determining whether DNA or RNA are recovered. At pH 5.0–6.0, DNA is selectively retained in the organic phase leaving RNA in the aqueous phase (hence water saturated phenol is useful for RNA extraction). At pH 8.0 or higher, both DNA and RNA are retained in the aqueous phase. The phenol used for biochemistry comes as a water-saturated solution with Tris buffer, as a Tris-buffered 50% phenol, 50% chloroform solution, or as a Tris-buffered 50% phenol, 48% chloroform, 2% isoamyl alcohol solution (25:24:1). Phenol is naturally somewhat water-soluble, and gives a fuzzy interface that is sharpened by the presence of chloroform. Phenol, when oxidized, may damage and degrade DNA, and thus phenol showing a pink or brown color (which indicates strong oxidation) should be avoided. This also applies to commercial phenol preparations that are similarly pink (such as Trizol). The isoamyl alcohol reduces foam, which commonly occurs with phenol–chloroform. Most solutions also have an antioxidant, as oxidized phenol will damage the DNA. Pure phenol crystals are no longer common. These had to be equilibrated into the buffer and then melted and dissolved, with due care taken to avoid inhalation of the fumes or fine aerosolized powders.

Chloroform is a commonly applied solvent as it is relatively unreactive, miscible with most organic liquids, and conveniently volatile. Being light sensitive when pure, chloroform is kept in brown bottle. Chloroform is often used in conjunction with phenol (in a 1:1 mixture) to form a biolayer with extraction buffer (e.g., Tris) since deproteinization is more effective when two different organic solvents are applied simultaneously, in which DNA will remain in the supernatant while proteins and nonsoluble cell materials will precipitate between the buffer chloroform layers. Chloroform can be stabilized with small quantities of amylene or ethanol because storage of pure chloroform solutions in oxygen and ultraviolet (UV) light tends to produce phosgene gas. Chloroform sometimes also comes as a 96% chloroform, 4% isoamyl alcohol solution that can be mixed with an equal volume of phenol to make 25:24:1. In addition to denaturing proteins, chloroform is useful in removing lipids and a final chloroform extraction helps to remove the last traces of phenol.

Isoamyl alcohol (also known as isopentyl alcohol) is an antifoaming agent, whose vapors are poisonous. Isoamyl alcohol helps with the phase separation, decreases the amount of material found at the aqueous and organic interface, and helps reduce foaming. For RNA extraction, isopropanol (or 2-propanol) is often used in lieu of isoamyl alcohol.

Cesium chloride has been also used in the earlier days for purification of DNA using high-speed centrifugation (>100,000 rpm) in a technique known as isopycnic centrifugation. Under centrifugal and diffusive forces, cesium chloride solution will establish a concentration gradient (and thus a density gradient) within the centrifuge tube. When DNA is centrifuged in this solution, fragments of DNA will migrate down the tube until they reach a zone where the density of the DNA is equal to the density of the solution. At this point, the DNA will stop migrating. This allows separation of DNA of differing densities (e.g., DNA fragments with differing A-T or G-C contents). By adding ethidium bromide to the solution, it will allow easy detection of the band and ethidium bromide can then be removed with butanol. As this technique requires use of the ultracentrifuge, it has been superseded by other simpler procedures.

To avoid the exposure of hazardous chemicals such as organic solvents, silica oxide is often applied nowadays for separation of lipopolysaccharide and other materials from DNA [15]. Nucleic acids show affinity for silica under chaotropic conditions, which can be created with high concentrations of sodium iodide, guanidine hydrochloride, or guanidine thiocyanate. One widely used silica oxide is diatomaceous earth (DE), which will bind double-stranded DNA but not RNA or proteins in the presence of highly concentrated chaotropic agents. DNA can then be eluted with low-ionic-strength buffer or water [16]. Further refinement of silica oxide-based approach led to the development of silica-coated beads (or glass beads, usually 3–10 μm in diameter). Similarly, nucleic acids bind to silica-coated beads in the presence of a high-ionic solution, which can be reversed in the presence of a low-ionic solution. This platform has been extended to the use of paramagnetic beads that are made of silica impregnated with ion. Because these beads display magnetic quality only upon exposure to an external magnetic force, they can be conveniently separated from other substances in the solution [17].

Another useful way to separate nucleic acids from degraded proteins and cell debris is through the use of anion exchange columns. The positively charged diethylaminoethyl (DEAE) groups in the columns interact with the negatively charged phosphate groups in the DNA backbone. The columns are designed so that chromosomal DNA, 20–150 kb in size, can be isolated. To purify DNA by DEAE column, bacterial lysate is filtered and loaded onto the column, and nonbinding materials are washed away. After elution from the column, the DNA is precipitated using isopropanol.

Because silica oxide and its derivatives (e.g., paramagnetic beads) and anion exchange column obviate the need to use hazardous chemicals (e.g., phenol–chloroform), and are easy to perform, they have gained increasing acceptance. Many in-house and commercial nucleic acid purification systems have been designed on these principles and widely applied to generate high-purity nucleic acids from bacteria and other biological organisms.

5.2.4 DNA PRECIPITATION AND CONCENTRATION

Following removal of degraded proteins and insoluble cell debris from the bacterial lysate by centrifugation, the DNA can be concentrated and precipitated with alcohol, which occurs in the presence of a monovalent cation (salt). Either 2 volumes of ethanol or 0.6 volumes of isopropanol can be used. Isopropanol has the advantage of precipitating DNA at lower concentrations and smaller volume. This is useful when volume of the tube does not allow the addition of 2 volumes of ethanol. Isopropanol is often used for the first precipitation, but not for final precipitations because it tends to bring down salts more readily and is less volatile than ethanol. Alcohol precipitation occurs very rapidly except when DNA content is low (<100 ng). Many earlier protocols indicate that precipitation should be allowed to occur in the cold (e.g., −20°C). This has been shown to be unnecessary. Typically, centrifugation is conducted at 12,000 g for 10 min. The DNA pellet is air-dried on the lab bench for a few minutes. It is best not to overdry the DNA, which makes resolution more difficult and can cause denaturation of DNA.

The most common monovalent cations used for DNA precipitation are sodium acetate (stock solution 2.5 M, pH 5.2–5.5; final concentration 0.25–0.3 M), sodium chloride (stock solution 5.0 M; final concentration 0.10–0.15 M or 1/50 volume), and ammonium acetate (stock solution 4.0 M, filter sterilized; final concentration 1.3 M). Ammonium ions help keep free nucleotides in solution. The choice of salt is usually determined by the nature of the sample and the intended use of the DNA. For instance, ammonium acetate should not be used if the DNA is going to be phosphorylated with polynucleotide kinase. If the DNA preparation contains SDS, sodium chloride is the choice as it allows SDS to remain soluble in 70% ethanol. Samples with greater than 10 mM EDTA should not be ethanol precipitated since the salts will come down with the DNA.

DNA can also be recovered (or concentrated) from dilute solution by extraction with 2-butanol. The water from the sample moves into the butanol, which is discarded, thus leaving a higher DNA concentration in the aqueous phase. Water saturated butanol can also be used to remove residual ethidium bromide from samples obtained via cesium chloride centrifugation or agarose gel purification. Sometimes, residual phenol, chloroform, or butanol can be removed by ether extraction without going through ethanol precipitation.

5.2.5 CURRENT TECHNIQUES FOR BACTERIAL DNA ISOLATION

Over the years, numerous DNA isolation protocols based on in-house reagents have been developed for bacterial DNA isolation from cultured isolates. Apart form being economical, the in-house methods have offered many ingenious and purpose-designed approaches for the purification of DNA from a diversity of bacteria including Gram-negative, Gram-positive, and acid-fast bacteria. In recent years, a diverse array of commercially manufactured DNA purification kits has come to the market. These kits provide a level of convenience and consistence at an increasingly reasonable cost that is not easily obtainable with the in-house reagents.

McOrist et al. [18] compared the relative efficacy of extraction of bacterial DNA (both Gram negative and positive origin) using four commercial kits (FastDNA kit, Bio 101; Nucleospin C + T kit, Macherey-Nagel; Quantum Prep aquapure genomic DNA isolation kit, Bio-Rad; QIAamp DNA stool mini kit, Qiagen) and a noncommercial guanidine isothiocyanate/silica matrix method of Boom et al. [16]. The Boom method involved lysis of the bacterial pellet in 5 M guanidine isothiocyanate followed by absorption (binding) of DNA with diatoms earth (DE). After washing in 5.5 M guanidine thiocyanate and 70% ethanol, DNA was eluted from the DE in TE buffer. While the Nucleospin method was the most sensitive procedure for the extraction of DNA from a pure bacterial culture of Gram-positive *Lactobacillus acidophilus*, detecting 10^3 bacteria by polymerase chain reaction (PCR), QIAamp and the guanidine method were most sensitive for cultures of Gram-negative *Bacteroides uniformis* also detecting 10^3 bacteria by PCR (Table 5.1).

In another report, Veloso et al. [19] evaluated three DNA extraction methods for *Leptospira* bacteria. These included (1) proteinase K followed by phenol–chloroform, (2) a plant proteinase (E6870) followed by phenol–chloroform, and (3) boiling of leptospires in 0.1 mM Tris, pH 7.0 for 10 min at 100°C, with no phenol treatment. While the proteinase K or E6870 methods gave positive PCR results (with a detection limit of 10^2 leptospres) on all samples, the boiling method was somewhat inconsistent. Moreover, the E6870 method (which contained EDTA in the extracting buffer) appeared to remove nucleases at a higher efficiency than the other two methods.

Bahador et al. [20] examined the performance of five DNA extraction methods for *M. tuberculosis* cells, which comprised Triton, Chelex, Nonidet, SDS/lysozyme, and silica-based methods. DNA extracted from these procedures was diluted from 10^{-1} to 10^{-7} for use as templates in a single-round PCR. The sensitivity of extraction by the silica-based was 1–10 cells (or 10^{-5} dilution), which was followed by the SDS/lysozyme method at 10^2 cells (or 10^{-4} dilution), the Nonidet method at 10^3 cells (or 10^{-3} dilution), and the Triton and Chelex methods at 10^4 cells (or 10^{-2} dilution). The silica-based method appeared to offer the most effective extraction with the most sensitive detection, and least labor and completion time. The end-point titrations were identical for both bacterial cells and spiked sputum (Table 5.1).

TABLE 5.1

Comparison of In-House Reagents and Commercial Kits for Bacterial DNA Extraction from Cultured Isolates

Bacterial Type	Example	DNA Isolation Method	Sensitivity	Reference
Gram-positive bacteria	*L. acidophilus* isolate	Nucleospin C+T kit (Macherey-Nagel)	10^3 CFU (based on a single-round PCR)	[18]
		QIAamp DNA stool minikit (Qiagen)	10^4 CFU	[18]
		Guanidine isothiocyanate/silica matrix [16]	10^6 CFU	[18]
		Quantum Prep aquaculture genomic DNA isolation kit (Bio-Rad)	10^7 CFU	[18]
		FastDNA kit (BIO 101)	10^7 CFU	[18]
Gram-negative bacteria	*B. uniformis* isolate	Guanidine isothiocyanate/silica matrix [16]	10^3 CFU (based on a single-round PCR)	[18]
		QIAamp DNA stool minikit (Qiagen)	10^3 CFU	[18]
		Nucleospin C + T kit (Macherey-Nagel)	10^4 CFU	[18]
		Quantum Prep aquaculture genomic DNA isolation kit (Bio-Rad)	10^4 CFU	[18]
		FastDNA kit (BIO 101)	10^6 CFU	[18]
Mycobacteria	*M. tuberculosis* isolate	Silica (diatoms) [16]	1–10 cells (based on a single-round PCR)	[20]
		SDS/lysozyme	10^2 cells	[20]
		Nonidet	10^3 cells	[20]
		Triton X-100	10^4 cells	[20]
		Chelex 100	10^4 cells	[20]

Gomes et al. [21] described a simple DNA isolation method using routine chemicals that yielded high-quality and integrity preparations when compared to some of the most well-known protocols. Bacterial broth culture in 1.5 mL tube was centrifuged at 13,000 g for 5 min and the pellet was suspended in 200 μL of 0.1 mol Tris and added with 200 μL of lysis solution (0.2 N NaOH and 1% SDS), mixed and deproteinized with 700 μL of phenol/chloroform/isoamyl alcohol (25:24:1), homogenized and centrifuged 10 min at 13,000 g. To precipitate DNA, 700 μL of cold 95% ethanol was added and spun, washed in 70% ethanol and centrifuged. Precipitated DNA is dried at room temperature and suspended in 100 μL of water. This method did not require the use of lysing enzyme and the DNA was obtained within 40 min. The amount of nucleic acid extracted (as measured at 260 nm) from *Xanthomonas*, *Pseudomonas*, and *Erwinia* strains was two to five times higher than that of the most commonly used method.

In case DNA bacterial isolates is intended only for PCR amplification and detection, several rapid processing procedures can be utilized. Recently, four such procedures (i.e., boiling, Triton X-100 and heat, alkali and heat, and alkaline poly ethylene glycol [PEG] and heat) were evaluated for preparation of *L. monocytogenes* isolates prior to PCR [22]. It is clear that PCR-ready DNA can be obtained by boiling *L. monocytogenes* isolates in distilled water or in 1% Triton X-100 followed by a brief spin. On the other hand, to produce consistent PCR test results,

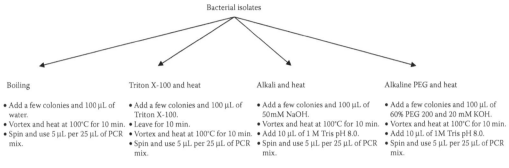

FIGURE 5.1 Rapid preparation of bacterial isolates for PCR amplification.

neutralization is required for alkali and heat, or alkaline PEG and heat procedures (Figure 5.1). Besides *L. monocytogenes*, these rapid sample handling procedures have the potential to be applied to other bacterial isolates as well as enriched (semienriched) broth cultures. While it is known that the enrichment broth cultures often contain other microbial organisms, this will not be a problem if the molecular diagnostic tests targeting organism-specific genes are used, because other microbial organisms present will be unlikely to cross-react with specific gene primers in the tests.

5.3 METHODS

General reagents, supplies, and equipment that are required for isolation of bacterial DNA are listed in Table 5.2.

5.3.1 SMALL-SCALE ISOLATION OF BACTERIAL GENOMIC DNA USING PHENOL–CHLOROFORM

The classic phenol–chloroform protocol represents a robust and reliable technique for isolation of DNA from a vast range of biological organisms including Gram-positive and Gram-negative bacterial isolates [23–35]. When applied to bacterial genomic DNA extraction, it is often necessary to

TABLE 5.2
General Reagents, Supplies, and Equipment Required for Bacterial DNA Isolation

Reagents	Supplies and Equipment
DNA lysis buffer: 30 mM Tris-HCl pH 8.0, 30 mM EDTA pH 8.0, 150 mM NaCl, 0.5% SDS, 0.5% Triton X-100, 0.5% Tween 20; sterilize using 0.22 µm filter	Pipettes: 0.5–10, 20–200, and 100–1000 µL
Lysozyme: 10 mg/mL in water freshly prepared	Tubes: 1.5, 15, and 50 mL
Proteinase K: 10 mg/mL in water and store at −20°C	Waterbath and heating block
RNase A: 10 mg/mL in water, boil for 15 min, and store at −20°C	Microfuge and centrifuge, room temperature and refrigerated
EDTA: 0.5 M pH 8.0, autoclaved	Refrigerator and freezer
Phenol	Vortex
Chloroform	Stir plates and stir bars
Isoamyl alcohol	pH meter, Shindengen Isfet KS723, Japan
Ethanol	Sterilizing filters, 0.22 µm
Isopropanol	Shaking oven
	Autoclave
1x TE buffer: 10 mM Tris pH 8 and 1 mM EDTA pH 8	Gloves

incorporate a lytic enzyme (e.g., lysozyme, lysostaphin, and mutanolysin) to disrupt the bacterial cell wall and a protease (e.g., proteinase K) to digest cellular proteins. Following separation of degraded proteins by phenol–chloroform, DNA is then precipitated with isopropanol (or ethanol). The contaminating RNA in the DNA preparation can be removed with RNase, and DNA is re-extracted with phenol–chloroform and precipitated with isopropanol (or ethanol). Alternatively, RNase can be added along with lysozyme during the cell-wall lysis step. Bacterial genomic DNA prepared by this method can be used for restriction digestion, cloning, PCR, and other molecular applications. For acid-fast bacteria, it may be helpful to include a bead beating step to efficiently break up the cell wall.

Procedure

1. Inoculate 5 mL of Luria Bertaini (LB) broth (or other media) into a 15 mL tube with a single colony of bacteria (from a previously inoculated agar plate), and incubate at 37°C for 16–18 h with shaking at 250 rpm (with OD at 600 nm of >1.0).
2. Pellet the bacteria at 3000 rpm for 10 min, and decant the supernatant.
3. Resuspend the cell pellet in 500 μL of DNA lysis buffer (30 mM Tris-HCl pH 8.0, 30 mM EDTA pH 8.0, 150 mM NaCl, 0.5% SDS, 0.5% Triton X-100, and 0.5% Tween 20) by vortexing; add 10 μL of freshly made lysozyme (10 mg/mL) and 5 μL of RNase A (10 mg/mL), and incubate at 37°C for 30 min.
   Note: A combination of detergents is included in this DNA lysis buffer to allow efficient lysis of all bacterial categories. While most Gram-negative and Gram-positive bacteria are readily lysed in the presence of 1% SDS, some Gram-positive bacteria such as *Listeria* tend to lyse more easily in the presence of 0.5%–1% Triton X-100.
4. Add 10 μL of proteinase K (10 mg/mL), and incubate at 56°C for 1–2 h (when the lysate becomes clear).
   Note: Protease (Sigma Cat. # P6911) can be used in place of proteinase K. As protease has a mixture of proteolytic activities, it may offer a higher efficiency for those hard-to-lyse bacteria such as Gram-positive organisms. Additionally, other lytic enzymes (e.g., lysostaphin for *Staphylococci*, mutanolysin for *Listeriae* or *Streptococci*) can be used individually or in combination if lysozyme does not give satisfactory result.
5. Transfer the lysate to a 1.5 mL tube, add 1 volume (500 μL) of phenol/chloroform/isoamyl-alcohol (25:24:1), and mix by inversion.
6. Centrifuge at 13,000 rpm for 10 min (repeat Steps 6 and 7 once for increased DNA purity if desired).
7. Transfer supernatant to a new 1.5 mL tube, add 1 volume (about 500 μL) of isopropyl alcohol and 20 μL of 5 M NaCl (to a final concentration of 0.10–0.15 M), mix by inversion, and centrifuge at 13,000 rpm for 10 min at 4°C.
8. Discard supernatant, add 300 μL of 70% ethanol (stored at −20°C), and centrifuge at 13,000 rpm for 2 min at 4°C.
9. Discard supernatant, invert the tube on a paper towel, and air-dry for 10 min.
10. Dissolve the DNA pellet in 200 μL of 1x TE (10 mM Tris-HCl pH 8.0 and 1 mM EDTA pH 8.0), determine the DNA concentration at UV 260/280 nm in a spectrophotometer, and store at −20°C.

5.3.2 LARGE-SCALE ISOLATION OF BACTERIAL GENOMIC DNA USING CTAB AND PHENOL–CHLOROFORM

Similar to the classic phenol–chloroform protocol, the CTAB and phenol–chloroform method involves lysis of bacterial cells with lytic enzyme and digestion of cellular proteins with proteinase K. A cationic surfactant CTAB further assists separation of DNA from degraded proteins, which are subsequently removed with phenol–chloroform. DNA is then precipitated with isopropanol (or ethanol). Again, RNase A treatment can be incorporated during the cell lysis stage (with lysozyme) or after DNA is precipitated (as outlined in Section 5.3.1). This method enables large-scale extraction of genomic DNA with high purity, which is especially useful for genome sequencing and other experiments [36,37].

Specific reagents

CTAB/NaCl solution: 10% CTAB in 0.7 M NaCl (Dissolve 4.1 g NaCl in 80 mL H$_2$O and slowly add 10 g CTAB [Sigma Cat. # 52365] while stirring on a hot plate. Adjust final volume to 100 mL. Add 200 μL [0.2%] of β-mercaptoethanol before use.)

Procedure

1. Inoculate 100 mL of Luria Bertani (LB) broth (or other liquid media) in a 500 mL flask with a single colony of bacteria (from a previously inoculated agar plate), and incubate at 37°C for 16–18 h with shaking at 250 rpm (with OD at 600 nm of >1.0).
2. Pour the broth culture into two 50 mL tubes, centrifuge at 3000 rpm for 15 min, and decant the supernatant.
3. Resuspend the cell pellets in each tube with 4.75 mL of 1x TE buffer (10 mM Tris-HCl pH 8.0, 1 mM EDTA pH 8.0) by vortexing, and combine the contents from the two 50 mL tubes into one.
4. Add 200 μL of freshly made lysozyme (10 mg/mL), and 50 μL of RNase A (10 mg/mL), and incubate at 37°C for 30 min.
5. Add 100 μL of proteinase K (10 mg/mL) and 500 μL of 10% SDS, and incubate at 56°C for 1 h (until the lysate becomes clear).
6. Add 1.8 mL of 5 M NaCl and 1.5 mL CTAB/NaCl solution, and incubate at 65°C for 10 min.
7. Add 1 volume (about 14 mL) chloroform/isoamyl alcohol (24:1) to the tube, mix, transfer the content to a 30 mL Sorvall centrifuge tube, and centrifuge at 10,000 rpm for 15 min. (Note: Steps 6 and 7 can be repeated once more if extra clean DNA is desired.)
8. Transfer the upper layer into a new 30 mL Sorvall centrifuge tube, add 1 volume (about 14 mL) phenol/chloroform/isoamyl alcohol, mix, and centrifuge at 10,000 rpm for 15 min.
9. Transfer the upper layer into a new 30 mL Sorvall centrifuge, add 0.7 volumes isopropanol, mix, and centrifuge at 10,000 rpm for 10 min.
10. Discard the solution and wash the pellet with 5 mL of 70% ethanol, and centrifuge at 10,000 rpm for 2 min.
11. Discard the ethanol, air-dry the DNA, and dissolve the DNA in 2–4 mL of 1x TE.
12. Determine the DNA concentration at UV 260/280 nm in a spectrophotometer, and store at −20°C. Quantify the DNA on 1% agarose gel using a DNA weight marker if preferred.

5.3.3   Isolation of Bacterial Genomic DNA Using UltraClean Microbial DNA Kit

The UltraClean microbial DNA kit (MO BIO) is designed to isolate high-quality genomic DNA from microorganisms including bacteria and fungi. Microbial cells are added to a bead beating tube containing beads, bead solution, and lysis solution, and the cells are lysed by a combination of heat, detergent, and mechanical force (using a specially designed vortex adapter on a standard vortex) against specialized beads. The released DNA in the cell lysate is then bound to a silica spin filter. After washing, the DNA is recovered in Tris buffer (or distilled water). The whole procedure takes about 20 min and yields up to 20 μg DNA (possibly of <20 kb in size) per filter, and without the conventional lysozyme and proteinase K treatments.

Specific reagent

UltraClean microbial DNA kit components (MO BIO Cat. # 12224-50), which consists of the following components:

Microbial DNA MicroBead tubes (50 units each containing 250 mg MicroBeads)
Microbial DNA MicroBead solution (16.5 mL)
Microbial DNA solution MD1 (2.75 mL)
Microbial DNA solution MD2 (5.5 mL)
Microbial DNA solution MD3 (50 mL)
Microbial DNA solution MD4 (16.5 mL)
Microbial DNA solution MD5 (3 mL)
Microbial DNA spin filters (50 units each in a 2 mL tube)
Microbial DNA 2 mL collection tubes (200)

Specific equipment

Vortex adapter (MO BIO Cat. # 13000-V1)

Procedure

(Please wear gloves at all times)

1. Add 1.8 mL of bacterial culture (with OD at 600 nm of >1.0) to a 2 mL collection tube and centrifuge at 10,000 g for 30 s at room temperature. Decant the supernatant and spin the tubes at 10,000 g for 30 s at room temperature and completely remove the media supernatant with a pipette tip. (Note: Some microbial cultures may require centrifugation longer than 30 s).
2. Resuspend the cell pellet in 300 µL of MicroBead solution and gently vortex to mix. Transfer the cell suspension to MicroBead tube. (The MicroBead solution contains salts and a buffer that stabilizes and homogeneously disperses the microbial cells.)
3. Add 50 µL of solution MD1 to the MicroBead tube. (Solution MD1 contains SDS and other disruption agents required for cell lysis.)
4. Optional: (1) To increase yields, heat the tubes at 65°C for 10 min and continue with Step 5. (2) To minimize DNA shearing, heat the tubes at 65°C for 10 min with occasional bump vortexing for a few seconds every 2–3 min. Skip Step 5 and go to Step 6. (3) To lyse the recalcitrant bacteria, heat the tubes at 70°C for 10 min and continue with Step 5.
5. Secure bead tubes horizontally using the MO BIO Vortex Adapter tube holder for the vortex or secure tubes horizontally on a flatbed vortex pad with tape. Vortex at maximum speed for 10 min.
6. Make sure the 2 mL MicroBead tubes rotate freely in the centrifuge without rubbing. Centrifuge the tubes at 10,000 g for 30 s at room temperature.
7. Transfer the supernatant to a clean 2 mL collection tube. Note: Expect 300–350 µL of supernatant.
8. Add 100 µL of solution MD2 to the supernatant. Vortex for 5 s. Then incubate at 4°C for 5 min. (Solution MD2 contains a reagent to precipitate non-DNA organic and inorganic materials.)
9. Centrifuge the tubes at room temperature for 1 min at 10,000 g.
10. Avoiding the pellet, transfer the entire volume of supernatant to a clean 2 mL collection tube. Expect approximately 450 µL in volume.
11. Add 900 µL of solution MD3 to the supernatant and vortex for 5 s. (Solution MD3 is a highly concentrated salt solution.)
12. Load about 700 µL into the spin filter and centrifuge at 10,000 g for 30 s at room temperature. Discard the flow through, add the remaining supernatant to the spin filter, and centrifuge at 10,000 g for 30 s at room temperature. (Note: A total of two to three loads for each sample processed are required.) Discard all flow-through liquid.
13. Add 300 µL of solution MD4 and centrifuge at room temperature for 30 s at 10,000 g. (Solution MD4 is an ethanol-based wash solution used to further clean the DNA that is bound to the silica filter membrane in the spin filter).

14. Discard the flow through.
15. Centrifuge at room temperature for 1 min at 10,000 g.
16. Be careful not to splash liquid on the spin filter basket, place the spin filter in a new 2 mL collection tube.
17. Add 50 μL of solution MD5 to the center of the white filter membrane. (Solution MD5 is 10 mM Tris pH 8.0 and does not contain EDTA.)
18. Centrifuge at room temperature for 30 s at 10,000 g.
19. Discard spin filter. The DNA in the tube is now ready for any downstream application. Store DNA at −20°C.
20. DNA may be further concentrated by adding 5 μL of 5 M NaCl to the 50 μL preparation and inverting three to five times to mix. Next, add 100 μL of 100% cold ethanol and invert three to five times to mix. Incubate at −20°C for 30 min and centrifuge at 10,000 g for 15 min at room temperature. Decant all liquid, air-dry, and resuspend the DNA in sterile water or solution MD5.

5.4 FUTURE DEVELOPMENT TRENDS

Isolation of bacterial DNA from cultures is a relatively straightforward process compared with that from non cultured specimens (e.g., clinical, food, and environmental samples). Few interfering substances remain in bacterial cultures, and therefore, a 10 min heating in water (or Triton X-100, or NaOH followed by neutralization for hardy bacteria) and a brief spin are all that is required to prepare bacterial isolates before PCR experiments (Figure 5.1). However, for DNA cloning and other molecular analysis (e.g., DNA–DNA hybridization, Southern blotting, genome sequencing), the availability of reasonable quantity of purified DNA is essential. Although a vast array of in-house protocols together with continued improvement of commercial kits (e.g., user's friendliness and per unit cost) are more than adequate to meet this demand, further refinement in the key steps of bacterial DNA isolation will help propel the bacterial DNA isolation to the next level of sophistication.

During the DNA isolation from bacteria (and indeed other biological organisms as well), one utmost important issue concerns the proper lysis of cell wall. Given the diversity of the cell-wall structures among Gram-positive, Gram-negative, and acid-fast bacteria, DNA isolation protocols (or kits) developed for one specific group of bacteria may not be ideal for the other. The fact that individual Gram-positive species displays remarkable variations in their sensitivity and susceptibility to lytic enzymes highlights the need to design tailor-made cell-lysis approaches. While use of bead-beating may possibly fulfill this role, its widespread adoption is hampered by the finding that bead-beating often contributes to DNA shearing (<20 kb), which is undesirable in certain experiments. Therefore, optimization of a lytic enzyme-based mix (e.g., lysozyme, lysostaphin, and mutanolysin) or development of other lysis approaches will help address the current deficiency in this area. In addition, identification and application of suitable bacteriophages will offer another option for efficient lysis of different bacterial species.

Once the bacteria are lysed, there is a necessity to digest and eliminate cellular proteins and other contaminating materials present in the lysate. Use of proteinase K (or protease) is certainly helpful in this regard, but at the moment, such treatment is unduly lengthy. Thus, future innovation in the cell-lysis step that speeds up the process without endangering the DNA quality will further streamline the DNA isolation process. Additionally, the development of more efficient procedures to remove the degraded proteins and other components will also contribute to the improved recovery of bacterial DNA from cultures. Currently, the commercial column-based systems offer an efficient and reliable way to separate proteins and other contaminants when a specified quantity of bacteria is applied. However, their efficiency with smaller or larger quantities of bacterial input is unknown. This limitation hinders their application in the quantitative studies of bacterial populations from noncultured clinical, food, and environmental specimens. The existing in-house methods are useful under the circumstances, but their reliance on toxic solvents and their lack of interlab standardization remain a concern.

REFERENCES

1. Erdos, G. et al., Construction and characterization of a highly redundant *Pseudomonas aeruginosa* genomic library prepared from 12 clinical isolates: Application to studies of gene distribution among populations, *Int. J. Pediat. Otorhinolaryngol.*, 70: 1891, 2006.
2. Shen, K. et al., Extensive genomic plasticity in *Pseudomonas aeruginosa* revealed by identification and distribution studies of novel genes among clinical isolates, *Infect. Immun.*, 74: 5272, 2006.
3. Hiller, N.L. et al., Comparative genomic analyses of seventeen *Streptococcus pneumoniae* strains: Insights into the pneumococcal supragenome, *J. Bacteriol.*, 189: 8186, 2007.
4. Hogg, J.S. et al., Characterization and modeling of the *Haemophilus influenzae* core and supragenomes based on the complete genomic sequences of Rd and 12 clinical nontypeable strains, *Genome Biol.*, 8: R103, 2007.
5. Glaser, P. et al., Comparative genomics of *Listeria* species, *Science*, 294: 849, 2001.
6. Nelson, K.E. et al., Whole genome comparisons of serotype 4b and 1/2a strains of the food-borne pathogen *Listeria monocytogenes* reveal new insights into the core genome components of this species, *Nucleic Acids Res.*, 32: 2386, 2004.
7. Hain, T. et al., Whole-genome sequence of *Listeria welshimeri* reveals common steps in genome reduction with *Listeria innocua* as compared to *Listeria monocytogenes, J. Bacteriol.*, 188: 7405, 2007.
8. Liu, D., *Listeria*-based anti-infective vaccine strategies, *Rec. Patents Anti-infect. Drug Disc.*, 1: 281, 2006.
9. Yeung, V.P. et al., Heat-killed *Listeria monocytogenes* as an adjuvant converts established murine Th2-dominated immune responses into Th1-dominated responses, *J. Immunol.*, 161: 4146, 1998.
10. Yamamoto, K. et al., Listeriolysin O, a cytolysin derived from *Listeria monocytogenes* inhibits generation of ovalbumin-specific Th2 immune response by skewing maturation of antigen-specific T cell into Th1 cells, *Clin. Exp. Immunol.*, 142: 268, 2005.
11. Herbert, S. et al., Molecular basis of resistance to muramidase and cationic antimicrobial peptide activity of lysozyme in staphylococci, *PLoS Pathog.*, 3: e102, 2007.
12. Boneca, I.G. et al., A critical role for peptidoglycan N-deacetylation in *Listeria* evasion from the host innate immune system, *Proc. Natl. Acad. Sci. U S A*, 104: 997, 2007.
13. Mujahid, S., Pechan, T., and Wang, C., Improved solubilization of surface proteins from *Listeria monocytogenes* for 2-DE, *Electrophoresis*, 28: 3998, 2007.
14. Loessner, M.J., Schneider, A., and Scherer, S., A new procedure for efficient recovery of DNA, RNA, and proteins from *Listeria* cells by rapid lysis with a recombinant bacteriophage endolysin, *Appl. Environ. Microbiol.*, 61: 1150, 1995.
15. Neudecker, F. and Grimm, S., High-throughput method for isolating plasmid DNA with reduced lipopolysaccharide content, *Biotechniques*, 28: 107, 2000.
16. Boom, R. et al., Rapid and simple method for purification of nucleic acids, *J. Clin. Microbiol.*, 28: 495, 1990.
17. Rudi, K. et al., Rapid, universal method to isolate PCR-ready DNA using magnetic beads, *Biotechniques*, 22: 506, 1997.
18. McOrist, A.L, Jackson, M., and Bird, A.R., A comparison of five methods for extraction of bacterial DNA from human faecal samples, *J. Microbiol. Methods*, 50: 131, 2002.
19. Veloso, I.F. et al., A comparison of three DNA extractive procedures with *Leptospira* for polymerase chain reaction analysis, *Mem. Inst. Oswaldo Cruz.*, 95: 339, 2000.
20. Bahador, A. et al., Comparison of five DNA extraction methods for detection of *Mycobacterium tuberculosis* by PCR, *J. Med. Sci.* (Pakistan), 4: 252, 2004.
21. Gomes, L.H. et al., A simple method for DNA isolation from *Xanthomonas* spp, *Sci. Agric.*, 57: 3, 2000.
22. Liu, D., Preparation of *Listeria monocytogenes* specimens for molecular detection and identification, *Int. J. Food Microbiol.*, 122: 229, 2008.
23. Liu, D. and Webber, J., A polymerase chain reaction assay for improved determination of virulence of *Dichelobacter nodosus*, the specific causative pathogen for ovine footrot, *Vet. Microbiol.*, 43: 197, 1995.
24. Liu, D. et al., Identification of *Listeria innocua* by PCR targeting a putative transcriptional regulator gene, *FEMS Microbiol. Lett.*, 223: 205, 2003.
25. Liu, D. et al., Characterization of virulent and avirulent *Listeria monocytogenes* strains by PCR amplification of putative transcriptional regulator and internalin genes, *J. Microbiol. Methods*, 52: 1065, 2003.
26. Liu, D. et al., Identification of a gene encoding a putative phosphotransferase system enzyme IIBC in *Listeria welshimeri* and its application for diagnostic PCR, *Lett. Appl. Microbiol.*, 38: 151, 2004.

27. Liu, D. et al., PCR detection of a putative *N*-acetylmuramidase gene from *Listeria ivanovii* facilitates its rapid identification, *Vet. Microbiol.*, 101: 83, 2004.

28. Liu, D. et al., Specific PCR identification of *Pasteurella multocida* based on putative transcriptional regulator genes, *J. Microbiol. Methods*, 58: 263, 2004.

29. Liu, D. et al., Species-specific PCR determination of *Listeria seeligeri*, *Res. Microbiol.*, 155: 741, 2004.

30. Liu, D. et al., Isolation and PCR amplification of a species-specific, oxidoreductase-coding gene region in *Listeria grayi*, *Can. J. Microbiol.*, 51: 95, 2005.

31. Liu, D. et al., PCR amplification of a species-specific putative transcriptional regulator gene reveals the identity of *Enterococcus faecalis*, *Res. Microbiol.*, 156: 944, 2005.

32. Liu, D. et al., Rapid identification of *Streptococcus pyogenes* with PCR primers from a putative transcriptional regulator gene, *Res. Microbiol.*, 156: 564, 2005.

33. Liu, D. et al., Evaluation of PCR primers from putative transcriptional regulator genes for identification of *Staphylococcus aureus*, *Lett. Appl. Microbiol.*, 40: 69, 2005.

34. Liu, D. et al., Use of a putative transcriptional regulator gene as target for specific identification of *Staphylococcus epidermidis*, *Lett. Appl. Microbiol.*, 43: 325, 2006.

35. Liu, D. et al., PCR detection of pathogenic *Leptospira* genomospecies targeting putative transcriptional regulator genes, *Can. J. Microbiol.*, 52: 272, 2006.

36. Ausubel, F.M. et al. (Eds.), *Current Protocols in Molecular Biology*, John Wiley & Sons, New York, 2002.

37. Woo, T.H., Cheng, A.F., and Ling, J.M., An application of a simple method for the preparation of bacterial DNA, *Biotechniques*, 13: 696, 1992.

6 Isolation of Bacterial RNA from Cultures

Som Subhra Chatterjee and Trinad Chakraborty

CONTENTS

6.1 INTRODUCTION

6.1.1 GENERAL CONSIDERATION

RNA (ribonucleic acid) is a nucleic acid that plays important roles in transcribing genetic information from DNA (deoxyribonucleic acid). RNA is a polymer of ribose sugar and phosphate backbone and nucleotide bases. Unlike DNA, RNA is usually single stranded (although these single-stranded molecules can form secondary structures giving rise to double-stranded stretches within a RNA molecule by forming hydrogen bonds). In addition, RNA nucleotide contains ribose sugar instead of deoxyribose sugar present in the DNA, and the nucleotide in RNA is uracil instead of thymine as present in DNA. The hydroxyl group present in the 2' position of the pentose sugar of the RNA molecules makes it more prone to hydrolysis and as a result RNA is less stable than DNA.

There are four major kinds of RNA present in a cell: messenger RNA (mRNA), ribosomal RNA (rRNA), transfer RNA (tRNA), and small RNAs (sRNAs). Constituting only 5% of the total RNA in a cell, the mRNAs are the information-carrying molecules, which serve as an intermediate between DNA and protein synthesis. The mRNA synthesis is constantly under regulation and the mRNA is the only RNA type that is vulnerable to cellular nucleases. As a result, the half-life of mRNAs is very short, which ranges from seconds to minutes. The rRNA is the catalytic component of ribosomes (the protein synthesis machinery of a cell), which are complex assemblies of rRNAs and are made of more than 50 different proteins. The rRNA molecules are extremely abundant in a cell and account for more than 80% of the total cellular RNA. tRNAs are sRNA chains that transfer amino acids to a growing polypeptide chain at the ribosomal site of protein synthesis. In addition, certain RNAs can function as enzymes and are referred to as ribozymes. Recent studies have led to the identification of a number of noncoding sRNAs in bacteria. These sRNAs change the stability of a specific mRNA by base pairing with the target mRNA. In recent years, the noncoding RNAs have been shown to regulate bacterial virulence and stress responses [1].

RNA is synthesized from DNA with the help of RNA polymerase. The initiation of RNA synthesis begins with the binding of the RNA polymerase to the promoter sites of the DNA. The RNA polymerase of bacteria consists of different subunits: α, β, β', and σ. The α component is present in two copies in the RNA polymerase holoenzyme while the rest of the components are present in a single copy. The σ is not tightly bound like the other subunits in the RNA polymerase. The sigma factor recognizes the appropriate sites on the DNA for the initiation of RNA synthesis and once a small portion of the RNA has been formed, the sigma factor dissociates from the holoenzyme and the RNA polymerase can complete the transcription of the RNA.

The process of transcription by RNA polymerase is similar in bacteria and eukaryotes. However, the processing of mRNA in bacteria and eukaryotes varies greatly. In most of the cases, bacterial mRNAs do not require any processing upon production. Unlike the bacterial mRNAs, eukaryotic mRNAs have to undergo a number of maturation steps such as 5' capping, splicing, editing, and polyadenylation before protein synthesis can take place by the translational machinery of the cell. Polyadenylation is the covalent linkage of the polyadenyl nucleotides to the 3' end of an mRNA molecule, often called as poly(A) tail. The poly(A) tail of the eukaryotic RNA is often manipulated for the isolation of mRNAs from the eukaryotic total RNA. It is also utilized as a prime target for the production of complementary DNA (cDNAs) in eukaryotes.

Following transcription of mRNA, translation takes place in order to make proteins. Ribosome is the central component of the protein manufacturing machinery in the cell. The genes that code for rRNAs are considered to be the most conserved among organisms and are often considered to be a landmark to measure genetic distances among various species of an organism. Both in bacteria and eukaryotes, ribosome is composed of two different subunits (small and large subunits) based on their sedimentation rates upon centrifugation. The sedimentation rate is a measure of shape and mass of a particular subunit of ribosome and is denoted by Svedberg units (S). The bacterial rRNA measures 70S, which comprises a large subunit of 50S (consisting of 5S and 23S rRNA species) and a small subunit of 30S (16S rRNA species); whereas eukaryotic rRNA measures 80S, which contains a large subunit of 50S (5S, 5.8S, and 23S rRNA species) and a small subunit of 40S (18S rRNA species).

6.1.2 Utility of Bacterial RNA in Research

The use of bacterial RNA in contemporary research has seen an overwhelming increase during the past few years. Just by searching for the key word "bacterial RNA isolation" in the PubMed (http://www.ncbi.nlm.nih.gov/sites/entrez) between the years 2002 and 2007 will retrieve around 1700 scientific publications, which is notably higher than the years 1997 and 2001 with around 800 papers and the years 1992 and 1996 with around 600 papers. The bacterial RNA has been exploited in scientific research for a number of purposes. These include the determination of the size and abundance of a particular RNA transcript (quantitative real-time PCR [polymerase chain reaction], northern hybridization, dot/slot blotting), determination of the fine structure of a particular transcript and the 5' and 3' ends of a RNA transcript (RNase protection assay, S1 mapping, and primer extension), determination of the precursors and processing intermediates of mRNAs (RNase protection and S1 mapping), and assessment of bacterial viability.

The availability of whole genome sequences of bacterial pathogens in the late 1990s marked the development of DNA microarrays, which has the ability to measure gene expression levels on a global basis at a faster rate. DNA microarrays can recapitulate the change in bacterial transcription upon a particular condition. Whole genome transcriptional profiling of pathogenic bacteria permits new insights into adaptive responses activated by the bacteria when growing in the niche of the host [2–5]. For example, RNA profiling can be used to identify novel vaccine candidates, as bacterial transcriptional response to infection offers insight into the physiological state of infecting bacteria and the mechanisms required by bacteria to successfully survive infection [6]. In addition, studies on bacterial noncoding sRNAs are also valuable for understanding the quorum sensing systems and pathogenesis processes in bacteria as these activities are often regulated by noncoding sRNAs, which generally ensure such a regulation by pairing to mRNAs of effector or regulatory genes, or by binding to proteins, thus directly and indirectly controlling gene expression [7,8].

RNA isolation from bacterial cultures is practiced routinely in many microbiology laboratories currently, and the use of bacterial RNA in research is likely to increase further in future. Below we discuss key issues relating bacterial RNA isolation, including specific considerations on the purification of RNA from cultures of different backgrounds (Gram negative, Gram positive, and acid-fast bacteria) and from intracellular bacterial pathogens. We also examine the features of current commercial kits for bacterial RNA isolation. We then provide step by step protocols for bacterial RNA isolation. This is followed by a discussion of the future development that will facilitate improved recovery of bacterial RNA from cultures.

6.2 BASIC PRINCIPLES OF BACTERIAL RNA ISOLATION

6.2.1 LYSIS OF BACTERIA AND INHIBITION OF RNASES

The ribose sugars present in RNA carry hydroxyl group at the 2' position in contrast to sugars present in DNA, which renders RNA more reactive chemically than DNA and can be easily cleaved by contaminating RNases. RNases are released virtually from all living cells upon lysis and are present on the skin, tabletops and aerosols, etc. Because of the presence of intrachain disulfide bonds, RNases are resistant to mild detergents and to autoclaving. In addition to the exogenous RNase sources, bacteria also have its endogenous RNase, which can degrade even its own RNA during isolation process. Thus, the key to successful purification of intact RNA from cells is speed and RNases should be inactivated at the very first stage of extraction. Thus, the following are good laboratory practices that must be strictly followed before RNA isolation. First, RNase buffers contaminated with bacteria and other microorganisms should be discarded; pipetting devices and centrifuges should be free from RNases and if possible use a separate set of pipetting device for RNA isolation only. Second, use of hand gloves helps preventing RNase contamination but gloves must be changed every time a piece of apparatus is touched, refrigerator opened, etc. Third, prepare solutions with RNase free glassware and DEPC (diethylpyrocarbonate) treated water. (DEPC is an alkylating agent used to inactivate RNases in buffers and glassware.) Finally, use disposable tips and treat solutions with 1% DEPC for 1 h at 37°C followed by autoclaving for 15 min [9].

Lysis of bacteria is often achieved with the use of a lytic enzyme (e.g., lysozyme) to disrupt the bacterial cell wall. This is also assisted with the addition of a detergent. After cell lysis, RNase inhibitors (e.g., RNasin, proteinase K, guanidinium hydrochloride, and guanidinium isothiocyanate) are added immediately to destroy RNases. Following the RNase inhibition step, organic solvents such as phenol and chloroform are used to remove ribonucleases.

6.2.2 SEPARATION OF BACTERIAL RNA FROM OTHER MOLECULES

It is clear from the discussion above that the isolation of high-quality RNA depends on efficient lysis of the cells and the early and efficient neutralization of endogenous and exogenous ribonucleases. Among the most frequently used RNase inhibitors, guanidinium hydrochloride and guanidinium isothiocyanate are strong denaturants, and have the capability to disrupt cells, solubilize their components, and denature endogenous RNases and other proteins simultaneously. These reagents are much more efficient than acid phenol in inactivating RNases. However, they do not physically separate RNA from proteins and DNA in one step, which need to be removed by other procedures (e.g., ultracentrifugation and chloroform treatment). Furthermore, some cellular polysaccharides and proteoglycans may remain in the preparation, which may sometimes affect downstream processing. Guanidinium isothiocyanate (4 M) was first used by Chirgwin et al. to homogenize cultured cells and the lysate was subjected to cesium chloride ultracentrifugation [10]. Because of the fact that RNA has a greater buoyant density compared to that of other cellular components, RNA migrates to the bottom of the tube during ultracentrifugation. This technique has been used extensively during 1980s for isolation of high molecular weight RNA. In 1997, Chomczynski and Sacchi introduced a single-step technique of RNA isolation in which the guanidinium thiocyanate homogenate of the cultured cells is extracted with phenol and chloroform at a reduced pH [11]. Nearly all of the RNA collects into the aqueous phase made up of a mixture of water-saturated phenol, chloroform, and a chaotropic denaturing solution (guanidinium), while the DNA and protein collect in the interphase and the organic phase made up of phenol. Use of this new technique eliminated the need for subsequent ultracentrifugation steps and introduced a quick method of isolating good quality of RNA. This method (as also known as TRIzol method or TRI reagents) still remains the protocol of choice to isolate RNA [9].

Another useful technique to separate bacterial RNA from other molecules is the hot phenol method described by von Gabain et al., in which bacteria are lysed in sodium dodecyl sulfate (SDS) solution followed by extraction with preheated phenol [12]. This technique is also known as acid phenol method as the phenol is adjusted to an acidic pH (equilibrated with 0.1 M sodium acetate pH 5.2). Besides assisting cell lysis and protein denaturation, acid phenol neutralizes the negative charge of DNA and causes it to partition into the organic phase, whereas RNA is unaffected at the acid pH and remains charged and partitions into the aqueous phase. After centrifugation, the RNA in the aqueous phase is recovered and further extracted. Although it represents an economic way to extract RNA, the procedure takes more time, and is prone to DNA contamination. In addition, it does not inactivate RNases immediately and can leave residual phenol in the sample inhibiting downstream reactions and introducing error into RNA quantitation.

Anion-exchange matrices are also useful for purification of RNA and DNA. They are essentially based on the interaction between negatively charged phosphates of nucleic acids and positively charged matrices. Under defined salt conditions, RNA binds to the matrices, while other molecules such as DNA and proteins are eluted sequentially with buffers of differing salt concentrations. Through this selective elution process, RNA, DNA, and sRNA can be isolated from the same sample. The eluted RNA is then precipitated with ethanol, which has a much higher purity than that prepared with cesium chloride centrifugation.

Silica matrices are also used to isolate RNA and DNA, as nucleic acids show a tendency to bind silica under chaotropic conditions (e.g., high salts) and then come off with lower salt buffers. Compared with anion-exchange matrices, silica matrices cannot hold very sRNA species (<200 nt), which include many types of noncoding RNAs including small interfering RNA (siRNA) and micro RNA (miRNA).

6.2.3 Precipitation of RNA

RNA can be precipitated (generally done if a chemical contamination is suspected, which may inhibit downstream processing of the RNA) from a solution by using alcohol. Usually, RNA is precipitated by adding 0.1 volume of 3 M sodium acetate at pH 5.2 and 3 volume of absolute ethanol to the RNA solution, followed by mixing and precipitation at −20°C overnight. The RNA can be pelleted by centrifugation at >10,000 g for 30 min at 4°C. Also, lithium chloride can be used at 0.5 M in place of sodium acetate for preferential precipitation of RNA. The major advantage of lithium chloride precipitation over other RNA precipitation reagents is that it does not precipitate DNA, protein, or carbohydrate efficiently [13]. As a consequence, it represents a method of choice for removing inhibitors of translation or cDNA synthesis from RNA preparations [14], in addition to providing a simple rapid method for recovering RNA from *in vitro* transcription reactions. Another approach to the isolation of RNA from bacterial lysates is to employ hexamine cobalt and spermidine for selective precipitation of RNA species. At a concentration of 3.5 mM hexamine cobalt, total RNA can be precipitated from a cell lysate; at 2 mM hexamine cobalt, rRNA can be fractionated selectively from low molecular weight RNA. The resulting RNA mixture can be further resolved to pure 5S and mixed 16S/23S rRNA by nondenaturing anion-exchange chromatography. Use of a second precipitation with 8 mM hexamine cobalt allows precipitation of the low molecular weight RNA fraction [15].

After the RNA precipitation, two washes with 80% (v/v) ethanol wash is recommended for removing salt and organic solvents, which may be still present in the RNA sample in trace amounts. The RNA sample must be dried (either by air-drying or by using speed vac) and resuspended in water. The resultant RNA can be stored at −20°C for short-term storage or at −70°C for extended storage.

6.2.4 QUANTIFICATION OF RNA AND DETERMINATION OF ITS QUALITY

RNA can be accurately quantified by measuring its absorbance in a spectrophotometer. RNA has an absorption maximum at 260 nm wavelength of light. One optical density (OD) unit of RNA is equivalent to 40 ng/μL of RNA concentration. Usually, a small aliquot of RNA is used to measure its concentration and then discarded. Contamination of RNA with traces of protein can be identified by the ratio of the absorbance of the RNA at 260 and 280 nm. Proteins have an absorption maximum at 280 nm. Thus, a 260–280 nm absorption ratio of 1.8–2.0 of an RNA sample is considered to be pure. Nonetheless, the exact absorption ratio may vary with the buffer used to resuspend the RNA. Namely, if RNA is dissolved in DEPC water (pH 5.0–6.0), an $A260/280$ nm value of 1.6 suggests the purity of RNA is high; if RNA is dissolved in nuclease free water (pH 6.0–7.0), an $A260/280$ nm value of 1.85 is a good indicator; and if RNA is dissolved in 1x TE buffer (pH 8.0), an $A260/280$ nm value of 2.14 is acceptable. An absorption ratio at 260–280 nm below 1.6 of an RNA sample indicates considerable protein contamination.

RNA preparations can also be contaminated with DNA, although this is rare. Both RNA and DNA have an absorption maximum at 260 nm wavelength of light and as a result DNA cannot be distinguished from RNA by spectrometry. In case of a suspected DNA contamination in a RNA sample, an aliquot of the RNA can be subjected to a PCR of a housekeeping gene. DNA polymerases used in PCR usually amplify from a DNA template, not RNA. If a PCR product is detected upon agarose gel electrophoresis and ethidium bromide staining, it confirms a DNA contamination. To remove the contaminating DNA in the preparation, RNA can be subjected to a DNase digestion. Alternatively, lithium chloride can be used to reprecipitate the RNA, while leaving DNA in the solution. Re-extraction with chloroform followed by ethanol precipitation is also helpful.

As RNA is liable to degradation by contaminated RNases during its isolation procedure, determination of the quality of RNA isolated is very important for its various downstream applications. Historically, mRNA quality has been assessed by denaturing agarose gel electrophoresis of total RNA followed by staining with ethidium bromide. This method relies on the fact that rRNA (16S and 23S subunits) quality and quantity reflect that of the underlying mRNA population. In other words, if the rRNA bands in an isolated RNA following agarose gel electrophoresis and staining with ethidium bromide appears to be intact, then the mRNA population in that RNA sample is intact too.

Recently, the Agilent 2100 bioanalyzer has become the standard in RNA quality assessment and quantitation. Using electrophoretic separation on microfabricated chips, RNA samples are separated and subsequently detected via laser-induced fluorescence detection. The bioanalyzer software generates an electropherogram and gel-like image (Figure 6.1). The electropherogram provides a detailed visual assessment of the quality of an RNA sample. The main advantage of Agilent 2100 bioanalyzer over the gel-based determination of RNA quality is that the former is able to detect the

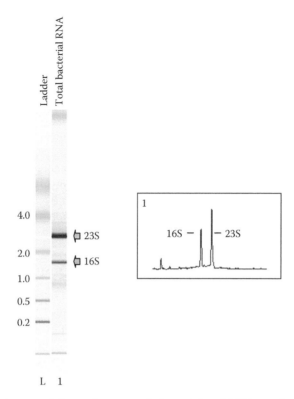

FIGURE 6.1   Gel-like image and electropherogram of a bacterial total RNA sample developed using Agilent 2100 bioanalyzer.

quality of RNA sample in nanogram level, whereas the latter needs at least 1 μg of total RNA to visualize the rRNA bands clearly with ethidium bromide. This is particularly important as in some cases the amount of RNA isolated from a sample may be limiting.

6.2.5   ISOLATION OF RNA FROM GRAM NEGATIVE, GRAM POSITIVE, AND ACID-FAST BACTERIA

Bacteria can be broadly divided into two groups depending upon their ability to retain a crystal violet-iodine dye complex following a treatment with acetone or alcohol. This reaction is called as Gram staining (named after Christian Gram). Bacterial cells that retain the Gram stain after treating with acetone are called as Gram-positive bacteria. On the other hand, those bacteria that cannot retain the Gram stain following acetone treatment are called as Gram-negative bacteria. The basis of the staining capability of the Gram-positive and Gram-negative bacteria underlies in the organization of the structures outside the plasma membrane of the bacteria. In Gram-negative bacteria, these structures constitute the cell envelope, whereas in Gram-positive bacteria, they comprise the cell wall. The Gram-positive cell wall is characterized by the presence of a very thick peptidoglycan layer, which is covalently attached to other cell wall polymers, such as teichoic acids and lipoteichoic acids. There may be up to 40 layers of this polymer conferring enormous mechanical strength on the cell wall.

In contrast to the Gram-positive cell wall, the Gram-negative cell envelope contains a thin layer of peptidoglycan layer adjacent to the cell membrane. Gram-negative bacteria are therefore mechanically much weaker than Gram-positive cells. In addition to this, the outer portion of the Gram-negative bacteria contains lipopolysaccharide, which forms the outer leaflet of the cell envelope.

A few medically important bacteria do not stain easily using Gram staining and need to be heated near boiling point in a dye. Having taken the stain, these bacteria resist decolorization with both acids and alcohol and are known as acid-fast bacteria. These types of bacteria include *Mycobacterium tuberculosis* (the causative agent of TB) and deserve special attention because of the unique nature of its cell wall structure in prokaryotes. The cell wall complex of acid-fast bacteria contains peptidoglycan and complex glycolipids. Over 60% of the cell wall of the mycobacterial cell wall is lipid, which leads to the impermeability to chemicals, resistance to killing by many antibiotics, acidic and alkaline compounds, as well as osmotic lysis.

The key to the successful RNA isolation is to lyse the bacterial cells prior to RNA isolation. The peptidoglycan layer of bacteria may be cleaved by the bacteriolytic enzyme—lysozyme. Predigestion of the bacterial cell wall with lysozyme improves lysis efficiency. This is essential for isolating RNA from Gram-positive bacteria. Preincubation of bacteria for 30 min at 37°C in a solution of 50 mM Tris pH 6.5 containing 0.4 mg/mL of lysozyme for Gram-negative bacteria and 50 mg/mL of lysozyme for Gram-positive bacteria is widely recommended. However, extraction of intact RNA from acid-fast bacteria is difficult owing to the physical strength of the cell wall. Thus, many methods were developed to physically disrupt the bacteria such as sonication [16], the French pressure cell [17], and the bead-beating device [18]. Among these, the bead-beating technique that uses 0.1 mm zirconium beads that permit rapid lysis of the bacteria is available commercially and can be used in conjunction with guanidinium-based RNA isolation [19].

6.2.6 ISOLATION OF RNA FROM BACTERIA UPON INFECTION OF HOST CELLS

Protective measures against pathogenic bacteria are based on detailed knowledge of its pathogenesis, which in turn is dependent on techniques that elucidate the underlying genetic and biochemical mechanisms. The availability of whole genome sequences of bacterial pathogens and the development of microarray technology in the late 1990s now offers the fastest way to generate gene expression profiles. Microarray technology is a powerful high-throughput tool for the analysis of host–pathogen interactions at the whole genome level. The availability of whole genome-based gene expression profiles from several bacterial pathogens during infection is beginning to represent a completely new type of resource for the investigation of the microbiology of infection. Large-scale analysis of gene expression during invasive disease provides not only transcriptional data with regard to virulence determinants, but also information about the status of the bacteria during infection and response of the bacteria to the host environment [2–5].

Obtaining the transcriptome profile of bacteria following infection is directly dependent on the isolation of high quality and adequate quantity of bacterial RNA from infected host. Collection of bacterial RNA from infected cells is limited first, by the inherent difficulties of getting rid of the large quantities of host-cell RNA, whereby the yield of bacterial RNA in such cases is also usually poor, and second, by the isolation of sufficient bacterial RNA from infected host cells.

It is therefore necessary to first choose the correct host model and bacterial load for the experiment. The host cells should be able to cope with the infecting bacterial load, which in turn allows the researcher to isolate sufficient amount of bacterial RNA. For example, in our laboratory, we infected *Listeria monocytogenes* to two murine macrophage cell lines, P388D1 and J774. Initially, both the cell lines were treated with different MOI (multiplicity of infection) of *L. monocytogenes* and bacterial count after 2, 4, 6, and 8 h postinfection was evaluated (Figure 6.2). P388D1 cells were chosen as they were more robust than J774 cells in surviving *L. monocytogenes* infection. Additionally, an MOI of 10 bacteria per cell was chosen for studies on infection of P388D1 cells as it gave the most reproducible results. At this point, the method of separating the bacteria from the infected cells before the RNA isolation was established. For the lysis of the eukaryotic cell membrane, a number of reagents were tried: water alone; 10% SDS, 1% Triton X 100 in water for 10 min at room temperature; and a mixture of 0.1% SDS, 1% acidic phenol, and 19% ethanol mixture in water for 5 min on ice. The lysate was centrifuged and the resultant pellet was subjected to RNA isolation. Of all these combinations, the yield of RNA derived from treating the sample with a mixture of SDS, phenol, and ethanol treated sample was satisfactory and at the same time the amount of host RNA contamination was

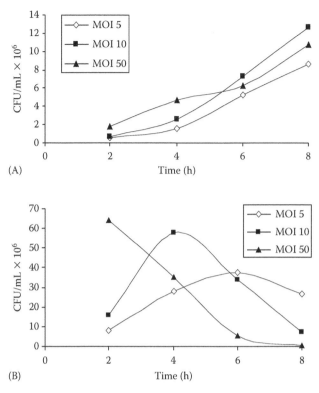

FIGURE 6.2 Survival of *L. monocytogenes* strain (EGD-e) intracellularly in P388D1 (A) and J774 (B) cells. Murine macrophage cells were infected with the wild type bacteria EGD-e with an MOI of 5, 10, and 50 in six-well tissue culture plate. After 2, 4, 6, and 8 h, postinfection eukaryotic cells were lysed with cold water and plated on agar plates for colony forming unit calculation.

negligible (Table 6.1 and Figure 6.3), as observed through electrophoretic run of the RNA samples with Agilent bioanalyzer. Eriksson et al. first described the cold phenol–ethanol method of isolating intracellular bacterial total RNA. This method allows the disruption of the eukaryotic cell and stabilization of the bacterial RNA at the same time and is applicable to Gram-negative as well as Gram-positive bacterial pathogens. The process of isolating intracellular bacteria upon infection of host cells using the cold phenol method is depicted in Figure 6.4.

6.2.7 ISOLATION OF sRNAs FROM BACTERIA

Small RNAs or noncoding RNAs are RNA molecules ranging in size from 50 to 250 bases and are present in the intragenic regions of the bacterial genomes [20]. The sRNAs have remained undetected

TABLE 6.1

Comparison of Cell Lysis Reagents for Bacterial RNA Isolation from
***L. monocytogenes* Infected P388D1 Cells**

Lysis Procedure	Total RNA Recovered (ng)	Comments
1% SDS treatment	556.5	Insufficient RNA yield
1% Triton X 100	5593	RNA contaminated with host-cell RNA
0.1% SDS + 1% acidic phenol + 19% ethanol	5624.5	Sufficient RNA yield and negligible contamination with host-cell RNA

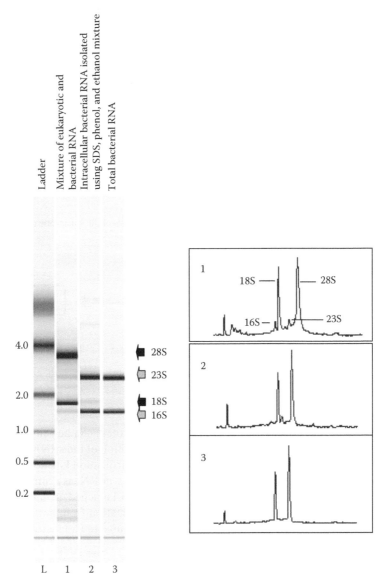

FIGURE 6.3 Quality of RNAs isolated from *L. monocytogenes* following infection to eukaryotic cells using cold phenol method.

for many years due to their small sizes, which show few detectable signals for biochemical assays and are poor targets for mutational screens. Furthermore, the sRNA genes do not encode proteins and, thus, are resistant to frameshift and nonsense mutations. Recent studies have led to the identification of a number of noncoding sRNAs in bacteria whose expressions are growth-phase dependent, stress related, and are shown to regulate bacterial virulence by changing the stability of a specific mRNA by base pairing with it [1]. In *Escherichia coli*, the first group of the sRNA genes identified consists of 4.5S, 6S, transfer-messenger RNA, RNase P, and Spot 42 sRNAs. The second group comprises OxyS. The third group includes MicF, DicF, and DsrA. All of these sRNAs have subsequently been assigned important regulatory and housekeeping functions [21].

There are presently many approaches to identify sRNAs in bacteria. Among them, use of DNA arrays containing the intragenic region of bacterial genomes and RNAs isolated by co-immunoprecipitation with the RNA binding protein Hfq is popular [21]. The RNA-binding protein Hfq is essential for the function of many sRNAs, which act as an antisense regulator.

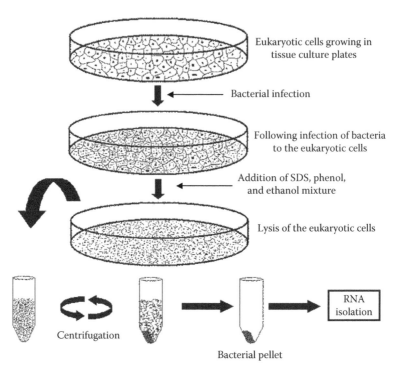

Eukaryotic cells growing in
tissue culture plates

Bacterial infection

Following infection of bacteria
to the eukaryotic cells

Addition of SDS, phenol,
and ethanol mixture

Lysis of the eukaryotic cells

Centrifugation

Bacterial pellet

RNA
isolation

FIGURE 6.4 Process of isolating intracellular bacteria upon infection of host cells using the cold phenol method.

Thus, co-immunoprecipitation of the sRNAs bound to Hfq protein can be achieved by antibody specific for Hfq protein [22]. Considering the small size, isolation of the sRNAs molecules need some special considerations. Most of the commercial RNA purification kits do not recover RNA molecules smaller than ~200 nucleotides. Recently, Qiagen's miRNeasy mini kit and Norgen Biotek Corporation's microRNA purification kit have been launched in the market and are capable of isolating RNA molecules <200 bases. RNA purification from both the kits is based on spin column chromatography, which uses standard procedures to isolate total bacterial RNA using commercial kits.

6.2.8 Isolation of mRNAs from Bacteria

Whereas eukaryotic mRNA can be separated from stable RNAs by virtue of its polyadenylated 3'-termini, bacterial mRNA is not coherently polyadenylated. Wendish et al. [23] developed a method to isolate *E. coli* mRNA by polyadenylating it in crude cell extracts with *E. coli* poly(A) polymerase I and purifying it by oligo(dT) chromatography. The mRNA thus purified gave comparable signal intensities in genome-wide expression study with only 1% as much oligo(dT)-purified mRNA as total RNA, providing evidence that *in vitro* poly(A) tailing works universally for *E. coli* mRNAs.

6.3 KEY FEATURES OF CURRENT COMMERCIAL KITS FOR BACTERIAL RNA ISOLATION

Due to their reliability and time effectiveness, many commercial kits for bacterial RNA isolation are developed and widely used in contemporary RNA research. These bacterial total RNA isolation kits are mostly based on the lysis of the bacteria by guanidine-based salts or by mechanical disruption and finally RNA extraction from the bacterial lysate either by organic extraction method (chloroform

TABLE 6.2

Features of Several Commonly Used Commercial Bacterial RNA Isolation Kits

Kit	Vendor	Isolation Method	Starting with	Expected Yield (Total RNA)
RNeasy mini kit	Qiagen	Guanidine thiocyanate and spin column	$<1 \times 109$ bacterial cells	Up to 100 µg
RiboPure-bacteria kit	Ambion	Zirconia beads, phenol, and glass-fiber filter	$0.5–1 \times 10^9$ bacterial cells	Between 40 and 90 µg depending on the bacterial species
TRIzol Max	Invitrogen	Lysis with Max reagent, Trizol, and chloroform	$<1 \times 108$ bacterial cells	>20 µg for *E. Coli* and 3 µg for *Lactococcus lactis*
SV total RNA	Promega	Guanidine thiocyanate and silica surface spin column	1×10^9 bacterial cells	>30 µg
UltraClean microbial isolation kit	MO BIO Laboratories	Guanidine thiocyanate, bead beating, and silica spin filter column	1.8 mL of bacterial culture	Up to 60 µg
FastRNA Pro Blue kit	qBiogene	Lysis with RNApro solution, chloroform, and ethanol	10^{10} bacterial cells	>50 µg

or phenol) or by silica surface-based glass-fiber filter columns. Table 6.2 summarizes some of the RNA isolation kits available commercially, the method utilized for RNA isolation, and the expected yield of RNA. The RNA isolated with these kits generally produce high-quality bacterial total RNA suitable for most of its downstream applications. Nevertheless, additional care should be taken during RNA isolation for the following matters:

1. All good laboratory practices must be strictly followed during RNA isolation.
2. The initial amount of starting material (i.e., number of bacterial cells used for the RNA isolation) must be strictly followed according to the manufacture's recommendation. Use of higher number of starting material may affect the efficiency of RNA isolation.
3. Considering the short half-life of RNA and that handling of the sample (centrifugation/ washing) can alter the bacterial gene expression, it is recommended to stabilize the bacterial RNA and freeze the bacterial gene expression at an early step of bacterial RNA isolation. RNAprotect (Qiagen) and RNAlater (Ambion) can be used for this purpose.
4. The peptidoglycan layer of bacteria may be cleaved by the bacteriolytic enzyme—lysozyme. Predigestion of the bacterial cell wall with lysozyme improves lysis efficiency. This is essential for isolating RNA from Gram-positive bacteria (see Section 6.2.5 for further details). Efficient extraction of RNA from acid-fast bacteria may be achieved by physically disrupting the bacteria prior to RNA isolation.
5. Some of the commercial kits integrate DNase digestion step to reduce the DNA contamination in the isolated RNA sample. Contamination of genomic DNA in the RNA sample may affect the downstream processing of the RNA and may influence final results. Thus, if DNA contamination is suspected, it is highly recommended to use an additional DNase treatment step prior to further processing of the RNA sample.

TRI Reagent.   This is a complete and ready-to-use reagent from Molecular Research Center, Inc., which represents the improved version of the popular single-step method of total RNA isolation developed by Chomczynski et al. [11]. It can be used for the isolation of total RNA or the simultaneous isolation of RNA, DNA, and proteins from samples of human, animal, plant, yeast, bacterial, and viral origin. TRI Reagent combines phenol and guanidine thiocyanate in a mono-phase solution to facilitate the immediate and most effective inhibition of RNase activity. A biological sample is homogenized or lysed in TRI Reagent and the homogenate is separated into aqueous and organic phases by bromochloropropane or chloroform addition and centrifuga-tion. RNA remains exclusively in the aqueous phase, DNA in the interphase, and proteins in the organic phase. RNA is precipitated from the aqueous phase by addition of isopropanol. DNA and proteins are sequentially precipitated from the interphase and the organic phase with ethanol and isopropanol. The entire procedure can be completed in 1 h. The isolated RNA can be used for northern analysis, dot blot hybridization, poly(A)$^+$ selection, *in vitro* translation, RNase protec-tion assay, molecular cloning, and reverse transcriptase polymerase chain reaction (RT-PCR).

Qiagen kit. The RNA isolation kit from Qiagen is relatively efficient in isolating bacterial RNA (Table 6.2). The company markets three different kits: RNeasy mini, midi, and maxi based on the yield of RNA quantity. Nevertheless, all of them utilize the same technique for the RNA isolation. Bacterial samples are first lysed and homogenized in the presence of a highly denaturing guanidine thiocyanate containing buffer (included in the kit). Following lysis of the bacteria, ethanol is added to provide appropriate binding conditions of RNA to the RNeasy spin column, where the total RNA binds to the membrane and contaminants are efficiently washed away. The total RNA bound to the spin column is then eluted in water. With the RNeasy procedure, all RNA molecules longer than 200 nucleotides are purified (excluding 5S rRNA and tRNAs).

RiboPure-Bacteria kit. This is a rapid RNA isolation kit marketed by Ambion, which combines disruption of bacterial cell walls with zirconia beads, phenol extraction of the lysate, and glass-fiber filter purification of the RNA. It can be used to isolate total RNA from a variety of gram-negative and gram-positive bacteria. This method disrupts bacterial cell walls by beating cells mixed with RNA WIZ (included in the kit) and 0.1 mm zirconia beads on a vortex adapter. The lysate is then mixed with chloroform and centrifuged to form three distinct phases. The upper aqueous phase contains RNA, which is then diluted with ethanol and bound to a silica filter. The RNA bound to the filter is washed to remove contaminants and eluted. The 5S rRNAs and tRNAs are not recovered using the RiboPure-Bacteria kit.

In addition to the bacterial total RNA isolation kits, several kits are available in the market that are designed to isolate mRNAs from total bacterial RNA, to isolate bacterial RNA from a mixture of bacterial and eukaryotic total RNA, and to amplify bacterial RNA, which are discussed below in details.

MICROBEnrich. This kit from Ambion is designed to rapidly enrich bacterial RNA from mixture containing eukaryotic and bacterial RNA. The kit captures and removes polyadenylated mRNAs, 18S rRNA, and 28S rRNA from a host–bacterial RNA mixture and thus is able to remove >90% of eukaryotic RNA from complex eukaryotic–bacteria RNA mixtures. Purified RNA (mixture of eukaryotic–bacterial RNA) is incubated with capture oligonucleotide mix. Magnetic beads, deriva-tized with an oligonucleotide that hybridizes to the capture oligonucleotide and to the polyadeny-lated 3' ends of eukaryotic mRNAs, are then added to the mixture. The magnetic beads with 18S rRNA, 28S rRNA, and polyadenylated mRNAs attached are pulled with a magnet. The enriched bacterial RNA in the supernatant can be precipitated with ethanol and subsequently resuspended in RNase free water. The resulting RNA contains total bacterial RNA and other sRNAs (5S, 5.8S rRNAs, and tRNAs) from the eukaryotic RNA population (depending on the RNA isolation procedure used to acquire the eukaryotic and bacterial RNA mixtures).

MessageAmp II-Bacteria kit. This kit from Ambion employs an *in-vitro* transcription-mediated linear amplification method optimized for use with 100–1000 ng of bacterial RNA. At first, bacterial RNA is polyadenylated using *E. coli* poly(A) polymerase. Following this, the tailed RNA is reverse transcribed in a reaction primed with an oligo(dT) primer bearing a T7 promoter. The resulting cDNA is then transcribed with T7 RNA polymerase to generate hundreds to thousands of antisense RNA copies of each RNA molecules in the sample. This reaction can also be conjugated with labeled nucleotides for microarray expression studies. This kit is very useful for the analysis of bacterial gene expression that often requires high amount of RNA that is virtually impossible.

MICROBExpress bacterial mRNA isolation kit. This kit from Ambion enables isolation of mRNA from eukaryotic sources using oligo(dT) selection. Bacteria, however, lack the poly(A) tails found on eukaryotic mRNAs and thus making it difficult to purify mRNA from bacterial total RNA. The MICROBExpress kit uses a technology to remove >95% of the 16S and 23S rRNAs from total RNA of a broad spectrum of bacteria. In the first step of the MICROBExpress procedure, bacterial total RNA is mixed with oligonucleotides that bind to the bacterial 16S and 23S rRNAs. Next, the rRNA is removed from the solution using magnetic microbeads, derivatized with an oligonucleotide that hybridizes to the capture oligonucleotide bound to 16S and 23S rRNAs. The mRNA remains in the supernatant and can be recovered by ethanol precipitation and subsequent resuspension in RNase free water. The resulting RNA will contain mRNA, tRNA, 5S rRNA, and other sRNAs (depending on the procedure used to isolate the total RNA).

mRNA-ONLY prokaryotic mRNA isolation kit with poly(A)-tailing. This kit from Epicenter provides a method for isolation of bacterial mRNA that is substantially free of rRNA and 3'-end poly(A)-tailing of the isolated prokaryotic mRNA for subsequent study. This kit includes a 5'-phosphate-dependent exonuclease that digests RNA having a 5'-monophosphate. Bacterial rRNAs usually contains a 5'-monophosphate and are cleaved by the enzyme. Thus, the kit can be used to isolate prokaryotic mRNA substantially free of 16S and 23S rRNAs. Further the kit also includes a poly(A) polymerase, which uses adenosine triphosphate (ATP) as a substrate for template-independent addition of adenosine monophosphates to the 3'-hydroxyl termini of RNA molecules. The manufacturer claims that the removal of rRNA from the total RNA greatly increases the efficiency of the poly(A) tailing reaction to bacterial mRNAs.

In terms of the relative performance of in-house reagents and commercial kits for bacterial RNA isolation, Huyts et al. [24] undertook a comparative study of three methods: an in-house acid–phenol extraction protocol of Gerischer et al. [25], SV total RNA isolation system (Promega, Madison, Wisconsin), and the RNeasy mini kit (Qiagen, Valencia, California) for Gram-positive bacterium species *Clostridium*. The acid–phenol protocol involved lysis of bacteria in SDS and acid–phenol (preheated to 60°C). The aqueous phase was further extracted three times with acid phenol to remove residual proteins, and the nucleic acids were precipitated with ethanol. The pellet was dissolved in nuclease-free water and treated with RNase-free DNase I. The RNA was extracted again with phenol/chloroform/isoamyl alcohol, precipitated with 0.3 M sodium acetate pH 6.0 and isopropanol, washed with ethanol, and resuspended in nuclease-free water. The Promega protocol achieved the cell lysis with 4 M guanidinium thiocyanate and 0.97% β-mercaptoethanol. After centrifugation to clear the precipitated proteins and cellular debris, nucleic acids in the lysate were precipitated with ethanol and bound to the silica surface of the glass fibers on the membrane. RNase-free DNase I was directly applied to the silica membrane to digest contaminating genomic DNA. The bound RNA was washed and eluted from the membrane with nuclease-free water. The Qiagen protocol involved lysis of bacteria with 3 mg/mL lysozyme. Addition of a guanidinium-isothiocyanate-containing buffer immediately inactivated proteins including RNases, and the sample was loaded onto a silica-based membrane on the RNeasy column. Contaminants were washed away and the sample was then treated with DNase I, while the nucleic acids were bound to the membrane. DNase was removed by a second wash and the RNA was eluted with nuclease-free water. As summarized in Table 6.3, the Qiagen RNeasy mini kit recovered 54 µg, the Promega kit gave 18 µg, and the acid–phenol protocol 17 µg of total RNA from 10 mL of *Clostridium* broth culture as assessed by spectrophotometry. The purity of RNA isolated by the two commercial kits appeared to be higher than that by the acid–phenol protocol as estimated by the ratio $A260/A280$ (Table 6.3). The integrity of the RNA species prepared by all three methods was all good as examined in a 1.3% agarose-formaldehyde gel in 3-(N-Morpholino)-propane sulfonic acid (MOPS) running buffer [24].

In a recent report, Werbrouck et al. [26] comparatively analyzed the efficiency of four commercial kits for RNA isolation from Gram-positive bacterium *L. monocytogenes*. These included RNeasy Mini and the RNeasy Micro kits from Qiagen and RNAqueous and the RNAqueous-Micro kits from

TABLE 6.3
Comparison of Three RNA Isolation Protocols
for Gram-Positive Bacterium Clostridium

Method	Source	RNA Yield (μg/ 10 mL Culture)	OD (A260/A280)
RNeasy mini kit	Qiagen	54	1.9
SV total RNA isolation system	Promega	18	2.0
Acid phenol	Gerischer et al. [25]	17	1.4

Source: Adapted from Huyts, S. et al., *J. Microbiol. Methods*, 44, 235, 2001.
Note: One $A260$ unit = 40 μg/mL of RNA.

Ambion. *L. monocytogenes* cell pellet (containing 10^1 to 10^9 cells from broth culture) was resuspended in 100 μL TE buffer containing 5 mg/mL of lysozyme (Roche Diagnostics) and 13 mg/mL of proteinase K (Promega) and subsequently incubated at 37°C for 10 min. After completion of other recommended steps, the total RNA was eluted in 50 μL DEPC-treated deionized water for the RNeasy Mini and the RNAqueous kit, and in 20 μL for the RNeasy Micro and the RNAqueous-Micro kits. It was noted that the RNAqueous-Micro kit was more sensitive than the other three, when low numbers (10^3 to 10^8) of *L. monocytogenes* bacteria were used for RNA isolation. However, the RNAqueous or the RNeasy Mini kits were more efficient for RNA isolation from high cell numbers (10^9) of *L. monocytogenes* bacteria, which have higher binding capacity of the column, resulting in more RNA recovered.

6.4 METHODS

Reagents

Acetic acid (glacial)
Ammonium acetate
Chloroform
Ethanol
Ethidium bromide
Ethylene diamine tetraacetic acid (EDTA) disodium salt
Isoamyl alcohol
Phenol (TE-saturated)
Sodium acetate
Sodium dodecyl sulfate (SDS)
Tris(hydroxymethyl)aminomethane (Tris)

Equipment

Centrifuge (microfuge)
Centrifuge tubes (1.5 and 15 mL)
Gloves (disposable)
Pipettors (20, 200, and 1000 μL) and pipette tips (20, 200, and 1000 μL)

6.4.1 GUANIDINIUM–PHENOL–CHLOROFORM METHOD

The in-house protocol for preparation of total cellular RNA presented below is adapted from that of Wilkinson [27], which is a modified version of the original Chomczynski's method [11]. Several commercial kits (e.g., TRI Reagents from Molecular Research Center, Inc. and TRIzol from Invitrogen) based on similar principles are available, which can provide a higher level of convenience and reproducibility than in-house reagents. This method allows recovery of all RNA from the samples, which can be further processed for the isolation of specific RNA species such as mRNA and sRNA. It remains a technique of choice for RNA isolation.

Specific Reagents

- Lysis buffer: 5 M guanidinium isothiocyanate, 0.5% (w/v) sarkosyl, 25 mM sodium citrate, 8% (v/v) β-mercaptoethanol. Dissolve 59 g guanidinium isothiocyanate in 50 mL of water by heating to 60°C. Add 5 mL of 10% (w/v) sarkosyl and 3.3 mL of 750 mM sodium citrate pH 7.0, filter the solution, and store at 4°C. Add β-mercaptoethanol to a concentration of 8% (v/v) just before use.
- 2 M sodium acetate pH 4.0, prepared by adding glacial acetic acid until the desired pH is obtained.
- Water-saturated phenol.

Procedure

1. Grow bacterial culture in broth overnight (with $OD_{600} = 0.5$).
2. Transfer $<1 \times 10^9$ bacterial cells (approximately 3 mL with $OD_{600} = 0.5$) to a 15 mL tube.
3. Spin at 5000 rpm for 10 min.
4. Wash the cells with a physiological buffer (e.g., phosphate buffer saline [PBS] or SET buffer [50 mM NaCl, 5 mM EDTA, and 30 mM Tris–HCl pH 7.0]).
5. Treat the cells with desired concentration of lysozyme or use mechanical disruption of the isolated bacteria (this choice depends on the nature of the bacteria and is discussed in Section 6.2.5 in detail).
6. Add lysis buffer and quickly vortex the cells vigorously to completely lyse the cells.
7. Add 0.1 volume of 2 M sodium acetate pH 4.0 and 1 volume of water-saturated phenol.
8. Mix and add 0.1 volume of 49:1 (v/v) chloroform and isoamyl alcohol.
9. Mix and incubate at 4°C for 15 min.
10. Centrifuge for 20 min at 10,000 g at 4°C and transfer the supernatant to a new tube.
11. Add 1 volume of isopropanol and precipitate for 30 min at −20°C.
12. Centrifuge for 10 min at 10,000 g at 4°C and discard the supernatant.
13. Add half the volume of the lysis buffer (i.e., half the volume of lysis buffer previously used) to the pellet.
14. Add equal amounts of 100% isopropanol and repeat Steps 11 and 12.
15. Precipitate the RNA with 2 volumes of absolute ethanol and salt and wash with 70% ethanol.
16. Resuspend the RNA in RNase-free water.

6.4.2 Hot Phenol Method

The following hot phenol method for bacterial total RNA is based on von Gabain et al. [12].

Specific Reagents

- Lysis buffer: 0.15 M sucrose, 10 mM sodium acetate, pH 5.2, 1% (w/v) SDS
- Phenol equilibrated with 0.1 M sodium acetate pH 5.2
- 3 M sodium acetate, pH 5.2, prepared by adding glacial acetic acid until pH is obtained

Procedure

1. Grow bacterial culture in broth overnight (with $OD_{600} = 0.5$).
2. Transfer $<1 \times 10^9$ bacterial cells (approximately, 3 mL with $OD_{600} = 0.5$) to a 15 mL tube.
3. Spin at 5000 rpm for 10 min.

4. Wash the cells with a physiological buffer (e.g., PBS).
5. Treat the cells with desired concentration of lysozyme or use mechanical disruption of the isolated bacteria (this choice depends on the nature of the bacteria and is discussed below in Section 6.2.5 in detail).
6. Add 10 volumes of lysis buffer (prewarmed at 65°C) to 1 volume of bacterial cells.
7. Add 2 volumes of phenol (prewarmed at 65°C) to the cell lysate.
8. Gently invert the tube several times and incubate for 10 min at 65°C.
9. Centrifuge the tubes at $10,000\,g$ for 5 min at 4°C.
10. Transfer the upper phase to a new tube and add 2 volumes of hot phenol.
11. Gently invert the tube and repeat Steps 5 and 6.
12. Transfer the upper phase to a tube containing equal volume of 25:24:1 (v/v/v) phenol:chloroform:isoamyl alcohol, vortex briefly, and centrifuge as in Step 9.
13. Transfer the upper phase in a tube containing equal volume of 24:1 (v/v) chloroform:isoamyl alcohol, vortex, and centrifuge as in Step 9.
14. Transfer the upper phase to another tube and ethanol precipitates the RNA (as described in Section 6.2.3).
15. Resuspend the RNA in RNase-free water.

Note: Considering the short half-life of RNA and that handling of the sample (centrifugation/washing) can alter the bacterial gene expression, it is recommended to stabilize the bacterial RNA and freeze the bacterial gene expression at an early step of bacterial RNA isolation. RNAprotect (Qiagen) and RNA later (Ambion) can be used for this purpose.

6.4.3 QIAGEN RNEASY KIT

The protocol shown below is based on the instruction booklet for RNeasy kit (Qiagen) and used to prepare total RNA from bacteria.

Procedure

1. Grow bacterial culture in broth overnight (with $OD_{600} = 0.5$).
2. Transfer $< 1 \times 10^9$ bacterial cells (approximately 3 mL with $OD_{600} = 0.5$) to a 15 mL tube.
3. Spin for 10 min at 5000 rpm.
4. Discard the supernatant carefully by decanting and gently tapping to remove any remaining liquid.
5. Resuspend the pellet in 100 μL of lysozyme (3 mg/mL in TE, not treated with DEPC) by vortexing.
6. Transfer to an Eppendorf tube and incubate at room temperature for 5 min with vortexing for 10 s after 2 and 4 min.
7. Add 1 μL of β-mercaptoethanol per 100 μL to buffer RLT from Qiagen RNeasy kit (350 μL per preparation).
8. Add 350 μL of the RLT buffer with β-mercaptoethanol to the lysed cells and vortex thoroughly.
9. Centrifuge at 13,000 rpm for 2 min.
10. Transfer supernatant to a fresh Eppendorf tube.
11. Add 250 μL of ethanol (96%) and vortex.
12. Transfer everything (about 700 μL) including any precipitate to an RNeasy column in a 2 mL collection tube.
13. Centrifuge for 15 s at 10,000 rpm and discard flow through but keep the collection tube.
14. Add 350 μL buffer RW1 from RNeasy kit to the column.
15. Centrifuge for 15 s at 10,000 rpm and discard flow through but keep the collection tube.

16. Add 10 μL RNase-free DNase to 70 μL buffer RDD from same and mix gently without vortexing.
17. Add 80 μL DNase mix to the membrane taking care to avoid any of the mix sticking to the sides.
18. Incubate at room temperature for 15 min.
19. Add 350 μL buffer RW1 from RNeasy kit to the column.
20. Centrifuge for 15 s at 10,000 rpm.
21. Transfer spin column into new 2 mL collection tube.
22. Add 500 μL buffer RPE from RNeasy kit to the column.
23. Centrifuge for 15 s 10,000 rpm and discard flow through but keep the collection tube.
24. Add 500 μL buffer RPE from RNeasy kit to the column.
25. Centrifuge for 15 s at 10,000 rpm.
26. Transfer spin column into 1.5 mL Eppendorf tube with the lid removed.
27. Centrifuge for 2 min at 13,000 rpm.
28. Transfer spin column to 1.5 mL collection tube.
29. Add 30 μL of RNase-free water from RNeasy kit to the membrane again taking care to avoid any sticking to the sides.
30. Centrifuge for 1 min at 10,000 rpm.
31. Add another 30 μL of RNase-free water from RNeasy kit to the membrane again taking care to avoid any sticking to the sides.
32. Centrifuge for 1 min at 10,000 rpm.
33. Transfer 50 μL of the supernatant to a fresh Eppendorf and label both tubes, the remaining 10 μL can be used for quality control.

The yield should be between 15 and 60 μg of RNA, giving a concentration of between 0.25 and 1.00 μg/μL. The RNA can be stored in a −70°C freezer.

6.5 FUTURE DEVELOPMENT TRENDS

Although major advances have been made in the recent past in relation to the isolation of bacterial RNA, critical issues remain. For example, the current technologies do not allow efficient recovery of bacterial RNA when the number of bacteria is small. Even the highly sensitive RNAqueous-Micro kit barely manages to generate sufficient RNA from 10^3 *L. monocytogenes* bacteria for consistent reverse transcription PCR analysis [26]. Thus, further development is required to enhance the sensitivity of RNA isolation procedures, so that isolation and analysis of bacterial RNA from small biopsies or individual cells will become possible in future.

While many studies have been described concerning the relative performance of in-house reagents and commercial kits for bacterial DNA extraction (see Chapters 5 and 8), few such reports are available that deal with the comparative efficiency of RNA isolation reagents and kits for bacterial RNA. These studies are invaluable as the ability of these reagents and kits to purify RNA from bacteria is best evaluated using the identical set of samples and under identical laboratory testing regimens. The availability of this information will help users to select most suitable reagents and kits that meet their specific requirement. This is certainly true through examination of DNA extraction reagents and kits that for recovery of DNA from Gram-positive, Gram-negative, and acid-fast bacteria from clinical specimens, different kits tend to yield varying amounts of DNA from the same type of samples.

Since many commercial RNA isolation kits are designed with eukaryotes in mind, the lysis conditions specified in the instruction booklets for Gram-positive, Gram-negative, and acid-fast bacteria may not be appropriate. Thus, further optimization in this area will help pinpoint the exact lysis conditions (e.g., different lytic enzymes, incubation temperature, and length) that may be demanded by Gram-positive, Gram-negative, and acid-fast bacteria for rapid lysis, which is critical for prompt inactivation of endogenous RNases with the addition of RNase inhibitors and increasing the overall rate of RNA recovery from the samples.

Som Subhra Chatterjee completed his masters degree in biophysics and molecular biology from the University of Calcutta, Calcutta, India; he then obtained a doctoral degree with summa cum laude from the Justus Liebig University, Giessen, Germany where he performed research on *Listeria monocytogenes* in 2006. He is currently a visiting postdoctoral fellow at the National Institutes of Health, Bethesda, Maryland.

Trinad Chakraborty is a professor and director of the Institute of Medical Microbiology at the Justus Liebig University, Giessen, Germany. He obtained his first degree from the University of London, London, United Kingdom and a doctorate from the Free University of Berlin, Berlin, Germany. He was the recipient of a Heisenberg professorship from the German Science Research Council prior to his appointment to the chair of medical microbiology in Giessen. His research interests are focused on host–pathogen interactions during infection.

REFERENCES

1. Gottesman, S., Micros for microbes: Non-coding regulatory RNAs in bacteria, *Trends Genet.*, 21: 399, 2005.
2. Eriksson, S. et al., Unravelling the biology of macrophage infection by gene expression profiling of intracellular *Salmonella enterica*, *Mol. Microbiol.*, 47: 103, 2003.
3. Chatterjee, S.S. et al., Intracellular gene expression profile of *Listeria monocytogenes*, *Infect. Immun.*, 74: 1323–1338, 2006.
4. Gaynor, E.C. et al., The *Campylobacter jejuni* stringent response controls specific stress survival and virulence-associated phenotypes, *Mol. Microbiol.*, 56: 8, 2005.
5. Staudinger, B.J. et al., mRNA expression profiles for *Escherichia coli* ingested by normal and phagocyte oxidase-deficient human neutrophils, *J. Clin. Invest.*, 110: 1151, 2002.
6. Waddell, S.J., Butcher, P.D., and Stoker, N.G., RNA profiling in host-pathogen interactions, *Curr. Opin. Microbiol.*, 10: 297, 2007.
7. Bejerano-Sagie, M. and Xavier, K.B., The role of small RNAs in quorum sensing, *Curr. Opin. Microbiol.*, 10: 189, 2007.
8. Toledo-Arana, A., Repoila, F., and Cossart, P., Small noncoding RNAs controlling pathogenesis, *Curr. Opin. Microbiol.*, 10: 182, 2007.
9. Sambrook, J., Fritsch, E.F., and Maniatis, T., *Molecular Cloning: A Laboratory Manual*, 2nd ed., Cold Spring Harbor Laboratory Press, Cold Spring Harbor, NY, 1989.
10. Chirgwin, J.M. et al., Isolation of biologically active ribonucleic acid from sources enriched in ribonuclease, *Biochemistry*, 18: 5294, 1979.
11. Chomczynski, P. and Sacchi, N., Single-step method of RNA isolation by acid guanidinium thiocyanate-phenol-chloroform extraction, *Anal. Biochem.*, 162: 156, 1987.
12. von Gabain, A. et al., Decay of mRNA in *Escherichia coli*: Investigation of the fate of specific segments of transcripts, *Proc. Natl. Acad. Sci. U S A*, 80: 653, 1983.
13. Barlow, J.J. et al., A simple method for the quantitative isolation of undegraded high molecular weight ribonucleic acid, *Biochem. Biophys. Res. Commun.*, 13: 61, 1963.
14. Cathala, G. et al., A method for isolation of intact, translationally active ribonucleic acid, *DNA*, 2: 329, 1983.
15. Murphy, J.C. et al., Fractionation with compaction agents, *Anal. Biochem.*, 295: 143, 2001.
16. Patel, B.K., Banerjee, D.K., and Butcher, P.D., Characterization of the heat shock response in *Mycobacterium bovis* BCG, *J. Bacteriol.*, 173: 7982, 1991.
17. Kinger, A.K., Verma, A., Tyagi, J.S., A method for the isolation of pure intact RNA from mycobacteria, *Biotechniques*, 14: 724, 1993.
18. Cheung, A.L., Eberhardt, K.J., and Fischetti, V.A., A method to isolate RNA from gram-positive bacteria and mycobacteria, *Anal. Biochem.*, 222: 511, 1994.
19. Mahenthiralingam, E., Extraction of RNA from Mycobacteria, Chapter 6, *Methods in Molecular Biology*, Vol. 101: *Mycobacteria Protocols*, Humana Press, Totowa, NJ, 1998.

20. Altuvia, S., Identification of bacterial small non-coding RNAs: Experimental approaches, *Curr. Opin. Microbiol.*, 10: 257, 2007.
21. Hu, Z. et al., An antibody-based microarray assay for small RNA detection, *Nucleic Acids Res.*, 34: e52, 2006.
22. Wassarman, K.M. et al., Identification of novel small RNAs using comparative genomics and microarrays, *Genes Dev.*, 15: 1637, 2001.
23. Wendisch, V.F. et al., Isolation of *Escherichia coli* mRNA and comparison of expression using mRNA and total RNA on DNA microarrays, *Anal. Biochem.*, 290: 205, 2001.
24. Huyts, S. et al., Efficient isolation of total RNA from *Clostridium* without DNA contamination, *J. Microbiol. Methods*, 44: 235, 2001.
25. Gerischer, U. and Dürre, P., mRNA analysis of the *adc* gene region of *Clostridium acetobutylicum* during the shift to solventogenesis, *J. Bacteriol.*, 174: 426, 1992.
26. Werbrouck, H. et al., Quantification of gene expression of *Listeria monocytogenes* by real-time reverse transcription PCR: Optimization, evaluation and pitfalls, *J. Microbiol. Methods*, 69: 306, 2007.
27. Wilkinson, M., Purification of RNA, Chapter 4, *Essential Molecular Biology*, Vol. 1, 2nd ed., Oxford University Press, New York, 2000.

7 Isolation of Plasmids

Armelle Bigot and Alain Charbit

CONTENTS

7.1 INTRODUCTION

7.1.1 GENERAL CONSIDERATION

7.1.1.1 Definition

Plasmid is a circle of self-replicating DNA, which is distinct from the chromosomal DNA and which contains genes that are generally not essential to the growth or survival of the cell. However, some plasmids can integrate into the host genome. The term "plasmid" was first introduced by Joshua Lederberg in 1952 and defined as the generic term for any extra-chromosomal genetic particle that is separate from the chromosomal DNA with a capacity of autonomous (self) replication [1]. Later, Jacob and Wollmann introduced the concept of episome to refer to accessory genetic elements transmissible from cell to cell and present either free in the cytoplasm or integrated into the bacterial chromosome [2]. The term "episome," which persisted in the literature for more than a decade, was somewhat confusing since it referred to both plasmid and phage genetic elements. Indeed, it included the fertility factor F, a large plasmid present either free in the cytoplasm of *Escherichia coli* (strains designated F+) or integrated at specific sites of the bacterial chromosome (strains designated Hfr for high frequency of homologous recombination), and the temperate bacteriophage lambda, which during its life cycle can be either integrated in the chromosome or extra-chromosome.

The term "replicon" first appeared in 1963 and is still currently used to designate self-replicating DNA molecules, such as chromosomal and plasmid DNAs. Replication of circular or linear plasmids is controlled by specific mechanisms whose role is to maintain their rate of replication synchronized with that of the host chromosome. Plasmids carry specific sequences and functions required for their replication, but they also involve the replication machinery of the host cell. This control is ensured by regulatory mechanisms occurring in the replication origin area.

Although more than 30 different replicons have been identified, almost all plasmids currently used in molecular biology carry a replicon derived from pMB1/ColE1 [3]. These plasmids are present at 15–20 copies per cells. A huge number of pMB1/ColE1 variants have been engineered over the years, some of which became very high copy number plasmids (e.g., pUC series). In plasmids of the pMB1/ColE1 family, an RNA molecule (also designated RNA II) positively regulates plasmid copy numbers. In other replicons, it is a *cis*-acting protein (designated RepA) that accomplishes this regulatory function. Plasmids regulated by an RNA molecule have generally high copy numbers. These plasmids, said to replicate in a relaxed fashion, can continue to replicate even when protein synthesis is stopped. This characteristic is sometimes used to increase the yield of plasmid DNA per cell by treatment of the cultures with chloramphenicol. By contrast, plasmids requiring the continuous synthesis of the plasmid-encoded RepA protein cannot be amplified by this process and are said to replicate under stringent control.

7.1.1.2 From Natural to Recombinant Plasmids

Most plasmids are double-stranded circular DNA molecules whose size can vary from 1 to more than 400 kb. Plasmids are naturally found in a wide variety of bacterial species, and often account for 3%–20% of the bacterial chromosomes, which help determine a number of bacterial properties including resistance to antibiotics and the ability to produce toxins. In addition, some plasmids are also present in eukaryotic organisms. Being an independent life-form and capable of autonomous replication in suitable (host) environments, plasmid may resemble virus to a certain extent. However, the plasmid tends to have a more symbiotic than parasitic relationship with its host than virus, considering that plasmids often contribute useful packages of DNA to their hosts for mutual survival under arduous conditions. This is exemplified by the ability of plasmids to convey antibiotic resistance to host bacteria, enabling bacterial growth and reproduction in the presence of otherwise lethal antibiotic concentrations.

Plasmids can be classified by their ability to be transferable to other bacteria. Conjugative plasmids contain *tra* genes, encoding the complex machinery responsible for their transfer to a recipient strain by conjugation. Some of them can only transfer within the same species (narrow host-range plasmids); while other have the capacity to transfer to other and sometimes distantly related species (broad host-range plasmids) [4,5]. On the other hand, nonconjugative plasmids are incapable of initiating conjugation. However, some of them, designated as mobilizable plasmids, have such specific features (i.e., an appropriate origin of transfer), which allow their mobilization by conjugation via another resident conjugative plasmid.

Alternatively, plasmids can be grouped according to the functions they carry. Five main groups are classically defined: (1) fertility plasmids, which contain *tra* (for transfer) genes and are capable of conjugation; (2) resistance plasmids, which carry antibiotic-resistance genes; (3) col plasmids, which contain genes that determine the production of bacteriocins (bactericidal proteins); (4) degradative plasmids, providing the capacity to grow on normally toxic compounds (e.g., pesticides and toluene, etc.); and (5) virulence plasmids, which carry genes essential for the pathogenicity of the recipient strain (e.g., the virulence plasmid of *Shigella flexneri*). Of course, some plasmids can belong to more than one of these functional groups and carry, for example, both virulence and antibiotic resistance genes.

In addition, plasmids have been assigned into a series of compatibility groups, referring to their capacities to coexist within the same host cell. Hence, plasmid from different incompatibility (or Inc) groups can persist stably within the same host, while plasmids belonging to the same Inc group are mutually exclusive possibly due to the regulation of vital plasmid functions (see Ref. [6] for a recent review).

Plasmid DNA usually assumes one of the five conformations (i.e., nicked open-circular, linear, relaxed circular, supercoiled, and supercoiled denatured), which tend to run at different speeds (from the slowest to the quickest), and thus show distinct band patterns in electrophoretic gel. Nicked open-circular plasmid DNA has one strand cut; linear plasmid DNA has free ends; relaxed circular plasmid DNA is fully intact with both strands uncut, but is enzymatically relaxed (supercoils removed); supercoiled (or covalently closed-circular) plasmid DNA is fully intact with both strands uncut, and with a twist built in, resulting in a compact form; and supercoiled denatured plasmid DNA is like supercoiled plasmid DNA, but has unpaired regions that make it slightly less compact; this can result from excessive alkalinity during plasmid preparation.

Plasmids have the capacity to accommodate foreign gene fragments (of 1–20 kb), and thus are often used in genetic engineering as vectors for transferring genes of interest into host cells, with the goal of making any multiple copies of a gene or large amounts of a particular protein. To multiply gene copies, the gene of interest is inserted into a plasmid containing selection genes (e.g., antibiotic resistance gene and lacZ gene) and a multiple cloning site (MCS or polylinker). The MCS is a short stretch of nucleotides consisting of some commonly used restriction enzyme sites, which permit the convenient insertion of DNA fragment of interest. Some plasmids have the MCS situated within a β-galactosidase gene, whose product digests galactose (X-gal, a substrate added in the plate) to display blue color for nonrecombinant clones and white color for recombinant clones (see below). The plasmid harboring a gene of interest (the so-called recombinant plasmid) is then inserted into bacteria by a process called transformation. In the presence of a particular antibiotic, only bacteria containing the plasmid will be resistant to the antibiotic and survive. The bacteria containing recombinant plasmid can be grown in large amounts, harvested, and purified. Besides making multiple copies of specific DNA fragments, plasmid can also be constructed to produce large amounts of proteins. Here, the underlying gene of interest is inserted into the plasmid containing an efficient, inducible promoter (a transcriptional driver of the vector's cloned insert or transgene), which is then transformed into bacteria. When the bacteria harboring the recombinant plasmid are grown in the presence of an inducing agent, large amounts of proteins are expressed from the inserted gene.

Examples of popular cloning plasmids consist of pBR322 and pUC18. pBR322 is a 4.3 kb plasmid containing an ampicillin resistance gene and a tetracycline resistance gene. Furthermore, it has a relaxed origin of replication for accumulation to high numbers in *E. coli* in addition to possessing 21 common enzymes that occur once within. pUC18 is a derivative of pBR322, and made up of only about 2.7 kb in size. Apart from an ampicillin resistance gene and an origin of replication both from pBR322, pUC18 comprises a multiple cloning site, which also codes for a small peptide. When transformed into a specific *E. coli* strain that lacks β-galactosidase activity, pUC18 makes the peptide for expression of an active enzyme, which can turn the bacteria blue in the presence of certain substrates (e.g., galactose or commonly known as X-gal). If, however, a foreign gene is inserted into one of the restriction endonuclease sites in the multiple cloning site, the peptide is no longer produced. Without β-galactosidase activity, the bacteria will appear white. Thus, it allows convenient discrimination of pUC plasmid containing a foreign insert of DNA from plasmids without an insert. Nonetheless, as a normal plasmid only has the capacity to handle inserts of about 1–20 kb, it is necessary to modify it to accommodate longer lengths of DNA. Two common plasmid derivatives for this purpose are cosmid and bacterial artificial chromosome (BAC).

A cosmid is a hybrid plasmid that contains *cos* sequences, which are ~200 bp long DNA sequences originally from the lambda phage. Cosmids are able to contain 37–52 kb of DNA, and replicate as plasmids if they have a suitable origin of replication: SV40 ori in mammalian cells, ColE1 ori for double-stranded DNA replication, or f1 ori for single-stranded DNA replication in bacteria. Similar to normal plasmids, cosmids also contain antibiotic resistance gene for selection.

However, because of their possession of *cos* sequences, cosmids can also be packaged in phage capsid, which allows the foreign genes to be transferred into or between cells by transduction. *Cos* sequences contain a *cosN* site where circular cosmid DNA is nicked by terminase to produce two cohesive or sticky ends of 12 bp for packaging purpose. The *cosB* site holds the terminase while it is nicking and separating the strands. The *cosQ* site of next cosmid is held by the terminase after the previous cosmid has been packaged, to prevent degradation by cellular DNases. Because of the fixed size of the phage head, terminase can only package cosmids that are between 75% and 105% of the length of the normal phage. Thus, the practical upper limit of the insert size is around 40 kb, since there will also need to be origins of replication, selection genes, and MCS.

A BAC is based on a fertility plasmid (or F-plasmid), and used for transforming and cloning in bacteria such as *E. coli*. F-plasmid contains partition genes to promote the even distribution of plasmids after bacterial cell division. BAC can handle insert ranging from 100 to 350 kb. A similar cloning vector, called a plasmid P1 derived artificial chromosome (PAC), has also been produced from the bacterial P1-plasmid. BACs are often used to sequence the genetic code of organisms. A short piece of the organism's DNA is amplified as an insert in BACs, and then sequenced. Finally, the sequenced parts are rearranged in silico (i.e., performed on computer or via computer simulation), resulting in the genomic sequence of the organism. Common gene components in BACs include *oriS* and *repE–F* for plasmid replication and regulation of copy number, *parA* and *parB* for partitioning F-plasmid DNA to daughter cells during division and ensuring stable maintenance of the BAC, a selectable marker for antibiotic resistance, lacZ at the cloning site for blue/white selection in some BACs, and *T7* and *Sp6* phage promoters for transcription of inserted genes.

Another plasmid derivative is phagemid or phasmid, which is a hybrid of the filamentous phage M13 and plasmids. Phagemids can grow as a plasmid, and also be packaged as single-stranded DNA in viral particles. In addition to phagemids with an ori for double-stranded replication, there are phagemids with an f1 ori to enable single-stranded replication and packaging into phage particles. Similar to a plasmid, a phagemid can be used to clone DNA fragments and be introduced into a bacterial host by a range of techniques (transformation or electroporation). However, infection of a bacterial host containing a phagemid with a helper phage (e.g., VCSM13 or M13K07) provides the necessary viral components to facilitate single-stranded DNA replication and packaging of the phagemid DNA into phage particles. Although filamentous phages retard bacterial growth, they are not generally lytic in comparison with lambda and T7 phages. Helper phage is usually engineered to package less efficiently than the phagemid so that the resultant phage particles contain predominantly phagemid DNA. F1 filamentous phage infection requires the presence of a pilus, so only bacterial hosts containing the F-plasmid or its derivatives can be used to generate phage particles. Phagemids were used to generate single-stranded DNA template for sequencing purposes prior to the development of cycle sequencing technique. Nowadays, phagemids are often used to generate templates for site-directed mutagenesis. The phage display technology is based on phagemid, in which a range of peptides and proteins can be expressed as fusions to phage coat proteins and displayed on the viral surface. The displayed peptides and polypeptides are associated with the corresponding coding DNA within the phage particle and so this technique is valuable to the study of protein–protein interactions and other ligand/receptor combinations.

7.1.2 Application of Plasmids in Molecular Biology and Other Fields

Plasmids are used by molecular biologists in all fields of research as vectors for cloning and amplifying double-stranded DNA (obtained by enzymatic digestion or from polymerase chain reaction (PCR) amplification) from many different organisms. With the rapid expansion of recombinant DNA technologies, a wide variety of recombinant forms have been created to fulfill all sorts of functions and tools, including cloning vehicles, expression vectors, mutagenic tools and DNA vaccines, etc.

7.1.2.1   From Cloning Vectors to Expression Vectors

An ideal, general-purpose cloning plasmid possesses the following features:

1. It is small so that it contributes only a minimal amount of extraneous DNA to the plasmid/passenger construct, thereby making it easier to prepare large amounts of the passenger DNA and to get into bacteria. Additionally, small plasmids replicate faster, require less energy for replication, and are less fragile and easier to purify.
2. Its DNA sequence is known.
3. It is a relaxed replication plasmid so that it can grow to high copy number in the host cell with increased yield (e.g., pUC-based vectors can yield between 500 and 700 copies per bacterial cell).
4. It should contain a selectable marker such as antibiotic resistance (e.g., ampicillin resistance or tetracycline resistance) for easy identification and isolation.
5. It should also contain a second selectable gene that is inactivated by insertion of the passenger for improved differentiation of bacteria containing plasmid only from bacteria with insert-containing plasmid.
6. There exist a large number of unique restriction sites (or MCS) within one of the two selectable markers described above. The presence of many unique sites permits maximum flexibility and ease in cloning as MCS allows for targeted insertion of DNA in a chosen orientation.

An expression vector (or expression construct) is a plasmid used to introduce and express a specific gene into a target cell. Once inside the target cell, the expression vector utilizes host cellular transcription and translation machinery to make multiple copies of stable messenger RNA (mRNA) transcripts from the gene of interest located within the plasmid, which are then translated into proteins. To be able to produce large amounts of mRNA transcripts, the plasmid is specifically designed so that the target DNA is inserted into a site that is under the control of a highly active promoter (e.g., T7 promoter and lac promoter). Upon completion of expression, the proteins of interest need to be separated from the proteins of the host cell. To enable easy purification, the cloned gene often contains a tag, which can be histine (His) tag or other marker protein.

Expressing a gene from another species in *E. coli* requires more specialized conditions. In particular, promoters of genes from other organisms are not likely to be recognized by the *E. coli* RNA polymerase. Thus, the gene to be expressed must be cloned in such a way that the transcription start site is spaced properly and is under the control of an *E. coli* promoter (transcriptional fusions). If the cloned DNA is from a eukaryote, protein expression in *E. coli* requires removal of introns prior to cloning, which is accomplished by isolating mRNA from the eukaryotic cell and reverse transcribing it into DNA (thus creating complementary DNA [cDNA]). In biotechnology and medicine, this process is used to isolate large quantities of proteins. In some cases, the produced recombinant proteins may need to be of high purity and in their active conformation and appropriate purification procedure must be adapted. In other cases, for example, to produce specific polyclonal antibodies, a native conformation may not be necessary.

7.1.2.2   From Transcriptional Fusion to Translational Fusion

Transcriptional fusion is a technique that is commonly used to trap a promoter and evaluate its efficiency under various conditions. Here, a plasmid vector containing a reporter gene (e.g., β-galactosidase gene [*lacZ*], β-glucuronidase [GUS] gene, or the chloramphenicol acetyl transferase [CAT] gene from *E. coli*; the green fluorescent protein [GFP] gene from the jelly fish *Aequorea victoria*; and the luciferase gene from the firefly *Photinus pyralis*) is modified in such a way that the promoter of this reporter gene is removed, rendering it nonfunctional. To restore the normal function and proper expression of the reporter gene, an external promoter has to be inserted (trapped) in the region just in front of the initiation codon of the promoterless reporter gene. This is often done by cloning random short genomic DNA fragments of an organism of interest into this region, and the resulting clones are selected by the signals generated by the reporter gene. A successful expression of the reporter gene indicates the functionality of a promoter cloned (trapped). For example, a promoter-less listeriolysin (*hly*) gene in *Listeria monocytogenes* has been used to directly screen chromosomal fragments for promoters induced during intracellular infection. Listeriolysin is a member of a family of bacterial pore-forming cytolysins that is central to *L. monocytogenes* pathogenesis. By placing the *hly* gene under the control of random promoter elements allows the selection of clones expressing listeriolysin during residence in the host phagosome. A further extension of this technique led to the development of the pGAD-HLY dual reporter system, which is a plasmid containing a promoter-less copy of a gene essential for *L. monocytogenes* acid survival (*gadB*, encoding glutamate decarboxylase) and a second gene required for *in vivo* virulence (*hly*, encoding listeriolysin). Application of this reporter system enabled trapping and identification of a number of genes involved in *L. monocytogenes in vivo* resistance to acid stress.

Translational fusion is another valuable technique that is exploited to monitor the activity of a given gene under a controllable heterologous promoter. Here, a plasmid vector containing a reporter gene is so designed that the promoter region together with part of the reporter gene is removed. Restriction or PCR fragments (containing their own promoter, ribosome binding site and initiation codon, and part of putative protein-encoding gene) can then be cloned into the deleted region. A successful expression of the reporter gene indicates the generation of an active translational gene fusion (hybrid protein). Generally, the fused protein or protein domain provides a technical advantage to the protein such as (1) easy purification of the protein (e.g., maltose-binding protein fusions purified on amylose columns or His-tag fusions purified on nickel column) or (2) easy detection (e.g., cMyc tags for specific immunodetection of proteins). Translational fusions are of value for monitoring the fate of a protein such as degradation, subcellular localization, secretion into the extracellular space, or translocation into other cells. Translational fusions at a given position of a chromosomal gene can be prepared by fusing reporter genes to regulatory elements of genes of interest. Most reporter genes mentioned above are also useful for construction of translational fusions, resulting in the synthesis of hybrid proteins. In particular, the *lacZ* gene from *E. coli* encoding the β-galactosidase has been widely applied to fuse heterologous genes or transcriptional units to produce either translational or transcriptional gene fusions, whose activity can be monitored by simple β-galactosidase assays. Thus, transcriptional and translational gene fusions can be utilized to identify unique regulatory mechanisms of gene expression at a given locus by comparing the data from a transcriptional fusion with the data from a translational fusion of the same genetic locus.

7.1.2.3 Mutagenesis

Plasmids can be employed for sequential deletions of a cloned DNA, which are useful in the subsequent sequencing and other analysis. In a commercial plasmid-based mutagenesis system marketed by Promega (Erase-a-Base system), exonuclease III (Exo III) is utilized to specifically digest insert DNA from a 5' protruding or blunt-end restriction site, while the adjacent sequencing primer binding site is protected from digestion by a 4-base 3' overhang restriction site or by an α-phosphorothioate-filled end. This allows the rapid construction of plasmid or M13 subclones containing progressive unidirectional deletions of any inserted DNA.

In addition, plasmids are useful as a means to alter a gene of interest by a single crossing-over, which may result in the inactivation of the gene concerned. A recombinant plasmid is introduced into host bacteria through conjugation or electroporation. Chromosomal integration of the recombinant plasmid is then obtained by homologous recombination after growth at restrictive temperature (between 37°C and 42°C). This leads to two separate truncated copies of the disrupted gene. This technique provides an easy and rapid way to inactivate target genes for screening purposes. However, a possible drawback of such a method is the high probability of generating polar effect on the expression of the downstream genes, when the insertion is not located at the end of an operon.

One valuable approach to avoid or reduce polar effects on downstream gene is to replace an allele (or a region) of the targeted gene in the chromosome with an inactive version (in general, a deletion). Allelic replacement is usually carried out by the following steps. The initial step involves the construction of an in-frame deletion of a gene from host bacteria by cloning the deleted region in an appropriate thermosensitive shuttle plasmid. For this, the upstream and downstream regions immediately flanking the target gene are amplified by PCR. Typically, 600–900 bp fragments are amplified to ensure efficient homologous recombination. The amplified products, flanked by suitable restriction sites, are then ligated together and cloned into the shuttle vector. In the next step, the recombinant plasmids are introduced into the host bacteria at permissive temperature. The resulting transformants are incubated at restrictive temperature with antibiotic pressure. This leads to chromosomal integration of the recombinant plasmid by homologous recombination between the plasmid-borne region and the corresponding chromosomal region (this recombination can occur either in the upstream or downstream region of the target). Upon integration, a mutated and a wild-type copy of the gene coexist in the chromosome. Finally, by shifting the bacteria back to permissive temperature in the absence of selective pressure, it allows the spontaneous excision of the integrated plasmid via a second crossing-over after a variable number of replication cycles. Thus, an allelic replacement is complete.

This approach permits the replacement of any chromosomal gene by any other gene either through in-frame deletion of the entire gene or through substitution with an antibiotic resistance gene. Most vectors currently used for this purpose are based on the thermosensitive mutation of the replication origin of plasmid pE194, and include pKSV7, pAUL-A, and pG + host5 or its derivatives. With plasmids like pAUL-A or pKSV7, the first step of allelic replacement (leading to chromosomal integration of the recombinant plasmid) is generally easy. Since the thermosensitivity of these plasmids is stringent, the plasmid is integrated in nearly all growing bacteria after one or two passages at restrictive temperature (with antibiotic pressure). On the other hand, plasmid excision can be time-consuming as this event cannot be directly selected. To ensure consistent outcome, the following screening procedure is usually applied. After five to ten passages at permissive temperature in broth (without antibiotic pressure), the recombinant bacteria are plated onto solid medium without antibiotic and then replica-plated onto the same medium with antibiotic. The antibiotic-sensitive clones (corresponding to plasmid loss) are then checked by PCR to identify the mutated clones, given that statistically 50% of the second crossing-over leads to a wild-type genotype.

A number of high copy number plasmids (e.g., pAT18, pAT19, pAT28, and pAT29) have also been constructed and applied for mutagenesis and complementation studies. pAT18 and pAT19 are 6.6 kb mobilizable shuttle cloning vectors, which contain the replication origins of pUC and pAMβ1, an erythromycin-resistance-encoding gene expressed in Gram-negative and Gram-positive bacteria, the transfer origin of plasmid RK2, and the multiple cloning site and the *lacZ*α reporter gene of pUC18 (pAT18) and pUC19 (pAT19). The latter feature permits selection of recombinant clones through a blue and white screening in *E. coli* carrying the *lacZ* delta M15 deletion by α-complementation. Plasmids pAT18, pAT19, and recombinant derivatives have been successfully transferred by conjugation from *E. coli* to various Gram-positive bacteria (including *Bacillus subtilis*, *L. monocytogenes*, and *Staphylococcus aureus*) at frequencies ranging from 10^{-6} to 10^{-9}. Plasmids pAT28 and pAT29 are pAT derivatives carrying spectinomycin-resistance gene instead of erythromycin-resistance gene.

Sometimes, low copy number plasmids may be preferred under certain circumstances. pTCV-lac represents one such example. This plasmid is a 12 kb derivative of the mobilizable shuttle vector pAT187, which contains the origin of replication of pBR322 and that of pAMβ1 as well as a promoterless *lacZ* gene preceded by a multiple cloning site. Plasmid pTCV-lac replicates at a low copy number (3–5) in Gram-positive hosts (*Bacillus*, *Clostridia*, *Listeria*, *Enterococcus*, *Staphylococcus*, and *Streptococcus*), and is useful either for complementation studies or for the quantification of promoter activities using the *lacZ* gene as a reporter (β-agalactosidase assay). Since it also contains the transfer origin of the IncP plasmid RK2, pTCV-lac can be transferred from an *E. coli* mobilizing donor to Gram-positive bacteria.

Additionally, plasmids have also been constructed to study gene *cis* complementation and to facilitate the chromosomal integration of the rescuing allele. pPL1 and pPL2 are site-specific shuttle integration vectors with two different chromosomal bacteriophage integration sites to facilitate strain construction in *L. monocytogenes*. pPL1 utilizes the listeriophage U153 integrase and attachment site within the *comK* gene for chromosomal insertion. pPL1 can be directly conjugated from *E. coli* into *L. monocytogenes*, forming stable, single-copy integrants at a frequency of approximately 10^{-4} per donor cell in both ½ and 4b serogroups. pPL2 utilizes the listeriophage phage of Scott A (PSA) integration site in the 3' end of an arginine tRNA gene. pPL2 demonstrates the same frequency of integration as pPL1 in both ½ and 4b serogroups. Furthermore, plasmid pLIV1 contains a ColE1 origin of replication and an ampicillin-resistance gene, for cloning and selection in *E. coli*; an origin of transfer, allowing conjugal transfer from *E. coli* to *L. monocytogenes*; a chloramphenicol-resistance gene, for plasmid selection in *L. monocytogenes*; and a temperature-sensitive origin of replication. With a unique *Xba*I restriction site, immediately downstream of the IPTG-inducible promoter SPAC/lacOid (designated pSPAC), this plasmid allows chromosomal integration of IPTG-inducible genes in *L. monocytogenes* (at nonpermissive temperature).

7.1.2.4   DNA Vaccines

Plasmid DNAs encoding specific proteins can be expressed in cells of a recipient host, which can elicit or enhance host's immune responses against infective organisms and cancers just like conventional vaccines do. The development of DNA vaccine technology was made possible through the earlier experimental findings that plasmid DNAs encoding β-galactosidase, luciferase, or acetylcholine transferase were able to express these enzymes in murine muscle cells upon inoculation. The first evidence of protective immunity provided by DNA vaccine obtained in animal models in 1993, in which plasmids encoding influenza virus hemagglutinin and influenza virus nucleoprotein stimulated specific immunity against challenge infection with influenza virus. Apart from introducing transgenes into cells that inherently lack the ability to produce the protein, plasmid DNA vaccines can correct genetic errors that produce functionally incompetent copies of a given protein. In addition to disease treatment, plasmids can be used as DNA vaccines for genetic immunization. DNA vaccines function through induction of immune response by introducing genes encoding antigens for specific pathogens. Indeed, NA vaccines have been shown to generate protective immunity against a number of viral, bacterial, and parasitic agents in animal models. Several DNA vaccines for human infectious diseases are currently in clinical trials, and DNA immunization has been proven to be safe and generally tolerated.

The mechanisms of plasmid DNA vaccines appear to involve two processes: (1) the plasmid produces antigen (protein) of interest in either professional antigen presenting cells (APCs) for direct priming of immune responses or in nonprofessional cells from which the antigen is subsequently transferred to APCs leading to cross-priming; and (2) being derived from bacteria, plasmid DNAs also stimulate the innate immune system by interacting with Toll-like receptor 9. This nonspecific immune response further augments the antigen-specific immune response.

Because plasmid DNA vaccines utilize the DNA transcription and translation apparatus in the host cell to biosynthesize the therapeutic protein, it is necessary for plasmid molecules to gain access into the nucleus after entering the cytoplasm. The entry of plasmid molecules into the nucleus through the nuclear pores can be a challenging and difficult process. Therefore, the plasmids used for DNA vaccines are engineered so that they contain the elements for efficient expression in eukaryotic cells. However, attention should be paid to the vector designs that do not contain a mammalian origin of replication (ORI), in order to minimize the risk of vector integration into the human chromosome and its related safety concerns. Besides the transgene of interest, plasmid DNA molecules typically possess several regulatory signals such as promoter and enhancer sequences that contribute to the regulation of gene expression. Promoter sequences provide recognition sites for the RNA polymerase to initiate the transcription process. By engineering, the plasmid with strong tissue- or tumor-specific promoters often results in higher efficiency of transcription and subsequent expression. Many promoter sequences are derived either from viral origins such as cytomegalovirus and roux sarcoma virus, or from human origins such as α-actin promoter. On the whole, human promoters have been shown to have greater resiliency against immune response activation than viral promoters. Promoter sequences often determine the immune response of the cell to the gene product, and tissue-specific promoters can also improve the efficacy of suicide gene therapy. Enhancers are regions in the plasmid DNA that enhance and maximize the production of the gene of interest, and they can be present on the plasmid locus either upstream or downstream from the promoter region. As some enhancers are tissue specific, judicious selection of suitable enhancers can improve transcription efficiency substantially. Additionally, plasmid DNA vaccines also contain a transcriptional terminator to terminate transcription in mammalian cells and a selectable marker to facilitate production of the plasmids in transformed bacterial cells. Further, splicing and polyadenylation sites are incorporated in the transgene construct to enable the correct processing of the mRNA generated after transcription. Some vectors also have introns that may increase pre-mRNA processing and nuclear transport.

In general, DNA vaccines are administered via intramuscular, intradermal, or intravenous injection by needle. Alternatively, intranasal inoculation of naked DNA or DNA–liposome complexes, use of live bacterial vectors, and oral delivery of DNA vaccines encapsulated in microspheres have been applied. More recently, particle-mediated epidermal delivery or gene gun has been shown to reduce the amount of plasmid DNA needed to induce immune responses. As the particle-mediated delivery puts plasmid DNA directly into the cells of the epidermis, leading to the transfection of Langerhans cells, it enables induction of immune responses with very small doses of plasmid DNA (i.e., 1–10 μg).

The plasmid DNA vaccines have several obvious advantages. As these vaccines do not require the use of purified proteins or viral vectors, they are much cost-effective and easy to produce and maintain. Additionally, expression of the immunizing proteins in host cells facilitates the presentation of normally processed proteins to the immune system for raising immune responses against the native forms of proteins. Specifically, the expressed immunogen has access to class I major histocompatibility complex presentation, which is critical for eliciting CD8+ CTL responses. Furthermore, plasmid DNA-based products are safe with no risk for infection and cause few side effects in relation to conventional vaccines, since they can be administered repeatedly without inducing anti-vector immunity. Moreover, DNA vaccines offer long-term persistence of immunogen. Another benefit is that maternal antibodies do not interfere with DNA vaccination, overcoming a potential problem that relates to live pediatric vaccines.

A major disadvantage of plasmid DNA vaccines is that they tend to induce poor immunogenicity when administered as an unformulated intramuscular injection. Often large quantities (5–10 mg) are required to elicit only modest immunogenicity. Fortunately, by altering the routes of administration together with the use of adjuvants, the immunogenicity and efficacy of DNA vaccines can be improved. The commonly used chemical adjuvants for enhancing plasmid DNA expression include liposomes, polymers, and microparticles. Use of nonparticulate polymeric adjuvants has also led to the enhancement of immune responses. The adjuvants can not only stimulate the immune system directly, but also potentiate the magnitude, localization, or duration of plasmid DNA expression as well as target delivery to specific cells. Other means to enhance the efficacy of DNA vaccines include (1) use of nuclear proteins (e.g., histones and polycationic reagents) to condense DNA and transfect cultured cells, as plasmids are large polyanions that require condensation into small particles (known as polyplexes) and masking of the negative charge for their efficient uptake by eukaryotic cells; due to their net positive charge, polyplexes can bind to the cell surface via electrostatic interactions and facilitate DNA entry; (2) use of structural proteins of DNA viruses to spontaneously assemble with plasmid DNA and form transfection-competent pseudocapsids; and (3) use of chimeric fusion proteins to incorporate in a single polypeptide chain heterologous protein domains, which facilitate binding to plasmid DNA, specific recognition of target cells, induction of receptor-mediated endocytosis, and DNA transport through intracellular compartments.

7.1.3   BASIC PRINCIPLES OF PLASMID ISOLATION

7.1.3.1   Critical Aspects in Plasmid Isolation

Isolation and purification of plasmid DNA from bacteria involves three major steps: growth of the bacterial culture, harvesting and lysis of the bacteria, and purification of plasmid DNA. The first two steps can be optimized to obtain the highest yield of plasmid recovery. The third step will be adapted to the requirement of the envisaged utilization of the plasmid preparation (transformation, transfection, and vaccination). These different aspects will be detailed below.

We shall review briefly below the major parameters that can influence, sometimes drastically, the yield and, thus, efficacy of recovery of a plasmid preparation, independently of the method used for extraction.

1. Recipient bacterial host strain: The choice of the most suitable host strain depends on the specific utilization of a plasmid preparation. For example, factors like presence of methylation systems should be taken into consideration.

Dam and Dcm methylation deficient strains.   Most laboratory strains of *E. coli* contain both Dam and Dcm methylases. Dam methylase transfers a methyl group to the adenine in the sequence GATC. Dcm methylase methylates the internal cytosine residues in the sequences CCAGG and CCTGG. Several restriction endonucleases will not cleave sites with these modified bases. The *dam dcm* strain allows growth and purification of DNA free of Dam and Dcm methylation. Such host strains are recommended when (1) the plasmid DNA (grown in *E. coli*) must be cut with a restriction endonuclease sensitive to *E. coli* K12 methylation patterns (such as *Bcl* I, TGATCA) and (2) a shuttle plasmid DNA grown in *E. coli* must be transferred into a strain of eubacteria or archaea that restricts DNA with this methylation.

Endonuclease 1-deficient (endA1⁻) strains.   The periplasmic space of wild type *E. coli* cells contains a nonspecific endonuclease. Endonucleasic activity of the host strain can be one parameter that may lead to significant plasmid degradation. Therefore, it is generally recommended to use bacterial host strains that are devoid of endonuclease 1 activity (*endA1⁻*). The *endA1⁻* mutation deletes this endonuclease activity and can significantly improve the quality of plasmid preparations. Bacterial strains such as JM109, DH5α, and XL1-Blue are *endA1⁻* and are appropriate hosts. Other characteristics of the host strain may alter plasmid yield and purity. For example, strain HB101 and its derivatives, such as TG1 and the JM100 series, in addition to high levels of endonuclease activity, also contain large amounts of carbohydrates that are released during lysis and can inhibit enzyme activities.

Restriction-deficient (hsdR2) strains.   Wild type *E. coli* K12 strains carry the EcoK type I restriction endonuclease, which cleaves DNA with sites AAC(N6)GTGC and GCAC(N6)GTT. While *E. coli* DNA is protected from degradation by a cognate methyl transferase, foreign DNA will be cut at these sites. The *hsdR2* mutation eliminates the endonuclease activity.

Partially methyl restriction-deficient (mcrA, mcrB1) strains.   *E. coli* has a system of enzymes, *mcrA*, *mcrB*, and *mrr*, which will cleave DNA with methylation patterns found in higher eukaryotes, as well as some plant and bacterial strains. Of note, DNA derived from PCR fragments, cDNA or DNA previously propagated in *E. coli*, will not be methylated at these sites and will not be cleaved.

2. Growth medium and growth conditions: By using richer media than Luria broth (LB), such as 2YT or terrific broth (TB), it is possible to reach very high cell densities and thus the recovery of higher amounts of plasmids. Generally, in standard growth conditions, it is recommended to harvest the bacteria after 12–16h of incubation at 37°C, which corresponds to the entry into stationary growth phase.

3. Plasmid copy number and size: Plasmids vary in copy number, depending on the origin of replication they contain, their size, and the size of insert. Thus, the yield of plasmid recovery will be directly dependent on the copy number per cell. For example, with a high copy number plasmid such

as pUC18, ca. 1–10 μg of plasmid DNA can be expected from a 1–3 mL of standard culture (e.g., LB), in a standard bacterial host (e.g., JM109 and DH5α). The yield can be at least ten fold lower with a low copy number plasmid such as pACYC184.

7.1.3.2 Scale

The scale of the extraction protocol should be adapted to the need. Most commercial kits propose three different scales designated: mini-, midi-, and maxi-preparations (which correspond to plasmid extraction from 1 to 10 mL, 10 to 100 mL, and 100 mL to 1 L of bacterial culture, respectively). Some companies also propose mega (from 0.5 to 2.5 L of culture) and giga (from 2.5 to 5 L of culture) scale purifications allowing isolation of milligrams of high-purity plasmid DNA using anion-exchange technology.

7.1.3.3 Separation of Plasmid DNA from Cellular Components

The separation of plasmid DNA from chromosomal DNA is based on the covalently closed form of plasmid DNA and its relatively small size compared to chromosomal DNA. Originally, the separation protocol used an equilibrium centrifugation in a cesium chloride–ethidium bromide (CsCl–EtBr) gradient, which relies on the difference of EtBr binding to covalently closed circular DNA and linear molecules. This method, which is very efficient to obtain pure plasmid preparation, is relatively time consuming as compared to the other methods and requires a subsequent step of dialysis to remove the EtBr (see Ref. [7] for additional information on this method).

Nowadays, the methods used worldwide for separation of plasmid DNA are based on the protocol of Birnboim and Doly [8] and use a selective denaturation of chromosomal DNA by alkali treatment. This method relies on a narrow pH range (12.0–12.5) to selectively denature linear, but not covalently closed circular DNA.

Plasmid preparations are carried out in three steps: (1) lysis of the bacteria, (2) separation of plasmid DNA, and (3) purification of plasmid DNA (Figure 7.1).

1. Lysis of the bacteria: The lysis step is critical since it can dramatically affect the yield of plasmid recovery. The amount of plasmid DNA in the lysate will directly depend on the amount of bacteria properly lysed during the procedure. Therefore, the concentration of bacteria should be taken into account before evaluating the volume of lysis buffer to be used. With *E. coli* strains, an overnight culture with agitation at 37°C in LB produces generally ca. $1–3 \times 10^9$ bacteria/mL. A standard miniprep procedure should be carried on a 1–5 mL culture sample. With a low copy number plasmid, the number of bacteria can be increased, but the volume of lysing buffer should also be adapted consequently. The bacterial culture is centrifuged and culture medium carefully removed before resuspension in a lysis buffer.

Two major procedures are routinely used to disrupt bacteria before plasmid preparation: alkaline lysis with sodium dodecyl sulphate (SDS) or the boiling method.

Alkaline lysis with SDS. In this method, SDS solubilizes the phospholipid and protein components of the cell membrane, leading to bacterial lysis and release of the cell contents. Bacterial pellet is resuspended in lysis buffer I containing 50 mM glucose, 25 mM Tris (pH 8.0), 10 mM ethylene diamine tetraacetic acid (EDTA) (pH 8.0). This buffer can be supplemented with DNase-free RNase A. Addition of lysing buffer II (0.2 N NaOH, 1% SDS) allows, in one step, lysis of the bacteria and denaturation of proteins and DNA.

Boiling lysis method. In this method, the bacterial pellet is resuspended in a lysis buffer (10 mM Tris, 0.1 M NaCl, 1 mM EDTA, and 5% Triton X-100) at pH 8.0 containing 250 μg of freshly added lysozyme. Bacteria are then incubated at 100°C for 40 s. The heat-induced disruption of the cell wall and the cytoplasmic membrane is facilitated by the lysozyme and the detergent. The heat also permits the denaturation of chromosomal DNA and proteins without denaturing the covalently closed circular plasmid DNA (boiling should not be prolonged more than 40 s to prevent plasmid denaturation).

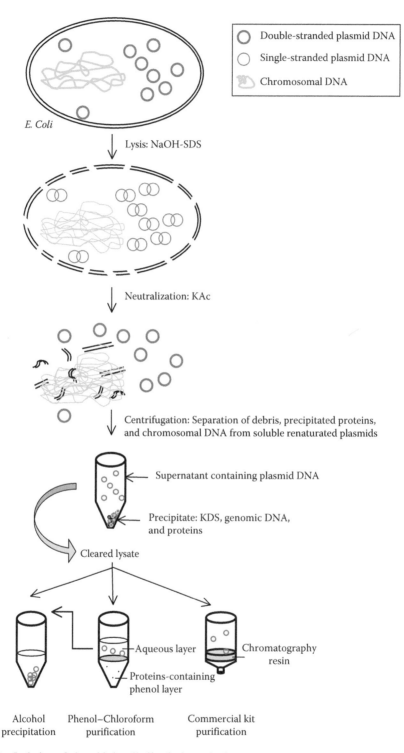

FIGURE 7.1   Isolation of plasmids by alkaline lysis method.

Lysis of Gram-positive bacteria.   The thickness of the cell wall of Gram-positive bacteria renders the classical methods less efficient and adaptations are often helpful. First, using a culture in exponential phase (OD_{600} of 0.6) gives bacteria with a thinner peptidoglycan layer than overnight grown bacteria. Moreover, exponential growth prevents the secretion of large amounts of nucleases in the culture medium (which occur mainly during postexponential growth). Cell disruption can be obtained by mechanical treatment, using silica or glass beads (e.g., bacteria can be broken in a Fastprep apparatus). This procedure allows a very efficient lysis of the bacteria, but can sometimes also break partially the chromosomal DNA and affect, thus, the subsequent step of separation of the chromosome from plasmid DNA molecules. A milder treatment, using an incubation with lysozyme before the alkaline lysis, has been successfully used in our laboratory for plasmid minipreps from *L. monocytogenes* (bacterial pellets are resuspended in 10 mg/mL lysozyme in TE buffer pH 8.0 containing RNase A and incubated at 37°C for 15 min before carrying over alkaline lysis treatment).

2. Separation of plasmid DNA: After alkaline lysis, the suspension is neutralized with a 5 M potassium acetate solution. During renaturation, chromosomal DNA forms an insoluble precipitate, while closed circular plasmid DNA remains soluble. A centrifugation step allows separation of the supernatant (which contain the solubilized plasmid) from the precipitate (which contains denatured chromosomal DNA, proteins, and cell-envelope debris). After boiling lysis, the centrifugation is also the method of choice for separating denatured chromosomal DNA and proteins from plasmid DNA.

3. Purification of plasmid DNA: This can be achieved by various ways.

Precipitation with ethanol. Plasmid DNA can be purified by ethanol (or isopropanol) precipitation. For this, the soluble fraction is first adjusted to a 0.1 M NaCl final concentration (or 0.3 M sodium acetate or 0.5 M ammonium acetate). Then 2 volumes of ultrapure ethanol (100% solution) are added. After gentle mixing, the suspension is frozen (three cycles of freeze-thaw in liquid nitrogen, or 30 min in dry ice). After 30 min centrifugation, the pellet containing precipitated plasmid DNA is washed twice with 70% ethanol (to remove salts). An RNase A treatment can be added after this step in order to remove coprecipitated RNA.

Precipitation with polyethylene glycol (PEG). DNA can be specifically precipitated and gives a maximum recovery yield with 13% PEG in the presence of 10 mM $MgCl_2$ at room temperature [9,10]. Contrary to ethanol, PEG cannot efficiently precipitate small DNA fragments (<150 bp) and can, thus, be used for removal of small DNA fragments. After precipitation, PEG is removed by two washes with 70% ethanol.

Phenol–chloroform extraction. Before precipitation with ethanol, removal of contaminating proteins can be achieved with a simple phenol–chloroform extraction. This step could be useful in order to prevent DNase contamination of the plasmid preparation (see protocol in Box 7.2).

Chromatography. Manufacturers sell purification plasmid kit using chromatography resins that selectively adsorb DNA. Different types of resins are available: silica resin or anion-exchange resin (e.g., DEAE-silicate). Nucleic acids are highly negatively charged linear polyanions. They bind efficiently to the resin while proteins and polysaccharides do not and are easily eliminated by washes of the resin with ethanol. Resin-bound DNA can be then eluted. Silica resins allow elution of nucleic acids with water or a low-salt buffer. Anion-exchange resins allow specific separation of different types of nucleic acid depending on the pH and the salt concentration and therefore give higher purity.

7.1.4 Comparison of In-House and Commercial Techniques for Isolation of Plasmids

Since the 1990s, the use of commercial kits for plasmid isolation has been generalized to almost all the laboratories worldwide. These kits often reproduce the basic alkaline lysis protocol and are generally coupled to a purification step on a chromatography column. The success of kit utilization has led to an increase of the number of manufacturers on the market as well as to the constant development of small adaptations and improvements of the basic protocol (sometimes referred to as new technologies).

The main inconvenience of kit utilization, as compared to homemade solutions and recipes, is the cost. However, kits ensure reproducibility and often help to gain time (which may at least in part compensate their cost). Of note, utilization of kits has led the students or new users to forget the biochemical principles underlying each step of the protocol. It is thus recommended to consult the commercial Web sites, which generally provide complete information about principles of plasmid extraction and helpful tips (see recommended web sites).

The whole-genome sequencing projects have open a new challenge in increasing the number of samples treated at the same time (see protocols provided in Refs. [11–13]). Manufacturers have developed their products in order to adapt to the demand of high throughput and many companies now sell kit for simultaneous 96-plasmid extraction.

7.2   METHODS

Buffers and Reagents

The recipes for stock solution preparation are given in Box 7.1
Appropriate growth medium (e.g., LB)
Appropriate antibiotic solution (the antibiotic and the concentration depend on the strain and the plasmid)
Lysis buffer I: 50 mM glucose, 25 mM Tris-HCl, 10 mM EDTA pH 8.0
Lysis buffer II: 200 mM NaOH, 1% SDS
Neutralization buffer: 5 M potassium acetate pH 4.8
Isopropanol 100% or ethanol 100%
Ethanol 70%
TE buffer: 10 mM Tris-HCl, 1 mM EDTA, pH 7.5
Optional (for phenol–chloroform purification): Phenol/chloroform/isoamyl alcohol (PCIA) mix (25:24:1, v/v)
7.5 M ammonium acetate solution

Supplies and Equipment

Standard equipment for bacterial culture (plate, inoculating loop, culture flask, a shaking incubator at 37°C)
Automatic pipettes and tips (1000 and 200 μL)
Tubes 1.5 and 15 mL (for low copy plasmid only)
Refrigerated tabletop centrifuge for 1.5 mL tubes (Note: A miniprep with an on-column purification does not require a cooling system since all the centrifugation can be carried out at room temperature)
Disposable gloves
pH meter or pH test strips to adjust the pH of buffer solution

7.2.1   PLASMID DNA MINIPREP WITH AN IN-HOUSE PROTOCOL

A detailed annotated in-house protocol for plasmid miniprep is presented in Box 7.2.

7.2.2   PLASMID DNA MINIPREP WITH A COMMERCIAL KIT

A detailed protocol for plasmid miniprep using a commercial kit is given in Box 7.3 [20].

BOX 7.1   BUFFERS AND COMMON STOCK SOLUTION RECIPES

LB medium

To make 1 L, use 10 g tryptone, 5 g yeast extract, 10 g NaCl. Adjust pH to 7.0. Sterilize by autoclaving.

2 M glucose

Glucose 36.04 g
Distilled water 100 mL. Filter and sterilize.

1 M Tris-HCl

To make 1 L, dissolve 121 g of Tris base in 800 mL of H_2O. Adjust pH to the desired value by adding approximately the following:
pH = 7.5, about 65 mL of concentrated HCl
pH = 8.0, about 42 mL of concentrated HCl
Sterilize by autoclaving.

0.5M EDTA solution pH 8.0

Add 186.1 g of disodium EDTA-$2H_2O$ to 800 mL of H_2O. Stir vigorously on a magnetic stirrer. Adjust the pH to 8.0 with NaOH (approximately 20 g of NaOH pellets). Dispense into aliquots and sterilize by autoclaving. The disodium salt of EDTA will not solubilize until the pH of the solution is adjusted to 8 by the addition of NaOH.

For tetrasodium EDTA, use 226.1 g of EDTA and adjust pH with HCl.

NaOH

The preparation of 10 N NaOH involves a highly exothermic reaction, which can cause breakage of glass containers. Prepare this solution with extreme care in plastic beakers. To 800 mL of H_2O, slowly add 400 g of NaOH pellets, stirring continuously. As an added precaution, place the beaker on ice. When the pellets have dissolved completely, adjust the volume to 1 L with H_2O. Store the solution in a plastic container at room temperature. Sterilization is not necessary.

SDS stock

To prepare a 20% (w/v) solution, weight 200 g of electrophoresis-grade SDS (using a mask) and dissolve it in 900 mL of H_2O. Heat to 68°C and stir with a magnetic stirrer to assist dissolution. If necessary, adjust the pH to 7.2 by adding a few drops of concentrated HCl. Adjust the volume to 1 L with H_2O. Store at room temperature. Sterilization is not necessary. Do not autoclave.

5 M potassium acetate

For a 100 mL solution, dissolve 49.07 g of potassium acetate (MW: 98.14 g/mol) in distilled water. Filter and sterilize by autoclaving.

7.5 M ammonium acetate

For a 100 mL solution, dissolve 57.81 g of ammonium acetate (MW: 77.08 g/mol) in distilled water. Filter and sterilize by autoclaving.

(continued)

BOX 7.1   (CONTINUED) BUFFERS AND COMMON STOCK SOLUTION RECIPES

TE buffer

10 mM Tris-Cl (pH 8.0)

1 mM EDTA (pH 8.0)

Use concentrated stock solutions to prepare. If sterile water and sterile stocks are used, there is no need to autoclave. Otherwise, sterilize solutions by autoclaving for 20 min. Store the buffer at room temperature.

Note: Always make sure solutions are at room temperature before making final pH adjustments.

BOX 7.2   PLASMID DNA MINIPREP IN-HOUSE PROTOCOL

1. Pick single colony and inoculate 5 mL of LB containing appropriate antibiotic. Incubate at 37°C with shaking (100–250 rpm) overnight.
2. Centrifuge 1.5 mL cells in 1.5 mL Eppendorf tube at maximum speed (10,000–15,000 rpm) for 1 min. Aspirate supernatant.
   For purification of low copy plasmid, increase the volume of bacteria (up to 10 mL) and increase the amount of lysis and neutralizing buffer in the next steps.
3. Resuspend cell pellet in 100 µL of lysis buffer I (50 mM glucose, 25 mM Tris-Cl, 10 mM EDTA pH 8.0). Vortex gently if necessary.
   This step ensures that all the cells will be exposed to the lysis reagent. For purification of low copy plasmid, double the lysis buffer volume.

   Note: RNase A can be added in lysis buffer I (20 mg/mL) in order to permit RNA digestion during lysis.
4. Add 200 µL of NaOH/SDS lysis solution II (0.2 M NaOH, 1% SDS). Invert tube 6–8 times.
   SDS solubilizes the phospholipids and the protein components of the cell membrane leading to lysis and release of the cell contents. NaOH denatures chromosomal DNA, plasmid, and proteins.
5. Immediately, add 150 µL of 5 M potassium acetate solution (pH 4.8). Spin at full speed for 10 min.
   This solution neutralizes NaOH in the previous lysis step. Plasmid DNA, being circular and relatively small compared to chromosomal DNA, renatures correctly while the high salt concentration causes potassium dodecylsulfate to coprecipitate with denatured proteins, chromosomal DNA, and cellular debris. Centrifugation separates soluble plasmid DNA (and RNA) from the precipitate.
6. Transfer supernatant to new tube, being careful not to pick up any white flakes. Precipitate the nucleic acids with 0.5 mL of isopropanol on ice for 1 min and centrifuge at maximum speed for 10 min.
   Isopropanol precipitation of plasmid DNA can be carried out at room temperature since this minimizes coprecipitation of salt.
7. Aspirate off all the isopropanol supernatant. Add 500 µL of ethanol 70%. Centrifuge for 10 min at 4°C at maximum speed. Remove carefully the supernatant. Air-dry the pellet for 5–15 min before dissolving it in 50 µL of TE buffer (10 mM Tris-Cl, 1 mM EDTA, pH 7.5).

BOX 7.2   (CONTINUED) PLASMID DNA MINIPREP IN-HOUSE PROTOCOL

ADDITIONAL PURIFICATION STEP (PHENOL–CHLOROFORM EXTRACTION)

7. Aspirate off all the isopropanol supernatant. Dissolve the pellet in 0.4 mL of TE buffer (10 mM Tris-Cl, 1 mM EDTA, pH 7.5).
8. Add 0.3 mL of PCIA mix. Vortex vigorously for 30 s. Centrifuge at full speed for 5 min at room temperature.

Note: Organic PCIA layer will be at the bottom of the tube.
At this step, remaining proteins are separated of DNA in the organic phenol layer.

9. Remove upper aqueous layer containing the plasmid DNA carefully avoiding the white precipitated protein layer above the PCIA layer and transfer it to a clean 1.5 mL Eppendorf tube.
10. Add 100 µL of 7.5 M ammonium acetate solution and 1 mL of absolute ethanol to precipitate the plasmid DNA on ice for 10 min. Centrifuge at full speed for 5 min at room temperature.

Note: Absolute ethanol can be replaced by isopropanol and precipitation carried out at room temperature.

11. Aspirate off all the ethanol solution. Add 500 µL of 70% ethanol. Centrifuge 10 min at 4°C at maximum speed. Remove carefully the supernatant. Air-dry the pellet for 5–15 min before dissolving it in 50 µL of TE buffer (10 mM Tris-Cl, 1 mM EDTA, pH 7.5).
12. Measure DNA concentration: dissolve 5 µL of DNA solution in 995 µL of water, and measure absorbance at 260 nm. The absorbance at 260 nm (for a 1 cm pathlength cuvette) multiplied by 10 is the DNA concentration in milligram per milliliter.

BOX 7.3   PLASMID DNA MINIPREP PROTOCOL USING A COMMERCIAL KIT (e.g., QIAPREP SPIN)

1. Pick single colony and inoculate 5 mL of LB containing appropriate antibiotic. Incubate at 37°C with shaking (100–250 rpm) overnight.
Do not exceed 16 h of growth since cells begin to lyse and plasmid yield may be reduced.
2. Centrifuge 1–5 mL cells at >8000 rpm for 3 min at room temperature (or at 5000 rpm for 10 min at 4°C). Discard supernatant.
For purification of low copy plasmid, increase the volume of bacteria (up to 10 mL) and double the amount of lysis and neutralizing buffer in the next steps.
3. Resuspend pelleted bacteria in 250 µL of buffer P1 (50 mM Tris-Cl pH 8.0, 10 mM EDTA, 100 µg/mL RNase A) and transfer to a microcentrifuge tube.
This step insures that all the cells will be exposed to the lysis reagent. For purification of low copy plasmid, double the buffer P1 volume.

Note: LyseBlue reagent can be added to buffer P1 in order to check proper lysis of the bacterial cells.

(continued)

BOX 7.3   (CONTINUED) PLASMID DNA MINIPREP PROTOCOL USING A COMMERCIAL KIT (e.g., QIAPREP SPIN)

4. Add 250 μL of buffer P2 (0.2 M NaOH, 1% SDS) and mix thoroughly by inverting the tube 4–6 times.

   Do not vortex, as this will result in shearing genomic DNA. Do not allow lysis reaction to proceed for more than 5 min.

   If LyseBlue reagent has been added, invert the tube until the blue color of the suspension appears homogenous light blue.

5. Add 350 μL of buffer N3 and mix immediately and thoroughly by inverting the tube 4–6 times. Centrifuge for 10 min at 13,000 rpm in a tabletop microcentrifuge.

   The solution became cloudy. If LyseBlue reagent has been added, mix until the solution became colorless and all trace of blue has gone.

   The exact composition of buffer N3 is proprietary (it contains acetic acid and guanidine hydrochloride); however, its acidic pH (around 4.5) allows neutralization of NaOH and renaturation of plasmid DNA, while genomic DNA, proteins, and cellular debris form a cloudy precipitate.

6. Apply the supernatant from Step 5 to the QIAprep spin column by decanting or pipetting. Centrifuge for 30–60 s at 13,000 rpm. Discard the flow through.

   At this step, plasmid DNA from the cleared lysate binds to the silica matrix of the column, while proteins and polysaccharides are eluted.

7. Wash the QIAprep spin column by adding 0.5 mL of buffer PB and centrifuging for 30–60 s. Discard flow through.

   This step is optional but allows removing of any trace of nuclease activity and can be useful when dealing with *endA*1⁺ strain like HB101 of the JM series or any strain, which have high level of nuclease and carbohydrate. Host strains such as DH5α do not require this additional wash.

8. Wash QIAprep spin column with 0.75 mL of buffer PE and centrifuge for 30 s at 13,000 rpm. Discard the flow through and centrifuge for an additional minute to remove residual of PE buffer.

   This washing PE buffer that contains ethanol and residual of ethanol can inhibit enzymatic reaction. At this step, plasmid DNA is still bound to the silica matrix.

9. Place the column in a clean 1.5 mL microcentrifuge tube. Add 50 μL of buffer EB (10 mM Tris-HCl, pH 8.5) or water to the center of each QIAprep spin column. Let stand for 1 min and centrifuge for 1 min.

   The conditions of binding to the silica resin allow elution in a low, or no, salt buffer. Water can be used to elute plasmid DNA. However, deionized water can sometimes be too acidic to obtain best yield and therefore use of EB buffer is recommended for plasmid elution.

   Elution can be improved by warming the EB buffer at 70°C prior to use. For large plasmid (>10 kb), this could increase significantly the yield.

Source: Adapted from *QIAprep Miniprep Handbook*, Qiagen, Gmbh, Germany, June 2005.

7.3   FUTURE DEVELOPMENT TRENDS

The development of research in the field of gene therapy and DNA vaccination has increased the demand for efficient large-scale methods to produce high-quality plasmid DNA. The potential medical use of plasmid DNA increased the need of a controlled production with safety guidelines like any other pharmaceutical production.

Enlarging the scale of production can dramatically modify the result of a simple protocol like alkaline lysis [14]. For example, after alkaline lysis of bacteria containing a high copy number plasmid, plasmid DNA does not represent more than 3% of the cleared lysate, while RNA can represent 21% of the content [15]. This low concentration is not in itself a problem in a miniprep protocol where the plasmid DNA can be efficiently purified and concentrated using a small chromatography column. In contrast, in a large batch (which may treat up to 100-L culture), a rapid concentration and separation procedure are required. Different protocol adaptations have been engineered (see Ref. [15] for review). For example, a continuous protocol for large-scale extraction based on the alkaline lysis has been very recently proposed [16,17]. This new procedure integrates the continuous control of reagent flow and the replacement of all centrifugation steps by filtering which is a faster process. Thus, 4 L of *E. coli* culture can be processed in less than 90 min, giving a yield of purified plasmid of 90 mg/L.

Plasmid DNA can be found in different topological states: linear form if the double strand is broken at one position, open circle form if only one strand is nicked, and covalently closed circular form. This last form is the most compact form and the most active topology. In this form, the DNA structure is intact. In large-scale protocols, each step of the purification process (e.g., filtration for the concentration of the cleared lysate) can alter DNA structure and properties. Therefore, a method for controlling plasmid DNA structure and quality during the production process is required in order to obtain a homogenous production of supercoiled covalently closed DNA. The control process of plasmid topology can be achieved by capillary gel electrophoresis [18]. Besides topology, many criteria must be controlled before using of a batch of plasmid DNA for vaccination, such as absence of proteins, absence of LPS or contaminating bacteria, and absence of contamination by genomic DNA or other plasmid DNA (see Ref. [19] for review). The development of common guidelines and analytical methods for quality control is an important issue for large-scale medical-grade plasmid production.

Armelle Bigot graduated in biological sciences from the University Paris 11, Paris, France and École Normale Supérieur, Paris. During her PhD training at the Necker Faculty of Medicine in Paris in the laboratory of pathogenesis of systemic infections, she focused on secretion pathways in *Listeria monocytogenes*. Her first interests were on the flagellar system, the cytosolic chaperones, and their potential roles in pathogenesis. She then worked on an RNase H associated with the signal peptidase of *L. monocytogenes* and virulence. More recently, as a postdoctoral associate, she undertook the study of the amino acid acquisition pathways and the relevance of amino acid uptake systems in the virulence of *L. monocytogenes*. For the past 3 years, she also has taught biochemistry and microbiology practical courses for bachelor students at the University Paris 7, Paris.

Alain Charbit received his bachelor degree in bacterial genetics at the Pasteur Institute, Paris, France. After his MS in microbiology at the University Paris 78, Paris, he spent a year and half of scientific military service in Marseille working on yellow fever transcription. He then joined the laboratory of Maurice Hofnung at the Pasteur Institute and obtained his third cycle thesis in microbiology in 1984. He was recruited as a *chargé de recherche* by the Centre National de la Recherche Scientifique (CNRS) in 1991. He is currently a senior scientist of the CNRS (*directeur de recherche* since 1997). His earlier research interests at the Pasteur Institute (1982–1998) were mostly focused on fundamental aspects of maltose uptake and bacteriophage binding mediated by the multifunctional pore-forming protein LamB of *Escherichia coli*. More recently, he has focused on the physiopathology of intracellular bacterial pathogens such as *Listeria monocytogenes* and *Francisella tularensis*. He has published over 100 scientific papers, invited reviews, and book chapters, and is a team leader in an Institut national de la santé et de la recherche médicale (INSERM) laboratory of the Faculty René Descartes Paris 5, Paris.

REFERENCES

1. Lederberg, J., Plasmid (1952–1997), *Plasmid*, 39: 1, 1998.
2. Jacob, F. and Wollman, E. L., Episomes, a proposed term for added genetic elements, *CR. Hebd. Seances Acad. Sci.*, 247: 154, 1958.
3. Hashimoto-Gotoh, T. and Timmis, K. N., Incompatibility properties of Col E1 and pMB1 derivative plasmids: Random replication of multicopy replicons, *Cell*, 23: 229, 1981.
4. Thomas, C. M. and Nielsen, K. M., Mechanisms of, and barriers to, horizontal gene transfer between bacteria, *Nat. Rev. Microbiol.*, 3: 711, 2005.
5. Dionisio, F. et al., Plasmids spread very fast in heterogeneous bacterial communities, *Genetics*, 162: 1525, 2002.
6. Bouet, J. Y., Nordstrom, K., and Lane, D., Plasmid partition and incompatibility—the focus shifts, *Mol. Microbiol.*, 65: 1405, 2007.
7. Sambrook, J. and Russell, D. W., *Molecular Cloning: A Laboratory Manual,* Vol. 1, 3rd ed., Cold Spring Harbor Laboratory Press, Cold Spring Harbor, NY, 2001, Chapter 1.
8. Birnboim, H. C. and Doly, J., A rapid alkaline extraction procedure for screening recombinant plasmid DNA, *Nucleic Acids Res.*, 7: 1513, 1979.
9. Lis, J. T., Fractionation of DNA fragments by polyethylene glycol induced precipitation, *Methods Enzymol.*, 65: 347, 1980.
10. Paithankar, K. R. and Prasad, K. S., Precipitation of DNA by polyethylene glycol and ethanol, *Nucleic Acids Res.*, 19: 1346, 1991.
11. Elkin, C. J. et al., High-throughput plasmid purification for capillary sequencing, *Genome Res.*, 11: 1269, 2001.
12. Engelstein, M. et al., An efficient, automatable template preparation for high throughput sequencing, *Microb. Comp. Genomics*, 3: 237, 1998.
13. Dederich, D. A. et al., Glass bead purification of plasmid template DNA for high throughput sequencing of mammalian genomes, *Nucleic Acids Res.*, 30: e32, 2002.
14. Lyddiatt, A. and O'Sullivan, D. A., Biochemical recovery and purification of gene therapy vectors, *Curr. Opin. Biotechnol.*, 9: 177, 1998.
15. Stadler, J., Lemmens, R., and Nyhammar, T., Plasmid DNA purification, *J. Gene Med.*, 6 (Suppl. 1): S54, 2004.
16. Li, X. et al., A continuous process to extract plasmid DNA based on alkaline lysis, *Nat. Protoc.*, 3: 176, 2008.
17. Li, X. et al., An automated process to extract plasmid DNA by alkaline lysis, *Appl. Microbiol. Biotechnol.*, 75: 1217, 2007.
18. Schmidt, T. et al., Quantitative analysis of plasmid forms by agarose and capillary gel electrophoresis, *Anal. Biochem.*, 274: 235, 1999.
19. Schleef, M. and Schmidt, T., Animal-free production of ccc-supercoiled plasmids for research and clinical applications, *J. Gene Med.*, 6 (Suppl. 1): S45, 2004.
20. *QIAprep Miniprep Handbook*, Qiagen, Gmbh, Germany, June 2005.

Recommended Web Sites

http://www1.qiagen.com/resources/info/
http://www.neb.com/nebecomm/tech_reference/default.asp
http://www.promega.com/tbs/dna_rna.htm

8 Preparation of Bacterial Samples for Direct Molecular Applications

Knut Rudi, Robert C. Wilson,
Kjetill S. Jakobsen, and Dongyou Liu

CONTENTS

8.1 INTRODUCTION

8.1.1 NECESSITY TO ACCESS NONCULTURED SPECIMENS

Prokaryotes represent a most diverse biomass on the Earth. It has been estimated that the number of bacterial taxa in oceans may approach 2 million (or 160/mL), in soil at least 4 million (or 6,400–38,000/gm), and in the atmosphere at least 4 million [1]. This estimation puts the total number of bacterial species in the order of 10 million and perhaps up to a billion. In contrast, there are probably between 10 and 30 million of animal species, the vast majority of which are insects. At the moment, the number of scientifically recognized animal species is about 1,250,000, and the number of recognized plant species amounts to almost 300,000. Given that the emergence and evolution of microbial organisms greatly preceded plants and animals, and that bacteria have relatively short generation times and are adapted to fine-scale microenvironments in all ecosystems, it is surprising that the total number of characterized bacterial species only stands at about 5000.

Several reasons may account for the discrepancy between the estimated number of bacterial taxa and the actual number that is currently known. First of all, traditional classification techniques based on morphology and microscopic examination are not appropriate for identification of bacteria, as relatively few bacteria can be determined by morphological characteristics alone. To facilitate improved characterization of bacteria, emphasis has been placed on their metabolic and physiological features, which are dependent on the availability of a pure culture of an organism. Unfortunately, *in vitro* culture is a time-consuming process. In addition, there are estimates that cultivatable microorganisms only represent a minor portion of all microbial species on Earth. Indeed, with many bacteria growing poorly in laboratory media, frequently less than 1% of the bacterial organisms present in the environmental samples are recovered. To access genetic information from the vast majority of microbial species that we have not yet managed to culture, one needs to analyze these organisms in their natural habitats, and from nonenriched specimens [2–4].

It is only in the past few decades that molecular techniques (especially nucleic acid amplification procedures) have been developed and utilized for bacterial classification and identification. These techniques usually target the small ribosomal RNA subunit (16S rRNA) and other genetic elements in bacteria. Since these techniques require minute amounts of starting materials, they have the potential to be applied to studies of both pure isolates and nonculturable organisms. The first microbial genome was published in the autumn of 1995 [5], and a few hundred bacterial genomes have become available in the public domain (www.ncbi.nlm.nih.gov) since then. The information gained from these model organisms is invaluable to the phylogenetic and molecular classification of bacteria; however, it is far from being representative of the real biodiversity [6–8].

Besides their value for taxonomic investigation, molecular detection and identification of bacteria from noncultured specimens have relevance in medicine, food science, and environmental microbiology. This is attributable to the fact that many different bacterial pathogens undergo unique life cycles and display contrasting antibiotic susceptibility profiles, which demand tailor-made strategies for effective control and prevention. Speedy and correct identification of bacterial species concerned is not only vital to the design and adoption of appropriate measures against the bacterial pathogens but also important to the evaluation of the efficacy of treatment undertaken, which forms a critical component in the overall management of infectious disease control and prevention campaign. Additionally, the ability to detect and quantify the bacteria from noncultured specimens provides useful means to gain new insights on the diversity and potential roles of bacterial taxa in the environment, which have been previously infeasible using traditional phenotype-based techniques.

8.1.2 Characteristics of Molecular Detection Technology

Conventional culture-based methods for determination of bacteria are noted for their time consumption and tediousness. In addition, as these methods assess the phenotypic characteristics of bacteria, their performance can be affected by external factors that influence microbial growth and metabolic processes. New generation, genotype-based techniques have been designed specifically to overcome the drawbacks of the phenotype-based procedures. By focusing on the nucleic acids (DNA or RNA) of bacteria, which (particularly DNA) are intrinsically more stable than proteins, the genotypic (or molecular) identification techniques are much more precise and less variable than the phenotype-based methods. Furthermore, as many of the genotypic techniques involve *in vitro* template and/ or signal amplification, they are also more sensitive, much faster, and more amenable to automation. Also, owing to their ability to generate consistent results from small quantity of template, the genotypic methods allow detection and identification of "viable but noncultivatable" cells that are metabolically active but nondividing, as well as dead cells.

From the initial, unsophisticated nonamplified DNA hybridization procedures, molecular detection technology has moved toward the all-encompassing *in vitro* nucleic acid amplification procedures along with the capacity for real-time monitoring and detection. These improvements have made nucleic acid amplification and detection technology one of the most widely adopted and applied techniques in both research and clinical laboratories worldwide. Among several elegant and distinct nucleic acid amplification approaches, polymerase chain reaction (PCR) is the first and most commonly utilized technique due to its simplicity, robustness, and versatility. In its classic form, PCR uses two single-stranded oligonucleotide primers (measuring 20–30 bases in length) to flank the front and rear ends of a specific DNA target, a thermostable DNA polymerase that is capable of synthesizing (duplicating) the specific DNA, and double-stranded DNA to function as a template for DNA polymerase. The PCR amplification process begins at a high temperature (e.g., 94°C) to convert the double-stranded template DNA into single strands. This is followed by a relatively low temperature (e.g., 55°C), which permits annealing of the single-stranded primer to the single-stranded template. Then temperature is changed to 72°C, which is optimal for DNA polymerase extending (copying) along the template. The whole denaturing, annealing, and extending process is repeated 25–30 times so that one single copy of DNA template is turned into billions of copies within 3–4 h. Because PCR is capable of selectively amplifying specific targets present in low concentrations (theoretically down to a single copy of DNA template), it not only demonstrates high specificity and extreme sensitivity but also has a rapid turnover time and is highly amenable to automation for high-throughput testing, in addition to its capacity for identifying both cultured and noncultivatable organisms. The resulting PCR products can be separated by gel electrophoresis and detected with a DNA stain, or alternatively by automated, real-time monitoring and identification.

However, as versatile and dependable as an identification technique can be, PCR has shortcomings of its own. The first relates to the risk of carryover contamination from the previously amplified products, leading to potential false positive results, due to the ability of PCR to generate product from a single copy of template. While separation of sample preparation, PCR amplification, and product detection rooms are valuable strategies to avoid such contamination, another approach involves replacement of dTTP with dUTP in the PCR mixture, which leads to the production of uracil-containing DNA (U-DNA). The PCR amplicons that are synthesized with dUTP can be eliminated by an enzyme called uracil-DNA-glycosylase prior to actual PCR amplification, and thus only nonamplified test DNA with resistance to uracil-DNA-glycosylase digestion, remains in the tube for subsequent PCR experiment. Another notable shortcoming of PCR-based technique is that it is prone to inhibition by substances that are present in noncultured, clinical, food, and environmental samples.

Other factors limiting effective molecular analyses relate to sampling methods, separation of cells from the sample matrix, and subsequent nucleic acid purification; despite this fact, the development of new strategies in the field of sample preparation has been relatively limited. In their natural habitats, microorganisms may occur in limited numbers, often in coexistence with nucleic acid degrading/modifying substances and/or inhibitors of the enzymes that are used in downstream analyses [9]. Methods for preparing samples for molecular analyses are often designed for defined biological sources such as tissues and cultures [10]. In the absence of culturing, nucleic acid concentrations may be very low in samples that are often difficult, if not impossible to define, due to their extreme heterogeneity. Finally, concerning the analysis of environmental samples, the origin of the nucleic acids is paramount. The determination of whether the nucleic acids derive from living or dead organisms, or if contaminating organisms or nucleic acids have been introduced during sample processing, is of utmost importance [11].

The true origin of the DNA purified from environmental sources has been a topic under increased scrutiny. Particularly, this relates to whether or not the DNA originates from viable or dead cells in the sample [12,13]. The limited cultivatability of the microorganisms in natural environments (about 0.1%–1%) renders it impossible to determine cell viability by standard techniques [6,14]. Soil samples, for example, often contain high quantities of free DNA, including DNA from dead microorganisms. The determination of viability is also important when investigating pathogenic microorganisms in the environment and also in clinical settings, for example, with respect to therapeutic regimes. Nonetheless, in most cases, DNA is too stable to be applied as a viability marker. Intact DNA has even been recovered from fossil material [15]. Furthermore, both the strains and manner of death may help determine the stability of sample DNA [16]. The current consensus is therefore that DNA cannot be used as a viability marker. The use of DNA as an indirect viability marker has been the focus of recent advances [11,17]. Living cells protect their DNA by an intact cell wall/membrane, while these barriers are compromised in dead cells. Exploiting this fact, samples are treated with an agent that modifies the exposed DNA, rendering it inapplicable for subsequent PCR analysis; positive PCR amplification is only possible from viable cells [18]. Alternative methods are also being developed, such as comparing differences in physical properties between viable and dead cells, or differences in DNA exposure [19]. Differential density or dielectric properties between viable and dead cells could be utilized for their separation [20]. Better description of the different DNA fractions in environmental and clinical samples using newly developed methodologies will foster increased understanding of microbial communities.

8.1.3 INHIBITORS IN NONCULTURED SPECIMENS

Apart from microorganisms of interest, clinical, food, and environmental specimens contain multiple substances, some of which may interfere with nucleic acid amplification processes (possibly through degradation and sequestration of available nucleic acids and inhibition of thermostable DNA polymerase). Substances impeding DNA polymerase activity have been identified in body fluids and feces (e.g., heme, hemoglobulin, lactoferrin, immunoglobulin G, leukocyte DNA, polysaccharides, and urea); in foods (e.g., phenolics, glycogen, calcium ions, fat, and other organic substances); in environmental specimens (e.g., phenolics, humic acids, and heavy metals); and in added anticoagulants (e.g., EDTA and heparin). Several nucleic acid extraction reagents (e.g., detergents, lysozyme, NaOH, alcohol, EDTA, and EGTA) have also been shown to hinder PCR amplification (Table 8.1).

Interestingly, many interfering substances in noncultured specimens or among nucleic acid extraction reagents appear to be involved in competition with Mg^{2+} ions that are essential for proper functioning of DNA polymerase. For example, being one of the major constituents of soil organic matter humus, humic acid has the ability to chelate (bind) positively charged multivalent ions (e.g., Mg^{2+}, Ca^{2+}, Fe^{2+}, Fe^{3+}). Thus, the presence of humic acid in the sample reduces the amount of Mg^{2+} ions available in a PCR mixture that are needed for optimal activity of DNA polymerase.

TABLE 8.1
Analysis of Microorganisms in Various Environmental Matrixes

	Separation from Matrix	Major PCR Inhibitors	Special Consideration
Liquids	Centrifugation, filtration, affinity binding, or dielectrophoresis	All possible—depending on liquid; however, relatively easy to define	Heterogeneous low amount of particles
Soil	Ion exchange, affinity binding, or density gradients	Organic polymers, humic acids and ions	Heterogeneous, strong binding of microorganisms to particles
Sediments	Centrifugation	Similar to soil	Potential high content of dead cells
Feces	Affinity binding or density gradients	Proteases, nucleases, and polysaccharides	High content of PCR inhibitors
Plant and animal tissues	Mechanical or enzymatic disruption in combination with affinity binding or density gradients	Proteins, ion complexes, proteases, polysaccharides and polyphenols	Very heterogeneous
Biofilms	Mechanical release from surface in combination with centrifugation, filtration, or affinity binding	Polysaccharides	Difficult to obtain representative sample due to biofilm formation, and binding to the solid surface

Similarly, as a deep red iron-containing prosthetic group $C_{34}H_{32}N_4O_4Fe$ of hemoglobin and myoglobin, the Fe^{2+} ions in heme may compete directly with Mg^{2+} ions in the PCR process. Other known chelators include EDTA (ethylene diamine tetraacetic acid), which can bind tightly to divalent cations such as Mg^{2+} and remove them from solution. Indeed, this property of EDTA has led to its use as a treatment for patients with extreme, life-threatening hypercalcemia. Due to their small sizes, these chelating molecules are neither easily nor effectively eliminated by conventional nucleic acid purification procedures such as ethanol precipitation.

In this chapter, we address the particular challenges associated with obtaining nucleic acids suitable for direct molecular (PCR) analyses of bacteria from real ecosystems containing recalcitrant environmental (soil, water), gastrointestinal (GI) tract (digests, feces) matrixes and foods, with special emphasis on how these problems are currently being overcome, and possible future solutions. We discuss the issues of sampling, sample preservation, separation of the microorganisms from the sample matrix, and nucleic acid purification (as a summary, see Figure 8.1) with regard to prokaryotic microorganisms. Postsampling treatments are our central focus, since the

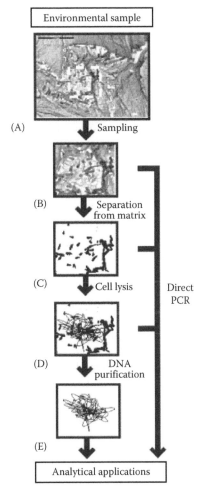

FIGURE 8.1 Schematic representation of the process of analyzing environmental samples. The environmental sample could have a heterogeneous composition (A). It is important to obtain a representative sample (B) in the analysis of microbial communities. The bacteria are separated from the matrix (C) after the sampling. Then the microorganisms are disrupted, and the DNA released (D). Finally, the DNA is purified (E), and is ready for downstream applications such as PCR. Steps B and D can be omitted in special cases.

various sampling procedures are highly dependent on the specific application (e.g., soil, water, or feces). Finally, issues important for implementing PCR in high-throughput clinical or environmental monitoring are covered.

8.2 BASIC PREPARATION PRINCIPLES

In addition to microorganisms of interest, specimens intended for bacterial testing harbor a variety of other substances that may degrade or sequester nucleic acids, or hamper subsequent detection by molecular methods. Therefore, obtaining representative environmental samples and keeping them sufficiently intact are crucial to their analysis. Precautions have to be taken at the site of sampling to avoid modification and/or degradation of the nucleic acid in the samples. For practical reasons, the sample pretreatment in the field or clinic should be kept to a minimum. However, prevention of nucleic acid modification or degradation is paramount, and any enzymatic activities that could compromise nucleic acid quality should be inactivated. The aim is to stabilize microbial nucleic acids and/or cells until samples reach the analytical laboratory for further treatment [21]. Upon arrival at the laboratory, bacterial samples can be preserved by drying, freezing, and fixing. Bacteria of interest can be physically separated from large particles via centrifugation and filtration, and further purified through selective capturing and binding with purpose-designed magnetic beads and columns.

8.2.1 SAMPLE PRESERVATION

Uncultured samples can be preserved with alcohol, drying, freezing, fixation in formaldehyde, or combinations of the above, and collectively, represent the most frequently utilized methods for sample pretreatment [10,22]. In many cases, isopropanol or ethanol is preferable as a preservative that is easy to use, relatively nontoxic, kills most organisms, and keeps nucleic acids stable. Using alcohol preservation also reduces risks associated with clinically infectious material. A simple alternative for sample pretreatment is drying (lyophilization). During dehydration, however, the sample is not immediately preserved, leaving the nucleic acid susceptible to damage or chemical modification by enzymes or chemical agents. Furthermore, the growth of microorganisms during the early stages of the preservation phase when the samples are still relatively hydrated is another significant problem with drying. However, once completely dehydrated, samples are relatively inert, and can be stored for prolonged periods. A successful sample preparation method involves the application of samples onto a special paper (FTA paper), followed by drying [23].

Snap freezing in liquid nitrogen is widely considered to be the optimal way to preserve a sample [24]. Aside from sample preservation, liquid nitrogen offers an additional advantage of grinding the frozen material to ease downstream nucleic acid purification steps [24]. This method of sample preservation is not without its disadvantages; the sample processing is quite laborious and the sampling site must be close to the laboratory. However, when immediate preservation and stability are important issues, liquid nitrogen is a conservation method worth serious consideration. Storage in glycerol at −80°C is also useful for preservation of unprocessed samples (e.g., soil) for subsequent nucleic acid isolation.

Formaldehyde, although widely used, is not a preferred choice as a fixative in sample preparation. Nucleic acids are nearly immediately destroyed by unbuffered acidic formaldehyde [25], while the effect is not as rapid when the formaldehyde is buffered. In either case, nucleic acids are not stable in formaldehyde over long periods of time [26]. The iodine-containing microscope fixation solution Lugol, unlike formaldehyde, does not interfere with nucleic acids, and has been employed for preservation of environmental samples, including sediments containing algae and protozoa with successful subsequent DNA analysis [27,28]. Although not yet tested, similar approaches can probably also be used for bacteria.

8.2.2 CENTRIFUGATION AND FILTRATION

The process of separating bacterial cells from a sample matrix is a crucial step, because the matrix can harbor major enzymatic inhibitors that affect nucleic acids or their downstream processing, and contain other organisms whose nucleic acids may compromise the sensitivity and specificity of assays targeting bacteria of interest. Common techniques to separate bacterial organisms from the contaminating materials (such as food particles) include centrifugation and filtration, which also help reduce (concentrate) the sample volume to a workable level. Centrifugation and filtration can be employed as stand-alone techniques, or in combination, and they often form as an integral part of the comprehensive sample preparation scheme for noncultured specimens. During the monitoring or diagnosis of harmful or pathogenic bacteria, sensitivity issues are of particular importance. Biofilms, where microorganisms are tightly attached to a surface [6], make cell–matrix separation difficult. The separation of cells from the matrix in soil samples is also problematic and quite critical. Microbial cells may be ionically bound to particles, especially clay, in soil matrices [29]. Direct lysis therefore underlies most methods for sample preparation from soil [30,31]. Binding to particles seems less pronounced for aquatic sediments, exemplified by the separation of cyst-forming dinoflagellates by centrifugation-based approaches [27]. Due to its small size, the direct particle binding for bacteria is difficult to evaluate.

Other ingenious physical separation and concentration techniques have also been described for preparation of noncultured clinical, food, and environmental samples prior to molecular application. For example, an aqueous two-phase system containing polyethylene glycol (PEG) 4000 and dextran 40 was used to restrict the PCR inhibitors to the top PEG phase and concentrate most *Listeria monocytogenes* cells (or DNA) in the bottom dextran phase, which can then be applied for molecular detection [32].

8.2.3 SELECTIVE BINDING AND ABSORPTION

Under chaotropic conditions (e.g., at high concentrations of sodium iodide, guanidine hydrochloride or guanidine thiocyanate, etc), nucleic acids show an affinity for silica-coated beads (or glass beads, usually 3–10 μm in diameter). The binding between nucleic acids and silica-coated beads in the presence of a concentrated ionic solution can be later reversed in the presence of a dilute ionic solution. This unique characteristic of nucleic acids has underscored a variety of in-house and commercial nucleic acid purification systems, in which beads are used to selectively bind and capture nucleic acids, unbound materials are washed away, and nucleic acids are then eluted with a low ionic strength buffer (or distilled water) to produce high purity DNA and RNA for molecular applications. Subsequent modification of this technological platform has led to the use of paramagnetic beads that are made of silica impregnated with iron. As these paramagnetic beads demonstrate magnetic qualities only upon exposure to an external magnetic force, they can be conveniently separated from other substances in the solution. By coupling paramagnetic beads with specific oligonucleotides complimentary to the target gene sequence, the nucleic acids from organisms of interest can be selectively absorbed and eluted. Immobilization of specific oligonucleotides on paramagnetic beads provides an effective approach for capturing bacterial nucleic acids from clinical, food, and environmental specimens. Apart from in-house paramagnetic beads based nucleic acid purification systems, commercial kits based on this principle are available that offer added convenience.

Alternatively, by covalently linking specific monoclonal antibodies (MAb) or polyclonal antibodies (against surface protein of target organism) and other ligands (for interaction with bacterial surface proteins) to paramagnetic beads, target microorganisms can be also isolated from other materials. This offers a valuable means to capture bacteria from clinical, food, and environmental samples and to eliminate PCR-inhibitory factors therein. Due to their relative low cost and high efficiency, immunocapture based on paramagnetic beads has been increasingly utilized for the separation of target bacteria from a matrix [33]. After the paramagnetic beads and matrix are mixed, the beads complex with the target microorganisms, which can then be separated from the matrix through the application of a magnetic force [34,35]. Sample complexity can be reduced through the enrichment of specific organisms by bead capture but may also reduce matrix interference in environmental or clinical samples. Immunomagnetic separation has been coupled directly to PCR or combined with DNA isolation and PCR detection with notable examples from the diagnosis of clinical pathogens such as *Mycobacterium* and various food pathogens [36–38]. Potential problems with immunomagnetic separation, however, are that only a few organisms can be identified simultaneously, and the antigens may not be expressed under given environmental conditions. Finally, secreted compounds, such as polysaccharides or highly viscous substances [39], may mask the antigens.

Microorganisms in water and other hydrophilic liquids have been isolated and/or concentrated by nonspecific adsorption onto polymer beads. Such approaches have been successfully applied in the analyses of cyanobacterial communities in water [40,41]. General binding properties or common affinities among whole groups of microorganisms can provide the basis for physical separation [42]. By coating surfaces with lecithin, carbon, or metal hydroxides [43–45], nonspecific adsorption can be achieved. While the isolation of a wide range of cells is a clear advantage with these strategies, a main disadvantage is that any given coating type may not be entirely selective with respect to cell binding. Undesirable compounds or compounds that prevent microbial binding could be co-purified, presenting potential problems. For bacteria in water, it is often necessary to process large volumes in order to concentrate enough cells. For instance, several thousand liters can be processed using the application tangential filtration [46]. The challenge with filtration approaches is that execution of the method is complicated and sometimes results in the co-purification of other particulate materials. Thus, this strategy is only applicable for processing relatively small number of samples that do not contain too much particulate material. The key advantage of this method is the increased sensitivity obtained by concentrating cells from very large volumes.

Generally, bacterial cells are dense relative to most tissues and other biological material. Density gradient centrifugation can be employed to exploit this fact to separate microbial cells from a biological matrix [32]. Benefits include removal of inhibitory compounds, and removal of DNA from other organisms that may also be inhibitory on PCR. The approach is, however, quite technically challenging. Dielectrophoresis can be also utilized to separate microorganisms present in liquid samples. By inducing an uneven charge distribution within a cell through the use of an oscillating electrical field, separation is possible [47]. The technique is limited by the small size of the electrophoresis unit, which results in sensitivity to the conductivity of the medium and to particulate contaminants.

8.2.4 Lysis of Bacteria

Regardless of whether sample analysis is performed directly, without DNA purification, or not (see below), bacterial cells must be lysed in order to release their nucleic acids. Although cultured bacterial colonies can often be lysed during the initial denaturing step of PCR thermocycling, analyses of complex microbial samples require a rigid lysis procedure that does not introduce errors due to differential lysis of different bacteria in the sample [41,48]. Commonly, mechanical, chemical, and/or enzymatic approaches are applied. Mechanical disruption of fresh, freeze dried, or material frozen in liquid nitrogen is achieved by grinding [10]. The addition of alumina or glass beads can facilitate the mechanical grinding process. Grinding has the advantage that any type of material can be processed, while disadvantages include the possibility of cross-contamination and difficulties with automating routines. The treatment of environmental samples with ultra sound (sonication) has also been successfully applied to release nucleic acids [49]. Certain types of biological matrixes can be selectively degraded by enzymes; proteases, for example, can be used to achieve effective lysis of tissue matrixes consisting mainly of proteins. The addition of detergents or chaotropic salts that denature biomolecules, accompany nearly all cell disruption and lysis strategies [50,51].

There are special circumstances when it is not necessary to purify the DNA from the samples to achieve effective analysis. In such cases, the presence of PCR inhibitors in the samples may be too low to interfere with the PCR [9], or, alternatively, the samples can be diluted to prevent the inhibition of enzymatic reactions [52]. When the amount of target material is extremely limited, such as single cells or bacteria concentrated by immunomagnetic separation, loss of DNA during the purification steps may occur [28,53].

Most environmental and clinical samples contain compounds that are potent inhibitors of downstream nucleic acid analyses. Proteases that degrade the polymerase or nucleases that break down nucleic acids are common examples. Substances like chaotropic salts that destabilize the enzymes, or polysaccharides that can interact with both the nucleic acids and/or enzymes, are also examples of potentially potent inhibitors [54]. Compounds that modify nucleic acid or that directly interfere with DNA polymerase activity in the PCR reaction should also be avoided [9].

Given the presence of inhibitors in the sample, two relatively simple and quick remedies may be implemented to ameliorate their effects. By adding substances that facilitate the PCR in the presence of inhibitors, or by selectively removing inhibitors from the sample, effective amplification can be achieved [9,55]. Protocol standardization, however, can be challenging due to the diverse nature of environmental samples.

8.2.5 PURIFICATION OF BACTERIAL NUCLEIC ACIDS

The rationale for purifying nucleic acids for molecular applications is to remove substances that may interfere (even marginally) with the enzymatic reactions in order to generate a DNA/RNA preparation yielding reproducible analytic results. Use of organic solvents such as phenol represents the classical way of extracting nucleic acids from complex, inhibitor-, and protein-containing solutions. Chloroform or ether can also be employed to separate other undesirable components, for instance, fat from the nucleic acid-containing aqueous phase. Plant or algal polysaccharides that co-purify with nucleic acids can be selectively precipitated with cetyltrimethyl ammonium bromide (CTAB) [24]. However, organic solvents are toxic, and given the complex handling routines involving centrifugations and aqueous phase transfer, their use in nucleic acid extraction is not ideal.

It is well documented that nucleic acids bind to glass, silica particles, or other polymer surfaces in the presence of alcohol, high salt, or chaotropic agents, and can be subsequently released from them using low salt buffers [50]. Both detergents [10,51] and PEG [56] have also been employed to bind nucleic acids onto polymer surfaces. The solid-phase (polymer) nucleic acid binding principle has been applied in several formats, including cartridges, filters, columns, and paramagnetic beads. Paramagnetic beads can easily be manipulated by a magnet, eliminating the need for centrifugations and thus accelerating the washing steps, and thus offering a clear advantage over other solid-phase materials. For these reasons, most automated platforms for sample and nucleic acid preparation are based on a magnetic solid phase.

The yield and the purity of the isolated nucleic acids are important parameters. Nucleic acid quality can be measured empirically simply by evaluating the amplification efficiency of the subsequent PCR reaction. However, such measurements do not provide information about the kinds of inhibitors present; such information is crucial for optimizing nucleic acid extraction protocols. Nucleic acid purity has largely been measured by the degree of protein contamination, most often determined by comparing the absorption of UV light with a wavelength of 260 nm ($OD_{260\,nm}$), and with that of 280 nm ($OD_{280\,nm}$). An $OD_{260\,nm}/OD_{280\,nm}$ ratio of 1.7 indicates a pure nucleic acid preparation, however, a ratio of 1.5 may indicate 99% protein contamination. Several pigments can also interfere with the absorption measurements [57]. OD determination does not provide sufficient information to identify potential PCR inhibitors in environmental samples. There are more sensitive and specific methods to evaluate nucleic acid purity in order to understand more about the purification method employed, and the presence of potential inhibitors. Standard analytical chemistry methods such as MALDI-TOF [58], high-pressure liquid chromatography (HPLC) [59], multispectral analyses, and LC MS yield detailed and accurate information regarding the different components in a sample [60]. However, the use of these methods is restricted to the optimization of sample preparation methods, as they are not suited for routine application.

8.3 PRACTICAL APPROACHES

8.3.1 CLINICAL SPECIMENS

Clinical specimens cover a wide range of materials that include blood, urine, feces, sputum, bronchial lavages, milk, cerebrospinal fluid, semen, tissues, bones, etc. As discussed above, these materials contain a number of known and uncharacterized substances with the capacity to interfere with the functionality of DNA polymerase and sequester the available nucleic acids for primer and probe recognition. Therefore, the ability to detect and identify bacterial organisms depends on how successful the inhibiting components are eliminated from these materials and the bacteria/nucleic acids of interest are retained for subsequent application. Toward this end, numerous sample handling protocols using both in-house reagents and commercial kits have been developed for improved recovery of bacteria/nucleic acids from clinical specimens. While these protocols have offered useful stand-alone techniques to handle various types of clinical samples prior to molecular analysis, their efficiency and overall performance often become more discernable upon side-to-side comparison. For this reason, we focus on a number of comparative studies that deal with common yet challenging types of clinical samples, with the aim to assist readers in the selection of appropriate sample preparation methods for their particular needs.

Undoubtedly, one of the most challenging clinical samples to work with is stool (feces), due not only to its odor, but also to its complex nature. Several DNA polymerase inhibitors (e.g., heme, polysaccharides, and urea) are known to be present in stool specimens, and they are highly recalcitrant to conventional nucleic acid purification procedures including phenol/chloroform extraction and ethanol precipitation. McOrist et al. [61] conducted an assessment of four commercial kits (FastDNA kit, Bio 101; Nucleospin C+T kit, Macherey-Nagel; Quantum Prep Aquapure Genomic DNA isolation kit, Bio-Rad; QIAamp DNA stool mini kit, Qiagen) and a noncommercial guanidine isothiocyanate/silica matrix method of Boom et al. [50] for extraction of both Gram-positive and Gram-negative bacterial DNA (i.e., *Lactobacillus acidophilus* and *Bacteroides uniformis*) from stool samples. They showed that of the five methods, QIAamp DNA stool mini kit appeared to be highly effective in obtaining *B. uniformis* and *L. acidophilus* DNA from stools, with a single-round PCR sensitivity of 1–10 colony forming units (cfu) per gram of *B. uniformis*- or *L. acidophilus*-spiked human stools (Table 8.2). The QIAamp DNA stool mini kit utilizes lysis buffer with high strength chaotropic guanidine salts and detergents and washing buffer with initial low strength chaotropic salts, and Tris/alcohol/acid buffers for DNA elution. It also includes a commercial polysaccharide mixture, which is added to the lysis buffer for the described purpose of removing PCR inhibitors of fecal origin.

In a separate study, Trochimchuk et al. [62] examined four different methods (i.e., phenol/chloroform purification, phenol/chloroform/Sepharose B4 spin columns, phenol/chloroform/polyvinylpolypyrrolidone (PVPP) spun columns and Mo Bio UltraClean kit) for their effectiveness in extracting and purifying Gram-negative *Escherichia coli* O157:H7 DNA and cells from cattle manure. They noted that the PVPP spun columns and the Mo Bio UltraClean kit demonstrated a high sensitivity, detecting 20 pg of *E. coli* DNA (about 2×10^3 cells) per 100 mg of manure, and 3×10^4 *E. coli* cfu per 100 mg of spiked manure (Table 8.2). By adding a brief enrichment step (tryptic soy broth at 37°C for 5 h), both the PVPP spun column and the UltraClean kit methods enabled detection of initial inocula of 6 cfu *E. coli* per 100 mg manure. Interestingly, the PVPP spun column method relies on a guanidine thiocyanate and phenol/chloroform extraction of the manure sample followed by the PVPP spun-column chromatography; and the UltraClean kit involves sample disruption with beads and DNA purification on a silica-based filter.

TABLE 8.2

Comparison of In-House Reagents and Commercial Kits for Recovery of Bacterial DNA from Nonenriched Specimens

Bacterium	Isolation Method	Outcome	References
Gram-positive			
Lactobacillus acidophilus–spiked human stool	QIAamp DNA stool mini kit (Qiagen)	1–10 cfu (based on a single-round PCR)	[61]
	Guanidine isothiocyanate/silica matrix [50]	10^7 cfu	
	Quantum Prep aquaculture Genomic DNA isolation kit (Bio-Rad)	$>10^7$ cfu	
	Nucleospin C+T kit (Macherey-Nagel)	$>10^7$ cfu	
	FastDNA kit (Bio 101)	$>10^7$ cfu	
Streptococcus pneumoniae–spiked human blood	Bilatest bead DNA 2 extraction kit (Bilatec)	1 genome equivalent (based on a single-round PCR)	[63]
	Wizard SV96 system (Promega)	2–3 genome equivalents	
	Wizard Magnesil bead kit (Promega)	2–3 genome equivalents	
	Nucleospin robot 96 (plasmid) kit (Macherey-Nagel)	5 genome equivalents	
	Montage plasmid Miniprep96 kit (Millipore)	6 genome equivalents	
Gram-negative			
Bacteroides uniformis–spiked human stool	QIAamp DNA stool mini kit (Qiagen)	1–10 cfu (based on a single-round PCR)	[61]
	Quantum Prep aquaculture genomic DNA isolation kit (Bio-Rad)	10^4 cfu	
	Nucleospin C+T kit (Macherey-Nagel)	$>10^7$ cfu	
	Guanidine isothiocyanate/silica matrix (Boom et al., 1990)	$>10^7$ cfu	
	FastDNA kit (Bio 101)	$>10^7$ cfu	
E. coli O157:H7–spiked cattle manure	UltraClean kit (Mo Bio)	20 pg DNA or 3×10^4 cfu per 100 mg manure (based on a single-round PCR)	[62]
	PVPP spin column	20 pg DNA or 3×10^4 cfu per 100 mg manure	
	Phenol/chloroform extraction	$>3 \times 10^4$ cfu per 100 mg manure	
	Phenol/chloroform/Sepharose B4 spin column	$>3 \times 10^4$ cfu per 100 mg manure	
Brucella melitensis–spiked serum	UltraClean DNA bloodSpin kit (Mo Bio)	10^2 fg DNA (based on a single-round PCR)	[64]
	QIAamp DNA blood mini kit (Qiagen)	10^2 fg DNA	
	High Pure PCR template preparation kit (Roche)	10^4 fg DNA	
	Nucleospin tissue kit (Macherey-Nagel)	10^4 fg DNA	
	Wizard Genomic DNA purification kit (Promega)	10^5 fg DNA	
	UGFX genomic DNA purification kit (Amersham)	10^5 fg DNA	
	Puregene DNA purification system (Gentra)	10^6 fg DNA	

(continued)

TABLE 8.2 (continued)
Comparison of In-House Reagents and Commercial Kits for Recovery
of Bacterial DNA from Nonenriched Specimens

Bacterium	Isolation Method	Outcome	References
Legionella pneumophila–spiked sputum	MagNA Pure (Roche)	526,200 cfu/mL (based on PCR quantification)	[66]
	NucliSens (bioMérieux)	171,800 cfu/mL	
	Roche High Pure kit (Roche)	133,900 cfu/mL	
	QIAamp DNA mini kit (Roche)	46,380 cfu/mL	
	ViralXpress kit (Chemicon)	13,635 cfu/mL	
Mycobacteria			
Mycobacterium tuberculosis–spiked sputum	IDI lysis tube (Infection Diagnostics, Inc.)	Recovering 42 pg DNA from 200 μL sample (based on a single-round PCR on a Smart Cycler)	[65]
	PrepMan Ultra extraction (Applied Biosystems)	Recovering 30 pg DNA from 200 μL sample	
	QIAamp DNA mini kit (Qiagen)	Recovering 28 pg DNA from 200 μL sample	
	Tris-EDTA buffer boil extraction	Recovering 7 pg DNA from 200 μL sample	
Mixed bacteria			
Soil	Direct lysis with glass beads plus SDS and protein removal with cesium chloride [79]	Recovering 5.94 μg DNA per gram of oven-dried soil as determined by OD_{260nm} reading	[78]
	Direct lysis with lysozyme and precipitation with isopropanol [83]	Recovering 2.29 μg DNA per gram of oven-dried soil	
	Direct lysis with 1% SDS and precipitation with 15% PEG 6000 [82]	Recovering 0.71 μg DNA per gram of oven-dried soil	
	Direct lysis with lysozyme and proteinase K and precipitation with ethanol [84]	Recovering 0.26 μg DNA per gram of oven-dried soil	
	Bead beating	$OD_{260nm} = 1.82$ (with an OD_{260nm}/OD_{280nm} ratio of 1.69)	[80]
	Sonication	$OD_{260nm} = 1.20$ (with an OD_{260nm}/OD_{280nm} ratio of 1.41)	
	Proteinase K digestion	$OD_{260nm} = 1.06$ (with an OD_{260nm}/OD_{280nm} ratio of 1.31)	
	Differential centrifugation and lysozyme	$OD_{260nm} = 0.83$ (with an OD_{260nm}/OD_{280nm} ratio of 1.10)	
	Bead beating/lysozyme/SDS	$OD_{260nm} = 1.71$	[81]
	Lysozyme/SDS	$OD_{260nm} = 1.68$	
	Bead beating/SDS	$OD_{260nm} = 1.57$	
	SDS	$OD_{260nm} = 1.40$	
	Bead beating	$OD_{260nm} = 1.33$	
	Lysozyme	$OD_{260nm} = 1.21$	
	Bead beating/lysozyme	$OD_{260nm} = 1.00$	
	Grinding in liquid N_2, extaction with CTAB/phenol–chloroform, separation of DNA from RNA with Qiagen Tip 100 RNA-DNA purification system	Recovering 23–435 μg DNA per gram of dry weight soil	[85]

More recently, Nechvatal et al. [27] analyzed the effects of preservative/extraction methods suitable for self-collection and shipping of fecal samples at room temperature for epidemiological investigations of bacterial DNA and RNA markers. They showed that DNA was successfully extracted after room temperature storage for 5 days from Whatman FTA cards, RNAlater or Paxgene preserved, silica gel dried, or liquid N_2 frozen samples. In particular, high amounts of PCR-amplifiable *Bacteroides* DNA with relatively little PCR inhibition was obtained with Qiagen Stool kit applied to RNAlater preserved the samples in comparison with other DNA extraction techniques (i.e., Whatman FTA cards and MoBio Fecal).

Blood (or serum) is another common type of clinical specimens that harbors a number of inhibitory components (e.g., heme, hemoglobulin, immunoglobulin G, leukocyte DNA, and polysaccharides) for DNA polymerase. Smith et al. [63] compared five commercially available kits based on 96-well binding plate, 96-well filter plate, and metallic bead formats for the extraction of Gram-positive *Streptococcus pneumoniae* DNA from whole-blood samples. The result indicated that the Bilatest bead DNA 2 extraction kit was more sensitive than the other four, achieving a detection limit of 1 genome equivalent using one round PCR amplification (Table 8.2). The Bilatest bead DNA 2 extraction kit uses magnetic particle suspension in the presence of guanidine thiocyanate and ethanol to capture bacterial DNA from blood sample followed by washing and DNA elution. Similarly, Queipo-Ortuno et al. [64] evaluated seven commercial kits (i.e., UltraClean DNA BloodSpin kit, Puregene DNA purification system, Wizard genomic DNA purification kit, High Pure PCR template preparation kit, GFX genomic blood DNA purification kit, NucleoSpin tissue kit, and QIAamp DNA blood mini kit) for recovery of Gram-negative *Brucella melitensis* Rev 1 DNA and cells from human serum samples. They found that although both the UltraClean DNA BloodSpin kit and the QIAamp DNA blood mini kit displayed a sensitivity down to 100 fg DNA as assessed by PCR, the UltraClean DNA BloodSpin kit was markedly better for generating *Brucella* DNA of high purity from human serum specimens (Table 8.2).

Sputum samples are frequently submitted to clinical laboratories for detecting *Mycobacterium*, an acid-fast bacterium notoriously resistant to commonly used cell lysis reagents and techniques. Aldous et al. [65] assessed five extraction methods (i.e., Tris-EDTA [TE] buffer, PrepMan Ultra, 2% sodium dodecyl sulfate [SDS]-10% Triton X with and without sonication, Infection Diagnostics, Inc. [IDI] lysing tubes, and Qiagen QIAamp DNA mini kit) for *M. tuberculosis*-spiked sputum samples followed by quantitative PCR. They observed that the IDI lysing tubes provided the greatest recovery of mycobacterial DNA, which was also the least time-consuming procedure (taking less than 1 h versus 2.5–3 h for other methods) (Table 8.2). The IDI extraction method employs a glass bead matrix and boiling treatment. In a separate study, Wilson et al. [66] evaluated the efficiencies of five commercially available nucleic acid extraction methods for the recovery of a standardized inoculum of Gram-negative bacterium *Legionella pneumophila* in respiratory specimens (sputum and bronchoalveolar lavage [BAL] specimens). These authors found that the automated MagNA Pure method (Roche) recovered 526,200 *Legionella* cfu per milliliter of sputum, which was followed by the automated NucliSens method (bioMérieux) at 171,800 cfu/mL, the manual Roche High Pure template preparation kit (Roche) at 133,900 cfu/mL, the manual QIAamp DNA mini kit (Qiagen) at 46,380 cfu/mL, and the manual ViralXpress kit (Chemicon) at 13,635 cfu/mL.

Among the many innovative procedures described in the literature for RNA purification from clinical specimens, a differential lysis protocol reported by Di Cello et al. [67] is notable. In comparison with a coextraction protocol in which human brain endothelial cells containing *E. coli* K1 were processed directly, the differential lysis protocol utilized Qiagen RLT lysis buffer to lyse the brain endothelial cells before proceeding with bacterial DNA extraction using Ambion RiboPure-Bacteria kit followed by cleaning with Qiagen RNeasy mini kit. This differential lysis protocol permitted isolation of microarray-grade *E. coli* RNA, which was free of human RNA contamination, eliminating bias that could be introduced in the gene expression pattern analysis using bacterial RNA preparation containing host RNA (Table 8.3).

TABLE 8.3

**Comparison of In-House Reagents and Commercial Kits for Recovery
of Bacterial RNA from Nonenriched Specimens**

Bacterium	Isolation Method	Outcome	References
Gram-negative			
E. coli K1 in human brain endothelial cells	Coextraction protocol (RiboPure-Bacteria kit, Ambion)	*E. coli* RNA contaminated with human RNA (unsuitable for microarray analysis)	[67]
	Differential lysis protocol (lysis of brain cells with Qiagen RLT lysis buffer followed by extraction with RiboPure-Bacteria kit, Ambion, and cleaning with RNeasy mini kit, Qiagen)	*E. coli* RNA free of human RNA contamination (suitable for microarray analysis)	
Mixed bacteria			
Soil	Mix soil with water, beat with glass beads, extract with phenol/chloroform plus LiCl, precipitate with isopropanol, and treat with DNase [87]	Recovering 3.08 μg RNA per gram of dry weight soil	[86]
	Mix soil with CTAB extraction buffer, freeze in liquid N_2, grind in a mixer mill, beat with glass beads, extract with chloroform plus LiCl, precipitate with ethanol and treat with DNase [88]	Recovering 1.43 μg RNA per gram of dry weight soil	
	Release ribosomes with glass beads, recover ribosomes by centrifugation, extract RNA by phenol, precipitate with ethanol, treat with DNase and pass through CL-6B spin column [89]	Recovering 0.64 μg RNA per gram of dry weight soil	
	Grinding in liquid N_2. extraction with CTAB/phenol-chloroform, separation of RNA from DNA with Qiagen Tip 100 RNA–DNA purification system	Recovering 1.4–56 μg DNA per gram of dry weight soil	[85]

8.3.2 FOOD SPECIMENS

Food specimens encompass meat, fish, milk, cheese, vegetables, fruits, juices, etc., that, in general, offer ideal nutritional substrates for many microorganisms and harbor multiple inhibitory components for nucleic acid amplifying enzymes. Thus, these samples (in particular meat, fish, and cheese) have presented considerable challenges for molecular detection and identification of bacterial pathogens, and often demand specific pretreatment protocols designed for each of the food types [68].

In dealing with meat products, the use of surface washings alone may generate insufficient numbers of bacteria for subsequent testing. Therefore, it is often necessary to homogenize the sample for maximal recovery of target organisms. By combining homogenization with filtering and multiple centrifugation steps, Wang et al. [69] detected 4–40 cfu *Listeria monocytogenes* in spiked meat using PCR primers derived from 16S rRNA gene. Similarly, Rodriguez-Lazaro et al. [70] optimized a sample preparation procedure for beef involving filtration and DNA purification with Chelex-100 resin, which led to sensitive detection of *L. monocytogenes* bacteria. Use of membrane filtration followed by transferring the filters on ALOA agar facilitated detection of 10 cfu *L. monocytogenes* per gram of smoked salmon [71].

To process cheese specimens, a multistep protocol was designed by Uyttendaele et al. [72] for efficient recovery of *L. monocytogenes*. This protocol utilizes centrifugation to get rid of large particles in cheese homogenate, sieving to further eliminate particles and fat from the supernatant, centrifugation again to concentrate the bacteria, and pronase to degrade the residual small food particles. Use of this procedure allowed recovery of 0.5–1.5 *L. monocytogenes* cfu per gram of cheese without enrichment. An alternative sample preparation procedure for cheese was described by Stevens and Jaykus [73]. This procedure employs high-speed centrifugation ($9700\,g$) followed by DNA extraction, PCR amplification, and detection by hybridization, which resulted in the detection of 10–10^3 *L. monocytogenes* cfu per $11\,g$ of cheese and yogurt.

A two-step method for processing potato salad was reported by Isonhood et al. [74]. This method includes a low-speed centrifugation step ($119\,g$ for $15\,min$ at $5°C$) to eliminate large food solids in potato salad filtrate and a second high-speed centrifugation step ($11,950\,g$ for $10\,min$ at $5°C$) to concentrate the bacterial cells in the supernatant. This procedure reduced sample volume by fivefold and test sensitivity by 1000 folds (from 10^6 cfu/g [no sample processing] to 10^3 cfu/g of salad) as examined by PCR and Southern blot analysis. In addition to the in-house procedures, commercial kits have also been applied to the preparation of salads and other food types for bacterial detection and identification. For example, using a High Pure *Listeria* sample preparation kit (Roche Diagnostics), Berrada et al. [75] obtained PCR-ready DNA from salad for use in quantitative PCR, with a sensitivity of 10 *L. monocytogenes* cfu per gram of salad.

8.3.3 ENVIRONMENTAL SPECIMENS

Environmental specimens comprise a diversity of materials such as wastewater, sludge, soil, rock, etc., which can provide additional hurdles for efficient recovery of bacterial cells and nucleic acids for molecular applications (Figure 8.1). In particular, wastewater and sludge samples usually come in large quantities (sometimes measured in liters), which contain very low numbers of microorganisms. The biomasses from these samples need to be recovered by centrifugation and/or filtration before further purification can take place. Shannon et al. [76] employed low-speed centrifugation (at $2600\,g$ for $30\,min$ at $4°C$) to collect biomass from wastewater, primary effluents, and sludge; and high-speed centrifugation (at $20,000\,g$ using a Westfalia Separator) to process large volume of final effluent wastewater. The resulting biomass was subject to DNA extraction process with a commercial kit (Wizard genomic DNA purification kit, Promega), and a sensitivity of 3.26 copies of *L. monocytogenes* genomes per $100\,mL$ in raw wastewater was achieved via quantitative PCR.

More elaborate sample-preprocessing procedures are often needed to prepare soil samples for molecular detection as microbe cells tend to bind tightly to soil colloids [77]. After comparative assessment of four direct lysis methods and three cell-extraction methods, Tien et al. [78] showed that the direct lysis method of Holben et al. [79] was superior to other techniques for the isolation of bacterial DNA from oven-dried soil. This method involves lysis of bacteria in soil with glass beads in the presence of SDS followed by removal of proteins with cesium chloride, and it enabled recovery of $5.95\,\mu g$ DNA per gram of soil as measured by $OD_{260\,nm}$ reading (Table 8.2). Additionally, Yeates et al. [80] examined four DNA extraction methods for soil, which included bead beating, sonication, proteinase K lysis, and differential centrifugation plus lysozyme. They showed that bead beating resulted in a higher quantity and purity of DNA from soil than sonication, proteinase K digestion or differential centrifugation plus lysozyme based on $OD_{260\,nm}/OD_{280\,nm}$ readings (Table 8.2).

In another more detailed investigation, Krsek and Wellington [81] demonstrated that by using a combination of bead beating, lysozyme, and SDS lysis in Crombach buffer, more DNA was recovered from soil than from lysozme and SDS lysis or other procedural permutations as assessed by reading at $OD_{260\,nm}$ (Table 8.2). The authors also noted that use of lysozyme improved the purity of DNA extracted in comparison with procedures containing no lysozyme treatment. While DNA isolated by bead beating or by lysozyme and SDS lysis is of higher molecular weight (around 20 kb or 40 kb, respectively), which is suitable for a wide range of molecular biological applications. In bead beating any other vigorous shaking must be omitted in order to maintain the integrity of higher molecular weight DNA (up to 80 kb). The authors recommended that the most reliable method of DNA isolation from the soil involves direct extraction in Crombach or TE buffer consisting of bead beating followed by lysozyme and SDS lysis and vigorous shaking with glass beads and following purification steps: potassium acetate purification, overnight PEG precipitation, phenol–chloroform purification, overnight isopropanol precipitation, and spermine–HCl precipitation.

Additionally, Hurt et al. [85] described a method for simultaneous recovery of bacterial DNA and RNA from soil, which was based on the use of an extraction buffer containing CTAB (i.e., 100 mM sodium phosphate pH 7.0, 100 mM Tris-HCl pH 7.0, 100 mM EDTA pH 8.0, 1.5 M NaCl, 1% hexadecyltrimethylammonium bromide [CTAB] and 2% SDS). By grinding the soil sample in a denaturing solution at a temperature below 0°C to inactivate nuclease activity and by passing the purified RNA through an anion exchange resin, the integrity, yield, and purity of RNA was much improved. The authors were able to obtain 1.4–56 µg of RNA and 23–435 µg of DNA consistently from 1 g of dried soil.

In another study, Sessitsch et al. [86] compared three in-house protocols for extracting bacterial RNA from soil, and noted that a method previously described by Fleming et al. [87] gave much higher yield, 3.08 µg of RNA per gram of dried weight soil (Table 8.3). This method involves mixing soil with water, beating with glass beads, extracting with phenol/chloroform in the presence of LiCl, precipitating with isopropanol, and treating the resulting RNA with DNase. The authors also found that lyophilization, storage at −20°C as well as storage in glycerol stocks at −80°C proved to be equally effective for the storage of soils for subsequent RNA isolation.

Taken together, the above findings suggest that some nucleic acid purification methods may work well with certain noncultured specimens, but are less effective with other sample types. In addition, different methods and kits may be preferred for maximal recovery of nucleic acids from Gram-positive, Gram-negative, or mycobacteria, even from an identical sample type (e.g., stools). Furthermore, the combined use of magnetic beads, chaotropic reagents, and/or CTAB often helps yield nucleic acids of high quantity and quality from uncultured clinical, food, and environmental specimens.

8.4   FUTURE DEVELOPMENT TRENDS

The number of direct nucleic acid methods that have been adapted in high-throughput platforms for environmental or clinical analyses is currently limited due to the technical complexity and high cost associated with these testing formats [90]. The development of automated protocols will likely be the key to the much wider adoption of the high-throughput molecular testing platforms for all kinds of routine diagnostic or detection purposes (particularly relating to harmful or pathogenic microbes). Process automation is a necessity for all large-scale screenings, and/or when eliminating human error to obtain reproducible results is a main concern (Figure 8.2). There are several commercially available pipetting robots adapted to DNA purification; most of these instruments consist of a liquid handling platform and an automated magnetic device for handling paramagnetic beads. Such instruments are, for example, available from Tecan, Roche, Qiagen, Hamilton, and NorDiag, all of which provide protocols for microorganisms. For some of the most important human bacterial pathogens, such as *Chlamydia*, successful automated DNA purification approaches have been demonstrated [91]. However, even with such defined samples, challenges with low or variable reliability are often encountered due to inhibitors in blood and urine or sample variation (e.g., urine and vaginal swabs) [91–93]. One promising automated approach for clinical samples for improving reproducibility is to

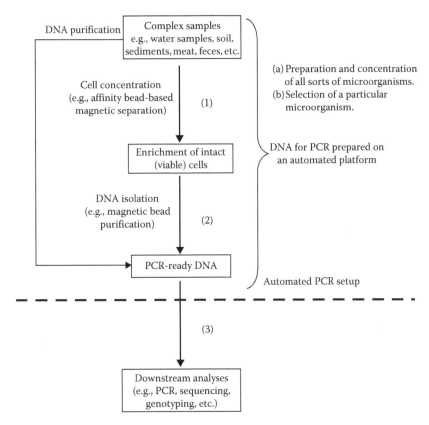

FIGURE 8.2 Platform for automated analyses of environmental samples. The figure illustrates different strategies for obtaining PCR-ready DNA from complex environmental samples. In an automated system, it is important to simplify the process and to avoid steps that cannot easily be automated. For instance, the application of paramagnetic beads both for cell concentration (1) and DNA purification (2) may enable automated preparation of environmental samples. Finally, the purified DNA is transferred to the downstream detection analyses (3).

first purify and concentrate the bacteria prior to the nucleic acid isolation step. This is automated in a robot that first binds the bacteria to paramagnetic beads and subsequently isolates the nucleic acid using the same beads. A large and comparative study on *Chlamydia* from urine has shown that such an integrated sample-preparation approach (i.e., where bacterial purification and DNA isolation are automated on the same platform) improves reproducibility and reliability (as well as increases speed) of *Chlamydia* diagnostics [91]. Still the challenge with handling even more complex environmental samples by automated approaches is that these samples are likely to be very diverse and difficult to define, making automation arduous.

Regarding the analysis of environmental samples, handheld equipment that can readily be brought into the field is under development [94]. The U.S. army is a driving force in this area due to its applicability in counteracting biological warfare [95]. Pathogen control in animals used for food production is also a field heralding recent advances [95]. Integration of all steps into a single apparatus, conceptually similar to lab-on-a-chip, will be one focus of future developments. Instead of the expensive silica used in current lab-on-a-chip production, cheap plastic chips will likely be increasingly employed in the future [96]. Acceptance of plastic chips is rising, mainly because of their low cost, and because they can process liquid volumes well within the practical range for most applications.

We foresee two main areas in the future development of automated analyses for environmental/clinical diagnostics and monitoring. One will be for routine environmental monitoring purposes. Here, the samples will be preserved on site and transported to a test laboratory for further processing.

The aim is to obtain relatively large amounts of accurate data. The other area will be rapid screening to confirm the absence of harmful microorganisms in relation to bioterrorism, pathogen outbreaks, or routine clinical applications where expediency is crucial [97]. Here, it is important to literally move the analyses into the field (including the physicians' offices), and out of the specialized molecular or clinical laboratories. If positive samples are detected, they can then be analyzed more thoroughly in centralized laboratories. We believe the future of nucleic acid based environmental analyses will become an interplay between both centralized and mobile test laboratories. Furthermore, with the emergence of new analytic tools, it is likely that the focus for routine analyses will be changed from single organisms to monitoring or detection of multiple microbial species or even entire communities.

In conclusion, only a minor fraction of all microorganisms on Earth has yet been characterized in culture. Direct nucleic acid based detection methods, without preparatory culturing, are crucial for future environmental monitoring, clinical diagnosis, and control of biological warfare, and to promote a basic understanding of microbial life. The most critical step in direct analyses, although often neglected, is sample preparation. It is extremely important to ensure that the results obtained really represent the microbial diversity in the original sample. A particular challenge is the heterogeneity of the different noncultured environmental or clinical samples to be analyzed, and that the target organisms may be present in very low concentrations. An overview is presented here on the many variations of sample preparation approaches for bacteria from challenging source matrixes such as soil and the gastrointestinal tract, including feces. Most current methods utilize binding (and release) of DNA in the final steps by binding to a solid phase (membrane, column, plate, particles, or magnetic beads) having properties suitable for automated handling. Automation is another crucial issue for reproducibility, in addition to high-throughput sample processing. The implementation of automated routine applications will likely boost the utilization of PCR in clinical microbiology, environmental monitoring, food safety in addition to basic research on microbial communities and the relatively undiscovered world of uncultivable microorganisms.

ACKNOWLEDGMENTS

We thank Ellen Tronrud, Siri Mathiesen, and Unn Hilde Refseth for carefully reading and commenting on this paper. This work has been supported by a Research Levy on Certain Agricultural Products and Hedmark Sparebank.

Knut Rudi received his education at the University of Oslo, Oslo, Norway, and his PhD training in molecular genetics and diagnostics of cyanobacteria in the laboratory of Kjetill S. Jjakobsen. He continued with a postdoctoral stint in the same laboratory. In 2000, he moved to the MATFORSK Norwegian Food Research Institute, Ås, Norway as a research scientist. In 2006, he received a full professorship at the Hedmark University College, Hamar, Norway. His current research interests include analyses and understanding of the gut microbiota, and coevolution of the microbiota and the host. He has published more than 60 research papers and book chapters, and holds 8 patents and patent applications.

Robert C. Wilson earned his BA in genetics at the University of California at Berkeley, California; he received his PhD training in molecular genetics of the Rhizobium–legume symbiosis at the University of California, Santa Barbara, California. He then received his postdoctoral training in the laboratory of Professor Zinmay Renee Sung at the University of California, Berkeley, working on the root meristemless mutants of the model plant Arabidopsis. In 1997, he moved to Norway, working as a researcher at the Agricultural University of Norway. Since 1999, he has been employed as an associate professor of molecular biology at Hedmark University College in Hamar, Norway. His current research interests include automated isolation and PCR-based analysis of DNA from animals, plants, and bacteria, and the molecular genetics of plant development, plant–microbe interactions, and plant response to biotic and abiotic stress.

Kjetill S. Jakobsen received his PhD in molecular genetics in 1989 at the University of Oslo, Oslo, Norway in the field of plant biology. Since then he has been working with a wide array of organisms ranging from large mammals to unicellular eukaryotes and bacteria. Methods development has been an important area of interest for him. He has been CEO of the biotech company Genpoint (now Nordiag ASA), developing automated microbial sample preparation protocols and equipment from 2000 to 2002. Currently, he is a professor of evolutionary genetics at the Centre of Ecological and Evolutionary Synthesis at the University of Oslo, Oslo, Norway. His current interests include endosymbiosis and the biology of organelle genomes, bacterial genome dynamics, and high-throughput sequencing approaches utilized for metagenomic and microbial research.

REFERENCES

1. Curtis, T.P., Sloan, W.T., and Scannell, J.W., Estimating prokaryotic diversity and its limits, *Proc. Natl. Acad. Sci. USA*, 99: 10494, 2002.
2. Torsvik, V., Ovreas, L., and Thingstad, T.F., Prokaryotic diversity–magnitude, dynamics, and controlling factors, *Science*, 296: 1064, 2002.
3. Acinas, S.G. et al., Fine-scale phylogenetic architecture of a complex bacterial community, *Nature*, 430: 551, 2004.
4. Chen, K. and Pachter, L., Bioinformatics for whole-genome shotgun sequencing of microbial communities, *PLoS Comput. Biol.*, 1: e24, 2005.
5. Fleischmann, R.D. et al., Whole-genome random sequencing and assembly of *Haemophilus influenzae* Rd, *Science*, 269: 496, 1995.
6. Amann, R.I., Ludwig, W., and Schleifer, K.H., Phylogenetic identification and in situ detection of individual microbial cells without cultivation, *Microbiol. Rev.*, 59: 143, 1995.
7. Oldach, D.W. et al., Heteroduplex mobility assay-guided sequence discovery: Elucidation of the small subunit (18S) rDNA sequences of Pfiesteria piscicida and related dinoflagellates from complex algal culture and environmental sample DNA pools, *Proc. Natl. Acad. Sci. USA*, 97: 4303, 2000.
8. Pace, N.R., A molecular view of microbial diversity and the biosphere, *Science*, 276: 734, 1997.
9. Abu Al-Soud, W. and Radstrom, P., Effects of amplification facilitators on diagnostic PCR in the presence of blood, feces, and meat, *J. Clin. Microbiol.*, 38: 4463, 2000.
10. Rudi, K. et al., Rapid, universal method to isolate PCR-ready DNA using magnetic beads, *Biotechniques*, 22: 506, 1997.
11. Rudi, K. et al., Use of ethidium monoazide and PCR in combination for quantification of viable and dead cells in complex samples, *Appl. Environ. Microbiol.*, 71: 1018, 2005.
12. Caron, G.N., Stephens, P., and Badley, R.A., Assessment of bacterial viability status by flow cytometry and single cell sorting, *J. Appl. Microbiol.*, 84: 988, 1998.
13. Klein, P.G. and Juneja, V.K., Sensitive detection of viable *Listeria monocytogenes* by reverse transcription-PCR, *Appl. Environ. Microbiol.*, 63: 4441, 1997.
14. Torsvik, V. et al., Novel techniques for analysing microbial diversity in natural and perturbed environments, *J. Biotechnol.*, 64: 53, 1998.
15. Lindahl, T., Facts and artifacts of ancient DNA, *Cell*, 90: 1, 1997.
16. Nogva, H.K. et al., Application of the 5'-nuclease PCR assay in evaluation and development of methods for quantitative detection of *Campylobacter jejuni*, *Appl. Environ. Microbiol.*, 66: 4029, 2000.
17. Nogva, H.K. et al., Ethidium monoazide for DNA-based differentiation of viable and dead bacteria by 5'-nuclease PCR, *Biotechniques*, 34: 804, 2003.
18. Marx, G. et al., Covalent attachment of ethidium to DNA results in enhanced topoisomerase II-mediated DNA cleavage, *Biochemistry*, 36: 15884, 1997.
19. Nishino, T., Nayak, B.B., and Kogure, K., Density-dependent sorting of physiologically different cells of *Vibrio parahaemolyticus*, *Appl. Environ. Microbiol.*, 69: 3569, 2003.
20. Lapizco-Encinas, B.H. et al., Dielectrophoretic concentration and separation of live and dead bacteria in an array of insulators, *Anal. Chem.*, 76: 1571, 2004.

21. Rogers, C. and Burgoyne, L., Bacterial typing: Storing and processing of stabilized reference bacteria for polymerase chain reaction without preparing DNA—an example of an automatable procedure, *Anal. Biochem.*, 247: 223, 1997.

22. Steffan, R.J. and Atlas, R.M., Polymerase chain reaction: Applications in environmental microbiology, *Annu. Rev. Microbiol.*, 45: 137, 1991.

23. Nechvatal, J.M. et al., Fecal collection, ambient preservation, and DNA extraction for PCR amplification of bacterial and human markers from human feces, *J. Microbiol. Methods*, 72: 124, 2008.

24. Snead, M., Kretz, P., and Short, J., Methods for generating plant genomic libraries, *Plant Mol. Biol. Manual*, H1: 1, 1994.

25. Inoue, T. et al., Feasibility of archival non-buffered formalin-fixed and paraffin-embedded tissues for PCR amplification: An analysis of resected gastric carcinoma, *Pathol. Int.*, 46: 997, 1996.

26. Hamazaki, S. et al., The effect of formalin fixation on restriction endonuclease digestion of DNA and PCR amplification, *Pathol. Res. Pract.*, 189: 553, 1993.

27. Bowers, H.A. et al., Development of real-time PCR assays for rapid detection of *Pfiesteria piscicida* and related dinoflagellates, *Appl. Environ. Microbiol.*, 66: 4641, 2000.

28. Jakobsen, K.S. et al., Discovery of the toxic dinoflagellate *Pfiesteria* in northern European waters, *Proc. Biol. Sci.*, 269: 211, 2002.

29. Hardarson, G. and Brough, W., eds. *Molecular Microbial Ecology of the Soil*. 1999, Kluwer Academic, the Netherlands.

30. McGregor, D.P. et al., Simultaneous detection of microorganisms in soil suspension based on PCR amplification of bacterial 16S rRNA fragments, *Biotechniques*, 21: 463, 1996.

31. Zhou, J., Bruns, M.A., and Tiedje, J.M., DNA recovery from soils of diverse composition, *Appl. Environ. Microbiol.*, 62: 316, 1996.

32. Lantz, P. et al., Removal of PCR inhibitors from human fecal samples trough the use of an aqueous two-phase sample preparation prior to PCR, *J. Microbiol. Methods*, 28: 159, 1997.

33. Swaminathan, B. and Feng, P., Rapid detection of food-borne pathogenic bacteria, *Annu. Rev. Microbiol.*, 48: 401, 1994.

34. Chandad, F., Guillot, E., and Mouton, C., Detection of *Bacteroides forsythus* by immunomagnetic capture and a polymerase chain reaction-DNA probe assay, *Oral Microbiol. Immunol.*, 12: 311, 1997.

35. Enroth, H. and Engstrand, L., Immunomagnetic separation and PCR for detection of *Helicobacter pylori* in water and stool specimens, *J. Clin. Microbiol.*, 33: 2162, 1995.

36. Lowery, C.J. et al., Detection and speciation of *Cryptosporidium* spp. in environmental water samples by immunomagnetic separation, PCR and endonuclease restriction, *J. Med. Microbiol.*, 49: 779, 2000.

37. Roberts, B. and Hirst, R., Immunomagnetic separation and PCR for detection of *Mycobacterium ulcerans*, *J. Clin. Microbiol.*, 35: 2709, 1997.

38. Olsvik, O. et al., Magnetic separation techniques in diagnostic microbiology, *Clin. Microbiol. Rev.*, 7: 43, 1994.

39. Tallgren, A.H. et al., Exopolysaccharide-producing bacteria from sugar beets, *Appl. Environ. Microbiol.*, 65: 862, 1999.

40. Rudi, K., Larsen, F., and Jakobsen, K.S., Detection of toxin-producing cyanobacteria by use of paramagnetic beads for cell concentration and DNA purification, *Appl. Environ. Microbiol.*, 64: 34, 1998.

41. Rudi, K. et al., Application of sequence-specific labeled 16S rRNA gene oligonucleotide probes for genetic profiling of cyanobacterial abundance and diversity by array hybridization, *Appl. Environ. Microbiol.*, 66: 4004, 2000.

42. Rudi, K. et al., Direct real-time PCR quantification of *Campylobacter jejuni* in chicken fecal and cecal samples by integrated cell concentration and DNA purification, *Appl. Environ. Microbiol.*, 70: 790, 2004.

43. Duffy, G. et al., The development of a combined surface adhesion and polymerase chain reaction technique in the rapid detection of *Listeria monocytogenes* in meat and poultry, *Int. J. Food Microbiol.*, 49: 151, 1999.

44. Lucore, L.A., Cullison, M.A., and Jaykus, L.A., Immobilization with metal hydroxides as a means to concentrate food-borne bacteria for detection by cultural and molecular methods, *Appl. Environ. Microbiol.*, 66: 1769, 2000.

45. Payne, M.J. et al., The use of immobilized lectins in the separation of *Staphylococcus aureus, Escherichia coli, Listeria* and *Salmonella* spp. from pure cultures and foods, *J. Appl. Bacteriol.*, 73: 41, 1992.

46. Giovannoni, S.J. et al., Tangential flow filtration and preliminary phylogenetic analysis of marine picoplankton, *Appl. Environ. Microbiol.*, 56: 2572, 1990.

47. Cheng, J. et al., Preparation and hybridization analysis of DNA/RNA from *E. coli* on microfabricated bioelectronic chips, *Nat. Biotechnol.*, 16: 541, 1998.
48. Field, K.G. et al., Diversity and depth-specific distribution of SAR11 cluster rRNA genes from marine planktonic bacteria, *Appl. Environ. Microbiol.*, 63: 63, 1997.
49. Millar, B.C. et al., A simple and sensitive method to extract bacterial, yeast and fungal DNA from blood culture material, *J. Microbiol. Methods*, 42: 139, 2000.
50. Boom, R., et al., Rapid and simple method for purification of nucleic acids, *J. Clin. Microbiol.*, 28: 495, 1990.
51. Deggerdal, A. and Larsen, F., Rapid isolation of PCR-ready DNA from blood, bone marrow and cultured cells, based on paramagnetic beads, *Biotechniques*, 22: 554, 1997.
52. Grevelding, C.G. et al., Direct PCR on fruitflies and blood flukes without prior DNA isolation, *Nucleic Acids Res.*, 24: 4100, 1996.
53. Tengs, T. et al., Phylogenetic analyses indicate that the 19'hexanoyloxy-fucoxanthin-containing dinoflagellates have tertiary plastids of haptophyte origins, *Mol. Biol. Evol.*, 17: 718, 2000.
54. Monteiro, L. et al., Complex polysaccharides as PCR inhibitors in feces: *Helicobacter pylori* model, *J. Clin. Microbiol.*, 35: 995, 1997.
55. Arbeli, Z. and Fuentes, C.L., Improved purification and PCR amplification of DNA from environmental samples, *FEMS Microbiol. Lett.*, 272: 269, 2007.
56. Hawkins, T.L. et al., DNA purification and isolation using a solid-phase, *Nucleic Acids Res*, 22: 4543, 1994.
57. Glasel, J.A., Validity of nucleic acid purities monitored by 260nm/280nm absorbance ratios, *Biotechniques*, 18: 62, 1995.
58. Hung, K.C. et al., Use of paraffin wax film in MALDI-TOF analysis of DNA, *Anal. Chem.*, 70: 3088, 1998.
59. Rubsam, L.Z. and Shewach, D.S., Improved method to prepare RNA-free DNA from mammalian cells, *J. Chromatogr. B Biomed. Sci. Appl.*, 702: 61, 1997.
60. Mock, K., Routine sensitive peptide mapping using LC/MS of therapeutic proteins produced by recombinant DNA technology, *Pept. Res.*, 6: 100, 1993.
61. McOrist, A.L., Jackson, M., and Bird, A.R., A comparison of five methods for extraction of bacterial DNA from human faecal samples, *J. Microbiol. Methods*, 50: 131, 2002.
62. Trochimchuk, T., A comparison of DNA extraction and purification methods to detect *Escherichia coli* O157:H7 in cattle manure, *J. Microbiol. Methods*, 54: 165, 2003.
63. Smith, K., Diggle, M.A., and Clarke, S.C., Comparison of commercial DNA extraction kits for extraction of bacterial genomic DNA from whole-blood samples, *J. Clin. Microbiol.*, 41: 2440, 2003.
64. Queipo-Ortuno, M.I. et al., Comparison of seven commercial DNA extraction kits for the recovery of Brucella DNA from spiked human serum samples using real-time PCR, *Eur. J. Clin. Microbiol. Infect. Dis.*, 27: 109, 2008.
65. Aldous, W.K. et al., Comparison of six methods of extracting *Mycobacterium tuberculosis* DNA from processed sputum for testing by quantitative real-time PCR, *J. Clin. Microbiol.*, 43: 2471, 2005.
66. Wilson, D. et al., Comparison of five methods for extraction of *Legionella pneumophila* from respiratory specimens, *J. Clin. Microbiol.*, 42: 5913, 2004.
67. Di Cello, F. et al., Approaches to bacterial RNA isolation and purification for microarray analysis of *Escherichia coli* K1 interaction with human brain microvascular endothelial cells, *J. Clin. Microbiol.*, 43: 4197, 2005.
68. Liu, D., Preparation of *Listeria monocytogenes* specimens for molecular detection and identification, *Int. J. Food Microbiol.*, 122: 229, 2008.
69. Wang, R.F., Cao, W.W., and Johnson, M.G., 16S rRNA-based probes and polymerase chain reaction method to detect *Listeria monocytogenes* cells added to foods, *Appl. Environ. Microbiol.*, 58: 2827, 1992.
70. Rodríguez-Lázaro, D. et al., Rapid quantitative detection of *Listeria monocytogenes* in meat products by real-time PCR, *Appl. Environ. Microbiol.*, 70: 6299, 2004.
71. Besse, N.G. et al., A contribution to the improvement of *Listeria monocytogenes* enumeration in cold-smoked salmon, *Int. J. Food Microbiol.*, 91: 119, 2004.
72. Uyttendaele, M., Van Hoorde, I., and Debevere, J., The use of immuno-magnetic separation (IMS) as a tool in a sample preparation method for direct detection of *L. monocytogenes* in cheese, *Int. J. Food Microbiol.*, 54: 205, 2000.
73. Stevens, K.A. and Jaykus, L.A., Direct detection of bacterial pathogens in representative dairy products using a combined bacterial concentration-PCR approach, *J. Appl. Microbiol.*, 97: 1115, 2004.

74. Isonhood, J., Drake, M., and Jaykus, L.A., Upstream sample processing facilitates PCR detection of *Listeria monocytogenes* in mayonnaise-based ready-to-eat (RTE) salads, *Food Microbiol.*, 23: 584, 2006.

75. Berrada, H. et al., Quantification of *Listeria monocytogenes* in salads by real-time quantitative PCR, *Int. J. Food Microbiol.*, 107: 202, 2006.

76. Shannon, K.E. et al., Application of real-time quantitative PCR for the detection of selected bacterial pathogens during municipal wastewater treatment, *Sci. Total Environ.*, 382: 121, 2007.

77. Herrera, A. and Cockell, C.S., Exploring microbial diversity in volcanic environments: A review of methods in DNA extraction, *J. Microbiol. Methods*, 70: 1, 2007.

78. Tien, C.C., Chao, C.C., and Chao, W.L., Methods for DNA extraction from various soils: Comparison, *J. Appl. Microbiol.*, 86: 937, 1999.

79. Holben, W.E., Isolation and purification of bacterial DNA from soil. *In Methods of Soil Analysis, Part 2. Microbiological and Biochemical Properties*, Eds. Weaver, R.W. et al., pp. 727–752. SSSA Book Series No. 5. Madison, WI: Soil Science Society of America.

80. Yeates, C. et al., Methods for microbial DNA extraction from soil for PCR amplification, *Biol. Proced. Online*, 1: 40, 1999.

81. Krsek, M. and Wellington, E.M.H., Comparison of different methods for the isolation and purification of total community DNA from soil, *J. Microbiol. Methods*, 39: 1, 1999.

82. Selenska, S. and Klingmuller, W., DNA recovery and direct detection of Tn5 sequences from soil, *Lett. Appl. Microbiol.*, 13: 21, 1991.

83. Tsai, Y.L. and Olson, B.H., Rapid method for separation of bacterial DNA from humic substances in sediments for polymerase chain reaction, *Appl. Environ. Microbiol.*, 58: 2292, 1992.

84. Porteous, L.A. and Armstrong, J.L., Recovery of bulk DNA from soil by a rapid, small scale extraction method, *Curr. Microbiol.*, 22: 345, 1991.

85. Hurt, R.A. et al., Simultaneous recovery of RNA and DNA from soils and sediments, *Appl. Environ. Microbiol.*, 67: 4495, 2001.

86. Sessitsch, A. et al., RNA isolation from soil for bacterial community and functional analysis: Evaluation of different extraction and soil conservation protocols, *J. Microbiol. Methods*, 51: 171, 2002.

87. Fleming, J.T., Yao, W.H., and Sayler, G.S., Optimization of differential display of prokaryotic mRNA: Application to pure culture and soil microcosms, *Appl. Environ. Microbiol.*, 64: 3698, 1998.

88. Chang, S., Puryear, J., and Cairney, J., A simple and efficient method for isolating RNA from pine trees, *Plant Mol. Biol. Rep.*, 11: 113, 1993.

89. Felske, A. et al., Direct ribosome isolation from soil to extract bacterial rRNA for community analysis, *Appl. Environ. Microbiol.*, 62: 4162, 1996.

90. Palmer, C.J. et al., Detection of *Legionella* species in reclaimed water and air with the EnviroAmp Legionella PCR kit and direct fluorescent antibody staining, *Appl. Environ. Microbiol.*, 61: 407, 1995.

91. Angles d'Auriac, M. et al., A new automated method for isolation of *Chlamydia trachomatis* from urine eliminates inhibition and increases robustness for NAAT systems, *J. Microbiol. Methods.*, 70: 416, 2007.

92. Shafer, M.A. et al., Comparing first-void urine specimens, self-collected vaginal swabs, and endocervical specimens to detect *Chlamydia trachomatis* and *Neisseria gonorrhoeae* by a nucleic acid amplification test, *J. Clin. Microbiol.*, 41: 4395, 2003.

93. Verkooyen, R.P. et al., Reliability of nucleic acid amplification methods for detection of *Chlamydia trachomatis* in urine: Results of the first international collaborative quality control study among 96 laboratories, *J. Clin. Microbiol.*, 41: 3013, 2003.

94. Belgrader, P. et al., A battery-powered notebook thermal cycler for rapid multiplex real-time PCR analysis, *Anal. Chem.*, 73: 286, 2001.

95. Iqbal, S.S. et al., A review of molecular recognition technologies for detection of biological threat agents, *Biosens. Bioelectron.*, 12: 549, 2000.

96. Bruin, G.J., Recent developments in electrokinetically driven analysis on microfabricated devices, *Electrophoresis*, 21: 3931, 2000.

97. Belgrader, P. et al., A reusable flow-through polymerase chain reaction instrument for the continuous monitoring of infectious biological agents, *Anal. Chem.*, 75: 3446, 2003.

Part III

*Purification of Nucleic Acids
from Fungi*

9 Isolation of Nucleic Acids from Yeasts

Lakshman P. Samaranayake, Nipuna B. Parahitiyawa, and C. Jayampath Seneviratne

CONTENTS

9.1 INTRODUCTION

9.1.1 GENERAL CONSIDERATION

Yeasts are unicellular eukaryotic microorganisms that are classified in the kingdom Fungi rather than form a specific taxonomic or phylogenetic grouping. The kingdom Fungi also includes another group of eukaryotes called true fungi (or filamentous fungi, see Chapter 10). Yeasts are distinguished from other fungi in the kingdom Fungi by their characteristic cell structure, modes of growth, and reproduction. Unlike other fungi, yeasts do not possess the sexual appendages or covered by fruiting bodies and they achieve vegetative reproduction predominantly through fission or budding [1]. In everyday language, the term "yeast" is often used as a synonym for *Saccharomyces cerevisiae*; however, the phylogenetic diversity of yeasts is reflected by their placement in both divisions Ascomycota and Basidiomycota. Furthermore, the budding yeasts (true yeasts) are classified in the order Saccharomycetales.

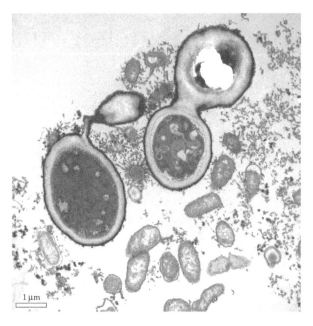

FIGURE 9.1 Transmission electron micrograph of the yeast *Candida albicans* and the Gram negative coliform bacillus *Escherichia coli* depicting the relative thickness of the eukaryotic yeast cell wall and the bacterial cell wall. Also seen is the budding phenomenon of yeasts with a large mother cell (blastospore) and the small daughter cell.

Yeasts represent a growth form, of which about 1500 species (representing only 1% of all yeasts) have been described. Most yeasts reproduce asexually by budding, but a few do so by binary fission (Figure 9.1). Although yeasts are predominantly unicellular, some species with yeast forms may become multicellular through the formation of a string of connected budding cells known as pseudohyphae or true hyphae as seen in most molds (Figure 9.2). Yeasts typically measure 3–4 μm in diameter, but their sizes can vary enormously depending

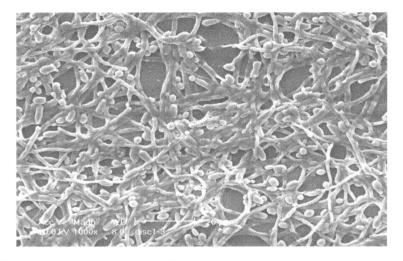

FIGURE 9.2 Scanning electron micrograph of the yeast (blastospore) phase and hyphal phase cells of the human pathogenic yeast *Candida albicans* growing on a denture acrylic surface.

on species, with some yeasts measuring over 40 μm. The morphological characteristics of yeast cell, ascospore, and colony are often used in combination with the physiological characteristics (e.g., the ability to ferment sugars for the production of ethanol) to identify yeast species.

9.1.2 Biological Aspects of Yeasts

Being chemoorganotrophs, yeasts can use organic compounds as a source of energy and do not require sunlight to grow. The main source of carbon for yeasts consists of hexose sugars (e.g., glucose and fructose) and disaccharides (e.g., sucrose and maltose), although some yeast species are capable of metabolizing pentose sugars, alcohols, and organic acids. Yeast species are either obligate aerobes or facultative anaerobes that can obtain oxygen from aerobic cellular respiration or have aerobic methods of energy production. In contrast to bacteria, there are no known obligate anaerobes yeast species that grow only anaerobically.

Yeasts can grow at temperatures ranging from 10°C to 37°C, with most species being adapted to temperatures between 30°C and 37°C. In fact, the optimal temperature for *S. cerevisiae* growth is at about 30°C. Yeasts are largely inactive in the temperature range of 0°C–10°C, and yeast cells can survive freezing under certain conditions, with decreasing viability over time. Yeast cells tend to become stressed and are unable to divide properly at above 37°C, and most of them die above 50°C.

While yeasts are distributed ubiquitously in the environment, they are most frequently isolated from sugar-rich specimens such as fruits and berries (grapes, apples, peaches, etc.) and exudates from plants (plant saps, cacti, etc.). Further, some yeasts are found in association with soil, insects, animals, and humans. In the laboratory, yeasts are usually cultured on solid growth media or liquid broths, which include potato dextrose agar (PDA) or potato dextrose broth, Wallerstein Laboratories Nutrient agar, yeast peptone dextrose (YPD) agar, and Yeast Mould agar or broth. The antibiotic cycloheximide is sometimes added to yeast growth media to help inhibit the growth of *Saccharomyces* yeasts and select for wild/indigenous yeast species.

Even though yeasts have the capacity to undergo asexual and sexual reproductive cycles, the most common mode of vegetative growth in yeast is asexual reproduction by budding or fission. During the budding process, a small bud (or daughter cell) forms on the parent cell, with the nucleus of the parent cell splitting into a daughter nucleus and migrating into the daughter cell. Eventually, the growing bud separates from the parent cell to become a new cell. Yeasts such as *S. cerevisiae* can stably exist as either a haploid or a diploid, the latter is usually formed under stressful condition such as nutrient depletion (especially in special media, such as potassium acetate medium). (Ploidy is the number of homologous sets of chromosomes in a biological cell, with a haploid cell containing one homologous set of chromosomes, and a diploid cell containing two homologous sets of chromosomes.) Both haploid and diploid yeast cells are capable of reproducing by mitosis (i.e., asexual reproduction cycle), with daughter cells budding off of mother cells. Haploid cells can also mate with other haploid cells of the opposite mating type (an "**a**" cell can only mate with an α cell, and vice versa) in sexual reproduction cycle to produce a stable diploid cell. In addition, diploid cells can undergo sporulation, entering meiosis (sexual reproduction) to produce four haploid spores (or ascospores, which are sac-like structures commonly known as asci or singular ascus): two **a** spores and two α spores, which constitute essential components for subsequent mating by yeasts. However, haploid cells cannot undergo meiosis.

In terms of sizes, diploid cells typically measure 5 × 6 μm ellipsoids and haploid cells are 4 μm diameter spheroids. Morphologically, haploid cells form buds that appear adjacent to one another, while diploid cells have buds that appear at the opposite pole. Each mother cell usually forms no more than 20–30 buds, and its age can be ascertained by the number of bud scars left on the cell wall. Furthermore, when grown on agar medium limiting for nitrogen sources, some diploid strains of *S. cerevisiae* may display a markedly different cell and colony morphology commonly known as pseudohyphae, which are notably elongated, with mother–daughter pairs remaining attached to each other. This characteristic pseudohyphal growth extends the branched chains outward from the center of the colony, and invades under the surface of agar medium.

Interestingly, wild type haploid yeast cells can switch mating type between **a** and α, which may result in the formation of both **a** and α mating types in a population previously made up of a single mating type. A combination of mating type switching and a drive for two haploid mating types to form diploid cells will lead to a predominantly diploid colony, irrespective of whether a haploid or diploid cell founded the colony in the first place. However, most laboratory yeast strains do not go through mating type switching as these strains have been altered by deletion of a *HO* gene encoding a DNA endonuclease, which cleaves DNA at the *MAT* locus that governs the sexual behavior of both haploid and diploid cells. This facilitates the stable propagation of haploid yeast, as haploid cells of the **a** mating type will remain **a** cells (and α cells will remain α cells), and will not form diploids.

Most laboratory haploid yeast strains can double in about 90 min when cultured in YPD medium (1% yeast extract, 2% peptone, and 2% glucose) during the exponential phase of growth at the optimum temperature of 30°C, to a maximum density of 2×10^8 cells/mL. The doubling time increases to about 140 min in synthetic media. To achieve a higher growth density, special conditions such as pH control, continuous additions of balanced nutrients, filtered-sterilized media, and extreme aeration that can be delivered in fermenters are required.

9.1.3 ECONOMIC IMPACT OF YEASTS

Yeasts have a number of physiological properties that can be usefully exploited for beverage and bread making. In winemaking, yeast converts the sugars present in grape juice or must into alcohol. Most widely used wine yeasts are strains of *S. cerevisiae*, which can yield up to 18% ethanol. During the fermentation of sugars in the grape juice by yeast, carbon dioxide is generated, which can be trapped to produce so-called sparkling wines. In brewing, *Saccharomyces carlsbergensis* is used in the production of several types of beers including lagers. Yeast, in particular *S. cerevisiae*, is utilized in baking as leavening agent, where it converts the fermentable sugars present in the dough into carbon dioxide, which sets in pockets when baked, giving the baked product a soft and spongy texture. While addition of potatoes, eggs, or sugar in bread dough accelerates the growth of yeasts, salt and fats (e.g., butter) slow down yeast growth. Additionally, *Saccharomyces exiguus* (also known as *S. minor*) is a wild yeast that is occasionally used for baking and is found on plants, fruits, and grains.

Some yeasts can be applied in the field of bioremediation. Yeast species *Yarrowia lipolytica* is known to degrade palm oil mill effluent, TNT (an explosive material), and other hydrocarbons such as alkenes, fatty acids, fats, and oils. *Saccharomyces* yeasts have also been engineered to ferment xylose, which is a major fermentable sugar present in cellulose biomasses such as agriculture residues, paper wastes, and wood chips. This has the potential to make cellulosic ethanol fuel efficiently from more inexpensive feedstock. Yeast (usually deactivated *S. cerevisiae*) is also useful as nutritional supplements, for which it is often referred to as "nutritional yeast" because it contains 50% protein and provides a rich source of B vitamins, niacin, and folic acid. Some probiotic supplements incorporate the yeast *Saccharomyces boulardii* to maintain and restore the natural flora in the large and small gastrointestinal tract. *S. boulardii* has been shown to alleviate the symptoms of acute diarrhea in children, prevent reinfection of *Clostridium difficile*, and reduce bowel movements in diarrhea patients.

As yeasts are tolerant of a relatively low pH (5.0 or lower), they can grow on many types of food, taking advantage of the available sugars, organic acids, and other easily metabolized carbon sources. The growth of yeasts causes the physical, chemical, and sensory properties of a food to change, and spoils the food as a consequence. The yeast of the *Zygosacchromyces* genus is a well-known spoilage yeast within the food industry, as these yeasts can grow in the presence of high sucrose, ethanol, acetic acid, sorbic acid, benzoic acid, and sulfur dioxide concentrations, which represent some of the commonly applied food preservation reagents.

In medicine, the yeast *Cryptococcus neoformans* causes a disease called cryptococcosis, which is a significant debilitating disease in immunocompromised individuals. The cells of this yeast are surrounded by a rigid polysaccharide capsule, which prevent human white blood cells to recognize and engulf the yeast. The yeast-like fungus species *Candida* forms part of commensal flora in the mucus membranes of humans and other warm-blooded animals. However, in physiologically and immunologically stressed individuals, *Candida* yeast cells can sprout a hyphal outgrowth, penetrating the local mucosal membrane and causing irritation and shedding of the tissues. The following *Candida* species are pathogenic to humans (in a descending order of virulence): *C. albicans*, *C. tropicalis*, *C. stelatoidea*, *C. glabrata*, *C. krusei*, *C. parapsilosis*, *C. guilliermondi*, *C. viswanathii*, *C. lusitaniae*, and *Rhodotorula muciaginosa*; where *C. glabrata* is the second most common *Candida* pathogen after *C. albicans*, causing infections of the urogenital tract, and of the bloodstream. Besides being the causative agent in vaginal yeast infections, *Candida* is also a cause of diaper rash and thrush of the mouth and throat. Nonpathogenic yeasts such as *S. cerevisiae* may also be involved in disease processes, as anti-*S. cerevisiae* antibodies (ASCA) have been found in familial Crohn's disease and other forms of colitis.

9.1.4 APPLICATION OF YEASTS IN MOLECULAR BIOLOGY RESEARCH

Compared with other higher level eukaryotes (e.g., mammals), the yeast genome is relatively simple and compact. The size of *S. cerevisiae* genome measures 12,052 bp, which is clustered into 16 chromosomes (ranging in size from 200 to 2,200 kb). A total of 6,183 open-reading frames (ORFs) of over 100 amino acids long are identified, and 5,800 of them are predicted to be protein-coding genes, which situate at 2 kb segments apart. The sizes of protein-coding genes in yeast range from 40 to 4910 codons, with 1.45 kb (or 483 codons) being the average.

Over 95% of yeast RNA are noncoding, most of which are ribosomal RNAs (rRNAs) consisting of 18S, 5.8S, and 25S–28S subunits. Other noncoding RNAs include small nucleolar RNAs (snoRNAs), small nuclear RNAs (snRNAs), transfer RNAs (tRNAs), telomerase RNA, signal-recognition-particle RNAs, and the RNA components of the RNase P and RNase MRP endonucleases. Yeast ribosomal RNA is coded by about 120 copies of a single tandem array on chromosome XII. Besides 262 tRNA genes, yeast chromosomes also possess movable DNA elements and retrotransposons, which amount to 30 copies in most laboratory strains. Yeast mitochondrial DNA encodes components of the mitochondrial translational machinery and about 15% of the mitochondrial proteins. Most *S. cerevisiae* strains harbor dsRNA viruses that account for about 0.1% of total nucleic acid. In addition to three families of RNA viruses (L-A, L-BC, and M), yeast also contains a 20S circular single-stranded RNA with three features: encoding an RNA-dependent RNA polymerase, acting as an independent replicon, and being inherited as a non-Mendelian genetic element.

In spite of having a greater genetic complexity and containing 3.5 times more DNA than bacteria such as *Escherichia coli* cells, yeasts share some important properties with bacteria, which include rapid growth, dispersed cells, the ease of replica plating and mutant isolation, a well-defined genetic system, and most important, a highly versatile DNA transformation system, making them valuable for biological studies. On the other hand, being nonpathogenic, yeasts are much safer than many bacteria to be used as a laboratory tool for biochemical studies. Furthermore, unlike most other microorganisms, strains of *S. cerevisiae* have both a stable haploid and diploid state. This allows recessive mutations to be conveniently isolated and manifested in haploid strains, and complementation tests to be conducted in diploid strains. Additionally, in contrast to most other organisms, integrative recombination of transforming DNA in yeast proceeds exclusively via homologous recombination.

Coupled with yeasts' high levels of gene conversion, homologous recombination has enabled the development of techniques for the direct replacement of genetically engineered DNA sequences into their normal chromosome locations. Also unique to yeast, transformation can be carried out directly with synthetic oligonucleotides, facilitating the production of many altered forms of proteins. Because the cell cycle in yeast is remarkably similar to that in humans, the basic cellular mechanics of DNA replication, recombination, cell division, and metabolism are comparable. Not surprisingly, yeast, especially *S. cerevisiae*, has been the model system for much of molecular genetic research. Indeed, many proteins with important roles in human biology (e.g., cell cycle proteins, signal proteins, and protein-processing enzymes) were first discovered by studying their homologs in yeast.

Three most widely used molecular tools involving yeast are (1) the two-hybrid screening systems for the general detection of protein–protein interactions; (2) the yeast artificial chromosomes (YACs) for cloning large fragments (200–800 kb) of DNA; and (3) expression systems for heterologous proteins. One version of the two-hybrid systems exploits the properties of certain eukaryotic transcription factors (e.g., Gal4p) that have one domain for DNA binding and another domain for transcriptional activation. These two domains are normally on the same polypeptide chain. However, the transcription factor can also function if these two domains are brought together by noncovalent protein–protein interactions. By linking DNA-binding domain to one protein, Yfg1p, and the activation domain to another protein, Yfg2p through gene fusions, the interaction between Yfg1p and Yfg2p brings the DNA-binding and activation domains close together, resulting in the expression of a reporter gene that is regulated by the transcription factor. Another version of the two-hybrid system makes use of the *lexA* operator sequence and the DNA-binding domain from the *E. coli lexA* repressor protein. Namely, the activator domain being a segment of *E. coli* DNA expresses an acidic peptide, which acts as a transcriptional activator in yeast when fused to a DNA-binding domain. The *lexA* transcriptional activator contains a nuclear localization signal that directs the protein into the nucleus. Yeast strains having *lexA* operators upstream of both the *E. coli lacZ* and yeast *LEU2* gene serve as reporter genes. The two-hybrid systems have been frequently used for the following three applications: testing proteins that are believed to interact on the basis of other criteria; defining domains or amino acids critical for interactions of proteins that are already known to interact; and screening libraries for proteins that interact with a specific protein.

YACs are a cloning system that can take up large DNA fragments (ranging from 200 to 800 kb) as compared to bacterial artificial chromosomes or plasmid P1 derived artificial chromosome (BAC or PAC) that can handle 100–200 kb fragments. YAC cloning systems utilize yeast linear plasmids, YLp, that contain homologous or heterologous DNA sequences to function as telomeres (TEL) *in vivo*, in addition to possessing yeast ARS (origins of replication) and CEN (centromeres) segments. As YLp linear plasmids *in vitro* are unable to propagate in *E. coli*, specially developed circular YAC vectors can be used. For instance, a circular YCp vector, containing a head-to-head dimer of *Tetrahymena* or yeast TEL, is resolved *in vivo* after yeast transformation into linear molecules with the free ends terminated by functional TEL. The YLp is maintained at high copy numbers, but it can be lost at high frequency because of its small size. Increasing the size of the YLp by homologous integration *in vivo* or by ligation *in vitro* increases the stability of the plasmid and reduces the copy number to approximately one per cell. The benefits of YAC technology have been demonstrated by the recently developed methods for transferring YACs to cultured cells and to the germ line of experimental animals.

The yeast *S. cerevisiae* is also useful for production of heterologous proteins, and it has some attractive features in comparison with *E. coli*–based protein expression system. Unlike the proteins produced in *E. coli*, proteins produced in yeast lack endotoxins. In some cases such as hepatitis B core antigen, the proteins produced in yeast display a higher activity than those produced in *E. coli*. Furthermore, in contrast with using *E. coli*, several posttranslational processing mechanisms available in yeast allow the expression of human pathogen-associated proteins with appropriate authentic modifications. Such posttranslational modifications include particle assembly, amino terminal acetylation, myristylation, and proteolytic processing. Additionally, heterologous proteins secreted from specially engineering strains are correctly cleaved and folded and are easily harvested from yeast culture media.

The use of either homologous or heterologous signal peptides has facilitated authentic maturation of secreted products by the endogenous yeast apparatus. The utility of yeast-based protein expression systems is highlighted by the fact that the first approved human vaccine, hepatitis B core antigen, and the first food product, rennin, were generated in yeast.

9.1.5 PRINCIPLES AND CURRENT METHODS FOR NUCLEIC ACID ISOLATION FROM YEASTS

Most, if not all, molecular biological procedures involving yeasts including a significant proportion of diagnostic workups in clinical microbiology are dependent on successful isolation of yeast nucleic acids. The starting material in biotechnological methods is often pure and authentic cultures of yeast are grown in specified media. However, in diagnostic procedures this is often contaminated with other microorganisms and host tissue debris. This is of particular concern when there is a scanty sample to start with and cultivation is deemed impossible. Such situations present the worker with the challenge of isolating and purifying the nucleic acid from the little material available.

Isolation of genomic DNA is sufficient for most procedures related to clinical microbiology. In contrast, more advanced biotechnological applications often require isolation of extranuclear DNA such as mitochondrial DNA and yeast plasmids. In addition, isolation of yeast RNA is a vital first step for many applications such as cDNA synthesis, reverse transcription PCR (RT-PCR), quantitative PCR (qPCR), microarray differential display, RNase protection assay, primer extension, and Northern blot. Total RNA may be used for isolation of various subtypes of RNAs. In recent years, real-time PCR has emerged as a powerful tool to identify and quantify gene expression from small amount of RNA. Important uses of different nucleic acids from yeasts are summarized in Table 9.1.

Of the yeasts, the most studied organisms are *Saccharomyces cerevisiae* and *Candida albicans*. Being eukaryotic, these organisms are ideal for the study of much higher other eukaryotes such as plants and mammals. In particular, completion of genome sequences of *S. cerevisiae* [2] and *C. albicans* [3,4] has paved the way for many downstream applications of yeast nucleic acids. A good understanding of the organization of nucleic acids of these organisms is useful for those who are embarking on molecular studies of these floras.

This chapter focuses on the yeast *S. cerevisiae*, and related interbreeding species. The fission yeast *Schizosaccharomyces pombe*, which is only distantly related to *S. cerevisiae*, has equally

---

TABLE 9.1
Application of Yeast Nucleic Acids in Medicine and Molecular Biology

Nucleic Acid	Application	References
Genomic DNA	Serotype identification of *Cryptococcus neoformans* by multiplex PCR	[26]
	Identification of medically important *Candida* and non-*Candida* yeast species by oligonucleotide array	[27]
Mitochondrial DNA	A yeast model of the neurogenic ataxia retinitis pigmentosa (mutation in a mitochondrial gene)	[28]
Plasmid	As expression vector for *Canavalia brasiliensis* lectin: a model for the study of protein splicing	[29]
Total RNA	Gene expression analysis of real-time PCR	[17]
mRNA	Microarray analysis of gene expression	[12,13]
tRNA	Identifying amino acids attached to RNA	[30]
Total RNA	Immune response dendritic cells pulsed with RNA	[14]

important features, but is not as well characterized. The general principles of the numerous classical and modern approaches for investigating *S. cerevisiae* are described, and the explanation of terms and nomenclature used in current yeast studies is emphasized. This chapter should be particularly useful to the uninitiated who are exposed for the first time to experimental studies of yeast. Detailed protocols are described in the primary literature and in a number of reviews in the books listed in the references. The original citations for the material covered in this chapter can also be found in these comprehensive reviews.

9.1.5.1 Principles of Yeast Nucleic Acid Isolation

The successful recovery of the nucleic acids devoid of contaminating material requires four essential steps; effective disruption of cells, denaturation of nucleoprotein complexes, inactivation of endo genous DNase/RNase activity, and removal of contaminating proteins.

The major challenge of nucleic acid extraction from yeasts is their rigid cell wall. Yeast cell wall is composed of two main layers: an inner layer of 1,3-β-glucan, 1,6-β-glucan, chitin, and an outer layer of densely packed mannoproteins and polysaccharides [6]. This architecture makes the yeast cell wall a robust structure, much more than that of the lipid bilayered mammalian cell membranes or the peptidoglycan walls of bacteria. Therefore, any successful nucleic acid isolation protocol warrants a combination of strategies to break this robust yeast cell wall first (Table 9.2). For example, enzymes such as lyticase, chitinase, zymolase, and gluculase, which partially disrupt the cell wall are often used to generate spheroplasts as the first step of cell breakage. Subsequent treatment with denaturing agents like sodium dodecyl sulfate (SDS) denatures cytosolic proteins and lipid membranes to

TABLE 9.2
Digestion of the Fungal Cell Walls

Method	Brief Description (Remarks)	Kits/In-House Protocol Subsequent to the Cell Wall Disruption Method [References]
Proteinase K	Fungal cells are treated with proteinase K (1 mg/mL) in a sorbitol buffer and incubated for 30 min at 56°C	QIAamp DNA Mini Kit [21,22]
Lyticase	20 mg/mL lyticase solution is added to the cell pellet (~10^8 cells) and incubated for 2 h at 37°C	Promega Wizard genomic DNA extraction kit/UltraClean microbial DNA isolation kit [23]
Zymolyase	Alkali treated cell pellets are incubated with 500 μL of zymolyase solution (300 μg/mL zymolyase) for 45 min at 37°C	[24]
Acid treatment	300 μL of conc. HCl added to the yeast suspension and incubated overnight at room temperature	[21]
Alkali treatment	300 μL of 5 M NaOH added to the yeast suspension and incubated overnight at room temperature	[21]
Sonication	Specimens sonicated on ice for 90 s at 150 Hz in 0.5 mL of the DNA extraction buffer (low efficacy compared to enzymatic and chemical methods)	Phenol chloroform method [21,25]
Glass beads and liquid N₂ freezing	Crushing cells with glass beads in liquid nitrogen before RNA extraction by the hot phenol method	Hot acid phenol method [17]
Glass beads	0.3 g of 500 μm acid-washed glass beads is added to the cell pellet and vortexed at high speed for 2 min	NucleoSpin RNA II kit [18]
Mortar and pestle grinding	Cells are homogenized through grinding with mortar and pestle (comparable or even better efficacy than enzymatic and chemical methods)	[21]

facilitate the purification. After the cell wall is dealt with, attempts are made to purify the nucleic acids from a homogenous mixture of cellular debris, which primarily consists of proteins. There are two alternative approaches that may be employed to isolate the nucleic acid thereafter. The first is to separate the proteinaceous debris using organic extractants such as phenol/chloroform which leaves the nucleic acids in an aqueous compartment. The second method is to trap the nucleic acid onto a membrane to be eluted thereafter by a buffer, yielding the pure product. Precipitation methods are the cornerstone of most of the in-house purification protocols as well as commercially available extraction kits. The latter uses either centrifugation or density gradients separations. Membrane or columns made of silica or cellulose are almost exclusively employed in commercial kits.

Both DNA and RNA tend to lose some features of their functionality when contaminated by one another. Thus, it is important to have a DNA extract devoid of RNA or vice versa. Therefore, RNase is used in DNA extraction while DNase is used in RNA extractions, in order to get rid of contaminating RNA or DNA, respectively.

Extraction of mitochondrial DNA demands a more meticulous approach than that of genomic DNA isolation. This is because a relatively abundant amount of yeast cells is required to fractionate and enrich the mitochondrial portion through fractional centrifugation [7]. The quality of the DNA extracted with commercial kits is shown to be superior to the quality of DNA extracted with conventional methods such as boiling and mechanical disruption [8].

Yeast RNA has been traditionally extracted using hot acid phenol method [9,10]. However, several modifications of the original method are available with marginal variations of the protocol among different groups. Phenol is toxic and care must be taken to avoid spoilage and direct contact. Due to these and related environmental constraints, use of phenol-based extraction methods has become less popular. An alternative protocol suggested for successful RNA extraction from the yeast *S. cerevisiae* is to boil the organisms in SDS and subsequent precipitation with NaCl [11].

For the purpose of mechanical disruption of yeast cell walls for RNA extraction, a bead mill is necessary especially for large-scale sample processing. However, for medium- or small-scale extractions, acid-washed glass beads in the presence of chaotropic agents have shown to be effective [12,13]. Acid-washed glass beads are prepared by soaking in concentrated nitric acid for 1 h, followed by extensive washing with deionized water, and oven drying. The glass beads should be chilled on ice prior to use.

Alternatively, repeated cycles of freezing in liquid nitrogen and thawing have also been used as a method to disrupt the cell wall [14]. RNA extraction protocol developed for *S. cerevisiae* [15,16] has also been successfully adopted for *C. albicans* [17].

Commercially available kits from a number of manufacturers such as Promega, QIAGEN RNeasy, Trizol, and Ambion use either enzymatic lysis or mechanical disruption of the cell walls singly or in combination. Moreover, most commercial kits use spin columns to isolate RNA. Aforementioned kits have been extensively used for yeast RNA extraction procedures covering a wide range of downstream applications.

Protocols involving prior incubation with enzymes or chemicals for an extensive period of time may impede quality of mRNA for downstream applications because of its short half-life. To counter this, a new protocol has been developed recently incorporating two purification strategies, i.e., acid phenol extraction and binding to a silica matrix into one shortening the whole procedure to be completed in less than 90 min [18].

Purified total RNA can subsequently be used to isolate mRNA, rRNA, or tRNA. For instance, there are commercially available kits for mRNA isolation that can be used in applications such as expression profiling [12,13,19].

A few useful notes:

1. Phenol is used in many extraction protocols for deproteinization of nucleic acids. Most proteins are relatively soluble in phenol than in the aqueous phase and conversely

nucleic acids are more soluble in aqueous phase. Phase partitioning of nucleic acids is also pH dependent. At pH 4–6, DNA will be retained in the organic phase and inter-phase leaving the RNA in the aqueous phase. Therefore, isolation of RNA is often done with acid–phenol. The pH dependence of DNA phase partitioning makes it necessary to raise the pH prior to extraction. Therefore, alkaline pH buffered to pH 7–9 is used in DNA isolation.

2. Addition of chloroform increases the efficiency of extraction due to its ability to denature protein and keeping them in the organic phase. Furthermore, it aids the removal of lipids.

3. Isoamyl alcohol is added to the phenol:chloroform to prevent foaming.

4. Do not let the RNA pellet dry completely, as this greatly decreases its solubility. Avoid drying the pellet by centrifugation under vacuum.

5. For larger cell pellets doubling reagent volumes yield better results.

6. It is necessary to alter temperature point and cycles empirically rather adhere to what protocols site.

7. Centrifugation at 4°C allows better precipitation of the proteins, thus increasing the recovery of DNA. However, bear in mind that columns are not compatible with lower temperatures.

9.2 METHODS

The general reagents, equipment, and media that are needed for nucleic acid isolation from yeasts are summarized in Table 9.3.

TABLE 9.3

General Reagents, Equipment, and Media for Nucleic Acid Isolation from Yeasts

Reagents	Equipment	Media
1 M sorbitol	Microcentrifuge tubes 200 and 500 and 1500 µL and tube racks and storage boxes	YPD broth
20 mg/mL lyticase	37°C and 55°C water bath	Sabouraud's dextrose agar
β-mercaptoethanol	Spectrophotometer	YNB broth and agar
0.15 M NaCl/ 0.1 M EDTA solution	Hemocytometer	
RNase A (10 mg/mL)	Orbital shaker	
Lyticase (20 mg/mL)	Incubator 37°C	
Ethanol and isopropanol	Autoclave	
Proteinase K (20 mg/mL)	Wire loops	
Phenol:chloroform:isoamyl alcohol	Bunsen burner	
10% SDS	Refrigerated centrifuge	
SE buffer (1.2 M sorbitol; 0.1 M EDTA pH 7.5)	Micropipettes and tips (10, 20, 100, 200, and 1000 µL)	
Chloroform	DNA and RNA workstations with laminar flow	
Phosphate buffered saline pH 7.2	Refrigerators (−20°C and −70°C)	
Ice cubes	Biological and chemical waste disposal bins	
50 mM EDTA (pH 8.0)	Thermal blocks	
Molecular weight markers and bromophenol blue	Ultraviolet illuminator/gel visualization system	
Agarose	Gel electrophoresis tanks	
Ethidium bromide	Quartz cuvettes	
Tris EDTA buffer		
Absolute isopropanol	Vortex mixer	

9.2.1 Basic Laboratory Requirements, Equipments, and Organization

It is imperative to organize the basic laboratory features and hardware prior to attempting nucleic acid isolation. Biosafety and contamination prevention go hand in hand and it is best to have a dedicated work area for nucleic acid extractions. In particular, RNA extraction demands more meticulous housekeeping and therefore, it is advisable to segregate DNA and RNA workstations. Apart from the specific reagents mentioned in different protocols below, basic laboratory equipments such as two dedicated sets of micropipettes for RNA and DNA are necessary. At least two water baths with temperature control, and $-20°C$ and $-70°C$ freezers, thermal blocks, and a spectrophotometer comprise the bare necessities.

We outline below widely used DNA and RNA extraction protocols for *C. albicans* that have yielded good results in our hands. These protocols can be successfully employed for other yeast species such as *S. cerevisiae* with minor modifications.

9.2.2 Preparation of Yeast Suspensions

In this chapter, we use *Candida albicans* as a model for isolation of yeast DNA and RNA.

1. Subculture the selected *C. albicans* isolate on Sabouraud's dextrose agar at $37°C$ for $18 h$.
2. Harvest a loopful of the growth and inoculate into yeast nitrogen base (YNB) medium supplemented with $50 mM$ glucose and, incubate at $37°C$ in an orbital shaker at $75 rpm$ for $18–24 h$.
3. Harvest the late exponential phase yeast cells through two rounds of washing with $20 mL$ of $0.1 M$ phosphate buffered saline (PBS pH 7.2).
4. Density of the yeast cell suspensions are then adjusted using a spectrophotometer (McFarland standards) or counting in a counting (hemocytometric) chamber.

9.2.3 Extraction of Genomic DNA from Yeasts

9.2.3.1 Wizard Genomic DNA Extraction Kit

The following procedures are based on Wizard genomic DNA extraction kit (Promega Corporation, Madison, Wisconsin):

1. Prepare the *Candida* cell suspensions as previously described.
2. Centrifuge at $13,000–16,000 g$ for $5 min$ to pellet the cells. Remove the supernatant.
3. Resuspend the cells thoroughly in $293 \mu L$ of $50 mM$ EDTA.
4. Add $7.5 \mu L$ of $20 mg/mL$ lyticase and gently pipette, four times, to mix.
5. Incubate the sample at $37°C$ for $30–60 min$ to digest the cell wall. Cool to room temperature.
6. Centrifuge the sample at $13,000–16,000 g$ for $2 min$ and then remove the supernatant.
7. Add $300 \mu L$ of nuclei lysis solution to the cell pellet and gently pipette to mix.
8. Add $100 \mu L$ of protein precipitation solution and vortex vigorously at high speed for $20 s$.
9. Let the sample sit on ice for $5 min$.
10. Centrifuge at $13,000–16,000 g$ for $3 min$.
11. Transfer the supernatant containing the DNA to a clean $1.5 mL$ tube containing $300 \mu L$ of room temperature isopropanol.
12. Gently mix by inversion until the thread-like strands of DNA form a visible mass.
13. Centrifuge at $13,000–16,000 g$ for $2 min$.
14. Carefully decant the supernatant and drain the tube on a clean absorbent paper. Add $300 \mu L$ of 70% ethanol at room temperature and gently invert the tube several times to wash the DNA pellet.

15. Centrifuge at 13,000–16,000 g for 2 min and carefully aspirate all of the ethanol.
16. Drain the tube on clean absorbent paper and allow the pellet to air-dry for 10–15 min.
17. Add 50 µL of DNA rehydration solution.
18. Add 1.5 µL of RNase solution to the sample. Vortex the sample for 1 s. Centrifuge briefly in a microfuge for 5 s and incubate at 37°C for 15 min.
19. Rehydrate the DNA by incubating at 65°C for 1 h, periodically mixing the solution by gently tapping the tube. Alternatively, rehydrate the DNA by incubating the solution overnight at room temperature or at 4°C.
20. Store the DNA at 2°C–8°C.

9.2.3.2 Gentra Puregene Yeast/Bact Kit

The following procedures are based on Gentra Puregene Yeast/Bact kit (QIAGEN, GmBH, Germany):

1. Prepare cell suspensions as described earlier.
2. Transfer 1 mL of the cell suspension to a 1.5 mL tube on ice.
3. Centrifuge at 13,000–16,000 g for 5 min to pellet cells. Carefully discard the supernatant by pipetting or pouring.
4. Add 300 µL of cell lysis solution, and pipette up and down.
5. Add 1.5 µL of lytic enzyme solution, and mix by inverting 25 times. Incubate at 37°C for 30 min.
6. Centrifuge at 13,000–16,000 g for 5 s, and carefully discard the supernatant by pipetting or pouring.
7. Add 300 µL of cell lysis solution, and pipette up and down gently to lyse the cells.
   (Note: Vigorous pipetting can damage the DNA.)
8. Add 100 µL of protein precipitation solution, and vortex vigorously for 20 s at high speed.
9. Centrifuge at 13,000–16,000 g for 3 min.
   (Note: The precipitated proteins should form a tight, white pellet. If the protein pellet is not tight, incubate on ice for 5 min and repeat the centrifugation.)
10. Pipette 300 µL of 100% isopropanol into a clean 1.5 mL tube and add the supernatant from the previous step by pouring carefully. Pay attention not to dislodge the protein pellet during pouring.
11. Mix by gently inverting 50 times.
12. Centrifuge for 1 min at 13,000–16,000 g.
13. The DNA will be visible as a small, white pellet.
14. Incubate at 65°C for 1 h to dissolve the DNA.
15. Incubate at room temperature overnight with gentle shaking. Ensure tube cap is tightly closed to avoid leakage. Samples can then be centrifuged briefly and transferred to a storage tube.

9.2.3.3 QIAamp DNA Mini Kit for Yeasts

The following procedures are based on QIAamp DNA Mini Kit for yeasts (QIAGEN, Germany):

1. Prepare the yeast cell suspension as mentioned.
2. Harvest the cells by centrifuging at 13,000–16,000 g for 5 min.
3. Resuspend the pellet in 600 µL sorbitol buffer. (1 M sorbitol, 100 mM EDTA, 14 mM β-mercaptoethanol). Add 200 U zymolase or lyticase and incubate at 30°C for 30–60 min.
4. Pellet the spheroplasts by centrifuging for 5 min at 13,000–16,000 g.
5. Resuspend the spheroplasts in 180 µL buffer ATL.

6. Add 20 μL Proteinase K, mix by vortexing, and incubate at 56°C until the cells are completely lysed. Vortex occasionally during incubation to disperse the cells, or place in a shaking water bath.

7. Briefly centrifuge the tube to remove drops from inside the lid. Continue with Step a, or if RNA-free genomic DNA is required, continue with Step b below:

   (a) Add 200 μL buffer AL to the sample, mix by pulse-vortexing for 15 s, and incubate at 70°C for 10 min. Briefly centrifuge the 1.5 mL microcentrifuge tube to remove drops from inside the lid.

   (b) First add 4 μL RNase A (100 mg/mL), mix by pulse-vortexing for 15 s, and incubate for 2 min at room temperature. Briefly centrifuge the 1.5 mL tube to remove drops from inside the lid before adding 200 μL buffer AL to the sample. Mix again by pulse-vortexing for 15 s, and incubate at 70°C for 10 min. Briefly centrifuge the 1.5 mL microcentrifuge tube to remove drops from inside the lid.

8. Add 200 μL ethanol (96%–100%) to the sample, and mix by pulse-vortexing for 15 s. After mixing, briefly centrifuge the 1.5 mL tube to remove drops from inside the lid.

9. Carefully apply the mixture from Step 8 (including the precipitate) to the QIAamp spin column (in a 2 mL collection tube) without wetting the rim. Close the cap, and centrifuge at 6000 g for 1 min. Place the QIAamp spin column in a clean 2 mL collection tube and discard the tube containing the filtrate.

10. Carefully open the QIAamp spin column and add 500 μL buffer AW1 without wetting the rim. Close the cap, and centrifuge at 6000 g for 1 min. Place the QIAamp spin column in a clean 2 mL collection tube and discard the collection tube containing the filtrate.

11. Carefully open the QIAamp spin column and add 500 μL buffer AW2 without wetting the rim. Close the cap and centrifuge at full speed for 3 min. Continue directly with Step 12, or to eliminate any chance of possible buffer AW2 carryover, perform Step 11a.

    (a) Optional: Place the QIAamp spin column in a new 2 mL collection tube and discard the collection tube containing the filtrate. Centrifuge at 20,000 g for 1 min.

12. Place the QIAamp spin column in a clean 1.5 mL microcentrifuge tube and discard the collection tube containing the filtrate. Carefully open the QIAamp spin column and add 200 μL buffer AE or distilled water. Incubate at room temperature for 1 min, and then centrifuge at 6000 g for 1 min to elute the purified DNA.

13. Repeat the centrifugation once to fully recover the DNA.

14. Store the DNA at −20°C.

9.2.3.4 In-House Method

Procedure

1. Prepare the *Candida* cell suspensions as previously described.

2. Centrifuge at 13,000–16,000 g for 5 min to pellet the cells. Remove the supernatant and resuspend in 1 mL 1 M sorbitol.

3. Repeat Step 2 once.

4. Centrifuge at 16,000 g for 5 min.

5. Resuspend the pellet in 1 mL SE buffer (1.2 M sorbitol; 0.1 M EDTA, pH 7.5).

6. Add 1 μL β-mercaptoethanol and 15 μL lyticase (20 mg/mL).

7. Incubate in a water bath at 37°C for 1 h with inverting at each 10 min intervals.

8. Collect the pellet (spheroplasts) by centrifugation at 16,000 g for 1 min, wash once with 1 mL SE, and resuspend in 0.4 mL of 0.15 M NaCl/0.1 M EDTA solution.

9. Add 10 μL of Proteinase K (20 mg/mL), 50 μL SDS (10%), and 40 μL of RNase A (10 mg/mL) and incubate in a water bath at 55°C for 1.5 h.

10. Centrifuge at 16,000 g for 15 min and transfer 400 µL of supernatant to a new microcentrifuge tube.
11. Add 400 µL of phenol:chloroform:isoamyl alcohol (25:24:1) vortex mix and centrifuge at 16,000 g for 3 min.
12. Transfer the upper aqueous layer to a new 1.5 mL tube taking care not to disturb the interface between the two layers.
13. Repeat Steps 11 and 12 once.
14. Add 400 µL of chloroform, vortex mix, and centrifuge at 16,000 g for 3 min.
15. Transfer the upper aqueous layer to a new microcentrifuge tube taking care not to disturb the interface between the two layers.
16. Add 180 µL of absolute isopropanol, vortex mix, and let stand at room temperature for 30 min.
17. Centrifuge at 16,000 g for 15 min.
18. Discard the supernatant taking care not to disrupt the DNA pellet and wash with 70% ethanol.
19. Centrifuge at 16,000 g for 5 min, remove supernatant, and dry the tube in a heat block at 55°C for 5 min.
20. Dissolve the DNA in 50 µL of TE buffer and store at 4°C.

9.2.4 Extraction of RNA from Yeasts

9.2.4.1 Hot Acid Phenol Extraction

Hot acid phenol extraction has been used traditionally to extract RNA [20].

Materials

1. Acid phenol: Heat 100 g of crystalline phenol to 65°C in fresh bottle, and add 20 mL of buffer containing 10 mM sodium acetate (pH 5.1), 50 mM NaCl, and 1 mM EDTA (pH 8.0). Mix the content of the bottle by stirring. Let the liquefied phenol cool to room temperature and add another 10 mL of buffer. Wrap the bottle in aluminium foil and store at 4°C. Note: Acid phenol should be freshly prepared each time.
2. AE buffer: 50 mM sodium acetate and 10 mM EDTA (pH 5).
3. ANE: 10 mM sodium acetate, 2 mM EDTA, and 100 mM NaCl (pH 6).
4. Phenol/AE: Equilibrate liquefied phenol with an equal volume of AE buffer. Store at 4°C.
5. Phenol:Chloroform/ANE: Mix 50% phenol/AE with 50% $CHCl_3$ and 0.25% 8-hydroxy-quinoline equilibrated with ANE buffer. Store at 4°C.

Procedure

1. Prepare the *Candida* cell suspensions as described above. Centrifuge at 13,000–16,000 g for 5 min to pellet the cells. Remove the supernatant.
2. Resuspend the cell pellet in appropriate volume of AE buffer.
   (Note: For a small pellet in a 1.5 mL tube 400 µL of buffer will be sufficient.)
3. Add 40 µL 10% SDS. Vortex briefly. Immediately add 500 µL hot phenol/AE. Vortex for 15 s.
4. Incubate at 65°C for 5 min. Vortex for 5 s at every 30 s.
5. Cool to room temperature by placing the tube in ice.
6. Centrifuge for 10 min at 6000 g at room temperature.
7. Use pipette to remove lower phenol (organic) layer leaving behind pellet, interphase layer, and aqueous supernatant.
8. To remove the proteins, the aqueous phase should be treated at least three times with acid phenol (repeat Steps 3–7).
9. Transfer aqueous supernatant to fresh 1.5 mL tube. Estimate volume.

10. Extract with equal volume of phenol:CHCl₃/ANE. Vortex for 1 min at room temperature. Spin for 30 s.

11. Transfer aqueous supernatant to a 1.5 mL tube. Extract with CHCl₃:isoamyl alcohol (24:1). Spin the tube.

12. Transfer aqueous supernatant to fresh 1.5 mL tube. Volume should be nearly 400 μL.

13. Add 1/10 volume of 3 M sodium acetate, pH 5. Vortex. Add 2.5 volumes of ethanol. Vortex. Hold at −20°C for 30 min or more.

14. Spin in 1.5 mL at 4°C at top speed for 20 min.

15. Wash the pellet with 75% diethylenepyrocarbonate (DEPC)-treated ethanol and dry.

16. Resuspend RNA in nuclease-free water and store at −20°C.

9.2.4.2 Promega SV Total RNA Isolation Kit

A commercially available system, SV Total RNA Isolation System (Promega) can be utilized for isolation of total RNA from *C. albicans*, although the system is not specific for fungi. In this system, breakage of *Candida* cell walls is achieved through a chemical method. Dilution of yeast cells in the presence of guanidine thiocyanate inactivates RNase and causes selective precipitation of cellular proteins, and RNA remains intact in the solution. After separation of lysate containing proteins and cellular debris by centrifugation, RNA can be selectively absorbed into a silica membrane. Contaminating genomic DNA is eliminated by applying RNase-free DNase I.

Procedure

1. Prepare the cell suspensions as described above and harvest the cell pellet by centrifugation.

2. Resuspend the pellet in 100 μL of the solution prepared from 1M sorbitol, 0.1 M EDTA, mix gently by pipetting up and down several times. Note: If the cell pellet is large, it is advisable to double the volume of each solution to make it 200 μL.

3. Add 0.1 μL of 0.1% β-mercaptoethanol and 3 μL of lyticase (20 mg/mL) and mix gently. Note: If the volume in the first step is doubled these volumes should also be doubled. Pipette the chemicals into the solution.

4. Incubate at 30°C for 30 min until the solution appears clear. Note it is best to keep the tubes for 45 min, in particular when the cell density is higher.

5. Add 75 μL of RNA lysis buffer and mix gently. Volumes should be doubled accordingly. RNA lysis buffer is prepared by adding 1 mL of β-mercaptoethanol to 50 mL of SV RNA lysis buffer and stored at 4°C.

6. Add 350 μL of RNA dilution buffer (blue colour), volume should be doubled accordingly. Mix by inversion and centrifuge at 13,000–16,000 g for 10 min.

7. Transfer the cleared lysate solution to fresh 1.5 mL tube, avoid disturbing the pellet.

8. Add 200 μL of 95% ethanol to the cleared lysate, and mix by pipetting three to four times. There is no need to double the volume from this step onward.

9. Transfer 700 μL of the mixture to the spin column assembly each time. If the volume is more than 700 μL, do this in two steps.

10. Centrifuge at 13,000–16,000 g for 1 min.

11. Take the spin basket out of the spin column assembly and discard the solution. Add 600 μL of RNA wash solution to the spin column assembly and centrifuge at 13,000–16,000 g for 1 min.

12. Discard the solution and add 50 μL of DNase solution directly into the membrane inside the spin basket. Care must be taken not to touch the membrane with pipette tip.

13. Incubate for 15 min at room temperature.

14. Add 200 μL of DNase stop solution. DNase stop solution is prepared by adding 8 mL of 95% ethanol to the bottle containing 5.3 mL concentrated SV DNase stop solution provided by the manufacturer.

15. Add 600 μL of RNA wash solution and centrifuge at 13,000–16,000 g for 1 min.
16. Empty the tubes and add 250 μL of RNA wash solution and centrifuge at 13,000–16,000 g for 2 min.
17. Transfer the spin basket to new 1.5 mL tube and add 100 μL of nuclease-free water making sure that the membrane is completely covered with water.
18. Centrifuge at 13,000–16,000 g for 1 min.
19. Remove the spin basket and discard the column.
20. Purified RNA should be stored in −70°C.

9.2.4.3 Single-Step RNA Isolation

A single-step RNA isolation method developed by Chomczynski and Sacchi has been used frequently for total RNA isolation of both cultured and clinical samples [12,13]. A few commercial suppliers such as Invitrogen (Carlsbad, CA, USA) (TRIzol) and Peqlab, Erlangen, Germany (peq-GOLD RNA-Pure) produce commercial kits based on this method.

Presented below is an outline of the extraction of *C. albicans* total RNA by using the single-step RNA isolation method. In brief, harvested yeast cell pellet is homogenized in a denaturing solution containing guanidine thiocyanate. The homogenate is mixed sequentially with sodium acetate, phenol, and finally chloroform/isoamyl alcohol. The resulting mixture is centrifuged, yielding an upper aqueous phase containing RNA. Following isopropanol precipitation, the RNA pellet is redissolved in denaturing solution (containing guanidine thiocyanate), reprecipitated with isopropanol, and washed with 75% ethanol.

Reagents

1. Denaturing solution
   Stock solution: Mix 293 mL water, 17.6 mL of 0.75 M sodium citrate (pH 7.0), and 26.4 mL of 10% (w/v) N-lauroylsarcosine (Sarkosyl). Add 250 g guanidine thiocyanate and stir at 60°C–65°C to dissolve. Store up to 3 months at room temperature.
   Working solution: Add 0.35 mL β-mercaptoethanol to 50 mL of stock solution. Final concentrations are 4 M guanidine thiocyanate, 25 mM sodium citrate, 0.5% Sarkosyl, and 0.1 M β-mercaptoethanol.
2. Sodium acetate 2 M (pH 4.0): Add 16.42 g sodium acetate (anhydrous) to 40 mL water and 35 mL glacial acetic acid. Adjust solution to pH 4 with glacial acetic acid and dilute to a volume of 100 mL with water.
3. Water saturated phenol: Dissolve 100 g phenol crystals in water at 65°C. Aspirate the upper water phase and store up to 1 month at 4°C. Note: Buffered phenol should not be used in place of water-saturated phenol.
4. 49:1 (v/v) chloroform/isoamyl alcohol.
5. 100% isopropanol.
6. 75% ethanol (prepared with DEPC-treated water).
7. DEPC-treated water.

Procedure

1. Prepare the *Candida* suspension as mentioned above.
2. Harvest the cells by maximum centrifugation for 10 min.
3. Add 1 mL of denaturing solution to the cell pellet and mix by repetitive pipetting.
4. Transfer the homogenate into a 1.5 mL tube.
5. Add 0.1 mL of 2 M sodium acetate pH 4, and mix thoroughly by inversion. Add 1 mL water-saturated phenol, mix thoroughly, and add 0.2 mL of 49:1 chloroform/isoamyl alcohol and mix thoroughly and incubate the suspension for 15 min at 4°C.

6. Centrifuge at 13,000–16,000 g in 4°C for 20 min. Transfer the upper aqueous phase to a clean tube.
7. Precipitate the RNA by adding 1 mL of isopropanol. Incubate the samples for 30 min at −20°C. Centrifuge at 13,000–16,000 g in 4°C for 10 min and discard supernatant.
8. Resuspend the RNA pellet in 75% ethanol, vortex, and incubate for 10–15 min at room temperature.
9. Centrifuge at 13,000–16,000 g, in 4°C for 5 min, and discard supernatant.
10. Dissolve the RNA pellet in 100 μL of DEPC-treated water. Store RNA dissolved in water at −70°C.

9.2.5 ASSESSING THE INTEGRITY, CONCENTRATION, AND PURITY OF EXTRACTED NUCLEIC ACIDS

Genomic DNA extracts from yeasts generally show up in an ethidium bromide stained 1% agarose gel as a sharp and clear band below the loading well. Any other band would suggest the presence of RNA and smears mean the contamination with proteins or degraded DNA. Integrity of the extracted total RNA is also evaluated similarly. Intact RNA should be visible as three sharp, clear bands, representing 28S and 18S, 5–5.8S species. Intensity of the 28S rRNA band should be approximately as twice as the 18S rRNA band and this is a good indication of RNA integrity. An example of an agarose gel electrophoresis of total RNA extracted from *C. albicans* is shown in Figure 9.3.

Accurate determination of concentration and purity is indispensable for downstream applications of both DNA and RNA. Concentration is commonly obtained by spectrophotometric measurement of the optical absorbance at 260 nm wavelength (A_{260}). DNA concentration (μg/mL) is estimated by multiplying the A_{260} value by appropriate dilution factor and 50. On the other hand, RNA concentration (μg/mL) is calculated by multiplying A_{260} value by 40. The purity of the sample is estimated by the A_{260}/A_{280} ratio. Pure DNA should have an A_{260}/A_{280} ratio of more than 1.7 while RNA should have an A_{260}/A_{280} absorbance ratio in excess of 2.0. Cuvettes made of quartz are traditionally being used for spectrophotometry, but novel standalone spectrophotometers such as "Nanodrop" yield accurate results while negating the cumbersome procedure of loading and cleaning the cuvettes.

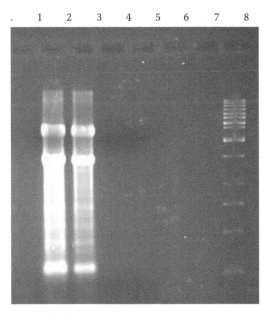

FIGURE 9.3 Agarose gel electrophoresis of the RNA extracted from *C. albicans* at different cell densities: lane 1, negative control; lanes 2–6, *C. albicans* cell suspensions at 1×10^7, 1×10^6, 1×10^5, and 1×10^4 cells/mL, respectively; lanes 6–7, blank wells; and lane 8, molecular weight marker (GeneRuler®, MBI Fermentas).

9.2.6 Storage of Extracted DNA and RNA

DNA, being relatively more robust than RNA, can be safely stored at 4°C. However, storage at −20°C prolongs the shelf life considerably. Alternatively, freeze dried (lyophilized) DNA have at least 1 year stability in room temperature provided that these are stored away from exposure to light.

RNA may be stored in a number of ways. However, it should be borne in mind that frequent freeze–thaw cycles tend to degrade the RNA even with best possible care. Therefore, it is better to preplan the experiments and use the sample as fresh as possible. Furthermore, for short-term storage, nuclease-free water either with or without chelating agents such as 0.1 mM EDTA or 1 mM Tris HCl buffer may be used. For long-term storage of RNA, addition of chelating agents such as EDTA is necessary to prevent inducing nonspecific cleavages by ions like Mg^{2+}. In this way, RNA is generally stable at −80°C for up to a year without degradation.

9.3 FUTURE DEVELOPMENT TRENDS

Successful isolation of DNA or RNA could be achieved through a number of commercial kit-based or in-house protocols. The choice of method is largely governed by the nature of downstream application cascades. In most diagnostic workups, the objective is to demonstrate a pathognomonic DNA band in an agarose gel. These qualitative applications only require a little amount of DNA devoid of any inhibitory contaminants. However, more meticulous applications such as cloning and gene expression studies necessitate the presence of an ample quality of highly specific DNA or RNA devoid of contaminants. This can be achieved by paying careful attention to the detail of the protocol. Commercial kits that are widely used and that have been mentioned in the preceding sections possess good efficacy of DNA or RNA recovery. In addition to being low cost, in-house methods offer the user with a degree of flexibility that can be exploited depending on the outcome.

In general, good results are directly an outcome of good laboratory practices. Prior to getting the hands on the laboratory procedure, careful planning is of utmost importance. Poor results could often be attributed to the nonadherence to the protocols and equipment failures. Routine quality control measures such as temperature regulation of water baths and centrifuges, and timing of different steps need special attention. In the multiuser laboratory, it is always better to reserve such equipment and adjust the temperatures before the start of the experiment. In addition, correct pipetting techniques also deserve a mention since most protocols advocate gentle pipetting at certain steps to prevent damage to the nucleic acids being extracted. The adage "garbage in garbage out" applies to molecular biology similar to any other laboratory procedure. Thus, the quality of the starting material is a central consideration.

Lakshman P. Samaranayake, PhD, is the chair of the oral microbiology department and is dean of the Faculty of Dentistry at the University of Hong Kong, Pokfulam, Hong Kong. His expertise is in oral fungal infections and he has published more than 300 articles on the subject.

Nipuna B. Parahitiyawa received his MBBS from the University of Peradeniya, Peradeniya, Sri Lanka. He obtained his postgraduate diploma in medical microbiology from the University of Colombo, Colombo, Sri Lanka, and is currently pursuing his PhD in oral microbiology in relation to molecular detection of the oral microbiome.

C. Jayampath Seneviratne received his dental education in the Faculty of Dental Sciences, University of Peradeniya, Peradeniya, Sri Lanka. He did his MPhil on the periodontal health of renal transplant recipients at the Faculty of Dental Sciences, the University of Peradeniya. He pursued his PhD on the molecular microbiology of *Candida* biofilms at the Faculty of Dentistry, the University of Hong Kong, Pokfulam, Hong Kong.

REFERENCES

1. Kurtzman, C.P. and Fell, J.W., Definition, classification and nomenclature of the yeasts. In *The Yeasts, A Taxonomic Study*, 4th ed.; Kurtzman, C.P. and Fell, J.W., Eds. Elsevier: Amsterdam, 1998; pp. 1–3.
2. Goffeau, A. et al., Life with 6000 genes, *Science*, 274: 563, 1996.
3. Braun, B.R. et al., A human-curated annotation of the *Candida albicans* genome, *PLoS Genet.*, 1: 36, 2005.
4. d'Enfert, C. et al., CandidaDB: A genome database for *Candida albicans* pathogenomics, *Nucleic Acids Res.*, 33: D353, 2005.
5. Granneman, S. and Baserga, S.J., Probing the yeast proteome for RNA-processing factors, *Genome Biol.*, 4: 229, 2003.
6. Deacon, J.W., *Fungal Biology*, 4th ed., Blackwell: Malden, MA/Oxford, 2006.
7. Querol, A. and Barrio, E., A rapid and simple method for the preparation of yeast mitochondrial DNA, *Nucleic Acids Res.*, 18: 657, 1990.
8. Yamada, Y. et al., Comparison of different methods for extraction of mitochondrial DNA from human pathogenic yeasts, *Jpn. J. Infect. Dis.*, 55: 122, 2002.
9. Schmitt, M.E., Brown, T.A., and Trumpower, B.L., A rapid and simple method for preparation of RNA from *Saccharomyces cerevisiae*, *Nucleic Acids Res.*, 18: 3091, 1990.
10. Manna, F. et al., A simple and inexpensive method for RNA extraction from yeasts, *Trends Genet.*, 12: 337, 1996.
11. Yamkovaya, V. et al., Isolation of total RNA from baker's yeast, *Appl. Biochem. Microbiol.*, 42: 84, 2006.
12. Fradin, C. et al., Granulocytes govern the transcriptional response, morphology and proliferation of *Candida albicans* in human blood, *Mol. Microbiol.*, 56: 397, 2005.
13. Fradin, C. et al., Stage-specific gene expression of *Candida albicans* in human blood, *Mol. Microbiol.*, 47: 1523, 2003.
14. Bacci, A. et al., Dendritic cells pulsed with fungal RNA induce protective immunity to *Candida albicans* in hematopoietic transplantation, *J. Immunol.*, 168: 2904, 2002.
15. Lopez de Heredia, M., and Jansen, R.P., RNA integrity as a quality indicator during the first steps of RNP purifications: A comparison of yeast lysis methods, *BMC Biochem.*, 5: 14, 2004.
16. Schultz, J.R. and Clarke, C.F., Characterization of *Saccharomyces cerevisiae* ubiquinone-deficient mutants, *Biofactors*, 9: 121, 1999.
17. Uppuluri, P., Perumal, P., and Chaffin, W.L., Analysis of RNA species of various sizes from stationary-phase planktonic yeast cells of *Candida albicans*, *FEMS Yeast Res.*, 7: 110, 2007.
18. Mutiu, A.I. and Brandl, C.J., RNA isolation from yeast using silica matrices, *J. Biomol. Tech.*, 16: 316, 2005.
19. Nantel, A. et al., Transcription profiling of *Candida albicans* cells undergoing the yeast-to-hyphal transition, *Mol. Biol. Cell*, 13: 3452, 2002.
20. Kohrer, K. and Domdey, H., Preparation of high molecular weight RNA, *Methods Enzymol.*, 194: 398, 1991.
21. Karakousis, A. et al., An assessment of the efficiency of fungal DNA extraction methods for maximizing the detection of medically important fungi using PCR, *J. Microbiol. Methods*, 65: 38, 2006.
22. Wahyuningsih, R. et al., Simple and rapid detection of *Candida albicans* DNA in serum by PCR for diagnosis of invasive candidiasis, *J. Clin. Microbiol.*, 38: 3016, 2000.
23. Chien, C.T. et al., The two-hybrid system: A method to identify and clone genes for proteins that interact with a protein of interest, *Proc. Natl. Acad. Sci. USA*, 88: 9578, 1991.
24. Loffler, J. et al., Comparison of different methods for extraction of DNA of fungal pathogens from cultures and blood, *J. Clin. Microbiol.*, 35: 3311, 1997.
25. van Burik, J.A. et al., Comparison of six extraction techniques for isolation of DNA from filamentous fungi, *Med. Mycol.*, 36: 299, 1998.
26. Ito-Kuwa, S. et al., Serotype identification of *Cryptococcus neoformans* by multiplex PCR, *Mycoses*, 50: 277, 2007.
27. Leaw, S.N. et al., Identification of medically important *Candida* and non-*Candida* yeast species by an oligonucleotide array, *J. Clin. Microbiol.*, 45: 2220, 2007.
28. Rak, M. et al., A yeast model of the neurogenic ataxia retinitis pigmentosa (NARP) T8993G mutation in the mitochondrial ATP synthase-6 gene, *J. Biol. Chem.*, 282: 34039, 2007.
29. Bezerra, W.M. et al., Establishment of a heterologous system for the expression of *Canavalia brasiliensis* lectin: A model for the study of protein splicing, *Genet. Mol. Res.*, 5: 216, 2006.
30. Suzuki, T., Ueda, T., and Watanabe, K., A new method for identifying the amino acid attached to a particular RNA in the cell, *FEBS Lett.*, 381: 195, 1996.

10 Isolation of Nucleic Acids from Filamentous Fungi

Margaret E. Katz and Brian F. Cheetham

CONTENTS

10.1 INTRODUCTION

10.1.1 CLASSIFICATION OF FUNGI

Based on morphological criteria, fungi are often separated into two groups: yeasts and filamentous fungi. Unicellular fungi are known as yeasts. The extraction of nucleic acids from yeast cells is described in Chapter 9. Filamentous fungi form a mycelia consisting of multinucleate, tubular hyphae, which may be separated into compartments by septa. Some fungi are dimorphic, that is, they can exist in a unicellular or multinucleate form. The switch from filamentous to pathogenic

191

yeast form is temperature dependent in the human dimorphic pathogens *Coccidioides immitis*, *Blastomyces dermatitidis*, *Histoplasma capsulatum*, and *Paracoccidioides brasiliensis*. In the dimorphic plant pathogen, *Ustilago maydis*, the filamentous form, which is produced by mating, is infectious while the fungus grows in culture as a yeast. Some groups of fungi belonging to the phylum Chytridiomycota do not form a true mycelium. Unlike other fungi, chytrids produce motile spores possessing a flagellum.

Analysis of DNA sequence data has been crucial to understanding the evolutionary relationship of fungi to other eukaryotes and the relationships between different groups of fungi. It has also revealed the groups of organisms that are true fungi.

A large body of molecular data shows that fungi are more closely related to animals than they are to plants [1]. Based on morphological characteristics and mechanism of sexual reproduction, the fungal kingdom has been divided into four phyla, Ascomycota, Basidiomycota, Zygomycota, and Chytridiomycota. Older classification schemes included a fifth subdivision, the Deuteromycota, for fungal species that had no known sexual cycle. The ability to perform molecular analyses to define phylogenetic relationships made this classification obsolete. Recent molecular studies suggest that neither Zygomycota nor Chytridiomycota are monophyletic and Hibbett et al. [2] have proposed replacing these two phyla with four new phyla and four additional unplaced subphyla.

Analysis of DNA sequence data indicates that microsporidia (obligate intracellular parasites that lack mitochondria) also belong to the fungal kingdom. Several other organisms that were considered to be protozoa have also been shown to be fungi, whereas organisms that were considered to be filamentous fungi (e.g., Oomycota) are placed elsewhere [1].

10.1.2 IMPORTANCE OF FILAMENTOUS FUNGI

In addition to their essential environmental role in decomposition and recycling of organic matter, filamentous fungi have many other characteristics that have made them the focus of scientific research. These studies have used molecular genetic techniques (involving extraction of DNA and RNA) to characterize, manipulate, and detect fungi that are important in medicine, agriculture, food production, and industry.

Environment.   Many filamentous fungi are saprotrophs that decompose plants, animals, and microorganisms (Figure 10.1). Some fungi are parasitic and obtain their nutrients by infecting living plants, animals, or microorganisms. Fungal predators, such as nematode-trapping fungi, are also known. A variety of fungi form mutualistic associations with other living things. These include mycorrhizal fungi, endophytes, lichens, termitomycetes, and rumen fungi.

Agriculture.   The many fungi that cause disease in crop plants are of major importance to human welfare and global food security. Fungi are capable of causing catastrophic famines as well as chronic damage to annual food production. The Great Bengal Famine of 1943 in which 2 million people died was caused by the fungal pathogen of rice, *Cochliobolus miyabeanus*, and 10%–30% of the rice crop, which supports almost half the world's population, is lost to the fungus *Magnaporthe grisea* every year [3]. Fungal associations with plants can also have a beneficial effect on crop production (e.g., increased nutrient uptake due to mycorrhizal fungi). Some fungi are under investigation as potential biological control agents such as entomopathogenic fungi for the control of insects, nematode-trapping fungi for the control of parasitic nematodes, and fungal parasites for the control of weeds. Endophytic fungi that produce toxic alkaloids have been used in grasses to improve resistance to insect herbivores in grasses.

Medicine.   Filamentous fungi that are of medical significance include species that produce antibiotics (e.g., penicillin and cephalosporin) and other pharmaceuticals (e.g., statins), fungi capable of causing disease in humans, fungi that produce mycotoxins and, of course, poisonous mushrooms. Systemic fungal infections are not common in healthy individuals although some dimorphic fungi are capable of causing systemic infections (e.g., *C. immitis*). Superficial infections (e.g., tinea and toenail infections)

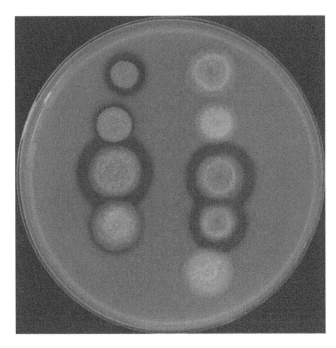

FIGURE 10.1   Growth of *Aspergillus nidulans* wild type and mutant strains on medium containing milk as a carbon source. The halos surrounding the colonies are due to extracellular protease activity.

by dermatophytes are much more common. In immunocompromised patients, opportunistic fungal infections (e.g., cryptococcosis, aspergillosis, and mucormycosis) are frequently fatal. Fungi are also associated with a number of allergic disorders. Contamination of food with mycotoxins (e.g., aflatoxin and fumonisin) is a significant health problem, affecting up to 25% of the world's food crops [4]. Ingestion of poisonous mushrooms, while rare, is very serious.

Industry.   Filamentous fungi are used for a variety of industrial applications including the production of food (e.g., soy sauce), enzymes (e.g., amylases for starch processing), and acids (e.g., citric acid). The use of fungi and fungal enzymes for a more environmentally friendly pulp and paper industry is an ongoing area of research. Genetically modified fungi can be used to produce recombinant proteins (e.g., rennet for cheese making).

Scientific research.   Filamentous fungi have played an important role since the beginning of molecular genetic research. The "one gene one enzyme hypothesis" of Beadle and Tatum was based on research in *Neurospora crassa*. *N. crassa*, *Aspergillus nidulans*, and many other filamentous fungi have continued to make important contributions to the study of eukaryotic molecular and cellular biology as model genetic organisms (Figure 10.2).

The information in this chapter has in part been collected by surveying the methods used in the Fungal Genetics Community. Thus, it represents techniques that have gained widespread use.

10.2   KEY ASPECTS OF NUCLEIC ACID PURIFICATION FROM FILAMENTOUS FUNGI

10.2.1   SOURCE OF FUNGAL TISSUE

Fungal DNA can be prepared from mycelia, vegetative spores, and fruiting bodies. Most methods are for extraction of nucleic acids from mycelia or fruiting bodies. Even dried herbarium samples of basidiomata can be used [5,6].

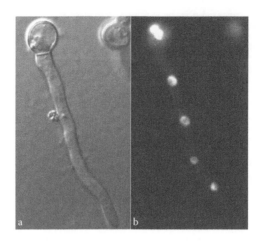

FIGURE 10.2 (a) *A. nidulans* germling viewed with Normarski differential interference contrast microscopy and (b) protein tagged with green fluorescent protein in the same germling viewed with fluorescence microscopy.

Many scientists report that they are unable to achieve consistent results when attempting to extract DNA from spores, which are resistant to disruption, and contain polyphenolic pigments, which can inhibit enzyme activity (Section 10.2.2). The use of glass beads [7] or other particles (Section 10.2.3.2) may be necessary to achieve physical disruption of resistant spores. To overcome the problem of disruption, the spores can be allowed to germinate before DNA extraction [8]. High-quality DNA can be isolated from even highly melanized spores [9].

For laboratory cultures, the fungus can be grown on liquid, semisolid, or solid medium, depending on the organism and experimental protocol. Commonly, mycelia are grown in submerged liquid cultures but they can also be grown on the surface of liquid medium in Petri dishes. For organisms that secrete large quantities of polysaccharides in liquid medium, it may be preferable to culture the fungus on solid or semisolid medium (Section 10.2.2). The preparation of nucleic acids from mycelia or spores grown on plates can also save time and allow the rapid screening of multiple strains. To prevent contamination of the DNA/RNA with agar, the fungus can be grown on cellophane [10,11], cellulose [12], or nylon membranes placed on the solid or semisolid medium. "Reverse agar," which is solid at room temperature but liquid at 4°C has also been used to culture fungi on solid medium for DNA extraction [13].

Methods for the analysis of clinical specimens are covered in Chapter 11. Samples that have been prepared for histological analysis (e.g., for detection of *Batrachochytrium dendrobatidis* in frogs) can be used for DNA extraction [14].

For extraction of fungal DNA from plant tissue, foods, and soils, techniques that are designed for tough fibrous materials (e.g., bark and roots) and which remove plant/soil compounds that can inhibit enzymatic reactions must be used [15–17]. This is especially important when preparing DNA from seeds that are very high in polysaccharides [18]. Commercial kits are available for extraction of DNA from food (e.g., Wizard Magnetic DNA Purification System for Food from Promega) and soil (e.g., UltraClean Soil DNA Isolation Kit from MO BIO Laboratories and Soil Microbe DNA Kit from Zymo Research). For extraction of fungal DNA from these sources, a suitable method of cell wall disruption (Section 10.2.3) is required.

10.2.2 SPECIFIC ISSUES FOR NUCLEIC ACID PURIFICATION FROM FILAMENTOUS FUNGI

Some difficulties are commonly encountered in the extraction of DNA and RNA from filamentous fungi. One of the most important obstacles of obtaining a high yield of nucleic acids from filamentous fungi is the fungal cell wall. A variety of methods have been developed to disrupt the fungal cell wall and these are reviewed in Section 10.2.3.

Secreted or cell wall polysaccharides and polyphenolic fungal pigments can cause problems during the extraction of nucleic acids and can interfere with the activity of enzymes such as polymerases and restriction enzymes that are used in downstream applications. Contamination with polyphenolic compounds can make it difficult to quantify nucleic acid preparations using spectrophotometry. In some cases, the levels of these undesirable compounds can be limited by changing the growth conditions of the fungus. In other cases, the nucleic acid purification procedure must be modified to allow the removal of these compounds.

Polysaccharides that are secreted during growth in liquid culture can prevent filtration of mycelia and interfere with DNA and RNA purification. Production of secreted polysaccharides can be limited by harvesting the mycelium before maximal growth is achieved. Alternatively, the fungus can be cultured on solid medium using one of the strategies mentioned in Section 10.2.1 to prevent contamination of the nucleic acid preparation with agar.

The detergent hexadecyltrimethylammonium bromide (CTAB) is commonly used to remove polysaccharides during the extraction of nucleic acids from plants and this method has been applied to filamentous fungi [19]. High salt can also be used to remove polysaccharides during DNA extraction [20].

In some fungi, pigment production is related to the age of the culture and nutrient depletion, both of which can be avoided by limiting the growth period. If pigment production is restricted to vegetative spores, mycelia can be harvested before the spores are formed. However, this strategy may not be an option for all species or experimental situations. The addition of polyvinylpyrrolidone (PVP) has been used to overcome problems with phenolic compounds in plants [21] and has been used during DNA [22] and RNA [23] extraction in filamentous fungi.

10.2.3 Disruption of the Fungal Cell Wall

10.2.3.1 Manual Grinding

The most commonly used method for disruption of the fungal cell wall is by grinding lyophilized or fresh mycelia in liquid nitrogen with a mortar and pestle. As this is a time-consuming and laborious process, the number of samples that can be processed at one time is limited. Cross-contamination of samples is also a possibility and thus this method is not ideal for screening large numbers of strains using polymerase chain reaction (PCR).

Methods that are based on the use of fresh tissue bypass the requirement for liquid nitrogen and/or lyophilization. Fresh tissue can be ground with sand [24] or glass beads [25] prior to the extraction of nucleic acids. For small-scale preparations of nucleic acids, fresh mycelia can be pulverized with sand [26] or glass beads [27] using a vortex mixer.

10.2.3.2 Instruments for Cell Disruption

A number of techniques have been devised to avoid the labor involved in grinding by hand. These include using a coffee grinder [28], a mechanical grinder [29], or a bead beater [30]. Fresh, frozen, or lyophilized tissues can be disrupted using glass beads with a homogenizer [25].

Devices designed to disrupt multiple samples of spores or mycelia are available. The Mini-BeadBeater (Biospec Products) can be used with a variety of beads including glass, ceramic, zirconium, and steel. A version that uses 96-well plates is available. For fungi, beads of 0.5 mm are recommended. Glass beads are suitable for the disruption of mycelia while denser beads such as zirconium or zirconium silicate are recommended for spores and other tough tissues. The FastPrep Instrument (Qbiogene) rapidly homogenizes samples in tubes containing a variety of special lysing matrices. The MagNA Lyser (Roche Diagnostics) uses ceramic beads and can simultaneously process 16 samples in 2 mL tubes. These instruments all process samples within sealed tubes, which minimizes the risk of cross-contamination (and release of pathogenic fungi).

10.2.3.3 Enzymatic Methods

An alternative to physical methods for the disruption of the fungal cell wall is the use of cell wall degrading enzymes [31]. This method has the advantage of yielding DNA of very high molecular weight, which can be used in the construction of gene libraries [32]. Lytic enzymes have also been used for preparation of template for PCR [33–35].

10.2.3.4 Other Methods

Some methods of cell disruption that are suitable for extraction of small amounts of DNA for PCR have been described. These include boiling mycelia and/or conidiospores scraped from Petri plates [36,37], cycles of freezing mycelia in lysis buffer using liquid nitrogen followed by thawing at 70°C [10], and incubation of mycelia in lysis buffer at room temperature for 10 min [38]. Further details about these methods are provided in Section 10.2.4.3.

10.2.4 DNA Extraction

In choosing a method for genomic DNA extraction (Table 10.1), a number of factors must be considered. The amount and purity of DNA that are required for a particular application will in many cases determine which method is chosen. Many DNA extraction procedures are designed specifically for PCR applications and the yield of DNA is insufficient for other applications.

Cost is an important issue for many laboratories. DNA extraction using homemade reagents is inexpensive, reliable, and yields large quantities of DNA of relatively high purity. The fact that most commercially available kits do not circumvent the time-consuming step of cell disruption means that the advantages of using a kit are relatively limited. The yield from commercial kits can also be disappointing.

For laboratories that are preparing DNA from many samples, the amount of labor and time involved is an important consideration. Disruption of the fungal cell wall is the most laborious part of many DNA extraction procedures. Thus, procedures that eliminate manual grinding of samples are most suitable for high throughput and have the advantage of eliminating a potential source of cross-contamination.

10.2.4.1 Homemade Reagents

Most fungal DNA extraction protocols are based on variations of the method of Raeder and Broda [39] and involve heating ground mycelia in a cell lysis buffer containing a detergent (e.g., sodium dodecyl sulphate (SDS), CTAB, N-lauroylsarcosine). Proteins and carbohydrates are precipitated by the addition of potassium acetate [40] or ammonium acetate [26] and are removed by phenol:chloroform or chloroform:isoamyl alcohol extraction. The DNA is then precipitated with isopropanol.

Two very popular methods for DNA extraction from filamentous fungi are the methods of Lee and Taylor [5] and Zolan and Pukkila [19]. In both methods, powdered lyophilized mycelia are incubated at 65°C in lysis buffer followed by extraction with phenol:chloroform or chloroform:isoamyl alcohol and precipitation with isopropanol. In the method described by Lee and Taylor [5], the lysis buffer contains Tris, EDTA, 2-mercaptoethanol and the detergent SDS. In the method of Zolan and Pukkila [19], SDS is replaced with the detergent CTAB and the lysis buffer also contains 0.7 M NaCl. The use of CTAB is recommended for samples that contain large amounts of polysaccharide (Section 10.2.2). Modifications of the CTAB method that include proteinase K [41], RNase [42], or PVP [22] in the lysis buffer have also been reported.

Depending on the organism, extraction with organic solvents may not be necessary for all applications. For example, a modified version of Yelton et al. [43] and Andrianopoulos and Hynes [40], which

TABLE 10.1
Methods for DNA Purification

Fungus	Phylum	Method[a]	Application
Absidia spp.[*]	Zygomycota	CTAB/chloroform [25]	PCR
Alternaria spp.	Ascomycota	Modified FastPrep System (BIO 101) [55]	PCR
Amanita spp.	Basidiomycota	CTAB/phenol:chloroform/PVP [22]	PCR
		CTAB/phenol:chloroform [41]	Southerns, libraries
Ascochyta spp.	Ascomycota	DNeasy Plant Kit (QIAGEN)	AFLP, inverse PCR, TAIL
		UltraClean Soil DNA Isolation Kit	PCR
		(MO BIO Laboratories) for soil and	PCR, qPCR
		chickpea seeds	
Aspergillus spp.	Ascomycota	Modified FastPrep System (BIO 101) [55]	PCR
		PreMan Ultra (Applied Biosystems) +	PCR
		10 min boiling step after MagNA Lyser	
		(Roche)/ceramic bead disruption	
A. flavus	Ascomycota	DNeasy Plant Kit (QIAGEN)	Southerns
		CTAB method	
A. fumigatus	Ascomycota	SDS/phenol:chloroform [40]	PCR, Southerns
		(phenol:chloroform can be omitted for	
		PCR and most Southerns)	
A. nidulans	Ascomycota	SDS/phenol:chloroform [56]	PCR, mtRFLP, Southern
			blots, viral dsRNA
			detection
		SDS/phenol:chloroform method of Chow	PCR, qPCR, Southerns,
		and Kafer [27] except use lyophilized,	dot blots
		ground mycelia	
		SDS/phenol:chloroform [40]	PCR, Southerns, libraries
		(phenol:chloroform can be omitted for	
		PCR and most Southerns)	
		SDS/phenol:chloroform [5]	PCR, Southerns, libraries
		CTAB/phenol:chloroform (K.S. Bruno,	PCR, Southerns
		personal communication)	
		Enzymatic disruption of cell wall [31]	PCR, Southerns, libraries
		Boiling lysis [36]	PCR screening
		DNeasy Plant Kit (QIAGEN)	PCR
A. niger	Ascomycota	SDS/phenol:chloroform [56]	PCR, mtRFLP, Southern
			blots, viral dsRNA
			detection
		Triisopropylnapthalene sulfonic	PCR, Southerns
		acid/*p*-aminosalicylic acid (TNS/PAS)	
		phenol:chloroform [57]	
		CTAB/phenol:chloroform (K.S. Bruno,	PCR, Southerns
		personal communication)	
A. terreus	Ascomycota	CTAB/phenol:chloroform (K.S. Bruno,	PCR, Southerns
		personal communication)	
B. dendrobatidis[*]	Chytridiomycota	CTAB/chloroform [19] followed by	PCR
		QIAEX II purification (QIAGEN) [42]	
Botrytis spp.	Ascomycota	PrepMan Ultra (Applied Biosystems) +	PCR
		10 min boiling step after MagNA Lyser	
		(Roche)/ceramic bead disruption	

(continued)

TABLE 10.1 (continued)
Methods for DNA Purification

Fungus	Phylum	Method[a]	Application
		CTAB/PVP/phenol:chloroform [18] after Fast Prep Instrument/ceramic spheres (or mortar and pestle for seeds) disruption	PCR, qPCR
		DNeasy Plant Kit (QIAGEN) after Fast Prep Instrument/ceramic spheres (or mortar and pestle for seeds) disruption	PCR, qPCR, AFLP
Botrytis cinerea	Ascomycota	SDS/Triton-X/phenol:chloroform [44]	PCR
Chrysosporium spp.	Ascomycota	Modified FastPrep System (BIO 101) [55]	PCR
Claviceps purpurea	Ascomycota	SDS, variation of Cenis [29] using lyophilized mycelia and 5 M potassium acetate instead of 3 M sodium acetate	PCR, Southerns
Colletotrichum spp.	Ascomycota	Mycelial tips in Milli-Q H_2O heated to 91°C before PCR (no centrifugation) [34]	PCR <1 kb
Coprinus cinereus	Basidiomycota	CTAB/chloroform [19]	PCR, Southerns
		Enzymatic disruption of cell wall [32]	BAC library
Cryptococcus neoformans	Basidiomycota	CTAB/phenol:chloroform [41]	PCR, Southerns, library construction
		SDS/Triton-X/phenol:chloroform [44]	PCR
Epidermophyton spp.	Ascomycota	DNeasy Plant Kit (QIAGEN)	PCR
Fusarium spp.	Ascomycota	Modified FastPrep System (BIO 101) [55]	PCR
		Wizard Magnetic DNA Purification System for Food (Promega)	qPCR
		CTAB/phenol:chloroform (K. S. Bruno, personal communication)	PCR, Southerns
Fusarium graminearum	Ascomycota	CTAB/phenol:chloroform/PVP [22]	PCR
		DNeasy Plant Kit (QIAGEN)	PCR
		CTAB/phenol:chloroform [41]	Southerns, libraries
Leptosphaeria biglobosa	Ascomycota	DNeasy Plant Kit (QIAGEN)	PCR
Leptosphaeria maculans	Ascomycota	DNeasy Plant Kit (QIAGEN)	PCR
Magnaporthe grisea	Ascomycota	CTAB/phenol:chloroform (K. S. Bruno, personal communication)	PCR, Southerns
Microsporum spp.	Ascomycota	DNeasy Plant Kit (QIAGEN)	PCR
Mucor spp.[*]	Zygomycota	CTAB/chloroform [25]	PCR
Mycosphaerella spp.	Ascomycota	Wizard Magnetic DNA Purification System for Food (Promega)	qPCR
N. crassa	Ascomycota	SDS/proteinase K/phenol:chloroform (Yeadon, www.fgsc.net/neurospora protocols)	PCR, Southerns, libraries
		N-lauroylsarcosine [26] (+ RNase + phenol:chloroform)	PCR, Southerns
		LETS buffer methods (van Diepeningen, personal communication)	PCR, Southerns, RFLP, plasmid extraction
		CTAB/phenol:chloroform (K. S. Bruno, personal communication)	PCR, Southerns

TABLE 10.1 (continued)
Methods for DNA Purification

Fungus	Phylum	Method[a]	Application
Penicillium spp.	Ascomycota	Modified FastPrep System (BIO 101) [55]	PCR
Penicillium marneffei	Ascomycota	SDS/phenol:chloroform [5]	PCR, Southerns
		DNAzol (Molecular Research Centre) for DNA from conidia	PCR
Phycomyces blakesleeanus[*]	Zygomycota	CTAB/phenol:chloroform [41]	PCR
Podospora anserina	Ascomycota	LETS buffer method (van Diepeningen, personal communication)	PCR, Southerns, RFLP, plasmid extraction
		Chelex (Bio-Rad) (van Diepeningen, personal communication)	PCR
		SDS/phenol:chloroform [10]	Southerns (not all enzymes), PCR < 3 kb
		Fungal DNA Mini Kit (PEQLAB Biotechnologie)	PCR
Rhizomucor spp.[*]	Zygomycota	CTAB/chloroform [25]	PCR
Rhizomucor tauricus[*]	Zygomycota	LiCl/SDS/phenol:chloroform [45,58]	PCR
Rhizopus spp.	Zygomycota	CTAB/phenol:chloroform (K. S. Bruno, personal communication)	PCR, Southerns
Schizophyllum commune	Basidiomycota	DNAzol or DNAzol ES (Molecular Research Center) after grinding in liquid nitrogen or a bead beater disruption	PCR, Southerns
Sclerotinia sclerotiorum	Ascomycota	DNeasy Plant Kit (QIAGEN)	PCR
Termitomyces spp.	Basidiomycota	DNeasy Plant Kit (QIAGEN)	PCR and other applications
		Chelex (Bio-Rad)	PCR
Trichoderma reesei	Ascomycota	High salt/phenol:chloroform [59]	PCR, Southerns, libraries
		Homogenization of fresh mycelia in QIAGEN PCR purification buffer + centifugation (Monika Schmoll, personal communication)	PCR
Trichophyton spp.	Ascomycota	DNeasy Plant Kit (QIAGEN)	PCR
Trichophyton rubrum	Ascomycota	SDS/phenol [39,60]	PCR, Southerns, libraries
Ustilago maydis	Basidiomycota	SDS/Triton-X/phenol:chloroform [44]	PCR, Southerns, library construction
Xylaria spp.	Ascomycota	DNeasy Plant Kit (QIAGEN)	PCR and other applications
		Chelex (Bio-Rad)	PCR

Note: RFLP = Restriction fragment length polymorphism.

[a] The methods used for DNA extraction were obtained through a survey of the Fungal Genetics Community except methods for fungal species marked with an asterisk, which were sourced directly from journal articles.

involves heating of powdered mycelia in 50 mM EDTA pH 8, 2% SDS; centrifugation to remove cell debris; precipitation of the proteins and SDS with 0.2 volumes of 5 M potassium acetate pH 4.8; and precipitation of the DNA with an equal volume of isopropanol yields DNA from *A. nidulans*, which is suitable for PCR and digestion with most restriction enzymes (Section 10.3.1). The method of Irelan [13] for preparation of DNA from *N. crassa* also does not require the use of organic solvents.

Two additional methods that are commonly used to prepare fungal DNA are the 10 min procedure of Hoffman and Winston [44], which was developed for the yeast *Saccharomyces cerevisiae*, and the method of Leach et al. [45]. In the former method, which is used for the dimorphic fungi, a vortex mixer or a mini-bead beater is used to disrupt cells in a lysis buffer containing the detergents SDS and Triton-X, phenol:chloroform, and glass beads. The latter method uses a lysis buffer containing LiCl (LETS buffer).

10.2.4.2   Commercial Kits

One of the most popular kits for the preparation of genomic DNA from filamentous fungi is the QIAGEN DNeasy Plant Kit. Other commercial products that are used by fungal biologists are the UltraClean Microbial DNA Kit (Mo Bio Laboratories), DNAzol (available from several suppliers), Fungal DNA Mini Kit (PEQLAB Biotechnologie), and Chelex 100 (Bio-Rad Laboratories).

10.2.4.3   Rapid Preparation of DNA for PCR

A large number of rapid methods have been reported for the extraction of fungal DNA for PCR. De Maeseneire et al. [46] tested different methods of cell disruption (liquid nitrogen, glass beads, lytic enzymes, boiling, microwave oven, salt extraction LiCl, and acetone) for yield and reproducible amplification of short and long DNA fragments from *A. nidulans* and *Myrothecium gramineum*. Use of mycelia without prior DNA extraction was also tested. They found that the highest yield of DNA was obtained with a protocol using cell wall degrading enzymes [33]. Good yields were also obtained with a boiling procedure [47]. When it came to reproducible amplification, most procedures showed some species specificity. A method using liquid nitrogen [48] worked well for both species and was suitable for long PCR.

Most protocols for the preparation of fungal DNA for PCR are tested on relatively few species, though there are many exceptions [49]. A single method that is suitable for all fungal species is especially important for diagnostic laboratories. The development of such a method is still in progress. Karakousis et al. [35] tested physical, chemical, and enzymatic methods for cell wall disruption in 16 fungi of medical importance. They found that the use of enzymes (lyticase or proteinase K) was required to obtain good yields of DNA from all 16 species. They also found that manual grinding in a mortar and pestle was more effective than glass bead milling indicating that there is room for improvement in the nonmanual methods.

10.2.5   RNA Extraction

The same issues that arise in choosing a method for extraction of DNA (yield, quality, cost, labor, and speed) are relevant to the choice of method for RNA extraction. The use of commercial kits and reagents is far more widespread due to the greater difficulty in obtaining good quality RNA. For some applications, RNA preparations may require treatment with RNase-free DNase to remove DNA contamination. Commercial kits are available for the purification of mRNA from total RNA (e.g., the Promega PolyATtract mRNA Isolation System).

10.2.5.1   Homemade Reagents

Relatively few fungal biologists report that they are using homemade reagents for the extraction of total RNA (Table 10.2). Most methods involve lysing the cells and inhibiting endogenous RNase activity by suspending ground mycelium in a solution containing phenol and either guanidine thiocyanate or SDS. The single step acid guanidine thiocyanate–phenol–chloroform extraction method of Chomczynski and Sacchi [50] is the most popular and is also the basis for the commercial products, TRIzol Reagent (Invitrogen) and TRI Reagent (Molecular Research Center or Sigma). A modified

TABLE 10.2
Methods for RNA Purification

Fungus	Phylum	Method[a]	Application
Ascochyta spp.	Ascomycota	TRIzol Reagent (Invitrogen)	cDNA libraries
A. flavus	Ascomycota	TRIzol Reagent (Invitrogen)	qRT-PCR, cDNA labeling for microarrays
A. fumigatus	Ascomycota	TRIzol Reagent (Invitrogen) + purification with QIAGEN spin column	qRT-PCR, cDNA synthesis, cDNA labeling for microarrays
		RNeasy Plant Kit (QIAGEN)	qRT-PCR, cDNA synthesis, cDNA labeling for microarrays
A. nidulans	Ascomycota	RNeasy Plant Kit (QIAGEN) after grinding lyophilized mycelia with sand	cDNA synthesis
		Guanidine hydrochloride method [61] after grinding frozen mycelia ground with sand	Northerns
		TRIzol Reagent (Invitrogen) after FastPrep Instrument (Qbiogene Inc.)/glass beads disruption	Northerns, RT-PCR
		RNeasy Plant Kit (QIAGEN)	RT-PCR
		TRIzol Reagent (Invitrogen)	RT-PCR, qRT-PCR, cDNA labeling for microarrays, Northerns
		SDS/phenol:chloroform [53] + PolyATtract RNA System (Promega) for arrays	Northerns, RT-PCR cDNA labeling for microarrays
A. niger	Ascomycota	TRIzol Reagent (Invitrogen)	RT-PCR, cDNA libraries, cDNA labeling for microarrays, Northerns
A. terreus	Ascomycota	TRIzol Reagent (Invitrogen)	RT-PCR, Northerns
Botrytis cinerea	Ascomycota	Acid guanidine thiocyanate–phenol–chloroform method + LiCl precipitation [51]	Northern blots, cDNA synthesis
Claviceps purpurea	Ascomycota	RNAgents Total RNA Isolation Kit (Promega)	RT-PCR, qRT-PCR, Northerns, cDNA libraries,
Coprinus cinereus	Basidiomycota	SDS/phenol:chloroform [28]	cDNA synthesis, Northerns
		RNAiso (Takara Co.)	cDNA synthesis, Northerns
		RNeasy Plant Kit (QIAGEN)	RT-PCR, cDNA libraries, Northerns
Cryptococcus neoformans	Basidiomycota	TRIzol Reagent (Invitrogen)	Northerns
Epidermophyton spp.	Ascomycota	RNeasy Plant Kit (QIAGEN)	RT-PCR
Fusarium spp.	Ascomycota	TRIzol Reagent (Invitrogen)	RT-PCR, cDNA libraries, Northerns
		RNeasy Plant Kit (QIAGEN)	RT-PCR, cDNA libraries
Fusarium graminearum	Ascomycota	TRIzol Reagent (Invitrogen) + CTAB + RNeasy Kit (QIAGEN) [23]	cDNA labeling, Northerns, cDNA synthesis
		RNeasy Plant Kit (QIAGEN)	cDNA labeling, Northerns
Magnaporthe grisea	Ascomycota	TRIzol Reagent (Invitrogen)	RT-PCR, Northerns
Microsporum spp.	Ascomycota	RNeasy Plant Kit (QIAGEN)	RT-PCR

(*continued*)

TABLE 10.2 (continued)
Methods for RNA Purification

Fungus	Phylum	Method[a]	Application
Mucor circinelloides	Zygomycota	RNeasy Plant Kit	RACE PCR, primer extension, slot blots
Mycosphaerella spp.	Ascomycota	TRIzol Reagent (Invitrogen)	RT-PCR, cDNA libraries
		RNeasy Plant Kit (QIAGEN)	RT-PCR, cDNA libraries
N. crassa	Ascomycota	TRIzol Reagent (Invitrogen)	RT-PCR, Northerns
Penicillium marneffei	Ascomycota	TRIzol Reagent (Invitrogen)	RT-PCR, Northerns
Phycomyces blakesleeanus[*]	Zygomycota	Perfect RNA Eukaryotic Kit (Eppendorf) after Mini-BeadBeater (Biospec) zirconium bead disruption [30]	RACE PCR, Northerns
Podospora anserina	Ascomycota	Acid guanidine thiocyanate–phenol–chloroform + CsCl centrifugation [52]	RT-PCR, Northerns, cDNA libraries
		RNeasy Plant Kit (QIAGEN)	RT-PCR
Rhizopus spp.	Zygomycota	TRIzol Reagent (Invitrogen)	RT-PCR, Northerns
Schizophyllum commune	Basidiomycota	RNeasy Plant Kit (QIAGEN)	RT-PCR, cDNA libraries, Northerns
		RNApure Reagent (GenHunter) + phenol:chloroform extraction + LiCl precipitation	RT-PCR, qRT-PCR, cDNA library construction, Northerns
Trichoderma reesei	Ascomycota	Acid guanidine thiocyanate–phenol–chloroform [50] (+ RNA purification (QIAGEN) for qRT-PCR)	Northerns, RT-PCR, RACE, qRT-PCR
Trichophyton spp.	Ascomycota	RNeasy Plant Kit (QIAGEN)	RT-PCR
Trichophyton rubrum	Ascomycota	TRIzol Reagent (Invitrogen)	Northern blots
		Illustra RNAspin RNA Isolation Kit (GE Healthcare)	RT-PCR, cDNA synthesis
Ustilago maydis	Basidiomycota	SDS/hot phenol method [54]	Northerns, RT-PCR
		TRIzol Reagent (Invitrogen)	qRT-PCR, cDNA labeling for microarrays

[a] The methods used for RNA extraction were obtained through a survey of the Fungal Genetics Community except methods for fungal species marked with an asterisk, which were sourced directly from journal articles.

version of this method which includes a LiCl precipitation step was developed by Lichter et al. [51] to minimize polysaccharide contamination. The RNA can also be purified by ultracentrifugation through a CsCl cushion [52]. A number of methods in which the extraction buffer contains SDS and phenol are used [28,53,54].

10.2.5.2 Commercial Products

Two commercial products are widely used to prepare RNA from filamentous fungi: TRIzol Reagent (Invitrogen) and the RNeasy Plant Kit (QIAGEN). Users of these products report that TRIzol Reagent provides good inhibition of RNase activity and a good yield of RNA. The yield from the RNeasy Plant Kit is lower but the purity of the preparation is very high. The RNeasy Kit, like TRIzol Reagent, uses an extraction buffer containing guanidine isothiocyanate but no phenol, and includes a silica gel-membrane purification step. These products may not perform well for material that is very high in polysaccharides. Hallen et al. [23] used TRIzol Reagent, followed by a CTAB–chloroform extraction step and final purification with RNeasy Mini Kit (QIAGEN) to extract RNA from polysaccharide-rich samples.

10.3 METHODS

10.3.1 EXTRACTION OF FUNGAL DNA

10.3.1.1 Large-Scale Preparation of Genomic DNA

The following protocol for extraction of genomic DNA from *Aspergillus* is based on Andrianopoulos and Hynes [40]. This method yields a large quantity of DNA that is suitable for PCR and digestion with most of restriction enzymes without further purification.

Reagents: Sterile deionized H_2O, liquid nitrogen, lysis buffer (0.5 M EDTA pH 8, 0.2% SDS), 5 M potassium acetate pH 4.8 (For 100 mL, 29.44 g potassium acetate, 11.5 mL glacial acetic acid, pH adjusted to 4.8 with HCl), isopropanol, 70% ethanol, TE buffer (10 mM Tris pH 8, 1 mM EDTA).

Disposable items: Sterile 10 mL centrifuge tubes, 1.5 mL microfuge tubes, blue and yellow micropipette tips, and glass Pasteur pipettes.

Equipment: −20°C freezer, freeze-drier, unglazed porcelain mortars and pestles, benchtop centrifuge with a swinging bucket rotor, water bath, ice maker, microcentrifuge, micropipettes (1 mL, 200 μL, 20 μL).

Procedure

1. Harvest mycelium and wash with sterile deionized H_2O to remove growth medium. Blot mycelium dry by pressing it between sheets of paper toweling and wrap in foil. The mycelium can be stored at −20°C until ready for processing.
2. Drop mycelium in liquid nitrogen for 5 min. Lyophilize mycelium in a freeze-drier.
3. Pulverize the lyophilized mycelium with an unglazed mortar and pestle. Transfer to a 10 mL disposable centrifuge tube. There should be less than 3 mL of powdered mycelium.
4. Add 3 mL of 50 mM EDTA pH 8, 0.2% SDS to the powdered mycelium. Use a vortex to mix thoroughly, making sure that all the mycelium is suspended in the lysis buffer.
5. Incubate at 65°C for 15 min in a water bath.
6. Spin for 15–20 min in a centrifuge with a swinging bucket rotor at 1700 g.
7. Decant supernatant to a fresh 10 mL tube and discard the tube containing the mycelial debris. From this point onward, a vortex mixer should not be used as it can lead to shearing of genomic DNA.
8. Add 0.6 mL of 5 M potassium acetate pH 4.8, mix by inversion, and incubate on ice for at least 1 h.
9. Spin for 15–20 min in the swinging bucket rotor at 1700 g.
10. Carefully transfer supernatant to a fresh 10 mL tube, leaving a little behind above the pellet.
11. Slowly add an equal volume of isopropanol to the tube by letting it slide down the wall of the tube to create a layer of isopropanol above the solution containing the DNA.
12. Swirl the tube and slowly rock it to precipitate the DNA in a mass at the interface. The aim is to have the DNA in one big blob.
13. Fish out the DNA with a Pasteur pipette and transfer it into a 1.5 mL microfuge tube containing 1 mL of cold 70% ethanol. If the DNA cannot be removed from the isopropanol with a Pasteur pipette, the DNA can be pelleted by centrifugation but the DNA preparation may then contain higher levels of polysaccharides.

14. Pellet the DNA by spinning in a microcentrifuge for 30 s. Tip off the 70% ethanol.
15. Wash the DNA pellet by adding 0.5 mL cold 70% ethanol to the microfuge tube and spinning briefly. Tip off ethanol and repeat the washing step.
16. Remove all traces of ethanol and dry pellet under vacuum or by leaving the tube open.
17. Add 100–200 μL of TE buffer (10 mM Tris pH 8, 1 mM EDTA) to the tube and leave at room temperature overnight to resuspend the DNA. The volume of TE can be adjusted if the amount of mycelium was smaller than usual.
18. After ensuring that the DNA is resuspended, run 1 μL on a gel to estimate concentration.

10.3.1.2 Small-Scale Preparation of Genomic DNA

A modified version of the same method can be used for small-scale preparations for PCR. To avoid cross-contamination, gloves and plugged tips should be used throughout the procedure. The solutions should be freshly autoclaved or used exclusively with plugged tips.

Reagents: 20 mM EDTA pH 8, acid-washed sand, ethanol, lysis buffer (0.5 M EDTA pH 8, 0.2% SDS), 5 M potassium acetate pH 4.8 (For 100 mL, 29.44 g potassium acetate, 11.5 mL glacial acetic acid, pH adjusted to 4.8 with HCl), 70% ethanol, TE buffer (10 mM Tris pH 8, 1 mM EDTA).

Disposable items: Gloves, sterile 1.5 mL microfuge tubes, blue micropipette tips, and plugged tips (1 mL, 200 μL, 20 μL).

Equipment: Water bath, ice maker, microcentrifuge, micropipettes (1 mL, 200 μL, 20 μL).

Procedure

1. Place a small amount of mycelium into a 1.5 mL microfuge tube containing 1 mL of 20 mM EDTA pH 8.
2. Pellet mycelium in a microcentrifuge by spinning for 5 min. Tip off the supernatant.
3. Add 1 mL of 100% ethanol to the mycelium. Mix with a vortex mixer.
4. Spin the tube in microcentrifuge for 1 min to pellet the mycelium. Tip off ethanol and remove remaining ethanol with a micropipette tip.
5. Dry the pellet under vacuum. It is important to make sure that the mycelium is totally dry.
6. Add a small amount of sterile acid-washed sand to the tube with a sterile spatula. Grind mycelium to a powder with a blue micropipette tip that has been rounded off using a flame.
7. Add 0.3 mL of 50 mM EDTA pH 8.0, 0.2% SDS to pulverized mycelium. Mix thoroughly with a vortex mixer.
8. Incubate at 65°C for 15 min.
9. Spin the tube at top speed in a microcentrifuge for 10 min.
10. Transfer the supernatant to a new tube and discard the tube containing the pellet.
11. Add 60 μL of ice-cold 5 M potassium acetate pH 4.8 to the tube and mix. Incubate tube on ice for at least 1 h.
12. Spin for 20 min in a microcentrifuge.
13. Carefully remove the supernatant to a fresh tube and discard the tube containing the pellet.
14. Add an equal volume of isopropanol to the supernatant and mix by inverting the tube.
15. Pellet the DNA by spinning in a microcentrifuge for 5 min. Tip off the supernatant.
16. Wash the DNA pellet by adding 0.5 mL 70% ethanol. Spin for 5 min.
17. Remove all ethanol and dry the pellet under vacuum or by leaving the tube open.
18. Resuspend the DNA in 50 μL TE buffer. Leave overnight at room temperature to resuspend.
19. Run 5 μL on a gel to estimate concentration.

10.3.2 EXTRACTION OF FUNGAL RNA

10.3.2.1 QIAGEN RNeasy Mini Plant Kit

The following protocol, for extraction of fungal RNA using the RNeasy Mini Plant Kit (QIAGEN), is based on the manufacturers' instructions. Read the "Important notes before starting" in the RNeasy Mini Protocol for Isolation of Total RNA from Plant Cells and Tissues and Filamentous Fungi (*RNeasy Mini Handbook,* pp. 75–76). Gloves and RNase-free micropipette tips, and tubes should be used for this procedure.

Reagents: Sterile deionized H_2O, liquid nitrogen, RNeasy Plant Mini Kit (QIAGEN), β-mercaptoethanol (for the RLC buffer), and ethanol (for the RPE buffer).

Disposable items: Gloves, RNase-free 1.5 mL microfuge tubes, and blue and yellow micropipette tips (handled with gloves at all times).

Equipment: −70°C freezer, unglazed porcelain mortars and pestles, heating block, microcentrifuge, micropipettes (1 mL, 200 μL, 20 μL), and spectrophotometer.

Procedure

1. Harvest the mycelium and wash with sterile deionized H_2O to remove growth medium. Blot mycelium dry by pressing it between sheets of paper toweling and wrap in foil. The mycelium can be stored at −70°C until ready for processing.
2. For each sample, prepare a 1.5 mL microfuge tube containing 450 μL RLC buffer.
3. Quickly weigh out 120 mg of frozen mycelium. (The manufacturer suggests using a maximum of 100 mg, but some is lost during grinding and transfer of the powdered mycelium.) Place frozen mycelium in an unglazed mortar and pestle, add liquid nitrogen, and grind to powder.
4. Allow most of the liquid nitrogen to evaporate, then quickly scrape the powdered mycelium into the tube containing RLC buffer using a spatula that was precooled in liquid nitrogen.
5. Mix the contents of the tube with a vortex mixer and place the tube at 56°C. Leave the tube at 56°C until all samples are ground.
6. Follow steps 4–9 of the RNeasy Mini Protocol for Isolation of Total RNA from Plant Cells and Tissues and Filamentous Fungi (*RNeasy Mini Handbook,* pp. 77–78).
7. As described in step 10 of the RNeasy Mini Protocol, place the RNeasy column in a fresh 1.5 mL collection tube and apply 30 μL of RNase-free H_2O to the membrane. Spin the column and tube in a microcentrifuge for 1 min.
8. Repeat Step 7, leaving the column in the same collection tube. Apply a second 30 μL volume of RNase-free H_2O to the membrane. Spin the column and tube in a microcentrifuge for 1 min.
9. Store the RNA at −70°C. The concentration of the RNA should be measured using a spectrophotometer. To check that the RNA is not degraded, run 3 μg on a formaldehyde-agarose gel.

10.3.2.2 TRI Reagent or TRIzol Reagent

TRI Reagent (Molecular Research Center or Sigma) and TRIzol Reagent (Invitrogen) are improved versions of the single-step total RNA isolation reagent developed by Chomczynski and Sacchi [50]. As it contains phenol and guanidine thiocyanate, the reagent should be handled with gloves and safety goggles and used in a fume cupboard.

Reagents: TRI Reagent (Molecular Research Center or Sigma) or TRIzol Reagent (Invitrogen), sterile deionized H_2O, liquid nitrogen, chloroform, isopropanol, and ethanol.

Disposable items: Gloves, RNase-free 1.5 mL microfuge tubes, and blue and yellow micropipette tips (handled with gloves at all times).

Equipment: −70°C freezer, unglazed porcelain mortars and pestles or a bead beater that processes samples in microfuge tubes, heating block, microcentrifuge, micropipettes (1 mL, 200 μL and 20 μL), and spectrophotometer.

Procedure

1. Harvest the mycelium and wash with sterile deionized H_2O to remove growth medium. Blot mycelium dry by pressing it between sheets of paper toweling and wrap in foil. The mycelium can be stored at −70°C until ready for processing.
2. Quickly weigh out 100 mg of frozen mycelium and add to a screw-cap microfuge tube containing glass, ceramic, silica, or zirconium beads and 1 mL TRI Reagent or TRIzol Reagent. Process in a bead beater. Alternatively, quickly weigh out 120 mg of frozen mycelium, place frozen mycelium in an unglazed mortar and pestle, add liquid nitrogen, and grind to a powder as in Section 10.3.2.1. Powdered lyophilized mycelia can also be used. The sample volume should not exceed 10% of the volume of TRI Reagent.
3. For TRI Reagent, follow steps 2–6 of the Isolation of RNA Protocol recommended by Molecular Research Center (http://www.mrcgene.com/tri.htm) or for TRIzol Reagent, follow the procedure described by the manufacturers (Invitrogen).

10.4 CONCLUSIONS

DNA extraction using homemade reagents and manual grinding is inexpensive, reliable, and yields large quantities of DNA of relatively high purity. A number of instruments are available for the efficient disruption of fungal cells without the labor-intensive and time-consuming step of manual grinding. Commercial kits for the extraction of DNA from filamentous fungi are also available and may be of particular use for laboratories that process a large number of specimens or for samples containing plants, food, or soil. Two commercial products are widely used to prepare RNA from filamentous fungi: TRIzol Reagent (Invitrogen) or TRI Reagent (Molecular Research Center, or Sigma) and the RNeasy Plant Kit (QIAGEN). Special procedures may be required for extraction of nucleic acids from fungal tissues containing high levels of polysaccharide or pigments.

Although numerous reports are available on the comparative performance of various nucleic acid isolation procedures and kits for preparation of DNA or RNA from other organisms (e.g., bacteria), relatively few such studies have been described to date concerning filamentous fungi. Clearly, this type of investigation is necessary to help determine and select highly efficient and cost-effective procedures for purification of DNA and RNA from individual fungal species. In addition, future development and optimization of improved procedures for the efficient disruption of the fungal cell wall will further streamline the nucleic acid extraction process for filamentous fungi.

ACKNOWLEDGMENTS

We gratefully acknowledge the help of David Backhouse and the following members of the Fungal Genetics Community: Duur Aanen, Kirk Bartholomew, Michael Boelker, Arnaud Bottin, Kenneth Bruno, Margi Butler, David Catcheside, Martin Chilvers, Robert Cramer, Fons Debets, Ineka de Vries, Maurizio Di Stasio, James Fraser, Allen Gathman, Heather Hallen, Andrea Hamann, Michael Hynes, Malgorzata Jedryczka, Regine Kahmann, Amnon Lichter, David Lubertozzi, Nilce Martinez-Rossi, Michelle Momany, Hajime Muraguchi, Greg O'Brian, Susan Rahman, Arthur Ram, Gabriela Roca, Yvonne Rolke, Monika Schmoll, Heather Sealy-Lewis, Hayley Smith, Nai Tran-Dinh, Martin Urban, Anne van Diepeningen, Oded Yarden, and Jane Yeadon.

Margaret E. Katz received her PhD in biological sciences from the University of California at Santa Barbara, California for research in microbial genetics. Her research, as a postdoctoral research fellow at the University of Melbourne (Parkville, Victoria, Australia) and Monash University (Clayton, Victoria, Australia) and, subsequently, as a lecturer in molecular genetics at the University of New England in Armidale, New South Wales, Australia, has focused on gene regulation in filamentous fungi, diagnosis of bacterial and fungal diseases, and bacterial and fungal genes involved in virulence. She is currently an associate professor at the University of New England.

Brian F. Cheetham completed his BSc (honors) and PhD at the Australian National University, Acton, Australia. During his thesis research, he focused predominantly on the effects of adenoviruses on the mammalian cell cycle. Following his doctoral work, he obtained an American Cancer Society (California division) postdoctoral fellowship at the University of California, Santa Barbara, California and carried out research on the yeast cell cycle. After 3 years in California, he took a position as lecturer (level A) at Monash University, Clayton, Victoria, Australia and broadened his research areas into the study of human interferons and their role in hepatitis B, multiple sclerosis, rheumatoid arthritis, and AIDS. He then worked as a lecturer at the University of New England, Armidale, New South Wales, Australia, where he is now an associate professor. His research interests include the genetic control of virulence in *Dichelobacter nodosus*, the anaerobic bacterium that causes foot rot in sheep; gene regulation in *Aspergillus*; and the development of molecular techniques for the diagnosis of Marek's disease and infectious bronchitis in chicken.

REFERENCES

1. Blackwell, M. et al., Fungi. Eumycota: Mushrooms, sac fungi, yeast, molds, rusts, smuts, etc. Version July 13, 2007. Available at http://tolweb.org/Fungi/2377/2007.07.13, in Tree of Life Web Project; http://tolweb.org/, 2007.
2. Hibbett, D.S. et al., A higher-level phylogenetic classification of the Fungi, *Mycol. Res*, 111: 509, 2007.
3. Strange, R.N. and Scott, P.R., Plant disease: A threat to global food security, *Annu. Rev. Phytopathol.*, 43: 83, 2005.
4. WHO, Basic food safety for health workers. World Health Organization, 1999.
5. Lee, S.B., Taylor, J.W., Isolation of DNA from fungal mycelia and single spores, in *PCR Protocols. A Guide to Methods and Applications*, Innis, M.A. et al., Eds., Academic Press, San Diego, CA, 1990, pp. 282–287.
6. Goes-Nato, A., Loguercio-Leite, C., and Guerrero, R., DNA extraction from frozen field-collected and dehydrated herbarium fungal basidiomata: Performance of SDS and CTAB methods, *Biotemas*, 18: 19, 2005.
7. Dobinson, K.F., Genetic transformation of the vascular wilt fungus *Verticillium dahliae*, *Can. J. Bot.*, 73: 710, 1995.
8. Cassago, A. et al., Cellophane based mini-prep method for DNA extraction from the filamentous fungus *Trichoderma reesei*, *BMC Microbiol.*, 2: 14, 2002.
9. Moller, E.M. and Peltola, J., Isolation of high-molecular-weight DNA from mycelium and the recalcitrant and heavily pigmented spores of *Stachybotrys chartarum*, *Anal. Biochem.*, 297: 99, 2001.
10. Lecellier, G. and Silar, P., Rapid methods for nucleic acids extraction from Petri dish-grown mycelia, *Curr. Genet.*, 25: 122, 1994.
11. Wendland, J., Lengeler, K.B., and Kothe, E., An instant preparation method for nucleic acids of filamentous fungi, *Fungal Genet. Newslett.*, 43: 54, 1996.
12. Ikeda, K.-I., Nakamura, H., and Matsumoto, N., Hypovirulent strain of the violet root fungus *Helicobasidium mompa*, *J. Gen. Plant. Pathol.*, 69: 385, 2003.
13. Seifert, K.A., A novel method of growing fungi for DNA extraction, *Fungal Genet. Newslett.*, 41: 79, 1994.
14. Simoncelli, F. et al., Evidence of *Batrachochytrium dendrobatidis* infection in water frogs of the *Rana esculenta* complex in central Italy, *EcoHealth*, 2: 307, 2005.
15. Bahnweg, G. et al., DNA isolation from recalcitrant materials such as tree roots, bark, and forest soil for the detection of fungal pathogens by polymerase chain reaction, *Anal. Biochem.*, 262: 79, 1998.

16. Vazquez-Marrufo, G. et al., DNA isolation from forest soil suitable for PCR assays of fungal and plant rRNA genes, *Plant Mol. Biol. Rep.*, 20: 379, 2002.
17. Malvick, D.K. and Grunden, E., Isolation of fungal DNA from plant tissues and removal of DNA amplification inhibitors, *Mol. Ecol. Notes*, 5: 958, 2005.
18. Chilvers, M.I., A real-time, quantitative PCR seed assay for *Botrytis* spp. that cause neck rot of onion, *Plant Dis.*, 91: 599, 2007.
19. Zolan, M.E. and Pukkila, P.J., Inheritance of DNA methylation in *Coprinus cinereus*, *Mol. Cell. Biol.*, 6: 195, 1986.
20. Fang, G., Hammar, S., and Grumet, R., A quick and inexpensive method for removing polysaccharides from plant genomic DNA, *Biotechniques*, 13: 52, 1992.
21. Loomis, W.D., Overcoming problems of phenolics and quinones in the isolation of plant enzymes and organelles, *Methods Enzymol.*, 31: 528, 1974.
22. Hallen, H.E., Watling, R., and Adams, G.C., Taxonomy and toxicity of *Conocybe lactea* and related species, *Mycol. Res.*, 107: 969, 2003.
23. Hallen, H.E. et al., Gene expression shifts during perithecium development in *Gibberella zeae* (anamorph *Fusarium graminearum*), with particular emphasis on ion transport proteins, *Fungal Genet. Biol.*, 44: 1146, 2007.
24. Weiland, J.J., Rapid procedure for the extraction of DNA from fungal spores and mycelia, *Fungal Genet. Newslett.*, 44: 60, 1997.
25. Schwarz, P. et al., Molecular identification of zygomycetes from culture and experimentally infected tissues, *J. Clin. Microbiol.*, 44: 340, 2006.
26. Irelan, J., Small scale DNA preps for *Neurospora crassa*, *Fungal Genet. Newslett.*, 40: 24, 1993.
27. Chow, T.Y.K. and Kafer, E., A rapid method for isolation of total nucleic acids from *Aspergillus nidulans*, *Fungal Genet. Newslett.*, 40: 25, 1993.
28. Stassen, N.Y. et al., Isolation and characterization of *rad51* orthologs from *Coprinus cinereus* and *Lycopersicon esculentum*, and phylogenetic analysis of eukaryotic *recA* homologs, *Curr. Genet.*, 31: 144, 1997.
29. Cenis, J.L., Rapid extraction of fungal DNA for PCR amplification, *Nucleic Acids Res.*, 20: 2380, 1992.
30. Rodriguez-Romero, J. and Corrochano, L.M., The gene for the heat-shock protein HSP100 is induced by blue light and heat-shock in the fungus *Phycomyces blakesleeanus*, *Curr. Genet.*, 46: 295, 2004.
31. Bainbridge, B.W. et al., Improved methods for the preparation of high molecular weight DNA from large and small scale cultures of filamentous fungi, *FEMS Microbiol. Lett.*, 54: 113, 1990.
32. Muraguchi, H., Kamada, T., and Yanagi, S.O., Construction of a bacterial artificial chromosome (BAC) library of *Coprinus cinereus*, *Mycoscience*, 46: 49, 2005.
33. van Zeijl, C.M. et al., An improved colony-PCR method for filamentous fungi for amplification of PCR-fragments of several kilobases, *J. Biotechnol.*, 59: 221, 1997.
34. Roca, M.G., Davide, L.C., and Wheals, A., Template preparation for rapid PCR in *Collectotrichum lindemuthianum*, *Brazil. J. Microbiol.*, 34: 8, 2003.
35. Karakousis, A. et al., An assessment of the efficiency of fungal DNA extraction methods for maximizing the detection of medically important fungi using PCR, *J. Microbiol. Methods*, 65: 38, 2006.
36. Chiou, C.H. et al., Chromosomal location plays a role in regulation of aflatoxin gene expression in *Aspergillus parasiticus*, *Appl. Environ. Microbiol.*, 68: 306, 2002.
37. Henderson, S.T., Eariss, G.A., and Catcheside, D.E.A., Reliable PCR amplification from *Neurospora crassa* DNA obtained from conidia, *Fungal Genet. Newslett.*, 52: 24, 2005.
38. Saitoh, K.-I., Togashi, K., and Arie, T.T.T., A simple method for a mini-preparation of fungal DNA, *J. Gen. Plant. Pathol.*, 72: 348, 2006.
39. Raeder, U. and Broda, P., Rapid preparation of DNA from filamentous fungi, *Lett. Appl. Microbiol.*, 1: 17, 1985.
40. Andrianopoulos, A. and Hynes, M.J., Cloning and analysis of the positively acting regulatory gene *amdR* from *Aspergillus nidulans*, *Mol. Cell. Biol.*, 8: 3532, 1988.
41. Pitkin, J.W., Panaccione, D.G., and Walton, J.D., A putative cyclic peptide efflux pump encoded by the TOXA gene of the plant-pathogenic fungus *Cochliobolus carbonum*, *Microbiology*, 142: 1557, 1996.
42. Annis, S.L. et al., A DNA-based assay identifies *Batrachochytrium dendrobatidis* in amphibians, *J. Wildl. Dis.*, 40: 420, 2004.
43. Yelton, M.M., Hamer, J.E., and Timberlake, W.E., Transformation of *Aspergillus nidulans* by using a *trpC* plasmid, *Proc. Natl. Acad. Sci. U. S. A.*, 81: 1470, 1984.
44. Hoffman, C.S. and Winston, F., A ten-minute DNA preparation from yeast efficiently releases autonomous plasmids for transformation of *Escherichia coli*, *Gene*, 57: 267, 1987.

45. Leach, J., Finkelstein, D.B., and Rambosek, J.A., Rapid miniprep of DNA from filamentous fungi, *Fungal Genet. Newslett.*, 33: 32, 1986.
46. De Maeseneire, S.L. et al., Rapid isolation of fungal genomic DNA suitable for long distance PCR, *Biotechnol. Lett.*, 29: 1845, 2007.
47. Vanittanakom, N., Vanittanakom, P., and Hay, R.J., Rapid identification of *Penicillium marneffei* by PCR-based detection of specific sequences on the rRNA gene, *J. Clin. Microbiol.*, 40: 1739, 2002.
48. DuTeau, N.M. and Leslie, J.F., A simple, rapid procedure for the isolation of DNA for PCR from *Gibberella fujikuroi* (*Fusarium* section *Liseola*), *Fungal Genet. Newslett.*, 38: 72, 1991.
49. Liu, D. et al., Rapid mini-preparation of fungal DNA for PCR, *J. Clin. Microbiol.*, 38: 471, 2000.
50. Chomczynski, P. and Sacchi, N., Single-step method of RNA isolation by acid guanidinium thiocyanate–phenol–chloroform extraction, *Anal. Biochem.*, 162: 156, 1987.
51. Lichter, A. et al., Survival responses of *Botrytis cinerea* after exposure to ethanol and heat, *J. Phyto-pathol.*, 151: 553, 2003.
52. Borghouts, C. et al., Copper-modulated gene expression and senescence in the filamentous fungus *Podospora anserina*, *Mol. Cell. Biol.*, 21: 390, 2001.
53. Reinert, W.R., Patel, V.B., and Giles, N.H., Genetic regulation of the *qa* gene cluster of *Neurospora crassa*: Induction of *qa* messenger ribonucleic acid and dependency on *qa*-1 function, *Mol. Cell. Biol.*, 7: 427, 1981.
54. Schmitt, M.E., Brown, T.A., and Trumpower, B.L., A rapid and simple method for preparation of RNA from *Saccharomyces cerevisiae*, *Nucleic Acids Res.*, 18: 3091, 1990.
55. Smith-White, J.L., Gunn, L.V., and Summerell, B.A., Analysis of diversity within *Fusarium oxysporum* populations using molecular and vegetative compatibility grouping, *Aust. Plant Pathol.*, 30: 153, 2001.
56. Van Diepeningen, A.D. et al., Efficient degradation of tannic acid by black *Aspergillus* species, *Mycol. Res.*, 108: 919, 2004.
57. Kolar, M. et al., Transformation of *Penicillium chrysogenum* using dominant selection markers and expression of an *Escherichia coli* lacZ fusion gene, *Gene*, 62: 127, 1988.
58. Vagvolgyi, C. et al., *Rhizomucor tauricus*: A questionable species of the genus, *Mycol. Res.*, 103: 1318, 1999.
59. Schmoll, M. et al., Cloning of genes expressed early during cellulase induction in Hypocrea jecorina by a rapid subtraction hybridization approach, *Fungal. Genet. Biol.*, 41: 877, 2004.
60. Fachin, A.L. et al., Role of the ABC transporter TruMDR2 in terbinafine, 4-nitroquinoline N-oxide and ethidium bromide susceptibility in *Trichophyton rubrum*, *J. Med. Microbiol.*, 55: 1093, 2006.
61. Deeley, R.G. et al., Primary activation of the vitellogenin gene in the rooster, *J. Biol. Chem.*, 252: 8310, 1977.

11 Preparation of Fungal Specimens for Direct Molecular Applications

Françoise Irlinger, Nicolas Berthet, Tatiana Vallaeys, Valérie Vasseur, Renaud Ioos, Marc Buée, and Jérôme Mounier

CONTENTS

11.1   INTRODUCTION

Direct DNA extraction methods, as first performed by Torsvik et al. in 1980, have opened the route to new molecular-based applications including the direct detection of pathogens in biological fluids or the characterization of the diversity of uncultivable microflora. These methods have been combined with signal amplification procedures such as polymerase chain reaction (PCR) to characterize fungi in particular habitats without the need for enrichment or isolation. Being recognized as a rapid, sensitive, and specific molecular diagnostic tool, PCR can be extremely effective with pure nucleic acids, with the capability of generating detectable signal from a single copy of target template. However, its sensitivity may be reduced dramatically when applied directly to biological (clinical, environmental, and food) samples. This is due mainly to the fact that many foodstuff, clinical, and environmental samples harbor substances that inhibit or reduce the amplification capacity of PCR. The PCR inhibitors may act on one or more of the following ways by inactivation of the thermostable

DNA polymerase, by degradation or capture of the nucleic acids, or by interfering with the cell lysis step. The optimization of PCR testing conditions has been used to improve the amplification capacity of the DNA polymerase, but in most cases, a sample preparation step is required prior to PCR. Thus, reliable and sensitive detection of the target fungal organisms from the complex samples is dependent on the abililty of sample processing procedures to recover and extract nucleic acids of adequate quantity and quality from these samples. Consequently, much effort is being devoted to the development of sample preparation methodologies that yield PCR-compatible templates from fungal samples; and from the vast number of procedures and articles being published on this topic to date, it appears that the problems associated with molecular detection of fungal pathogens directly from uncultured samples are still far from being solved.

The requirements for direct detection and quantification of fungi in different complex matrices include (1) reproducible rate of target nucleic acids and (2) removal of inhibitors to aid alternative molecular methods for downstream analysis. In this chapter, we review the general strategies for improved lysis of fungal organisms and subsequent extraction and precipitation of fungal nucleic acids for the detection and characterization of fungal species in food, clinical, and environmental samples. We also present various in-house reagents and commercial kits that have been shown to generate PCR-ready templates from uncultured fungal samples. A constantly recurring theme emerging from these data is that direct nucleic acid extraction methods must be optimized for each application, as different fungal samples (fluid, solid, and biochemical characteristics), target organisms (yeast and molds), nucleic acid template yield and integrity, and potential existence of PCR inhibitors must all be considered in the sample preparation prior to PCR experiments.

11.1.1 Fungi in Food Samples

Fungi including yeasts and molds are able to grow in a diverse range of foods such as milk, cheese, fermented products (wine, beer, and meat), cereals, vegetables, fruits, and related products. The impact of yeasts and molds on the production, quality, and safety of foods and beverages is strongly linked to their ecology and biological activities [1].

Fungi can influence the quality and safety of foods and beverages either positively or negatively. On the positive side, fungi contribute to the maturation and stabilization of fermented products such as wine, bread, beer, and cheese [1]. It is well established that many fungal species other than *Saccharomyces cerevisiae* are involved in the maturation of wine, bread, and beer. On the negative side, yeasts and filamentous fungi may be responsible for food spoilage such as toxin production, off-flavors, and color defects, mostly through their enzymatic activities. The most important aspect of mold spoilage in food is the production of mycotoxins, which could have adverse effects on animal and human health. More than 400 mycotoxins are known today, which are produced in large quantities by different fungal genera such as *Penicillium*, *Aspergillus*, and *Fusarium* spp. [2].

Consequently, it is of importance for the food industry to be able to clearly identify and quantify the yeasts and molds present in food products in order to ensure their production, quality, and safety. Until recently, the identification and quantification of fungi in food products has relied solely on cultivation-dependent methods, which involve a preliminary isolation of the microorganisms on specific and nonspecific media prior to identification using phenotypic or genotypic methods. Although these methods are effective, they are also time consuming because culturing of yeasts and molds and subsequent identification require at least several days. Moreover, it is well known that culture-dependent approaches may lead to biased results because universal media do not exist to cultivate all species. The presence of viable but noncultivable yeasts and molds has also been revealed in different food products such as wine [3]. For these reasons, culture-independent approaches are extensively utilized alone or in combination with culture-dependent techniques to study fungi diversity in food products. These approaches target the nucleic acids (DNA and RNA) of fungal organisms for both the assessment of community structure and the quantification of individual constituents. An overview of the different food products, for which DNA and RNA extraction has been developed for direct detection and quantification of fungi, is shown in Table 11.1. Indeed,

TABLE 11.1

Examples of Fungal DNA and RNA from Food Samples Used for Molecular Assessment

Food Samples	Target Taxon	Application	References
Grains of wheat and barley	*Fusarium* spp.	PCR	[117]
Grains of malt and barley	*Fusarium* spp.	Quantitative real-time PCR	[154]
Grains of wheat	*Fusarium* spp.	Quantitative real-time PCR	[116,118]
	Saccharomyces cerevisiae	PCR	[87]
Powdered pepper, paprika, or ground maize kernels	*Aspergillus flavus*	Quantitative real-time PCR	[103]
Grape berries	*Aspergillus carbonarius*	Quantitative real-time PCR	[94]
	Non-*Saccharomyces*	PCR–DGGE	[65]
	Penicillium spp., *Aspergillus* spp., and *Botrytis* spp.	PCR- temporal temparature gradient gel electrophoresis (TTGE)	[64]
Figs	*A. flavus*	PCR	[120]
Green coffee beans	*Aspergillus ochraceus*	Quantitative real-time PCR	[155]
Cocoa fermentation	*S. cerevisiae*/non-*Saccharomyces*	PCR–DGGE	[66]
Bakery products	*S. cerevisiae*	PCR	[87]
Sourdough	*S. cerevisiae*/non-*Saccharomyces*	PCR–DGGE	[115]
Beer	*S. cerevisiae*	PCR	[122]
Wine	*S. cerevisiae*/non-*Saccharomyces*	PCR–DGGE PCR-TTGE	[53,121,156,157]
	S. cerevisiae	PCR-TTGE	[158]
	Dekerra bruxellensis	Quantitative real-time PCR	[159,160]
	S. cerevisiae	Quantitative real-time PCR	[161]
	Saccharomyces spp., *Hanseniaspora* spp.	Quantitative real-time PCR	[162]
	Zygosaccharomyces bailii	Quantitative real-time PCR	[163]
Camembert–Roquefort	*Penicillium camemberti* *Penicillium roqueforti*	PCR	[111]
Livarot	*Geotrichum candidum, Debaryomyces hansenii,* and *Yarrowia lipolytica*	Quantitative real-time PCR	[164]
Livarot	Yeast	PCR	[145,165]
Salers	Yeast	PCR–single-strand conformation polymorphism (SSCP)	[166]
Smeared soft cheese	Yeast	PCR–SSCP	[112]
Raw milk	Yeast	PCR–DGGE	[53]
Yoghurt	*Kluyveromyces marxianus*	PCR	[56]
Salami	*D. hansenii, Rhodotorula mucilaginosa,* and *Trichosporon brassicae*	PCR–DGGE	[113]
Fermented sausages	*D. hansenii, Candida* spp., and *Willopsis*	PCR–DGGE	[114]
Fruit juices, fruit preserves, milk, yoghurt	*S. cerevisiae*/non-*S. cerevisiae* mRNA (actin)	Real-time RT–PCR	[51]
Wine	Total yeast rRNA (26 rRNA)	Real-time RT–PCR	[138]
Wheat	*Penicillium nordicum* mRNA (ochratoxin polyketide synthase)	Real-time RT–PCR	[137]
Milk	Yeasts–Molds mRNA (elongation factor)	RT–PCR	[54]
Livarot	Yeasts rRNA (26 rRNA)	RT–PCR	[145]

the development of culture-independent approaches provides an interesting tool for the food industry because it enables fast and reliable identification and quantification of fungi present in food products compared with culture-dependent approaches. Furthermore, during the last few years, a concerted effort has been made to characterize gene expression profiles during food maturation, related to stress response and carbohydrate metabolism, mainly in *S. cerevisiae* [4,5].

11.1.2 FUNGI IN CLINICAL SAMPLES

Fungal diseases in humans, commonly known as mycoses, include a range of infections of contrasting severity. While superficial skin and nail colonizations are commonly encountered by otherwise healthy people without life-threatening consequences, fungal contamination of deep tissues including lungs, esophagus, brain, etc. is a cause of significant health concern in humans, with pneumonia, septicemia, skin, and systemic disease being common outcomes [6]. Fungal contamination can also affect biological fluids, either remaining locally (such as in lachrymal canals in cases of eye infections) or affecting the whole body in extreme cases of fungal-induced septicemia. Methods to extract nucleic acids directly from human samples have thus to take into consideration the variety of inner characteristics of the infected tissues or biological fluids. Mycoses are caused by a variety of fungi from a morphological, ecological, and clinical point of view [6]. Indeed, to date, more than 100 yeast species and filamentous fungi have been identified as opportunistic or well-recognized human pathogens. Among these, *Candida*, *Microsporum*, *Trichophyton*, *Epidermophyton*, and *Blastomyces* cause several common mycoses (e.g., skin rash and skin mycoses); *Candida*, *Aspergillus*, and *Fusarium* are responsible for invasive fungal infections (e.g., sinusitis, skin lesions, and endophthalmitis); and *Histoplasma*, *Coccidioides*, *Aspergillus*, *Pneumocystis*, *Blastomyces*, and *Cryptococcus* produce severe invasive fungal infections (e.g., pneumonia and febrile illness, persistent fever, meningo-encephalitis and septicemia) [9,62,167,168].

Candidoses.   Among the >200 species of *Candida*, only a small proportion (around 20) are reported to infect humans. Of these, *Candida albicans* is the most common fungal contaminant that causes a variety of infections [7], contracted mainly as superficial skin and nail contaminations. Intestinal, vaginal, and oral mucosal contaminations involving *C. albicans* are also commonly encountered. However, outbreaks of systemic infections as well as the severity of nosocomial *C. albicans* infections seem to have increased in recent years [8,9].

Aspergilloses.   *Aspergillus fumigatus* is the causative agent of invasive aspergillosis, which represents the second common cause of death resulting from fungal infections in hospitals. Invasive aspergillosis is associated with a high mortality. This is mostly due to the poor sensitivity of the currently available diagnostic tests and the overreliance on amphotericin B therapy, which has deleterious side effects.

Cryptococcoses.   Cutaneous cryptococcosis usually appears in acne-form pustules, granuloma-like ulcers, deep-seated abscesses, or tumor-like lesions on the skin. These tumor-like lesions mimic myxomas and are formed of pure culture of *Cryptococcus neoformans*. Under rare circumstances, secondary infections can reach the nervous system [10].

Fungal infections in immunocompromised and HIV patients.   The epidemiology of fungal infections has changed over recent decades with the rise in the number of immunocompromised patients and the pressures of antifungal treatment and prophylaxis. This dramatic increase in immunocompromised individuals is attributable to the current widespread application of new technologies and therapies such as bone-marrow or solid-organ transplantation and use of chemotherapeutic agents of broad-spectrum antimicrobial agents along with the AIDS epidemic. These patients are highly susceptible to nosocomial infections caused by organisms such as fungi that were previously considered to be of low virulence or nonpathogenic [11]. Indeed, fungal infections in these patients are often severe, rapidly progressive, and difficult to diagnose or treat [8,12,13].

This results in a sudden increase in the list of potential species to be tracked. For instance, recent infections caused by less common yeast species such as *Pichia*, *Rhodotorula*, *Trichosporon*, and *Saccharomyces* spp. have been reported [14–16]. Finally, cases of mixed contaminations are now observed [17]. These phenomena underlie the urgent need for growth-independent techniques allowing direct, rapid routine diagnosis, identification of emergent pathogenic fungal species, prediction of antibiotic resistance patterns, and, finally, delineation of strains of medically important fungal species for epidemiological tracking.

Traditionally, definitive diagnosis of invasive mold infections requires demonstration of either tissue invasion by fungal hyphae or growth of a mold from a sample obtained by sterile procedure [18]. Suspective fungal specimens are first cultured on Sabouraud dextrose agar (without tissue grinding), and fungal colonial morphology and microscopic structures are then examined. Biochemical tests are also available to aid in identification such as tissue staining using periodic acid Schiff and Grocott's stain for detection of fungal elements. This detection and identification scheme has long been (and still remained) the commonly used method for fungal diagnosis in hospitals. However, phenotypic methods can in some case take weeks, a time frame that is clinically irrelevant. In addition, culture results from biopsies are often negative and hyphal morphology permits only limited distinction of different fungal species [19–21].

Poor diagnostic sensitivities, occurrence of emerging opportunistic species as novel fungal contaminants in human samples, and long turnaround times associated with cultivation-based identification have led clinicians to apply direct, nucleic acid-based alternative methods for detection, identification, and molecular typing of pathogenic and opportunistic fungi. These rapid identification tests of molds causing invasive diseases are mainly based on PCR amplification of nucleic acids extracted from biological samples with or without prior cultivation-enrichment step. The detection of fungal nucleic acids directly extracted from tissue specimens has been shown to be more sensitive than culture for the diagnosis of invasive fungal infections in animal models [22,23] and human biopsies [17].

However, nucleic acid extraction from uncultured fungal specimens remains, in many cases, the bottleneck of this new diagnostic approach. Relatively, few studies have focused on the critical nucleic acid extraction stage of sample processing [24] in contrast to the multitude of reports on the fungal PCR assay methods. Optimization of nucleic acid-based detection and identification methods requires taking into consideration both the diversity of contaminant fungi and the variety of biological conditions in which these fungi could be potentially found. Additionally, as analytic methods of direct nucleic acid-based fungal detection are currently evolving toward new applications such as micro/oligo-arrays, the yield and quality of extracted nucleic acids have to be taken into consideration as they may affect downstream array-based applications.

11.1.3 FUNGI IN ENVIRONMENTAL SAMPLES

Fungal plant pathogen species. Approximately, half of the plant diseases are caused by fungal species and among the 100,000 fungal species described up to now, more than 10% are reported to be pathogenic for cultivated plants [25]. Fungal plant pathogens encompass true fungi, classified as members of the kingdom Fungi, but also include fungal-like organisms such as members of the kingdom Protista (phylum *Myxomycota* or *Plasmodiophoromycota*, or slime molds) or members of the recently described Kingdom Stramenopila (phylum *Oomycota*) that comprises plant pathogens causing infection known as downy mildew. Examples of famous plant diseases and the fungal species involved are briefly summarized in Table 11.2.

Ectomycorrhizal (ECM) fungi. Fungi are among the most ecologically and evolutionarily diverse organisms and are classified within three eukaryotic kingdoms (Fungi, Stramenopila, and Protista).

TABLE 11.2

Examples of Well-Known Plant Diseases Caused by Fungi or Fungal-Like Organisms

Kingdom/Phylum	Target Taxon	Disease	Molecular Detection	References
Protista/ Plasmodiophoromycota	*Plasmodiophora brassicae*	Clubroot of crucifers	PCR	[169]
	Spongospora subterranea	Powdery scab of potato	RT-PCR	[170]
Stramenopila/*Oomycota*	*Plasmopara halstedii*	Downy mildew of sunflower	PCR	[171]
	Bremia lactucae	Downy mildew of lettuce	PCR	[172]
	Peronospora	Downy mildew of crucifer	PCR	[173]
	Phytophthora infestans	Tomato and potato late blight	RT-PCR	[174]
	Phytophthora cinnamomi	Chestnut ink	PCR	[175]
	Phytophthora spp.	Root and stem rots on numerous plants	PCR	[88]
	Pythium spp.	Seed rot, root rot, and damping-off on numerous plants	RT-PCR	[176]
Fungi/*Chytridiomycota*	*Sychytrium endobioticum*	Potato wart	RT-PCR	[177]
Fungi/*Zygomycota*	*Mucor* spp.	Storage rot of fruits and vegetables	RT-PCR	[178]
	Rhizopus spp.	Storage rot of fruits and vegetables	RT-PCR	[179]
Fungi/*Ascomycota*	*Taphrina deformans*	Peach leaf curl	PCR	[180]
	Venturia inaequalis	Apple scab	PCR	[181]
	Erysiphe spp.	Powdery mildew	PCR	[182]
	Ophiostoma ulmi	Dutch elm disease	RT-PCR	[183]
	Ceratocystis fagacearum	Oak wilt	PCR	[184]
Fungi/*Basidiomycota*	*Tilletia* spp.	Smut on cereal	RT-PCR	[185]
	Puccinia spp.	Rust on numerous plants	PCR	[186]
Fungi/*Deuteromycota*	*Fusarium* spp.	Head blight on cereals and wilt of tomato	RT-PCR	[187]
	Magnaporte grisea	Rice blast	RT-PCR	[188]
	Botrytis spp.	Gray mod disease on fruits and vegetables	RT-PCR	[189]
	Monilia spp.	Brown rot on stone fruits	PCR	[190]
Mycelia sterilia	*Rhizoctonia* spp.	Root rot on several plants	RT-PCR	[191]

In the boreal and temperate forests, fungal species have traditionally been divided into two distinct functional groups, which correspond to different strategies of carbon acquisition: saprotrophic fungi and symbiotic (mutualistic, commensalistic, and parasitic) fungi. Saprobes are free-living decomposers, which acquire carbon and nutrients exclusively from dead and decaying organic matter: litter, wood, insects, animals, fungi, etc. [26,27]. In the second category, the ECM symbiosis is a mutualistic interaction between soil fungi and the roots of majority of temperate and boreal forest trees allowing formation of ectomycorrhizas. Ectomycorrhizas have a beneficial impact on plant growth in natural and agroforestry ecosystems. Central to the success of these symbioses is the exchange of nutrients between the symbionts [28]. The fungus gains carbon from the plant while plant nutrient uptake is mediated via the fungus [29]. Ectomycorrhizas are characterized structurally by the presence of a dense mass of fungal hyphae forming a pseudoparenchymatous tissue ensheathing the root. This is the Hartig net of intercellular hyphae, characterized by labyrinthine branching and an outward network of hyphae prospecting the soil and gathering nutrients. Numerous ECM fungi are characterized by mycelial structures that are similar to those of saprobes, including rhizomorphs or mycelial cords [30–32]. The richness and diversity of ECM fungi contrast with the low number of tree species.

The number of ECM fungi can reach hundreds of species at the stand scale [33–36] and over 5000 species of ECM fungi have been described [37]. Fungal ecology and ECM fungal community studies based on sporocarp surveys failed to reveal the actual number and distribution of ECM fungi, because many species form inconspicuous resupinate or hypogeous fruit bodies. In addition, some species never form sporocarps. Correspondence between above- and below-ground views of species composition, spatial frequency, and abundance is usually very low [38–41]. Therefore, it is essential to study below-ground aspects with molecular methods to get more precise and reliable results. During the last 10 years, considerable progress has been made in ECM fungi molecular ecology and fungal taxonomy [35,42], in particular with the development of electrophoretic techniques [43,44] and array technology [45,46], PCR primers for molecular diversity studies [47,48], enrichment of databases (http://www.ncbi.nlm.nih.gov/), or creation of new rDNA sequence databases [49]. Without exception, these techniques all require nucleic acid extracts free of the numerous inhibitory contaminants commonly found in the soil environmental samples or in the hardwood and coniferous tree roots, such as proteins, polysaccharides, and phenolic compounds.

11.2 EXTRACTION OF FUNGAL NUCLEIC ACIDS FROM FOOD, CLINICAL, AND ENVIRONMENTAL SAMPLES

In this section, we undertake a review of general aspects and detailed protocols for the direct extraction and purification of DNA and RNA from all major type of samples.

11.2.1 SAMPLE PREPARATION METHODS

Depending on the type of samples from which nucleic acids are extracted and the type of molecular application, different sample preparation methods may be used or optimized. To assist selection of a suitable sample processing procedure, the distinct characteristics of these methods have to be considered. For RNA extraction, special care must be taken such as maintaining RNase-free workspace including glassware, plasticware and solutions, etc. in order to ensure successful extraction. It is also important to work as fast as possible because RNA is very rapidly degraded by RNases that are extremely ubiquitous and resilient proteins [50]. Therefore, if RNA is not extracted immediately after sampling, it is recommended to resuspend the sample in extraction buffer followed by flash freezing in liquid nitrogen and storage at −80°C. For DNA extraction, the collected sample can be stored at −20°C.

11.2.1.1 Liquid Sample Preparation

11.2.1.1.1 Food Liquid Samples

In liquid samples such as wine or fruit juices, a single centrifugation step is sufficient to collect the cell pellets and spores [51,52]. For the recovery of yeast cells from yoghurt and milk samples without fats, proteins, and polysaccharides, specific additional steps should be introduced such as washing with saline solutions [53,54], ammonium hydroxide, diethyl ether, and petroleum ether [51] or with a milk clearing solution containing iminodiacetic acid and detergent [55,56]. After washing, the fat layer, separated by centrifugation, is removed from the tubes.

Filtration may also be used to collect the cell pellets and spores and may have the advantage of eliminating potential PCR inhibitors such as organic and inorganic compounds that may be present in the sample. Following filtration, the filter can be cut into fine pieces and directly used for DNA or RNA extraction [54].

While methods for extracting RNA have been optimized, not much attention has been given to the method used for sampling. Usually, the cells can be recovered by centrifugation at 500 rpm for ~3 min at room temperature [57], but in some protocols, the cells are collected on ice [58]. However, when collecting cells on ice, changes in mRNA levels due to cold-shock are possible [57]. The

Qiagen protocols using RNeasy columns advise to collect the yeast cells by 5 min centrifugation at 4°C, while the 2003 edition of current protocols [59] involves 3 min centrifugation at 4°C. Belinchón et al. showed that rapid filtration under vacuum through a nitrocellulose membrane followed by flash freezing in liquid nitrogen could be used as an alternative to centrifugation for collecting the cells [60]. According to Belinchón et al., rapid filtration improved the yield of mRNAs (2–10 folds) in *S. cerevisiae* compared with centrifugation at room temperature or at 4°C. Recovery of total RNA was similar when using centrifugation or filtration. However, in *Yarrowia lipolytica*, depending on the medium used for cultivation, yields of mRNAs were higher using centrifugation suggesting that each method should be tested in order to maximize mRNA recovery.

11.2.1.1.2 Clinical Liquid Samples

Clinical samples include mainly throat swabs, nasal swabs, nasopharyngeal swabs, nasal washes, bronchoalveolar lavage fluid, sputum, blood, lachrymal fluids, bone marrow, etc. Standard protocols and regulations must be followed (including seeking for authorization from local ethical committee, etc.). Readers are also encouraged to consult Chapter 16.

Blood sampling and methods derived from hemoculture [61]. The fungal pellet is resuspended and incubated in 50 mM NaOH and neutralization with 1 M Tris-HCl (pH 7.0) is not necessary, and the rates of centrifugation for the leukocyte lysis and spheroplast genesis steps are reduced to 1500 and $2000 \times g$, respectively.

Collection and handling of whole blood prior to RNA extraction. For stabilization of RNA in blood samples prior to RNA extraction from whole blood, 300 μL of RLT lysis buffer (Qiagen, Valencia, CA) is added for each 100 μL of blood, followed by immersion in liquid nitrogen. Samples are kept at −80°C until use. Then, 20 μL of RNA secure (Ambion, Austin, TX) are added to the whole blood mixture followed by incubation for 20 min at 60°C.

Sampling aqueous humor or vitreous fluid from patient with clinical diagnosis of presumed endophthalmitis. A volume of 50–150 μL of aqueous humor is collected aseptically in a tuberculin syringe after application of topical anesthesia. Vitreous fluids are aspirated by syringe connected to suction port of vitreous cutter at the beginning of vitrectomy. Sterile disposable needle fixed to syringe is capped with sterile rubber bung after collection [62].

11.2.1.1.3 Environmental Liquid Samples

Water-borne oomycetes such as *Phytophthora* or *Pythium* spp. may also be isolated from culture irrigation systems. To avoid the centrifugation of liters of liquid to pellet the pathogenic zoospores, an alternative consisting of a size-selective filtration of the water through a membrane is now frequently used [63]. The zoospores trapped onto the membrane are subsequently recovered by washing and their nucleic acid content may be extracted through one of the regular protocol described below.

11.2.1.2 Solid Sample Preparation

11.2.1.2.1 Solid Food Samples

Washing. To harvest yeast cells and fungal spores from the surface of plant material such as grape berries, a washing step may be used to remove the organisms from the surface of the berries [64,65]. This washing solution should contain 0.01%–0.1% Tween 80. The rinse is then centrifuged to collect the cell pellets.

Grinding. Solid samples may be grinded in small fragments with, for example, mortar and pestle (in the presence or absence of liquid nitrogen or not), blender, coffee grinder, stomacher, and Ultra-Turrax depending on the food sample and directly used for nucleic acid extraction. Alternatively, a defined amount of the sample may be mixed with a defined volume of sterile distilled water and subsequently homogenized using one of the methods cited above. Part of this solution is then centrifuged to collect the cell pellets. It is important to get the smallest particles as possible to release the organisms from the food matrix and to allow access of the nucleic acid extraction reagents to the microbial cells.

Freeze-drying. It can be used for DNA extraction from plant materials that are difficult to grind. It makes the matrix brittle and therefore facilitates subsequent DNA extraction from the food product. For example, it has been used for DNA extraction from cocoa beans [66].

11.2.1.2.2 Clinical Solid Samples
The sample preparation methods have to be adapted to the characteristics of the samples to be analyzed, typically, soft or hard tissues. Solid samples include particularly soft muscular tissues [67], biopsies, and corneal scrapings [62]. Analysis of muscular tissues, usually freshly collected or fixed by formalin, embedded in paraffin [17] or frozen [67], requires specific removal of these adjuvants that may otherwise interfere with further enzymatic amplification or labeling of purified DNA. Extraction of DNA from hard tissues including hair, nails, and skin samples becomes a difficult task due to sample characteristics. The inherent characteristics of nails require a specific dissolution of nail material. We also advise the reader to see Chapter 18 for nucleic acid extraction from miscellaneous samples. We further propose a set of methods to deal with the samples according to the type of clinical material provided.

Preparation of samples from biopsies and soft muscular tissues. (1) Immediate freezing of collected samples: Sample tissues collected from paranasal sinuses, orbital, nasal mucosa, palate biopsies, lung biopsies, biopsies of the central nervous system, and liver biopsies are immediately frozen upon collection at −20°C for subsequent DNA extraction and −80°C for RNA extraction. (2) Collection of formalin (or paraffin embedded, or untreated) biopsies and soft tissue sample [68–70]: Clinical samples infected by invasive fungi include potentially a large variety of sample types, such as paranasal sinuses, orbita, nasal mucosa, palate biopsies, lung biopsies, biopsies of the central nervous system, liver biopsies, and skin biopsy. Most often samples are immediately fixed using formalin and then paraffin-wax embedded. Biopsies are then placed in sterile tubes for further analysis and kept for up to 2 days as prolonged storage plays a negative effect on the quality of extracted DNA and thus further PCR amplification or labeling. The wax embedding allows an increased storage time and decreases cross-contamination rate. However, the use of wax requires the addition of xylene (1 mL) to each Eppendorf tube containing two 5 mm sections of paraffin wax embedded tissue section to remove wax before proceeding to nucleic acid extraction. It was reported that the type of formalin used to fix the sample may have a varying effect on further amplification of the extracted DNA. Zsikla et al. reported that buffered formalin had a positive effect on DNA preservation. Freezing remains however far more advisable when RNA extraction is planned [71].

Preparation of samples from hard tissues. (1) Nail preparation [72,73]: Whole nails and relatively large nail fragments are cut into small pieces with a surgical blade. Nail shavings are processed directly. (2) Collection of skin sample [74]: Dermatological clinical specimens are taken from patients by scarification from the active edges of the lesion. About 2 mg of clinical material, after mincing and crumbling on a microscope slide, is diluted in 25 μL of Tris-based buffer and further homogenized by shearing several times through an 18-gauge syringe needle in 100 μL of Tris-base buffer (Tris-HCl 50 mM pH 8.0, ethylenediaminetetraacetic acid [EDTA] 25 mM, and NaCl 75 mM) in 1.5 mL sterile Eppendorf tubes.

Preparation of samples from intestine and stools. Collection and storage of mixed samples from intestine and stools in guanidine thiocyanate [75]: Fresh stool samples are collected preferentially directly into stomacher bags and immediately blended 2 times at 4°C for 1 min using a stomacher blender in a solution of 4 M guanidine thiocyanate—0.1 M Tris-HCl (pH 7.5) (w/v) and 150 μL of 10% *N*-lauroyl sarcosine per gram of stool sample (1% final concentration). Aliquots containing 250 μL of the grounded material are immediately transferred to a 2 mL screw-cap polypropylene microcentrifuge tube, and directly used for extraction of their total nucleic acids or frozen at −20°C for further DNA extraction or −80°C for further RNA extraction.

11.2.1.2.3   Environmental Solid Samples

Since plant pathogenic fungi are either biotroph or nonobligate parasite and consequently not cultivable on a synthetic medium, the only pure fungal samples available for nucleic acid extraction are taken directly from parasitized plant. Hence, caution must be observed during the sampling, since adhering plant cells may also be harvested along with fungal cells.

A second type of fungi only grows between plant cuticle and epidermic cells (e.g., *Spilocaea* sp. or *Fusicladium* sp., anamorphic stages of *Venturia* spp. causing apple or pear scab). Nevertheless, these fungi can be isolated from the plant and subsequently grown on agar medium, enabling pure fungal nucleic acids to be extracted. When the fungus is not cultivable (e.g., rust fungi), nucleic acid extraction can be carried out directly starting from disrupted [76] or previously germinated spores [77].

The last type of fungi colonizes extensively the plant cells (e.g., *Phytophthora* diseases) or will move upward and downward the plant through the vessels (e.g., *Fusarium* or *Verticillium* wilts). In this case, the fungi can be isolated from the plant by plating on universal or selective agar media prior to nucleic acid extraction. When it is necessary to recover a fungal plant parasite during its saprophytic stage in soil or in culture substrates, the latter may be diluted in water serial dilution and plated on selective agar media in order to isolate the fungi.

Another way to recover the plant fungal pathogen is to carry out a biological baiting with sensitive plants that will attract the pathogen. The fungal pathogen potentially present in the soil sample will parasitize the baiting plant cells and may be easily isolated in pure culture or detected directly by testing the baiting plant tissue by PCR. This procedure is widely used for *Phytophthora* spp. [78] or several other genera of true fungi [79].

11.2.2   DNA Extraction Protocols

11.2.2.1   Lysis and Purification Variants

11.2.2.1.1   Cell Disruption

The lysis step is a critical step in DNA extraction because it must lead to the lysis of as many yeasts and filamentous fungi species as possible. It is absolutely essential when conducting fungal community assessment. Complete destruction of the cell wall and the release of all intracellular components require destruction of the strength-providing components of the cell wall, that is, glucan in yeast. For yeasts and fungi, this lysis step can be separated into three types: (1) physical, (2) chemical, and (3) enzymatic methods. These three methods are usually combined in order to optimize DNA yield recovery and examples of different lysis methods used for nucleic acid extraction from food, clinical, and environmental samples are summarized in Tables 11.3 through 11.5, respectively.

Physical methods.   Physical disruption may be achieved by freezing/thawing, sonication, and bead beating homogenization treatments. Mechanisms of disruption are cavitation, shear, impingement, or their combination. These methods are effective to disrupt fungal cell walls and spores but they often result in significant DNA shearing. Intensive cooling of the cell suspension subjected to the treatment is necessary to remove the heat generated by dissipation of the mechanical energy. The boiling technique was also successfully used directly with plant material infected by *Eutypa lata* [80], *Gremmeniella abietina* [81], or *Tilletia tritici* [82] to disrupt tissues and cells.

Chemical methods.   The outer wall of fungi can be permeabilized by a large variety of chemical compounds, which differ in selectivity and efficiency toward different species. Chemical permeabilization could be accomplished by reducing agents (β-mercaptoethanol), chelating agents (EDTA), chaotropes (urea, guanidine, ethanol), detergents (sodium dodecyl sulfate [SDS], Triton X, Sarcosyl,

TABLE 11.3

Examples of Lysis Methods Used to Extract DNA from Food Samples

Sample	Chemical Lysis	Enzymatic Lysis	Physical Lysis	References
Must, wine	SDS, Triton X-100, and phenol		Bead beating and Freezing/thawing	[3,65,121,158, 160,192]
Wine	SDS	Zymolyase	Heat (65°C)	[193]
Berries	CTAB		Heat (65°C)	[94]
Grains of wheat	CTAB	Proteinase K	Heat (65°C)	[87]
Sourdough	CTAB, SDS, and phenol	Proteinase K, lyticase, and lysing enzymes from *Trichoderma harzianum*		[115]
Camembert–Roquefort	EDTA, SDS, CTAB, and chloroform	Proteinase K	Heat (65°C)	[111]
Milk	Triton X, SDS, EDTA, phenol, and chloroform		Freezing/thawing and bead beating	[53]
Yoghurt	EDTA, SDS, phenol, and chloroform	Lyticase	Bead beating	[194]
Salers and Livarot	SDS, phenol, and chloroform		Bead beating and heat (80°C)	[145,164–166]
Smeared soft cheese	Laurylsarcosine, SDS, phenol, and chloroform	Proteinase K	Heat (65°C) and bead beating	[112]
Salami, fermented sausages	Petrol ether, hexane, SDS, EDTA, Triton X, phenol, and chloroform	Proteinase K	Heat (65°C) and bead beating	[113,114]

cetyltrimethyl-ammonium bromide [CTAB]), solvents (toluene, chloroform, acetone), or hydroxides and hypochlorites. Detergent causes the cell membrane to breakdown by emulsifying the lipids and proteins of the cell and disrupting the polar interactions that hold the cell membrane together. Use of chaotropes leads to the solubilization of membrane proteins and the disruption of hydrogen bonds. Phenol is a strong denaturant agent for proteins.

Enzymatic methods. Enzymatic lysis has the advantage of being specific and gentle. Disruption requires the use of protease and glucanase to attack, at first the mannoprotein complex of the cell wall and then the glucan backbone. To date, lyticase (also called zymolyase) is the most efficient system for fungal cell lysis. Lyticase is in fact a mixture of enzymes found in *Arthrobacter luteus* [83] or in *Oerskovia xanthineolytica* [84]. The main lytic enzyme, a β-1,3-glucan laminaripentaohydrolase or a β-glucanase from *A. luteus* or *O. xanthineolytica*, respectively, hydrolyzes glucose polymers linked by β-1,3-bonds, which are present in fungi cell walls. Other enzymes may be present depending on the purity of the product such as protease, mannanase, and β-1,3-glucanase. Lyticase from *A. luteus* is effective on a large number of yeast species including *Ashbya, Candida, Debaryomyces, Eremothecium, Endomyces, Hansenula, Hanseniaspora, Kloeckera, Kluyveromyces, Lipomyces, Metschikowia, Pichia, Pullularia, Torulopsis, Saccharomyces, Saccharomycopsis, Saccharomycodes,* and *Schwanniomyces* spp. Lyticase is also effective on filamentous fungi such as *Thanatephorus cucumeris, Ulocladium botrytis, Penicillium roqueforti, Trichoderma harzianum, Trichoderma longibrachiatum, Verticillium albo-atrum,* and *Verticillium dahliae* [85].

It should be noted that fungal spores can be very difficult to lyse depending on fungal species. For example, using a commercial kit for DNA extraction (FastDNA spin kit for soil, Qbiogene, Irvine, CA), Doaré-Lebrun et al. could only obtain sufficient DNA for PCR amplification with spore concentration of *Aspergillus carbonarius* above 10^6 spores/mL [64]. This detection limit could not be improved despite the use of lytic enzymes such as lyticase and chitinase. The initiation of germination

TABLE 11.4

Examples of Lysis Methods Used to Extract DNA from Clinical Samples

Sample	Chemical Lysis	Enzymatic Lysis	Physical Lysis	References
Organs (lung and eyes)		Recombinant lyticase/ proteinase K	Freezing/thawing	[67]
Organs (liver, lung, and brain)		Proteinase K	Sonication	[67]
Bone marrow		Proteinase K	Sonication	[67]
Skin and facial soft tissues		Proteinase K	Sonication	[67]
Corneal scraping	Biogene (Texas) kit lysis solution			[62]
Ocular specimens and soft tissues (paranasal sinuses and polypous tissue biopsies)	EDTA and SDS			[105,108,195]
Blood samples and plasma			Freezing/thawing	[23,196]
Blood samples	RLT lysis buffer (Qiagen)		Freezing with liquid nitrogen	[23]
Serum	KCl, Tris-HCl, and Tween 20	Proteinase K		[197]
Bronchoalveolar lavage fluid	Triton X-100, NaOH, Tween 20, and Nonidet P-40		Heat	[198]
Cerebrospinal fluid	Guanidine thiocyanate, phenol, and chloroform/isoamyl alcohol		Boiling	[199,200]
Nails	Bicarbonate solution, potassium chloride, and BSA		Boiling	[72]
Nail specimens	GenomicPrep cells and tissue DNA isolation kit (Amersham Biosciences)	Proteinase K	Bead beating	[201]

during an incubation step in culture medium was used in order to increase the detection limit of 3–4 log spores/mL. Alternatively, in order to quantify spores of *A. carbonarius* in grapes, Selma et al. used a 10 min incubation step at 95°C to initiate breakdown of conidia [86]. This was followed by a 10 min cooling step on ice and DNA extraction using the EZNA fungal DNA kit (Omega Bio-teck, Doraville, GA). With this boiling step and kit, Selma et al. detected 1×10^2 spores/mL using quantitative real-time PCR.

11.2.2.1.2 Nucleic Acid Purification

DNA isolated using the traditional methods is highly contaminated with proteins and other enzyme-inhibiting components and the yields and purity can vary according to the fungi present and sample type. Moreover, in food, clinical, and environmental constituents, a multitude of substances (organic

TABLE 11.5

Examples of Lysis Methods Used for Nucleic Acid Extraction from Environmental Samples

Sample/Fungi Species	Chemical Lysis	Enzymatic Lysis	Physical Lysis	References
Material plant/ *Eutypa lata*, *Gremmeniella abietina*, and *Tilletia tritici*			Boiling and crushing	[80–82]
Filamentous plant pathogenic fungi/ *Phytophthora* spp.	β-mercaptoethanol		Grinding with liquid nitrogen	[202,203]
Spruce, Alnus trees, and sunflower seeds/ filamentous plant pathogenic fungi			Bead beating	[88,171,204–206]
Plant leaves and soft roots/ *Plasmopara halstedii* and *Phytophthora ramorum*	Guanidine isothiocyanate		Grinding	[204,207]
Soil and ligneous plant tissue/*Verticillium dahliae* and *Phytophthora alni*	Guanidine isothiocyanate		Bead beating	[88,100]
Cereal and seed/*Fusarium* spp.	Guanidine isothiocyanate		Sonication and grinding	[117,187,208]
Neurospora crassa	Guanidine isothiocyanate			[209,210]
Cereal and seed/*Fusarium* spp.	Guanidine hydrochloride			[211,212]
Plant pathogenic fungi	Phenol, chloroform			[110]
Conifer roots and needles/ plant pathogenic fungi	CTAB and SDS	Proteinase K		[213]
Fruiting bodies and ECM root tips	SDS and EDTA		Freeze/thawing, and grinding	[214]
ECM fungi root tips	CTAB			[124–126]
Arbuscular mycorrhizal roots			Grinding	[215]
Arbuscular mycorrhizal roots	Phenol and chloroform		Heating	[216,217]
Endomycorrhizal soil and root samples			Bead beating	[132–134]
ECM fungi from soil			Sonication	[218]

compounds, polyphenols, heparin, glycogen, fat, salt, polysaccharides, amino acids, casein-hydrolysate, chelators, heme and humic acid, etc.) can affect PCR efficiency by inhibiting DNA polymerase activity. Therefore, nucleic acid purification is an essential step in providing a good quality template for subsequent applications such as PCR, cloning, and restriction enzyme digestions.

Removal of proteins and lipids.   Proteins and lipids are present in wide range of food products and can be coextracted during nucleic acid extraction. Chloroform/phenol/isoamyl alcohol is a reagent commonly used in nucleic acid purification for the removal of proteins and lipids. Phenol is a strong denaturant of proteins that leads to the partition of the proteins into the organic phase (and interface), whereas nucleic acid partition into the aqueous phase. Usually, phenol is used in a 1:1 mixture with chloroform since deproteinization is more effective when two different organic solvents are used simultaneously. In addition to denaturing proteins, chloroform is useful in removing lipids and a final chloroform extraction helps to remove the last traces of phenol. The isoamyl alcohol helps with the phase separation, decreases the amount of material found at the aqueous and organic interface, and helps reduce foaming.

Proteinase K can also be useful in the removal of proteins. For example, in *Triticum*-related products, the presence of gluten proteins such as gliadins–glutenins complex can inhibit DNA amplification. The method employed for DNA cleanup used the addition of proteinase K (20 mg/mL) to the lysis buffer in order to optimize the lysis of the plant and fungal cell walls and to improve the partial digestion of proteins particularly in foods [87].

Removal of polyphenols.   The presence of polyphenols, which are powerful antioxidizing agents present in many plants species including grape berries and also in wine, can reduce the yield and purity of extracted DNA. Polyphenols can bind covalently to proteins and nucleic acids while in their oxidized forms. This irreversible binding makes the extracted DNA unavailable for most research applications, including restriction enzyme digestion, amplification, and cloning. This inhibition effect may lead to false-negative results following PCR detection of a fungal pathogen. To prevent the false-negative results, it is recommended to include an internal amplification control [88,89] in each PCR or to test the DNA extract with a universal Plant PCR primer [90] to identify possible false-negative samples that could be due to inhibition. However, different protocols enable a significant removal of these inhibitory components. For example, in wine, polyphenols can be removed by a single centrifugation step of the wine prior to DNA extraction and subsequently, the cell pellets are resuspended in water. In some cases, it may not be sufficient especially in the presence of ropy strains that may bind more polyphenols than non-ropy strains [91]. Moreover, a series of chemicals such as polyvinyl polypyrrolidone (PVPP), skim milk, bovine serum albumin (BSA), sodium sulfite, or dimethyl sulfoxide [92] added during the extraction process or in the PCR mixture are valuable for overcoming inhibitory effect that may be encountered with fungi-infected plant DNA. In DNA extraction from wine, 1% PVPP is added after lysis [91] or after precipitation of cellular fragments [52] followed by vortexing for 10 s (this step can be repeated for highly tannic wines) and centrifugation (10,000×g for 5 min at 4°C) to remove PVPP. Spinning the final DNA extracts through a column filled with PVPP showed a good efficiency for the removal of polyphenols and humic acid (see below) in fungal-infected roots [93]. PVPP adsorbs polyphenols thereby preventing their interaction with DNA. To prevent oxidation of polyphenols, 1% mercaptoethanol can be added in the extraction buffer [94].

Removal of humic acid.   Purification protocol is a necessary step in nucleic acid extraction from soil. Several methods for separating and purifying nucleic acids from soil components have been investigated. A major soil component, humic acid, inhibits restriction enzyme digestion of DNA and PCR and also alters the results of membrane hybridizations [95].

Nucleic acid binding on hydroxyapatite columns was successful for extracting DNA and RNA [96,97]. Crude DNA extracts were also loaded onto PVPP minicolumns (e.g., Bio-Rad laboratories, Hercules, CA), centrifuged and the eluate were passed through a sepharose 4B (Sigma, St. Louis, MO) spin minicolumn by centrifugation [98].

Introduction of prelysis buffer in samples from forest soil with high organic matter or addition of PVPP or milk powder during the lysis step can strongly improve the nucleic acid extraction [99,100]. Moreover, the addition of BSA to the PCR mixtures overcomes the inhibitors present in fungal and root cells after nucleic acid extraction, and enables the amplification of DNA/RNA regions directly from ECM tissues [101].

Cesium chloride density gradient centrifugation was also shown to allow further enzymatic restriction of purified nucleic acids. Less time-consuming purification protocols have been used individually or in association: agarose gel electrophoresis to facilitate separation of DNA from humic contaminants coextracted, gel filtration resins including, for example, sephadex G200 [102] or Sepharose 4B [98], and other commercial purification columns (e.g., Wizard DNA clean-up system or Tip-500 columns from Qiagen).

Removal of polysaccharides. CTAB is a common reagent used to remove polysaccharides that can be coextracted from plant material. Indeed, negatively charged DNA/RNA might bind to positively charged polysaccharides and become trapped in polysaccharides. CTAB binds polysaccharides and therefore releases DNA/RNA.

Several methods to remove the contaminating substances from nucleic acid preparations have been described. However, these molecules can still be present in the final DNA extract and interfere with PCR amplification. One of the easiest procedures to circumvent the inhibition of PCR is to dilute the DNA extracted before amplification. For example, Mayer et al. diluted 20-fold of the DNA extracted from powdered pepper, paprika, and ground maize, to minimize the influence of possibly co-isolated inhibitory substances on PCR [103].

Removal of inhibitors from clinical samples. Clinical material usually contains numerous contaminant molecules that are inhibitory for the enzymes used in nucleic acid labeling and amplification such as the *Taq* polymerase. Some well-known inhibitory compounds include hemoglobin in whole blood, or compounds introduced to the clinical material (such as heparin) or nucleic extraction reagents (such as phenol traces), which might not be removed completely from the extracted nucleic acids [23,104,219]. Indeed, contaminating substances in blood sample have been found especially troublesome for PCR [105]. One study showed that molecular-based fungal detection was more often positive when serum was used for testing than when whole blood was analyzed [106]. A number of studies have also shown that vitreous and aqueous biological fluids may exhibit inhibitory properties when added directly to PCR [107–109]. Many protocols have been developed to extract nucleic acid from clinical liquid specimens, but no clear consensus has been achieved for optimally extracting, purifying, and concentrating fungal nucleic acids from such samples, especially during the purification step. However, many of the most popular and widely applied purification methods for nucleic acids include the posttreatment with spin columns, because they are quick and efficient in terms of recovery and adaptable to most standard laboratory equipments. For instance, although inhibitors have not been clearly identified, the sample purification step, using QIAamp system (Qiagen), has been shown to eliminate this problem [105]. Other inhibitors essentially include not only adjuvant introduced to the clinical of solid material such as formalin and paraffin but also nucleic extraction reagents such as phenol traces. Thus, the use of formalin and paraffin in the preparation of clinical sample requires the addition of xylene to remove these components [69,71].

11.2.2.1.3 Nucleic Acid Precipitation

Nucleic acid precipitation is the final step in DNA/RNA extraction. The two most known methods are ethanol and isopropanol precipitation. Ethanol precipitation is the commonest but isopropanol can alternatively be used because smaller volumes are needed [110]. Ethanol or isopropanol addition may be combined with the use of a salt such as sodium acetate or sodium chloride. The negative charge on the PO_4^{3-} groups on the nucleic acids is neutralized by the positively charged sodium ions. Ethanol facilitates the electrostatic attraction between Na^+ and PO_4^{3-} and makes the nucleic acid less hydrophilic. This leads to the precipitation of the nucleic acid.

11.2.2.2 In-House Protocols for Fungal DNA Extraction and Applications

11.2.2.2.1 Food Samples

Milk [53].   This method was used to prepare DNA for PCR amplification of ribosomal RNA (rDNA) genes and subsequent examination by denaturing gradient gel electrophoresis (DGGE) to study the natural distribution of yeast species in raw milk. Milk samples of 2 mL were centrifuged at $14,000 \times g$ for 10 min at 4°C, put at −80°C for 30 min to solidify the sample and the fat layer, separated by centrifugation, and was removed from the tube. After thawing, the supernatant was discarded and precipitated cells were resuspended in 1 mL of 8 g/L NaCl solution and transferred to a microcentrifuge tube containing 0.3 g of 0.5 mm diameter glass beads. The mixture was centrifuged at $14,000 \times g$ for 10 min at 4°C and the supernatant discarded. The cell/bead pellet was resuspended in 300 μL of breaking buffer (2% Triton X-100, 1% SDS, 100 mM Tris pH 8.0, and 1 mM EDTA pH 8.0) and 300 μL of phenol/chloroform/isoamyl alcohol (25:24:1) were added. The cells were then homogenized in a bead beater three times, each for 30 s at maximum speed at room temperature. TE of 300 μL (10 mM Tris and 1 mM EDTA pH 7.6) was added and the tubes were centrifuged at $12,000 \times g$ for 10 min at 4°C. The aqueous phase was collected and the DNA was precipitated with 1 mL ice-cold absolute ethanol. After centrifugation at $14,000 \times g$ for 10 min at 4°C, the pellet was dried under vacuum at room temperature and resuspended in 50 μL of sterile distilled water containing 2 IU DNase-free RNase. The samples were then incubated at 37°C for 30 min before storage at −20°C.

Cheese—Protocol 1 [111].   Fungi DNA of *Penicillium camemberti* or *P. roqueforti* was extracted from two types of soft cheese, a blue mould cheese and a white mould cheese, both with 60% fat. PCR was then performed for the detection of *Penicillium* spp. in cheese by testing two sets of primers (internal transcribed spacer [ITS] region sequences), one which specifically identifies all members of *Penicillium*, and one which specifically recognizes *P. roqueforti* and *Penicillium carneum*. To obtain cheese samples containing *P. roqueforti*, Roquefort cheese (25 g) was homogenized in 250 mL water using stomacher 400 (Struers, Denmark) for 120 s at medium speed. A cheese sample containing *P. camemberti* was obtained scrapping fungal tissue (2 g) from the surface of Camembert cheese and stomaching for 120 s at medium speed in 20 mL water. Camembert and Roquefort cheese homogenates (4 mL) were digested with proteinase K in a 10 mL volume with 0.01 M Tris-HCl pH 8.0, 0.005 M EDTA, 0.5% SDS, and 250 μg/mL proteinase K. The mixture was incubated for 1 h at room temperature [110]. Two volumes of 2% CTAB, 0.1 M Tris-HCl pH 8.0, 20 mM EDTA, and 1.4 M NaCl were added to a 500 μL aliquot and the mixture was incubated at 65°C for 10 min and centrifuged at $20,000 \times g$ for 10 min. The supernatant was extracted in an equal volume of chloroform by inversion and centrifuged at $20,000 \times g$ for 10 min. An equal volume of 1% CTAB, 50 mM Tris-HCl pH 8.0, and 10 mM EDTA was added to the supernatant, which was held at room temperature for 30 min, centrifuged for 10 min, and the supernatant was discarded. To the pellet, 1 M NaCl (450 μL) and 96% ethanol (900 μL) were added. After 5 min, the mixture was centrifuged for 5 min at $20,000 \times g$ and the supernatant was discarded. The DNA was washed with 80% ethanol, and dissolved in 15 μL of TE buffer (10 mM Tris-HCl pH 8.0 and 1 mM EDTA).

Cheese—Protocol 2 [112].   This simple fast method was used to isolate total RT-PCR quality microbial (yeast and bacteria) DNA from smeared soft cheese. Cheese smear of 1 g was homogenized in 400 μL of TE buffer containing 25 μL of 10% *N*-lauroyl sarcosine. Suspensions were treated with 50 μL of proteinase K (40 mg/mL) for 2 h at 55°C and then were transferred to a 2 mL tube containing 20 μL of 10% SDS, 0.2 g of acid-washed glass beads (200–300 μm), and 1 volume of phenol/chloroform/isoamyl alcohol (25:24:1). Tubes were shaken twice for 40 s in Savant FastPrep instrument (Savant, Farmingdale, NY) and then centrifuged at 14,000 rpm for 20 min. The upper phase containing DNA was precipitated with absolute ethanol at −20°C. After centrifugation, pellets were washed with 70% ethanol, dried and dissolved in 200 μL of TE buffer. The DNA extracted was then purified using Qiaquick PCR purification kit (Qiagen) prior to PCR amplification of the 16S rRNA gene.

Fermented sausages [113,114].   The total microbial community (yeast and bacteria) was profiled by the DGGE method without cultivation by analyzing the DNA that was directly extracted from the salami and naturally fermented sausage samples. Samples of 10 g were homogenized in a stomacher bag with 20 mL of saline–peptone water for 1 min. Each preparation had settled for 1 min and 1 mL of supernatant was transferred into a screw-cap tube containing 0.3 g of glass beads with a diameter of 0.5 mm and centrifuged at 4°C for 10 min at $14,000 \times g$. The resulting pellet was treated with 1 mL of petrol ether/hexane (1:1) for 10 min at room temperature to extract lipids. A second centrifugation was performed, as described above, and the pellet was resuspended in 150 μL of proteinase K buffer (50 mM Tris-HCl, 10 mM EDTA pH 7.5, and 0.5% SDS). Proteinase K of 25 μL (25 mg/mL) was added, and treatment at 65°C for 1 h was performed. After this step, 150 μL of 2x breaking buffer (4% Triton X-100, 2% SDS, 200 mM NaCl, 20 mM Tris pH 8.0, and 2 mM EDTA pH 8.0) was mixed in tubes, and 300 μL of phenol/chloroform/isoamyl alcohol (25:24:1, pH 6.7) was added. Then, three 30 s treatments at maximum speed, with an interval of 10 s, were performed in a bead beater. The tubes were centrifuged at 12,000 × g at 4°C for 10 min, and the aqueous phase was collected and mixed with 1 mL of ice-cold absolute ethanol. The DNA was precipitated at 14,000 × g at 4°C for 10 min, and the pellets were dried under vacuum at room temperature. Sterile water of 50 μL was added, and a 30 min incubation at 45°C facilitated the nucleic acid solubilization. DNase-free RNase of 1 μL was added to digest the RNA by incubation at 37°C for 1 h.

Sourdough [115]. This method was used to monitor the dynamics of yeasts in sourdough fermentation processes by PCR–DGGE. For the extraction of total DNA from sourdough samples, 10 g was homogenized for 5 min in a stomacher bag containing 90 mL of saline–tryptone diluent. An aliquot (50 mL) was centrifuged at 4°C for 5 min at 200 × g (dough made with rye flour) or for 5 min at 1500 × g (dough made with rye bran). Finally, to harvest the cells, the supernatant was centrifuged for 15 min at 5000 × g, and the cell pellet was stored at −20°C. The frozen pellet was thawed on ice and washed three times with 1 mL of phosphate-buffered saline and once with 1 mL of water. The pellet was resuspended in 130 μL of lysis buffer (6.7% sucrose, 50 mM Tris-HCl pH 8.0, and 1 mM EDTA) and 6 μL of an enzyme mixture of zymolyase (Seikagaku America, Falmouth, MA) (12 mg/mL), lysing enzyme from *Trichoderma harzianum* (Sigma) (40 mg/mL), and lyticase (Sigma) (20 mg/mL). After incubation for 1 h at 37°C, 30 μL of NaCl (5 M) and 25 μL CTAB (10% CTAB in 0.7 M NaCl) were added, and the mixture was incubated for 10 min at 65°C. Afterward, 20 μL of proteinase K solution (15 mg/mL) and 10 μL of SDS (20%) were added, and the mixture was incubated for 30 min at 60°C. Finally, 200 μL of phenol (65°C, pH 7.0) was added and mixed, and the mixture was incubated for 6 min at 65°C. After the mixture was cooled on ice, 220 μL of Tris-HCl (10 mM, pH 8.0) was added, and the mixture was extracted twice with equal volumes of phenol/chloroform/isoamyl alcohol (25:24:1) and twice with chloroform. After ethanol precipitation, the DNA was dissolved in 100 μL of Tris-HCl (10 mM, pH 8.0).

Bread, thin breadsticks, crackers, cookies, and cakes [87]. This method was applied to detect and identify *S. cerevisiae* in some *Triticum*-related bakery products by PCR amplification of the ITS region of rDNA. Samples of 50–100 mg of food were extracted with 600 μL of lysis buffer (2% CTAB, 0.1 mM Tris-HCl pH 9.0, 1.4 mM NaCl, and 20 mM EDTA pH 8.0) and 5 μL of 20 mg/mL proteinase K was added. The samples were incubated for 1 h at 65°C with occasional shaking and centrifuged at 8000 × g for 15 min. The supernatant was then transferred to a clean tube. DNA was extracted with phenol/chloroform/isoamyl alcohol in a 25:24:1 ratio, again with chloroform and precipitated with cold isopropanol at −20°C. After centrifugation at 8000 × g for 15 min, the pellets were washed twice with 70% ethanol. The dried pellets were resuspended in 100 μL of TE buffer and the concentration of DNA estimated by absorbance at 260 nm.

Grape berries [94]. This method was used to monitor and quantify *A. carbonarius*, the main species responsible for the production of ochratoxin A, in wine grapes by quantitative real-time PCR assay. Extraction of DNA from grape berries was performed by using conventional extraction and cleanup through EZNA Hi-bond spin columns (Omega Bio-teck). A portion of fresh and frozen grape berries (300 mg) was incubated with 1.5 mL extraction buffer (1 M Tris-HCl pH 8.0, 1.4 M NaCl, 20 mM EDTA, and 3% CTAB) and 15 μL β-mercaptoethanol for 90 min at 65°C under constant shaking on an orbital shaker. After incubation, samples were centrifuged at 6500 rpm for 5 min at 4°C and the supernatant was collected in a 2 mL Eppendorf tube. One volume of chloroform/isoamyl alcohol (24:1) was added and samples mixed and centrifuged at 6500 rpm for 20 min at 4°C. The upper aqueous phase was transferred into another tube, adding 0.1 volume of 10% CTAB. Again, 1 volume of chloroform/isoamyl alcohol (24:1) was added and samples mixed and centrifuged at 6500 rpm for 20 min at 4°C. The upper aqueous phase was transferred into another tube, adding 0.1 volume of cold 2-propanol. Samples were then incubated for 60 min at −80°C and centrifuged for 20 min at 13,000 rpm. The pellet was dissolved in 300 μL of sterile water and processed according to the EZNA fungal DNA miniprep kit protocol (Omega Bio-tek, Doraville, Georgia), starting from step 8 of protocol B that involves DNA cleanup through Hi-bond spin column. In the final step, DNA was eluted in 100 μL of deionized water.

Fermented cocoa beans [66].   This method was used to identify and monitor the dynamics of yeasts associated with cocoa fermentations using PCR–DGGE. Pulp was scraped off the beans with a sterile scalpel, transferred to a freeze-drying vial, and stored at −20°C for at least 24 h before treatment in a Heto FD3 freezedrier (Heto Lab Equipment, Allerød, Denmark). The freeze-dried cocoa pulp was ground thoroughly with a sterile pestle. Homogenized freeze-dried pulp of 30 mg was transferred to a FastPrep vial (Bio 101, Vista, CA) containing a 1/4 in. ceramic sphere and a garnet matrix, as supplied by the manufacturer. DNA extraction buffer I of 400 μL (150 mM EDTA and 225 mM NaCl, pH 8.5) was added and the mixture homogenized for 10 s (speed 4) in a FastPrep instrument (Bio 101). Following homogenization, 180 μL lysozyme (50 mg/mL, Sigma) was added and the samples are incubated at 37°C for 30 min with rigorous whirli mixing every 10 min. Subsequently, 12 μL of 25% SDS (Pharmacia Diagnostics, Uppsala, Sweden) and 12 μL proteinase K (20 mg/mL, Merck, Darmstadt, Germany) were added. After incubation for 60 min at 37°C (mixing every 10 min), 200 μL hot (90°C) DNA extraction buffer II (100 mM EDTA, 400 mM Tris-HCl, 400 mM Na_2HPO_4 buffer, pH 8.0, 5.55 M NaCl, and 4% CTAB [Sigma] pH 8.0] and 36 μL 25% SDS were added. After mixing, the samples were incubated for 5 min at room temperature and subsequently subjected to bead beating in a FastPrep instrument (3 × 30 s, speed 5.5). Following bead beating, the samples were subjected to 3 freeze/thaw cycles (−80°C for 20 min, 65°C for 20 min) and subsequently centrifuged (5000× g, 2 min, room temperature) to pellet the debris and the CTAB complex. The supernatant was transferred to a 2 mL vial, 800 μL chloroform/isoamyl alcohol (24: 1) was added, and the solution was mixed for 30 s. The aqueous phase was recovered by centrifugation (14,000× g, 5 min, room temperature) and the DNA in the supernatant further purified with a QIAamp DNA mini kit (Qiagen), following the instructions of the manufacturer. Purified DNA was stored at −20°C.

Grains—Protocol 1 [116].   This method was used in order to develop a quantitative real-time PCR assay to quantify trichothecene-producing *Fusarium* spp. CTAB buffer of 30 mL (sorbitol, 23 g; *N*-laurylsarcosine, 10 g; CTAB, 8 g; NaCl, 87.7 g; PVPP, 10 g; and water, 1 L) was added to a 10 g sample of crushed wheat grain in a 50 mL centrifuge tube, mixed and incubated at 65°C for 16 h. Potassium acetate of 10 mL (5 M) was added and mixed and the tube was frozen for 1 h at −20°C. The tubes were thawed and the contents were mixed and centrifuged (3000× g, 15 min). A 1.3 mL aliquot of supernatant was removed and added to 0.6 mL of chloroform in a 2 mL Eppendorf tube. The contents of the tubes were mixed by gentle inversion for 1 min and then centrifuged (12,000× g, 15 min). A 1 mL aliquot of the aqueous phase was removed to a fresh tube containing 0.8 mL of 100% isopropanol. The contents of the tubes were mixed by gentle inversion for 1 min and the tubes incubated at 18°C for 30 min and then centrifuged (6000× g, 15 min). The resulting DNA pellets were washed twice with 44% isopropanol and then air-dried. Pellets were resuspended in 200 μL of TE buffer (10 mM Tris-HCl and 1 mM EDTA pH 8.0) at 65°C for 1 h before storage at 4°C. Total DNA was quantified by spectrophotometry.

Grains—Protocol 2 [117,118].   This method was used in order to detect *Fusarium* spp. by PCR. Two different ultrasonic processors were used throughout the study. For cereal samples exceeding 10 g, a model UP 200S processor (Dr. Hielscher GmbH, Stannsdorf, Germany) equipped with a 14 mm diameter steel sonotrode (model S14, energy density = 105 W/cm²) was applied. Ultrasonification of pure DNA and a single infected kernels was performed in 2 mL Eppendorf tubes using a model 50S processor equipped with a 3 mm diameter sonotrode (model MS3, energy density = 460 W/cm²). All sonications were done at maximum amplitude. Sonotrodes were rinsed with ddH_2O and absolute ethanol after each sample. Preparation of fungal DNA from cereal samples was achieved by ultrasonification for 1 min with the model S14 sonotrode in 25 mL lysis buffer (1 L contained 66 mM EDTA, 33 mM Tris, 3.3% Triton X-100, 1.65 M guanidinium-HCl, 0.825 M NaCl, 6% PVPP-40T, and ddH$_2$O of 1 L, adjusted to pH 7.9) added to 55 mL ddH$_2$O in a sterile 50 mL conical screw-cap plastic tube. Following ultrasonification, 0.8 mL was taken from the supernatant and 0.4 volume (stored at −20°C) absolute ethanol was added. This mixture was spun through DNA extraction columns supplied with High pure PCR template preparation kit (Roche, Mannheim, Germany) by sequentially applying two portions of 600 μL. All centrifugation steps were done at 25°C with a bench-top centrifuge at 5900×g. The extraction was performed according to the manufacturer's recommendation. DNA was eluted with two rinses of 100 μL elution buffer, preheated to 72°C, into a sterile 1.5 mL Eppendorf tube. In an alternative lysis protocol, 10 g samples of naturally infected wheat or wheat spiked with infected kernels were rigorously shaken by hand for 60 s, instead of ultrasonification, in the same buffer. Extraction of DNA was performed principally as described above. This DNA preparation technique was also used successfully to prepare DNA from wheat artificially infected with *Aspergillus flavus* [119]. It was used to detect *Alternaria alternata* in tomato products and *Zygosaccharomyces rouxii* in honey and marzipan (unpublished results).

Grains—Protocol 3 [87].   This method was used to identify *S. cerevisiae* by PCR amplification of the ITS region of rDNA in order to understand the fermentation of bread and bakery products. Samples of 50–100 mg of food were extracted with 600 μL of lysis buffer (2% CTAB, 0.1 mM Tris-HCl pH 9.0, 1.4 mM NaCl, and 20 mM EDTA pH 8.0) and 5 μL of 20 mg/mL proteinase K was added. The samples were incubated for 1 h at 65°C with occasional shaking and centrifuged at 8000×g for 15 min. The supernatant was then transferred to a clean tube. DNA was extracted with phenol/chloroform/isoamyl alcohol in a 25:24:1 ratio, again with chloroform and precipitated with cold isopropanol at −20°C. After centrifugation at 8000×g for 15 min, the pellets were washed twice with 70% ethanol. The dried pellets were resuspended in 100 μL of TE buffer and the concentration of DNA estimated by absorbance at 260 nm.

Figs [120].   This method was used to detect aflatoxinogenic *A. flavus* by a PCR reaction. Fungal material from the surfaces of the figs was isolated, frozen in liquid nitrogen, and grinded, and then resuspended in lysis buffer (50 mM EDTA and 0.2% SDS pH 8.5). This suspension was heated at 68°C for 15 min and then centrifuged for 15 min at 15,000×g. After centrifugation, 1 volume of 7 mL of the supernatant was transferred to a new centrifuge tube and 1 mL of 4 M sodium acetate was added. This solution was placed on ice for 1 h and centrifuged for 15 min at 15,000×g. After centrifugation, 6 mL of the supernatant was transferred to a fresh tube. The solution was phenol extracted and the isolated DNA was precipitated by the addition of 2.5 volumes of ethanol. The concentration of DNA was determined by a densitometer (Bio-Rad, Model GS-670).

Wine and wine-related samples (berry washing solution, equipment washing solution).   To our knowledge, there is not any in-house protocol published for DNA extraction from fungal spores found in wine and berry washing solution. The DNA extraction method described below has only been used to study yeast diversity in wine [121] and in berry washing solution [65] using PCR–DGGE. For yeast DNA extraction from wine, must, and winery equipment cleaning solution, the cell pellets were obtained using 1–100 mL of the sample, depending on the cell density, by centrifugation at 10,000×g for 10 min at 4°C, and the cell pellets were stored at −20°C until extraction. Yeasts from the surface of grape berries can be harvested as follows prior to DNA extraction:

1. Harvesting yeast at the surface of grape berries: To harvest yeast at the surface of grape berries, Prakitchaiwattana et al. used 50 g of berries that were randomly and aseptically removed from the bunches and combined to give 50 g samples [65]. These samples were rinsed in 450 mL of 0.1% peptone water with 0.01% Tween 80 by orbital shaking in a flask at 150 rpm for 30 min. The rinse was then centrifuged at 16,000×g for 15 min at 4°C. The

sediment microbial cells were taken up in a small volume of 0.1% peptone water, transferred to a 1.5 mL cryogenic tube, centrifuged at 16,000×*g* for 15 min at 4°C, and the cell
pellet is stored at −20°C until DNA extraction.

2. Extraction [121]:   For the DNA preparation, the cell pellet samples were resuspended in
1 mL of an 8 g/L NaCl solution and transferred to a microcentrifuge tube containing 0.3 g
of 0.5 mm diameter glass beads (BioSpec Products Inc., Bartlesville, OK). The cell–bead
mixture was centrifuged at 11,600×*g* for 10 min at 4°C, and the supernatant was
discarded. The cell–bead mixture was resuspended in 300 μL of breaking buffer (2%
Triton X-100, 1% SDS, 100 mM NaCl, 10 mM Tris pH 8.0, and 1 mM EDTA pH 8.0) and
300 μL of phenol/chloroform/isoamyl alcohol (50:48:2). The cells were then homogenized
in a bead beater instrument (FastPrep; Bio 101) 3 times for 45 s each at a speed setting of
4.5. The mixture was then centrifuged at 11,600×*g* for 10 min at 4°C, and the aqueous
phase was removed to another microcentrifuge tube.

3. Purification: In the method described above [121], DNA was further purified using
DNeasy plant mini kit (Qiagen). An in-house method for purification [52] is described
below. Residual polyphenols were precipitated after addition of a 10% PVPP (Sigma) solution to reach a 1% PVPP concentration in the aqueous phase and vortexing at high speed
for 10 s. For highly tannic wines, this step can be repeated. After centrifugation (10,000×*g*
for 5 min at 4°C), the supernatant was transferred to a 1.5 mL microcentrifuge tube containing 500 μL of isopropanol. The tube was gently mixed by inversion until a visible mass of
DNA could be seen and left at −20°C for 3 h. After centrifugation (10,000×*g* for 20 min at
4°C), 300 μL of a room temperature 70% ethanol solution was added to the pellet before a
final centrifugation (10,000×*g* for 5 min at 4°C). Ethanol was carefully removed and the
tube dried. DNA was resuspended in 50 μL of sterile water containing 2 U of RNase and
rehydrated overnight at 4°C before storage at −20°C.

Beer [122].   This method was used to identify, by PCR, genetically engineered yeast strains that
may have been used in the brewing process.

Protocol 1:   Beer of 2 mL was centrifuged at 14,000 rpm for 15 min. The pellet was resuspended
in 0.5 mL of extraction buffer containing 100 mM Tris-HCl pH 8.0, 1.4 M NaCl, 20 mM EDTA,
and 20 g/L CTAB, incubated at 65°C for 30 min, and centrifuged at 16,000×*g* for 10 min. The
supernatant was mixed with an equal volume of Tris-EDTA (TE: 100 mM Tris and 1 mM EDTA)-
saturated phenol, vortexed and centrifuged at 16,000×*g* for 10 min. The aqueous (upper) phase was
treated with 1 volume of phenol/chloroform (1:1, v/v), vortexed and centrifuged as above, and
further extracted by adding 200 μL chloroform. For DNA precipitation, 1 volume of isopropanol
was added to the aqueous phase, and after 10 min incubation at room temperature, the pellet was
collected by centrifugation at 16,000×*g* for 10 min, washed with 70% (v/v) ethanol, air-dried and
finally redissolved in 100 μL TE buffer.

Protocol 2:   Beer of 2 mL was boiled, cooled at room temperature, and passed through a DNA-
binding membrane (BioTrace HP 0.45 μm, diameter 15 mm, Gelman Sciences, Ann Arbor, MI). For
membrane filtration, a self-made Plexiglas manifold was used, which consisted of two plates having
six drill-holes of 12 mm diameter that were screwed together with membrane filters between them.
The membranes were first conditioned by passing 5 mL of prewarmed 0.5% (w/v) Triton X-100
solution through each well-using vacuum. Beer samples treated as described were passed through
the filters followed by washing with 3 mL of prewarmed Triton X-100 solution. Filter disks were
removed and transferred into 2 mL Eppendorf tubes, where they were boiled with 200 μL water for
15 min and cooled to room temperature. For solubilization, 500 μL chloroform was added and the
tubes centrifuged at 14,000×*g* for 5 min. The aqueous phase was transferred into another tube and
phenol extracted. Finally, the DNA was precipitated, collected by centrifugation, washed with 70%
(v/v) ethanol, air-dried, and redissolved in 20 μL of water.

11.2.2.2.2 Clinical Samples

Whole blood sample [61].   Maaroufi et al. developed a Taqman-based detection assay of *C. albicans* in blood samples. Using a species-specific probe and the following whole blood DNA extraction protocol, they showed that the sensitivity and specificity of their assay were of 100% and 97%, respectively, compared with the results of blood culture. The fungal pellet was resuspended and incubated in 50 mM NaOH and neutralization with 1 M Tris-HCl (pH 7.0) were not necessary, and the rates of centrifugation for the leukocyte lysis and spheroplast genesis steps were reduced to 1500 and 2000 × *g*, respectively. Briefly, after the hypotonic lysis of erythrocytes in 5 mL of EDTA-anticoagulated blood samples with erythrocyte lysis buffer (10 mM Tris pH 7.6, 5 mM $MgCl_2$, and 10 mM NaCl) and the enzymatic lysis of leukocytes with leukocyte lysis buffer (10 mM Tris pH 7.6, 10 mM EDTA, 50 mM NaCl, 0.2% SDS, and 200 µg/mL of proteinase K [Roche Diagnostics, Indianapolis, IN]), the spheroplasts were directly generated by incubation of the pellets with 500 µL of lyticase buffer (10 mM Tris pH 7.5, 1 mM EDTA, 0.2% β-mercaptoethanol, and 1 U of recombinant lyticase [ICN Biomedicals, Aurora, OH] per 100 µL) for 45 min at 37°C. Finally, spheroplasts lysis and DNA extraction were accomplished with the QIAmp tissue kit (Qiagen), according to the instructions of the manufacturer. The DNA recovered in 100 µL of elution buffer was immediately analyzed or stored at −20°C until testing.

Nail specimens [72,73].   Whole nails and relatively large nail fragments are cut into small pieces with a surgical blade. DNA from nail samples is extracted by a 10 min incubation of the sample in 100 µL of extraction buffer (60 mM $NaHCO_3$, 250 mM KCl, and 50 mM Tris pH 9.5) at 95°C and by subsequent addition of 100 µL anti-inhibition buffer (2% BSA). In the original patent description, after vortex mixing, the DNA-containing solution is directly used for downstream molecular analysis. It can also be stored at −20°C. However, the protein content should not exceed 3% (w/v). It is advisable otherwise to proceed with a standard phenol/chloroform extraction, followed by an ethanol precipitation of the nucleic acids, as mentioned above. Purification of extracted nucleic acids using Qiaquick columns (Qiagen) or the QIAmp system (Qiagen) may also improve further PCR yields.

Fresh stool samples modified [75,123].   The following method has been successfully used to extract total microbial DNA from stool samples. It can be applied to recover DNA from fungal contaminants. A preincubation step of stool samples with lyticase or other fungi lysing enzyme may increase the yield. Stool sample of 1 g was concentrated by centrifugation at 4000 × *g* for 10 min and resuspended in 1 mL of 4 M guanidine thiocyanate—0.1 M Tris pH 7.5 and 150 µL of 10% *N*-lauroyl sarcosine. Mix was then transferred to a 2 mL screw-cap tube (Sarstedt, Newton, NC) containing 0.2 g of acid-washed glass beads (212–300 µm; Sigma). After the addition of 500 µL of 5% *N*-lauroyl sarcosine and 0.1 M phosphate buffer pH 8.0, the 2 mL tube was incubated at 70°C for 1 h. Tubes were then shaken twice for 40 s in the Savant FastPrep instrument. PVPP (15 mg) was added to the tube, which was vortexed and then centrifuged for 3 min at 12,000 × *g*. After recovery of the supernatant, the pellet was washed with 500 µL of TENP (50 mM Tris pH 8.0, 20 mM EDTA pH 8.0, 100 mM NaCl, and 1% PVPP) and centrifuged for 3 min at 12,000 × *g*. The new supernatant was added to the first supernatant. The washing step was repeated three times. The pooled supernatants (about 2 mL) were briefly centrifuged to remove particles and then split into two 2 mL tubes. Nucleic acids were precipitated by the addition of 1 volume of isopropanol for 10 min at room temperature and centrifuged for 15 min at 20,000 × *g*. Pellets were resuspended and pooled in 450 µL of 100 mM phosphate buffer pH 8.0, and 50 µL of 5 M potassium acetate. The tube was placed on ice for 90 min and centrifuged at 16,000 × *g* for 30 min. The supernatant was transferred to a new tube containing 20 µL of RNase (1 mg/mL) and incubated at 37°C for 30 min. Nucleic acids were precipitated by the addition of 50 µL of 3 M sodium acetate and 1 mL of absolute ethanol. The tube was incubated for 10 min at room temperature, and nucleic acids were recovered by centrifugation at 20,000 × *g* for 15 min. The DNA pellet was finally washed with 70% ethanol, dried, and resuspended in 400 µL of TE buffer.

11.2.2.2.3 Environmental Samples
ECM fungi DNA extraction by the modified CTAB protocol [124–126].   Fungal material of 10–50 mg
was collected and transferred to an Eppendorf plastic tube and dipped in liquid nitrogen. The frozen
sample was crushed with a plastic micropestle specially designed to grind small samples in 1.5 mL
Eppendorf tubes. Immediately, 300–500 µL CTAB buffer (100 mM Tris-HCl pH 9.0, 20 mM EDTA
pH 8.0, 1.4 M NaCl, 2% CTAB, 0.2% β-mercaptoethanol, and freshly added guanidinium isothiocyanate buffer [110]) containing 0.1 mg proteinase K (Bioprobe PROK02, Montreuil-sous-Bois,
France) was added to ground mycelium. The mixture was vortexed briefly and incubated for 1 h at
65°C. A plastic or Styrofoam floating rack was useful for this and subsequent manipulations.
The extract was centrifuged for 5–10 min at 13,000×g to remove cell debris. The supernatant was
removed to a new 1.5 mL Eppendorf tube. The proteins were removed from the suspension by
sequential extractions with 500 µL of Tris-saturated phenol/chloroform/isoamyl alcohol (25:24:1).
The emulsion was centrifuged for 5–10 min at 13,000×g. The upper aqueous layer was transferred
to a new 1.5 mL Eppendorf tube. The aqueous supernatant was extracted with 500 µL of chloroform/
isoamyl alcohol (24:1) and centrifuged as above. The upper aqueous layer was removed to a new
1.5 mL Eppendorf tube as above. A volume of 1 mL of isopropanol was added to the aqueous phase,
mixed gently, and incubated for 1 h at −70°C or overnight at −20°C. The mixture was centrifuged
at 13,000×g for 30 min to pellet the DNA. The supernatant was removed and the pellet retained.
The DNA pellet was washed with 150 µL of 70% ethanol and dried at room temperature. Finally,
the DNA pellet was solubilized in 30–50 µL of sterile ultrapure water or Tris-EDTA buffer
(10 mM Tris-HCl pH 8.0 and 1 mM EDTA). The DNA solution should be stored at −20°C if not
used immediately.

11.2.2.3 Commercial Kits

Most commercially available DNA or RNA purification kits use spin columns containing silica
resin, gel membrane, or magnetic beads that bind the nucleic acids when *ad hoc* binding buffers
are used.

Commercial kits using a silica-gel membrane technology are the most popular technology in the
recent literature. Several companies have developed a wide range of products that selectively bind
either RNA or DNA and separate nucleic acids within certain size parameters. Nucleic acids contained in a lysate are adsorbed to the silica-gel membrane in the presence of chaotropic salts, which
remove water from hydrated molecules in solution. Thus, polysaccharides and protein present in the
lysate do not adsorb and are removed. After washing steps, pure nucleic acids are eluted under low-
or no-salt conditions in small volumes, ready for immediate use.

The magnetic-bead technology allows an even faster purification process. Beads or particles
with high affinity to DNA or RNA molecules bind to them, then are washed, and the target molecules are finally eluted. This technology does not require centrifugation steps, but the magnetic
particles and the bound target molecules are captured and concentrated using magnetic sticks or
magnetic holders.

These protocols are quicker and allow serial analyses, and several companies have now developed automatic workstation that are able to complete the whole extraction step in sometimes less
than 1 h. Thanks to their robustness and celerity, some of these commercial extraction kits may be
adapted for on-site detection of regulated or quarantine fungi [127].

11.2.2.3.1 Food Samples
A large number of kits have been used in the published literature for DNA extraction and purification from food samples. They are used either for DNA extraction, DNA purification, or both.
These methods are rapid, relatively inexpensive, and are quite safe because they do not use
organic solvents for DNA purification. A list of kits that have been used successfully is shown
in Table 11.6.

TABLE 11.6

Commercial Kits Used for Fungal DNA Extraction from Food Samples

Sample	Kit	Extraction	Purification	References
Grains of wheat and barley	High pure PCR template preparation kit (Roche, Mannheim, Germany)	Yes	Yes	[117,118]
Grains of malt and barley	FastDNA spin kit for soil (Qbiogene, Irvine, CA)	Yes	Yes	[154]
Grains of wheat	GenomicPrep (Pharmacia Biotech, Uppsala, Sweden)	Yes	Yes	[87]
Bakery products	GenomicPrep (Pharmacia Biotech)	Yes	Yes	[87]
Powdered pepper, paprika, or ground maize kernels	DNeasy plant mini kit (Qiagen, Valencia, CA)	Yes	Yes	[103]
Grape berries	DNeasy plant kit (Qiagen)	No	Yes	[65]
	EZNA fungal DNA miniprep kit (Omega Bio-teck, Inc., Doraville, GA)	Yes	Yes	[86]
	EZNA fungal DNA miniprep kit (Omega Bio-teck, Inc.)	No	Yes	[94,220]
Green coffee beans	High pure PCR template preparation kit (Roche)	Yes	Yes	[155]
Wine	Lysing matrix B and C (Qbiogene), cell lysis, and protein precipitation solutions (Promega, Madison, WI)	Yes	Yes	[159]
	FastDNA spin kit for soil (Qbiogene)	Yes	Yes	[64]
	Prepman kit (PE Applied Biosystems, Foster City, CA)	Yes	Yes	[161]
	Masterpure yeast DNA purification kit (Epicenter, Madison, WI)	Yes	Yes	[163]
Wine	Cell lysis and protein precipitation solutions (Promega)	Yes[a]	Yes[a]	[52]
	Geneclean kit (Qbiogene)	No	Yes	[3]
	DNeasy plant kit (Qiagen)	No	Yes	[138,158,162,221]
Beer	InViSorb genomic DNA kit III (InVitek, Berlin, Germany)[b]	Yes	Yes	[122]
Milk and yoghurt	Wizard genomic DNA purification kit (Promega)	Yes[a]	Yes[a]	[55,56]

[a] With modifications and supplementary steps.

[b] Comparing the extraction procedures in terms of DNA yield and subsequent amplifiability of DNA, the InViSorb kit and membrane binding (described in Section 11.2.2.2.1) performed best, while CTAB extraction (described in Section 11.2.2.2.1) was less efficient. The procedures using the kit and the membrane yielded DNA amounts in the order of <10 ng DNA from 2 mL of beer, which were sufficient to serve as a template for PCR. The amplifiability of the DNA also indicates the absence of DNA polymerase inhibition, and the selective DNA binding is very efficient in terms of removal of inhibitory substances.

11.2.2.3.2 Clinical Samples

Numbers of nucleic acid methods developed for extraction from natural environments have been successfully applied to biological fluids (Table 11.7). Fredricks et al. used quantitative PCR assays to measure the recovery of DNA of two important fungal pathogens (*A. fumigatus* and *C. albicans*) from bronchoalveolar lavage fluid, subjected to six DNA extraction methods [24]. Differences among the DNA yields from the six extraction methods were highly significant. An extraction method based on

TABLE 11.7

Commercial Kits Used for Fungal DNA Extraction from Clinical Samples

Sample	Kit	Extraction	Purification	References
Bronchoalveolar lavage fluid	MPY, MasterPure yeast DNA purification kit (Epicenter, Madison, WI)	Yes	Yes	[24]
	FDNA, FastDNA kit (Qbiogene, Irvine, CA)	Yes	Yes	[24]
	UCS, UltraClean soil DNA isolation kit (MoBio, Inc., Solana beach, CA)	Yes	Yes	[24]
	MPPL, MasterPure plant leaf DNA purification kit (Epicenter)	Yes	Yes	[24]
	YL-GNOME, Yeast cell lysis kit + GNOME kit (Qbiogene)	Yes	Yes	[24]
	SM, SoilMaster DNA extraction kit (Epicenter)	Yes	Yes	[24]
Blood and soft tissues	DNeasy Blood and tissue kit (Qbiogene)	Yes	Yes	[222]
Soft tissues	DNeasy Plant mini kit (Qiagen, Valencia, CA)	Yes	Yes	[223]
Soft tissue and purulent specimens	QIAamp DNA mini kit (Qiagen), lysis solution	Yes	Yes	[62]
Nails	GenomicPrep cells and tissue DNA isolation kit (Amersham Biosciences, Piscataway, NJ)	Yes	Yes	[201]

enzymatic lysis of fungal cell walls (yeast cell lysis plus the use of Gnome kits) produced high levels of fungal DNA with *C. albicans*, but low levels of fungal DNA with *A. fumigatus* conidia or hyphae. Extraction methods employing mechanical agitation with beads produced the highest yields with *Aspergillus* hyphae (FastDNA kit and UltraClean soil DNA isolation kit). The MasterPure yeast method produced high levels of DNA from *C. albicans* but only moderate yields from *A. fumigatus*. In conclusion, the six extraction methods produce markedly differing yields of fungal DNA and thus can significantly affect the results of fungal PCR assays. No single extraction method was optimal for all organisms.

11.2.2.3.3 Environmental Samples

DNA or RNA can be extracted from dirty roots, rhizospheric soil, or bulk soil using a variety of purpose-designed kits (e.g., MoBio laboratories soil DNA and RNA isolation kits, Q-BIOgene FastDNA spin kit for soil, Qiagen DNeasy plant mini kits) as well as other protocols [100,128]. These different protocols have been applied to field or forest soils for mycorrhizal fungi nucleic acid extraction. Addition of CTAB or PVPP has also been reported [129] and optimizes results for recalcitrant samples, collected in forest soils or dead woody debris [130,131].

Landeweert et al. report fungal DNA extraction from soil samples and molecular identification of ECM fungi using a bead-beater, as described by Smalla et al. (1993) and DNA cleanup system (Promega) [132,133]. A number of authors compared DNA extraction methods from arbuscular mycorrhizal roots [134] and concluded that bead-beating was a rapid and simple root crushing method, which can be coupled with microbial DNA isolation kit (e.g., MoBio laboratories) or Qiagen DNeasy plant mini kit.

A squash blot method initially described by Langridge et al. for plant tissue may be used for fungal infected plant tissue [135]. The protocol consists in crushing plant tissue on a NaOH prewetted membrane. The membrane is subsequently rinsed with NaCl and Tris-EDTA buffers, then allowed to dry before storage or may be used immediately. FTA (Flinders Technology Associates Whatman Inc., Clifton, NJ) paper is a commercially available paper specially treated to bind and protect, from degradation, nucleic acids from a large scale of different tissues, including plant and fungal tissues. Mycelium of the fungal pathogen or potentially parasitized plant tissue is directly squashed on the FTA paper without any preliminary treatment and will be allowed to soak in it. The piece of paper is subsequently dried and can be stored at room temperature until utilization [136]. To release DNA, a small disk is punched from the FTA paper on which the DNA is bound and is washed then dried, and directly used for PCR applications.

11.2.3 RNA EXTRACTION PROTOCOLS AND APPLICATIONS

A possible bias with the direct use of microbial DNA for microbial quantification from food, clinical, and environmental samples is the quantification/identification of dead cells. Therefore, the detection of RNA by using, for example, real-time RT-PCR is a better indicator of cell viability than the detection of DNA since, conversely to RNA, DNA remain stable after cell death. However, gene expression of a specific RNA (mRNA or rRNA) may vary depending on external factors or growth phases resulting in under- or overestimation of a microbial population. Despite this, real-time RT-PCR has a great potential for the quantification of fungi in food [51,137] and beverages [138].

Moreover, yeasts and filamentous fungi starter cultures are used in many food industrial processes such as fermentation processes. In order to improve starter culture properties such as flavor production or resistance to stress, the knowledge of interrelated regulatory and metabolic processes within cells, through RNA analysis during food maturation, will provide new insights for the selection of starter cultures for the food industry [139]. To our knowledge, there are very few studies that deal with expression profiles of starter cultures directly in food. Indeed, most of the studies have been undertaken in model synthetic media mimicking food composition because of the need of standard and reproducible medium composition [140,141].

11.2.3.1 RNA Isolation Steps

In this section, detailed protocols for total RNA extraction and mRNA purification from fungi present in different samples are provided. In general, they are the same methods used for DNA extraction. They involve disruption and lysis of the starting material, followed by removal of proteins, DNA, and other contaminants. Nevertheless, the enzymatic digestion step is not recommended since lysis with proteinase K or lyticase is relatively slow and it is difficult to prevent endogenous RNases from degrading the RNA. Organic extraction is a classical technique that is often combined with use of strong denaturants, followed by alcohol precipitation.

Typically, the sample is mixed with phenol at acid pH. The phenol lyses cells and denatures the proteins in the sample. At acid pH, DNA in the sample is protonated, neutralizing the charge and causing it to partition into the organic phase. RNA remains charged and partitions into the aqueous phase. The two phases are separated by centrifugation, and the aqueous phase is reextracted with a mixture of phenol and chloroform, and then with chloroform to extract the remaining phenol.

11.2.3.2 Special Consideration for Different Sample Sources

Some sample sources have specificities in their RNA or contain substances that can cause problems in RNA isolation and analysis. These samples require special considerations, which are not generally necessary when working with standard material.

Influence of the pH on the specific RNA recovery from smeared soft cheese.   The ripened cheese is characterized by a high buffering capacity and pH values at the end of ripening can reach 7.8–8.0. Buffering capacity depends on the manufacturing technology used. The buffering capacity of a cheese matrix is higher when curd is rennet-type (gel with a strong cohesion) mainly [142]. In this curd type, the important mineralization, especially phosphate ions, generates the strong buffering capacity [143]. The smeared soft cheeses such as Livarot cheese are mixed-type (lactic and rennet-type mixed) but with rennet prevalence and consequently, buffering capacity is important. Another factor has to be considered in the chemical balance between alkaline and acid compounds that influences pH: during the ripening, alkaline compounds, such as aroma compounds (nitrogen-containing compounds, pyrazines) or ammonia are often produced by the smear microflora and can be effective in the chemical balance. Consequently, the extraction environment is not acid enough to remove proteins, lipids, and DNA by using phenol and chloroform separation and leaving RNA in the liquid phase [144]. DNA is always extracted with the RNA. It is difficult in such cheese medium to obtain acid condition for the specific RNA extraction.

11.2.3.3 In-House Protocols for RNA and mRNA Extraction

11.2.3.3.1 RNA Extraction from Food Samples

Cheese [145].   This study developed an effective method for the direct and simultaneous isolation of yeast and bacterial rRNA and genomic DNA from the same cheese samples. DNA isolation was based on a protocol used for nucleic acid isolation, with the combined use of the action of chaotropic agent (acid guanidinium thiocyanate), detergents (SDS and *N*-lauroyl sarcosine), chelating agent (EDTA), and a mechanical method (bead-beating system). The DNA purification was carried out by two washing steps of phenol/chloroform. RNA was isolated successfully after the second acid extraction step by recovering it from the phenolic phase of the first acid extraction (Figure 11.1). The novel method yielded pure preparation of undegraded RNA accessible for RT-PCR.

Wine.   This method of extraction in the presence of hot phenol [146] has been developed for preparation of high molecular weight RNA and used in transcriptional profiling and northern analysis in *S. cerevisiae* wine yeast in fermenting grape juice [5]. A full-detailed procedure is available online at http://web.wi.mit.edu/young/expression/ for total RNA preparation.

Fruit juices, yoghurt, fruit preserves, and milk.   This method was developed for the quantification, using real-time reverse-transcriptase PCR, of yeasts and molds contaminating yogurts and pasteurized food products [51]. Each sample was initially diluted 1:3 with sterile 0.9% NaCl and centrifuged at $5500 \times g$ for 5 min. Cells were resuspended in LETS buffer (200 mM LiCl, 20 mM EDTA, 20 mM Tris-HCl pH 8.0, and 0.4% SDS). Then 300 μL of acid phenol (pH 4.3; Sigma)/chloroform/isoamyl alcohol (25:24:1), 1 μL of diethyl pyrocarbonate (DEPC; Sigma), and about 60 mg of acid-washed glass beads were added. The preparations were treated by alternating 1 min cycles of vortexing with incubation on ice for about 5 min. Extracts were then centrifuged at $15,000 \times g$ for 10 min at 4°C. An equal volume of 2x CTAB buffer (2% CTAB 1% PVPP, 100 mM Tris-HCl pH 8.0, 20 mM EDTA pH 8.0, 1.4 mM NaCl, and DEPC-purified water) was added to the supernatant before extraction with an equal volume of chloroform/isoamyl alcohol (24:1). These steps were repeated until a clear interface between aqueous and organic layers was obtained after centrifugation. Total RNA was precipitated with 2 volumes of ice-cold 100% ethanol and 0.1 volume of 3 M potassium acetate and was left at −80°C for 1 h before the nucleic acids were pelleted at $15,000 \times g$ for 15 min at 4°C. The pellet was washed with 70% ethanol and resuspended in 30 μL of sterile DEPC-treated water. The sample was stored at −80°C until use.

Contaminating genomic DNA was removed from total RNA by incubation with 10 U of RNase-free DNase I (Boehringer Mannheim, Germany) in a 20 μL reaction mixture containing 10 μL of RNA, 4 U of RNase inhibitor (RNaseOUT; Invitrogen S. Giuliano Milanese, Italy), DNase assay buffer (100 mM sodium acetate and 0.5 mM $MgSO_4$), and DEPC-treated water. The reaction mixture was incubated first for 30 min at 37°C and then for 5 min at 60°C. A PCR was performed to check for any contaminating DNA as described below. When necessary, the DNase treatment was repeated. RNA concentrations were determined spectrophotometrically.

Beer.   This method was used to examine global gene expression profile of *S. cerevisiae* brewer's yeast subjected to brewing stress [4]. The fermented wort was transferred to a prechilled 3 L conical flask. Particular attention was given not to overexpose the fermented wort to air. In some cases, an overlay of

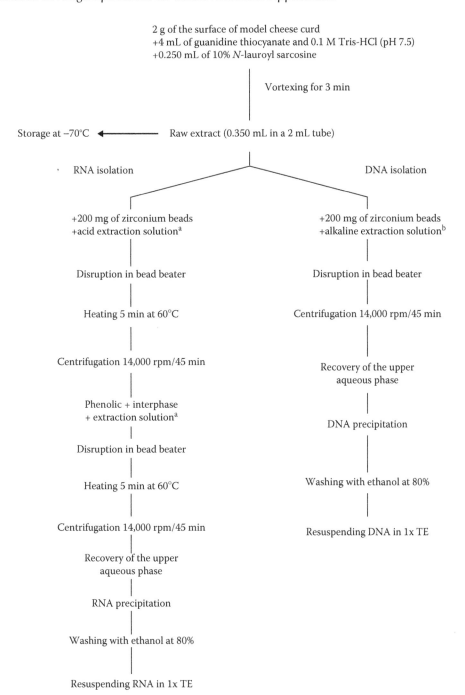

FIGURE 11.1 Flowchart for DNA and RNA extraction from smeared cheese.

[a] Acid extraction solution: 0.05 mL of 20% SDS solution + 0.30 mL of 0.1 M phosphate buffer pH 6.0 + 0.30 mL of 50 mM sodium acetate/10 mM EDTA pH 4.7 + 0.40 mL phenol/chloroform/isoamyl alcohol (25:24:1) pH 4.7.

[b] Alkaline extraction solution: 0.05 mL of 20% SDS solution + 0.30 mL of 0.1 M phosphate buffer pH 8.0 + 0.30 mL of 50 mM sodium acetate/10 mM EDTA pH 5.5 + 0.40 mL phenol/chloroform/isoamyl alcohol (25:24:1) pH 8.0.

mineral oil was used to ensure anaerobic conditions. At the start of fermentation (prior to pitching), one end of a long sterile rubber tubing (3 mm internal diameter) was inserted into the bottom of the fermentation vessel, while the other end was attached to a locked 50 mL syringe outside the vessel through an airtight stopper. This was used to remove 50 mL samples during fermentation. The aliquot was transferred to a prechilled centrifuge tube and centrifuged at $5000 \times g$ at 4°C for 10 min and the yeast pellet was collected. The yeast pellet was quickly resuspended in cold distilled water (50 mL) and centrifuged again. The cells were aliquoted into $2\,cm^3$ microfuge tubes and quick–frozen in dry ice and stored at −70°C.

Total RNA was extracted from the yeast pellet as described previously [147]. Approximately, 100–200 μL cell pellet was mixed with 2 volumes of phenol-saturated AE buffer (300 mM sodium acetate pH 5.5, 10 mM EDTA, and 1% SDS), phenol at 80°C (hot phenol), and 1 volume of acid-washed glass beads (size 80 μm). The cells were vortexed for 5 min. AE buffer of 1 volume was added and the process repeated. The tubes were transferred to an 80°C water bath and incubated for 5 min. The process was repeated once more. The cell lysate was centrifuged to separate the two phases and the aqueous phase was re-extracted with an equal volume of hot phenol. The aqueous phase was extracted at room temperature, once with a mixture of phenol/chloroform (1:1) and once with chloroform alone. The final aqueous phase was precipitated with 2.5 volumes of alcohol after adjusting the sodium acetate (pH 5.5) concentration to 300 mM. The total RNA pellet (ca. 200 μg) was dissolved in TNM buffer (10 mM Tris-HCl pH 7.5, 10 mM $MgCl_2$, and 50 mM NaCl) and any residual DNA contamination was removed by digestion with RNase-free DNase (Roche Biochemicals) at a concentration of 100 U/mL for 30 min at 37°C using buffer conditions recommended by the supplier. Following digestion, a commercially available kit (RNeasy, Qiagen) was used to reclaim the RNA in sterile distilled water.

11.2.3.3.2   mRNA Extraction from Food Samples

A procedure for the removal of tRNAs, rRNAs, and contaminated traces of DNA is given below as described by Körher and Domdey (1991) using oligo(dT)-cellulose [148]. For preparation of the oligo(dT)-cellulose column, 0.6–0.8 mL (wet volume in water) of oligo(dT)-cellulose is pipetted into a 15 mL column, washed with several column volumes of sterile water, washed with 10 mL of 0.1 M NaOH, and subsequently washed with sterile water until the pH of the effluent is neutral. The column is equilibrated with 50 mL high-salt buffer (HSB) (10 mM Tris-HCl pH 7.5, 0.2% SDS, 5 mM EDTA, and 0.5 M NaCl). The ethanol-precipitated total RNA (up to 20 ng) is dissolved in 2.5 mL low-salt buffer (LSB) (10 mM Tris-HCl pH 7.5, 0.2% SDS, and 1 mM EDTA), incubated at 65°C for 3 min, and cooled in an ice bath to room temperature. One-tenth volume of 5 M NaCl is added, and the RNA solution is applied to the column. For good binding of the poly(A)$^+$ RNA, the flow rate has to be adjusted to around 200 d/min (or 1 drop/5 s) and should be controlled from time to time. The effluent is collected, heated again to 65°C, cooled, and reapplied to the column. Subsequently, the column is washed with at least 100 mL HSB to remove most of the rRNA and tRNA. The bound poly(A)$^+$ RNA is eluted with 8 mL LSB, which is added in 0.5 mL portions. For efficient removal of rRNA and tRNA, a second passage through the column is recommended. To do so, NaCl is added to a final concentration of 0.5 M, and the sample is loaded on the freshly equilibrated (HSB) column. The eluted poly(A)$^+$ RNA is precipitated at −20°C overnight with 1/10 volume of 3 M sodium acetate (pH 5.3) and 3 volumes ethanol. A second ethanol precipitation helps to get rid of traces of SDS. The RNA is dissolved to a final concentration of 2 μg/μL in DEPC-treated water and stored at −70°C.

11.2.3.3.3   RNA Extraction from Environmental Samples

Ectomycorrhizal tips [149].   Tissues (100 mg) were ground in a mortar with liquid nitrogen, glass beads (106 μm, # G4649, Sigma), and diatomaceous earth suspension (# D3877, Sigma). The resulting powder was homogenized in an extraction buffer (100 mM Tris-HCl pH 8.0, 20 mM EDTA, 0.5 M NaCl, 0.5% SDS, 0.1 M β-mercaptoethanol, and phenol [aquaphenol, 5:1 (v:v)]), followed by incubation at 65°C. After the addition of chloroform (1:2 [v:v]), the extract was maintained on ice and then centrifuged at 4°C. The upper layer of the supernatant was collected and RNA was then precipitated by the addition of LiCl to a final concentration of 2 M and incubated on ice. After centrifugation, the

RNA pellet was resuspended in Tris-EDTA buffer (10 mM Tris-HCl pH 8.0 and 1 mM EDTA, pH 7.5) and purified twice by the phenol–chloroform procedure. RNA was finally precipitated at −20°C in 3 M sodium acetate (1:3 [v:v], pH 5.2) and 100% ethanol (2 volumes). The solution was centrifuged and the RNA pellet washed twice in 70% ethanol, and resuspended in diethyl pyrocarbonate-treated water (0.02%). To ensure that the RNA solution is clear from phenolics or humic acids, a supplementary cleaning step using the RNeasy plant mini kit (#74904, Qiagen) was added.

11.2.3.4   Commercial Kits

Table 11.8 summarizes the use of different commercial kits used for total RNA preparation and mRNA purification from food and clinical samples.

TABLE 11.8

Commercial Kits Used for Total RNA and mRNA Purification from Food and Clinical Samples

Food Samples	Total RNA Extraction	mRNA Purification	Target Taxon	Application	References
Beer	FastRNA RED kit (Qbiogene, Irvine, CA)	PolyATtract mRNA isolation system IV (Promega, Madison, WI)[a]	*Saccharomyces carlsbergensis*	DNA-based array	[224]
Wine	None	Oligotex kit (Qiagen, Valencia, CA)[a]	*Saccharomyces cerevisiae*	Microarray	[5]
Sake	None	RNeasy mini kit[b] (Qiagen)	*S. cerevisiae*	Microarray	[141]
Wheat	EZNA fungal RNA kit (Omega Bio-teck, Doraville, GA)	None	*Penicillium nordicum*	Real-time reverse-transcriptase PCR	[137]
Whole blood[c]	RLT lysis buffer (Qiagen, Hilden, Germany) + RNA secure (Ambion, Austin, TX)	RNeasy mini kit + QiaShredder spin columns (Qiagen) + TURBO DNA-freeTM kit (Ambion)	*Aspergillus*	NASBA-based on 18S rRNA gene	[23]
Soft clinical tissue	QIAzol lysis reagent (Qiagen) + Lysing matrix E (Qbiogene)	RNeasy MinElute cleanup kit or RNeasy mini, midi, or maxi kit (Qbiogene)		Amplification and microarray	Unpublished

[a] mRNA purification.

[b] Total RNA purification.

[c] Pretreatment: For RNA extraction from whole blood, 300 mL of RLT lysis buffer (Qiagen) was added to 100 mL of blood, followed by immersion in liquid nitrogen. Then, 20 mL of RNA secure (Ambion) was added, and the mixture was incubated for 20 min at 60°C. Isolation and purification of the RNA were performed according to the manufacturer's protocol by using the RNeasy mini kit and QiaShredder spin columns (Qiagen). Next, 80 mL of eluate was obtained and stored at −80°C until further use.

11.3 FUTURE DEVELOPMENT TRENDS

Advances in molecular technologies have provided new analytical tools as PCR technique for studying the diversity and biological activities of fungi associated with food production, environment, and also nosocomial outbreaks. Other technologies have been reported recently in the literature and are based on functional analyses: the search for specific mRNA, proteins, or metabolites has become increasingly popular. For well-known microorganisms, there are abundant investigations at the level of gene expression (transcriptomics), protein translation (proteomics), and more recently the metabolite network (metabolomics). The literature data are relatively scarce for other microorganisms, especially fungi from food, environmental, and clinical samples, despite the prevalence of these organisms in the biotechnology industry, and their importance as both human and plant pathogens. Some transcriptomic studies on filamentous fungi using the microarray technology allowed to detect a large diversity of pathogens in only one step [150,151], to estimate gene expression levels on a genome-wide level and to describe successful cross-species hybridizations in closely related filamentous fungal species such as *Fusarium* spp. [152] It is evident that these promising techniques offer advantages such as fast, specificity, and sensitivity analysis. However, they have weaknesses that are, for example, the presence of inhibitors and the small sample volume prior to labeling or PCR. It has been shown that nucleic acids extracted from complex biological samples could be directly hybridized on microarray but had a limited sensibility [153]. Although alternative solutions exist (protocol for amplification of all extracted nucleic acid; Berthet et al., unpublished data), a properly selected sample preparation step, as well as the utilization of an appropriate thermostable DNA polymerase or reverse transcriptase may circumvent some of these weaknesses and significantly increase the applicability of these methods.

Françoise Irlinger graduated from a national engineering school in Compiègne, France; she did her PhD in microbiology at the National Institute for Agronomic Research (INRA) in the AgroParisTech institute, Thiverval Grignon, France. She is currently a research engineer at the INRA. She has experience in the taxonomy of cheese coryneform bacteria and their detection and identification. Currently, her research focuses on the study of microbial community structure and functioning of surface-ripened cheeses (interactions between yeast and bacteria).

Nicolas Berthet received his pharmacy education at Grenoble, France. He obtained his PharmD in 2003 in medical mycology, where he worked on the determination of susceptibility of *Aspergillus* and *Candida* species to different benzimidazoles such as albendazole and triclabendazole and their use as novel antifungal agents. He then did his PhD training in molecular microbiology at the Institut Pasteur in Paris, France, where he worked on the development of high-density microarrays for the massive identification of both bacterial and viral pathogens and their pathogenicity determinants, that is, genes conferring antibiotic resistance and producing toxins. He developed different universal processes for total nucleic acid extraction and amplification directly from clinical samples. Currently, he is developing a different molecular strategy for the identification of emerging and unknown viruses from different clinical samples.

Tatiana Vallaeys is employed as a full-time researcher at the National Institute for Agronomic Research (INRA), France since 1993, with her education in microbiology at the École Normale Supérieure in Paris (France) and in statistics at the University of Lyon (Lyon, France). She obtained a PhD in molecular ecology at Lille University (Lille, France). She then specialized in environmental soil microbiology while working at the University of Cardiff (Cardiff, Wales, United Kingdom), the University of Adelaide (Adelaide, South Australia, Australia), Michigan State University (East Lansing, Michigan), and Centre d'Études Nucléaires (Mol, Belgium). She also spent 4 years at the Pasteur Institute in Paris (France), where she took part in genome sequencing projects and in a project aiming at the development of a multipathogen detection microarray and associated nucleic acid extraction methods. She later joined the team at the Laboratoire de Génie et Microbiologie des Procédés

Alimentaires at INRA, where she is developing microarray-based tools for investigating yeast–bacteria interactions during cheese ripening.

Valérie Vasseur is a microbiologist. She teaches microbiology, mycology, and molecular biology at the Brest Engineering School of Microbiology and Food Safety (University of West Brittany, Brest, France). Her main scientific research interest concerns stress in fungi. She has focused on the effect of temperature on the transcriptional response of *Penicillium glabrum*, a food spoilage filamentous fungus, by using subtractive suppressive hybridization associated with cDNA microarray in order to isolate potential molecular markers for thermal stress.

Renaud Ioos first graduated from a national agronomic engineering school in Bordeaux in France specializing in plant pathology. He did his PhD in plant pathology and mycology at the University of Nancy, Nancy, France and the National Institute for Agronomic Research, Nancy. He is currently a researcher in the mycology station of the National Plant Health Laboratory in Nancy. His early research was devoted to the interspecific hybridization process in the genus *Phytophthora*, with a special focus on the pathogen *Phytophthora alni*, which causes a large-scale and lethal alder disease in Europe. He is now in charge of the development of molecular tools for the detection and the characterization of numerous plant pathogenic fungi, with special interest in fungi listed as quarantined and regulated pests for the European Union.

Marc Buée is a research scientist in fungal ecology at the Institut National de la Recherche Agronomique of Nancy, France since 2003. He has 12 years of research experience in plant physiology and mycorrhizal fungi (endomycorrhizal and ectomycorrhizal fungi), with particular focus on molecular taxonomy, mycorrhizal ecology, and microbial functional diversity.

Jérôme Mounier completed his PhD training in microbiology in 2002 at the University College Cork (Cork, Ireland) and Moorepark Food Research Centre (Fermoy, Ireland). After a 2 year post-doctoral position at the Institut National de la Recherche Agronomique, he has continued his research at the University of Brest (Plouzané, France) since 2005. His interests are mainly in food microbial ecology and especially in the study of bacterial and fungal community structure and functioning of surface-ripened cheeses.

REFERENCES

1. Fleet, G.H., Yeasts in foods and beverages: Impact on product quality and safety, *Curr. Opin. Biotechnol.*, 18: 170, 2007.
2. Filtenborg, O., Frisvad, J.C., and Thrane, U., Moulds in food spoilage, *Int. J. Food Microbiol.*, 33: 85, 1996.
3. Divol, B. and Lonvaud-Funel, A., Evidence for viable but nonculturable yeasts in botrytis-affected wine, *J. Appl. Microbiol.*, 99: 85, 2005.
4. James, T.C. et al., Transcription profile of brewery yeast under fermentation conditions, *J. Appl. Microbiol.*, 94: 432, 2003.
5. Marks, D., van der Merwe, G.K., and van Vuuren Hennie, J.J., Transcriptional profiling of wine yeast in fermenting grape juice: Regulatory effect of diammonium phosphate, *FEMS Yeast Res.*, 3: 269, 2003.
6. Rippon, J.W., The changing epidemiology and emerging patterns of dermatophyte species, *Curr. Top. Med. Mycol.*, 1: 208, 1985.
7. Turenne, C.Y. et al., Rapid identification of fungi by using the ITS2 genetic region and an automated fluorescent capillary electrophoresis system, *J. Clin. Microbiol.*, 37: 1846, 1997.
8. Fridkin, S.K. and Jarvis, W.R., Epidemiology of nosocomial fungal infections, *Clin. Microbiol. Rev.*, 9: 499, 1996.
9. Jarvis, W.R., Epidemiology of nosocomial fungal infections, with emphasis on *Candida* species, *Clin. Infect. Dis.*, 20: 1526, 1995.
10. Droitcourt, C. et al., Primary cutaneous cryptococcosis in transplant recipients: A report of two cases, *Rev. Med. Interne*, 26: 157, 2005.

11. Galgiani, J.N. and Ampel, N.M., Coccidioidomycosis in human immunodeficiency virus-infected patients, *J. Infect. Dis.*, 162: 1165, 1990.
12. Edwards, J.E. Jr., Invasive candida infections-evolution of a fungal pathogen, *New Engl. J. Med.*, 324: 1060, 1991.
13. Eickhoff, T.C., Airborne nosocomial infection: A contemporary perspective, *Infect. Control Hosp. Epidemiol.*, 15: 663, 1994.
14. Cassone, M. et al., Outbreak of *Saccharomyces cerevisiae* subtype boulardii fungemia in patients neighboring those treated with a probiotic preparation of the organism, *J. Clin. Microbiol.*, 41: 5340, 2003.
15. Toscano, C.M. and Jarvis, W.R., Emerging issues in nosocomial fungal infections, *Curr. Infect. Dis. Rep.*, 1: 347, 1999.
16. Petrocheilou-Paschou, V. et al., *Rhodotorula* septicemia: Case report and minireview, *Clin. Microbiol. Infect.*, 7: 100, 2001.
17. Rickerts, V. et al., Diagnosis of invasive aspergillosis and mucormycosis in immunocompromised patients by seminested PCR assay of tissue samples, *Eur. J. Clin. Microbiol. Infect. Dis.*, 25: 8, 2006.
18. Ascioglu, S. et al., Defining opportunistic invasive fungal infections in immunocompromised patients with cancer and hematopoietic stem cell transplants: An international consensus, *Clin. Infect. Dis.*, 34: 7, 2002.
19. Denning, D.W., Invasive aspergillosis, *Clin. Infect. Dis.*, 26: 781, 1998.
20. Nosari, A. et al., Utility of percutaneous lung biopsy for diagnosing filamentous fungal infections in hematologic malignancies, *Haematologica*, 88: 1405, 2003.
21. Ribes, J.A., Vanover-Sams, C.L., and Baker, D.J., Zygomycetes in human disease, *Clin. Microbiol. Rev.*, 13: 236, 2000.
22. Bialek, R. et al., Diagnosis and monitoring of murine histoplasmosis by a nested PCR assay, *J. Clin. Microbiol.*, 39: 1506, 2001.
23. Loeffler, J. et al., Nucleic acid sequence-based amplification of *Aspergillus* RNA in blood samples, *J. Clin. Microbiol.*, 39: 1626, 2001.
24. Fredricks, D.N., Smith, C., and Meier, A., Comparison of six DNA extraction methods for recovery of fungal DNA as assessed by quantitative PCR, *J. Clin. Microbiol.*, 43: 5122, 2005.
25. Agrios, G.N., *Plant Diseases Caused by Fungi*, 4th ed., New York: Academic Press, 1997.
26. Dix, N.J. and Webster, J., *Fungal Ecology*, London, United Kingdom: Chapman & Hall, 1995.
27. Leake, J.R., Donnelly, D.P., and Boddy, L., *Interaction between Ecto-Mycorrhizal and Saprotrophic Fungi*, Berlin, Germany: Springer-Verlag, 2002.
28. Emmerton, K.S. et al., Assimilation and isotopic fractionation of nitrogen by mycorrhizal fungi, *New Phytol.*, 101: 503, 2001.
29. Smith, S.E. and Read, D.J., *Mycorrhizal Symbiosis*, 2nd ed., London, United Kingdom: Academic Press, 1997.
30. Agerer, R., Exploration types of ectomycorrhizae: A proposal to classify ectomycorrhizal mycelial systems according to their patterns of differentiation and putative ecological importance, *Mycorrhiza*, 11: 107, 2001.
31. Bending, G.D. and Read, D.J., The structure and function of the vegetative mycelium of ectomycorrhizal plants V. Foraging behaviour and translocation of nutrients from exploited litter, *New Phytol.*, 130: 401, 1995.
32. Cairney, J.W.G., Translocation of solutes in ectomycorrhizal and saprotrophic rhizomorphs, *Mycol. Res.*, 2: 135, 1992.
33. Buée, M., Vairelles, D., and Garbaye, J., Year-round monitoring of diversity and potential metabolic activity of the ectomycorrhizal community in a beech (*Fagus silvatica*) forest subjected to two thinning regimes, *Mycorrhiza*, 15: 235, 2005.
34. Dahlberg, A., Community ecology of ectomycorrhizal fungi: An advancing interdisciplinary field, *New Phytol.*, 150: 555, 2001.
35. Horton, T.R. and Bruns, T.D., The molecular revolution in ectomycorrhizal ecology: Peeping into the black-box, *Mol. Ecol.*, 10: 1855, 2001.
36. Jonsson, L.M. et al., Context dependent effects of ectomycorrhizal species richness on tree seedling productivity, *OIKOS*, 93: 353, 2001.
37. Molina, R. et al., Mycorrhizae: Ectomycorrhizal fungi. *Encyclopedia Environ. Microbiol.*, 2: 2124, 2002.
38. Gardes, M. and Bruns, T.D., Community structure of ectomycorrhizal fungi in a Pinus muricata forest: Above- and below-ground views, *Can. J. Bot.*, 74: 1572, 1996.
39. Jonsson, L. et al., Ectomycorrhizal fungal communities in late-successional Swedish boreal forests, and their composition following wildfire, *Mol. Ecol.*, 8: 205, 1999.

40. Peter, M. et al., Above- and below-ground community structure of ectomycorrhizal fungi in three Norway spruce (*Picea abies*) stands in Switzerland, *Can. J. Bot.*, 79: 1134, 2001.
41. Richard, F. et al., Diversity and specificity of ectomycorrhizal fungi retrieved from an old-growth Mediterranean forest dominated by *Quercus ilex*, *New Phytol.*, 166: 1011, 2005.
42. James, T.Y. et al., Reconstructing the early evolution of Fungi using a six-gene phylogeny, *Nature*, 443: 818, 2006.
43. Anderson, I.C. and Cairney, J.W., Diversity and ecology of soil fungal communities: Increased understanding through the application of molecular techniques, *Environ. Microbiol.*, 6: 769, 2004.
44. Muyzer, G., DGGE/TGGE a method for identifying genes from natural ecosystems, *Curr. Opin. Microbiol.*, 2: 317, 1999.
45. Cook, K.L. and Sayler, G.S., Environmental application of array technology: Promise, problems and practicalities, *Curr. Opin. Biotechnol.*, 14: 311, 2003.
46. El Karkouri, K. et al., Identification of internal transcribed spacer sequence motifs in truffles: A first step toward their DNA bar coding, *Appl. Environ. Microbiol.*, 73: 5320, 2007.
47. Martin, K.J. and Rygiewicz, P.T., Fungal-specific PCR primers developed for analysis of the ITS region of environmental DNA extracts, *BMC Microbiol.*, 5: 28, 2005.
48. Schüßler, A. et al., Analysis of partial Glomales SSU rRNA gene sequences: Implications for primer design and phylogeny, *Mycol. Res.*, 105: 5, 2001.
49. Kõljalg, U. et al., UNITE: A database providing web-based methods for the molecular identification of ectomycorrhizal fungi, *New Phytol.*, 166: 1063, 2005.
50. Purdy, K.J., Nucleic acid recovery from complex environmental samples, *Methods Enzymol.*, 397: 271, 2005.
51. Bleve, G. et al., Development of reverse transcription (RT)-PCR and real-time RT-PCR assays for rapid detection and quantification of viable yeasts and molds contaminating yogurts and pasteurized food products, *Appl. Environ. Microbiol.*, 69: 4116, 2003.
52. Renouf, V., Claisse, O., and Lonvaud-Funel, A., Inventory and monitoring of wine microbial consortia, *Appl. Microbiol. Biotechnol.*, 75: 149, 2007.
53. Cocolin, L. et al., An application of PCR-DGGE analysis to profile the yeast populations in raw milk, *Int. Dairy J.*, 12: 407, 2002.
54. Vaitilingom, M., Gendre, F., and Brignon, P., Direct detection of viable bacteria, molds, and yeasts by reverse transcriptase PCR in contaminated milk samples after heat treatment, *Appl. Environ. Microbiol.*, 64: 1157, 1998.
55. Garcia, T. et al., Enumeration of yeasts in dairy products: A comparison of immunological and genetic techniques, *J. Food Prot.*, 67: 357, 2004.
56. Mayoral, M.B. et al., Detection of *Kluyveromyces marxianus* and other spoilage yeasts in yoghurt using a PCR-culture technique, *Int. J. Food Microbiol.*, 105: 27, 2005.
57. Gash, A., Yeast genomic expression using DNA microarrays, *Methods Enzymol.*, 350: 393, 2002.
58. ter Linde, J.J. et al., Genome-wide transcriptional analysis of aerobic and anaerobic chemostat cultures of *Saccharomyces cerevisiae*, *J. Bacteriol.*, 181: 7409, 1999.
59. Collart, M. and Oliveiro, S., Preparation of yeast RNA, New York: Wiley, 1993.
60. Belinchón, M.M., Flores, C.L., and Gancedo, J.M., Sampling *Saccharomyces cerevisiae* cells by rapid filtration improves the yield of mRNAs, *FEMS Yeast Res.*, 4: 751, 2004.
61. Maaroufi, Y. et al., Rapid detection of *Candida albicans* in clinical blood samples by using a TaqMan-based PCR assay, *J. Clin. Microbiol.*, 41: 3293, 2003.
62. Bagyalakshmi, R., Therese, K.L., and Madhavan, H.N., Application of semi-nested polymerase chain reaction targeting internal transcribed spacer region for rapid detection of panfungal genome directly from ocular specimens, *Indian J. Ophthalmol.*, 55: 261, 2007.
63. Hong, C., Richardson, P., and Kong, P., Comparison of membrane filters as a tool for isolating *Pythiaceous* species from irrigation water, *Phytopathology*, 92: 610, 2002.
64. Doare-Lebrun, E. et al., Analysis of fungal diversity of grapes by application of temporal temperature gradient gel electrophoresis—Potentialities and limits of the method, *J. Appl. Microbiol.*, 101: 1340, 2006.
65. Prakitchaiwattana, C.J., Fleet, G.H., and Heard, G.M., Application and evaluation of denaturing gradient gel electrophoresis to analyse the yeast ecology of wine grapes, *FEMS Yeast Res.*, 4: 865, 2004.
66. Nielsen, D.S. et al., Yeast populations associated with Ghanaian cocoa fermentations analysed using denaturing gradient gel electrophoresis (DGGE), *Yeast*, 22: 271, 2005.
67. Imhof, A. et al., Rapid detection of pathogenic fungi from clinical specimens using LightCycler real-time fluorescence PCR, *Eur. J. Clin. Microbiol. Infect. Dis.*, 22: 558, 2003.

68. Bialek, R. et al., PCR based identification and discrimination of agents of mucormycosis and aspergillosis in paraffin wax embedded tissue, *J. Clin. Pathol.*, 58: 1180, 2005.
69. Bialek, R. et al., Nested PCR assays for detection of *Blastomyces dermatitidis* DNA in paraffin-embedded canine tissue, *J. Clin. Microbiol.*, 41: 205, 2003.
70. Cawkwell, L. and Quirke, P., Direct multiplex amplification of DNA from a formalin fixed, paraffin wax embedded tissue section, *Mol. Pathol.*, 53: 51, 2000.
71. Zsikla, V., Baumann, M., and Cathomas, G., Effect of buffered formalin on amplification of DNA from paraffin wax embedded small biopsies using real-time PCR, *J. Clin. Pathol.*, 57: 654, 2004.
72. Brillowska-Dabrowska, A., Saunte, D.M., and Arendrup, M.C., Five-hour diagnosis of dermatophyte nail infections with specific detection of *Trichophyton rubrum*, *J. Clin. Microbiol.*, 45: 1200, 2007.
73. Brillowska-Dabrowska, A.H., PCR diagnostic of dermatophytes and other pathogenic fungi, DK patent WO/2006/133701, 2006.
74. Turin, L. et al., Fast, simple and highly sensitive double-rounded polymerase chain reaction assay to detect medically relevant fungi in dermatological specimens, *Eur. J. Clin. Invest.*, 30: 511, 2000.
75. Suau, A. et al., Direct analysis of genes encoding 16S rRNA from complex communities reveals many novel molecular species within the human gut, *Appl. Environ. Microbiol.*, 65: 4799, 1999.
76. Pey, M.H. et al., Distinction between stem- and leaf-infecting forms of *Melampsora* rust on *Salix viminalis* using RAPD markers, *Mycol. Res.*, 101: 7, 1997.
77. Barrett, L.G. and Brubaker, C.L., Isolation and characterization of microsatellite loci from the rust pathogen, *Melampsora lini*, *Mol. Ecol. Notes*, 6: 930, 2006.
78. Erwin, D.C. and Ribeiro, O.K., *Phytophthora Diseases Worldwide*, 1st ed., St. Paul, MN: APS Press, 1996.
79. Wilhelm, S., A sand culture technique for the isolation of fungi associated with roots, *Phytopathology*, 46: 293, 1956.
80. Lecomte, P. et al., PCR assays that identify the grapevine dieback fungus *Eutypa lata*, *Appl. Environ. Microbiol.*, 66: 4475, 2000.
81. Hamelin, R.C. et al., PCR detection of *Gremmeniella abietina*, the causal agent of Scleroderis canker of Pine, *Mycol. Res.*, 104: 527, 2000.
82. Josefsen, L. and Christiansen, S.K., PCR as a tool for the early detection and diagnosis of common bunt in wheat, caused by *Tilletia tritici*, *Mycol. Res.*, 106: 1287, 2002.
83. Kitamura, K. and Yamamoto, Y., Purification and properties of an enzyme, zymolyase, which lyses viable yeast cells, *Arch. Biochem. Biophys.*, 153: 403, 1972.
84. Scott, J.H. and Schekman, R., Lyticase: Endoglucanase and protease activities that act together in yeast cell lysis, *J. Bacteriol.*, 142: 414, 1980.
85. Phalip, V., Hatsch, D., and Jeltsch, J.M., Application of a yeast method for DNA extraction associated with database interrogations for the characterization of various filamentous fungi from diseased hop, *Biotechnol. Lett.*, 26: 409, 2004.
86. Selma, M.V., Martinez-Culebras, P.V., and Aznar, R., Real-time PCR based procedures for detection and quantification of *Aspergillus carbonarius* in wine grapes, *Int. J. Food Microbiol.*, 122: 126, 2008.
87. Arlorio, M., Coïsson, J.D., and Martelli, A., Identification of *Saccharomyces cerevisiae* in bakery products by PCR amplification of the ITS region of ribosomal DNA, *Eur. Food Res. Technol.*, 209: 185, 1999.
88. Ioos, R. et al., SCAR-based PCR primers to detect the hybrid pathogen *Phytophthora alni* and its subspecies causing alder disease in Europe, *Eur. J. Plant Pathol.*, 112: 323, 2005.
89. Langrell, S.R.H., Molecular detection of *Neonectria galligena* (Syn. Nectria galligena), *Mycol. Res.*, 106: 280, 2002.
90. Tooley, P.W. et al., Real-time fluorescent polymerase chain reaction detection of *Phytophthora ramorum* and *Phytophthora pseudosyringae* using mitochondrial gene regions, *Phytopathology*, 96: 336, 2006.
91. Gindreau, E., Walling, E., and Lonvaud-Funel, A., Direct polymerase chain reaction detection of ropy *Pediococcus damnosus* strains in wine, *J. Appl. Microbiol.*, 90: 535, 2001.
92. Ma, Z. and Michailides, T.J., Approaches for eliminating PCR inhibitors and designing PCR primers for the detection of phytopathogenic fungi, *Crop Prot.*, 26: 145, 2007.
93. Bonants, P.J.M. et al., Combination of baiting and different PCR formats, including measurement of real-time quantitative fluorescence, for the detection of *Phytophthora fragariae* in strawberry plants, *Eur. J. Plant Pathol.*, 110: 689, 2004.
94. Mule, G. et al., Development of a quantitative real-time PCR assay for the detection of *Aspergillus carbonarius* in grapes, *Int. J. Food Microbiol.*, 111: 28, 2006.
95. Robe, P. et al., Extraction of DNA from soil, *Eur. J. Soil Biol.*, 39: 183, 2003.

96. Purdy, K.J. et al., Rapid extraction of DNA and rRNA from sediments by a novel hydroxyapatite spin-column method, *Appl. Environ. Microbiol.*, 62: 3905, 1996.

97. Steffan, R.J. et al., Recovery of DNA from soils and sediments, *Appl. Environ. Microbiol.*, 54: 2908, 1988.

98. Ranjard, L. et al., Sampling strategy in molecular microbial ecology: Influence of soil sample size on DNA fingerprinting analysis of fungal and bacterial communities, *Environ. Microbiol.*, 5: 1111, 2003.

99. He, J., Xu, Z., and Hughes, J., Pre-lysis washing improves DNA extraction from forest soil, *Soil Biol. Biochem.*, 37: 2337, 2005.

100. Volossiouk, T., Robb, E.J., and Nazar, R.N., Direct DNA extraction for PCR-mediated assays of soil organisms, *Appl. Environ. Microbiol.*, 61: 3972, 1995.

101. Iotti, M. and Zambonelli, A., A quick and precise technique for identifying ectomycorrhizas by PCR, *Mycol. Res.*, 110: 60, 2006.

102. Kuske, C.R. et al., Small-scale DNA sample preparation method for field PCR detection of microbial cells and spores in soil, *Appl. Environ. Microbiol.*, 64: 2463, 1998.

103. Mayer, Z. et al., Quantification of the copy number of nor-1, a gene of the aflatoxin biosynthetic pathway by real-time PCR, and its correlation to the cfu of *Aspergillus flavus* in foods, *Int. J. Food Microbiol.*, 82: 143, 2003.

104. Read, S.J., Recovery efficiences on nucleic acid extraction kits as measured by quantitative LightCycler PCR, *Mol. Pathol.*, 54: 86, 2001.

105. Jaeger, E.E. et al., Rapid detection and identification of *Candida*, *Aspergillus*, and *Fusarium* species in ocular samples using nested PCR, *J. Clin. Microbiol.*, 38: 2902, 2000.

106. Bougnoux, M. et al., Serum is more suitable than whole blood for diagnosis of systemic candidiasis by nested PCR, *J. Clin. Microbiol.*, 37: 925, 1999.

107. Carroll, N.M. et al., Detection of and discrimination between gram-positive and gram-negative bacteria in intraocular samples by using nested PCR, *J. Clin. Microbiol.*, 38: 1753, 2000.

108. Okhravi, N. et al., Polymerase chain reaction and restriction fragment length polymorphism mediated detection and speciation of *Candida* spp. causing intraocular infection, *Invest. Ophthalmol. Vis. Sci.*, 39: 859, 1998.

109. Wiedbrauk, D.L., Werner, J.C., and Drevon, A.M., Inhibition of PCR by aqueous and vitreous fluids, *J. Clin. Microbiol.*, 33: 2643, 1995.

110. Sambrook, J., Fritsch, E.F., and Maniatis, T., *Molecular Cloning: A Laboratory Manual*, 2nd ed., Cold Spring Harbor, NY: Cold Spring Harbor Laboratory Press, 1989.

111. Pedersen, L.H. et al., Detection of *Penicillium* species in complex food samples using the polymerase chain reaction, *Int. J. Food Microbiol.*, 35: 169, 1997.

112. Feurer, C. et al., Does smearing inoculum reflect the bacterial composition of the smear at the end of the ripening of a French soft, red-smear cheese? *J. Dairy Sci.*, 87: 3189, 2004.

113. Aquilanti, L. et al., The microbial ecology of a typical Italian salami during its natural fermentation, *Int. J. Food Microbiol.*, 120: 136, 2007.

114. Rantsiou, K. et al., Culture-dependent and -independent methods to investigate the microbial ecology of Italian fermented sausages, *Appl. Environ. Microbiol.*, 71: 1977, 2005.

115. Meroth, C.B., Hammes, W.P., and Hertel, C., Identification and population dynamics of yeasts in sourdough fermentation processes by PCR-denaturing gradient gel electrophoresis, *Appl. Environ. Microbiol.*, 69: 7453, 2003.

116. Edwards, S.G. et al., Quantification of trichothecene-producing *Fusarium* species in harvested grain by competitive PCR to determine efficacies of fungicides against *Fusarium* head blight of winter wheat, *Appl. Environ. Microbiol.*, 67: 1575, 2001.

117. Knoll, S. et al., Rapid preparation of *Fusarium* DNA from cereals for diagnostic PCR using sonication and an extraction kit, *Plant Pathol.*, 51: 728, 2002.

118. Schnerr, H., Vogel, R.F., and Niessen, L., Correlation between DNA of trichothecene-producing *Fusarium* species and deoxynivalenol concentrations in wheat-samples, *Lett. Appl. Microbiol.*, 35: 121, 2002.

119. Geisen, R., Mulfinger, S., and Niessen, L., Detection of aflatoxinogenic fungi in wheat by PCR, *J. Food Mycol.*, 1: 211, 1998.

120. Farber, P., Geisen, R., and Holzapfel, W.H., Detection of aflatoxinogenic fungi in figs by a PCR reaction, *Int. J. Food Microbiol.*, 36: 215, 1997.

121. Mills, D.A., Johannsen, E.A., and Cocolin, L., Yeast diversity and persistence in botrytis-affected wine fermentations, *Appl. Environ. Microbiol.*, 68: 4884, 2002.

122. Hotzel, H., Müller, W., and Sachse, K., Recovery and characterization of residual DNA from beer as a prerequisite for the detection of genetically modified ingredients, *Eur. Food Res. Technol.*, 2009: 192, 1999.

123. Godon, J.J. et al., Molecular microbial diversity of an anaerobic digestor as determined by small-subunit rDNA sequence analysis, *Appl. Environ. Microbiol.*, 63: 2802, 1997.

124. Gardes, M. and Bruns, T.D., ITS primers with enhanced specificity for basidiomycetes-application to the identification of mycorrhizae and rusts, *Mol. Ecol.*, 2: 113, 1993.

125. Gherbi, H. et al., High genetic diversity in a population of the ectomycorrhizal basidiomycete *Laccaria amethystina* in a 150-year-old beech forest, *Mol. Ecol.*, 8: 2003, 1999.

126. Henrion, B., Chevalier, G., and Martin, F., Typing truffle species by PCR amplification of the ribosomal DNA spacers, *Mycol. Res.*, 98: 37, 1994.

127. Tomlinson, J.A. et al., On-site DNA extraction and real-time PCR for detection of *Phytophthora ramorum* in the field, *Appl. Environ. Microbiol.*, 71: 6702, 2005.

128. Griffiths, R.I. et al., Rapid method for coextraction of DNA and RNA from natural environments for analysis of ribosomal DNA- and rRNA-based microbial community composition, *Appl. Environ. Microbiol.*, 66: 5488, 2000.

129. Izzo, A., Agbowo, J., and Bruns, T., Detection of plot-level changes in ectomycorrhizal communities across years in an old-growth mixed-conifer forest, *New Phytol.*, 166: 619, 2005.

130. Buée, M. et al., Soil niche effect on species diversity and catabolic activities in an ectomycorrhizal fungal community, *Soil Biol. Biochem.*, 39: 1947, 2007.

131. Tedersoo, L. et al., Fine scale distribution of ectomycorrhizal fungi and roots across substrate layers including coarse woody debris in a mixed forest, *New Phytol.*, 159: 153, 2003.

132. Landeweert, R. et al., Molecular identification of ectomycorrhizal mycelium in soil horizons, *Appl. Environ. Microbiol.*, 69: 327, 2003.

133. Smalla, K. et al., Rapid DNA extraction protocol from soil for polymerase chain reaction-mediated amplification, *J. Appl. Bacteriol.*, 74: 78, 1993.

134. Ishii, S. and Loynachan, T., Rapid and reliable DNA extraction techniques from trypan-blue-stained mycorrhizal roots: Comparison of two methods, *Mycorrhiza*, 14: 271, 2004.

135. Langridge, U., Schwall, M., and Langridge, P., Squashes of plant tissue as substrate for PCR, *Nucleic Acids Res.*, 19: 6954, 1991.

136. Lin, J.J. et al., Detection of plant genes using a rapid, nonorganic DNA purification method, *Biotechniques*, 28: 346, 2000.

137. Geisen, R. et al., Development of a real time PCR system for detection of *Penicillium nordicum* and for monitoring ochratoxin A production in foods by targeting the ochratoxin polyketide synthase gene, *Syst. Appl. Microbiol.*, 27: 501, 2004.

138. Hierro, N. et al., Real-time quantitative PCR (QPCR) and reverse transcription-QPCR for detection and enumeration of total yeasts in wine, *Appl. Environ. Microbiol.*, 72: 7148, 2006.

139. Perez-Ortin, J.E., Garcia-Martinez, J., and Alberola, T.M., DNA chips for yeast biotechnology. The case of wine yeasts, *J. Biotechnol.*, 98: 227, 2002.

140. Tanaka, F. et al., Functional genomic analysis of commercial baker's yeast during initial stages of model dough-fermentation, *Food Microbiol.*, 23: 717, 2006.

141. Wu, H. et al., Global gene expression analysis of yeast cells during sake brewing, *Appl. Environ. Microbiol.*, 72: 7353, 2006.

142. Salaün, F., Mietton, B., and Gaucheron, F., Buffering capacity of dairy products, *Int. Dairy J.*, 15: 95, 2005.

143. Mietton, B. et al., *Bactéries lactiques*, Uriage, France: Lorca, 1994.

144. Felske, A. et al., Direct ribosome isolation from soil to extract bacterial rRNA for community analysis, *Appl. Environ. Microbiol.*, 62: 4162, 1996.

145. Bonaiti, C., Parayre, S., and Irlinger, F., Novel extraction strategy of ribosomal RNA and genomic DNA from cheese for PCR-based investigations, *Int. J. Food Microbiol.*, 107: 171, 2006.

146. Holstege, F.C. et al., Dissecting the regulatory circuitry of a eukaryotic genome, *Cell*, 95: 717, 1998.

147. Bracken, A.P. and Bond, U., Reassembly and protection of small nuclear ribonucleoprotein particles by heat shock proteins in yeast cells, *RNA*, 5: 1586, 1999.

148. Körher, K. and Domdey, H., Preparation of high molecular weight RNA, *Methods Enzymol.*, 194: 398, 1991.

149. Bugos, R.C. et al., RNA isolation from plant tissues recalcitrant to extraction in guanidine, *Biotechniques*, 19: 734, 1995.

150. Leinberger, D.M. et al., Development of a DNA microarray for detection and identification of fungal pathogens involved in invasive mycoses, *J. Clin. Microbiol.*, 43: 4943, 2005.

151. Wiesinger-Mayr, H. et al., Identification of human pathogens isolated from blood using microarray hybridisation and signal pattern recognition, *BMC Microbiol.*, 7: 78, 2007.

152. Guldener, U. et al., Development of a *Fusarium graminearum* Affymetrix GeneChip for profiling fungal gene expression in vitro and in planta, *Fungal Genet. Biol.*, 43: 316, 2006.
153. Palacios, G. et al., Panmicrobial oligonucleotide array for diagnosis of infectious diseases, *Emerg. Infect. Dis.*, 13: 73, 2007.
154. Sarlin, T. et al., Real-time PCR for quantification of toxigenic *Fusarium* species in barley and malt, *Eur. J. Plant Pathol.*, 114: 371, 2006.
155. Schmidt, H. et al., Detection and quantification of *Aspergillus ochraceus* in green coffee by PCR, *Lett. Appl. Microbiol.*, 38: 464, 2004.
156. Cocolin, L. et al., Molecular detection and identification of *Brettanomyces/Dekkera bruxellensis* and *Brettanomyces/Dekkera anomalus* in spoiled wines, *Appl. Environ. Microbiol.*, 70: 1347, 2004.
157. Fernandez-Gonzalez, M. et al., Yeasts present during wine fermentation: Comparative analysis of conventional plating and PCR-TTGE, *Syst. Appl. Microbiol.*, 24: 634, 2001.
158. Manzano, M. et al., A PCR-TGGE (temperature gradient gel electrophoresis) technique to assess differentiation among enological *Saccharomyces cerevisiae* strains, *Int. J. Food Microbiol.*, 101: 333, 2005.
159. Delaherche, A., Claisse, O., and Lonvaud-Funel, A., Detection and quantification of *Brettanomyces bruxellensis* and "ropy" *Pediococcus damnosus* strains in wine by real-time polymerase chain reaction, *J. Appl. Microbiol.*, 97: 910, 2004.
160. Phister, T.G. and Mills, D.A., Real-time PCR assay for detection and enumeration of *Dekkera bruxellensis* in wine, *Appl. Environ. Microbiol.*, 69: 7430, 2003.
161. Martorell, P., Querol, A., and Fernandez-Espinar, M.T., Rapid identification and enumeration of *Saccharomyces cerevisiae* cells in wine by real-time PCR, *Appl. Environ. Microbiol.*, 71: 6823, 2005.
162. Hierro, N. et al., Monitoring of *Saccharomyces* and *Hanseniaspora* populations during alcoholic fermentation by real-time quantitative PCR, *FEMS Yeast Res.*, 7: 1340, 2007.
163. Rawsthorne, H. and Phister, T.G., A real-time PCR assay for the enumeration and detection of *Zygosaccharomyces bailii* from wine and fruit juices, *Int. J. Food Microbiol.*, 112: 1, 2006.
164. Larpin, S. et al., *Geotrichum candidum* dominates in yeast population dynamics in Livarot, a French red-smear cheese, *FEMS Yeast Res.*, 6: 1243, 2006.
165. Gente, S. et al., Development of primers for detecting dominant yeasts in smear-ripened cheeses, *J. Dairy Res.*, 74: 137, 2007.
166. Callon, C. et al., Application of SSCP-PCR fingerprinting to profile the yeast community in raw milk Salers cheeses, *Syst. Appl. Microbiol.*, 29: 172, 2006.
167. Kanbe, T. et al., Species-identification of dermatophytes *Trichophyton*, *Microsporum* and *Epidermophyton* by PCR and PCR-RFLP targeting of the DNA topoisomerase II genes, *J. Dermatol. Sci.*, 33: 41, 2003.
168. Hazen, K.C., New and emerging yeast pathogens, *Clin. Microbiol. Rev.*, 8: 462, 1995.
169. Cao, T.S., Tewari, J., and Strelkov, S.E., Molecular detection of *Plasmodiophora brassicae*, causal agent of clubroot of crucifers, in plant and soil, *Plant Dis.*, 91: 80, 2007.
170. Ward, L.I. et al., A real-time PCR assay based method for routine diagnosis of *Spongospora subterranea* on potato tubers, *J. Phytopathol.*, 152: 633, 2004.
171. Ioos, R. et al., Development of a PCR test to detect the downy mildew causal agent *Plasmopara halstedii* in sunflower seeds, *Plant Pathol.*, 56: 209, 2007.
172. Paran, I. and Michelmore, R.W., Development of reliable PCR-based markers linked to downy mildew resistance genes in lettuce, *Theoret. Appl. Genet.*, 85: 985, 1993.
173. Casimiro, S. et al., Internal transcribed spacer 2 amplicon as a molecular marker for identification of *Peronospora parasitica* (crucifer downy mildew), *J. Appl. Microbiol.*, 96: 579, 2004.
174. Prakob, W. and Judelson, H.S., Gene expression during oosporogenesis in heterothallic and homothallic *Phytophthora*, *Fungal Genet. Biol.*, 44: 726, 2007.
175. Kong, P., Hong, C.X., and Richardson, P.A., Rapid detection of *Phytophthora cinnamomi* using PCR with primers derived from the Lpv putative storage protein genes, *Plant Pathol.*, 52: 681, 2003.
176. Schroeder, K.L. et al., Identification and quantification of pathogenic *Pythium* spp. from soils in Eastern Washington using real-time polymerase chain reaction, *Phytopathology*, 96: 637, 2006.
177. van den Boogert, P.H.J.F. et al., Development of PCR-based detection methods for the quarantine phytopathogen *Synchytrium endobioticum*, causal agent of potato wart disease, *J. Plant Pathol.*, 113: 47, 2005.
178. Jiang, X.Y., Zou, S.M., and Zhou, P.G., Cloning and sequence analysis of complete cDNA of chitin deacetylase from *Mucor racemosus*, *Chin. J. Agri. Biotechnol.*, 4: 167, 2007.
179. Yoshida, S. et al., Cloning and characterization of a gene rpg1 encoding polygalacturonase of *Rhizopus oryzae*, *Mycol. Res.*, 108: 1407, 2004.

180. Tavares, S. et al., Direct detection of *Taphrina deformans* on peach trees using molecular methods, *Eur. J. Plant Pathol.*, 110: 973, 2004.
181. Tenzer, I. and Gessler, C., Subdivision and genetic structure of four populations of Venturia inaequalis in Switzerland, *Eur. J. Plant Pathol.*, 103: 565, 1997.
182. Falacy, J.S. et al., Detection of *Erysiphe necator* in air samples using the polymerase chain reaction and species-specific primers, *Phytopathology*, 97: 1290, 2007.
183. Tadesse, Y. et al., Real time RT-PCR quantification and Northern analysis of cerato-ulmin (CU) gene transcription in different strains of the phytopathogens *Ophiostoma ulmi* and *O.novo-ulmi*, *Mol. Genet. Genom.*, 269: 789, 2003.
184. Witthuhn, R.C. et al., PCR-based identification and phylogeny of species of *Ceratocystis* sensu stricto, *Mycol. Res.*, 103: 743, 1999.
185. Tan, M.K. and Murray, G.M., A molecular protocol using quenched FRET probes for the quarantine surveillance of *Tilletia indica*, the causal agent of Karnal bunt of wheat, *Mycol. Res.*, 110: 203, 2006.
186. Zhao, J. et al., A PCR-based assay for detection of *Puccinia striiformis* sp. *tritici* in wheat, *Plant Dis.*, 91: 1669, 2007.
187. Waalwijk, C. et al., Quantitative detection of *Fusarium* species in wheat using TaqMan, *J. Plant Pathol.*, 110: 481, 2004.
188. Suzuki, F., Arai, M., and Yamaguchi, J., Genetic analysis of *Pyricularia grisea* population by rep-PCR during development of resistance to scytalone dehydratase inhibitors of melanin biosynthesis, *Plant Dis.*, 91: 176, 2007.
189. Chilvers, M.I. et al., A real-time, quantitative PCR seed assay for *Botrytis* spp. that cause neck rot of onion, *Plant Dis.*, 91: 599, 2007.
190. Ioos, R. and Frey, P., Genomic variation within *Monilinia laxa*, *M. fructigena* and *M. fructicola*, and application to species identification by PCR, *Eur. J. Plant Pathol.*, 106: 373, 2000.
191. Sayler, R.J. and Yang, Y.N., Detection and quantification of *Rhizoctonia solani* AG-1 IA, the rice sheath blight pathogen, in rice using real-time PCR, *Plant Dis.*, 91: 1663, 2007.
192. Cocolin, L., Bisson, L.F., and Mills, D.A., Direct profiling of the yeast dynamics in wine fermentations, *FEMS Microbiol. Lett.*, 189: 81, 2000.
193. Ibeas, J.I. et al., Detection of *Dekkera-Brettanomyces* strains in sherry by a nested PCR method, *Appl. Environ. Microbiol.*, 62: 998, 1996.
194. Mayoral, M.B. et al., A reverse transcriptase PCR technique for the detection and viability assessment of *Kluyveromyces marxianus* in yoghurt, *J. Food Prot.*, 69: 2210, 2006.
195. Hendolin, P.H. et al., Panfungal PCR and multiplex liquid hybridization for detection of fungi in tissue specimens, *J. Clin. Microbiol.*, 38: 4186, 2000.
196. Loeffler, J. et al., Comparison between plasma and whole blood specimens for detection of *Aspergillus* DNA by PCR, *J. Clin. Microbiol.*, 38: 3830, 2000.
197. Yamakami, Y. et al., PCR detection of DNA specific for *Aspergillus* species in serum of patients with invasive aspergillosis, *J. Clin. Microbiol.*, 34: 2464, 1996.
198. Bretagne, S. et al., Detection of *Aspergillus* species DNA in bronchoalveolar lavage samples by competitive PCR, *J. Clin. Microbiol.*, 33: 1164, 1995.
199. Ralph, E.D. and Hussain, Z., Chronic meningitis caused by *Candida albicans* in a liver transplant recipient: Usefulness of the polymerase chain reaction for diagnosis and for monitoring treatment, *Clin. Infect. Dis.*, 23: 191, 1996.
200. Rappelli, P. et al., Development of a nested PCR for detection of *Cryptococcus neoformans* in cerebrospinal fluid, *J. Clin. Microbiol.*, 36: 3438, 1998.
201. Kardjeva, V. et al., Forty-eight-hour diagnosis of onychomycosis with subtyping of *Trichophyton rubrum* strains, *J. Clin. Microbiol.*, 44: 1419, 2006.
202. Raeder, U. and Broda, P., Rapid preparation of DNA from filamentous fungi, *Lett. Appl. Microbiol.*, 1: 17, 1985.
203. Lee, S.B. and Taylor, J.W., *Isolation of DNA from Fungal Mycelia and Single Spores*, San Diego, CA: Academic Press, 1990.
204. Ioos, R. et al., Distribution and expression of elicitin genes in the interspecific hybrid oomycete *Phytophthora alni*, *Appl. Environ. Microbiol.*, 73: 5587, 2007.
205. Langrell, S.R.H., Development of a nested PCR detection procedure for *Nectria fuckeliana* direct from Norway spruce bark extracts, *FEMS Microbiol. Lett.*, 242: 185, 2005.
206. van Vaerenbergh, B., Grootaert, B., and Moens, W., Validation of a method for the preparation of fungal genomic DNA for polymerase chain reaction (PCR) and random amplification of polymorphic DNA (RAPD), *J. Mycol. Med.*, 5: 133, 1995.

207. Hughes, K.J.D. et al., Development of a one-step real-time polymerase chain reaction assay for diagnosis of *Phytophthora ramorum*, *Phytopathology*, 96: 975, 2006.
208. Mulfinger, S., Niessen, L., and Vogel, R.F., PCR based quality control of toxinogenic *Fusarium* spp. in brewing malt using ultrasonication for rapid sample preparation, *Adv. Food Sci.*, 22: 38, 2000.
209. Grube, M. et al., DNA isolation from lichen ascomata, *Mycol. Res.*, 99: 1321, 1995.
210. Lucas, M.C., Jacobson, J.W., and Giles, N.H., Characterization and in vitro translation of polyadenylated messenger ribonucleic acid from *Neurospora crassa*, *J. Bacteriol.*, 130: 1192, 1977.
211. Doohan, F.M. et al., Development and use of a reverse transcription-PCR assay to study expression of Tri5 by *Fusarium* species in vitro and in planta, *Appl. Environ. Microbiol.*, 65: 3850, 1999.
212. Logemann, J., Schell, J., and Willmitzer, L., Improved method for the isolation of RNA from plant tissue, *Analyt. Biochem.*, 163: 16, 1987.
213. Moller, E.M. et al., A simple and efficient protocol for isolation of high molecular weight DNA from filamentous fungi, fruit bodies, and infected plant tissues, *Nucleic Acids Res.*, 20: 6115, 1992.
214. Griffin, D.W. et al., A rapid and efficient assay for extracting DNA from fungi, *Lett. Appl. Microbiol.*, 34: 210, 2002.
215. van Tuinen, D. et al., Characterization of root colonization profiles by a microcosm community of arbuscular mycorrhizal fungi using 25S rDNA-targeted nested PCR, *Mol. Ecol.*, 7: 879, 1998.
216. Edwards, S., Fitter, A., and Young, J.P.W., Quantification of an arbuscular mycorrhizal fungus *Glomus mosseae*, within plant roots by competitive polymerase chain reaction, *Mycol. Res.*, 101: 1440, 1997.
217. Simon, L., Lalonde, M., and Bruns, T.D., Specific amplification of 18S fungal ribosomal genes from vesicular-arbuscular endomycorrhizal fungi colonizing roots, *Appl. Environ. Microbiol.*, 58: 291, 1992.
218. Yeates, C. et al., Methods for microbiol DNA extraction from soil for PCR amplification, *Biol. Proced. Online*, 1: 40, 1998.
219. Akane, A. et al., Identification of the heme compound copurified with deoxyribonucleic acid (DNA) from bloodstains, a major inhibitor of polymerase chain reaction (PCR) amplification, *J. Forensic. Sci.*, 39: 362,1994.
220. Atoui, A., Mathieu, F., and Lebrihi, A., Targeting a polyketide synthase gene for *Aspergillus carbonarius* quantification and ochratoxin A assessment in grapes using real-time PCR, *Int. J. Food Microbiol.*, 115: 313, 2007.
221. Di Maro, E., Ercolini, D., and Coppola, S., Yeast dynamics during spontaneous wine fermentation of the Catalanesca grape, *Int. J. Food Microbiol.*, 117: 201, 2007.
222. Kano, R. et al., Isolation of *Aspergillus udagawae* from a fatal case of feline orbital aspergillosis, *Mycoses*, 51: 360, 2008.
223. Kumar, M. and Shukla, P.K., Use of PCR targeting of internal transcribed spacer regions and single-stranded conformation polymorphism analysis of sequence variation in different regions of rRNA genes in fungi for rapid diagnosis of mycotic keratitis, *J. Clin. Microbiol.*, 43: 662, 2005.
224. Olesen, K. et al., The dynamics of the *Saccharomyces carlsbergensis* brewing yeast transcriptome during a production-scale lager beer fermentation, *FEMS Yeast Res.*, 2: 563, 2002.

Part IV

Purification of Nucleic Acids
from Parasites and Insects

Vitaliano A. Cama and Lihua Xiao

CONTENTS

12.1 INTRODUCTION

12.1.1 General Consideration

Protozoa (singular protozoon) are small, unicellular eukaryotes belonging to the kingdom Protista, which once included plant-like algae and fungus-like water molds and slime molds. However, with autotrophic algae being now placed in the kingdoms Plantae and Chromista, heterotrophic protozoa are the remaining life forms in the kingdom Protista. In fact, protozoa are considered as the origin for the evolution of all multicellular organisms that comprise the plants, fungi, and animals. Based on their trophic preferences (i.e., either auto [photo]trophic [plants] or heterotrophic [fungi and animals]), protozoa can be of both purely auto- and heterotrophic, and fit easily into a plant or animal context. In addition, there exist some protozoan species (e.g., dinoflagellates) that demonstrate an intermediate trophic capability (thus called mixotrophic).

Being largely microscopic, most protozoa measure from 10 to 200 μm in size. Indeed, as mobile microorganisms, the most striking feature of protozoa at the light microscope level is the variation they show in their locomotory structures, which have formed the basis for the traditional classification system for this group of organisms. Many protozoa are covered by a skeletal structure (called the pellicle) that is made up of a plasma membrane and underlying cytoskeleton, and that is critical in maintaining the shape of the cell. The plasma membrane forms the outer surface and its associated cytoskeleton may contain additional membranes, microtubules, microfilaments, or plates of cellulose or protein. The cytoplasm is separated into a thin outer ectoplasm and an inner endoplasm. Inside the cytoplasm, vacuoles exist, with some being stomach-like and involved in food digestion and others being contractible for eliminating excess water. One by-product of protozoan digestion is nitrogen, which can be utilized by plants and other higher creatures. Cilia or flagella in protozoa provide mobility, and they often have the $9 \times 2 + 2$ axoneme and $9 \times 3 + 0$ basal body microtubular structure typical of eukaryotes. Protozoa can reproduce sexually and asexually, with each individual protozoon being male and for female. Unlike the open spindle of multicellular animals, cell division in protozoa usually involves a closed spindle, which is constructed inside the intact nuclear envelope.

Protozoa are both diverse and abundant. Their diversity is highlighted by the fact that in spite of the current description of over 92,000 species, the precise number of protozoan species remains unknown. Their abundance is shown by the fact that protozoa occupy all habitats including marine, freshwater, and soil, and make one of the largest biomasses on earth. Being heterotrophic, protozoa are predators for unicellular or filamentous algae, bacteria, and microfungi, assuming a role both as herbivores and as consumers in the decomposer link of the food chain, which is vital to the control of bacteria population and biomass, and leads to ecological balance. In addition, protozoa form an important food source for microinvertebrates, contributing the transfer of bacterial and algal production to successive trophic levels. Further, protozoa may be parasitic or symbiotic, living attached to or inside other organisms.

12.1.2 Classification of Protozoa

Protozoa in the kingdom Protista can be subclassified into three phyla: Apicomplexa, Sarcomastigophora, and Ciliophora [1] (Table 12.1). Microsporidia [2], *Pneumocystis* [3,4], and *Blastocystis* [5] are taxonomically no longer considered protozoa, but are still traditionally covered within medical and veterinary protozoology.

The Apicomplexa comprise the bulk of what used to be called the Sporozoa, and is still taken by some as a synonym. These protozoan parasites do not have flagella, pseudopods, or cilia, but most are motile. This is a diverse group of organisms that include several pathogens like *Plasmodium*, *Cryptosporidium*, *Toxoplasma*, *Eimeria*, *Cyclospora*, *Isospora* (*Cystoisospora*), *Sarcocystis*, and *Babesia*. Most apicomplexans have a complex life cycle, involving both asexual and sexual reproduction. Typically, a definitive host is infected by ingesting cysts or oocysts, which release mobile stages that invade host cells where they replicate asexually. Eventually, the infected cells burst,

TABLE 12.1

Common Parasitic Protozoa and Types of Samples for Nucleic Acid Extraction

Protozoa	Blood or Fluids	Organ or Mucosa	Stool	Brain	Culture	Water	Other Samples
Apicomplexa							
Plasmodium	++	+[a]	−	+[b]	++	−	*Anopheles* +
Babesia	++	−	−	−	−	−	Ticks +
Toxoplasma	−	++[c]	++[d]	+	+[e]	++	−
Isospora	−	−	++	−	−	+	−
Cryptosporidium	−	+[f]	++	−	−	++	Soil, produce +
Cyclospora	−	−	++	−	−	+	Produce, foods ++
Neospora	−	++[c]	++	−	+	+	−
Sarcocystis	−	++[c]	++	−	+	+	−
Sarcomastigophora							
Mastigophora = Flagellates							
Trypanosoma	++	+[g]	−	−	+	−	Tsetse flies,[h] triatomines ++[i]
Leishmania	++	+	−	−	+	−	Sand flies +
Giardia	−	−	++	−	++	++	Soil, produce ++
Trichomona	−	+ +[j]	−	−	+	−	−
Sarcodina = Amoebas							
Acanthamoeba	−	++[k]	−	+	++	++	−
Naegleria	−	−	−	++	++	++	−
Balamuthia	−	−	−	++	+	++	−
Entamoeba	−	+[l]	++	−	+	+	−
Ciliophora = Ciliates							
Balantidium coli	−	−	++	−	+	+	−
Organisms Traditionally Classified as Protozoa (Not True Protozoa)							
Blastocystis	−	−	++	−	+	−	−
Pneumocystis	++[m]	+	−	−	−	−	−
Microsporidia	+	+	++[n]	++	−	+	−

Note: Samples frequently used = ++, less frequently = +, not frequently = −.

[a] Liver.
[b] *P. falciparum* only.
[c] Muscles and organs from intermediate hosts.
[d] Cats.
[e] Tachizoites.
[f] Biliary duct biopsies in people with chronic infections.
[g] Heart in chronic infections with *T. cruzi*.
[h] *T. brucei*, sleeping sickness.
[i] *T. cruzi*, Chagas disease.
[j] Vaginal swabs.
[k] Cornea.
[l] Liver abscesses.
[m] Sputum, bronchio alveolar lavages (BAL).
[n] Enterocytozoon bieneusi.

merozoites are released, and infect new cells. This may occur several times, until sexual stages or gamonts are produced. These will form gametes that will fuse to create new oocysts, which are the infectious stages. There are many variations on this basic pattern, however. Several apicomplexa, such as *Cryptosporidium*, *Cyclospora*, and *Isospora*, complete their life cycle in a single host; while a second group forms tissue cysts and requires two mammalian hosts in a predator–prey relationship, such as *Toxoplasma*, *Sarcocystis*, and *Neospora*.

The Sarcomastigophora are divided into two subphyla: the Mastigophora and the Sarcodina. Members of the subphylum Mastigophora are also called flagellates, and can be divided into noninvasive luminal, and body and tissue flagellates. *Giardia* and *Trichomonas* belong to the first group, are elongated to oval, have sucking disks or axostyles adhering to the mucosa, replicate by binary fission, and have more than one flagellum. Among the body and tissue flagellates are *Trypanosoma* and *Leishmania*. These parasites are fusiform, and have a single polar flagellum, a mitochondria-like structure called kinetoplast, and a complex life cycle that requires an intermediate arthropod host [6].

The main characteristic of the Sarcodina is locomotion through amoebic motion. Human pathogens within this subphylum are *Entamoeba* and free-living amoeba such as *Naegleria*, *Acanthamoeba*, and *Balamuthia*. These parasites have a relatively simple life cycle, reproducing by binary fission. While *Entamoeba* are obligate parasites, the free-living amoeba are fortuitous parasites of humans or other mammals [7].

Ciliophora are characterized by locomotion by cilia. The most significant human pathogen of this group is *Balantidium coli*, which has a single host life cycle. It may produce ulcerative lesions in the colon. This parasite also infects pigs, which are considered the reservoir for human infections.

Most protozoa detected in blood are linked to vector-borne transmission, and require an arthropod to complete their life cycle. This is the case of *Plasmodium*, *Trypanosoma*, *Leishmania*, and *Babesia*. Among protozoa transmitted via the fecal-oral route, some complete their cycle within one host, such as *Cryptosporidium*, *Cyclospora*, *Giardia*, *Entamoeba histolytica*, *Isospora*, *Eimeria*, and *Balantidium*, while others require an intermediate host to complete their development. In the latter group, carnivores are the definitive hosts and excrete adult parasite stages through their feces. These are ingested by mammalian herbivores where the intermediate stages develop, usually in muscles although several solid organs can also be infected. The life cycle is completed when a susceptible carnivore ingests the intermediate stages. This is the case of *Toxoplasma*, *Sarcocystis*, and *Neospora*. Humans can contract toxoplasmosis when they ingest infectious oocysts and become accidental intermediate hosts. Among the flagellates, *Trichomonas* are sexually transmitted through direct contact with a susceptible host. Finally, free-living protozoa can cause infection when they are accidentally introduced into a susceptible host, like *Naegleria*, *Acanthamoeba*, and *Balamuthia*. These free-living amoeba does not require an animal host to survive, but can cause severe pathological damage or death when they accidentally infect humans or animals.

12.1.3 ECONOMIC IMPORTANCE OF PROTOZOA

While many of the 92,000 recognized species of protozoa are free-living or symbiotic living, at least 10,000 species are parasitic and bring more harm than benefit to their hosts. Protozoa are found in a variety of vertebrates, invertebrates, and plants, and remarkably, individual protozoan species have evolved to occupy certain ecological niches in the hosts. For example, in the human body, *Leishmania* can be found in the skin, spleen, and liver; *Acanthamoeba* infects the eye and the central nervous system; amoebae and flagellates are detected in the mouth (often nonpathogenic); *Giardia*, *Entamoeba*, *Cryptosporidium*, and *Isospora* are located in the gut; *Trichomonas* resides in the genital-urinary tract; *Plasmodium* and *Trypanosoma* lives in bloodstream and the central nervous system; etc. The pathogenic protozoa with greatest economical impact in the phylum Apicomplexa include *Plasmodium*, *Babesia*, *Eimeria*, *Cryptosporidium*, and *Toxoplasma*, and *Tryposoma*, *Leishmania*, *Giardia*, *Trichomonas*, and *Entamoeba* in the phylum Sarcomastigophora. Some protozoa of medical importance are shown in Figure 12.1.

Plasmodium is the causative agent for human malaria, which is the world's second biggest killer after tuberculosis. Malaria places 40% of the world population at risk, produces 300 million clinical

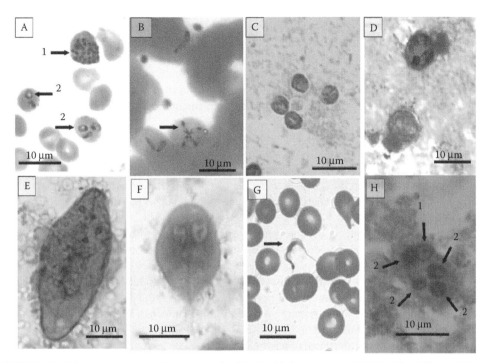

FIGURE 12.1 Photomicrographs of protozoa of public health importance. (A) Thin blood films of *Plasmodium falciparum* using Giemsa stain: 1 = schizonts, 2 = ring shaped trophozoites. (B) *Babesia* spp. (hetoxylin and eosin stain); arrow indicates tetrad of trophozoites in Maltese cross. (C) *Cryptosporidium parvum* oocysts (acid-fast stain). (D) *Cyclospora cayetanensis* unsporulated oocysts (acid-fast). (E) Unsporulated oocyst of *Isospora* (*Cystoisospora*) *belli* (acid-fast). (F) Fecal smear of a trophozoite of *Giardia intestinalis* (hematoxylin and eosin stain). (G) Blood smear of a trypomastigote of the kinetoplastidae flagellate *T. cruzi* (Giemsa stain). (H) Trophozoite of *Entamoeba histolytica* in a fecal smear (trichrome stain): 1 = nucleus, 2 = ingested red blood cells found only in *E. histolytica*. (*Giardia intestinalis*: Photograph courtesy of Y. Ortega.)

cases and causes 1.5–2.7 million deaths annually, the majority of which occur in sub-Saharan Africa. Among the four species (*Plasmodium falciparum*, *Plasmodium malariae*, *Plasmodium malariae*, and *Plasmodium vivax*) within the genus *Plasmodium*, *P. falciparum* is the most common and also the most lethal. The parasite chiefly infects red blood cells and causes them to develop surface knobs that stick to endothelial cells. In the severe form of malaria, this results in blood vessel blockages and subsequent brain and end organ damage, with death often occurring within a few days of infection. *P. falciparum* is especially dangerous to small children in highly endemic areas and to travelers from nonmalarious areas who, without prior exposure, have no partial immunity to the parasite. Chronic malaria also poses another threat to individuals, as it is possibly responsible for introducing several serious genetic disorders (e.g., sickle cell anemia and thalassemia) in populations stressed by endemic malaria. These disorders produce abnormal hemoglobin in the patients' red blood cells.

Babesia also targets red blood cells and causes a serious and fatal disease in cattle called babesiosis. The parasite breaks down the red blood cells, and the resulting hemoglobin renders the urine red. The disease is controlled by regular dipping of cattle in acaricide baths to get rid of the transmitting ticks and also by applying an antitick vaccine that targets components of the tick salivary glands to prevent their feeding. Occasionally, animal *Babesia* may cause severe infections in humans, which can be fatal especially in persons who have been splenectomized for other reasons.

Eimeria grows in the intestinal tracts of vertebrates, and it is a significant animal pathogen of economic importance, because of the economic resources spent on its prevention and treatment, and losses in many parts of the world every year. Animals (e.g., poultry and laming ewes) are particularly vulnerable to *Eimeria* infection (coccidiosis) when kept under high density or stressful conditions.

Different *Eimeria* species show predilections for specific sections of the digestive tract and positions within the host cell, with some developing below the nucleus, others above it, and in a few cases actually within the nucleus itself. As intestinal cells have a rapid turnover rate (with half-lives of between 24 and 48 h), *Eimeria* parasites have to (and are able to) complete their development within the intestinal cell before being discharged and cause diarrhea that can be bloody. Coccidiosis can be controlled by good farming practices, and the implementation of prevention programs using coccidiostats or vaccines.

Cryptosporidium is infective to a wide range of vertebrates including cattle, sheep, rodents, cats, dogs, and man. The parasite is usually located in the brush border of the gastrointestinal cell and appears to be extracellular in location. The chief route of infection is fecal-oral, and people acquire partial immunity after repeated exposures to the parasite. *Cryptosporidium* is a major health problem to immunocompromised population groups such as patients with AIDS or those undergoing transplantation surgery.

Toxoplasma normally infects the intestine in cats, which may shed vast numbers of oocysts within a few days of infection. The oocysts can reinfect a cat or infect other mammalian hosts when ingested. In the latter, the parasite does not stay in the gut, but migrates into the tissues, liver, lungs, and muscles resulting in a disseminated infection. The disease is usually asymptomatic in healthy individuals, who often acquire immunity after exposure. However, the parasite can cause considerable damage to the fetus when the gestating mother becomes infected. In fact, 40% of the infected infants may develop symptoms ranging from mild to severe. Additionally, *Toxoplasma* can cause a serious disease in immunocompromised individuals.

Trypanosoma is usually transmitted by insect vectors via stercorarian and salivarian routes. The stercorarian trypanosomes are taken up by an insect vector in the blood meal and grow in its hindgut. The trypanosomes gain entry into vertebrate the host through the feeding site or by the skin being scratched, when the insect feeds and defecates. The most important stercorarian trypanosome is *Trypanosoma cruzi*, which is spread by triatomid bugs. It causes Chagas disease that affects 12–24 million people in Central and South America. Being essentially a tissue parasite, *T. cruzi* spends a brief time in the blood, and then enters various cells in the body, particularly muscle and nerve cells, where they multiply. Infection of heart muscle and nerve cells can result in heart failure, cardiac dysrhytmias, megacolon or megaesophagus.

The salivarian trypanosomes develop in the midgut of blood-feeding flies and are injected via the salivary glands when the fly feeds. The two salivarian trypanosomes that cause human diseases are *Trypanosoma brucei rhodesiense*, which causes acute sleeping sickness, and *Trypanosoma brucei gambiensis* (for chronic sleeping sickness), which invade the nervous systems. Both of these parasites are spread by tsetse flies and both can infect a wide range of mammals including man.

Leishmania is an intracellular pathogen of the immune system targeting macrophages and dendritic cells. Leishmaniasis produces disfiguring cutaneous and mucocutaneous lesions as well as visceral disease affecting the hemopoietic organs. Among several pathogenic *Leishmania* species (e.g., *Leishmania donovani*, *Leishmania infantum*, and *Leishmania chagasi*), *L. donovani* causes the most severe and life-threatening visceral leishmaniasis. This disease is endemic in 88 countries, especially Bangladesh, Brazil, India, and Sudan, with a total of more than 350 million people at risk, an estimated 0.5 million new cases worldwide each year and tens of thousands of deaths. Whereas the typical infection is cutaneous that causes slow-to-heal sores, the most serious infection (known as kala azar) is visceral involving internal organs such as the liver and spleen. The fact that *Leishmania* can live in macrophages without getting killed highlights its ability to mitigate the two main killing mechanisms of macrophages: the production of reactive oxygen intermediates such as superoxide anions or hydrogen peroxide, which is down regulated by the parasite, and the production of lysosomal enzymes such as glycosidases and acid phosphatases, which is also inhibited by the parasite.

Giardia intestinalis (also known as *Giardia duodenalis* and *Giardia lamblia*) is commonly found in the small intestine, with overall prevalence rates ranging from 1% to 30%. The parasite attaches to the cells of the gut using its two suckers and divides rapidly by binary fission. Although the parasite does not break down host cells, its sheer numbers over the surface of the intestine probably interfere with absorption and trigger diarrhea, vomiting, and loss of weight. Typically, *Giardia* is noninvasive and often causes asymptomatic infections. It nonetheless produces over 200 million clinical cases per year due to its widespread nature, with acute or chronic diarrhea, or other gastrointestinal manifestations.

Trichomonads are generally nonpathogenic commensals and only a few species are of importance in animals and humans. Among these, only *Trichomonas vaginalis* is clearly pathogenic and is a common sexually transmitted disease found in the urogenital tract. *Trichomonas tenax* (also called *Trichomonas buccalis*) is a commensal of the human oral cavity. *Pentatrichomonas hominis* (formerly known as *Trichomonas hominis*) is a nonpathogenic commensal of the large intestine. *Dientamoeba fragilis* is a flagellate without flagella. A distinctive feature of the trichomonads is its axostyle (ax), which is a cytoskeletal element composed of concentric rows of microtubules and is believed to function in the attachment of the parasite to epithelial cells. The axostyle runs the length of the organism and appears to protrude from the posterior end. *Trichomonas foetus*, a related parasite found in bovines, causes sterility or abortion in cattle.

Entamoeba histolytica is an amoeba that usually grows in the lower small intestine or colon where it replicates by binary fission and produces characteristic four-nucleated cysts. Passing out in the feces, the cysts contaminate water and food. Following ingestion, the amoebae start to invade the mucosa causing ulcers, then get into the blood stream and are transported to other sites in the body such as the liver, causing amoebic abscesses. Three-quarters of the people infected with *Entamoeba* are asymptomatic carriers, and the rest are symptomatic. Several other *Entamoeba* spp. are present in humans and animals.

12.2 BASIC PRINCIPLES OF NUCLEIC ACID EXTRACTION FROM PROTOZOA

Nucleic acid information has been critical in understanding the evolution, genetic diversity, and phenotypic differences among protozoan parasites. Extraction and purification of nucleic acids from protozoa rely on the same techniques that are used in mammals, other eukaryotes, bacteria, and viruses. However, there are biological and technical differences to be considered. Among the most important are the relatively low number of organisms per unit of clinical or environmental samples, the hardiness of the life cycle stages in samples, and the presence of multiple substances that may interfere with the integrity of the nucleic acids or their successful use in downstream applications. Below we outline a few aspects of nucleic acid purification that have particular relevance to protozoa.

12.2.1 Release of Nucleic Acids from the Protozoa

The initial step for extraction of nucleic acids is their release from the parasite in the sample. This can be accomplished either by physical or chemical treatment of the samples.

Physical treatment includes homogenization by mechanical devices such as grinders, blenders, mortars, and bead beating. Cell lyses can also be accomplished by consecutive cycles of freezing and thawing [8,9], or a combination of freezing followed by homogenization [10].

Chemical treatment of samples is preferably used when the nucleic acids of interest cannot be physically released. Chemically induced release of nucleic acids, however, was frequently used for parasites enclosed within a sturdy cyst-type structure. Typical examples are *Cryptosporidium* or microsporidia, where the nucleic acids were released after chemically induced excystation using an alkaline treatment that was followed by a neutralization step [11].

Physical and chemical methods can be used in combination, for example, the use of ceramic or glass beads in conjunction with lysing buffers [12]. The samples to be lysed are placed in bead-loaded tubes with buffers that usually contain chaotropic salts or detergents. The tubes are then subjected to vigorous agitation in mechanical devices that generate high impact speed of the beads against the parasites. Bench-top vortexes can be retrofitted with tube holders to provide fast circular agitation. There are dedicated units that generate either very fast linear or sigmoidal motion to the sample. This latter technology has become very useful for the release of nucleic acids from protozoa in environmental samples [13] or food matrices [14].

In all cases, the released nucleic acids are accompanied by proteins, cell membranes, and other molecules that can alter the integrity of the nucleic acids of interest, or may interfere with downstream applications. Thus, further purification of the nucleic acids is accomplished by extraction methods.

12.2.2 SOLVENT-BASED EXTRACTION OF NUCLEIC ACIDS

One of the first protocols widely used for the extraction of DNA is based on protein digestion using proteinase K, followed by precipitation of the denatured proteins using a solution of phenol and chloroform (or phenol–chloroform and isoamyl alcohol). The purified DNA remains in the supernatant and it is recovered by precipitation using cold alcohol and centrifugation. This method has the advantages of low cost and flexibility for samples with large mass or volume. Some limitations of the phenol–chloroform extraction are the use of hazardous chemicals, the need of strict quality control of the reagents, primarily phenol, and its inability to remove certain inhibitors, especially those found in environmental and food samples [13].

12.2.3 SOLID-PHASE METHODS: SILICA-BASED FILTERS (MEMBRANES), WOOL OR SLURRY

Filter-based methods rely on the characteristics of silica to retain nucleic acids. Using the adequate buffers and salt concentrations, kits using silica-based membranes can be tailored for the recovery of DNA, RNA, proteins, or combinations thereof. Filter extraction offers the advantage of being simple, highly reproducible, and allowing the selective pass-through of nondesirable compounds. Silica membranes are also used in specially designed plates, allowing the automated and simultaneous extraction of multiple samples. Filter-based methods, however, are restricted to fixed volumes of samples, the retention capacity of the silica, and have a greater cost than solvent-based methods.

A variation of the silica membrane method is the use of either silica slurry or fibers (wool). In these methods, the nucleic acids are liberated from the parasites and exposed via direct contact to silica microbeads or wool. These matrices with attached nucleic acids are then washed from undesirable substances, either by centrifugation or elution through a membrane. The performance of silica beads or fibers is very similar to that of the filter-based methods, with the potential advantage of having an increased binding surface for nucleic acids. Nonetheless, a critical element that affects the performance of the solid-based methods is the chemical characteristics of the buffers used. They play a major role in the performance of the different silica-based purification methods.

Both silica membrane and slurry methods are among the technologies most frequently reported in the literature for nucleic acid extraction from protozoa. Historically, a major limitation has been its ability to only process small volumes per sample. Lately, most manufacturers are offering larger size kits that can accommodate volumes, in most cases from 1 up to 10 mL per sample.

Automated nucleic acid extraction relies on magnetic beads covered with silica. In these systems, a magnet retains the silica-magnetized beads with nucleic acids, while allowing the flow through of buffers and undesirable substances. The desired nucleic acids are released in the final step of the process. The automated methods allow the simultaneous processing of multiple samples, all with small or fixed volumes, and require the use of proprietary hardware and consumables, which are compatible with laboratories that routinely process large number of samples. Bench-top kits are also marketed by some companies. These methods, however, usually require high numbers of targeted organisms to prevent the saturation of bead surface with DNA from sample matrices.

12.2.4 CHELEX EXTRACTION

Chelex is a substance frequently used in the extraction of DNA. This method is based on the properties of Chelex to neutralize DNases [15]. It is a simple and fairly inexpensive process, but the nucleic acids are not purified from other substances, and its applications have been limited to DNA suitable for polymerase chain reaction (PCR) applications.

12.2.5 AFFINITY CHROMATOGRAPHY

These methods are used for the extraction and purification of mRNA from samples. They use T-oligonucleotide probes, either polyT or oligoT [16] that specifically bind to the polyA tails of the mRNA resulting in better yields and purity of this type of RNA. Although this method is restricted by the volume capacity of each extraction and its higher cost, it is the method of choice for purification of mRNA.

12.3 ASPECTS RELATING TO SPECIFIC SAMPLE MATRICES

Nucleic acids from protozoa can be used for diagnosis, molecular characterization, phylogenetic, metabolic, and drug resistance studies. The taxonomic diversity among protozoa and differences in life cycles and transmission patterns result in a wide range of samples that can be used as sources of nucleic acids. As a consequence, nucleic acids can be extracted from a highly diverse set of samples that also require different extraction methods. Below we discuss aspects relating to specific handling for various sample matrices: blood, organs or secretions, stools, environmental and food matrices, and vectors. Table 12.1 lists types of samples used for DNA extraction of common parasitic protozoa, and Tables 12.2 and 12.3 present common DNA and RNA extraction methods for different types of samples of parasitic protozoa.

12.3.1 BLOOD

Whole blood samples are frequently collected for the microscopic detection of protozoa. In this process, substances like ethylenediaminetetraacetic acid (EDTA), heparin, or acid citrate are frequently used as anticlotting agents. The stability of the nucleic acids in these samples will decrease over time, unless the samples are specifically preserved for further nucleic acid work. The presence of hemoglobin or EDTA is known to interfere with downstream applications, although multiple commercial kits now use buffers that are tailored to control them.

Protozoan DNA can be extracted from freshly or specially preserved blood derived samples. Whole blood samples collected in heparin can be processed by proteinase K digestion followed by phenol–chloroform extraction and cold ethanol precipitation [17]. For freshly collected samples (<24 h), DNA can be extracted by lysing the cells in 5 volumes of buffer containing 10 mM Tris-HCl pH 7.6, 10 mM EDTA, 0.1 M NaCl, 0.5% sodium dodecyl sulfate (SDS), and 300 mg/mL of proteinase K and incubated at 50°C for 2 h. Nucleic acids are extracted twice using phenol/chloroform/isoamyl alcohol (25:24:1) and precipitated by centrifugation with chilled ethanol [18]. Parasite nucleic acids can also be extracted from whole blood samples using membrane filters [19], or from the residual buffy coat and packed cells from samples collected with anticlotting agents [20].

There are cases when the low level of parasites in the blood sample yields a small or invisible pellet of DNA. When samples are extracted with phenol–chloroform and chilled alcohol precipitation, the DNA pellet can be better visualized by the addition of a soluble polyacrylamide (GenElute LPA, Sigma Chemical Co., St. Louis, MO.) during precipitation with isopropanol, thus improving the reproducibility of the precipitation process. This polymer does not interfere with PCR applications [21].

Parasite DNA that is ready for PCR amplification can be obtained from blood samples (250 μL) exposed to an equal volume of lysis buffer (0.31 M sucrose, 0.01 M Tris-Cl pH 7.5, 5 mM $MgCl_2$, and 1% Triton X-100). The resulting pellet can be directly resuspended in 1 × PCR buffer (10 mM Tris-HCl pH 8.4, 50 mM KCl, and 1% Triton X-100) and 1.5 μL of proteinase K (10 mg/mL) and incubated

TABLE 12.2

Some Common DNA Extraction Methods by Type of Samples and Protozoa

Type of Sample	Typical Protozoa	DNA Extraction	References
Blood, fresh	*Plasmodium, Babesia, Leishmania,* and *Trypanosoma*	Phenol–chloroform	[17]
		Phenol–chloroform	[18]
		Guanidine/phenol/chloroform	[21]
		Phenol–chloroform, and polyacrylamide additive	[21]
		Proteinase K digestion alone	[22]
		Silica membrane kits	[19,20]
Blood, frozen		Guanidine buffer and phenol–chloroform	[23]
Blood in filter membranes		Methanol elution	[25]
		Tris-EDTA elution	[26]
		Silica membrane kits	[27]
		Saponin/Chelex	[29]
		Water/Chelex	[30]
		None, directly used for PCR	[28]
		FTA membranes	[33]
Blood in slides or smears		Phenol–chloroform	[44]
		Phenol–chloroform and silica kits	[43]
		Silica membrane kits	[42]
Skin	*Leishmania*	Phenol–chloroform	[61]
		Imprint on paper, eluted with water	[62]
Urine	*Plasmodium,* microsporidia, and *Leishmania*	Phenol–chloroform	[35,38]
		Silica membrane kits	[36]
		Silica fiber fleece kit	[37]
Urine and saliva		Chelex or silica membrane kits	[39]
Sputum, BAL	*Pneumocystis*	Silica beads kits	[63]
		Silica membrane kits	[64]
		Phenol–chloroform	[65]
Sputum, BAL in membranes		FTA filter as template for PCR	[66,67]
Brain	*Naegleria, Balamuthia, Toxoplasma,* and microsporidia (*Encephalitozoon*)	Silica slurry and silica membrane	[68]
		Homogenization and grinding, and silica fiber fleece kit	[69]
		Homogenization and guanidine-free silica membrane kits	[70]
Cerebrospinal fluid	*Toxoplasma*	Silica membrane kits	[71]
Intestine	*Cryptosporidium*	Lysis buffer and phenol–chloroform	[72]
Liver	Piroplasmas and *Entamoeba*	Silica membrane kit[a]	[73]
Muscles, meat	*Toxoplasma,*[b] *Sarcocystis,*[b] and *Neospora*[b]	Lysis buffer SDS/proteinase K and phenol–chloroform	[74]

TABLE 12.2 (continued)

Some Common DNA Extraction Methods by Type of Samples and Protozoa

Type of Sample	Typical Protozoa	DNA Extraction	References
Mucosal swabs	*Trichomonas*	Magnetic silica beads, automated	[75]
Stools	*Cryptosporidium, Giardia, Cyclospora,*	Freeze–thaw proteinase K and phenol–chloroform	[76]
	Isospora, Balantidium, microsporidia,	Alkaline digestion and silica membrane	[11]
	Blastocystis, Toxopla	Silica membrane kits	[77]
	sma,[c] Sarcocystis,[c]	Bead beating and silica slurry kit	[12,40]
	and *Neospora[c]*	Culture and phenol–chloroform	[78]
		Silica membrane kits	[79]
Stools, from slides	*C. Giardia*	Lysis and proteinase K digestion	[8]
Environmental samples	*C. Giardia,*	Silica membrane kits	[80]
	Cyclospora, and microsporidia	Silica membrane slurry	[80]
Food samples	*C. Giardia, Cyclospora,* and microsporidia	FTA membrane	[50]
		Silica membrane slurry	[14]
Vectors, blood meal	*Plasmodium* and	Silica membrane kit	[47]
	Trypanosoma	Water elution	[24]
Vector, stool		Guanidine lysis and selective DNA precipitation	[48]

[a] GFX genomic blood now replaced with Illustra blood GenomicPrep mini kit.

[b] From herbivore intermediate hosts.

[c] Stools form carnivore definitive hosts.

at 56°C for 1 h. The proteinase K is heat inactivated, and the resulting supernatant with parasite DNA is ready for PCR amplification [22].

Stored blood samples require extra processing for the preservation of nucleic acids. Chaotropic agents such as guanidine have been reported to enhance the extraction of parasite nucleic acids, and most commercial kits currently incorporate guanidine in their lysis buffers. Guanidine can also be used to preserve blood specimens, by mixing the sample with an equal volume of 6 M guanidine–HCl and 0.1 M EDTA pH 8.0, followed by boiling for 15 min, and stored at 4°C until extracted using phenol–chloroform and cold ethanol precipitation [21]. Parasitized red blood cells can also be stored frozen when kept at −70°C in 2 volumes of 6 M guanidine–HCl, 50 mM Tris pH 8.0, and 20 mM EDTA [23].

Another alternative for long-term storage or shipping and transporting of blood samples is the use of dried spots, where a drop of whole blood is collected and dried onto a filter paper. Fractions of the filter can be either cut into strips or punched out. The DNA can be eluted by the following methods:

1. Soaking in water at 37°C for 30 min and centrifuged at 7600 g for 10 min to extract the proteins, followed by incubation in water at 99°C for 30 min and centrifugation to extract the DNA [24].
2. Soaking in methanol, letting the methanol to air-dry, and resuspending the DNA in water [25].
3. Soaking the filter fractions in Tris-EDTA buffer with heating at 97°C for 15 min [26].

TABLE 12.3

Some Common RNA Extraction Methods by Type of Samples and Protozoa

Type of Sample	Typical Protozoa	RNA Extraction Method	References
Blood samples, cultures	*Plasmodium, Babesia, Leishmania*, and *Trypanosoma*	Guanidine-based kit	[81]
		RNA preservative and guanidine-based kit	[82]
		Guanidine extraction	[83]
		Guanidine-based kit	[84]
Blood, frozen		RNA preservative and silica membrane kit	[34]
Skin	*Leishmania*	Silica particles	[85]
In-vitro culture	*Naegleria, Balamuthia, Toxoplasma*, and microsporidia (*Encephalitozoon*)	Guanidine-based kit	[86]
		Guanidine, urea, and lithium chloride centrifugation	[87]
		Guanidine-based kit	[88]
		Silica-based kits	[89]
	Cryptosporidium, Giardia, Entamoeba, Balantidium, Trichomonas, Toxoplasma, Sarcocystis, and *Neospora*[a]	Guanidine-based kit	[90]
		Guanidine-based kit	[91]
		Silica membrane kits	[92]
Purified or cultured parasites	*Cryptosporidium, Giardia, Entamoeba, Balantidium, Trichomonas, Toxoplasma, Sarcocystis*, and *Neospora*[a]	Silica membrane kit	[93]
		Oligo(dT)-magnetic beads	[94]
Vectors, blood meal	*Plasmodium*	Silica-based membranes	[95]

[a] Not a human pathogen.

4. Extraction using silica membrane kits [27].
5. In cases where the number of parasites is high, the filter punch can be used directly as the template for PCR amplification [28].
6. Soaking the filters in 0.5% saponin-phosphate buffered saline (PBS), followed by incubation in 20% Chelex-100 and water at 99°C for 15 min; the sample is centrifuged and the resulting supernatant has the DNA [29].
7. Soaking in 1 mL of water for 30 min, centrifuged and the supernatant mixed with 0.25 volume of 1% freshly prepared Chelex-100 solution. The mixture is incubated at 56°C for 30 min, boiled for 8 min, vortexed for 2 min, and finally centrifuged. The DNA is recovered in the supernatant and is suitable for PCR [30].

There are specialized membranes that have been designed to be the support matrix for whole blood or other cell-rich samples and contain specific chemicals to better preserve the DNA. Flinders Technology Associates (FTA) media cards are made of a cellulosic-based dry solid support that is coated with a proprietary media [31] and facilitates the use of the samples in clinical diagnostic methods. This media contains denaturants, chelating agent buffers, and free radical traps, thus eliminating most common organisms that may grow in the samples while preserving the integrity of the DNA during storage [32]. Punch fractions of the FTA matrix cards with samples are washed twice for 15 min in FTA wash buffer, followed by one rinse with TE buffer, after which they can be used for PCR amplification [33].

Extraction of parasite RNA from stored blood samples. RNA now can be saved for future extraction from blood samples with the use of preservation reagents. *Plasmodium*-infected blood samples can be stored frozen after pretreatment in an RNA preservation solution (RNAlater, Qiagen), and the RNA can be later extracted using silica-based membrane filtration kit. The resulting RNA can be used in reverse transcriptase PCR assays to test for *Plasmodium* mRNA [34].

Parasite infected erythrocytes can be saved for later RNA work by washing the red blood cells in 20–40 times volumes of phosphate buffered saline, followed by preservation at −70°C in five times pellet volumes of RNA extraction reagent (TRIzol, Invitrogen, Carlsbad, CA). Total RNA extraction is accomplished by double extraction using the same reagent following the manufacturer's instructions and the addition of a DNase digestion step [23].

12.3.2 Tissues

Protozoa can be found in diverse types of tissues including skin, solid organs, mucosal epithelium, etc. Parasite nucleic acid can be extracted from thin skin sections by direct phenol–chloroform extraction, but a lysis step prior to extraction will aid in the recovery.

The extraction of nucleic acids from muscle or other solid organs requires prior homogenization or grinding of the sample, followed by digestion in a lysis buffer containing a detergent. Alternatively, mechanical disruption by bead beating followed by silica-based kits can be used.

Mucosal tissues are generally collected in swabs from where the parasite nucleic acids have to be extracted, usually using silica-based extraction kits.

12.3.3 Urine and Saliva

Parasite DNA has been recently detected in body fluids, such as urine and saliva. If this approach is validated through all protozoa, it could simplify the collection of clinical specimens for the detection of these parasites, as these fluids can be collected through noninvasive procedures. The samples need to be preconcentrated by centrifugation and the pellet used for DNA extraction using conventional phenol–chloroform [35], Chelex, or silica-based kits [36–39].

12.3.4 Stools

The method of sample preservation is critical for further use in extraction of nucleic acids. Potassium dichromate 2.5% wt/vol used in 1:1 sample to solution ratio has proven effective in preserving *Cryptosporidium*, *Giardia*, *Cyclospora*, and microsporidia for DNA extraction and related applications. A limitation is that potassium dichromate is a research chemical not routinely used for preservation of diagnostic samples. Potassium dichromate needs to be removed prior to nucleic acid extraction, usually by two centrifugation washes with reagent water or 0.85% saline, at $>5000\,g$ for at least 5 min. DNA can be successfully extracted from clinical specimens fixed in polyvinyl alcohol (PVA) or specimens preserved frozen, although the success rates of the latter can be affected by accidental freeze/thaw or inability to wash the sample. Formalin fixation is frequently used in the fixation or processing of stool samples for coproparasitological analyses. However, it renders the sample unsuitable for molecular biologic work.

Parasites present in stools have a hard environmentally resistant cyst, oocyst, or spore membrane that needs to be disrupted for successful extraction of nucleic acids. Chemical treatment of samples using alkaline buffers [11], or mechanical disruption methods using either consecutive freeze–thaw cycles or bead beating [40] is frequently used. Alternatively, parasites can be subjected to partial purification using cesium chloride centrifugation followed by incubation in lysis buffer [41].

12.3.5 Slide Smears

The detection of blood protozoa usually involves the microscopic detection of the parasites in thin or thick smears. However, slides can also be considered as a source of DNA for molecular studies. The samples are recovered from the slides by scrapping with a scalpel, and DNA can be extracted by phenol–chloroform or solid-based methods [42–44]. One key consideration on this process is the potential cross-contamination of the slides that has been previously diagnosed by microscopy, as there may be carry over of DNA from one slide to another. The potential sources of contamination are stain-processing using jars and immersion oil from the objective lens or droppers used for immersion oil application [45].

Slides previously used for the detection of enteric parasites in stool smears, or the detection of *Cryptosporidium* and *Giardia* in environmental samples can be used for extraction of parasite DNA. The sample is scraped from the slide and incubated in a lysis buffer containing SDS, followed by 15 freeze–thaw cycles and proteinase K denaturation. The material can be subjected to further purification using silica-based kits or used directly for PCR applications [46].

12.3.6 Vectors

The analyses of gut contents or stools from vectors provide valuable information on several protozoa. However, there are challenges in the detection of parasite DNA due to the presence of inhibitory factors such as hemoglobin, other plasma proteins, and cell debris. Traditional phenol–chloroform and alcohol precipitation can be used for nucleic acid extraction from these samples, but may be impractical for large numbers of samples. Silica-based kits designed for the extraction of blood products can also be used. When collecting the gut contents from triatomes, it is recommended to avoid the stomach contents, as they may inhibit PCR amplification [47]. Smear preparations of gut contents can be also be placed on Whatman paper, treated with acetone, and stored at 4°C until used. The nucleic acids and proteins are eluted in separate fractions by incubation at different temperatures: 37°C, also called cold eluate, for proteins, and 100°C, or hot eluate, for DNA [24]. Stools from vectors can be used directly for the extraction of protozoan nucleic acids using a guanidine-based buffer to lyse the parasites and precipitation of DNA using ice-cold ethanol [48].

12.3.7 Environmental and Food Matrices

Purification of nucleic acids of protozoa found in environmental and food matrices shall be preceded by the recovery of parasites. *Giardia* cysts and *Cryptosporidium* oocysts in environmental samples or produce can be recovered through a two-step process: elution followed by parasite recovery.

Recuperation of these parasites from water samples is described in detail in the U.S. EPA Method 1623 [49]. Briefly, the method requires a sample size of at least 10 L of surface water or 100 L of finished water to be passed through certified filters. Parasites are then eluted from the filters by agitation using an elution buffer. The eluate is concentrated by centrifugation and the parasites are recovered from 0.5 mL of the pelleted eluate using parasite-specific immunomagnetic separation (IMS) beads. In most cases, the nucleic acid of interest is DNA, which is used for species and genotype determination of these protozoa. DNA can also be extracted directly from 0.5 mL of the concentrated pellet, without the use of IMS, using extraction kits designed for environmental samples [13].

The detection of *Cryptosporidium*, *Cyclospora*, or *Giardia* in fresh produce requires the elution of parasites from the produce surface. Samples of 10 g of produce are placed in stomacher bags with the addition of 250 mL of elution buffer described in Method 1623 [49], and agitated on an orbital shaker at 100 rpm for 5 cycles, 10 min each [14]. The wash is collected and concentrated by centrifugation or IMS. Parasite DNA can be extracted from the pellets obtained by either method.

In both environmental and food samples, the direct extraction of DNA without IMS leads to faster results, possibility to test for multiple organisms, and significantly reduced testing costs. Meanwhile, the use of IMS may be beneficial as it selectively captures parasites of interest and reduces the presence of substances that may interfere with nucleic acid analyses. However, the use of IMS requires reagents of much higher cost and is limited to the specificity of the beads, thus it is not feasible to detect DNA of other organisms that may be present in the same sample.

An alternative for food matrices is the retention of DNA in FTA cartridges [50]. This method was designed under the consideration that parasite densities are usually very low and the presence of multiple substances could interfere with downstream applications. One limitation was the difficulty in releasing DNA from the FTA membrane. However, an elution protocol has been recently developed [51].

12.4 METHODS

General reagents, supplies, and equipment that are needed for nucleic acid preparation from protozoa are listed in Table 12.4.

12.4.1 ISOLATION OF DNA FROM PROTOZOA

12.4.1.1 Phenol–Chloroform Extraction (Pilcher et al, *Nature Protocol*, 2007) [52]

1. Place 100–200 μL of sample with lysed parasites in a 1.5 mL microcentrifuge tube. Add 10 μL of proteinase K, 10 mg/mL, and incubate overnight at 55°C.
2. Add 300 μL of phenol/chloroform/isoamyl alcohol (25:24:1) to the tube and centrifuge for 10 min at 12,000 g. The upper layer will contain the DNA; carefully collect the supernatant and transfer to a new clean tube.
3. To extract any residual phenol, add 300 μL of chloroform, mix gently, and centrifuge for 5 min at 12,000 g. Carefully transfer the supernatant to a new tube.
4. To precipitate the DNA, add 750 μL of chilled 100% ethanol, shake gently to mix, and incubate for 5 min at room temperature. Centrifuge at 10,000 g for 5 min. Remove the ethanol, and wash the DNA pellet with 1 mL of 70% ethanol. Centrifuge at 12,000 g for 15 min at 4°C.
5. To remove RNases, carefully remove the supernatant and resuspend the DNA pellet in 100 μL of TE buffer pH 7.4, containing 10 mg/mL of RNase A. Incubate for 15 min at room temperature.

TABLE 12.4
General Reagents, Supplies, and Equipment Required for Nucleic Acid Purification from Protozoa

Reagents for DNA Extraction	Reagents for RNA Extraction	Supplies and Equipment
Proteinase K, 10 mg/mL	Guanidinium thiocyanate, 4 M	1.5 mL snap cap microcentrifuge tubes
Phenol/chloroform/isoamyl alcohol (25:24:1)	Sodium citrate, 25 mM pH 7.0	Polypropylene tubes, 4 and 15 mL
Chloroform	*N*-lautorosyl sarcosine (Sarkosyl) 0.05% (wt/vol)	Pipettes of 10 or 20, 100 or 200, and 1000 μL
Ethanol	2-mercaptoethanol, 0.1 M	Aerosol resistant tips (filter tips) for each pipettes
TE buffer pH 7.4 (10 mM Tris pH 7.5 and 1 mM EDTA pH 8.0)	Sodium acetate, 2 M pH 4.0	Microcentrifuges, room temperature, and refrigerated
RNase A	Water-saturated phenol	Freezer, −20°C
Sodium acetate, 3 M pH 5.2	Chloroform/isoamyl alcohol (49:1)	
	Isopropanol	
	Ethanol, 75%	
	DEPC-treated water	

6. Precipitate the DNA by adding 1/10 volume (10 μL) of 3 M sodium acetate pH 5.20 and 2.5 volumes (250 μL) of ice-cold 100% ethanol. Incubate the tube on ice for 15 min, and centrifuge it at 4°C for 15 min at 12,000 g and carefully collect the supernatant without disturbing the pellet.
7. Wash the DNA pellet with 2 volumes (200 μL) of ice-cold 70% ethanol and centrifuge for 2 min at 12,000 g. Collect the supernatant and air-dry the DNA. Resuspend the pellet in 50 μL of water or TE buffer pH 7.4.

12.4.1.2 QIAamp DNA Mini Kit (Qiagen, Valencia, California) [13]

1. Add 180 μL of buffer ATL from the kit to a 1.5 mL microfuge tube that contains the sample with concentrated parasites, such as IMS-isolated *Cryptosporidium* oocysts, and vortex for 30 s.
2. Freeze/thaw the tube for 5 cycles. Freezing may be accomplished by incubation at −70°C for 30 min, or by placing the tube 1–3 min in dry ice-chilled alcohol. Thaw the sample at 56°C using a water bath or heat block.
3. Add 20 μL of proteinase K to the tube, vortex for 10 s, and incubate at 56°C overnight. Try not to exceed 14 h incubation.
4. Add 200 μL of buffer AL to the sample, vortex, and incubate the tube at 70°C for 10 min.
5. Centrifuge at full speed to precipitate the undigested pellet.
6. Transfer the supernatant into a new 1.5 mL tube.
7. Add 200 μL of ethanol to the sample and vortex for 15 s.
8. Carefully transfer the mixture to a QIAamp spin column without wetting the rim, and centrifuge the column at 6000 g for 1 min.
9. Place the spin column in a clean 2 mL collection tube, and discard the tube containing the filtrate.
10. Add 500 μL of buffer AW1 without wetting the rim, and centrifuge at 6000 g for 1 min.
11. Place the spin column in a clean 2 mL collection tube and discard the tube containing the filtrate.
12. Add 500 μL of buffer AW2 without wetting the rim and centrifuge at full speed for 3 min.
13. Place the spin column into a clean 1.5 mL microfuge tube and discard the tube containing the filtrate.
14. Add 100 μL of buffer AE and incubate the tube at room temperature for 1 min.
15. Centrifuge the tube at 6000 g for 1 min.
16. Save the filtrate containing DNA and store the extraction at −20°C.

12.4.2 ISOLATION OF RNA FROM PROTOZOA

A method commonly used for RNA extraction from protozoa is the guanidinium thiocyanate/phenol/chloroform. It has been used with various types of specimens containing these parasites (Table 12.3). This method is based on the characteristics of guanidinium thiocyanate to denature proteins and RNases, which in conjunction with a single-step centrifugation in phenol–chloroform results in the precipitation of denatured proteins and DNA while total RNA remains soluble in the acidic upper phase.

12.4.2.1 Acid Guanidinium Thiocyanate/Phenol/Chloroform Extraction [53]

1. Incubate the samples in a denaturing solution composed of 4 M guanidinium thiocyanate, 25 mM sodium citrate pH 7.0, 0.05% (wt/vol) *N*-lautorosyl sarcosine (Sarkosyl) and 0.1 M β-mercaptoethanol, at 1 mL/100 mg (100 μL) of tissue or 1 mL/10^7 cells.
2. Transfer the lysate to 4 mL polypropylene tubes, and sequentially add (a) 0.1 mL of 2 M sodium acetate pH 4.0, thoroughly mixing by inversion; (b) 1 mL of water-saturated phenol,

mixing thoroughly by inversion; and (c) 0.2 mL chloroform/isoamyl alcohol (49:1), shaking vigorously by hand.

3. Cool the samples on ice for 15 min, and centrifuge for 20 min in a refrigerated centrifuge (4°C) at 10,000 g. Carefully collect the upper aqueous phase that contains the RNA, and transfer to a new clean tube.

4. Add 1 mL of isopropanol and incubate for 1 h or longer at −20°C. Precipitate the RNA by centrifugation in a refrigerated centrifuge for 20 min at 10,000 g. The RNA pellet has the appearance of a gel-like substance. Remove the supernatant carefully and use the pellet in a secondary precipitation.

5. Dissolve the pellet in 0.3 mL of the same denaturing solution (Step 1) and transfer the suspension to a new 1.5 mL polypropylene tube. Add an equal volume of isopropanol (0.3 mL) to the RNA, followed by incubation for at least 30 min at −20°C.

6. Precipitate the RNA by centrifugation for 10 min at 10,000 g. The supernatant is carefully discarded, and the RNA pellet is resuspended in 1 mL of 75% ethanol, mixed by gentle vortexing, and incubated for 10 min at room temperature to aid in the removal of residues of guanidinium. Precipitate the RNA by refrigerated centrifugation for 5 min at 10,000 g.

7. Discard the supernatant and allow the pellet to air-dry. The solubility of the RNA pellet will decrease dramatically if the pellet dries completely or if it is dried in a vacuum drier.

8. Dissolve the RNA in DEPC-treated water and incubate at 60°C for 15 min, and store at −80°C until used.

12.4.2.2   RNeasy Kit (Qiagen, Valencia, California)

1. Release the RNA from pure parasites by mechanical disruption, or a series of freeze–thaw cycles. Use 10^5 to 10^7 organisms per extraction. Remove any culture or storage media by centrifugation at 14,000 g for 5 min. Discard the supernatant and use the pelleted parasites (about 350 µL).

2. Prepare buffer RLT by adding 10 µL of β-mercaptoethanol per 1 mL of RLT.

3. Add 1 volume (~350 µL) of buffer RLT.

4. Lyse the parasites by 10 freeze–thaw cycles.

5. Centrifuge the lysate for 3 min at 14,000 g. Remove the supernatant with a pipette and transfer to a new clean tube.

6. Add 700 µL (about twice the original sample volume) of 70% ethanol. Mix by pipetting.

7. Transfer 700 µL of the mixture to an RNeasy spin column placed on a 2 mL collection tube.

8. Centrifuge for 15 s at 9000 g. Discard the flow through.

9. Repeat Steps 7 and 8 until all lysate is transferred to the spin column.

10. Add 700 µL of buffer RW1 to the column.

11. Centrifuge for 15 s at 9000 g. Discard the flow through.

12. Ensure that ethanol is added to RPE buffer prior to use.

13. Add 500 µL of RPE buffer.

14. Centrifuge for 15 s at 9000 g. Discard the flow through.

15. Add 500 µL of RPE buffer.

16. Centrifuge for 2 min at 9000 g. Discard the flow through.

17. Carefully place the RNeasy spin column to a new 2 mL discard tube.

18. Centrifuge for 1 min at 14,000 g. Discard the flow through.

19. Carefully place the RNeasy spin column to a new 1.5 mL microcentrifuge tube.

20. Add 30–50 µL of RNase-free water. Make sure to place it directly to the membrane.

21. Centrifuge for 1 min at 9000 g to elute the RNA from the filter. RNA will be in the collection tube.

Notes: (1) For higher yields, repeat Steps 20 and 21. (2) For higher RNA concentration, repeat Step 20 using the eluate from Step 21 and the same collection tube.

12.5 FUTURE DEVELOPMENT TRENDS

Recent advances in the extraction of nucleic acids are mostly in two areas: the increase use of solid-phase technologies, and the ability to process larger number of samples, primarily through automated extractions.

Using the principles of solid-based technologies, solid-phase reversible immobilization is frequently used in automated extraction of nucleic acids. Its key feature is the use of carboxyl-coated magnetic particles that can reversibly bind DNA in the presences of polyethylene glycol and salts [54], allowing the reduction in the number of processing steps and facilitating its use in manual or automated processes. Other processes that allow automation are based on silica-solid-phase membranes [55] or silica-coated magnetic beads [56]. Some considerations when using these technologies are the need of preparatory or lysing steps, the limited capacity of the magnetic beads, and costs, both of capital equipment and supplies.

A more recent innovation in solid-phase technologies is the development of microchips that use photoactivated polycarbonate [57] or silica monoliths [58] for nucleic acid extraction. These chips have two immobilization or capture beds, consisting of ordered arrays solid matrix. These beds are specifically UV-photoactivated for selective capture of DNA or RNA, which is later released with water. Some advantages are the potential of simple and fast extractions even with limited laboratory infrastructure, and significant reductions of cross-contamination. Important limitations when using microchips are the small capacity of the chips, which are currently designed for microscale volumes, their limitation to nonexisting compatibility with automated processing, and much higher cost per extraction.

Vitaliano A. Cama received his bachelor and veterinary degrees from San Marcos University in Lima, Peru. He spent the first 2 years of his veterinary career in poultry and avian pathology, with interest in coccidiosis research. Later he joined AB Prisma as a researcher where he worked in the production of diagnostic antibodies in poultry systems and serodiagnosis of porcine cysticercosis. Thereafter, he joined the University of Arizona as a research associate, earning his PhD in veterinary pathobiology. He worked for 5 years in the biotechnology industry at IgX Ltd./Synergy Pharmaceuticals, where he developed antibody-based therapies against enteric and viral pathogens. Since 2002, he has been working at the Centers for Disease Control and Prevention (CDC), in the Division of Parasitic Diseases, National Center for Zoonotic, Vector-Borne and Enteric Diseases, CDC, Atlanta, Georgia. His recent research focuses on the molecular epidemiology and transmission dynamics of *Cryptosporidium, Giardia*, microsporidia, and other protozoa that infect humans and animals. He has published over 40 scientific papers and book chapters. He is a guest lecturer at Emory University and the University of Georgia, and holds adjunct faculty positions at Johns Hopkins University and San Marcos University in Peru.

Lihua Xiao received his veterinary education in China. After his MS in veterinary parasitology and 2 years of teaching at the Northeast Agricultural University in Harbin, China, he undertook his PhD training in veterinary parasitology at University of Maine, Orono, Maine, and postdoctoral training at the Ohio State University College of Veterinary Medicine, Columbus, Ohio. In 1993, he moved to the Centers for Disease Control and Prevention (CDC), first as a guest researcher, then as a senior staff fellow. He is currently a senior scientist in the division of parasitic diseases, National Center for Zoonotic, Vector-Borne, and Enteric Diseases, CDC, Atlanta, Georgia. His earlier research interests were mostly in the epidemiology, pathogenesis, and control of gastrointestinal nematodes and epidemiology and biology of cryptosporidiosis and giardiasis of farm animals. More recently, he has focused on the taxonomy, molecular epidemiology, and environmental biology of *Cryptosporidium, Giardia*, microsporidia, and other enteric protists in humans and animals. He has published over 200 scientific papers, invited reviews, and book chapters, and holds adjunct faculty positions at Cornell University (Ithaca, New York), the Ohio State University, and several universities in China.

REFERENCES

1. Ravdin, J.I. and Stauffer, W.M., Introduction to protozoal diseases. In Mandell, B.D. (Ed.), *Principles and Practice of Infectious Diseases*, 6th ed., New York, Churchill Livingston, 2005.
2. Hibbett, D.S. et al., A higher-level phylogenetic classification of the Fungi, *Mycol. Res.*, 111: 509, 2007.
3. Redhead, S.A. et al., *Pneumocystis* and *Trypanosoma cruzi*: Nomenclature and typifications, *J. Eukaryot. Microbiol.*, 53: 2, 2006.
4. Sugiyama, J., Hosaka, K., and Suh, S.O., Early diverging Ascomycota: Phylogenetic divergence and related evolutionary enigmas, *Mycologia*, 98: 996, 2006.
5. Silberman, J.D. et al., Human parasite finds taxonomic home, *Nature*, 380: 398, 1996.
6. Plorde, J.J., Flagellates. In Sherris, J.C. (Ed.), *Medical Microbiology*, 2nd ed., New York: Elsevier, 1990, Chapter 54.
7. Plorde, J.J., Rhizopodes. In Sherris, J.C. (Ed.), *Medical Microbiology*, 2nd ed., New York: Elsevier, 1990, Chapter 53.
8. Nichols, R.A., Moore, J.E., and Smith, H.V., A rapid method for extracting oocyst DNA from *Cryptosporidium*-positive human faeces for outbreak investigations, *J. Microbiol. Methods*, 65: 512, 2006.
9. Cama, V. et al., Mixed *Cryptosporidium* infections and HIV, *Emerg. Infect. Dis.*, 12: 1025, 2006.
10. Ortega, Y.R. et al., Restriction fragment length polymorphism analysis of *Cryptosporidium parvum* isolates of bovine and human origin, *J. Protozool.*, 38: 40S, 1991.
11. Xiao, L. et al., Genetic diversity of *Cryptosporidium* spp. in captive reptiles, *Appl. Environ. Microbiol.*, 70: 891, 2004.
12. Da Silva, A.J. et al., Fast and reliable extraction of protozoan parasite DNA from fecal specimens, *Mol. Diagn.*, 4: 57, 1999.
13. Jiang, J. et al., Development of procedures for direct extraction of *Cryptosporidium* DNA from water concentrates and for relief of PCR inhibitors, *Appl. Environ. Microbiol.*, 71: 1135, 2005.
14. Ortega, Y. and Cama, V., Foodborne transmission. In Fayer, R. and Xiao, L. (Eds.), *Cryptosporidium and Cryptosporidiosis*, 2nd ed., Boca Raton, FL: CRC Press, 2008.
15. Giraffa, G., Rossetti, L., and Neviani, E., An evaluation of chelex-based DNA purification protocols for the typing of lactic acid bacteria, *J. Microbiol. Methods*, 42: 175, 2000.
16. Jacobsen, N. et al., Direct isolation of poly(A) + RNA from 4 M guanidine thiocyanate-lysed cell extracts using locked nucleic acid-oligo(T) capture, *Nucleic Acids Res.*, 32: e64, 2004.
17. Biswas, S. et al., Prevalence of point mutations in the dihydrofolate reductase and dihydropteroate synthetase genes of *Plasmodium falciparum* isolates from India and Thailand: A molecular epidemiologic study, *Trop. Med. Int. Health*, 5: 737, 2000.
18. Kirchhoff, L.V. et al., Comparison of PCR and microscopic methods for detecting *Trypanosoma cruzi*, *J. Clin. Microbiol.*, 34: 1171, 1996.
19. Mccollum, A.M. et al., Common origin and fixation of *Plasmodium falciparum* dhfr and dhps mutations associated with sulfadoxine-pyrimethamine resistance in a low-transmission area in South America, *Antimicrob. Agents Chemother.*, 51: 2085, 2007.
20. Herwaldt, B.L. et al., Use of polymerase chain reaction to diagnose the fifth reported US case of autochthonous transmission of *Trypanosoma cruzi*, in Tennessee, 1998, *J. Infect. Dis.*, 181: 395, 2000.
21. Virreira, M. et al., Comparison of polymerase chain reaction methods for reliable and easy detection of congenital *Trypanosoma cruzi* infection, *Am. J. Trop. Med. Hyg.*, 68: 574, 2003.
22. Clausen, P.H. et al., Use of a PCR assay for the specific and sensitive detection of *Trypanosoma* spp. in naturally infected dairy cattle in peri-urban Kampala, Uganda, *Ann. N. Y. Acad. Sci.*, 849: 21, 1998.
23. Rottmann, M. et al., Differential expression of *var* gene groups is associated with morbidity caused by *Plasmodium falciparum* infection in Tanzanian children, *Infect. Immun.*, 74: 3904, 2006.
24. Boid, R., Jones, T.W., and Munro, A., A simple procedure for the extraction of trypanosome DNA and host protein from dried blood meal residues of haematophagous diptera, *Vet. Parasitol.*, 85: 313, 1999.
25. Gil, J.P. et al., Detection of atovaquone and Malarone resistance conferring mutations in *Plasmodium falciparum* cytochrome b gene (cytb), *Mol. Cell Probes*, 17: 85, 2003.
26. Bereczky, S. et al., Short report: Rapid DNA extraction from archive blood spots on filter paper for genotyping of *Plasmodium falciparum*, *Am. J. Trop. Med. Hyg.*, 72: 249, 2005.

27. Zhou, Z. et al., Decline in sulfadoxine-pyrimethamine-resistant alleles after change in drug policy in the Amazon region of Peru, *Antimicrob. Agents Chemother.*, 52: 739, 2008.

28. Kaestli, M. et al., Longitudinal assessment of *Plasmodium falciparum* var gene transcription in naturally infected asymptomatic children in Papua New Guinea, *J. Infect. Dis.*, 189: 1942, 2004.

29. Plowe, C.V. et al., Pyrimethamine and proguanil resistance-conferring mutations in *Plasmodium falciparum* dihydrofolate reductase: Polymerase chain reaction methods for surveillance in Africa, *Am. J. Trop. Med. Hyg.*, 52: 565, 1995.

30. De Almeida, P.P. et al., Diagnostic evaluation of PCR on dried blood samples from goats experimentally infected with *Trypanosoma brucei*, *Acta Trop.*, 70: 269, 1998.

31. Fomovskaia, G. et al., FTA-coated media for use as a molecular diagnostic tool. US Patent number 6746841 United States, Whatman Inc., 2000.

32. Rogers, C. and Burgoyne, L., Bacterial typing: Storing and processing of stabilized reference bacteria for polymerase chain reaction without preparing DNA—an example of an automatable procedure, *Anal. Biochem.*, 247: 223, 1997.

33. Gonzales, J.L., Loza, A., and Chacon, E., Sensitivity of different *Trypanosoma vivax* specific primers for the diagnosis of livestock trypanosomosis using different DNA extraction methods, *Vet. Parasitol.*, 136: 119, 2006.

34. Mlambo, G. et al., A filter paper method for the detection of *Plasmodium falciparum* gametocytes by reverse transcription polymerase chain reaction, *Am. J. Trop. Med. Hyg.*, 78: 114, 2008.

35. Manna, L. et al., Urine sampling for real-time polymerase chain reaction based diagnosis of canine leishmaniasis, *J. Vet. Diagn. Invest.*, 20: 64, 2008.

36. Franceschi, A. et al., Occurrence of *Leishmania* DNA in urines of dogs naturally infected with leishmaniasis, *Vet. Res. Commun.*, 31: 335, 2007.

37. Solano-Gallego, L. et al., Detection of *Leishmania infantum* DNA by fret-based real-time PCR in urine from dogs with natural clinical leishmaniosis, *Vet. Parasitol.*, 147: 315, 2007.

38. Parija, S.C. and Khairnar, K., Detection of excretory *Entamoeba histolytica* DNA in the urine, and detection of *E. histolytica* DNA and lectin antigen in the liver abscess pus for the diagnosis of amoebic liver abscess, *BMC Microbiol.*, 7: 41, 2007.

39. Mharakurwa, S. et al., PCR detection of *Plasmodium falciparum* in human urine and saliva samples, *Malar. J.*, 5: 103, 2006.

40. Xiao, L. et al. Possible transmission of *Cryptosporidium canis* among children and a dog in a household, *J. Clin. Microbiol.*, 45: 2014, 2007.

41. Santin, M., Trout, J.M., and Fayer, R., Prevalence of *Enterocytozoon bieneusi* in post-weaned dairy calves in the eastern United States, *Parasitol. Res.*, 93: 287, 2004.

42. Brustoloni, Y.M. et al., Sensitivity and specificity of polymerase chain reaction in Giemsa-stained slides for diagnosis of visceral leishmaniasis in children, *Mem. Inst. Oswaldo Cruz*, 102: 497, 2007.

43. Ekala, M.T. et al. Evaluation of a simple and rapid method of *Plasmodium falciparum* DNA extraction using thick blood smears from Gabonese patients, *Bull. Soc. Pathol. Exot.*, 93: 8, 2000.

44. Kimura, M. et al., Amplification by polymerase chain reaction of *Plasmodium falciparum* DNA from Giemsa-stained thin blood smears, *Mol. Biochem. Parasitol.*, 70: 193, 1995.

45. Aubouy, A. and Carme, B., *Plasmodium* DNA contamination between blood smears during Giemsa staining and microscopic examination, *J. Infect. Dis.*, 190: 1335, 2004.

46. Nichols, R.A. and Smith, H.V., Optimization of DNA extraction and molecular detection of *Cryptosporidium* oocysts in natural mineral water sources, *J. Food Prot.*, 67: 524, 2004.

47. Pizarro, J.C., Lucero, D.E., and Stevens, L., PCR reveals significantly higher rates of *Trypanosoma cruzi* infection than microscopy in the Chagas vector, *Triatoma infestans*: High rates found in Chuquisaca, Bolivia, *BMC Infect. Dis.*, 7: 66, 2007.

48. Marcet, P.L. et al., PCR-based screening and lineage identification of *Trypanosoma cruzi* directly from faecal samples of triatomine bugs from northwestern Argentina, *Parasitology*, 132: 57, 2006.

49. Method 1623: *Cryptosporidium* and *Giardia* in water by filtration/IMS/FA. Office of Water, USEPA, Washington, DC, 2005.

50. Chu, D.M. et al., Detection of *Cyclospora cayetanensis* in animal fecal isolates from Nepal using an FTA filter-base polymerase chain reaction method, *Am. J. Trop. Med. Hyg.*, 71: 373, 2004.

51. Lema, C. et al., Optimized pH method for DNA elution from buccal cells collected in Whatman FTA cards, *Genet. Test.*, 10: 126, 2006.

52. Pilcher, K.E. et al., A reliable general purpose method for extracting genomic DNA from *Dictyostelium* cells, *Nat. Protoc.*, 2: 1325, 2007.

53. Chomczynski, P. and Sacchi, N., The single-step method of RNA isolation by acid guanidinium thiocyanate–phenol–chloroform extraction: Twenty-something years on, *Nat. Protoc.*, 1: 581, 2006.

54. Deangelis, M.M., Wang, D.G., and Hawkins, T.L., Solid-phase reversible immobilization for the isolation of PCR products, *Nucleic Acids Res.*, 23: 4742, 1995.

55. Selley, P. et al., Automated solutions for total RNA isolation from diverse sample types, *Clin. Lab. Med.*, 27: 155, 2007.

56. Nagy, M. et al., Optimization and validation of a fully automated silica-coated magnetic beads purification technology in forensics, *Forensic Sci. Int.*, 152: 13, 2005.

57. Witek, M.A. et al., Purification and preconcentration of genomic DNA from whole cell lysates using photoactivated polycarbonate (PPC) microfluidic chips, *Nucleic Acids Res.*, 34: e74, 2006.

58. Wen, J. et al., DNA extraction using a tetramethyl orthosilicate-grafted photopolymerized monolithic solid phase, *Anal. Chem.*, 78: 1673, 2006.

59. Garcia, L. et al., Culture-independent species typing of neotropical *Leishmania* for clinical validation of a PCR-based assay targeting heat shock protein 70 genes, *J. Clin. Microbiol.*, 42: 2294, 2004.

60. Garcia, J.L. et al., *Toxoplasma gondii*: Detection by mouse bioassay, histopathology, and polymerase chain reaction in tissues from experimentally infected pigs, *Exp. Parasitol.*, 113: 267, 2006.

61. Ergin, M. et al., Cutaneous leishmaniasis: Evaluation by polymerase chain reaction in the Cukurova region of Turkey, *J. Parasitol.*, 91: 1208, 2005.

62. Marques, M.J. et al., Simple form of clinical sample preservation and *Leishmania* DNA extraction from human lesions for diagnosis of American cutaneous leishmaniasis via polymerase chain reaction, *Am. J. Trop. Med. Hyg.*, 65: 902, 2001.

63. Robberts, F.J., Liebowitz, L.D., and Chalkley, L.J., Polymerase chain reaction detection of *Pneumocystis jiroveci*: Evaluation of 9 assays, *Diagn. Microbiol. Infect. Dis.*, 58: 385, 2007.

64. Beser, J., Hagblom, P., and Fernandez, V., Frequent in vitro recombination in internal transcribed spacers 1 and 2 during genotyping of *Pneumocystis jirovecii*, *J. Clin. Microbiol.*, 45: 881, 2007.

65. Saito, K. et al., Detection of *Pneumocystis carinii* by DNA amplification in patients with connective tissue diseases: Re-evaluation of clinical features of *P. carinii* pneumonia in rheumatic diseases, *Rheumatology* (Oxford), 43: 479, 2004.

66. Nuchprayoon, S. et al., Flinders Technology Associates (FTA) filter paper-based DNA extraction with polymerase chain reaction (PCR) for detection of *Pneumocystis jirovecii* from respiratory specimens of immunocompromised patients, *J. Clin. Lab. Anal.*, 21: 382, 2007.

67. Orlandi, P.A. and Lampel, K.A., Extraction-free, filter-based template preparation for rapid and sensitive PCR detection of pathogenic parasitic protozoa, *J. Clin. Microbiol.*, 38: 2271, 2000.

68. Lehmann, T. et al., Strain typing of *Toxoplasma gondii*: Comparison of antigen-coding and housekeeping genes, *J. Parasitol.*, 86: 960, 2000.

69. Flori, P. et al., Experimental model of congenital toxoplasmosis in guinea-pigs: Use of quantitative and qualitative PCR for the study of maternofetal transmission, *J. Med. Microbiol.*, 51: 871, 2002.

70. Hurkova, L. and Modry, D., PCR detection of *Neospora caninum*, *Toxoplasma gondii* and *Encephalitozoon cuniculi* in brains of wild carnivores, *Vet. Parasitol.*, 137: 150, 2006.

71. Khan, A. et al., Genotyping of *Toxoplasma gondii* strains from immunocompromised patients reveals high prevalence of type I strains, *J. Clin. Microbiol.*, 43: 5881, 2005.

72. Jenkins, M.C., Trout, J., and Fayer, R., Development and application of an improved semiquantitative technique for detecting low-level *Cryptosporidium parvum* infections in mouse tissue using polymerase chain reaction, *J. Parasitol.*, 84: 182, 1998.

73. Johnson, E.M. et al., Field survey of rodents for *Hepatozoon* infections in an endemic focus of American canine hepatozoonosis, *Vet. Parasitol.*, 150: 27, 2007.

74. Cummings, K.L. and Tarleton, R.L., Rapid quantitation of *Trypanosoma cruzi* in host tissue by real-time PCR, *Mol. Biochem. Parasitol.*, 129: 53, 2003.

75. Simpson, P. et al., Real-time PCRs for detection of *Trichomonas vaginalis* beta-tubulin and 18S rRNA genes in female genital specimens, *J. Med. Microbiol.*, 56: 772, 2007.

76. Strong, W.B., Gut, J., and Nelson, R.G., Cloning and sequence analysis of a highly polymorphic *Cryptosporidium parvum* gene encoding a 60-kilodalton glycoprotein and characterization of its 15- and 45-kilodalton zoite surface antigen products, *Infect. Immun.*, 68: 4117, 2000.

77. Fayer, R. et al., Detection of *Cryptosporidium felis* and *Giardia duodenalis* Assemblage F in a cat colony, *Vet. Parasitol.*, 140: 44, 2006.

78. Paris, D.H. et al., Loop-mediated isothermal PCR (LAMP) for the diagnosis of *Falciparum* malaria, *Am. J. Trop. Med. Hyg.*, 77: 972, 2007.

79. Stensvold, C.R. et al., Detecting *Blastocystis* using parasitologic and DNA-based methods: A comparative study, *Diagn. Microbiol. Infect. Dis.*, 59: 303, 2007.
80. Jiang, J., Alderisio, K.A., and Xiao, L., Distribution of *Cryptosporidium* genotypes in storm event water samples from three watersheds in New York, *Appl. Environ. Microbiol.*, 71: 4446, 2005.
81. Agop-Nersesian, C. et al., Functional expression of ribozymes in Apicomplexa: Towards exogenous control of gene expression by inducible RNA-cleavage, *Int. J. Parasitol.*, 38: 673, 2008.
82. Salanti, A. et al., Selective upregulation of a single distinctly structured var gene in chondroitin sulphate A-adhering *Plasmodium falciparum* involved in pregnancy-associated malaria, *Mol. Microbiol.*, 49: 179, 2003.
83. Perez, J.A. et al., Validation of a rapid method for extraction of total RNA applied to *Leishmania* promastigotes, *J. Parasitol.*, 85: 757, 1999.
84. Dahlback, M. et al., Changes in var gene mRNA levels during erythrocytic development in two phenotypically distinct *Plasmodium falciparum* parasites, *Malar. J.*, 6: 78, 2007.
85. Boom, R. et al., Rapid and simple method for purification of nucleic acids, *J. Clin. Microbiol.*, 28: 495, 1990.
86. Decatur, W.A., Johansen, S., and Vogt, V.M., Expression of the *Naegleria* intron endonuclease is dependent on a functional group I self-cleaving ribozyme, *RNA*, 6: 616, 2000.
87. Johnson, A.M., Mcdonald, P.J., and Illana, S., Characterization and in vitro translation of *Toxoplasma gondii* ribonucleic acid, *Mol. Biochem. Parasitol.*, 18: 313, 1986.
88. Lin, A. et al., Genetic characteristics of the Korean isolate KI-1 of *Toxoplasma gondii*, *Korean J. Parasitol.*, 43: 27, 2005.
89. Ronnebaumer, K. et al., Identification of novel developmentally regulated genes in *Encephalitozoon cuniculi*: An endochitinase, a chitin-synthase, and two subtilisin-like proteases are induced during meront-to-sporont differentiation, *J. Eukaryot. Microbiol.*, 53(Suppl 1): S74, 2006.
90. Chen, X.S. et al., Combined experimental and computational approach to identify non-protein-coding RNAs in the deep-branching eukaryote *Giardia intestinalis*, *Nucleic Acids Res.*, 35: 4619, 2007.
91. Bansal, D. et al., Multidrug resistance in amoebiasis patients, *Indian J. Med. Res.*, 124: 189, 2006.
92. Benitez, A.J., Mcnair, N., and Mead, J., Modulation of gene expression of three *Cryptosporidium parvum* ATP-binding cassette transporters in response to drug treatment, *Parasitol. Res.*, 101: 1611, 2007.
93. Mahbubani, M.H. et al., Detection of *Giardia* cysts by using the polymerase chain reaction and distinguishing live from dead cysts, *Appl. Environ. Microbiol.*, 57: 3456, 1991.
94. Stinear, T. et al., Detection of a single viable *Cryptosporidium parvum* oocyst in environmental water concentrates by reverse transcription-PCR, *Appl. Environ. Microbiol.*, 62: 3385, 1996.
95. Dessens, J.T. et al., CTRP is essential for mosquito infection by malaria ookinetes, *EMBO J.*, 18: 6221, 1999.

13 Isolation of Nucleic Acids from Helminthes

Munehiro Okamoto and Akira Ito

CONTENTS

13.1 INTRODUCTION

The term "helminth" (plural, helminthes) is commonly used to describe various organisms belonging to the phylum Platyhelminthes (flukes, tapeworms, and other flatworms) and the now-obsolete phylum Nemathelminthes (roundworms and their relatives) [1]. Platyhelminthes is one of the acoelomate phyla and includes several important classes, such as Cestoda and Trematoda. Nemathelminthes (or Aschelminthes) was formerly considered a phylum to represent the pseudocoelomate organisms. It is now obsolete and divided into several distinct phyla, such as Nematoda, Nematomorpha, Acanthocephala, Entoprocta, Rotifera, Gastrotricha, Kinorhyncha, etc. Among these, Nematoda (roundworms), Nematomorpha (horsehair worms), and Acanthocephala (thorny-headed worms) are important. In recognition of recent proposals based on molecular phylogenetic analysis, taxonomical status of these phyla is rapidly altered. In this chapter, the more traditional and common classification is embraced. The Annelida (earthworms, leeches) are not regarded as helminthes, though some (e.g., leeches) may be parasitic and others (e.g., earthworms) may serve as intermediate hosts of helminthes [1,2].

Besides distinct morphology, helminthes also demonstrate characteristic biology and life cycle, with some being free-living while others being parasitic for animals or plants [1]. Parasitic helminthes parasitize not only humans and domestic animals but also many other vertebrates and invertebrates. Many species of molluscs and arthropods are commonly utilized as their intermediate hosts. In most cases, parasitic helminthes are harmful and pathogenic to their hosts. Because there are many species of parasitic helminthes, and because their morphological features and host change with developing stage (Table 13.1), use of molecular tools is critical to the identification and control of parasitic helminthes. In this chapter, the key features of helminthic parasites are reviewed briefly, and practical methods for DNA and RNA extractions from helminthes are discussed.

13.1.1 ACOELOMATE HELMINTHES

Acoelomates have three well-defined germ layers: ectoderm, endoderm, and mesoderm. However, acoelomates lack a coelom—a fluid-filled cavity between the outer body wall and the gut. Being one of acoelomate phyla, Platyhelminthes comprises the following four important classes: Cestoda (tapeworms), Trematoda (flukes), Monogenea, and Turbellaria (Table 13.1). Platyhelminthes have a dorsoventrally flattened and bilaterally symmetrical body without a body cavity. Many are hermaphroditic (with both male and female sex organs) and practice self-fertilization. Other than those species in the class Turbellaria, most Platyhelminthes in classes Cestoda (tapeworms), Trematoda (flukes), and Monogenea are parasitic. The organs are embedded in tissue called the parenchyma and the excretory organs are flame cells. The life cycle is usually indirect.

13.1.1.1 Class Cestoda

Class Cestoda consists of two subclasses: Eucestoda and Cestodaria.

13.1.1.1.1 Subclass Eucestoda
All members in Eucestoda are parasitic and referred to as tapeworms (Figure 13.1). Tapeworms are endoparasitic worms with elongate flat body and without a body cavity and alimentary canal. The anterior end of the body is modified into a holdfast organ called a scolex, which may have a rostellum, suckers, bothria, bothridia, tentacles, hooks, and spines to aid attachment to the gut of the host. The scolex is usually followed by a short unsegmented portion called the neck and, in general, the remainder of the body or strobila consists of a number of segments or proglottids. With the exception of the members of the order Dioecocestidae, the tapeworms are protandric hermaphrodites, that is, each proglottid contains one or two complete copies of the male and female reproductive organs. The body is covered by a tegument composed of a syncytial outer layer formed by the tegumental cells. The outer cytoplasm is extended into microtriches (sometimes referred to as microvilli) and acts as the absorptive structures. The life cycles are usually indirect.

Eggs may be embryonated or unembryonated when passed from the definitive host and the fully embryonated egg contains an oncosphere, which has three pairs of hooks. An oncosphere is infectious to intermediate hosts and develops into larval stages (metacestodes). After all, cestode life cycles are indirect and require the development of metacestodes in one or more intermediate hosts. The common forms of metacestodes that occur in the life cycles of cestodes can be classified

TABLE 13.1
Primary Features of Helminthic Taxa

Phylum	Class	Digestive System	Reproductive System	Life Cycle	Definitive Hosts	Intermediate Hosts	Major Genera
Acoelomate							
Platyhelminthes	Cestoda (tapeworm)	Absent	Hermaphroditic[a]	Indirect	All classes of vertebrates	Vertebrates Invertebrates	Echinococcus Diphyllobothrium Hymenolepis
	Trematoda (fluke)	Present without anus	Hermaphroditic[a]	Indirect	All classes of vertebrates	Vertebrates Invertebrates	Clonorchis Paragonimus Schistosoma
	Monogenea	Present without anus	Hermaphroditic	Direct	Fishes, amphibians, and reptiles	—	Gyrodactylus Diplozoon
	Turbellaria	Present without anus	Hermaphroditic	Free living	—	—	Dugesia
Pseudocoelomate							
Nematoda (roundworm)	Two classes	Present	Dioecious	Free living, direct, or indirect	 Vertebrate Invertebrate Plant	— Vertebrate Invertebrate — 	Caenorhabditis Ascaris Trichinella Brugia Steinernema Globodera
Nematomorpha (horsehair worm)	Two classes	Present	Dioecious	Adult: free living Larva: parasitic	—	Invertebrate	Spinochordodes
Acanthocephala (thorny-headed worm)	Three classes	Absent	Dioecious	Indirect	Vertebrate, mainly fishes	Invertebrate	Macracanthorhynchus

[a] There are some exceptions.

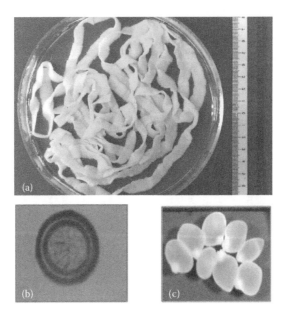

FIGURE 13.1 Photographs of eucestodes: (a) Adult tapeworms of *Taenia saginata* from human. (b) Eggs of *T. solium* from human. The egg cannot be distinguished from eggs of other taeniid cestodes morphologically. (c) Cysticerci of *T. saginata* developed in severe combined immunodeficient (*SCID*) mouse. *SCID* mice are highly useful for confirmation of the larval stages of human *Taenia* species.

as follows: (1) procercoid is the first metacestode stage in the life cycles of parasites such as the Pseudophyllidea (e.g., *Diphyllobothrium*); (2) plerocercoid follows the procercoid and occurs in the second intermediate host; (3) tetrathyridium is an elongate, solid-bodied metacestode with a deeply invaginated acetabular scolex (e.g., *Mesocestoides*); (4) cysticercoid is a metacestode with a single noninvaginated scolex withdrawn into a small vesicle with practically no cavity (e.g., *Dipylidium*); (5) cysticercus is a metacestode of Taeniid tapeworms, consisting of a single invaginated scolex enclosed in a fluid-filled cyst (e.g., *Taenia*); (6) hydatid is a large fluid-containing bladder that develops other cysts called brood capsules in which the scolices develop (e.g., *Echinococcus*). The metacestode is passively transferred to the definitive host when the latter ingests the infected intermediate host. The scolex excysts or evaginates and attaches to the mucosa of the intestine. The adult usually lives in the digestive tract of the host.

13.1.1.1.2   Subclass Cestodaria

Cestodaria are monozoic hermaphroditic worms that lack strobilation (proglottization) and include only a single set of male and female reproductive organs. There is no mouth or digestive tract. The body surface is covered by the syncytial tegument. A scolex is absent but there may be an anterior sucker for attachment. The ciliated larva in embryonated egg is called the lycophora and has 10 hooks. Cestodarians are all endoparasites in the intestine and coelomic cavities of various fishes and rarely in reptiles.

The life cycles of these parasites are incompletely known except for a very few species. Members of Amphilinidea are parasitic in the coelom of sturgeons, other primitive fish, and tortoises. The adult worms bore through the body wall of the fish host to lay their eggs. The egg consists of a lycophora. After the egg is ingested in the intermediate host, the lycophora hatches and enters the hemocoel. Inside the body cavity of the intermediate host, the lycophora attains the appearance of a procercoid and plerocercoid. When the definitive host ingests the intermediate host, the plerocercoid burrows into the intestine wall and enters the body cavity. It attains sexual maturity there.

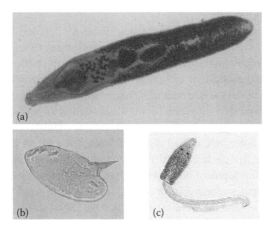

FIGURE 13.2 Photographs of digenean trematodes: (a) adult fluke of *Echinostoma hortense*, (b) egg of *S. haematobium*, and (c) Cercaria of *Metagonimus* sp.

13.1.1.2 Class Trematoda

Class Trematoda consists of two subclasses: Digenea and Aspidogastrea.

13.1.1.2.1 Subclass Digenea

The digenetic trematodes are usually dorsoventrally flattened, some being long and narrow, and some leaf-shaped (Figure 13.2). A small number of the paramphistomes have thick fleshy bodies. The schistosomes are long and worm-like shaped. Digeneans are parasitic, and adults are particularly common in the digestive tract, but occur throughout the organ systems of all classes of vertebrates. Adult trematodes have typically two suckers, an anterior oral sucker surrounding the mouth, and a ventral sucker, sometimes termed the acetabulum, on the ventral surface. Most digeneans have a forked digestive system that opens at the mouth, but there is no anus. Digeneans are also capable of direct nutrient uptake through the surface of the body, the tegument. The tegument is syncytial, that is, a mass of protoplasm containing many nuclei but not divided into cells. With the exception of the Schistomatidae and Didymozoidae, the digenetic trematodes are hermaphrodite. The life cycle requires one, two, or more intermediate hosts.

Five larval states may occur in the life cycle: miracidium, sporocyst, redia, cercaria, and metacercaria. The eggs of the digenea are usually passed in the feces of the host and under suitable conditions of moisture and warmth a larva, miracidium, hatches. Hatching is controlled by a number of factors such as light, temperature, and salinity. Miracidia do not feed and further development occurs after it enters a first intermediate host, a snail. They penetrate a snail actively probably by enzyme secretions from the apical gland. Following penetration, the ciliated coat is lost and it becomes a sporocyst. Within the sporocyst, the germinal cells multiply and produce either daughter sporocysts or rediae. One or more generations of rediae may occur. The next stage, the cercaria, is produced by the sporocyst or the redia. The cercaria leaves the snail host actively through an opening or through the tegument, or cercariae are expelled passively in masses. Cercariae usually encyst on a definitive host or in a second intermediate host such as a wide range of other invertebrates or even vertebrates, or on vegetation. The encysted form undergoes physiological maturation to produce the infective stage, the metacercaria. Metacercariae are orally ingested in a definitive host with second intermediate host. However, in the family Schistosomatidae, metacercaria is absent and cercariae actively penetrate the skin of the definitive host. The reproductive potential in the trematode life cycle is enormous and millions of cercariae may be produced from a single miracidium.

13.1.1.2.2 Subclass Aspidogastrea

Aspidogastrea is a small group of flukes comprising about 80 species. It is a sister group to the Digenea. Species range in length from approximately one millimeter to several centimeters. They are typically endoparasites of many mollusca, elasmobranches, teleosts, turtles, or decapod crustaceans. Maturation may occur in the mollusc or vertebrate host. A single large ventral sucker, known as an opisthaptor, takes up most of the surface area of its underbelly. The opisthaptor is divided into adhesive depressions (loculi) formed by muscular septa, which are useful in classification. A wide mouth has its opening at the anterior end of a flexible neck-like process. Unlike the related digenean worms, aspidobothreans have a simple, unbranched digestive tract that ends in the cecum, a digestive sac surrounded by muscle. The life cycle is usually direct and requires no intermediate host. However, larvae may continue their development in different hosts. Host specificity of most aspidogastreans is very low, that is, a single species of aspidogastrean can infect a wide range of host species. The larval form is called a cotylocidium and there are no multiplicative larval stages in the mollusc host.

13.1.1.3 Class Monogenea

Monogenea are small flatworms and parasites of cold-blooded aquatic or amphibious vertebrates (fishes, amphibians, and reptiles) and occasionally aquatic invertebrates. There are primarily ectoparasites, particularly of the gills, skin, fins, and buccal cavity. The body is usually flat and oval. Superficially, the monogeneids are not unlike the digenetic trematodes, except for the presence of a posterior adhesive structure, the opisthaptor, by which the parasite is attached to its host. They have a simple digestive system consisting of a mouth opening with a muscular pharynx and an intestine with no terminal opening (anus). The structure of the tegument is essentially the same as that of digenetic trematodes. The life cycles are, so far as is known, direct. Monogenea are mainly parasites on the surface of fish. They are usually hermaphrodites, and viviparous or oviparous. The life cycle is direct with no asexual reproduction. The ciliated larva in egg is called an oncomiridium. Once hatched in the water, the oncomiridia find a host in which they can reach sexual maturity.

13.1.1.4 Class Turbellaria

Class Turbellaria has approximately 3000 species in 12 orders. Some of the colorful marine polycladids are in demand in the aquarium trade, and the planarian *Dugesia* is a common laboratory animal, but the vast majority of turbellarians are little known. They usually have a mouth, a pharynx, and an intestine, but sometimes lack a pharynx or an intestine. The life cycles are usually simple. Most of the turbellarians are free-living, but some are symbiotic. They are marine and benthic organisms, but some also inhabit fresh water, moist temperate, and tropical terrestrial habitats. The order Temnocephalida is entirely commensal or parasitic, but some members of other orders are also commensal.

13.1.2 Pseudocoelomate Helminthes

Pseudocoelomate animals have a fluid-filled body cavity. This cavity surrounds the gut, may contain various other organs, and is called the pseudocoelom. It differs from the coeloms of true coelomates because it is not derived from or completely lined with tissue derived from the mesoderm. Each phylum was previously included in the now-obsolete phylum Nemathelminthes as one of several classes. In the following paragraphs, features of each phylum are highlighted.

13.1.2.1 Phylum Nematoda

Nematodes (roundworms) are one of the most common phyla of animals (Figure 13.3). The number of described species is around 12,000, and the true number may be closer to 500,000. They are

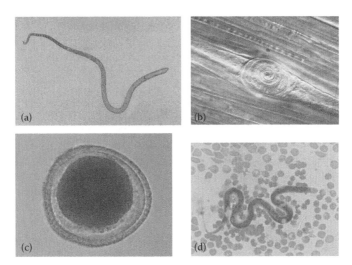

FIGURE 13.3 Photographs of nematodes: (a) adult female of *Trichinella spiralis* recovered from the small intestine of experimentally infected mouse, (b) encapsulated larva of *T. spiralis* in skeletal muscle, (c) egg of *Toxocara canis* from dog, and (d) microfilaria of *Dirofilaria immitis*.

ubiquitous in freshwater, marine, and terrestrial environments. Furthermore, they show a great many parasitic forms, including pathogens in most plants, animals, and also in humans. Nematodes are bilaterally symmetrical, both ends being usually somewhat pointed (Table 13.3). The body is surrounded by a strong, flexible, and noncellular layer called a cuticle. The cuticle is usually provided with circular annulations not readily visible to the naked eye, or it may be smooth or have longitudinal striations. The body cavity is a pseudocoelom (persistent blastula), which lacks the muscles of coelomate animals used to force food down the digestive tract. Nematodes have a complete digestive system, with a mouth at the anterior end of the worm, a muscular esophagus, and an intestine leading to anus. Most nematodes are dioecious. The life cycles are either direct or indirect. There are two classes: Adenophorea and Secernentea.

The larval worm is developed in the egg. After hatching, four molts or ecdyses usually take place before the adult stage is reached. In the case of parasitic nematodes, the larva becomes infective for the definitive host as a rule after the second molt. The infective stage may, in certain species, be reached in the eggshell. In species that use an intermediate host, the infective larva develops inside the intermediate host. Infection to the definitive host may be effected by (1) an active, nonparasitic third-stage larva that enters the host through its mouth or skin; (2) a passive infective egg containing an infective larva; and (3) an intermediate host in which the infective larva develops is either eaten by the definitive host or it carries the infective larva to the definitive host, and the infective larva then penetrates through the skin of the definitive host. After having entered the definitive host, many nematodes migrate through the body before settling down in their normal habitat. The third and fourth ecdyses take place in the definitive host.

13.1.2.2 Phylum Nematomorpha

Nematomorpha is a smallish phylum with about 320 known species, which are relatively long, thin worms (1–3 mm diameter and 10–100 cm in length). The adult worms are free living, but the larvae are parasitic on beetles, cockroaches, grasshoppers, and crustaceans. The adults have a nonfunctional gut and do not feed. The larvae have a better-developed digestive system than the adults, but it is likely they derive most of their nutrition from nutrients absorbed through their body wall. Nematomorpha is dioecious. The life cycles are indirect. Relationships within the phylum are still somewhat unclear, but two classes, Nectonematoida and Gordioidea, are recognized. The adult worms are free living. The female lays her eggs in long strings in water. After hatching, the larva penetrates an appropriate intermediate host, typically an insect. The adult nematomorpha emerges when the host is in or near water. Then it molts once after emerging and takes up its brief adult existence.

13.1.2.3 Phylum Acanthocephala

Phylum Acanthocephala covers a group of parasitic worms that are usually considered as being closely allied to the Nematoda, but lack a mouth or alimentary canal. The worms feed like cestodes, by absorbing their nourishment through the body wall. They are commonly called thorny-headed worms, because they have an evaginable proboscis, which is a cylindrical or oval structure armed with transverse or longitudinal rows of recurved hooks. The Acanthocephala is dioecious, and is divided into three classes, Archiacanthocephala, Palaeacanthocephala, and Eoacanthocephala, with indirect life cycles.

Major definitive hosts of acanthocephala are aquatic vertebrates, mainly fishes and birds. The eggs contain acanthor larvae, which are provided with an anterior circlet of hooks. They require to be ingested by an intermediate host, which is usually an arthropod. The acanthor larva in the egg hatches in the intermediate host and then encysts as a cystacanth in the hemocoel of the arthropod. The cystacanth may require several months for further development to the infective stage and often orange–red in color. Definitive hosts become infected by ingesting the arthropods. The cystacanth may re-encyst in vertebrates other than the definitive host following their ingestion. These act as paratenic hosts and may be important epidemiologically, acting as a link between the intermediate host and the definitive host.

13.1.3 IMPORTANCE OF MOLECULAR APPROACHES IN THE STUDY OF HELMINTHES

Helminthes parasitize not only humans and domestic animals but also many other vertebrates and invertebrates. Many species of molluscs and arthropods serve as intermediate hosts for their larval stage. In most cases, parasitic helminthes are harmful to their host. Therefore, identification of species is very important. In the case of dioecism, when an intact male adult is obtained, species may be identified based on morphology. However, identification of parasitic helminthes is particularly difficult and requires specialized training and relevant data. Moreover, since adult worms are parasitic in bodies of definitive hosts, they cannot easily be obtained in antemortem diagnosis. Therefore, identification of parasitic helminthes must be carried out using their eggs or larvae. The eggs and larvae share many morphological characteristics, and related species are similar to each other in their appearance. For example, eggs of three species of human *Taenia* cannot be distinguished by the conventional microscopic technique even by a trained parasitologist. In such case, molecular tools are very useful [3].

As identification of species using molecular tools is not reliant on specialized knowledge, molecular tools should be used whenever they are available. It is therefore no surprise that extraction of nucleic acids from parasitic worms and subsequent amplification by polymerase chain reaction (PCR) are widely applied. After isolation and purification of the adult or larva, nucleic acids can be extracted using methods that are similar to those for soft tissues. In the case of eggs, the eggshell must be crushed prior to extraction [4]. When extracting nucleic acids from eggs in the feces, efficient removal of PCR inhibitors is important [5].

The taxonomical groups of parasitic helminthes are diverse, and various intermediate hosts are utilized and their morphologies also differ. When the intermediate hosts are large animals, such as vertebrate animals, the larvae can be isolated. However, when the intermediate hosts are invertebrate animals, such as insects or molluscs, the worm is not as easily isolated or purified. In such a case, nucleic acids can be extracted from the worm together with the intermediate host. Since PCR inhibitors are often present in samples from invertebrate animals, removal of these inhibitive substances is essential [6]; the same applies to feces. In the case of an epidemiological survey, since many samples must be treated simultaneously, a quick and inexpensive method is required [7].

Occasionally, a pathological specimen and a formalin-fixed specimen are submitted for identification. As mentioned below, these specimens are barely dissolved, and DNA is remarkably fragmented [3]. As DNA to be used for diagnosis is not easily extracted from these specimens by the conventional methods, development of improved techniques is critical. Along with recent progress in molecular phylogeny and evolution, the systematics of organisms (including helminthes) have changed greatly. In the case of helminthes, since few fossils of helminthes have been found, molecular technique is indispensable for the phylogenetic analysis of helminthes, which often targets the mitochondrial genes and rDNA of the parasites. Type specimens, which have been previously described, are mainly preserved in universities and museums; many of them are fixed in formalin. The extraction of DNA from these formalin-fixed specimens for the phylogenetic analysis of helminthes demands extra attention [8,9].

Some parasitic helminthes are pathogenic and extremely harmful to their hosts, such as human and livestock. Against these parasitic helminthes, protective immunities and vaccines have been actively studied using molecular tools. In such studies, extraction of mRNA is necessary [10].

Although adult worms of *Ascaris suum* live in the anaerobic environment of the mammalian intestinal tract, they are exposed to the aerobic environment outside of their host at the larval stage. Since the mitochondrial respiratory chain of *A. suum* changes dramatically during life cycle, *A. suum* is used for studies on energy metabolism as a model [11]. In these studies, extraction of nucleic acid, mRNA in particular, is also essential.

Conversely, studies on nonparasitic helminthes have made little progress. *Caenorhabditis elegans* and some species of planaria (e.g., *Dugesia*), however, are used for research of embryology and metabolomics as model animals [12].

13.2 CURRENT TECHNOLOGIES FOR NUCLEIC ACID ISOLATION FROM HELMINTHES

Numerous nucleic acid extraction methods have been reported and many commercial kits are also available now. DNA extraction methods which are commonly used for helminthic worms are summarized in Table 13.2. The commonly applied methods for nucleic acid isolation from helminthes are discussed below.

13.2.1 Preservations of Helminthic Worms

The specimen of the helminthic worm has been stored in 10% formalin traditionally. Preservation in formalin should not be recommended, because it causes not only fragmentation of DNA but also difficulty in dissolving the worms [3]. On the other hand, preservation in 100% ethanol or at −80°C is convenient for DNA, but it is not necessarily suitable for morphological examination. Especially, preservation in 100% ethanol may destroy the structure of the worms. Judging from above, preservation in 70% ethanol is most practical. As a matter of first priority, 70% ethanol can be easy to obtain. However, DNA is gradually broken down even if the worm is stored in 70% ethanol. Then, preservation in 70% ethanol at −20°C may be recommended.

Recently, it was reported that eggs of trematodes [13] or microfilaria (MF) of nematodes [14] could be stored using fast technology for analysis of nucleic acids (FTA) cards (Whatman, England). FTA cards provide a safe, secure, and reliable method for the collection, transport, and safe room temperature storage of DNA. FTA is a chemical treatment, which allows for the rapid isolation of pure DNA. When samples are applied to FTA-treated paper, cell lysis occurs and high molecular weight (MW) DNA is immobilized within the matrix. To use FTA cards, simply apply sample, air-dry at room temperature, then remove a small piece of FTA card. This is then washed and used in PCR-based analysis.

TABLE 13.2
Common DNA Extraction Methods for Helminthic Worms

Material	Method	Taxon	Organism	Condition	Purpose	References
Purified adults or larvae	Phenol/chloroform extraction	Cestoda	*Taenia*	Stored at −80°C	RFLP	[15]
		Trematoda	Paramistome fluke	Stored at −20°C	RAPD	[17]
	CTAB precipitation	Cestoda	*Taenia*	Stored at −70°C	mtDNA, cloning	[19]
	GNOME	Cestoda	*Taenia*	Stored at −80°C	PCR	[21]
	Easy-DNA kit	Cestoda	*Hymenolepis* and *Taenia*	Stored at −80°C, or fixed in 70% ethanol	PCR	[22,24]
	AquaPure	Trematoda	*Schistosoma*	Fixed in ethanol	Southern hybridization	[23]
	DNeasy tissue kit	Cestoda	*Taenia* and *Echinococcus*	Fixed in 70% ethanol	PCR	[25,26]
		Trematoda	Several species of Digenea	Stored in 95% ethanol at −20°C	PCR	[28]
		Nematoda	*Trichinella*	Live larvae	PCR	[27]
	QIAamp DNA mini kit	Cestoda	*Spirometra*	Fixed in 70% ethanol	PCR	[32]
		Trematoda	*Calicophoron*	Fixed in 70% ethanol	PCR	[33]
		Nematoda	*Enterobius*	Fixed in 70% ethanol	PCR	[29]
			Phasmarhabditis	Stored at 4°C	PCR	[31]

Category	Method	Phylum	Species	Source	Detection	Reference
Eggs	NucleoSpin + inhibitors	Nematoda	Ascaris and Toxocara	Eggs in soil	PCR	[34]
	QIAamp DNA stool mini kit	Cestoda	Taenia	Eggs in feces	PCR	[35,37]
		Trematoda	Schistosoma	Eggs in feces	PCR	[36]
	Glass beads + QIAamp DNA stool mini kit	Cestoda	Taenia	Eggs in feces	PCR	[38]
	Autoclaved + QIAamp DNA stool mini kit	Trematoda	Opisthorchis	Eggs in feces	PCR	[39]
	Ceramic beads + DNeasy plant kit	Nematoda	Ostertagia	Purified eggs	PCR	[4]
Larvae in hosts	Phenol/chloroform extraction + GuSCN	Trematoda	Fasciola	Larvae in snail	PCR	[40]
	Dried, crushed, and boiled in TE	Nematoda	Brugia and Wuchereria	Microfilaria in mosquitoes	PCR	[7,41]
Formalin fixed specimens	Phenol/chloroform extraction	Cestoda	Ligura	Fixed in formalin	PCR	[9]
	DNeasy tissue kit (modified)	Nematoda	Several species	Fixed in formalin	PCR	[8]
Histopathological specimens	0.02N NaOH + Phenol/chloroform extraction	Cestoda	Taenia	Paraffin sections	PCR	[42]
	0.02N NaOH + DNeasy tissue kit		Taenia	Paraffin sections	PCR	[43]

13.2.2   DNA from Isolated Worms

13.2.2.1   In-House Reagents

As helminthic worms are essentially made of soft tissue, the conventional in-house isolation technique from soft tissues (Chapter 17) may be used. The most commonly applied conventional methods for DNA extraction from isolated worms include phenol/chloroform extraction and cetyl-trimethylammonium bromide (CTAB) precipitation. Azuma et al. [15] isolated intact DNA from taeniid cestode, *Taenia taeniaeformis*, by phenol/chloroform extraction. Although there are some disadvantages in phenol/chloroform extraction including the use of toxic organic solvent and sample loss during successive extraction, intact and large-sized DNA can be obtained. When a sufficient quantity of helminthic materials can be prepared, it seems that phenol/chloroform extraction is still applicable [9,16,17]. Isolation of DNA from taeniid cestodes by CTAB precipitation has been reported [18,19]. The precipitate formed is of high MW and does not require high-speed centrifugation to pellet the CTAB–nucleic acid complex. Nucleic acids can be easily recovered by redissolving the precipitate with a high salt buffer, and dissociating the CTAB plus residual protein into the organic phase by chloroform extraction. This technique is simple, rapid, reproducible, and yields undegraded DNA. This technique has also proved effective for mitochondrial DNA as well as chromosomal DNA from cestodes [19]. Previous attempts to separate cestode mitochondrial DNA from total DNA using gradient centrifugation have not been successful [20]. In addition to *Echinococcus* and *Taenia*, the procedure has been successfully used for *Hymenolepis diminuta*, *Spirometra erinaceieuropaei*, *Mesocestoides corti*, *Toxocara canis*, *Giardia intestinalis*, and *Plasmodium yoelii* [18].

13.2.2.2   Commercial Kits

Nucleic acids have also been extracted from isolated worms by using commercial kits. As DNA is often used as a template for PCR diagnosis of parasitic infections, no intact DNA is necessary under the circumstance. Important points are that purified DNA is free of contaminants and enzyme inhibitors and that procedure is simple and rapid. Many commercial DNA extraction kits have been launched, some of them are suitable for this purpose. DNA extraction methods can be generally classified into two types: affinity or nonaffinity methods. These principles have been adopted in the commercial kits. Nonaffinity methods, such as GNOME, Easy-DNA kit, AquaPure, are cheaper methods and sometimes used for helminthic worms [21–23]. However, extraction kits using affinity methods are overwhelmingly employed in recent studies. Most of them adopt the silica method. In these, a series of products by QIAGEN such as DNeasy tissue kit seems to be most common. Several protocols are usually designed for each kit, but the protocol for animal tissue is suitable in many cases.

GNOME DNA kit (Bio101, USA).   This kit is a rapid and efficient method for isolation of high MW genomic DNA from bacteria, yeast, and animal cells and tissues of all types. No organic extractions are required, and up to 100 µg of DNA can be isolated per preparation. The protocol, briefly, is as follows: suspension of tissue or cells (homogenized if necessary); lysis of cells in the presence of RNase; incubation with protease; precipitation of digested proteins and other debris by proprietary salting-out procedure, followed by centrifugation; and addition of ethanol to supernatant, followed by isolation of precipitated DNA by spooling or centrifugation. Each preparation starts with approximately 100 mg of cells or tissue and yields up to 100 µg of high MW genomic DNA. This kit was used for extraction of DNA from metacestodes of *T. taeniaeformis* [21]. By my impression, however, this method seems to be unsuitable for extraction from ethanol fixed specimens.

Easy-DNA kit (Invitrogen, USA).   Cells are lysed by the addition of solution A and subsequent incubation at 65°C. Proteins and lipids are precipitated and extracted by the addition of solution B and chloroform. The solution is then centrifuged to separate the solution into two phases with a solid interface separating the two phases. The DNA is in the upper, clear aqueous phase, the proteins and lipids are in the solid interface, and the chloroform forms the lower phase. The DNA is then removed, precipitated with ethanol, and resuspended in TE buffer. Purified DNA may be used for PCR or restriction digestion. This kit was used for extraction of DNA from adults of *Hymenolepis* spp. or *T. taeniaeformis* [22,24].

AquaPure genomic DNA purification system (Bio-Rad, USA).   DNA is isolated from cells such as cultured mammalian cells, white blood cells, animal tissue, or microbes by first lysing the cells with an anionic detergent in the presence of a DNA stabilizer. The DNA stabilizer in the lysis buffer works by limiting the activity of DNases that are contained in the cell and elsewhere in the environment. Contaminating RNA is then removed by treatment with an RNA-digesting enzyme. Other contaminants, such as proteins, are removed by salt precipitation. Finally, the genomic DNA is recovered by precipitation with alcohol and dissolved in a buffered solution containing the DNA stabilizer. This kit was used for DNA extraction from adults of *Schistosoma haematobium* [23].

DNeasy tissue kit (QIAGEN, the Netherlands).   It is designed for rapid purification of total DNA (e.g., genomic, mitochondrial, and pathogen) from a variety of sample sources including fresh or frozen animal tissues and cells, blood, or bacteria. DNeasy purified DNA is free of contaminants and enzyme inhibitors and is highly suited for PCR, Southern blotting, random amplified polymorphic DNA (RAPD), amplified fragment length polymorphism (AFLP), and restriction fragment length polymorphism (RFLP) applications. Purification requires no phenol or chloroform extraction or alcohol precipitation, and involves minimal handling. Principle and procedure of DNeasy tissue procedures are simple. Samples are first lysed using proteinase K. Buffering conditions are adjusted to provide optimal DNA binding conditions and the lysate is loaded onto the DNeasy mini spin column. During centrifugation, DNA is selectively bound to the DNeasy membrane as contaminants pass through. Remaining contaminants and enzyme inhibitors such as divalent cations are removed in two efficient wash steps and DNA is then eluted in water or buffer, ready for use in downstream applications. DNeasy purified DNA typically has an A260/A280 ratio between 1.7 and 1.9, and is up to 50 kb in size, with fragments of 30 kb predominating. The DNeasy procedure also efficiently recovers DNA fragments as small as 100 bp. This method is most commonly used for extraction from purified helminthic worms [25–28].

QIAamp DNA mini kit (QIAGEN).   This method seems to be not so different from DNeasy extraction kit as far as these handbooks are compared. However, the length of DNA extracted by QIAamp kit is little bit shorter than that by DNeasy kit. DNA purified using QIAamp kits is up to 50 kb in size, with fragments of approximately 20–30 kb predominating. DNA of this length denatures completely during thermal cycling and can be amplified with high efficiency. This method is also commonly used [29–33].

13.2.3   DNA FROM EGGS

Eggs of helminthic worms usually exist in feces or soil. If possible, isolation of eggs by suitable technique such as a modified Wisconsin sucrose flotation method should be employed before DNA extraction. In many cases, however, DNA is extracted from eggs together with feces or soil. There are large amounts of PCR-interfering substances in feces or soil. Krämer et al. [34] reported that inhibitors of PCR-interfering substances increased PCR sensitivity. The NucleoSpin tissue kit was used for extraction of DNA. The NucleoSpin columns contain special silica membranes designed to ensure a high DNA binding capacity. Animal tissues or cultured cells are incubated with a mixture of lysis buffer and enhanced proteinase K stock solution at 56°C. No further mechanical or organic extraction is necessary. After spinning down the lysis mixture, the clear lysates are transferred to the columns and centrifuged to bind the DNA to the column. This is followed by removal of the cellular contaminants with the provided wash buffers, and elution of the purified genomic DNA. According to the manufacturer's instructions, the resulting DNA preparation is suitable for PCR. In Krämer's report, genomic DNA from the eggs of *Ancylostoma caninum* or *T. canis* within soil was extracted after an overnight proteinase K digest (56°C) using NucleoSpin with enlarged buffer volumes. However, extracted DNA still contained PCR-interfering substances and was not suitable for PCR diagnosis as it was. Then, two different inhibitors of PCR-interfering substances, GeneReleaser (BioVentures Inc., USA) and Maximator (Connex GmbH, Germany), were tested. Although both of them increased PCR sensitivity, Maximator caused better results.

Recently, commercial DNA extraction kits, which enable isolation of DNA from stool samples, are available. The QIAamp DNA stool mini kit (QIAGEN) is one of them. This kit provides fast and easy purification of total DNA from fresh or frozen stool samples. Since the QIAamp DNA stool mini kit is one of QIAamp kits, its procedures are not so different from others. To ensure removal of PCR-interfering substances, however, the QIAamp DNA stool mini kit contains InhibitEX tablets, a unique reagent provided in a convenient tablet form. InhibitEX tablets efficiently adsorb these substances early in the purification process so that they can easily be removed by a quick centrifugation step. In addition, the kit contains buffer ASL, which is specially developed to remove inhibitory substances from stool samples. This kit is commonly used for DNA diagnosis of parasitic disease from fecal samples [35–37].

In addition to the problem of PCR-interfering substances, the eggshells of some kinds of helminthic worms are very strong and hard to destroy by proteinase K digestion. Furthermore, in the case of dangerous parasites for human, eggs are sometimes fixed by ethanol or boiling. When DNA is extracted from such samples, it is better to disrupt eggs prior to extraction. Several disruption procedures for nematode eggs have been tested including sonication, bead beating, boiling, microwaving, proteinase K/SDS digestion, freezing, and various combination of the above [4]. Results showed that egg disruption was best accomplished with the bead beater and ceramic beads. In the case of taeniid eggs, previous shaking with glass beads gave best results [38]. It was reported that incubation in 0.5 N NaOH for 60 min at room temperature and autoclaved for 60 min (121°C) are effective to break eggs of *Opisthorchis*, which is a fishborne liver fluke of human [39]. In every case, the QIAamp DNA stool mini kit was used for DNA extraction after the disruption of eggs.

13.2.4 DNA from Worms in Intermediate Hosts

Larvae of parasitic helminthes sometimes parasitize invertebrate intermediate hosts such as molluscs, insects, copepods, etc. When larvae can be purified from their hosts, conventional DNA extraction methods from animal tissue are generally applicable. In many cases, however, DNA must be extracted from worms together with their host. Since large amount of PCR-interfering substances often exist in host tissue, some device for extracting DNA is necessary. Of course, above-mentioned method for removing inhibitors may be applicable. The PCR assay for the sensitive and specific detection of *Fasciola hepatica* in field-collected *Lymnaea* sp. snail was reported [40]. In that report, DNA was extracted by standard phenol/chloroform extraction, and then contaminated inhibitors were removed by unique procedure.

The procedure of the removal step is as follows: DNA sample is resuspended in 1 mL of 60% guanidine thiocyanate, 50% Tris-HCl pH 6.4, 22 mM ethylenediaminetetraacetic acid (EDTA) and 1.2% (v/v) Triton X-100. A previously autoclaved diatomaceous earth suspension of 40 μL (20% [w/v] in 1% [v/v] HCl) is added to the resuspended pellet. After a 10 min incubation at room temperature with occasional vortexing, two washes with 0.5 mL of 60% guanidine thiocyanate, 50% Tris-HCl pH 6.4, and two washes with 0.5 mL of 70% ethanol were performed by 1 min centrifugation at 12,000 g. After drying the pellet, nucleic acids are eluted by 10 min incubation at 56°C in 50 μL of 5 mM Tris-HCl pH 8.5.

When field survey of filarial parasites in their vectors are carried out, large number of vector mosquito must be examined. Although conventional method of dissection and microscopy is widely used, PCR assay has been developed recently. Conventional methods and available commercial kits are acceptable for extraction of DNA from infected vectors. However, they are time consuming and involve hazardous and expensive chemicals. Then a unique method of DNA extraction from infected mosquitoes has been developed [41]. The procedure is as follows: Mosquitoes are pooled in appropriate size, dried at 95°C for 3 h, crushed into a fine powder using a micropestle and further homogenized in 30 μL of TE buffer. The pestle is washed with another 30 μL of TE buffer. The homogenate, after thorough vortexing, is held in boiling water bath for 10 min, centrifuged at 14,000 rpm for 10 min, and the supernatant is used as template DNA in PCR.

13.2.5 DNA FROM FORMALIN-FIXED SPECIMENS

Formalin fixation has historically been used to preserve specimens for morphological analysis. A large number of formalin-fixed helminthic specimens are available in museums and research laboratories; those could be exploited for molecular phylogenetic study. However, formalin-fixed specimens are hard to dissolve and DNA is remarkably fragmented. Recently, DNA extraction method from formalin-fixed nematodes was reported [8]. The protocol is as follows: DNA was extracted from formalin-fixed nematodes using an extended hot lysis protocol. Formalin-fixed nematodes were placed in 0.5 mL PCR tubes containing 200 μL of ATL lysis buffer from the DNeasy tissue kit. The tubes were initially incubated at 56°C for 24 h. Subsequently, 5 μL of proteinase K (50 mg/mL) and an additional 80 μL of the ATL were added to each tube and incubated for another 72 h at 55°C. The extraction procedure was then completed according to the DNeasy kit following the manufacturer's instruction. More recently, this method was applied for diphyllobothriid specimens, which were stored in formalin for less than 30 years; DNA that was suitable for PCR was extracted from about half the specimens.

13.2.6 DNA FROM HISTOPATHOLOGICAL SPECIMENS

Although histopathological examination of biopsy specimens is a useful method for confirmation of some disease, it is not always easy to make a definitive diagnosis. Therefore, DNA diagnosis from such specimen seems to be very useful. However, there are only a few reports about this. It was reported that two cases of neurocysticercosis in human, which is caused by *Taenia solium*, were confirmed by mitochondrial DNA analysis of biopsied lesion [42,43]. In the first case, mitochondrial DNA analysis was performed using a small piece of a formalin-fixed, paraffin-embedded specimen. The paraffin was melted in a heat block at 70°C, and a tiny amount of parasite material was separated. The parasite was lysed in 60 μL of 0.02 N NaOH containing proteinase K at 90°C for 15 min. After removal of the proteinase K by use of phenol/chloroform, the resulting solution was used directly as template DNA for PCR [42]. In the other case, DNA was extracted from 4–5 paraffin sections with a 5 μm thickness using 0.02 N NaOH containing proteinase K solution or DNeasy blood and tissue kit [43]. Yamasaki et al. reviewed molecular diagnosis using histopathological specimens in cestode zoonoses [44].

13.2.7 RNA FROM HELMINTHES

As for DNA extraction, most of RNA extraction methods for animal tissue are applicable to helminthic worms. Many commercial RNA extraction kits are also available. A series of RNeasy kits (QIAGEN), which utilizes the selective binding properties of a silica-based membrane, is a fast and simple method, and is indeed used in some researches [45]. In most studies on helminthic worms, however, one nonaffinity method, TRIzol reagent (Invitrogen, USA), has been selected for RNA extraction method [46–49]. Although TRIzol reagent includes toxic substances, such as phenol and guanidine isothiocyanate, nevertheless it seems to be one of the most superior methods for RNA extraction until now.

TRIzol (or TRI reagent) is a ready-to-use reagent for the isolation of total RNA from cells and tissues. The reagent, a monophasic solution of phenol and guanidine isothiocyanate, is an improvement to the single-step RNA isolation method developed by Chomczynski and Sacchi [50].

During sample homogenization or lysis, TRIzol maintains the integrity of the RNA, while disrupting cells and dissolving cell components. Addition of chloroform followed by centrifugation separates the solution into an aqueous phase and an organic phase. RNA remains exclusively in the aqueous phase. After transfer of the aqueous phase, the RNA is recovered by precipitation with isopropyl alcohol. After removal of the aqueous phase, the DNA and proteins in the sample can be recovered by sequential precipitation.

Precipitation with ethanol yields DNA from the interphase, and an additional precipitation with isopropyl alcohol yields proteins from the organic phase. Copurification of the DNA may be useful for normalizing RNA yields from sample to sample. This technique performs well with small quantities of tissue (50–100 mg) and cells (5×10^6), and large quantities of tissue (≥ 1 g) and cells ($>10^7$) of human, animal, plant, or bacterial origin. The simplicity of the TRIzol method allows simultaneous processing of a large number of samples. The entire procedure can be completed in 1 h. Total RNA isolated by TRIzol is free of protein and DNA contamination. It can be used for Northern blot analysis, dot blot hybridization, poly(A) + selection, *in vitro* translation, RNase protection assay, and molecular cloning.

13.3 METHODS

General reagents, supplies, and equipment that are required for preparation of nucleic acids from helminthes are listed in Table 13.3.

13.3.1 EXTRACTION OF DNA FROM HELMINTHES

13.3.1.1 Phenol/Chloroform Method

Specific reagents

Extraction buffer: 50 mM Tris-HCl pH 9.0, 100 mM EDTA, 200 mM NaCl, 0.5% (w/v) SDS, 200 μg/mL proteinase K, and 100 μg/mL RNase A

TABLE 13.3

General Reagents, Supplies, and Equipment Required for Nucleic Acid Purification from Helminthes

Reagents for DNA Extraction	Reagents for RNA Extraction	Supplies and Equipment
Proteinase K, 10 mg/mL	TRIzol reagent (containing phenol and guanidinium thiocyanate)	1.5 mL snap cap tubes
Phenol/chloroform (1:1)	Chloroform	Polypropylene tubes: 4, 15, and 50 mL
Chloroform/isoamyl alcohol (24:1)	Isopropyl alcohol	Pipettes: 10–20, 100–200, and 1000 μL
Phenol/chloroform/isoamyl alcohol (25:24:1)	Ethanol, 75%	Aerosol resistant tips (filter tips) for each pipettes
TE buffer, pH 7.4 (10 mM Tris pH 7.5 and 1 mM EDTA pH.8)	Diethylpyrocarbonate (DEPC)-treated water	Microfuges, room temperature, and refrigerated
CTAB solution (2% CTAB, 1.4 M NaCl, 100 mM Tris-HCl pH 8.0, and 20 mM EDTA).	0.5% SDS solution	Centrifuges, room temperature, and refrigerated
Triton X-100		Freezer: −20°C and −80°C
Sucrose		Homogenizer
RNase A		Rotator
10% sodium dodecyl sulfate		Pasteur pipette
Sodium chloride		
Sodium acetate, 3 M pH 5.20		
Ethanol		

Procedure [15]

1. Crush metacestodes of *T. taeniaeformis* in liquid nitrogen.
2. Digest in 10 mL of extraction buffer per 1 g of worms for 3 h at 65°C with occasional agitation.
3. Extract the resultant viscous solution several times with an equal volume of phenol that had been equilibrated with 1 M Tris-HCl pH 8.0 and then extracted 2 or 3 times with phenol/chloroform (1:1) and once with chloroform/isoamyl alcohol (24:1).
4. After an additional chloroform extraction, precipitate the nucleic acid with 2.5 volumes of cold absolute ethanol containing 300 mM sodium acetate and then chill at −80°C for 1 h.
5. Centrifuge at 12,000 rpm for 30 min at 0°C, and dissolve the pellet in 200 μL of TE buffer (10 mM Tris-HCl pH 7.6 and 1 mM EDTA).
6. Digest contaminating RNA with 100 μg/mL RNase A at 37°C for 3 h.
7. Add 5 mL of TE buffer to this solution and remove the RNase by extraction with chloroform/isoamyl alcohol.
8. Precipitate the DNA and dissolve in TE buffer as described above.

13.3.1.2 CTAB Method

Specific reagents [18]

Lysis buffer: 8% Triton X-100, 0.25 M sucrose, 50 mM Tris-HCl, 50 mM EDTA pH 7.5, and 1 mg/mL proteinase K
  CTAB solution: 2% CTAB, 1.4 M NaCl, 100 mM Tris-HCl pH 8.0, and 20 mM EDTA

Procedure

1. Crush the parasite in liquid nitrogen.
2. Suspend homogenized tissue (0.5 mL) in 1 mL of lysis buffer, and incubate at 65°C for 1–2 h.
3. Precipitate nucleic acid by adding 1.0 mL of a sterile 2% CTAB solution to the clear lysate, and pellet white precipitation at 1500 g.
4. Dissolve precipitate in 0.5 mL of 2.5 M NaCl and 10 mM EDTA pH 7.7, and dilute with 1.0 mL of 40 mM Tris-HCl and 2 mM EDTA pH 7.7.
5. Add 2 volumes of chloroform and mix gently, and centrifuge at 12,000 g for 5 min.
6. Add 2 volumes of ethanol (room temperature) to the aqueous phase. Recover nucleic acid by centrifugation at 12,000 g for 10 min at room temperature.

13.3.1.3 QIAGEN DNeasy Kit

Recently, DNeasy tissue kits have been renamed as DNeasy blood and tissue kits to reflect the range of sample types used with these kits. Protocols for animal tissues in the new DNeasy blood and tissue kits are as follows:

Procedure

1. Crush the helminthic worms in liquid nitrogen or cut into small pieces.
2. Add 180 μL buffer ATL.
3. Add 20 μL proteinase K, mix by vortexing, and incubate at 56°C until the samples are completely lysed. Vortex occasionally during incubation to disperse the cells, or place in a shaking water bath.
4. Briefly centrifuge the tube to remove drops from the inside of the lid. Add 200 μL buffer AL to the sample, and mix by pulse-vortexing for 15 s.

5. Add 200 µL ethanol (96%–100%) to the sample, and mix by pulse-vortexing for 15 s. After mixing, briefly centrifuge the 1.5 mL tube to remove drops from inside the lid. Buffer AL and ethanol can be premixed and added together in one step to save time when processing multiple samples.
6. Pipet the mixture from Step 5 (including any precipitate) into the DNeasy mini spin column placed in a 2 mL collection tube (provided). Centrifuge at 6000 g for 1 min. Discard flow-through and collection tube.
7. Place the DNeasy mini spin column in a new 2 mL collection tube (provided), add 500 µL buffer AW1, and centrifuge for 1 min at 6000 g. Discard flow-through and collection tube.
8. Place the DNeasy mini spin column in a new 2 mL collection tube (provided), add 500 µL buffer AW2, and centrifuge for 3 min at 20,000 g to dry the DNeasy membrane. Discard flow-through and collection tube.

13.3.2 EXTRACTION OF RNA FROM HELMINTHES

We present below a widely applied RNA isolation method, that is, TRIzol (Invitrogen), for preparation of helminthic RNA. Based on a monophasic solution of phenol and guanidine isothiocyanate, this reagent represents an improvement to the single-step RNA isolation method first described by Chomczynski and Sacchi in 1987 [50].

Procedure

1. Crush metacestodes of the helminthic worms in liquid nitrogen.
2. Add 1 mL TRIzol per 100 mg worms and homogenize using a glass-Teflon or power homogenizer. The lysate is transferred immediately to a sterile polypropylene tube and incubated for 5 min at room temperature to allow complete dissociation of nucleoprotein complexes.
3. Add 0.2 mL chloroform per 1 mL of TRIzol. After vigorous shaking (or vortexing) for 15 s and incubation for 2–3 min at room temperature, centrifuge the homogenate at 12,000 g for 15 min at 2°C–8°C.
4. Transfer the aqueous phase containing RNA to a sterile microfuge tube. Use 0.5 mL of isopropyl alcohol per 1 mL of TRIzol for the initial homogenization.
5. Incubate for 10 min at room temperature, and centrifuge the sample at 12,000 g for 10 min at 4°C.
6. Remove the supernatant, wash the RNA pellet once with 75% ethanol, and add at least 1 mL of 75% ethanol per 1 mL of TRIzol for the initial homogenization. Mix the sample by vortexing and centrifuge at 7500 g for 5 min at 2°C–8°C.
7. Air-dry for 5–10 min, dissolve RNA in DEPC-treated water, and store at −70°C or as an ethanol suspension at −20°C.

13.4 FUTURE DEVELOPMENT TRENDS

As described above, the methods for nucleic acid extraction from helminthic worms are not so different from those for mammalian tissues, as long as helminthic worms are purified. As novel commercial nucleic acid extraction kits of different variety and improving efficiency are constantly coming onto the market, they will be more commonly applied to the study of helminthic worms in future. On the other hand, the DNA extraction techniques from eggs still have room to improve. The QIAamp DNA stool mini kit is without doubt efficient in eliminating inhibitory substances. However, the process using InhibitEX tablets is not convenient by any standard. In the case of field survey, it is desirable to treat many samples simultaneously. The development of other methods, which are simple, rapid, safe, and inexpensive, is expected.

In recent years, population of some species have decreased rapidly. Additional samples of such organisms or helminthic parasites in such organisms are difficult to obtain. Therefore, use of formalin-fixed specimens kept in museums will be increasingly necessary. Given that the DNA extraction methods from formalin-fixed samples are not well standardized, further development in this area (especially design of easy-to-use commercial kits) will be extremely helpful.

ACKNOWLEDGMENTS

We would like to acknowledge support from the Japan Society for the Promotion of Science to M.O. (16500277, 18406008) and to A.I. (17256002, Asia/Africa Science Platform Fund 2006–2008). We also thank Dr. N. Nonaka, Hokkaido University, Japan, for his kind permission to use several parasite pictures.

Munehiro Okamoto received his MSc in veterinary medicine and his veterinary degree (VMD) from Hokkaido University, Sapporo, Japan in 1985. He then worked at the same university on the phylogeny and variation among taeniid cestodes. In 1990, he moved to the Institute of Experimental Animal Science, Osaka University Medical School, Osaka, Japan to study the molecular biology of nephritic mice. After completing his PhD in veterinary medicine at Hokkaido University in 1995, he went to the School of Veterinary Medicine, Faculty of Agriculture, Tottori University, Tottori, Japan, as an associate professor, and became the head of its Department of Parasitology in 2007. His research interests include molecular phylogeny, evolutionary biology, and biological geography, with special focus on the relationship between parasitic helminthes and their hosts.

Akira Ito received his MSc and PhD in biological sciences from Tohoku University, Sendai, Japan in 1971 and 1975, respectively, and DMedSci at Showa University, Tokyo, Japan in 1985. His research centers on the immunobiological aspects of host–parasite relationships using cestode infections in laboratory animals as models. His early work (1971–1985) dealt with *Hymenolepis nana* and other related species. From 1985, his studies extended to the immunochemical and molecular biological analyses of zoonotic cestodes such as *Taenia* spp. and *Echinococcus* spp. with an additional focus on the serodiagnosis of zoonotic cestodiases, cysticercosis, and echinococcosis. Having been the head of Department of Parasitology, Asahikawa Medical College, Asahikawa, Japan since 1998, he maintains a keen interest on the molecular and immunological diagnosis, epidemiology, and molecular evolution of zoonotic cestode infections.

REFERENCES

1. Soulsby, E.J.L., Helminths, *Arthropods and Protozoa of Domesticated Animals*, Lea & Febiger, Philadelphia, PA, 1982.
2. Cheng, T.C., *General Parasitology*, Academic Press, Orlando, FL, 1986.
3. Yamasaki, H. et al., Mitochondrial DNA diagnosis for taeniasis and cysticercosis, *Parasitol. Int.*, 55: S81, 2006.
4. Harmon, A.F. et al., Improved methods for isolating DNA from *Ostertagia ostertagi* eggs in cattle feces, *Vet. Parasitol.*, 135: 297, 2006.
5. Harmon, A.F., et al., Real-time PCR for quantifying *Haemonchus contortus* eggs and potential limiting factors, *Parasitol. Res.*, 101: 71, 2007.
6. Qvarnstrom, Y. et al., PCR-based detection of *Angiostrongylus cantonensis* in tissue and mucus secretions from molluscan hosts, *Appl. Environment. Microbiol.*, 73: 1415, 2007.
7. Vasuki, V. et al., A rapid and simplified method of DNA extraction for the detection of *Brugia malayi* infection in mosquitoes by PCR assay, *Acta Tropica*, 79: 245, 2001.
8. Bhadury, P. et al., Exploitation of archived marine nematodes—A hot lysis DNA extraction protocol for molecular study, *Zool. Scripta*, 36: 93, 2007.
9. Li, J. et al., Molecular characterization of a parasitic tapeworm (*Ligula*) based on DNA sequences from formalin-fixed specimens, *Biochem. Genet.*, 38: 309, 2000.

10. Chow, C. et al., *Echinococcus granulosus*: Oncosphere-specific transcription of genes encoding a host-protective antigen, *Exp. Parasitol.*, 106: 183, 2004.
11. Amino, H. et al., Stage-specific isoforms of *Ascaris suum* complex II: The fumarate reductase of the parasitic adult and the succinate dehydrogenase of free-living larvae share a common iron–sulfur sub-unit, *Mol. Biochem. Parasitol.*, 106: 63, 2000.
12. Sato, K. et al., Identification and origin of the germline stem cells as revealed by the expression of *nanos*-related gene in planarians, *Dev. Growth Differ.*, 48: 615, 2006.
13. Prasad, P.K. et al., PCR-based determination of internal transcribed spacer (ITS) regions of ribosomal DNA of giant intestinal fluke, *Fasciolopsis buski* (Lankester, 1857) Looss, 1899, *Parasitol. Res.*, 101: 1581, 2007.
14. Nuchprayoon, S., Random amplified polymorphic DNA (RAPD) for differentiation between Thai and Myanmar strains of *Wuchereria bancrofti*, *Filaria J.* (Online), July 30, 2007.
15. Azuma, H., et al., Intraspecific variation of *Taenia taeniaeformis* as determined by various criteria, *Parasitol. Res.*, 81: 103, 1995.
16. Steinauer, M.L., et al., First sequenced mitochondrial genome from the Phylum Acanthocephala (*Leptorhnchoides thecatus*) and its phylogenetic position within metazoa, *J. Mol. Evol.*, 60: 706, 2005.
17. Sripalwit, P. et al., High annealing temperature-random amplified polymorphic DNA (HAT-RAPD) analysis of three paramphistome flukes from Thailand, *Exp. Parasitol.*, 115: 98, 2007.
18. Yap, K.W. and Thompson, R.C.A., CTAB precipitation of cestode DNA, *Parasitol. Today*, 3: 220, 1987.
19. Yap, K.W. et al., *Taenia hydatigena*: Isolation of mitochondrial DNA, molecular cloning, and physical mitochondrial genome mapping, *Exp. Parasitol.*, 63: 288, 1987.
20. McManus, D.P. et al., Isolation and characterization of nucleic acids from the hydatid organisms, *Echinococcus* spp., *Mol. Biochem. Parasitol.*, 16: 251, 1985.
21. Okamoto, M. et al., Phylogenetic relationships within *Taenia taeniaeformis* variants and other taeniid cestodes inferred from the nucleotide sequence of the cytochrome *c* oxidase subunit I gene, *Parasitol. Res.*, 81: 451, 1995.
22. Okamoto, M., et al., Phylogenetic relationships of three hymenolepidid species inferred from nuclear ribosomal and mitochondrial DNA sequences, *Parasitology*, 115: 661, 1997.
23. Copeland C.S. et al., Identification of the *Boudicca* and *Sinbad* retrotransposons in the genome of the human blood fluke *Schistosoma haematobium*, *Mem. Inst. Oswaldo Cruz.* (Online), 101: 565, 2006.
24. Matoba, Y. et al., Detection of a taeniid species *Taenia taeniaeformis* from a feral raccoon *Procyon lotor* and its epidemiological significance, *Mamm. Study*, 28: 157, 2003.
25. Yamasaki, H. et al., DNA differential diagnosis of human taeniid cestodes by base excision sequence scanning thymine-base reader analysis with mitochondrial genes, *J. Clin. Microbiol.*, 40: 3818, 2002.
26. Nakao, M. et al., The complete mitochondrial DNA sequence of the cestode *Echinococcus multilocularis* (Cyclophyllidea: Taeniidae), *Mitochondrion*, 1: 497, 2002.
27. Hill, D.E. et al., *Trichinella nativa* in a black bear from Plymouth, New Hampshire, *Vet. Parasitol.*, 132: 143, 2005.
28. Olson, P.D. et al., Phylogeny and classification of the Digenea (Platyhelminthes: Trematoda), *Int. J. Parasitol.*, 33: 733, 2003.
29. Nakano, T. et al., Mitochondrial cytochrome *c* oxidase subunit 1 gene and nuclear rDNA regions of *Enterobius vermicularis* parasitic in captive chimpanzees with special reference to its relationship with pinworms in humans, *Parasitol. Res.*, 100: 51, 2006.
30. Le, T.H. et al., *Paragonimus heterotremus* Chen and Hsia (1964), in Vietnam: A molecular identification and relationships of isolates from different hosts and geographical origins, *Acta Tropica*, 98: 25, 2006.
31. MacMillan, K. et al., Quantification of the slug parasitic nematode *Phasmarhabditis hermaphrodita* from soil samples using real time qPCR, *Int. J. Parasitol.*, 36: 1453, 2006.
32. Okamoto, M. et al., Intraspecific variation of *Spirometra erinaceieuropaei* and phylogenetic relationship between *Spirometra* and *Diphyllobothrium* inferred from mitochondrial CO1 gene sequences, *Parasitol. Int.*, 56: 235, 2007.
33. Rinaldi, L. et al., Characterization of the second internal transcribed spacer of ribosomal DNA of *Calicophoron daubneyi* from various hosts and locations in southern Italy, *Vet. Parasitol.*, 131: 247, 2005.
34. Krämer, F. et al., Improved detection of endoparasite DNA in soil sample PCR by the use of anti-inhibitory substances, *Vet. Parasitol.*, 108: 217, 2002.
35. Yamasaki, H. et al., DNA differential diagnosis of taeniasis and cysticercosis by multiplex PCR, *J. Clin. Microbiol.*, 42: 548, 2004.

36. Gobert, G.N. et al., Copro-PCR based detection of *Schistosoma* eggs using mitochondrial DNA markers, *Mol. Cell. Probes*, 19: 250, 2005.
37. Li, T. et al., Taeniasis/cysticercosis in a Tibetan population in Sichuan Province, China, *Acta Tropica*, 100: 223, 2006.
38. Nunes, C.M. et al., Fecal specimens preparation methods for PCR diagnosis of human taeniosis, *Rev. Inst. Med. Trop, Sao Paulo*, 48: 45, 2006.
39. Stensvold, C.R. et al., Evaluation of PCR based coprodiagnosis of human opisthorchiasis, *Acta Tropica*, 97: 26, 2006.
40. Cucher, M.A. et al., PCR diagnosis of *Fasciola hepatica* in field-collected *Lymnaea columella* and *Lymnaea viatrix* snails, *Vet. Parasitol.*, 137: 74, 2005.
41. Vasuki, V. et al., A simple and rapid DNA extraction method for the detection of *Wuchereria bancrofti* infection in the vector mosquito, *Culex quinquefasciatus* by *Ssp* I PCR assay, *Acta Tropica*, 86: 109, 2003.
42. Yamasaki, H. et al., Solitary neurocysticercosis case caused by Asian genotype of *Taenia solium*: Confirmed by mitochondrial DNA analysis, *J. Clin. Microbiol.*, 42: 3891, 2004.
43. Yamasaki, H. et al., Molecular identification of *Taenia solium* cysticercus genotype in the histopathological specimens, *Southeast Asian J. Trop. Med. Public Health*, 36(Suppl. 4): 131, 2005.
44. Yamasaki, H. et al., Significance of molecular diagnosis using histopathological specimens in cestode zoonoses, *Trop. Med. Health*, 35: 307, 2007.
45. Adekunle, O.K. et al., Plant parasitic and vector nematodes associated with asiatic and oriental hybrid lilies, *Bioresour. Technol.*, 97: 364, 2006.
46. Morimoto, M. et al., *Ascaris suum*: cDNA microarray analysis of 4th stage larvae (L4) during self-cure from the intestine, *Exp. Parasitol.*, 104: 113, 2003.
47. Li, B.W. et al., Quantitative analysis of gender-regulated transcripts in the filarial nematode *Brugia malayi* by real-time RT-PCR, *Mol. Biochem. Parasitol.*, 137: 329, 2004.
48. Gauci, C.G. et al., *Taenia solium* and *Taenia ovis*: Stage-specific expression of the vaccine antigen genes, *TSOL18*, *TSOL16*, and homologues, in oncospheres, *Exp. Parasitol.*, 113: 272, 2006.
49. Fitzpatrick, J.M. et al., Dioecious *Schistosoma mansoni* express divergent gene repertoires regulated by pairing, *Int. J. Parasitol.*, 36: 1081, 2006.
50. Chomczynski, P. and Sacchi, N., Single-step method of RNA isolation by acid guanidinium thiocynate–phenol–chloroform extraction, *Anal. Biochem.*, 162: 156, 1987.

14 Isolation of Nucleic Acids from Insects

Kun Yan Zhu

CONTENTS

14.1 INTRODUCTION

14.1.1 INSECTS AND THEIR DIVERSITIES

Insects, such as beetles, wasps, flies, butterflies, and moths, are invertebrate animals in the class Insecta, which is one of the five major taxonomical classes within the phylum Arthropoda (arthropods). Other important arthropod classes include Arachnida (spiders, scorpions, ticks, and mites), Chilopoda (centipedes), Diplopoda (millipedes), and Crustacea (crabs, lobsters, shrimps, barnacles, woodlice, etc) (Table 14.1). Arthropods are characterized by possessing exoskeleton (a tough outer body-shell containing chitin: linear biopolymers composed of β-1,4 linked *N*-acetylglucosamines), segmented body, jointed limbs and jointed mouthparts, bilateral symmetry, ventral nerve cord (as opposed to dorsal nerve cord in vertebrate), and dorsal blood pump.

Like other arthropods, insects are characterized by their exoskeleton and various internal features, including open circulatory system, ventilatory tubules (tracheal system), and Malpighian tubules as a major organ involved in filtration of the hemolymph (insect blood). In the adult stage, insects show five distinct morphological characteristics: (1) three distinct body regions (head, thorax, and abdomen); (2) three pairs of segmented legs on the thorax; (3) often one or two pairs of wings

TABLE 14.1

Phylum Arthropoda and Its Five Major Classes

Class	Main Features	Common Names
Arachnida (arachnids)	Two segments; cephalothorax and abdomen; four pairs of legs; one pair of chelicerae; no antennae	Spiders, scorpions, ticks, mites, etc.
Chilopoda (chilopods)	Multiple body segments; one pair of legs per body segment; first pair of legs modified into venomous fangs; one pair of antennae	Centipedes
Diplopoda (diplopods)	Multiple body segments; two pairs of legs per body segment; one pair of antennae	Millipedes
Crustacea (crustaceans)	Multiple body segments; head, thorax, and abdomen, which may be fused; variable number of legs; two pairs of antennae	Crabs, shrimp, barnacles, sowbugs, woodlice, etc.
Insecta (insects)	Three body segments; head, thorax, and abdomen; six legs; one pair of antennae	Beetles, bugs, wasps, moths, flies, etc.

on the thorax; (4) a pair of segmented antennae on the head; and (5) a pair of compound eyes. These morphological features distinguish insects from other arthropod classes.

Insects have existed for about 350 million years, compared with less than 2 million years for humans, and are among the most diverse and abundant of all organisms on the Earth [1]. They vary in size from tiny gall midges that are only 80 μm long to very slender walking sticks that are 330 mm long. They have evolved to adapt to almost every type of habitats, including the arid deserts, hot springs up to 80°C, arctic temperatures below −20°C, and mountain peaks as high as 6096 m [2]. With over 1 million different species described, insects account for approximately 70%–75% of the known species of animals. As the estimated number of insect species could approach 30 million [3], insects may make up over 90% of the life forms on the planet.

Based mainly on the type of metamorphosis and the way of wing development they go through, insects are separated into three taxonomic groups: Apterygota, Exopterygota, and Endopterygota. Apterygota consists of five primitively wingless orders with simple metamorphosis; Exopterygota consists of 17 winged orders with incomplete metamorphosis; whereas Endopterygota consists of 9 winged (in adult stages) orders with complete metamorphosis (Table 14.2). Among these, the orders Orthoptera (grasshoppers, crickets, and katydids), Hemiptera (true bugs, cicadas, hoppers, psyllids, whiteflies, aphids, and scale insects), Coleoptera (beetles), Hymenoptera (bees, wasps, ants, sawflies, parasitic wasps), Lepidoptera (butterflies and moths), and Diptera (true flies and mosquitoes) are most abundant, each containing over tens of thousands of species (Table 14.2) [3].

14.1.2 Biology of Insects

Insects start their development as eggs produced by female adults. Although some insects such as aphids give birth to live young, the young actually hatch from eggs carried inside the mother. After the eggs hatch, insects grow in a series of distinct stages by periodically molting, which is a necessary process for the individual to escape the confines of the exoskeleton and to grow a new and larger outer covering. During the molting process, they shed old exoskeleton and expand the soft new exoskeleton by inhaling air. In a few hours, the new exoskeleton hardens and there is no further

TABLE 14.2

Insect Orders and Their Estimated Numbers of Described Species in the World

Taxonomic Group	Order	Common Name	Number of Species (World Estimates)
Apterygota: Primitively wingless insects with simple metamorphosis	Protura	Proturans	500
	Collembola	Springtails	>6,000
	Diplura	Diplurans	800
	Microcoryphia	Jumping bristletails	350
	Thysanura	Silverfish	370
Exopterygota: Insects with wings developing outside the body and undergoing incomplete metamorphosis, i.e., no pupal stage; the smaller young (nymphs) with underdeveloped wings resembling the adults	Ephemeroptera	Mayflies	2,000
	Odonata	Dragonflies and damselflies	5,000
	Orthoptera	Grasshoppers, crickets, and katydids	>20,000
	Phasmatodea	Walking sticks and leaf insects	>2,500
	Grylloblattodea	Rock crawlers	25
	Mantophasmatodea	Gladiators	3
	Dermaptera	Earwigs	1,800
	Plecoptera	Stoneflies	2,000
	Embiidina	Web-spinners	<200
	Zoraptera	Zorapterans, angel insects	30
	Isoptera	Termites	>2,300
	Mantodea	Mantids	1,800
	Blattodea	Cockroaches	<4,000
	Hemiptera	True bugs, cicadas, hoppers, psyllids, whiteflies, aphids, and scale insects	35,000
	Thysanoptera	Thrips	4,500
	Psocoptera	Psocids or booklice	>3,000
	Phthiraptera	Lice	>3,000
Endopterygota: Insects with wings developing inside the body and undergoing complete metamorphosis; the young (larvae) developing into the adults during a nonfeeding pupal stage	Coleoptera	Beetles	>300,000
	Neuroptera	Alderflies, dobsonflies, fishflies, snakeflies, lacewings, antlions, and owlflies	5,500
	Hymenoptera	Bees, wasps, ants, sawflies, parasitic wasps	115,000
	Trichoptera	Caddisflies	>7,000
	Lepidoptera	Butterflies and moths	150,000
	Siphonaptera	Fleas	2,380
	Mecoptera	Scorpionflies and hangingflies	500
	Strepsiptera	Twisted-wing parasites	550
	Diptera	Flies and mosquitoes	>150,000

change in body size until the following molt. The specialized molting process in which the insect undergoes a major morphological change is known as metamorphosis. Indeed, the division of insects into Apterygota, Exopterygota, and Endopterygota groups is largely based on differences in the type of metamorphosis.

The insects (e.g., silverfish) in Apterygota do not show distinct metamorphosis except for the change in their body size; and their immature stages closely resemble the adults (wingless). However, the insects (e.g., grasshoppers and cockroaches) in Exopterygota undergo an incomplete metamorphosis (hemimetabolous insects) which involves three basic life stages: egg, nymph, and adult. After the embryo develops into a nymph, the nymph undergoes several molts. The stages of the nymph between molts are called instars. Typically, a nymph undergoes three to five instars. Each molt attains the larger size of the next instar, and the adult finally emerges from the last instar nymph. Insects with incomplete metamorphosis do not have radical changes in morphologies and life styles between the nymph and adult. The principal changes occurring during metamorphosis are changes in body proportions, sexual maturity, and the development of wings. Furthermore, nymphs and adults often live in the same habitat and feed on the same food.

In contrast, the insects (e.g., butterflies, beetles, wasps, and flies) in Endopterygota undergo a complete metamorphosis (holometabolous insects), which involves four basic life stages: egg, larva, pupa, and adult. These insects have dramatically different body plans between their larval and adult stages; an example is the caterpillar larva and the butterfly adult. After a caterpillar hatches from an egg, there will be several molts during the larval stage (which may or may not have legs, and the wings are not visible). The stages of the larva between molts are also called instars. The caterpillar is an active and ferocious plant feeder. Under hormonal control, the caterpillar pupates to form a pupa in a cocoon. Although the pupal stage is a nonfeeding (inactive) stage, many physiological and morphological changes occur. Finally, the butterfly (with wings) emerges, its wings and cuticle harden, and it flies off. Complete metamorphosis allows the vast majority of insects to pursue radically different life styles as larvae and as adults and therefore to adapt to the foods and conditions present during different seasons of the year or stages of the insect life cycle.

Because insects do not possess lungs, they utilize internal tube and sac systems for respiration, through which gases either diffuse or are actively pumped to directly deliver oxygen to various tissues in the body. In addition, insects have no closed vessels (i.e., veins and arteries), and instead possess little more than a single, perforated dorsal tube that pulses peristaltically, and helps the hemolymph circulate inside the body cavity. Many insects are renowned for possessing very sensitive and specialized sensory organs. For example, bees have the capacity to perceive ultraviolet wavelengths, or detect polarized light, whereas male moths utilize their antennae to detect the pheromones of female moths over long distances (measured by kilometers).

Some insects can detect predators and avoid predation, as exemplified by the fact that some nocturnal moths can perceive the ultrasonic emissions of bats. Furthermore, certain predatory and parasitic insects are able to detect the characteristic sounds made by their prey/hosts. With special sensory structures, bloodsucking insects can detect infrared emissions to facilitate their homing-in on their hosts. Nonetheless, a trade-off exists between the insect's visual acuity and its chemical or tactile acuity, and many insects with well-developed eyes tend to have reduced or simple antennae, and vice versa. Some insects use mechanical action of appendages to produce sounds and to sense them. Sound-making represents a useful way for insects to communicate. Insects also use chemical means for communication, as they can detect semiochemicals that are often derived from plant metabolites that include those meant to attract, repel, and provide other kinds of information.

Insects are the only invertebrates that have developed flight, which relies heavily on turbulent aerodynamic effects. Besides powered flight, many small insects (e.g., the aphids) are dispersed over long distances by winds and by low-level jet streams. Many adult insects use their six legs to walk. All these have undoubtedly contributed to their spread to all parts of the world and their abundance and diversity.

14.1.3 Significance of Insects

The significance of insects is attributed to their tremendous diversity and success relative to other organisms, and their extreme importance from the human perspective [1]. Insects were among the earliest terrestrial herbivores and have acted as major selection agents on plants. Many insects are perceived as threats to agricultural production and human health because they are crop pests and disease vectors. For example, the European corn borer (*Ostrinia nubilalis*) is one of the most damaging pests of corn in North America and the Western world. In the United States alone, annual grain losses due to its infestations exceed $500 million [4]. The diamondback moth (*Plutella xylostella*) causes significant damages to the cruciferous vegetables, including cabbage, cauliflower, radish, and turnip, with crop yield losses ranging from 31% to 100% [5].

Many insects are also regarded as pests by humans. These include parasitic (lice and bedbugs), disease-transmitting (mosquitoes and flies), and structure-damaging (termites) insects. Mosquitoes are historically the most important insects affecting human health because they are vectors of human diseases such as malaria, filaria, dengue fever, and West Nile virus [6]. Even today, mosquito-borne diseases are still a major threat to human health and well-being in the world. Indeed, malaria is prevalent in over 100 countries and territories with more than 40% of the world population at risk [7]. Approximately 300–500 million cases of malaria occur, and between 1 and 1.5 million people worldwide die from it every year.

Nevertheless, many insects are beneficial to humans and our environment because they contribute to specific processes and functions, such as pollination (e.g., wasps, bees, butterflies, and ants), nutrient recycling, population regulation, and biological control in ecosystem. In the United States, for example, about 130 agricultural plants are pollinated by bees, and the annual value of honey bee pollination is estimated at over $9 billion [8]. Insects also produce many useful raw materials (e.g., honey, wax, lacquer, and silk). Many insects (e.g., beetles) are scavengers that feed on dead animals and fallen trees and recycle the biological materials into forms useful for other organisms, as well as contribute to the process of topsoil formation. Some insects feed on other insects as insectivores, which can be useful for controlling harmful insects.

14.1.4 Entomological Research and Impact of Molecular Biology

Insect science is one of the most active research areas in life sciences. It spans from fundamental to applied research which include behavior, biochemistry, developmental biology, ecology, evolutionary biology, genetics, insect pathology, molecular biology, morphology and systematics, pest management, plant–insect interactions, toxicology, and veterinary and medical entomology. Because insects make up the bulk of terrestrial species in almost every type of habitats on the Earth, they provide for numerous observations that are essential for testing fundamental hypotheses explaining patterns of species diversity, abundance, and distribution in ecological communities of life on the Earth. Because insects have a relatively short life cycle, often several generations per year, they allow researchers to make multiple observations of all life stages of a species in a relatively short period of time [9]. Many insect species have been used as models for various areas of biological research. The most commonly used insect is the fruit fly (*Drosophila melanogaster*). In fact, it represents one of the most intensively studied organisms in biology and serves as a model system for the investigation of many developmental and cellular processes common to higher eukaryotes, including humans [10].

Molecular techniques have been widely used in virtually all aspects of entomological research and have greatly benefited insect sciences. In recent years, the insect genetic manipulation techniques [11], the completions of genome sequences of several insect species [10,12,13], the DNA microarray [14–16] and RNA interference (RNAi) technologies [17,18], and bioinformatics [19–22] have played increasingly important roles in entomological research. For example, genetic manipulations of insects through molecular approaches have been used in the attempt to fight against insects and the diseases they carry [23]. Several effective methods of germline transformation have been developed and used in malaria mosquito vectors [24–26] and two different laboratories have developed genetic constructs that significantly reduce vector competence in experimental malaria models [27,28]. Although RNAi is a highly conserved cellular mechanism, it has been reported that certain insect species, such as the red flour beetle (*Tribolium castaneum*), can have a robust systemic RNAi response [18]. Thus, the red flour beetle can serve as an excellent model organism to study gene functions.

Furthermore, complete genome sequences are rapidly becoming available for a growing number of insects of great medical and agricultural importance. The first insect genome was sequenced from *D. melanogaster* in 2000 [10]. During the last 7 years, additional genomes have been sequenced from other insects, including the two important human disease vectors, Africa malaria mosquito (*Anopheles gambiae*) [12], and the yellow fever mosquito (*Aedes aegypti*) [29]; an important plant pollinator and honey producer, the honeybee (*Apis mellifera*) [30]; the silkworm (*Bombyx mori*) [13]; 11 *Drosophila* species (*D. ananassae, D. erecta, D. grimshawi, D. mojavensis, D. persimilis, D. pseudoobscura, D. sechellia, D. simulans, D. virilis, D. willistoni,* and *D. yakuba*) [31,32]; and an important stored product pest, the red flour beetle [33]. The sequencing of the genomes of three species of parasitic wasps (*Nasonia vitripennis, N. giraulti,* and *N. longicornis*) whose hosts include housefly (*Musca domesticus*) are underway [34]. These insect genome sequences provide researchers with exciting opportunities to tackle various research problems for an in-depth understanding of virtually all aspects of insect sciences. The insect genome sequences can also help researchers identify novel and insect-specific targets for developing new insecticides and transgenic crops for insect pest control.

14.2 PURIFICATION OF NUCLEIC ACIDS FROM INSECTS

14.2.1 General Consideration

The advance of genetic engineering and recombinant DNA techniques has significantly increased the need of the high-quality nucleic acids purified from insects. In fact, the quality of nucleic acids is the single most important factor determining optimal results in downstream analyses of the nucleic acids. Although the basic techniques for purification of nucleic acids from insects are not significantly different from those used for other organisms, several issues related to specific research purposes or unique to insects deserve special attention and are discussed below.

Because insects have several developmental stages, the first question is which stage of an insect should be used for purification of nucleic acids. For purification of genomic DNA, developmental stage may not be very crucial since genomic DNA purified from different stages is essentially the same. However, the amount of genomic DNA purified from different developmental stages could be very different. Obviously, the use of a later instar of an insect with its larger body size will result in a larger amount of genomic DNA than those of early instars of the same insect species. In contrast, it is important to choose an appropriate stage if total RNA or mRNA is to be purified from an insect, because the levels of different RNAs can vary significantly among different developmental stages of an insect species. Certain genes may express only in certain developmental stages. Therefore, depending on the specific research questions to be addressed, researchers may choose a specific developmental stage for purification of RNA or mRNA. If researchers are interested in profiling gene expressions during the embryo developmental stage, clearly insect eggs should be used as a starting material. On the other hand, if researchers are interested in studying molecular aspects of digestive enzymes, larvae or adults might be used for purification of total RNA or mRNA [35,36].

Generally, the nucleic acid contents are not homogeneous among different body parts, tissues, and even cell types of an insect. Certain genes may be expressed only in certain tissues. Therefore, the second question is whether the whole insect body or a specific tissue dissected from individual insects should be used for purification of nucleic acids. Again, depending on the specific research questions to be addressed, researchers may choose the whole insect body if they do not have a specific interest as to which body parts or tissues to be examined in the study [37,38]. Because most insect species are relatively small in their body size, it is relatively easy to use the whole body of an insect for nucleic acid purification. However, if a researcher anticipates addressing specific questions in a specific tissue, such a tissue is often dissected from individual insects and used for nucleic acid isolation. For example, Chen et al. [39] used salivary glands dissected from the Hessian fly (*Mayetiola destructor*) for isolation of RNA to study secreted salivary gland proteins, whereas Li et al. [35] used the guts dissected from the larvae of the European corn borer for isolation of RNA to characterize cDNAs encoding trypsin-like proteinases.

Finally, it is extremely important to use an appropriate technique to grind insects or insect tissues so that an appropriate purification procedure can be followed. Because the exoskeleton (body covering) of certain insects during their adult stage (e.g., beetles) is hard, such insects must be completely ground in the initial step of purification. Conventionally, researchers first grind insects into powder in liquid nitrogen in a porcelain mortar and then transfer the insect powder into a homogenizer followed by homogenization in an extraction buffer [40,41]. This method is very effective to break insects into a powder. However, if an insect sample is very small or precious, this method may not be a good choice because a high proportion of the insect powder may stick to the mortar, which makes the sample difficult to be recovered. More importantly, this grinding process is one of the most risky steps for RNA to be degraded by RNases because the RNA and RNases are mixed but RNases are not denatured. If this grinding step is absolutely necessary, it is very important to not let the sample or the ground powder thaw during this step, particularly when RNA and mRNA are to be purified from the sample. This can be achieved by repeatedly adding liquid nitrogen into the porcelain mortar during this grinding step.

A good strategy to avoid the sample loss and to minimize or prevent from the degradation of nucleic acids during the initial step is to directly homogenize insects or insect tissues in an appropriate extraction buffer by using a Potter–Elvehjem glass homogenizer coupled with a motor-driving Teflon pestle or using a rotor–stator tissue homogenizer. All the homogenizer parts exposed to the sample must be sterilized. For RNA purification, for example, a homogenizer can be treated with an RNase decontamination solution such as RNase*Zap* [42]. During the homogenization, as soon as RNA and RNases are released from cells, RNases are simultaneously denatured by guanidinium hydrochloride or guanidinium thiocyanate presented in the extraction buffer. This method not only can simplify the purification procedure but also often provides highly reproducible results. Once nucleic acids are released from the insect cells typically by homogenization as described above, the next step is to use basic enzymatic or chemical methods to remove contaminating proteins and RNA when genomic DNA is to be purified or to remove proteins and genomic DNA when RNA is to be purified. The basic methods to remove the contaminants are essentially the same as those used for purification of nucleic acids from other organisms.

There are many different methods available for isolation and purification of nucleic acids from insects [40,41]. Different methods use different technologies and often result in different qualities and yields of nucleic acids that may affect their downstream applications. When researchers make a choice as to which purification method to be used in their research, the following five factors are usually considered: (1) probability of the carryover of contaminants from the purification processes; (2) integrity of nucleic acids to be obtained; (3) yield of nucleic acids; (4) applicability of a method; and (5) cost of using a particular method or a commercial kit to purify nucleic acids.

The carryover of contaminants, such as salts, phenol, ethanol, and detergents to nucleic acids, are one of the most common factors affecting downstream applications of the nucleic acids because these contaminants can inhibit many enzymatic reactions. On the other hand, the degradation of nucleic acids, particularly when RNA and mRNA are to be purified, is often problematic for their downstream applications including cDNA library construction. Usually, the application of traditional methods based on the homemade extraction reagents is often inexpensive. However, these methods often rely on the use of phenol or other toxic reagents, and may not fulfill all the requirements due to the carryover of contaminants. In contrast, the application of new methods or commercial kits usually uses nonhazardous chemicals and is more costly than homemade kits. However, these methods render high yields and purity of the final product that can be used in almost all downstream applications, including highly sensitive applications such as polymerase chain reaction (PCR).

14.2.2 CURRENT TECHNIQUES FOR PURIFICATION OF GENOMIC DNA FROM INSECTS

Genomic DNA can be purified from insects or insect tissues using a simple salting-out method [41,43]. Briefly, a sample is first homogenized in an extraction buffer containing ethylenediaminetet-raacetic acid (EDTA) and sodium dodecyl sulfate (SDS) using a tissue grinder or a homogenizer as previously described. EDTA can inhibit DNase activity to reduce the degradation of DNA whereas SDS can denature various proteins. Proteins and other contaminants are then precipitated using high concentrations of a salt such as potassium acetate or ammonium acetate. After the precipitates are removed by centrifugation, DNA in the supernatant is recovered by alcohol precipitation. Genomic DNA can also be purified from insects or insect tissues by organic extraction methods [40,44]. Briefly, after the homogenate is incubated in the presence of proteinase K to degrade proteins, the contaminants are removed by organic extractions usually first using phenol, then phenol–chloroform/isoamyl alcohol, and finally chloroform/isoamyl alcohol. DNA remaining in the aqueous phase of the last extraction can be recovered by alcohol precipitation. These in-house DNA purification methods are typically simple and inexpensive, and the integrity of the resultant DNA is often sufficiently high [45]. However, the DNA may not be sufficiently pure or may contain carryover contaminants or organic solvents that may affect some of sensitive downstream applications such as PCR (Table 14.3).

High-quality genomic DNA can often be purified from insects and insect tissues using a commercial kit that relies on the binding of DNA to a solid-phase support of a column through either anion exchange or selective adsorption of DNA to a silica-gel membrane. After the contaminants are washed away from the column, DNA is eluted with a buffer into a collection tube. The DNA purified by these methods is suitable for all downstream applications, including Southern blotting, PCR, real-time PCR, random amplification of polymorphic DNA (RAPD), restriction fragment length polymorphism (RFLP), and amplified fragment length polymorphism (AFLP) analyses.

Colton and Clark [46] compared PCR results among the DNA templates that were isolated from *Drosophila* using three isolation techniques. The first method was a standard method for *Drosophila* DNA isolation using a lysis buffer containing proteinase K, phenol and chloroform extractions, followed by ethanol precipitation [47]. The second method was based on the user manual of the commercial kit, DNAzol, manufactured by Invitrogen. The DNAzol procedure is based on the use of a novel guanidine-detergent lysing solution that hydrolyzes RNA and allows the selective precipitation of DNA from a cell lysate [48]. The third method was modified from a procedure originally developed for isolation of DNA from blood samples [49]. All the three methods resulted in adequately high qualities of DNA for PCR amplifications.

Dillon et al. [50] examined the effects of insect preservation methods on the recovery of genomic DNA from two species of parasitic wasp, *Venturia canescens* and *Leptomastix dactylopii*. They noted that material stored at −80°C, 100% ethanol, air-drying in a desiccator, critical-point dried from alcohol, and ethylene glycol all yielded good PCR results after short- and long-term storage, but not formalin. Specimens of the insects killed in ethyl acetate vapor and air-dried resulted in highly degraded DNA, which was unsuitable for PCR. Carvalho and Vieira [51] compared the influence of six different storage methods on DNA extraction from *Atta* spp. (leaf-cutting ants). These included (1) −70°C; (2) 95% ethanol at −20°C; (3) 95% ethanol at 4°C; (4) 95% ethanol at room temperature; (5) silica gel at room temperature; and (6) buffer (0.25 M EDTA, 2.5% SDS, 0.5 M Tris–HCl, pH 9.2 at room temperature for 90, 210, and 360 days of storage. They found all methods were efficient to preserve *Atta* spp. DNA up to 210 days. At 360 days, DNA was degraded only in 95% ethanol at room temperature.

14.2.3 CURRENT TECHNIQUES FOR PURIFICATION OF TOTAL RNA
AND mRNA FROM INSECTS

All current RNA purification procedures basically rely on organic extraction followed by alcohol precipitation or adsorption of nucleic acid molecules on a glass fiber filter (GFF) or silicate matrix. Because endogenous RNases in a biological sample can rapidly degrade RNA, guanidinium

TABLE 14.3

Comparison of Common In-House and Commercial Kits for Purification of Genomic DNA

Method	Description	DNA Quality	Application Note	References or Representative Commercial Kit
Salting-out	Proteins and other contaminants are precipitated using high concentrations of salt such as potassium acetate or ammonium acetate. After the precipitates are removed by centrifugation, DNA in the supernatant is recovered by alcohol precipitation.	Low and highly variable	Simple in-house method, but repeated alcohol precipitation of DNA usually required for downstream applications	[41,43]
Organic extraction	Contaminants are separated into the organic phase in each of the sequential organic extractions usually using phenol, phenol–chloroform/isoamyl alcohol, and chloroform/isoamyl alcohol. DNA remaining in the aqueous phase of the last extraction is then recovered by alcohol precipitation.	Fair	Simple in-house method, but residual phenol and chloroform likely to inhibit enzyme reactions in downstream applications	[40,41]
Anion exchange	Interactions between the negatively charged phosphates of DNA and the positively charged surface molecules on a solid phase of a column lead to the binding of DNA to the solid phase of the column under low salt condition. After the contaminants are washed away from the solid phase using medium-salt buffers, DNA is eluted using a high-salt buffer and recovered by alcohol precipitation.	High	Commercial kits to be used for purification of DNA suitable for all downstream applications	Genomic-tips (Qiagen)
Silica-based method	DNA is selectively adsorbed to a silica-gel membrane on a solid phase of a column under optimal buffer conditions. After the contaminants are subsequently washed away, DNA is eluted from the silica-gel membrane using a low-salt buffer.	High	Commercial kits to be used for purification of DNA suitable for all downstream applications	DNeasy tissue kits (Qiagen); PureLink genomic DNA mini kit (Invitrogen)

hydrochloride or guanidinium thiocyanate has been commonly used to disrupt cells, solubilize their components, and simultaneously denature endogenous RNases in the initial step of RNA purification [41]. After a sample is homogenized in the extraction buffer containing guanidinium thiocyanate, RNA is extracted using phenol/chloroform at reduced pH [52]. To date, this single-step technique still is the method of choice to isolate RNA from insects.

Many commercial RNA purification kits are developed based on the same mechanisms. For example, TRIzol reagent, manufactured by Invitrogen [53], is a proprietary mixture of acidic phenol and guanidine isothiocyanate. After a tissue sample is lysed in TRIzol, total RNA is obtained by chloroform extraction followed by isopropanol (isopropyl alcohol) precipitation. The Qiagen RNeasy 96 kit [54] represents a new technology for high-throughput RNA preparation. The system combines the advantages of guanidinium thiocyanate lysis, selective binding properties of a silica-gel based membrane, and the speed of vacuum and/or spin technology. Briefly, after cells are lysed, ethanol is added to provide appropriate binding conditions. Samples are then applied to the wells of the RNeasy 96 plate. After the contaminants are washed away, RNA is eluted in water. This method is designed for purification of RNA from animal cells. However, it can be adapted for purifications of multiple RNA samples from individual insects or insect tissues.

Some studies require the use of mRNA or poly(A) RNA purified from total RNA. For example, mRNA is often used as a template to synthesize cDNA for the construction of a cDNA library. If the transcription level of a gene is very low, it is often necessary to use mRNA for Northern blot analysis. Nevertheless, because insect mRNA contains a poly(A) at its 3' termini, the mRNA can be conveniently separated from the bulk of cellular RNA by affinity chromatography using an oligo(dT)-cellulose column [55]. Purification of mRNA can also be achieved based on the binding of mRNA to biotinylated oligo(dT) in solution. The Promega PolyATtract mRNA isolation system is an example of using biotinylated oligo(dT) to bind mRNA [56]. Bound mRNA is then immobilized by MagneSphere streptavidin-coated paramagnetic particles in a magnetic separation stand. After several washes of the particles to remove the contaminants, mRNA is eluted from the particles.

Finally, if total RNA or mRNA is to be purified from insect tissues, a large number of the same tissue (e.g., salivary gland) or each of several different tissues is usually required for the purification because many insects are relatively small. It could take hours or even days to dissect out a large number of tissues from insects. To stabilize RNA during the dissection or storage after the dissection, the dissected tissues can be submerged in an RNA stabilization solution such as RNA*later* tissue collection solution manufactured by Ambion [57]. An RNA stabilization solution is an aqueous tissue storage reagent that rapidly permeates most tissues to stabilize and protect RNA in fresh specimens. Tissues preserved in the RNA stabilization solution can be processed after the completion of all dissection work. Alternatively, the tissues in the solution can be frozen, then thawed and processed like fresh tissues without concern for cell rupture and release of RNases since the RNases have already been inactivated.

14.3   METHODS

In the last decade, isolation and purification of nucleic acids from insects or insect tissues have become routine activities in many research laboratories. High qualities of nucleic acids are the most important factors determining optimal results in downstream applications. Currently, there are many excellent in-house methods and commercial kits available for purification of nucleic acids from various biological samples. Many of these methods and commercial kits are not only easy to use but also often produce high quality nucleic acids for virtually all types of downstream analyses. Furthermore, these methods and kits can be directly used or slightly modified for purification of nucleic acids from insects and insect tissues. This section describes four representative methods, including an in-house method for purification of genomic DNA and three commercial kits for purification of genomic DNA, total RNA, and mRNA from insects or insect tissues.

14.3.1 Purification of Genomic DNA Using an In-House Method

The genomic DNA purification method described below is a commonly used in-house method that combines the salting-out and organic extraction methods. Although the method is relatively tedious, it can produce high-quality genomic DNA samples for many downstream applications. Depending on the amount of genomic DNA required for a research project, the method can be conveniently scaled down or up by proportionally decreasing or increasing the amount of the starting material and the volume of each reagent, respectively.

Reagents

Chloroform
DNA isolation buffer: 100 mM Tris–HCl (pH 8.0), 1% SDS, 100 mM EDTA in autoclaved deionized distilled water; stored at room temperature
70% ethanol
95% ethanol
Isopropyl alcohol
Phenol: Equilibrate clear and colorless liquefied phenol by repeated extractions using an equal volume of 100 mM Tris–HCl (pH 8.0) at room temperature until the pH of the phenolic phase is greater than 7.8 as measured by pH paper, add 8-hydroxyquinoline as an antioxidant to a final concentration of 0.1%, store in a light-tight bottle at 4°C for up to 1 month
5 M potassium acetate: Mix 60 mL 5 M potassium acetate, 11.5 mL glacial acetic acid, and 28.5 mL water, and store at room temperature
Proteinase K (20 mg/mL): Dissolve 100 mg proteinase K in 5 mL TE buffer for 30 min at room temperature, aliquot and store at −20°C
RNase A (10 mg/mL): Dissolve 100 mg RNase A in 10 mL autoclaved distilled water, boil for 15 min, cool to room temperature, aliquot and store at −20°C
3 M sodium acetate: Dissolve 40.8 g sodium acetate with just enough deionized distilled water, adjust pH to 5.3 with glacial acetic acid, adjust volume to 100 mL, and then autoclave the solution
10x TE buffer (pH 7.6): Dissolve 1.21 g Tris-base in 60 mL deionized distilled water, add 2 mL 0.5 M EDTA (pH 8.0), adjust pH to 7.6 with concentrated HCl, bring the volume to 100 mL, and then autoclave
1x TE buffer (pH 7.6): Dilute the 10x TE buffer by tenfold using autoclaved deionized distilled water
Autoclaved deionized distilled water

Supplies and equipment

10 μL adjustable pipetter and 10 μL pipetter tips
100 μL adjustable pipetter and 100 μL pipetter tips
1000 μL adjustable pipetter and 1000 μL pipetter tips
1.5 mL microcentrifuge tubes
Microcentrifuges (4°C and room temperature)
Potter–Elvehjem glass homogenizer (4 mL) coupled with a motor-driving Teflon pestle or other types of homogenizer (e.g., Fisher PowerGen homogenizer)
Nitrile gloves
Vortex mixer
Water baths (37°C, 65°C)

Procedure

1. Gently homogenize 30–100 mg insects or insect tissues in 0.6 mL DNA isolation buffer in a homogenizer for 1–2 min or until the homogenate becomes relatively homogenous (Note: If whole insects are used, small insect cuticular particles may be still visible).

2. Transfer the homogenate into a 1.5 mL microcentrifuge tube.
3. Add 6 μL proteinase K to the tube, mix gently, and incubate the tube in a water bath at 65°C for 30 min.
4. Remove the tube from the water bath and allow the tube to cool to room temperature.
5. Add 0.3 mL of 5 M potassium acetate to the tube, mix the solution by inverting, and place the tube on ice for 30 min.
6. Pellet the precipitated protein/SDS complex by centrifuging the tube at 13,000 g for 10 min at 4°C.
7. Transfer the supernatant to a new tube using a pipetter to avoid any floating lipids on the surface of the supernatant.
8. Add 0.5 mL isopropyl alcohol to the tube, gently mix the solution, and pellet DNA by centrifuging the tube at 13,000 g for 5 min at 4°C.
9. Decant isopropyl alcohol, and rinse the DNA pellet with 1.0 mL 70% ethanol.
10. Briefly dry the DNA pellet in air and resuspend DNA in 200 μL 1x TE buffer.
11. Add 2 μL RNase A to the tube (final concentration: 100 μg/mL), mix, and incubate the tube at 37°C for 30 min.
12. Add 100 μL phenol and 100 μL chloroform to the tube, mix, and centrifuge at 4500 g for 10 min at room temperature.
13. Transfer the aqueous phase to a new tube and add the same volume of chloroform, mix, and centrifuge at 4500 g for 10 min at room temperature.
14. Transfer the aqueous phase to a new tube, add 0.1 volume of 3 M sodium acetate and 2.5 volume of 95% ethanol, invert to mix, and place the tube on ice for at least 10 min or −20°C overnight to precipitate DNA.
15. Centrifuge the tube at 13,000 g for 10 min at 4°C, and rinse the pellet with 1.0 mL 70% ethanol.
16. Decant the ethanol and pipet off remaining solution using a pipetter.
17. Air dry the DNA pellet by inverting the tube over a paper towel for 10–30 min.
18. Resuspend the DNA pellet in 50–100 μL 1x TE buffer or in autoclaved, deionized distilled water by leaving the tube on bench overnight or heating at 65°C for 1 h.
19. If the DNA sample is not used immediately, store it at either 4°C or −20°C.

14.3.2 PURIFICATION OF GENOMIC DNA USING AQUAGENOMIC REAGENT

AquaGenomic is a commercial genomic DNA purification reagent manufactured by MultiTarget Pharmaceuticals (Salt Lake City, Utah). It is a multifunctional aqueous solution that allows the completion of cell lysis, DNA extraction, and debris removal by the same solution, and has been known as one of the simplest and most powerful genomic DNA extraction solutions ever developed. The method can be conveniently scaled down or up by proportionally decreasing or increasing the amount of the starting material and the volume of each reagent, respectively. The following protocol is modified based on a manufacturer's recommended procedure for the purification of genomic DNA from *Drosophila* (http://www.aquaplasmid.com/Protocol.html#AquaGenomic%20Protocol).

Reagents

AquaGenomic kit (Cat. No. 2030; MultiTarget Pharmaceuticals, Salt Lake City, Utah)
Autoclaved deionized distilled water
70% ethanol
Isopropyl alcohol
10x TE buffer (pH 7.6): Dissolve 1.21 g Tris-base in 60 mL deionized distilled water, add 2 mL 0.5 M EDTA (pH 8.0), adjust pH to 7.6 with concentrated HCl, bring the volume to 100 mL, and then autoclave
1x TE buffer (pH 7.6): Dilute the 10x TE buffer by tenfold using autoclaved deionized distilled water

Supplies and equipment

100 µL adjustable pipetter and 100 µL pipetter tips
1000 µL adjustable pipetter and 1000 µL pipetter tips
1.5 mL microcentrifuge tubes
Microcentrifuge (room temperature)
Potter–Elvehjem glass homogenizer (4 mL) coupled with a motor-driving Teflon pestle or other types of homogenizer (e.g., Fisher PowerGen homogenizer)
Nitrile gloves
Vortex mixer
Water bath (60°C)

Procedure

1. Gently homogenize 40–120 mg insects or insect tissues in 0.8 mL AquaGenomic solution in a homogenizer for 1–2 min or until the homogenate becomes relatively homogenous (Note: If whole insects are used, small insect cuticular particles may be still visible).
2. Transfer the homogenate to a 1.5 mL microcentrifuge tube and incubate the tube at 60°C for 5 min.
3. Vigorously vortex the tube for about 30 s using a vortex mixer and centrifuge the sample at maximum speed (e.g., 16,000 g) for 5 min.
4. Transfer the supernatant to a new tube containing 0.56 mL (0.7 volume of the supernatant) isopropyl alcohol, invert to mix, and centrifuge the tube at maximum speed for 5 min.
5. Decant the supernatant and rinse the pellet with 1.0 mL 70% ethanol.
6. Decant the ethanol and pipet off remaining solution using a pipetter.
7. Air dry the DNA pellet by inverting the tube over a paper towel for 10–30 min at room temperature.
8. Resuspend the DNA pellet in 50–100 µL 1x TE buffer or in autoclaved deionized distilled water by leaving the tube on bench overnight or heating at 60°C for 1 h.
9. If the DNA sample is not used immediately, store it at either 4°C or −20°C.

14.3.3 ISOLATION OF TOTAL RNA FROM INSECTS USING TRIzol REAGENT

The methods for purification of RNA from biological samples are relatively simple and straightforward. However, every precaution must be taken to avoid or reduce degradation of the RNA sample because RNA molecules can be rapidly degraded by RNases during the purification process. There are many excellent commercial kits available nowadays for isolation of total RNA from various biological samples. However, the principles of purification are virtually the same among the kits as discussed previously. Many researchers choose commercial kits for purifications of total RNA from insects or insect tissues to obtain high-quality RNA samples for downstream applications. This section describes a step-by-step procedure for purification of total RNA from insects or insect tissues using a representative commercial reagent, TRIzol, manufactured by Invitrogen. The protocol is essentially adapted from the user manual of TRIzol reagent [53] and is suitable for small-scale preparations of total RNA. However, the procedure can be conveniently scaled down or up by proportionally decreasing or increasing the volume of each reagent used in the purification process.

Reagents

TRIzol reagent (Cat. No. 15596-018; Invitrogen, Carlsbad, California)
RNase-free water
Chloroform
70% ethanol (in RNase-free water)
Isopropyl alcohol

Supplies and equipment

100 μL adjustable pipetter and 100 μL RNase-free pipette tips
1000 μL adjustable pipetter and 1000 μL RNase-free pipette tips
1.5 mL RNase-free microcentrifuge tubes
Microcentrifuge (4°C)
Potter–Elvehjem glass homogenizer (4 mL) coupled with a motor-driving Teflon pestle or other types of homogenizer (e.g., Fisher PowerGen homogenizer)
Nitrile gloves
Vortex mixer

Procedure

1. Homogenize 50–150 mg insects or insect tissues in 1.0 mL of TRIzol reagent using a homogenizer.
2. Transfer the homogenate into a 1.5 mL RNase-free microcentrifuge tube.
3. Add 0.2 mL chloroform to the tube, mix, and incubate for 3 min at room temperature.
4. Centrifuge the tube in a microcentrifuge at 12,000 g for 15 min at 4°C.
5. Transfer the aqueous phase to a new RNase-free tube.
6. Add 0.5 mL isopropyl alcohol to the tube, mix, and incubate for 10 min at room temperature.
7. Centrifuge the tube at 12,000 g for 10 min at 4°C.
8. Decant the supernatant and wash the RNA pellet with 1.0 mL 70% ethanol by vortexing, and centrifuge the tube at 7500 g for 5 min at 4°C.
9. Air dry the RNA pellet for 5 min at room temperature.
10. Add 50–100 μL RNase-free water to the tube to dissolve RNA.
11. If RNA is not used immediately, adjust the RNA concentration with RNase-free water, aliquot RNA in several microcentrifuge tubes, and store them at −70°C.

In order to reduce potential degradation of RNA, each tube is thawed only a few times before it is empty. For long-term storage (e.g., longer than a couple of months) of the RNA sample, researchers can add 0.1 volume of 3 M sodium acetate and 3 volumes of 95% ethanol to the sample, and store it as an ethanol precipitation at −70°C. RNA is most stable in a sodium acetate/ethanol precipitation mixture at −70°C to −80°C.

14.3.4 PURIFICATION OF mRNA FROM TOTAL RNA USING OLIGOTEX mRNA KITS

Similar to the purification of total RNA from insects and insect tissues, every precaution must also be taken to avoid or reduce degradation of mRNA sample because mRNA molecules can be rapidly degraded by RNases during the purification process. There are many excellent commercial kits available nowadays for isolation of mRNA from various biological samples. However, many kits only allow researchers to purify mRNA from total RNA as a starting material. This means that researchers would need to purify total RNA with a total RNA purification kit and then use an mRNA purification kit to purify mRNA from the total RNA. This is convenient because some projects require the use of mRNA (e.g., construction of a cDNA library) but others require only total RNA (e.g., synthesis of first-strand cDNA for RT-PCR).

The protocol described below is essentially adapted from the user manual of Oligotex mRNA kits manufactured by Qiagen as an example to purify mRNA from total RNA as a starting material [54]. The company currently markets three different kits according to the amount of total RNA processed, including the mini prep for less than 250 μg total RNA, the midi prep for 250 μg/mg total RNA, and the maxi prep for 1–3 mg total RNA as starting material. Researchers can choose a desired prep size based on the amount of total RNA as a starting material. The following procedure is based on the use of Oligotex midi mRNA prep kit as recommended by the manufacturer [54].

Reagents

Oligotex midi mRNA prep kit (Cat. No. 70042, Qiagen, Valencia, California)

Supplies and equipment

10 µL adjustable pipetter and 10 µL RNase-free pipette tips
100 µL adjustable pipetter and 100 µL RNase-free pipette tips
1000 µL adjustable pipetter and 1000 µL RNase-free pipette tips
1.5 mL RNase-free microcentrifuge tubes
Microcentrifuge (room temperature)
Nitrile gloves
Vortex mixer
Water bath or heating block (37°C and 70°C)

Procedures

1. Heat Oligotex suspension to 37°C in a water bath or heating block, mix by vortexing, and then place at room temperature.
2. Transfer 0.75–1.0 mg of total RNA in solution to an RNase-free 1.5 mL microcentrifuge tube, and adjust its final volume to 500 µL with RNase-free water. (If the purification starts with precipitated RNA, dissolve the RNA pellet in RNase-free water by heating the tube for 5 min at 70°C followed by vortexing for 5 s, and sharply flicking the tube. Repeat the procedure at least twice.)
3. Add 500 µL buffer OBB and 55 µL well-mixed Oligotex suspension to the tube, mix the solution thoroughly by pipetting or flicking the tube.
4. Incubate the tube for 3 min at 70°C in a water bath or heating block.
5. Remove the tube from the water bath/heating block, and place it at room temperature for 10 min.
6. Pellet the Oligotex/mRNA complex by centrifuging the tube for 2 min at maximum speed (e.g., 16,000 g), and carefully remove the supernatant by pipetting.
7. Resuspend the Oligotex/mRNA pellet in 400 µL buffer OW2 by vortexing or pipetting.
8. Pipet the resuspension onto a small spin column placed in a 1.5 mL microcentrifuge tube and centrifuge for 1 min at maximum speed.
9. Transfer the spin column to a new RNase-free 1.5 mL microcentrifuge tube, and apply 400 µL buffer OW2 to the column, centrifuge for 1 min at maximum speed, and discard the flow through.
10. Transfer spin column to a new RNase-free 1.5 mL microcentrifuge tube, pipet 20–50 µL hot (70°C) buffer OEB onto the column, pipet up and down three or four times to resuspend the resin, and centrifuge for 1 min at maximum speed to collect the eluate containing mRNA.
11. To ensure maximal recovery of mRNA from the resin, pipet another 20–50 µL hot (70°C) buffer OEB onto the column, pipet up and down three or four times to resuspend the resin, and centrifuge for 1 min at maximum speed to collect mRNA.
12. If mRNA is not used immediately, adjust the mRNA concentration with RNase-free water, aliquot mRNA in several microcentrifuge tubes, and store them at −70°C.

In order to reduce potential degradation of mRNA, each tube is thawed only a few times before it is empty. For long-term storage (e.g., longer than a couple of months) of mRNA sample, researchers can add 0.1 volume of 3 M sodium acetate and 3 volumes of 95% ethanol to the sample, and store it as an ethanol precipitation at −70°C. RNA is most stable in a sodium acetate/ethanol precipitation mixture at −70°C to −80°C.

14.4 FUTURE DEVELOPMENT TRENDS

The trends for developing new technologies and reagents for purifications of nucleic acids from insects are not significantly different from those for purifications of nucleic acids from other organisms. The current trends may include (1) maximizing the recoveries of nucleic acids, and (2) automating nucleic acid purification methods to streamline the recovery of nucleic acids.

14.4.1 MAXIMIZATION OF THE RECOVERIES OF NUCLEIC ACIDS

High recoveries of nucleic acids are often desirable but may not be crucial if nucleic acids are purified from a relatively large organism or tissue, or from a large number of individuals or tissues for most of downstream applications. However, high recoveries of nucleic acids could be extremely important if they are purified from individual insects or tissues of small insects, such as aphids and whiteflies. Even if nucleic acids are to be purified from a number of small organs or tissues such as brains and Malpighian tubules, it is a very challenging task to dissect a large number of insects for nucleic acid purifications. Therefore, the use of protocols or reagents with increased recoveries of total RNA from insects or tissues, and mRNA from total RNA is extremely important to obtain sufficient amounts of nucleic acids for downstream applications working on insects. Although very little is known as to what maximum recoveries of nucleic acids should be expected for a given protocol or reagent, the different recoveries of DNA or RNA when different protocols or reagents are used for the same biological sample imply that there are great potentials to develop new protocols or reagents, or modify the existing protocols or reagents to increase the recoveries of nucleic acids [58]. The use of such protocols or reagents will greatly facilitate the purification of sufficient amounts of nucleic acids from individual insects or tissues for downstream applications.

14.4.2 AUTOMATION OF NUCLEIC ACID PURIFICATION METHODS

Recent research advances in insect molecular genetics, genomics, and functional analysis of genes is a major driving force for developing automated nucleic acid purification methods. Apparently, the major advantage of using an automated purification approach is the increased efficiencies of nucleic acid purification process. Because the purification of nucleic acids using magnetic particles is suitable for automation, such technologies have been used for the automation of nucleic acid purification in biological research. For example, MagAttract magnetic-bead technology in combination with a BioRobot M-Series workstation (e.g., BioRobot M48 and BioRobot M96) manufactured by Qiagen [59] is a fully automated system for purification of DNA and RNA from a variety of blood and tissue samples. The technology removes the need for centrifugation or vacuum processing, eliminating tedious and time-consuming processing steps.

However, current automated nucleic acid purification systems often use proteinase K to lyse samples as a first step of nucleic acid purification. Such a lysis step may be adequate if nucleic acids are to be purified from soft tissues or hemolymph of insects, but may not work well if nucleic acids are to be purified from whole bodies of insects because of their tough exoskeleton structures. To accommodate currently available automated system for purification of nucleic acids from insects, it is necessary to develop a multiple homogenizer that can be used to homogenize multiple samples (e.g., 96) simultaneously. ffrench-Constant and Devonshire [60] developed a multiple homogenizer for preparing enzyme samples of small invertebrates or tissue in a 96-well flat-bottom microplate. A similar homogenizer could be developed for nucleic acid purification from insects. Once such a homogenizer becomes available, it is relatively easy to adapt currently available automated nucleic acid purification systems for purification of nucleic acids from insects.

ACKNOWLEDGMENTS

The author thanks Troy D. Anderson and Yoonseong Park for reviewing an earlier version of this manuscript. The relevant research was supported by Arthropod Genomics Center and Ecological Genomics Institute, both funded by K-State Targeted Excellence Program, and Kansas Agricultural Experiment Station. The research was also supported by the NIH Grant no. P20 RR016475 from the INBRE Program of the National Center for Research Resources. However, mention of trade names or commercial products in this chapter is solely for the purpose of providing specific information and does not imply recommendation or endorsement by Kansas State University and NIH. This manuscript is contribution No. 08-307-B from the Kansas Agricultural Experiment Station, Kansas State University, Manhattan, Kansas.

Kun Yan Zhu is a faculty member of both the Department of Entomology and Ecological Genomics Institute and the Arthropod Genomics Center at Kansas State University, Manhattan, Kansas. He obtained his BS in plant protection (with entomology as specialty) from Zhejiang Agricultural University (presently Zhejiang University) in Hangzhou, China in 1982. After serving as a teaching assistant at the same university for 5 years, he undertook his graduate research in Professor William A. Brindley's laboratory in the Department of Biology at Utah State University in Logan, Utah, and received his MS and PhD in biology there in 1989 and 1992, respectively. He then joined Professor John M. Clark's laboratory in the Department of Entomology at the University of Massachusetts, Amherst, Massachusetts in 1992 as a postdoctoral research associate and later as a research associate professor. He became an assistant professor at Kansas State University in 1996, and was promoted to associate professor in 2002 and to full professor in 2007. His current research interests include molecular basis of insect resistance to insecticides, molecular mechanisms of insecticide actions, and insect toxicogenomics. Besides teaching molecular entomology, toxicology of insecticides, and insecticides' properties and laws at Kansas State University, he also serves as a subject editor of *Journal of Economic Entomology* and an editorial board member for both *Insect Science* and *Acta Entomologica Sinica*.

REFERENCES

1. Romoser, W.S. and Stoffolano, J.G., *The Science of Entomology* (4th ed.), WCB/McGraw-Hill, New York, 1998.
2. Elzinga, R., *Fundamentals of Entomology* (5th ed.), Prentice Hall, Upper Saddle River, NJ, 2000.
3. Triplehorn, C.A. and Johnson, N.F., *Borror and Delong's Introduction to the Study of Insects* (7th ed.), Thomson Brooks, Cole, Belmont, CA, 2005.
4. Ferriss, R., Research advances on European corn borer resistance, *Seed World*, 126: 36, 1988.
5. Lingappa S. et al., Threat to vegetable production by diamondback moth and its management strategies, in Mukerji, K.G. (Ed.), *Fruit and Vegetable Diseases*, Springer, Netherlands, 2004, pp. 357–396.
6. Campbell, J.B., Keith, D.L., and Kramer, W., Mosquito Control Guide (G74-154-A). Cooperative Extension, Institute of Agricultural and Natural Resources, University of Nebraska, Lincoln, NE, 2003.
7. CDC, Traveler's Health—Malaria: General Information. National Center for Infectious Diseases, Center for Disease Control and Prevention, 2000. http://www.cdc.gov/travel/malinfo.htm.
8. Delaplane, K.S. and Mayer, D.F., *Crop Pollination by Bees*, CAB International, New York, 2000.
9. Will, K.W., Insects as research tools, in Resh, V.H. and Cardé, R.T. (Eds.), *Encyclopedia of Insects*, Academic Press, New York, 2003, pp. 1000–1002.
10. Adams, M.D. et al., The genome sequence of *Drosophila melanogaster, Science*, 287: 2185, 2000.
11. O'Brochta, D.A., Introduction: Genetic manipulation of insects, *Insect Biochem. Mol. Biol.,* 35: 647, 2005.
12. Holt, R.A. et al., The genome sequence of the malaria mosquito *Anopheles gambiae, Science*, 298: 129, 2002.

13. Xia, Q. et al., A draft sequence for the genome of the domesticated silkworm (*Bombyx mori*), *Science*, 306: 1937, 2004.

14. David, J.P. et al., The *Anopheles gambiae* detoxification chip: A highly specific microarray to study metabolic-based insecticide resistance in malaria vectors, *Proc. Natl. Acad. Sci. U S A*, 102: 4080, 2005.

15. Marinotti, O. et al., Microarray analysis of genes showing variable expression following a bloodmeal in *Anopheles gambiae, Insect Mol. Biol.*, 14: 365, 2005.

16. Wang, J. et al., An annotated cDNA library and microarray for large-scale gene-expression studies in the ant *Solenopsis invicta, Genome Biol.*, 8: R9, 2007.

17. Turner, C.T. et al., RNA interference in the light brown apple moth, *Epiphyas postvittana* (Walker) induced by double-stranded RNA feeding, *Insect Mol. Biol.*, 15: 383, 2006.

18. Tomoyasu, Y. et al., Exploring systemic RNA interference in insects: A genome-wide survey for RNAi genes in *Tribolium, Genome Biol.*, 9: R10, 2008.

19. Eisen, M.B. et al., Cluster analysis and display of genome-wide expression patterns, *Proc. Natl. Acad. Sci. U S A*, 95: 14863, 1998.

20. Zhu, Q. et al., Computational identification of novel chitinase-like proteins in the *Drosophila melanogaster* genome, *Bioinformatics*, 20: 161, 2004.

21. Dana, A.N. et al., Gene expression patterns associated with blood-feeding in the malaria mosquito *Anopheles gambiae, BMC Genomics*, 6: 5, 2005.

22. Southey, B.R., Sweedler, J.V., and Rodriguez-Zas, S.L., Prediction of neuropeptide cleavage sites in insects, *Bioinformatics*, 24: 815, 2008.

23. Alphey, L. et al., Malaria control with genetically manipulated insect vectors, *Science*, 298: 119, 2002.

24. Coates, C.J. et al., Mariner transposition and transformation of the yellow fever mosquito, *Aedes aegypti, Proc. Natl. Acad. Sci. U S A*, 95: 3748, 1998.

25. Catteruccia, F. et al., Stable germline transformation of the malaria mosquito *Anopheles stephensi, Nature*, 405: 900, 2000.

26. Grossman, G.L. et al., Germline transformation of the malaria vector, *Anopheles gambiae*, with the piggyBac transposable element, *Insect Mol. Biol.*, 10: 597, 2001.

27. de Lara Capurro, M. et al., Virus-expressed, recombinant single-chain antibody blocks sporozoite infection of salivary glands in *Plasmodium gallinaceum*-infected *Aedes aegypti, Am. J. Trop. Med. Hyg.*, 62: 427, 2000.

28. Ito, J. et al., Transgenic anopheline mosquitoes impaired in transmission of a malaria parasite, *Nature*, 417: 452, 2002.

29. Nene, V. et al., Genome sequence of *Aedes aegypti*, a major arbovirus vector, *Science*, 316: 1718, 2007.

30. Honey Bee Genome Sequencing Consortium, Insights into social insects from the genome of the honey bee *Apis mellifera, Nature*, 443: 931, 2006.

31. Crosby, M.A. et al., FlyBase: Genomes by the dozen, *Nucleic Acids Res.*, 35 (Database issue): D486, 2007.

32. Michael, F. et al., Revisiting the protein-coding gene catalog of *Drosophila melanogaster* using 12 fly genomes, *Genome Res.*, 17: 1823, 2007.

33. *Tribolium* Genome Sequencing Consortium, The genome of the model beetle and pest *Tribolium castaneum, Nature,* 452: 949–955, 2008.

34. Sattelle, D.B., Jones, A.K., and Buckingham, S.D., Insect genomes: Challenges and opportunities for Neuroscience, *Invert. Neurosci.*, 7: 133, 2007.

35. Li, H. et al., Characterization of cDNAs encoding three trypsin-like proteinases and mRNA quantitative analysis in Bt-resistant and -susceptible strains of *Ostrinia nubilalis, Insect Biochem. Mol. Biol.*, 35: 847, 2005.

36. Prabhakar, S. et al., Sequence analysis and molecular characterization of larval midgut cDNA transcripts encoding peptidases from the yellow mealworm, *Tenebrio molitor* L., *Insect Mol. Biol.*, 16: 455, 2007.

37. Zhang, J. and Zhu, K.Y., Characterization of a chitin synthase cDNA and its increased mRNA level associated with decreased chitin synthesis in *Anopheles quadrimaculatus* exposed to diflubenzuron, *Insect Biochem. Mol. Biol.,* 36: 712, 2006.

38. Anderson, T.D. et al., Gene expression profiling reveals decreased expression of two hemoglobin genes associated with increased consumption of oxygen in *Chironomus tentans* exposed to atrazine: A possible mechanism for adapting to oxygen deficiency, *Aquat. Toxicol.*, 86: 148, 2008.

39. Chen, M.S. et al., A super-family of genes coding for secreted salivary gland proteins from the Hessian fly, *Mayetiola destructor, J. Insect Sci.*, 6: 15, 2005.

40. Loius, C., Isolation of genomic DNA, in Crampton, J.M., Beard, C.B., and Louis, C. (Eds.), *The Molecular Biology of Insect Disease Vectors: A Methods Manual*, Chapman & Hall, New York, 1997, pp. 159–163.
41. Sambrook, J. and Russell, D.W., *Molecular Cloning: A Laboratory Manual* (3rd ed.), Cold Spring Harbor Laboratory Press, Cold Spring Harbor, New York, 2001.
42. Ambion, RNaseZap RNase Decontamination Solution, Austin, TX, 2006.
43. Miller, S.A., Dykes, D.D. and Polesky, H.F., A simple salting-out procedure for extracting DNA from human nucleated cells, *Nucleic Acids Res.*, 16: 1215, 1988.
44. Ausubel, F.M. et al., *Current Protocols in Molecular Biology*, John Wiley & Sons, New York, 1994.
45. Cheng, S. et al., Template integrity is essential for PCR amplification of 20- to 30-kb sequences from genomic DNA, *PCR Methods Appl.*, 4: 294, 1995.
46. Colton, L. and Clark, J.B., Comparison of DNA isolation methods and storage conditions for successful amplification of *Drosophila* genes using PCR, *Dros. Inf. Serv.*, 84: 180, 2001.
47. Ashburner, M., *Drosophila: A Laboratory Manual*, Cold Spring Harbor Laboratory, Cold Spring Harbor, New York, 1989.
48. Chomczynski, P. et al., DNAzol: A reagent for the rapid isolation of genomic DNA, *Biotechniques*, 22: 550, 1997.
49. Gustincich, S. et al., A fast method for high-quality genomic DNA extraction from whole human blood, *Biotechniques*, 11: 298, 1991.
50. Dillon, N., Austin, A.D., and Bartowsky, E., Comparison of preservation techniques for DNA extraction from hymenopterous insects, *Insect Mol. Biol.*, 5: 21, 1996.
51. Carvalho, A.O.R. and Vieira, L.G.E., Comparison of preservation methods of *Atta* spp. (Hymenoptera: Formicidae) for RAPD analysis, *An. Soc. Entomol. Brasil.*, 29: 489, 2000.
52. Chomczynski, P. and Sacchi, N., Single step method of RNA isolation by acid guanidinium thiocyanate-phenol-chloroform extraction, *Analyt. Biochem.*, 162: 156, 1987.
53. Invitorgen, TRIzol Reagent, Invitrogen Corporation, Carlsbad, CA, 2007.
54. Qiagen, *RNeasy 96 Handbook: For RNA Isolation from Animal and Human Cells and for RNA Cleanup*, Valencia, CA, 2002.
55. Paladichuk, A., Isolating RNA: Pure and simple, *The Scientist*, 13: 20, 1999.
56. Promega, PolyATtract mRNA Isolation System, Madison, WI, 2006.
57. Ambion, RNAlater Tissue Collection: RNA Stabilization Solution, Austin, TX, 2007.
58. Noriega, F.G. and Wells, M.A., A comparison of three methods for isolating RNA from mosquitoes, *Insect Mol. Biol.*, 2: 21, 1993.
59. Qiagen, *QIAamp 96 DNA Swab BioRobot Kit Handbook*, Valencia, CA, 2008.
60. ffrench-Constant, R.H. and Devonshire, A.L., A multiple homogenizer for rapid sample preparation in immunoassays and electrophoresis, *Biochem. Genet.*, 25: 493, 1987.

15 Preparation of Parasitic Specimens for Direct Molecular Applications

Huw Vaughan Smith and Rosely Angela Bergamin Nichols

CONTENTS

15.1 INTRODUCTION

15.1.1 Parasitic Lifestyle: Implications for Sampling, DNA Extraction, and Purification

Parasites are organisms that live in a symbiotic relationship in or on other organisms (hosts), which are normally of a different species. They exhibit a high degree of specialization with respect to their hosts and influence host fitness. Parasites whose symbiotic relationships occur primarily within the body of a host are known as endoparasites, while those whose symbiotic relationships occur primarily at the outer surface of their hosts are known as ectoparasites. Although parasites affect most life forms, causing some harm, this chapter focuses on the extraction of nucleic acids from those eukaryotic parasites known to cause disease, morbidity, and mortality in human beings and livestock.

Eukaryotic parasites are transmitted by a variety of direct and indirect (e.g., environment, food, water, invertebrate intermediate hosts [vectors] found in the environment) routes and have distinct life cycle stages, which are required for transmission. Those with an environmental component to transmission require a sturdy transmissive stage to survive and are environmentally robust. These include the transmissive stages of intestinal endoparasites (e.g., oocysts, cysts, ova, larvae) and ectoparasites. While the robustness of transmissive stages poses distinct challenges to nucleic acid extraction and purification, as they are difficult to disrupt, other, less robust, life cycle stages resident within host tissues and fluids pose different challenges depending on their location. For those with limited knowledge of clinical parasitology, diagnostics, and the morphology and morphometry of endoparasites, particularly their transmissive stages, visit http://whqlibdoc.who. int/publications/2003/9241545305.pdf (Part II, Parasitology). Endoparasite transmissive stages (spores, cysts, oocysts, ova) can range from 1.0–1.6 μm × 0.9 μm (*Enterocytozoon bieneusi*, microsporidia) to 140–180 μm × 50–85 μm (*Schistosoma intercalatum*). Helminth ova can range from less than 30 to greater than 150 μm in length and an appreciation of their relative sizes can be obtained from viewing page 132.

Because of the diversity of parasites, their niches, and their numerous and varied life cycles, this chapter focuses primarily on parasitic protozoa of public and veterinary health importance, and the inclusion of some well recognized protozoan zoonoses (cryptosporidiosis, giardiasis, leishmaniasis, microsporidiosis) should increase the usefulness of this chapter (Table 15.1). As this chapter focuses on diagnosing parasites using polymerase chain reaction (PCR)-based methods, we review methods for extracting nucleic acid from protozoan parasites present in clinical (feces, blood, urine, respiratory tract, cerebrospinal fluid, and formalin-fixed specimens), food and environmental samples, including invertebrate intermediate hosts (vectors) found in the environment. Furthermore, we present in-house protocols for maximizing parasite DNA recovery from these different matrices for direct molecular applications, based primarily on specific protozoan parasites as model systems, and review the key features of some commercial DNA extraction and purification kits, highlighting, as far as possible, their convenience, efficiency, and applicability.

For intestinal parasites, the production of transmissive stages (cysts, oocysts, ova, larvae) can augment parasite (and nucleic acid) abundance, but their production and appearance in feces is dependent upon the life cycle of the parasites in question. Similarly, for those parasites present in the clinical matrices commonly examined (blood, feces, urine), their abundance will depend upon their life cycles. Here, a clear knowledge of parasite life cycles is a prerequisite for conventional and molecular diagnosticians. The sensitivities of morphological and molecular detection are expected to differ, with the molecular approach, using validated methods and primers, often being more sensitive. Animal models of human disease have demonstrated that PCR amplification of parasite DNA during early infection (e.g., *Trypanosoma cruzi*, *Toxoplasma gondii*, *S. mansoni*) [1–3] can assist the diagnosis of infection in humans earlier than when using conventional diagnostic techniques such as microscopy or serology. Amplifying naked DNA or DNA from morphologically unrecognizable life cycle forms is a distinct advantage in clinical samples, but can prove problematic in food and environmental samples, where often the demonstration of an intact (and preferably viable) transmissive stage is required. Transmissive stages might not be detectable in clinical samples and their absence in repeated submissions of samples from symptomatic individuals does not necessarily indicate the absence of infection. In recuperating immunocompetent cases and in immunosuppressed individuals (e.g., transplant patients, patients with deficient cell mediated immunity, primary immunodeficiency diseases such as single gene disorders of the immune system) who have sufficient immunity to downregulate the production of the transmissive stages, but insufficient immunity to down regulate the production of vegetative stages (e.g., cryptosporidiosis) [4], samples should be subjected to PCR-based detection, as sufficient DNA from other life cycle forms should be present, particularly when clinical suspicion is high. For PCR-based methods, nested PCR methods, being more sensitive than direct PCR methods, are likely to have a higher diagnostic index. Unlike prokaryotic parasites, the abundance of eukaryotic parasites cannot be increased readily by selective enrichment using *in vitro* culture techniques prior to their identification, therefore maximizing the numbers already present in a sample by physicochemical and immunological concentration methods, prior to DNA extraction, has obvious advantages.

TABLE 15.1

Some Parasite Zoonoses Transmitted by the Waterborne or Foodborne Routes

	Parasite	Transmission Route	Contaminated/Infected Matrix	Final Hosts
Microspora				
Enterocytozoonidae	*Enterocytozoon bieneusi*	?Water, food	?Spores in water and on uncooked or undercooked food	Humans, rhesus monkeys
Unikaryonidae	*Encephalitozoon cuniculi*	?Water, food	?Spores in water and on uncooked or undercooked food	Humans, pets/animals residing in and around human dwellings (e.g., rabbits, canines, mice, pigs, goats, cows)
	E. intestinalis			Parakeet, parrot
	E. hellem			
Pleistophoridae	*Pleistophora*-like organisms	Food	Uncooked or undercooked fish or crustacea	Humans, fish, crustacea
Protozoa				
Cryptosporidiidae	*Cryptosporidium parvum* (Genotype 2)	Water, food	Oocysts in water and on uncooked or undercooked food	Humans and other mammals
Hexamitidae	*Giardia duodenalis*	Water, food	Cysts in water and on uncooked or undercooked food	Humans, other mammals and birds
Sarcocystidae	*Toxoplasma gondii*	Food, water	Oocysts in water and on uncooked or undercooked food, tissue cysts in uncooked or undercooked meat	Felines
Balantiididae	*Balantidium coli*	Water, food	Cysts in untreated or minimally treated water and on uncooked or undercooked food	Humans, pigs, nonhuman primates, cats, rodents
Blastocystidae	*Blastocystis hominis*, *Blastocystis sp.*	Water, food	Cysts in untreated or minimally treated water and on uncooked or undercooked food	Humans and other mammals

Trematodes				
Opisthorchiidae	*Clonorchis* spp., *Opisthorchis* spp.	Meat, freshwater fish	Metacercariae in musculature	Humans, cats, dogs, etc.
Heterophyidae	*Metagonimus yokogawai*, *Heterophyes* spp.	Meat, freshwater fish (sweetfish) Brackish water fish	Metacercariae in musculature	Humans, cats, dogs, etc
Echinostomatidae	*Echinostoma* spp.	Meat: loach, frogs, snails	Intestinal submucosa of loach; kidney of frogs; head, mantle, and liver of snails	Humans, dogs, rats, birds, etc.
Fasciolidae	*Fasciola hepatica*	Waterplants (e.g., watercress, rice, dandelion, *Nasturtium* and *Mentha* spp.)	Metacercariae encysted on leaves (about 10% of metacercariae float in water)	Primarily ruminants
	Fasciolopsis buski	Water chestnut, water caltrop, water hyacinth	Metacercariae encysted on leaves	Humans, pigs
Troglotrematidae	*Paragonimus* spp.	Potamid and other crabs, crayfish, shrimp	Metacercariae in lungs and musculature of crabs	Humans, canines, felines, etc.
Schistosomatidae	*Schistosoma* spp.	Water, skin penetration	Cercariae in water	Humans, nonhuman primates, bovines, cats, dogs, pigs, rodents, etc.
	Schistosome dermatitis	Water, skin penetration	Cercariae in fresh and marine waters	Birds, nonhuman mammals
Cestodes				
Diphyllobothriidae	*Diphyllobothrium latum*	Salmonid and other fish	Plerocercoid in musculature, liver, roe	Humans, canines, felines, various land and marine mammals
	Marine diphyllobothriasis	Marine fish: ceviche (made of raw fish in Peru and Chile)	Plerocercoid in musculature	Marine mammals
Taeniidae	*Taenia saginata*	Meat: bovine and cervine	Cysticerci in musculature	Humans
	Taenia solium	Meat: pig, camel, rabbit, bear, etc.	Cysticerci in musculature	Humans
	Echinococcus spp.	Unfiltered water	Ova in water	Canines

(continued)

TABLE 15.1 (continued)
Some Parasite Zoonoses Transmitted by the Waterborne or Foodborne Routes

	Parasite	Transmission Route	Contaminated/Infected Matrix	Final Hosts
Nematodes				
Ascarididae	*Ascaris suum*	Contaminated vegetables	Infective ova on contaminated vegetables	Pigs, primarily
	Toxocara canis	Contaminated vegetables, liver, paratenic hosts such as snails	Infective ova on contaminated vegetables, infective larvae in tissues	Canines
	Toxascaris leonina	Contaminated vegetables	Infective ova on contaminated vegetables	Canines
	Toxocara cati	Contaminated vegetables	Infective ova on contaminated vegetables	Felines
	Lagochilascaris minor	Contaminated vegetables	Infective ova on contaminated vegetables	Felines, raccoons
Anisakidae	*Anisakis simplex* and *Pseudoterranova decipiens*	Intestine and musculature of marine fish, squid	Third stage larvae in tissues of marine fish and squid	Dolphins and toothed whales
Metastrongylidae	*Angiostrongylus* spp.	Contaminated vegetables	Third stage larvae on vegetables	Rodents, especially rats
		Infected frogs, prawns, crabs, etc.	Third stage larvae in frogs, prawns, crabs, etc.	
Gnathostomatidae	*Gnathostoma spinigerum* (and other species)	Meat, fresh water fish	Third stage larvae in musculature	Canines and felines
Trichinellidae	*Trichinella* spp.	Meat	Infective larvae in musculature	Humans, pigs, bears, wild boar, warthog, walrus, seal
Others				
Acanthocephalans	*Macracanthorhynchus hirudinaceus*	Beetles (as food/folk remedy)	Cystacanth in body cavity	Pigs
Pentastomids	*Armillifer armillatus* and *A. moniliformis*	Contaminated water or food: snake meat contaminated with eggs	Eggs in water or on vegetables Nymphs in snake meat	Python and other snakes
	Linguatula serrata	Organs (esp. liver) of infected herbivores	Nymphs in organs/tissues of herbivores (halzoun/marrara)	Canines

15.1.2 PARASITIC LIFESTYLE: INFLUENCE OF THE MATRICES

Clinical, food, environmental, and invertebrate vector matrices can contain different abundances of parasites. For example, clinically ill, immunocompetent hosts will harbor a high abundance of parasites, whereas foodstuffs and environmental samples normally contain parasites in low abundance, which will influence PCR outcomes if DNA extraction is not maximised. In addition, clinical, food, environmental, and invertebrate vector matrices contain many inhibitory substances in varying quantities, which will decrease the sensitivity of PCR detection. Two essential steps should be exploited to ensure that parasite nucleic acids are amplified maximally: the disruption of parasite cell membranes and proteins and the extraction of parasite nucleic acids from other sample contaminants such as host and parasite cell proteins and debris. For protozoan parasites, this process is complicated by the abundance of host nucleic acids, which can be 1 million-fold more abundant than parasite nucleic acids, and can reduce the efficiency of PCR amplification. This demands more effective methods both for neutralizing inhibitory effects and extracting nucleic acids. While methods for preparing DNA template from purified or partially purified parasite isolates may yield satisfactory PCR results, detection sensitivities can be influenced greatly by interferents derived from the sample matrix and its processing. Currently, for many PCR assays, there is a distinct difference between laboratory and field data. Neutralizing inhibitory effects in different matrices such as blood, feces, urine, food, and the environment have been a key feature of some commercial kits, which include proprietary reagents for this purpose. Reagents that have been shown to be effective in increasing nucleic acid extraction and reducing PCR interferents for the molecular detection of parasites of human and veterinary importance are identified in Section 15.3 and Table 15.2.

Nested PCR assays increase detection sensitivity, but can be a liability in the diagnostic laboratory because of the likelihood of carryover contamination from the first PCR amplification; however, the use of nested PCR cannot always be avoided. Protozoan parasites, as contaminants of water, food, and other environmental samples are often present in sample concentrates at very low abundance (1–10 organisms) due to low level contamination of the initial matrix, the inefficient methods used for their concentration, and the inability to increase their numbers *in vitro*. Here, nested assays are essential (e.g., particularly for environmental samples) [5], and stringent practices to limit carryover contamination in the laboratory must be applied. Alternatively, single-tube nested assays can be used and because both primary and secondary PCR reactions are set up initially in a closed tube system, the likelihood of carryover contamination is greatly reduced [6].

15.2 PREPREPARATION STEPS TO REDUCE PCR INTERFERENTS

Standardized concentration techniques are used regularly to increase parasite numbers in clinical samples such as feces, blood, and urine because we cannot expand them readily by selective enrichment in *in vitro* culture, with the exception of the protozoan parasites *Trichomonas vaginalis* and *Blastocystis* sp. Therefore, the abundance of the transmissive stage determines method sensitivity. Concentration techniques partially purify and separate parasites from particulate interferents that compromise examination by light microscope. The principle of increasing parasite numbers is also pertinent when extracting and purifying DNA for direct molecular applications, as it increases available nucleic acid, while also reducing PCR interferents. Different methods are required to concentrate parasites from feces, blood, and urine. Parasites can be concentrated by centrifugation or filtration in blood, urine, and cerebrospinal fluid (CSF), but, for blood parasites, whether the parasite is intracellular (e.g., malaria) or extracellular (e.g., *Trypanosoma*, microfilaria) must also be considered, which requires a working knowledge of parasite life cycles.

Concentrating parasites from feces is a routine procedure; however, stool consistency can vary from solid to fluid depending on disease state, diet, age, etc., which can influence parasite recovery. Typically, the stages of the parasite life cycle present in feces are the transmissive stages but, in instances of florid diarrhea, or following drug treatment, the vegetative stages of parasitic protozoa,

TABLE 15.2

Characteristics of Some DNA Extraction and Purification Methods Used for Parasites

Method	Parasite Disruption	DNA Extraction	DNA Purification	Advantages	Disadvantages
FTA® filter cards and FTA concentrator-PS filter	Yes, Chemical lysis	Yes	No, Filter used directly in PCR after filter washing	Simple to perform, Useful for field collection, transport, and archiving of dried material	PCR outcome dependent on the matrix; Large pore size of membrane (~25 μm) may allow passage of intact parasites through the matrix before they are lysed and their DNA bound; Uneven distribution of microorganisms on the filter paper may require testing of replicate sub-samples by PCR particularly when parasite abundance is low; Incomplete lysis with more resistant microorganisms, particularly environmentally robust parasites; Possibility of PCR inhibition when more than one disk is used; Not yet fully tested with a range of parasites
Whatman no. 3 filter paper	No	Boiling with Chelex™	Spin column (QIAamp DNA mini kit)	Simple to perform, Useful for field collection, transport, and archiving of dried material, Economical	PCR outcome dependent on the matrix; Uneven distribution of microorganisms on filter may require testing of replicate sub-samples by PCR particularly when parasite abundance is low; Not fully tested with transmissive stages of parasites which require harsh disruption treatments
Boiling with Chelex™ 100	Yes		No, Used directly after centrifugation	Rapid, Fairly economical	Dilution of sample; Requires partially purified material to start with

				In-house preparation	
PVPP	No	No	Yes	Rapid method for eliminating PCR inhibitors by pretreatment of feces and other matrices before DNA extraction; Purification of extracted DNA	Not fully tested with environmentally robust parasites; Can inhibit PCR if carried over; Dilution of sample; Requires partially purified material to start with; Not fully tested with environmentally robust parasites; Can inhibit PCR if carried over
Lysis buffer (Tris–EDTA–SDS) + pK digestion—low concentration formulation	No/yes Inclusion of mechanical disruption step necessary for resistant parasites	No	No Heat denaturation of pK	Rapid and inexpensive; Reagents can be prepared in-house, or kits can be purchased; Can be used directly in PCR for some applications	May require DNA purification; May be unsuitable for RNA extraction since the extraction procedure does not destroy all ribonucleases
Lysis buffer (Tris–EDTA–SDS) + pK digestion—high concentration formulation	No/yes Inclusion of mechanical disruption step necessary for resistant parasites	No	No	Applicable to DNA extraction from complex tissues	Time consuming; May require DNA purification
Guanidinium isothiocyanate (L6 buffer)	No/yes Inclusion of mechanical disruption step necessary for resistant parasites	Yes	Yes Centrifugation to separate silica, followed by two washes of silica particles with buffer L2	Yields purified DNA/RNA; Reagents can be prepared in-house, or kits can be purchased	More time consuming compared to column based methods for DNA purification; Not tested with a range of parasites; Not fully compared with other methods for extracting parasite nucleic acids

(continued)

TABLE 15.2 (continued)
Characteristics of Some DNA Extraction and Purification Methods Used for Parasites

Method	Parasite Disruption	DNA Extraction	DNA Purification	Advantages	Disadvantages
GeneReleaser® (BioVentures)	No	Includes proprietary polymeric materials to facilitate release of DNA from cells	The matrix separates inhibitors released during cell lysis	Yields PCR amplifiable nucleic acid from small amounts of material Cell lysis can be performed directly in the PCR amplification tube on a thermocycler within minutes Microwave protocol can shorten lysis time	Not fully investigated with parasites Proprietary kits can be expensive
InstaGene™ matrix (BioRad)	No	Sample is incubated with InstaGene matrix at 56°C for 15–30 min, then boiled for 8 min	InstaGene matrix absorbs cell lysis products and DNA is recovered by centrifugation	Rapidity, simplicity May be used as a secondary cleaning procedure producing an improved substrate for PCR	Not fully investigated with parasites Proprietary kits can be expensive
QIAamp® DNA mini kit (Qiagen)	No	pK in proprietary buffer	Yes Spin column	Rapidity Automation available	Lack of comparative studies with other methods for identifying the detection limit for parasites in different matrices Purified DNA can still contain PCR inhibitors Proprietary kits can be expensive
Wizard® genomic DNA purification kit (Promega)	No	Include cell lysis solution and nuclei lysis solution	Salt precipitation to remove proteins	DNA is concentrated and desalted by isopropanol precipitation	For the isolation of parasites from blood, provides initial step to lyse red blood cells with the cell lysis solution	Not fully investigated with parasites Proprietary kits can be expensive

High Pure PCR template preparation kit (Boehringer Mannheim-Roche)	Yes/No Applies a chemical/enzymatic approach to cell lysis and nuclease inactivation	pK in proprietary buffers	Yes Spin through glass fiber filter/with three different buffers	Can be used with whole blood, solid tissue, and mammalian cells	Not fully investigated with parasites Reports suggest that this kit works with a relatively limited range of sample types compared to competing systems Proprietary kits can be expensive
QIAamp® DNA stool mini kit (Qiagen)	No/Yes Inclusion of mechanical disruption step necessary for resistant parasites	pK in proprietary buffer and inhibitEX-tablets	Yes Spin column	Specially formulated to overcome PCR inhibition by substances in stool samples Use of proprietary kits can be advantageous in some settings (e.g., forensic and ancient material)	Not fully compared with other methods for DNA extraction from parasites Proprietary kits can be expensive
MasterPure™ DNA kits (Epicentre)	No/Yes Enzyme/salt/detergent-mediated cell lysis	Proprietary salt precipitation to remove proteins	Nucleic acid precipitation	Reported to be efficient in extracting nucleic acid from yeasts and filamentous fungal species Separate kits for the isolation of either DNA or RNA	Not fully investigated with parasites Proprietary kits can be expensive
NucliSens® isolation reagents and NucliSens™ lysis buffer (Biomérieux)	No/Yes NucliSens lysis buffer contains 5M guanidinium isothiocyanate	No Adherence of nucleic acid to silica particles under high concentration of chaotropic agent	Yes Centrifugation and several washes of silica-nucleic acid complex followed by nucleic acid elution in low salt buffer	Specimens can be stored in NucliSens lysis at −70°C for up to 1 year before extraction Automated platform available for multiple extractions	More time consuming compared to column purification Proprietary kits can be expensive
GENECLEAN® kits (I, II, and III) (Bio 101)	No	Yes	Yes Spin filters or Glassmilk® (silica-based matrix) purification	Nucleic acid purification from TAE/TBE buffered gels Bulk slurry form of the patented silica matrix allows for flexibility regarding the scale of purification	Not fully investigated with parasites Proprietary kits can be expensive

(continued)

TABLE 15.2 (continued)
Characteristics of Some DNA Extraction and Purification Methods Used for Parasites

Method	Parasite Disruption	DNA Extraction	DNA Purification	Advantages	Disadvantages
NucleoSpin® Trace, NucleoSpin XS (Clontech Laboratories)	No/Yes Proprietary lysis buffer	Yes	Yes NucleoSpin funnel columns	Used for forensic samples and suitable for trace amounts of DNA Specially designed columns to elute small volumes of DNA	Not suitable for all matrices Not fully investigated with parasites Proprietary kits can be expensive
FastDNA SPIN kit for soil (Q.BIOgene)	Yes FastPrep® instrument and proprietary Lysing matrix E, PPS solution, needed to lyse cells	Yes Proprietary components include Binding Matrix, SEWS-M solution, DES solution, Sodium Phosphate buffer, BBS solution, MT buffer	Yes GENECLEAN® procedure or Spin filters, Catch tubes	Used for purification of DNA from soil Can be adapted for other matrices such as environmental waters	Not fully investigated with parasites Requires dedicated lysing cell apparatus Proprietary kits can be expensive
UltraClean™ soil DNA isolation kit (Mo Bio Labs)	Yes Mechanical agitation (vortexing) with bead solution provided with the kit	Yes Proprietary solutions S1 to S5, inhibitor remover (IRS) solution	Yes Spin filters	Designed to extract DNA from soil	Not fully investigated with parasites Proprietary kits can be expensive

and larval and adult stages of helminths can also be voided. Protozoan cysts, oocysts, helminth ova, and larvae can withstand concentration procedures, but the reproductive (non transmissive) stages of protozoa cannot [7]. Two concentration procedures, sedimentation and flotation, or a combination of the two are effective [7].

15.2.1 Parasite Concentration Techniques

Centrifugation and filtration concentrates parasites and can reduce PCR interferents in blood, urine, and CSF, and repeating these procedures can increase these benefits as long as the removal of interferents far exceeds potential losses of the target organisms from repeated manipulations. Examples include the concentration of the extracellular protozoan blood parasite *Trypanosoma* spp. from blood (e.g., http://www.oie.int/eng/normes/mmanual/A_00066.htm), the intracellular protozoan parasite *T. gondii* from CSF by centrifugation, ova of the blood trematode *Schistosoma* spp. from urine by filtration, etc. Sufficient nucleic acid might be present in the sample such that repeated centrifugation and filtration steps as a pretreatment is not necessary, but this will depend on the parasite, its life cycle, and its abundance in the clinical sample submitted. As the benefits of stool concentration methods lie in being able to maximize the detection of a range of enteric parasites rather than individual parasites, the efficiency of these methods is dependent on the buoyant density and the robustness of the parasite life cycle stage sought. Commercial kits for extracting DNA from clinical and environmental samples (Table 15.2) incorporate proprietary reagents that reduce PCR interferents, but the impact of different disruption and extraction treatments on the quality of the extracted parasite nucleic acid has not been fully investigated. Selected methods for extracting and purifying parasite DNA are presented in Sections 15.4 through 15.6 and Tables 15.2 and 15.3.

15.2.1.1 Fecal Parasites: Sedimentation by Centrifugation

Denser parasites, such as helminth ova, and larvae, settle more rapidly than protozoan cysts, oocysts, and spores, and the ova of *Schistosoma* spp., *Clonorchis*, *Opisthorchis*, and heterophyid flukes, being the densest, will settle the most rapidly in suspension. Parasites settle more rapidly if the stool suspension is centrifuged, but food particles also sediment more rapidly and can mask the presence of parasites in the film examined. Larger food particles can be removed before centrifugation by filtering the emulsified stool through a sieve (425 μm aperture size; Endecotts [Filters] Ltd, London, United Kingdom) with an aperture size large enough for parasites to pass-through, but which retains larger food particles. A small fecal sample (500 mg–1 g: the size of a pea) is sufficient for examination. The efficiency of detection is increased by adding formalin for fixation and preservation of parasites, and ether to remove fats and oils (formalin ether (ethyl acetate) method). The use of both 10% formalin and ether renders the sample microbiologically safer; additionally, ether (or ethyl acetate) dilapidates the sample. After centrifugation, a fatty plug can be seen at the interface of the two liquids. The ether layer, the fatty plug, and the formalin below it are discarded and the whole pellet retained for examination. Many modifications to this procedure are used (e.g., Ref. [8]), but the method of Allen and Ridley [9] is typical of the formalin ether method used in diagnostic laboratories [10]. This method can achieve a concentration of 15- to 50-fold, dependent upon the parasite sought, and provides a good concentrate of protozoan cysts and helminth eggs, which are morphologically and morphometrically satisfactory. Less distortion of protozoan cysts occurs with this method than with $ZnSO_4$ flotation.

15.2.1.2 Fecal Parasites: Flotation

The specific gravity (sp. gr.) of helminth ova and larvae, and protozoan cysts and oocysts ranges from 1.050 to 1.150 and concentration by flotation utilizes a liquid suspending medium, which is denser than the parasites to be concentrated therefore, when mixed with the flotation medium, they

TABLE 15.3
Some Methods for Extracting *Cryptosporidium* Oocyst DNA from Feces, Water, and Food

Matrix	DNA Extraction and Purification Methods from Oocysts	PCR Facilitators	Target (Detection Limit)	Sensitivity (% Positives) Numbers Positives/Total	References
Unpreserved human stool (180–200 mg)	QIAamp® DNA stool mini kit[a], boil 5 min, freeze-thaw three times (liquid nitrogen 1 min, boil 2 min)	Inhibit-EX-tablet	Nested COWP (5×10^2 oocysts)	(97%) 86/89	[32]
Frozen pig and calf stools (diluted 1:4 w/v in PBS)	Freeze-thaw three times (liquid nitrogen for 2 min, 75°C for 2 min), centrifuge, and add lysis buffer (Tris-EDTA- SDS-proteinase K) to pellet, glassmilk[b]	BSA	Nested Laxer et al. (1991) (100 seeded oocysts)	NA	[33]
Human stool (0.3–0.5 g) preserved in 2.5% potassium dichromate	Modified FastDNA Prep kit[c] and QIAquick spin column	PVP in extraction method. 11.3% (24 of 213 samples) were inhibitory	Direct 18S rRNA	(100%) 53/53 plus 31 negatives by microscopy	[34]
Unpreserved solid (0.4–0.5 g) or liquid (200 μL) human stool	Water–ether extraction and washes in lysis buffer (Tris-EDTA-SDS)[d], freeze-thaw 15 times (liquid nitrogen for 1 min, 65°C for 1 min)	BSA, Tween 20, and PVP incorporated in PCR	Single tube nested COWP and nested 18S rRNA	(97.8%) 90/92 (98.9%) 91/92	[17]
Whole human stool (200 μL)	Oocyst disruption by shaking in guanidinium thiocyanide and zirconia beads with the addition of isoamyl alcohol to prevent foaming. DNA extraction following [24]	PVP in extraction method	Direct 18rRNA and Direct COWP and Direct TRAP-C1	(97%) 204/218 (91%) 191/218 (66%) 139/218	[35]

Storm waters (filtered and purified by sucrose–Percoll flotation)	Oocysts concentrated from a 0.5 mL pellet of storm water concentrated by IMS. Five cycles of freeze-thawing without dissociating oocysts from magnetizable beads, proteinase K digestion, and diluted with ethanol, QIAamp® DNA mini kit	None	Nested 18S rRNA	27/29 include 12 samples negative by microscopy	[36]
Raw milk (50 mL samples)	Centrifugation, IMS, five cycles of freeze-thawing (−80°C, 5 min; 95°C, 5 min), QIAGEN LB + proteinase K digestion, QIAmp spin column[e]	None	18S rRNA and COWP (<10 seeded oocysts)	NA	[37]
Apple juice (50 mL samples)	Centrifugation, sucrose gradient, IMS, proteinase K digestion, QIAmp blood kit	None	COWP (30–100 seeded oocysts)	NA	[38]

a   QIAamp® DNA stool mini kit is based on the use of proprietary buffers and DNA purification through a silica column.

b   Glassmilk (GENECLEAN® Bio 101) is based on DNA adsorption to silica suspensions that allows subsequent washing of bound DNA to remove inhibitors.

c   Modified FastDNA kit uses FP120 FastPrep Cell Disruptor to disrupt oocysts in a proprietary buffer that minimizes adsorption of DNA to fecal particles (Cell lysis/DNA Solubilizing Solution). Polyvinylpyrolidone (final concentration of 0.5% w/v) used in this step precipitates polyphenolic compounds and the solubilized DNA is bound to the Binding Matrix in the presence of chaotropic salt which is washed and then eluted. The final purification of DNA is performed in a QIAquick spin column.

d   This method uses semi purified oocysts and no DNA purification. The presence of 0.5% SDS in the oocyst lysate is inactivated by the addition of 2% Tween 20 to the PCR mixture.

e   DNA bound to spin columns is washed twice (instead of the recommended single wash) with two different wash buffers to improve the purity of the DNA.

rise to the surface and can be skimmed out of the surface film. For a flotation fluid to be useful in diagnostics, the suspending medium not only must be heavier than the object to be floated but also must not produce shrinkage sufficient to render the object undiagnosable. Originally, brine, a concentrated aqueous NaCl solution, which has a sp. gr. between 1.120 and 1.200 depending on the impurity of the salt used, was employed, and ova of the common intestinal helminths, such as *Ascaris*, *Trichuris*, and the hookworms, are not damaged by this process. It is especially useful for hookworm ova. *Schistosoma* ova, the large operculated ova of *Diphyllobothrium*, *Fasciola* and *Fasciolopsis*, hookworm and *Strongyloides* larvae, and protozoan cysts become badly shriveled or open up in this flotation fluid. Additionally, ova of *Clonorchis*, *Opisthorchis*, and heterophyid species have a sp. gr. higher than 1.200 and therefore, do not float in brine. The optimal time to examine specimens obtained from brine flotation is between 5 and 20 min after flotation. For these reasons, other flotation fluids have been advocated (see below). Sucrose solutions are more satisfactory than brine solutions for protozoa because the latter tends to plasmolyze most cysts.

15.2.1.3 Fecal Parasites: Centrifugal Flotation

The $ZnSO_4$ centrifugal flotation technique was developed both to overcome the inherent problems of using brine and to attempt to provide an efficient concentration method for protozoan cysts, helminth ova, and larvae from stool in an undistorted condition. This method combines the principles of centrifugation and flotation. Large particles are removed by passing a fecal suspension through a sieve (see above) and suspended particles by decanting the supernatant following centrifugation. Parasites present in the pellet are resuspended in the flotation medium, and the suspension is recentrifuged. During centrifugation, objects denser than the medium settle to the bottom of the tube while objects that are less dense than the medium will rise to the surface. The meniscus is sampled for the presence of parasites. Organisms that rise to the surface may begin to sink after about 1 h, therefore the meniscus should be sampled before then. Prolonged exposure of cysts to $ZnSO_4$ can cause them to distort, making identification difficult, therefore preparations should be examined as soon as possible. Centrifugal flotation provides a sample with high parasite abundance and which is relatively free of contaminating particulate material. The ability to detect the parasites sought depends on the sp. gr. of the flotation medium employed. Most parasites, except for operculate ova, and those heavier than the floating medium, can be recovered efficiently in a viable condition. The most useful concentration of $ZnSO_4$ for floating the commonly encountered parasites has a sp. gr. of 1.180. A sp. gr. of 1.20 is recommended for formalinized specimens. $MgSO_4$ (sp. gr. 1.1–1.4) has also been used successfully as a flotation medium.

15.2.2 IMMUNOMAGNETIC SEPARATION

In immunomagnetic separation (IMS), surface-exposed epitopes on parasites bind to their complementary antibody paratopes, which are covalently linked to magnetizable beads. Once bound, the bead–parasite complex can be separated from its matrix by successive concentration with a magnet, which reduces particulates and PCR interferents [4,7]. Where necessary (Sections 15.5 and 15.6), the bead–parasite complex can also be dissociated with acid (pH 2.75). Paramagnetic colloidal particles and iron-cored latex beads can be used as the antibody binding matrix, and IMS [11] has been used extensively to concentrate *Cryptosporidium* oocysts and *Giardia* cysts [(oo)cysts] from various matrices including feces, water, and food (see below) prior to DNA extraction [12] (Sections 15.5 and 15.6). High turbidities, low pH, and particulates can reduce the efficiency of IMS. IMS is more sensitive than the biophysical/biochemical methods used to concentrate (oo)cysts, particularly when they occur at low abundance in food and environmental matrices, but it is expensive. IMS provides an immunologically based method for separating parasites from contaminating matrices containing PCR inhibitors, but currently commercial kits can only be used for *Cryptosporidium* and *Giardia* detection. Focused development of monoclonal antibodies reactive to exposed epitopes on the transmissive stages of other parasites should enable rapid nucleic acid extraction from a greater range of parasites.

15.2.3 Parasite Ova in Feces, Soil, and Wastewater

Flotation, sedimentation, and centrifugal flotation/sedimentation techniques have been used most frequently for feces and soils. Inconsistency of recovery has been widely reported, with matrix effects and treatment of large particulates exerting major influences. Various methods, based upon sedimentation, centrifugal flotation, or centrifugal sedimentation, have been used to recover helminth parasites from raw and treated wastewater samples [13–15]. The modified Balinger method, based on centrifugal sedimentation [9] and $ZnSO_4$ flotation, reliably recovers *Ascaris*, *Trichuris*, and hookworm ova [15]. Although effective for recovering geohelminth ova, the method is not suitable for many operculate or large ova (e.g., *Clonorchis sinensis*, *Paragonimus westermani*, *P. pulmonalis*, *Fasciola hepatica*, *Fasciolopsis buski*, *Diphyllobothrium latum*, *Schistosoma* spp.).

15.2.4 Options for Concentrating *Cryptosporidium* Oocysts from Feces

Partial purification of oocysts from positive fecal samples can be achieved by a combination of water–ether treatment [16] followed by flotation in sucrose, NaCl, or $ZnSO_4$ solution; however, parasite losses occur through successive centrifugations. Protocols for oocyst concentration methods such as sucrose and saturated salt flotation and information regarding oocyst purification can be found in Smith [4]. The efficacy of concentrating *Cryptosporidium* oocysts (and probably other intestinal parasites) from feces varies depending on the consistency of the stool samples: liquid > semisolid > solid [17]. This approach can be used to concentrate the transmissive stages of other protozoan parasites from stools. Without knowledge of such information for other parasites that are sought, many investigators resort to using unprocessed, comminuted, or homogenized stools to extract parasite DNA. Alternatively IMS (Section 15.2.2) can be used with stools containing *Cryptosporidium* and *Giardia* (oo)cysts, particularly when the sample is important.

The viability of (oo)cysts can be affected by formalin–ether purification, and flotation (salt, sucrose, $ZnSO_4$, etc.), isopycnic centrifugation (Percoll, Ficol–Hypaque, etc.), or IMS methods should be used to purify (oo)cysts for RNA analysis. Viable organisms are not normally required for diagnostic applications (e.g., epidemiological and environmental) but knowledge of the presence and integrity of their nuclei can prove useful and, for *Cryptosporidium* and *Giardia*, this can be determined by staining (oo)cysts with the nuclear fluorogen 4'6-diamidino-2-phenyl indole (DAPI), which intercalates with their DNA, highlighting nuclei under the UV filters of a fluorescence microscope [18,19]. DAPI has been incorporated into the standardized methods for identifying *Cryptosporidium* and *Giardia* in environmental samples (Sections 15.5 and 15.6). DAPI has also proven useful for determining the presence or absence of nuclei within (oo)cysts prior to PCR, particularly when (oo)cyst abundance is low, or when (oo)cyst positive samples fail to amplify [5,20].

15.3 PARASITE DNA PURIFICATION OPTIONS: THE PRINCIPLES

The extraction of nucleic acids from non transmissive stages of protozoan parasites is far more readily accomplished than from the transmissive stages of intestinal parasites because the former does not possess the robust cell wall of the latter which is required for survival in the environment. For this reason, many approaches have been devised to extract DNA from the transmissive stages of intestinal protozoan parasites. The physicochemical and immunological concentration techniques for intestinal parasites outlined in Section 15.2 increases parasite abundance, and, of importance, these concentration techniques provide sufficient parasites in a medium containing depleted PCR interferents which maximizes DNA extraction and purification.

15.3.1 MECHANICAL DISRUPTION USING FREEZE-THAW LYSIS

The nontransmissive stages of protozoan parasites suspended in a buffer suitable for DNA extraction can be readily disrupted by a few consecutive freezing and thawing cycles that disrupt the cell membrane and liberate nuclei and DNA into suspension. The environmentally resistant parasite life forms that possess a thick wall protecting the organisms in the environment are more resistant to disruption, and require more freeze-thaw cycles. Freeze-thaw lysis of parasites suspended in a buffer containing a suitable detergent (sodium dodecyl sulphate [SDS] or another nonionic detergent) aids cell disruption by solubilizing cell membrane lipids and denaturing proteins, causing them to lose their native shape. As freeze-thawing liberates nucleic acids into the surrounding medium, a nucleic acid purification step is often performed subsequently, which can be time consuming. Mechanical disruption methods (freeze-thawing and bead beating; Section 15.3.2) are frequently used to disrupt and release nucleic acids from the transmissive stages of intestinal protozoan parasites. Effective methods for oocyst/cyst/ova/larval disruption must be developed prior to, or in conjunction with optimizing DNA extraction methods.

15.3.2 MECHANICAL DISRUPTION USING GLASS, ZIRCONIA–SILICA, OR CERAMIC BEADS

Environmentally robust transmissive stages can also be disrupted mechanically using ceramic, glass, or zirconia–silica beads. Glass beads are useful for general applications while zirconia beads, which have a greater density and thus a higher impact power, are useful for more specialist applications. The nature of the material to disrupt dictates the beads that should be used: glass has a density of 2.5 g/cc (most commonly used for general bead beating), zirconia–silica has a density of 3.7 g/cc (used for spores and most tissues), and zirconia has a density of 5.5 g/cc (100% denser than that of glass beads, and used for tough tissues) (http://www.biospec.com/Beads.htm). The FastPrep System is a rapid sample homogenization method (bead beating; http://www.mpbio.com/demo_request. php) that can be used to purify nucleic acids from various lysed matrices including plant material, frozen mammalian tissue, spores, bone, and archived specimens. Samples are added to impact-resistant tubes containing a lysing matrix and buffer and the rotation of the machine rotor in a figure of eight vertical, angular motion causes the beads to impact the sample from all directions, simultaneously, releasing nucleic acids and proteins into a protective buffer. A silica-based nucleic acid purifying step can be included as can a chloroform extraction and ethanol precipitation step. The FastPrep FP120 bead-beater apparatus uses silica or ceramic beads to mechanically disrupt samples, supports beating speeds (maximum speed of the tube during vertical movement) between 4.0 and 6.5 m s^{-1} corresponding to 4200–6800 rpm, and allows simultaneous processing of up to 12 samples (FastPrep Manual, Bio101/Savant). A similar commercial apparatus is the Mini-Bead-beater (Stratech Scientific, Luton, United Kingdom) that gives a maximum rotation of 5000 rpm. Bead beating has been used successfully to release nucleic acids from protozoan parasites, particularly in stools, but the released nucleic acids require further purification prior to use in PCR-based assays, which can be time consuming.

15.3.3 DISRUPTION IN CHELEX 100

Chelex 100 is a chelating resin (styrene-divinylbenzene resin-containing iminodiacetic acid groups) that has been used successfully for nucleic acid sample preparation from small numbers of cells and organisms, forensic specimens, and paraffin embedded tissues [21–23]. Chelex 100 chelates by ion exchange and has high binding capacity for transition metals. Transition metals are those elements in the d-block (groups 3 to 12) of the periodic table, including zinc, cadmium, and mercury. Chelex 100 chelates heavy metal ions that act as catalysts for the degradation of DNA at high temperatures in low ionic strength solutions. The polar resin beads bind polar cellular components in cell lysates while the nonpolar nuclear DNA and RNA remain in solution in the liquid above the Chelex 100.

Eluting the resin with a small volume of 2 M nitric acid, which protonates the iminodiacetate groups, results in concentrating chelated metals in solution. Methods vary as to whether samples are boiled or not in Chelex 100, but Chelex 100 can also aid cell lysis during boiling, and, as it is water insoluble, it can be removed effectively from the sample by centrifugation, unlike high concentrations of EDTA (which could inhibit PCR). Chelex 100 also binds other PCR interferents, particularly from blood. Both the chelation of PCR interferents onto an insoluble matrix which can be separated by centrifugation and the lysis of cells during boiling provide obvious benefits for releasing nucleic acids into a matrix more amenable to PCR amplification.

Methods can vary slightly but consist of suspending a sample in 50–100 μL of 5% Chelex 100 (BioRad) and heating at 60°C for 10 min, followed by boiling for 10 min in a water bath or by simply boiling for 20 min, depending on the sample. Samples are then centrifuged at $10,000 \times g$ for 3 min and the supernatant can be used directly for PCR [23].

Chelex 100 can also be used in combination with polyvinylpolypyrrolidone (PVPP) (Section 15.4.1.3). Benefits of using the combination include chelating heavy metal ions which degrade DNA during boiling and removing polyphenolics and polysaccharides (derived from plant tissues, humic materials in soil, or from the breakdown of hemoglobin) which co–purify with DNA and inhibit PCR reactions, possibly through an adverse effect on *Taq* polymerase.

15.3.4 POLYVINYLPOLYPYRROLIDONE AND POLYVINYLPYRROLIDONE

Polyvinylpolypyrrolidone (PVPP) is exceptionally good at adsorbing polyphenols by hydrogen bonding. Polyphenols are common in many plant tissues and can deactivate proteins if they are not removed, thus inhibiting reactions like PCR. PVPP binds these PCR-inhibiting polyphenolics and can be used either for pretreating clinical, food, and environmental samples prior to DNA extraction or for the final DNA purification of crude extracts by centrifuging the extract through prepprepared 10% w/v PVPP spin columns. As PVPP is water insoluble, it can also be removed effectively from the sample by centrifugation. Fecal samples can be pretreated with PVPP prior to DNA extraction. Water soluble polyvinylpyrrolidone (PVP) can also be used for DNA extraction as it also adsorbs polyphenols. PVP of different molecular sizes, either of 20–30 kDa or 360 kDa can be incorporated as soluble components directly into PCR reactions or incorporated into DNA extraction methods. While the monomer is carcinogenic, polymer PVP is safe. As it is inert to humans, it is used in many technical applications in the pharmaceutical, food (as a food additive, PVPP is E1202), and personal care industries. PVP binds to polar molecules exceptionally well.

15.3.5 SILICA PARTICLES, GLASS MILK, AND DIATOMS

Glass milk or silica particles can be used to purify extracted nucleic acids as these particles bind to nucleic acids in the presence of high concentrations of chaotropic agents. The chaotropic salts and cellular components, which are not adsorbed to the silica, are removed by washing and centrifugation, then the nucleic acids are eluted from the silica. These methods provide nucleic acids in water or buffer, such as TE buffer (10 mM Tris–HCl, 1 mM EDTA, pH 8.0), containing far fewer salt or macromolecule contaminants that can interfere with further processing or PCR analysis. This approach has found many applications in molecular parasitology diagnostics and also forms the basis of numerous methods and commercial kits.

Boom et al. [24] used the chaotrope guanidinium thiocyanate (GuSCN), which lyses cells and destroys nucleases simultaneously, to develop a method for nucleic acid extraction and purification, which is now widely used. The method consists of an extraction procedure that uses the L6 buffer (10 M GuSCN in 0.1 M Tris–HCl (pH 6.4)- 0.2 M EDTA (pH 8.0), 2% (w/v) Triton X-100) to lyse cells and release their contents, including nucleic acids. The nucleic acids are then purified by the addition of activated silica particles or preferably, diatoms (unicellular algae which have silicic acid cell walls) that, in the presence of high concentrations of GuSCN, bind nucleic acids. Subsequently, two washes of the silica-/diatom-nucleic acid complex with L2 buffer (10 M GuSCN in 0.1 M Tris-HCl, pH 6.4) are performed by centrifugation to ensure the removal of the chelating agent and detergent from the complex, followed by two washes with 70% ethanol and one with acetone. The washed silica-/diatom-nucleic acid complex is dried (56°C, 10 min) and the nucleic acids are eluted into TE buffer. Nucleic acids are recovered in the supernatant after centrifugation and the silica/diatom particles are discarded.

15.3.6 EXTRACTION-FREE DNA PREPARATION ON FTA FILTER PAPER

Flinders Technology Associates (FTA) filter paper is a cotton-based, cellulose membrane containing lyophilized chemicals that lyse most bacteria, viruses, and parasites on contact, denature proteins, and protect nucleic acids from nucleases, oxidative and UV damage. The DNA is entrapped and stabilized in the fibers of the patented matrix. The coating contains denaturing agents, a chelating agent, and a free radical trap that together permits the long-term storage of double-stranded DNA at room temperature (RT) [25]. As RNA is chemically less stable than DNA, it is best analyzed immediately upon return of samples to the laboratory. Frozen storage aids RNA preservation. FTA Cards (Whatman) combine the concentration and purification of nucleic acids in one step. FTA Cards also prevent the overgrowth of bacteria and fungi, thus immobilized nucleic acids are stabilized for transport, immediate processing, or long-term storage at RT. Immobilized nucleic acid can be ready for further applications in less than 30 min. Storage at RT, before and after sample application, reduces the need for laboratory freezer space, facilitates sample collection in remote and humid locations, and simplifies sample transport. A key benefit of FTA Cards for clinical samples is that both the sample and the user are protected.

One procedure [26] recommends spotting samples onto FTA card and drying either at RT for 1 h or at 56°C for 10 min. Disks of 1.2–2 mm in diameter are then punched out of the card, purified by washing with FTA purification reagent, and rinsed with TE buffer. Disks are dried at RT for 1 h or at 56°C for 10 min and can then be used immediately in PCR assays [26]. Genomic DNA that remains unsampled on the filter can be reused for further PCR applications [27].

The benefits of commercially produced FTA Cards, filter paper disks, and the FTA Concentrator-PS Parasite Purification apparatus for molecular parasitology diagnostics include standardization of the matrix, their ease of use, the sensitivity of the procedure, the reduced preparation time, and increased quality assurance. FTA Cards are being adopted more frequently for nucleic acid extraction, purification, and storage in diagnostic parasitology laboratories, particularly for clinical and food applications. Although standardization is important, currently there is no evidence that they maximize DNA extraction from the small numbers of parasites found in food and environmental samples, and dilutions of known numbers of intact oocysts and cysts should be included regularly with each batch of tests to ensure sufficient quality assurance.

15.3.7 CETYL TRIMETHYLAMMONIUM BROMIDE

Cetyl trimethylammonium bromide (CTAB) is a cationic detergent that has been used to purify DNA from plants and protozoan parasites. It is particularly useful for extracting DNA from polysaccharide-rich cells and can be incorporated into other methods, or used to further purify DNA prepared using different methods. CTAB binds polysaccharides under precise conditions and the salt concentration (NaCl) should be maintained above 0.5 M, otherwise a CTAB–DNA complex will form. This method was useful for extracting *Entamoeba* spp. DNA from feces [28]. Standard DNA extraction methods often yield *Entamoeba* DNA that is refractory to restriction enzymes digestion and PCR applications possibly because of the glycogen stored in these parasites. A fast CTAB DNA isolation method was developed by Ali et al. [29] and was used to detect *Entamoeba histolytica* in feces, pus from amoebic liver abscesses, and xenic or axenic cultured trophozoites.

15.4 DNA EXTRACTION FROM CLINICAL SAMPLES

This section reviews published methods and presents in-house protocols for preparing samples suspected of containing parasites from clinical specimens for direct molecular detection. The most common method for extracting DNA from lysed parasites uses a lysis buffer whose formulation can vary but is usually Tris-based, pH 8.0, containing a chelating agent (usually EDTA) and SDS. Proteinase K (pK) is widely used to digest proteins (56°C, 3h or overnight) in this buffer. Following proteolytic digestion, pK can be heat inactivated, which allows the DNA extract to be used immediately. pK is a broad spectrum serine protease discovered in 1974 in extracts of the fungus, *Tritirachium album*. As pK remains active in the presence of chemicals that usually inactivate proteins, it is used commonly to destroy nucleases during DNA extraction. pK retains its activity in the presence of SDS (0.5%–1%), urea (1–4 M), chelating agents (eg, EDTA), and trypsin or chymotrypsin inhibitors. pK is activated by denaturants that unfold the protein which enables greater access to its substrate. pK activity is maximum at temperatures ranging from 50°C to 60°C but can be denatured by heat (90°C–95°C). pK is also stable over a wide pH range (4–12), with a pH optimum of 7.5–12 [30,31].

Although there are many commercial DNA extraction kits available, kits do not offer specific procedures for extracting DNA from different parasites. PCR inhibition has been addressed and proprietary reagents that counteract the action of inhibitors present in different clinical matrices are incorporated into commercial kits (e.g., the QIAamp DNA Stool Mini Kit uses inhibitEX-tablets; Qiagen) (Table 15.2). Different laboratories can use different in-house techniques for disrupting parasites and extracting DNA, which may not be optimized for use with commercial kit protocols. This makes data difficult to compare.

15.4.1 FECES

Fecal samples can contain many PCR inhibitors, which can vary according to the diet of the host. In addition to bilirubin and bile salts, complex polysaccharides are also significant inhibitors. There is no standardized method for extracting DNA from fecal parasites. Prior to adopting specific DNA extraction techniques in clinical laboratories, both the variability between methods and the recognized difficulties in amplifying nucleic acids from fecal specimens by PCR must be overcome. This section focuses on extracting DNA from *Cryptosporidium* oocysts (Table 15.3) and microsporidial spores, although methods specific for other intestinal protozoa, and those utilizing the increasingly popular filter paper approach are also included. As *Cryptosporidium* oocysts are very robust, they have presented the greatest challenges in maximizing the liberation of nucleic acids, which is why these methods have been included. A brief historical overview of method development is also incorporated.

15.4.1.1 Bead Disruption of *Cryptosporidium* Oocysts

A variety of bead types have been used to disrupt *Cryptosporidium* oocysts in fecal samples, and as this approach has proved successful with oocysts (~5 μm diameter), it is expected to prove useful for disrupting larger and less robust transmissive stages of intestinal (and other) parasites. A method combining oocyst disruption by bead beating and DNA extraction in guanidinium buffer followed by silica purification [24] (Section 15.3.5) was used to extract and purify DNA, which was then used to determine the number of *Cryptosporidium* oocysts in human stools [35]. Approximately 200 μL of whole feces were added to 900 μL of L6 buffer [24] (Section 15.3.5) [24] together with zirconia (0.3 g, 0.5 mm diameter) beads (Stratech Scientific, Luton, United Kingdom) and 60 μL of isoamyl alcohol. The tubes were shaken (5000 rpm [maximum speed], 2 min) in a Mini-Beadbeater (Stratech Scientific), left at RT for 5 min and centrifuged. DNA was purified from the supernatant using silica [24] (Section 15.3.5) [24] and PCR-negative samples were further purified with PVP [39].

This consisted of mixing 50 μL of extracted DNA with 150 μL of PVP-TE (10% w/v PVP in TE buffer) and incubating (RT, 10 min). The DNA was then concentrated by precipitation with 100 μL of 2 M ammonium acetate and 600 μL of isopropanol at −20°C for 30 min. DNA was recovered by centrifugation (11,000 × g, 10 min), dried and reconstituted in water for use in PCR reactions. This method of DNA extraction and purification provided a sensitivity of 97.91% (n = 211) using the *Cryptosporidium* 18S rRNA gene locus on microscopy positive samples [35].

da Silva et al. [34] used the FP120 FastPrep Cell Disruptor to disrupt *Cryptosporidium* oocysts in human fecal samples, followed by extensive preparation and purification using PVPP and reagents from commercial kits (FastDNA kit; BIO 101) and QIAquick PCR purification kit (Qiagen). Of 213 samples analyzed, 153 were *Cryptosporidium* positive by microscopy using a modified acid-fast stain, of which 24 (11.3%) inhibited PCR as determined by spiking duplicate samples. No correlation of results with oocyst abundance, determined by microscopy, was provided however, the authors reported that PCR detected a further 31 *Cryptosporidium* positive samples that were negative by microscopy, indicating the usefulness of molecular detection methods in the diagnostic scenario.

Siliconized glass beads can also be used to mechanically disrupt *Cryptosporidium* oocysts by shaking a fecal slurry containing oocysts or a partially purified suspension with 0.5–0.75 mm diameter siliconized glass beads (Philip Harris Scientific, Glasgow, United Kingdom) using the Pulsifier (Kalyx Biosciences Inc., Ontario, Canada). Beads were siliconized by immersion in Sigmacoate solution (Sigma SL-2), drained immediately, then transferred to a Petri dish to dry overnight in a 50°C oven. Volumes of ~100 μL of beads (equivalent to 134 ± 20.8 mg [n = 5]) were placed into a 1.5 mL microcentrifuge tube and dispensed into presterilized tubes for DNA extraction. The sample, containing partially purified oocysts suspended in 100 μL of either PCR or lysis buffer (LB; 50 mM Tris-HCl, 1 mM EDTA and 0.5% SDS, pH 8.0) containing 1% antifoam A, was added to the tube containing the beads, capped, and subjected to pulses equivalent to 3500 rpm, for 10 min in the Pulsifier [40]. Bead beating with siliconized glass beads was as effective at extracting DNA from partially purified fresh or aged oocysts suspended in LB as using 14 cycles of freeze-thawing followed by pK digestion [20] and each method detected DNA equivalent to two oocysts, the minimum number tested [40]. Extracting DNA by freeze-thawing, for small numbers of oocysts, is preferable to glass bead disruption as it avoids surface-bound losses and shearing of DNA during extraction (Section 15.4.1.2). Shearing of *Cryptosporidium* oocyst DNA extracted by high speed homogenization with zirconia beads in a mini-bead beater was reported to result in poor PCR amplification when extracts were subjected to prolonged mechanical agitation [35]. However, DNA extraction by bead beating can prove advantageous, particularly when oocysts are attached to debris (e.g., in fecal, food, and environmental matrices), or when DNA is extracted from heat inactivated or formalized oocysts.

Fedorko et al. [41] developed a PCR–RFLP (restriction fragment length polymorphism) assay targeting the 18S rRNA microsporidial gene to detect *Encephalitozoon cuniculi*, *Encephalitozoon hellem*, *Encephalitozoon bieneusi*, and *Encephalitozoon (Septata) intestinalis* (following digestion of the PCR product separately with *Pst*I and *Hae*III) using parasites cultured in monkey kidney (E6) cells and microsporidia-positive stool samples from HIV-positive individuals. Mechanical disruption was by bead beating. DNA was readily extracted from cultured organisms, but extracting DNA from spores in fecal samples required a laborious 4 day procedure employing both mechanical and chemical disruption methods. Cultured organisms were washed twice in PBS, and the DNA was extracted using bead beating, PC, and ethanol precipitation. PCR inhibitors were removed by diluting the fecal sample with an equal volume of 0.5% sodium hypochlorite, which was then centrifuged (15,000 × g, 5 min) and the pellet washed (three times) with PBS, resuspended in 200 μL of lysis buffer (10 mM Tris–HCl, 100 mM NaCl, 20 mM dithiothreitol, 2 mg of pK mL^{-1}, 250 U of lyticase [pH 8.0] mL^{-1}), and incubated for 15 min. Spores were mechanically disrupted (425–600 mm diameter glass beads [Sigma] and a mini-beater [Biospec Products]) for 2 min. Extracts were incubated (37°C, 18 h), 150 μL of 2% SDS and 200 μL of pK (2 mg mL^{-1}) were added and then incubated further (50°C, 17 h). DNA was extracted by PC, purified (GENECLEAN II Kit, Bio 101) and resuspended in 30 mL of TE buffer.

15.4.1.2   Freezing and Thawing of *Cryptosporidium* Oocysts

Freezing and thawing has been used to disrupt *Cryptosporidium* oocysts (and *Giardia* cysts) in fecal samples directly in lysis buffer. Disruption of the *Cryptosporidium* oocyst wall using several consecutive cycles of freezing and thawing is the preferred method for the release of *Cryptosporidium* sporozoite DNA [22,42,43,44,45]. However, protocols developed by different laboratories vary, especially with respect to number of cycles of freezing and thawing, the temperature for thawing, and the medium for DNA extraction. As the ability to disrupt oocyst walls varies between isolates, the age of the isolate, and the temperature used for thawing, Nichols and Smith [20] proposed an optimized method for freezing (liquid nitrogen [LN$_2$]) and thawing (65°C) *Cryptosporidium* oocysts suspended in PCR or lysis buffer consisting of 15 consecutive freeze-thaw cycles of 1 min each duration for freezing and thawing. Samples were vortexed for 10 s following each set of five cycles in order to increase the separation of oocysts from fecal particles and aid DNA extraction [17,46] (Protocol 15.1).

This procedure proved particularly useful in the authors' laboratory, with small numbers (<10) of *C. hominis* oocysts obtained following IMS to help identify the indicator case of a waterborne outbreak, which had been asymptomatic for days. It is also useful with small numbers of oocysts obtained from the environment or foodstuffs with unknown disruption characteristics, where a maximized method for disruption is essential for efficient PCR. In addition, maximizing oocyst disruption and sporozoite nucleus release is important when analyzing clinical samples which may contain a mixture of *Cryptosporidium* species/genotypes with different oocyst disruption properties. Heat inactivated *Cryptosporidium* oocysts (70°C for 30 min) [47] suspended in PCR buffer and freeze-thawed for between 4 and 14 cycles with thawing at 37°C; 65°C; 90°C ,or 100°C could not be disrupted efficiently [20], and since heat treatment is commonly used to render specimens safe for handling and transport, freezing and thawing oocysts suspended in a lysis buffer containing SDS (50 mM Tris–HCl pH 8.5, 1 mM EDTA, 0.5% SDS) is more efficient than in PCR buffer [20].

15.4.1.3   Removal of PCR Interferents and Extraction of *Cryptosporidium* Oocyst DNA

To identify *Cryptosporidium* species in human stool samples, Morgan et al. [48] used a combination of pretreatment (10 min boiling in PVPP [20 μL of fecal sample added to 80 μL of 10% PVPP]) and DNA extraction using proprietary buffers (Qiagen) and glass milk to reduce PCR inhibitors and provide more sensitive detection by PCR. Of 511 samples, PCR detected 36 positives, while microscopy detected 29 positives, but the additional PCR positives were later confirmed by microscopy [48]. Morgan et al. [48] found that PCR was more sensitive and easier to interpret, but was more time consuming and expensive than microscopy. Importantly, unlike microscopy, it could differentiate between different *Cryptosporidium* species and genotypes.

To detect *C. parvum*, *Entamoeba histolytica*, and *Giardia duodenalis* DNA simultaneously using multiplex real-time PCR, Verweij et al. [49] extracted and purified parasite DNA from human fecal samples (200 μL) by first mixing with equal volumes of 4% PVPP (Sigma-Aldrich), boiling (10 min), digesting with pK (55°C, 2 h) and extracting and purifying the DNA with the QIAamp Tissue Kit spin columns (Qiagen). A similar sample preparation and PVPP treatment was used to isolate and extract DNA from *Giardia duodenalis* cysts in human feces. Extracted DNA was used in a real-time PCR assay to detect *G. duodenalis* 18S rRNA in 102 of 104 microscopy-positive human stools (98% sensitivity) and from 10 cases that were *Giardia* antigen-positive but microscopy negative [50]. The same protocol containing PVPP was used to extract *E. histolytica* DNA [51]. Similarly, PVPP was used to extract DNA from *Oesophagostomum bifurcum* and *Necator americanus* [52].

A combination of 10% (w/v) each of Chelex 100 and PVPP was shown to be more effective than either component alone for purifying *Cryptosporidium* DNA from solid, human fecal samples by direct PCR, using an 18S rRNA gene locus (Nichols and Smith, unpublished). Approximately 1.0 g of

Cryptosporidium-negative feces seeded with 10^5 oocysts mL^{-1} was comminuted in 700 µL of water and concentrated using the mini variant of the water–ether concentration method [17] (Protocol 15.2). The supernatant containing semi purified oocysts suspended in lysis buffer was subjected to 15 freeze-thaw cycles to disrupt oocysts [20] (Protocol 15.1). Following pK digestion (200 µg mL^{-1}; 55°C, 3 h) samples were heated to inactivate pK, then equal volumes of DNA lysate and Chelex 100/PVPP were mixed, boiled (10 min), and cooled (RT, 2 min). After centrifugation (10,000 × *g*, 5 min) the supernatant was used for direct PCR. Chelex 100/PVPP treated samples, tested in triplicate, yielded higher concentrations of amplicons for both genomic *Cryptosporidium* DNA and a coamplified PCR internal control [53], compared to Chelex 100 or PVPP alone [54]. Similarly, the combination of Chelex 100 and PVPP was used with partially purified *Cryptosporidium* oocysts isolated from human fecal samples to detect *Cryptosporidium* DNA and used in PCR reactions without further purification. The combination of reagents provided economy of time since both chelation and the purification of samples from PCR inhibitors were achieved simultaneously.

Alternatively, water soluble PVP can be incorporated directly into PCR mixtures. Koonjul et al. [55] added water soluble PVP 25 kDa directly into the PCR to reduce the inhibitory effects of plant-derived polyphenolics that copurified with RNA. Lawson et al. [39] used PVP 25 kDa to further treat human fecal DNA samples that had been extracted and purified by the Boom method [24]. Nichols et al. [17] incorporated PVP 25 kDa at 2 mg mL^{-1} final concentration (stock solution in water, pH 8.0 adjusted with 0.5M NaOH) into a PCR mixture containing the *Cryptosporidium*-specific primers CPB-DIAGF/R, and noted that (1) DNA human fecal samples that were previously negative generated visible amplicons and (2) the inclusion of PVP 25 kDa generated higher concentrations of amplicons with samples that previously gave poor PCR yields. It is imperative to optimize the concentration of PVP for each primer pair used in PCR assays using known positive controls. Whereas 2 mg mL^{-1} could be used with *Cryptosporidium*-specific primers CPB-DIAGF/R, a concentration of 1 mg mL^{-1} was the maximum that could be used with the *Cryptosporidium*-specific primers that amplify the *Cryptosporidium* oocyst wall protein (COWP) gene [6]. Higher concentrations gave adverse effects [17].

15.4.1.4 Extraction of *Cryptosporidium* and Microsporidial DNA from Filter Paper

Orlandi and Lampel [56] used FTA Cards to prepare *C. parvum* oocyst DNA, *Enterocytozoon bieneusi*, and *Encephalitozoon intestinalis* spore DNA from both purified oocysts and spores and from fecal samples. Using the Johnson et al. [22] *Cryptosporidium* primers, they detected the expected 435 bp product from filters seeded with as few as 10 oocysts. They achieved similarly sensitive detection limits (10 spores) when *E. intestinalis*-spotted FTA filter templates were amplified with primers SINTF1 and SINTR, detecting the 520 bp product [57]. In fecal samples, 10–50 *E. intestinalis* spores could be detected when seeded in a 100 µL stool sample, with similar outcomes when urine and sputum were tested [56]. Orlandi and Lampel [56] also used FTA Cards to prepare *Cyclospora cayetanensis* oocyst DNA from partially purified oocysts and extracts from raspberries (and other fruit extracts) seeded with partially purified *C. cayetanensis* oocyst DNA (Section 15.5.1.2). Multiplex PCR amplification was reported for *C. cayetanensis*, *C. parvum*, and microsporidial DNA. Benefits of using FTA Cards included the simplicity of template production, the sensitivity of the procedure, and the reduced preparation time.

Subrungruang et al. [58] compared FTA filter paper, a QIAamp Stool Mini Kit, and a conventional phenol–chloroform (PC) method for detecting *E. bieneusi* spores in human fecal samples, using known concentrations (100,000; 20,000; 4,000; 800; and 160) of spores mL^{-1} of sample. Five sets of primers were compared (MSP3-MSP4B [59], EBIEF1-EBIER1 [60], Primer set 2 [61], Eb.gc-Eb.gt [62], and V1-Mic3 [63]). The FTA filter paper and a QIAamp Stool Mini Kit (Qiagen) were the most sensitive using the primer pairs EBIEF1–EBIER1 and MSP3–MSP4B (which amplify the intergenic transcribed spacer sequences and is suitable for species determination following sequencing), detecting a minimum of 800 spores mL^{-1} with a 100% sensitivity and specificity. Permanent staining followed by light microscopy gave a sensitivity of 86.7% and a specificity of 100% [58].

Carnevale et al. [64] extracted *E. bieneusi* DNA from feces applied to Whatman No. 1 filter paper disks. Feces was preserved in four different solutions (5% formaldehyde [volume ratio of stool to formalin = 1:3], 0.05% saline solution, 2.5% potassium dichromate, and merthiolate–formalin [volume ratio of stool for each = 1:2]). Homogenized samples were spotted onto Whatman No. 1 filter paper disks (1 cm diameter) either as suspensions in their respective preservatives or as 100 μL resuspended pellets following ethyl ether extraction, and the filter papers were stored in individual plastic bags at 4°C for at least 6 months. Filter paper disks were incubated (56°C, 2 h) in 500 μL of lysis buffer (100 mM Tris-HCl [pH 8.0], 100 mM EDTA [pH 8.0], 2% SDS, 150 mM NaCl, and pK at 200 μg mL^{-1}) and the DNA purified by PC extraction and absolute ethanol precipitation. DNA was resuspended in 10 μL of sterile redistilled water. DNA of 5 μL purified from each filter paper disk was used for PCR with the primers Eb.gc and Eb.gt, which amplify a 210 bp fragment of the unique *E. bieneusi* rRNA intergenic spacer sequence. Only DNA from concentrated fresh samples amplified in the first round of PCR, but a second round of PCR resulted in amplifying all non-formalin-fixed specimens, highlighting known issues with formalin preservation [65] (Section 15.4.7). More PCR product was present in fecal concentrates, and specific amplicons were detected in the second round of PCR amplification using DNA prepared from formalin-fixed samples that had been ethyl ether extracted.

15.4.1.5 Comparison of *Giardia* Cyst DNA Extraction Using FTA FilterPaper, QIAamp Stool Mini Kit, and Phenol–Chloroform

Nantavisai et al. [66] determined the efficiencies of the three DNA extraction methods, FTA filter paper (Whatman), the QIAamp Stool Mini Kit (Qiagen), and the conventional phenol–chloroform method of Hopkins et al. [67], using known numbers of *G. duodenalis* cysts in phosphate buffered saline (PBS), which were concentrated using saturated sodium nitrate flotation. Seeded FTA filter paper disks were air-dried overnight and a quarter of the disk was washed twice (15 min) with 200 μL of FTA purification reagent (Life Technologies, Gaithersburg, Maryland), then washed twice (15 min) with 200 μL of TE buffer and air-dried overnight. The washed, air-dried filter paper was used directly as DNA template in PCR amplification. For the QIAamp Stool Mini Kit (Qiagen), 200 μL of each sample was used for DNA extraction, following manufacturer's instructions and the extracted DNA was kept frozen at −20°C until used. PC extraction was performed according to Hopkins et al. [67]. The most efficient extraction method (determined using the RH11/RH4-GiarF/GiarR primer set and PCR conditions described by Hopkins et al. [67]) was FTA filter paper (detecting 168 cysts mL^{-1}), whereas DNA extraction using either the QIAamp Stool Mini Kit or PC detected 674 cysts mL^{-1} of diluted stool. Although the authors stated that the FTA filter paper assay was simple to use, easy to handle and transport, and could be used for large-scale investigations, the major disadvantage was that some parts of the disk may contain more DNA template than others, which can affect PCR amplification. Nantavisai et al. [66] recommended that at least two PCR amplifications per disk of FTA filter paper were performed.

15.4.1.6 *Cryptosporidium* DNA Extraction for Outbreak and Epidemiological Investigations

Rapid oocyst isolation and extraction methods are often required for outbreak investigations, particularly when nested PCR-based methods are used. Oocysts can be scraped from air-dried unfixed, unstained smears and modified Ziehl Neelsen (mZN), auramine phenol (AP) [68], and immunofluorescence (IF) [17,69,70] stained microscope slides, responsible for the original diagnosis, for subsequent DNA extraction. Amar et al. [69] developed a microscope slide scraping method using fecal samples that had been stored at 4°C for up to 2 years without preservatives. Methanol fixed smears were stained with mZN, AP, or IF and 2 weeks later, immersed in L6 buffer [24] (Section 15.3.5) to extract DNA. The fecal sample containing oocysts was removed by rubbing the stained smear vigorously with a sterile cotton swab then the head of the swab was placed into a microcentrifuge tube containing zirconia (0.3 g, 0.5 mm diameter) beads and disrupted in a Beatbeater-8 (Stratech Scientific) for 1.5 min at maximum speed. DNA was extracted with activated silica, purified by washing the silica twice with L2 buffer [24] (Section 15.3.5), twice with ice-cold 80% ethanol, and once with ice-cold acetone. Once dried (55°C, 10 min), the silica–water mixture was incubated (55°C, 5 min), and the supernatant (DNA sample) recovered by centrifugation. This procedure is similar to that used for disrupting *Cryptosporidium* oocysts using the FastPrep bead-beater apparatus (Section 15.4.1.1).

A mini-variant of water–ether purification [16] was developed to partially purify *Cryptosporidium* oocysts from human stools for rapid throughput of samples during outbreak or large-scale epidemiological investigations [17] (Protocol 15.2). Small volumes of stool comminuted in water (200 μL) were concentrated and delipidated by water–ether concentration in microcentrifuge tubes (total volume 1.2 mL). The pellet, containing semipurified oocysts, was suspended in lysis buffer (50 mM Tris-HCl pH 8.5, 1 mM EDTA, 0.5% SDS). DNA was extracted directly using an optimized freeze-thaw lysis treatment [20] followed by pK digestion and its subsequent heat inactivation. No DNA purification was necessary because the buffer composition was suitable for PCR reactions as long as Tween 20 was used in the PCR mixture to instantaneously inactivate SDS [5,17,20,46,53,71]. The sensitivity of the method using *Cryptosporidium* negative stool samples (*n* = 9) seeded with ~60 oocysts per 500 mg of feces was 100% using a nested 18S rRNA PCR assay, and when this method was used with 92 *Cryptosporidium* positive fecal samples from outbreaks, a sensitivity of 98.9% was obtained [17].

Protocol 15.1   DNA Extraction from Small Numbers of Partially Purified Oocysts

The following method is effective for extracting DNA from small numbers (~10+) of partially purified oocysts, and is used in the author's laboratory [5,20,46]. DNA is extracted following 15 cycles of freeze-thawing. Some oocyst isolates are more resistant to disruption by freeze-thawing than others, and, in order to ensure that DNA extraction is maximized, 15 cycles of freeze-thawing are recommended [5,20]. Of importance, *C. parvum* (Iowa isolate) oocysts, which are used by many researchers to develop molecular methods, or for positive controls, are very susceptible to freeze-thawing [20], and this should be noted particularly when small numbers of fecally derived or environmentally derived oocysts, which may be more resistant to freeze-thawing, are used.

In the authors' laboratory, a partially filled, wide-necked cryogenic reservoir containing liquid nitrogen is placed in proximity to a 65°C water bath. Utmost attention should be given to local safety codes of practice when handling liquid nitrogen. Tubes containing the samples are placed into a cut out, expanded polystyrene, drilled tube rack insert ensuring that the meniscus of each sample protrudes below the base of the insert. The tube rack insert should be of a lesser diameter than the neck of the liquid nitrogen reservoir so that it can be floated on the surface of the liquid nitrogen (and water) thus ensuring that the samples become immersed in each fluid. Placing both liquid nitrogen reservoir and water bath in close proximity to each other enables the operator to transfer the tube rack insert containing the samples safely and rapidly between the two fluids.

Reagents, supplies, and equipment:

Cold resistant (for cryogenic temperatures) and disposable gloves, 1.5 mL microcentrifuge tubes, expanded polystyrene tube rack drilled insert suitable for 1.5 mL microcentrifuge tubes,* cryogenic reservoir containing liquid nitrogen, water bath set at 65°C and 55°C, oven set at 90°C, microcentrifuge (10,000 × *g*) lysis buffer reagents (1M Tris pH 8.5; 0.5M EDTA; 10% SDS and DNase/RNase

---

* SDS is inhibitory to *Taq* polymerase at concentrations as low as 0.01%, therefore, it is necessary to neutralize the SDS present in the extracted DNA prior to PCR. The addition of 2% Tween 20 will instantaneously neutralize up to 0.05% SDS and is added to the PCR mixture. This can neutralize a volume of lysate corresponding to one-tenth of the PCR total reaction volume.

free water from Sigma-Aldrich; to prepare 10×lysis buffer, mix 5 mL of 1M Tris, pH 8.5, 0.2 mL of 0.5M EDTA and 5 mL of 10% SDS), pK (at a final concentration of 200 µg mL^{-1}), ice. Always include a positive control.

Procedure:

1. Suspend the partially purified oocysts in 90 µL of DNase/RNase free water, then add 10 µL of 10× lysis buffer in a microcentrifuge tube.
2. Cap and insert each tube fully into the individual holes of the drilled tube rack insert.[*]
3. When all the samples have been inserted into the insert, gently lower it onto the surface of the liquid nitrogen. [†,‡]
4. Leave to float on the liquid nitrogen for 1 min.
5. Gently raise the insert out of the liquid nitrogen and float it immediately onto the water in the 65°C water bath. Leave for 1 min.
6. Repeat this process a further 14 times. Make a note of each cycle of freezing and thawing. Vortex the samples for 10 s every fifth cycle, immediately after removing them from the 65°C water bath, then continue with the following freeze-thaw cycle.
7. Add pK (at a final concentration of 200 µg mL^{-1}) to each sample, recap each tube, and incubate for 3 h in a 55°C water bath.
8. Transfer microcentrifuge tubes to an oven set at 90°C for 20 min to denature pK.
9. Chill microcentrifuge tubes on ice for 1 min.
10. Centrifuge microcentrifuge tubes at 10,000 × *g* for 5 min.
11. Uncap tubes and remove ~70 µL of supernatant.[§]
12. Store each extract in a clean, capped, and labeled tube at −20°C until used for PCR amplification.

Protocol 15.2   Partial Purification of *Cryptosporidium* Oocysts for Rapid Throughput of Samples (Nichols et al., 2006a) [17]

Reagents, supplies and equipment:

Disposable gloves, 1.5 mL microcentrifuge tubes, bench top microcentrifuge (fixed angle rotor), water/vacuum pump, wooden applicator sticks, lysis buffer reagents (1 M Tris pH 8.5; 0.5 M EDTA; 10% SDS, and DNase/RNase free water [Sigma-Aldrich]). To prepare 1× lysis buffer, mix 5 mL of 1M Tris pH 8.5, 0.2 mL of 0.5 M EDTA, and 5 mL of 10% SDS and make upto 100 mL with DNase/RNase free water.

Procedure:

1. Add 100 µL of reverse osmosis (RO) water to prelabeled 1.5 mL microcentrifuge tubes. For solid stools, place ~0.5 g of stool sample on a wooden applicator stick into 1.5 mL microcentrifuge tube and comminute in 500 µL of RO water. For liquid or semisolid fecal samples pipette 200 µL or 100 µL, respectively, of feces into a microcentrifuge tube using a 1 mL automatic micropipette with its tip cut at an angle to enlarge the tip aperture.

---

[*] Tubes containing the samples are pushed fully into the tube rack insert ensuring that the meniscus of each sample protrudes below the base of the insert. The insert should be of a lesser diameter than the neck of the liquid nitrogen reservoir so that it can be floated on the surface of the liquid nitrogen (and water) thus ensuring that the samples become immersed in each fluid. The expanded polystyrene tube rack insert can be readily cut to shape to fit into the aperture of the cryogenic liquid nitrogen reservoir.

[†] Observe local safety codes of practice when using liquid nitrogen.

[‡] A pair of long forceps is suitable for transferring samples between each fluid.

[§] For important samples, a larger volume can be withdrawn as long as the pellet remains stable after centrifugation.

2. Make up to a total volume of 700 μL with RO water, cap each tube, and vortex for 30s.
3. Add 300 μL of ether and shake the tubes vigorously for 30s. Invert the tubes a few times during this procedure.
4. Centrifuge the tubes in the microcentrifuge (10,000× g, 1 min).
5. Aspirate the fluid above the fatty plug and the fatty plug to waste, gently, under slow suction using a water pump or equivalent.
6. Cap each tube and vortex until all the sediment is resuspended. Add 1 mL of RO water.
7. Wash twice by centrifuging (10,000× g, 1 min) then gently aspirate the supernatant to waste. Leave ~100 μL after each centrifugation.
8. Cap each tube, vortex well, and place 5 μL on a well of a 12-well slide for direct immuno-fluorescence and DAPI staining.* Conversely, prepare a smear on a slide for auramine phenol staining (optional)*a*.
9. Wash with 1 mL of 1× lysis buffer. Mix gently, centrifuge (10,000× g, 1 min) and repeat the wash in LB and centrifugation step, then aspirate the supernatant to leave ~100 μL of fluid. Cap the tube and resuspend by vortexing. This can be stored at 4°C or frozen until required for freeze-thawing.
10. Vortex the sample and freeze-thaw to extract *Cryptosporidium* DNA as described in Protocol 15.1.

15.4.1.7 Extraction of *Entamoeba histolytica* DNA Using CTAB

As standard DNA extraction methods often yield *Entamoeba* DNA that is refractory to restriction enzymes digestion and PCR applications (possibly because of the glycogen stored), a fast CTAB DNA isolation method was developed by Ali et al. [72] and was used to detect *E. histolytica* in feces, pus from amoebic liver abscesses, and xenic- or axenic-cultured trophozoites. Fresh or frozen feces of 0.1 g, 50 μL of fresh or lyophilized pus, or 50 μL of culture pellet were dispersed in 250 μL of lysis buffer (0.25% SDS in 0.1 M EDTA pH 8.0) and 100 μg mL^{-1} pK were added and incubated (55°C, 20 min) then NaCl was added to a concentration of 0.7 M and CTAB to a concentration of 1%. After mixing, the sample was incubated (65°C, 10 min), then extracted with equal volumes of chloroform and PC–isoamyl alcohol followed by ethanol precipitation. The dried DNA was resuspended in sterile distilled water and further purified by passage through a spin column (Microspin S-200, HR column; Amersham Pharmacia Biotech, Inc, United Kingdom). Fecal samples were further purified using a QIAamp DNA Stool Mini Kit (Qiagen). The extracted DNA was used for PCR amplification of short tandem repeats that are linked with the tRNA genes in the *Entamoeba* genome. Species-specific primer pairs were designed to differentiate between *E. histolytica* from *E. dispar* and revealed intraspecies PCR product polymorphisms that could prove useful for strain typing and determining associations between symptomatology and genotype [72].

15.4.1.8 Proprietary DNA Extraction Kits

The QIAamp DNA Stool Mini Kit (Qiagen) was used to detect *Cryptosporidium* DNA in stools. Stools were boiled in the proprietary buffer for 5 min followed by three cycles of freezing and thawing (LN$_2$, 1 min; boiling, 2 min), centrifugation, and the addition of inhibitEX-tablets (provided with the kit) to the centrifuged lysate. Following pK digestion at 70°C for 10 min (1 mg mL^{-1} final concentration), the sample was applied to a silica column, centrifuged (10,000 × g, 10 min, RT), and DNA was recovered using a proprietary buffer. A minimum of 5 × 10^2 oocysts per sample were detected using two-step nested COWP and 18S rRNA assays specific for *Cryptosporidium* [32].

---

* This step provides information on the abundance of oocysts in the sample and whether they contain sporozoite DNA. Commercially available FITC-labeled anti-*Cryptosporidium* monoclonal antibodies stain the oocyst wall apple green and DAPI, which intercalates with sporozoite DNA, stains each of the four sporozoite nuclei sky blue, under epifluorescence microscopy [4,18,19].

Although QIAamp DNA kits are widely used for DNA extraction from parasites, the procedures described to disrupt oocyst walls in the above example have not necessarily been adopted by other laboratories, making comparisons difficult, if not impossible.

15.4.2   Parasite DNA Extraction from Blood

DNA can be extracted from blood parasites present in whole blood, pelleted erythrocytes, buffy coats, blood smears, and bone marrow aspirates, depending on the parasite in question (e.g., *Plasmodium* spp., *Trypanosoma* spp., *Leishmania* spp., *T. gondii*, microfilaria, etc.). A variety of procedures have been developed for extracting DNA from whole blood, but sampling and storing blood on a filter paper matrix offers practical advantages for analyzing both parasite and human genes. Blood spots dried and stored on filter paper are useful for diagnostics, population screening, drug monitoring, and genetic analysis, and especially for molecular epidemiologic studies in remote, tropical areas, where transport and storage conditions are suboptimal. FTA, Whatman (Nos. 1–5, 3MM), and Schleicher and Schuell filter papers have been used to immobilize blood parasites such as *Plasmodium* species, as dried in whole dried blood spots for detection using PCR-based methods (e.g., Refs. [73–77]). These filter papers are particularly useful for collecting field samples, and for transporting and archiving sample material. Of importance, samples deposited onto filter paper should be protected from direct sunlight, which degrades DNA (unless FTA filter paper is used).

Clearly, the volume of blood deposited onto the filter paper determines the sensitivity of the assay, particularly when the sensitivity of PCR can be as few as one organism. The use of filter paper for low parasitemia studies (e.g., Ref. [78]) brings with it certain limitations including the small volumes of blood analyzed (10–15 µL) compared with whole blood (250 µL), the uneven distribution of DNA on the filter paper resulting in different DNA concentrations being retrieved from each punched disk, and the fact that using more than one disk per PCR test can inhibit the reaction [66,78].

Natural PCR inhibitors (protein, hemoglobin, iron) present in Guthrie cards stored for 1–30 months were analyzed under nondenaturing conditions using sodium dodecyl sulphate-polyacrylamide gel electrophoresis (SDS-PAGE) and quantitated by Makowski et al. [79]. PCR inhibitors became increasingly resistant to elution over time, remaining "fixed" in the cards. For blood spots stored for 1 month, 600 µg protein, 1.87 arbitrary units (au) hemoglobin, and 374 ng iron were solubilized, whereas only 137 µg protein (22%), 0.34 au haemoglobin (18%), and 147 ng iron (39%) were solubilized from 30 month blood spots. PCR inhibitor "fixation" was not a result of excessive desiccation. Albumin and two erythrocyte metal-containing proteins, carbonic anhydrase, and haemoglobin, were the major proteins characterized by SDS-PAGE. Makowski et al. [79] amplified two regions (98 bp and 491 bp amplicons) encoding the ΔF508 cystic fibrosis mutation, despite the presence of these "fixed" PCR inhibitors from Guthrie cards stored for up to 30 months, and suggested that nucleic acid also became "fixed" to the filter paper matrix accounting, in part, for the low DNA yield following microextraction methods.

15.4.2.1   DNA Extraction from Blood Spotted onto Filter Paper with Chelex 100/TE/Methanol Extraction

Shigidi et al. [75] placed excised Whatman 3 MM filter paper disks containing dried blood into tubes containing a preheated (100°C, 5 min in a heating block) 5% Chelex100 solution (180 µL in a 1.5 mL microcentrifuge tube). Tubes were vortexed and heated for a further 10 min and the supernatant obtained after centrifugation was used to detect *Plasmodium* DNA diversity in cases of cerebral malaria.

Bereczky et al. [76] compared Tris–EDTA (TE), methanol [80], and Chelex [81] based DNA isolation methods for genotyping *Plasmodium falciparum* from 15 samples stored on Whatman 3 MM filter paper and 15 samples stored on 903 Schleicher and Schuell (Dassel, Germany) filter paper using archived (1–2 years) samples. In the TE-based extraction method, each 4 mm filter paper punched disk was placed in a microcentrifuge tube, soaked in 65 µL of TE buffer, and incubated (50°C, 15 min). Each disk was pressed gently on the bottom of the tube several times and heated (97°C, 15 min) to elute the DNA, which was kept at 4°C or stored at −20°C. Methanol-based DNA extraction [80] entailed soaking individual filter paper disks in 125 µL of methanol, incubating (RT, 15 min), then removing the methanol, and drying the samples before adding 65 µL of distilled water. Disks were then mashed using a pipette tip and heated (97°C, 15 min) to elute the DNA. Chelex extraction [81] involved incubating individual disks overnight at 4°C in 1 mL 0.5% saponin in PBS, washing in PBS (4°C, 30 min) then transferring into new tubes containing 25 µL of stock solution (20% Chelex 100 and 75 µL of distilled water), and vortexing (30 s). Tubes were heated (99°C, 15 min) to elute the DNA, vortexed, and centrifuged at 10,000 × g for 2 min, then supernatants were transferred into new tubes for use as template. For Whatman 3 MM filter paper, the sensitivity of the PCR with primers M2-FCR and M2FCF [82] was 100%, 73%, and 93% for the TE, methanol, and Chelex methods, respectively, whereas for the samples that were stored for a longer period of time on 903 Schleicher and Schuell filter paper, the sensitivity was 93%, 73%, and 0%, respectively, for the TE, methanol, and Chelex methods.

Gonzales et al. [78] compared four different DNA extraction methods and four different primer sets with microscopy on 75 blood samples for detecting the salivary trypanosome *T. vivax* in the blood of a laboratory reared sheep that was experimentally infected with a bovine isolate. Whole blood, blood dried on filter paper, or Whatman FTA Cards were used to evaluate four DNA extraction methods for detecting *T. vivax* DNA by PCR and PCR outcomes were compared to microscopy. Whole blood (250 µL) was mixed with an equal volume of lysis buffer (0.31 M sucrose, 0.01 Tris-HCl, pH 7.5, 5 mM MgCl$_2$, 1% Triton X-100) and the mixture washed (three times with 500 µL of lysis buffer) by centrifugation (13,000 × g, 20 s). The pellet was resuspended in 250 µL of 1× PCR buffer and pK (10 mg mL^{-1}) and incubated (56°C, 1 h). pK was heat inactivated (95°C, 10 min) and the sample stored at −20°C until used. Two methods for extracting DNA from blood spotted onto filter paper were tested [83,84]. The method of de Almeida et al. [83] consisted of eluting two (6 mm diameter) punched disks in 1 mL of distilled water for 30 min with occasional inversion, and discarding 850 µL of supernatant following centrifuging (7800 × g, 10 min). Freshly prepared Chelex 100 (200 µL of 1% solution in distilled water) was added to the pellet and disks and stirred, incubated (56°C, 30 min), boiled (8 min), vortexed (2 min), and centrifuged (7800 × g, 5 min), then 100 µL of supernatant was stored (−20°C). The method of Boid et al. [84] (Section 15.6.5) consisted of extracting DNA from one punched (6 mm dia.) disk twice, firstly in 200 µL of water (37°C, 30 min) followed by centrifugation (7,800 × g, 10 min) to collect the supernatant (cold eluate) and a second elution in water (99°C, 30 min), followed by centrifugation (7800 × g, 10 min) to collect the supernatant (hot eluate), which was stored at −20°C until used in PCR. Disks (2 mm diameter) were punched from blood spotted on Whatman FTA Cards into a tube, washed twice (15 min each wash) with FTA purification reagent, twice with TE buffer, air-dried and used for PCR.

Four sets of primers were tested: ILO 1264/ ILO 1265 [85], TVW A/ TVW B [86], TV 80.24/ TV 322.24 [87], and ITS 1BR/ ITS 1CF [88]. DNA extracted from whole blood gave an estimated sensitivity of >90% with all primer sets. Using DNA extracted from FTA Cards, the highest sensitivity (93.24%) was with primer pair TVW A/TVW B and the lowest was with primer pair ITS 1BR/ITS 1CF. The highest sensitivities for all the primer sets tested occurred with DNA extracted from whole blood, while the lowest sensitivities occurred when DNA was extracted from filter paper preparations. DNA extracted from FTA Cards, when used with primer set TVW A/TVW B was recommended for the surveillance and diagnosis of *T. vivax* in remote areas [78].

15.4.2.2 DNA Extraction from Buffy Coat and Peripheral Blood

Reithinger et al. [89] found that the choice of DNA extraction method heavily influenced their ability to detect *Leishmania* (*Viannia*) spp. in dog blood and bone marrow. DNA was extracted using standard protocols with either PC, Chelex 100 resin, or the DNeasy DNA extraction kit (Qiagen).

None of the seeded samples extracted with Chelex could be amplified and PCR with PC extracted samples was between 2- and $>10^4$-fold more sensitive than reactions with DNeasy and Chelex–ethanol-extracted samples, respectively. The authors chose to use PC in the DNA extraction protocol for their field samples, as it was almost as good as the DNeasy kit in extracting parasite DNA from blood, but at a significantly lower cost.

Lachaud et al. [90] compared DNA extraction methods to detect *L. infantum* seeded into buffy coat and peripheral blood and visceral leishmaniasis in humans and dogs. Samples of peripheral blood collected into EDTA-coated tubes (WB) or the buffy coats of peripheral blood (BC) were seeded with live *L. infantum* promastigotes corresponding to DNA equivalents of 10, 1, 0.1, 0.05, and 0.01 parasite per PCR tube, respectively. Samples were lysed with either guanidine-EDTA (GE) or pK. For GE lysis, WB samples were centrifuged ($1600 \times g$, 10 min), two-thirds of the plasma was removed, and 1 volume of GE (6 M guanidine hydrochloride–0.2 M EDTA pH 8.0) added. BC samples were centrifuged ($1600 \times g$, 10 min), 500 µL of BC was transferred to another tube, and 1 volume of GE was added. Preparations were then incubated (≥ 2 days, RT), boiled (10 min), left between 1 and 7 days at RT, then stored at 14°C. For pK lysis, two volumes of TNNT buffer (0.5% Tween 20, 0.5% Nonidet P-40, 10 mM NaOH, 10 mM Tris pH 7.2) and pK (320 mg mL^{-1}) were added to 500 µL of WB or BC and incubated (56°C, between 2 and 24 h), boiled (10 min), and then stored at 14°C. For dog blood, the pK was raised to 960 mg mL^{-1}.

DNA was extracted with PC, silica beads (S; Organon Teknika,) or the DNeasy tissue kit (K; Qiagen, according to manufacturer's instructions, where lysis is pK based). Sterile distilled water (300 µL) was added to WB-GE and BC-GE lysates (200 µL) which were subjected to PC followed by chloroform extraction (FC). DNA was ethanol precipitated and resuspended in 150 or 200 µL of sterile distilled water for WB-GE and BC-GE, respectively. Assay sensitivity was improved if DNA extraction was performed immediately after boiling. Similarly, 500 µL of WB-pK or BC-pK lysates were subjected to FC, except that the ethanol precipitated DNA was resuspended in 130 µL of sterile distilled water. For dog blood, a phenol step was added before the simplified FC. The silica bead method (S) was part of a commercial kit using guanidine at high concentrations, and only suitable for GE lysates. GE lysates (200 µL) were mixed with 10 µL of silica beads, and processed according to the manufacturer's instructions. The final elution volume was 150 µL for WB and 200 µL for BC. The DNA target for PCR amplification was the 18S rRNA gene, a 20- to 40-fold-repeated sequence specific for the genus *Leishmania*, using the R221 and R332 primers, which produce a 603 bp fragment.

Following PCR optimization, specificity was 100% with all methods tested, and differences between the eight methods only became significant at low parasite concentrations (≤ 100 mL^{-1}). Using WB as template, WB-GE-S was best, particularly at very low concentrations, but at a concentration of 100 mL^{-1}, PK-FC and K produced more intense banding patterns on gel. Using BC as template, BC-PK-FC and K proved best at all concentrations, and they reliably detected 10 parasites per milliliter of blood. The optimized, maximized DNA extraction methods were also tested on samples from four AIDS patients and seven dogs diagnosed with visceral leishmaniasis. BC-pK-FC generated better results (number of positive reactions and banding pattern intensity) in two of the four samples from AIDS patients, and with dog samples, BC-pK-FC and BC-K methods were most sensitive.

15.4.3 Urine

15.4.3.1 Extraction of *T. vaginalis* DNA from Urine and Vaginal Swabs

T. vaginalis is conventionally detected in urine by *in vitro* culture and/or microscopy of wet mounts. The sensitivity of wet mounts is 60% and culture is 85%–95%. Comparison between microscopy and PCR suggests that PCR is more sensitive. Wet mount microscopy of vaginal discharges detected 16% positives ($n = 155$) whereas urine samples from the same women subjected to a direct PCR of a 112 bp fragment of the β-tubulin gene detected 48% positives [91]. The urine sample was centrifuged, the supernatant decanted, and lysis buffer (100 mM sucrose,

10 mM Tris, % mM MgCl$_2$, 1% Triton X-100) added to the pellet, which was then boiled. DNA was recovered for PCR amplification following ethanol precipitation.

Chelex 100 was used to extract *T. vaginalis* DNA from vaginal swabs or urine and proved to be a good substitute for the conventional culture method for detecting infections in women [92]. To extract *T. vaginalis* DNA from vaginal swabs, samples ($n = 378$; women attending obstetrics or infertility clinics) were placed in a tube containing 500 μL of buffer (0.01M Tris–HCl pH 8.0) and, after vortexing, a 100 μL subsample was transferred to a 1.5 mL tube. Cells were sedimented by centrifugation (12,000 × g, 15 min, RT), the supernatant discarded, and 200 μL of 5% Chelex 100 (w/v; in buffer) was added to each tube. After vortexing, the tubes were incubated (56°C, 45 min), with vortexing every 15 min. The samples were then boiled (10 min) and centrifuged (12,000 × g, 30s, RT) and a 5 μL portion of the supernatant was used for PCR to amplify a 312 bp amplicon of the 18S rRNA gene. For urine, 5 mL samples were centrifuged (3500 × g, 5 min) and the sediment washed twice in 0.01 M Tris–HCl buffer by sequential centrifugation. The final pellets were resuspended in 500 μL of 0.01 M Tris–HCl buffer and a 200 μL subsample of each was treated with Chelex 100 as described for the swabs [92]. The PCR was compared with paired samples of vaginal swabs used for *in vitro* culture and had an overall sensitivity and specificity of 100% and 98%, respectively. Using urine, the PCR had a sensitivity and specificity of 100% and 99.7%, respectively, when compared with culture of vaginal swabs, but the sensitivity dropped to 83.3% when compared with the PCR for vaginal swabs.

15.4.3.2 Extraction of *Schistosoma* spp. DNA from Urine and Feces

Schistosomiasis (bilharzia) is caused by any of five species of schistosomes and is commonly found in Africa, Asia, and South America, in bodies of water inhabited with its freshwater snail intermediate hosts. Each species causes different clinical presentations as the parasites localize in different parts of the body. Urine samples are routinely investigated for the presence of *S. haematobium* ova, since the adult worms reside in the venous plexus of bladder. *S. japonicum* adult worms reside in the superior mesenteric veins draining the small intestine and their ova can be found both in feces and urine. *S. mansoni* adult worms reside in the superior mesenteric veins draining the large intestine and eggs are excreted in feces. Concentration of urine by filtration (e.g., nucleopore 12 μm, 47 mm polycarbonate membrane) is necessary to detect *Schistosoma* sp. ova at low abundance. Ova are confirmed under the ×40 (dry) objective of a bright field microscope, once the whole membrane has been scanned under the ×10 objective. The sensitivity is ≥1 egg L^{-1} of urine. PCR has also been used to detect *Schistosoma* ova in urine. Sandoval et al. [3] extracted DNA from the urine and feces of *S. mansoni*-infected 7 week old female BALB/c mice using the NucleoSpin Trace and the Nucleo-Spin Tissue kits (Macherey-Nagel, Germany), respectively, following the manufacturer's instructions. The authors showed that their PCR technique was more sensitive than the two conventional diagnostic clinical parasitology methods (Kato-Katz fecal smear and indirect ELISA) for diagnosing schistosomiasis and that PCR could potentially be applied to detect acute human *S. mansoni* infection using noninvasive samples such as urine.

15.4.4 Respiratory Tract Samples

Samples from the respiratory tract are examined for the presence of parasites that cause respiratory diseases. Parasites can be resident in the lungs (e.g., *Pneumocystis*, *Cryptosporidium* spp.) or can use the respiratory tree as part of their migratory route (e.g., *Strongyloides stercoralis*, *Ascaris lumbricoides*, *Toxocara canis*, etc.). The conventional microscopic methods used for sputum are not recognized as having high diagnostic indices, except when used to diagnose fulminant disease.

Pneumocystis carinii (*jirovecii*) pneumonia (PCP) is a common opportunistic parasitic infection of the respiratory tract in HIV/AIDS patients, and sputum, pulmonary secretions, and/or pulmonary tissue obtained by bronchoalveolar lavage (BAL), transbronchial biopsy (TBB), or lung biopsy specimens are used to identify organisms (trophozoites and cysts). The sensitivity of BAL using standard cytological stains is 86%–97%. Because BAL is an invasive procedure, with associated morbidity, sputum is often preferred, but the sensitivity of detection in sputum is lower (55%–78%). Sputum can be induced (IS) following the inhalation of 3% saline solution using an ultrasonic nebulizer.

PCR detection in sputum and induced sputum was reported to be more sensitive than cytology for detecting *P. carinii* in patients with, or without high risk of, HIV infection [93]. Respiratory samples were collected in sterile containers and diluted in 70% ethanol. Sputum and IS specimens were treated with Sputolysin, and BAL sputum and IS samples were centrifuged and washed in Hanks balanced salt solution. DNA was extracted from the pellet by digesting with pK (5 mg mL^{-1}) and 10 µL used for PCR amplification. PCR was more sensitive than cytology, and detected *P. carinii* DNA in more bronchoalveolar washing and expectorated sputum samples. Nuchprayoon et al. [94] used the less invasive samples of IS and BAL to detect *P. jirovecii* cysts/trophozoites by cytology and PCR using FTA filters for sample collection and preparation together with a one-step PCR method (FTA-PCR) amplifying a target on the mitochondrial 5S rRNA gene. The sensitivity and specificity of the FTA-PCR method compared to microscopic examination were 67% and 90% for IS, and 67% and 91% for BAL, respectively [94], indicating that, for these clinical samples, cytology was superior.

15.4.5 CEREBROSPINAL FLUID

PCR is used to detect *Toxoplasma* DNA in CSF in cases of *T. gondii* suspected encephalitis (TE), pretransplant donor and recipient screening and ocular toxoplasmosis. Constrains include the invasiveness of the method, the limited quantity of sample, and the lack of a standardized method for specimen preparation prior to PCR. CSF can be used directly for DNA extraction from fresh samples, but some protocols centrifuge CSF to recover the cells which are washed in buffer by centrifugation prior to DNA extraction. Frozen CSF specimens have also been used successfully for DNA extraction and PCR. The specificity of PCR detection of *Toxoplasma* DNA (primarily detection of the repetitive B1 gene) in CSF, and the positive predictive value, is reported to be very high, but the sensitivity of detection is low [95–99]. Parmley et al. [95] extracted DNA from CSF that was frozen immediately, then transported on dry ice to the laboratory. Frozen CSF was thawed on ice and centrifuged (10,000 × *g*, 4°C) to pellet cells. The pellet was resuspended in water and heated (94°C, 10 min) to lyse cells and the lysate was used directly in a PCR amplifying the B1 gene. Four of nine TE cases were PCR positive, and no false positives were observed [95].

PCR was used to detect *T. gondii* in CSF specimens from 26 Japanese HIV-positive individuals presenting with focal neurological signs and a possible diagnosis of TE. Five milliliters of CSF was collected by lumbar puncture and kept at 4°C until DNA was extracted. CSF was frozen and thawed three times to disrupt parasites and DNA was extracted from 100 µL of fluid using a commercial extraction kit (SMI test EX-R&D [Sumitomo Metal Industries Ltd, Tokyo, Japan]) or QIAmp DNA Mini Kit (Qiagen) according to the manufacturer's instructions. The DNA was resuspended in 10 µL of water. Of 8 cases diagnosed with TE using accepted diagnostic criteria, PCR amplified the B1 gene target in CSF samples from 5 patients, whereas CSF samples from 18 patients without TE were negative for *T. gondii* DNA. The sensitivity, specificity, and positive and negative predictive values for detecting *T. gondii* in CSF using PCR were 62.5%, 100%, 100%, and 85.7%, respectively [99]. Although improvements in sensitivity are required, PCR can be useful clinically for detecting *T. gondii* DNA in HIV positive patients presenting with focal neurological signs.

A higher sensitivity of detection was obtained by Vidal et al. [100]. Sample preparation consisted of an initial centrifugation (3000 × *g*, 10 min) of the specimen to sediment cells followed by two washes of cells in PBS by centrifugation. Pelleted cells were resuspended in 50 µL ultrapure water with 20 µL of RNase mL^{-1} and lysed by heating (100°C, 10 min). Lysed cells were digested with pK (100 µg mL^{-1}) in 50 mM Tris–HCl pH 8.0, 25 mM EDTA, 2% SDS for 2 h at 56°C. DNA was extracted using PC-isoamyl alcohol, precipitated with ethanol, and washed with 70% ethanol for 10 min at 5000 × g. The DNA pellet was resuspended in ultrapure water with 20 μL of RNase mL^{-1}. Using this protocol and fresh CSF specimens from AIDS patients who did not have highly active antiretroviral therapy (HAART) treatment, a direct PCR amplifying a fragment of the B1 gene gave 100% sensitivity (12/12 PCR-positives from AIDS patients with cerebral toxoplasmosis) and 94.4% specificity (1/18 PCR-positive from AIDS patients with other neurological disorders) [100].

Since a fundamental pathological event in the development of TE is the switch from the bradyzoite stage to the tachyzoite stage of *T. gondii*, Contini et al. [98] detected these stages by nested PCR using stage-specific primers to amplify target sequences expressed on bradyzoites (SAG4 and MAG1), tachyzoites (SAG1) or both stages (B1). CSF specimens were obtained from 46 patients with AIDS, of whom 27 had TE (16 first episode, 11 relapse) and 19 had other AIDS-related brain lesions (AIDS-OBL) in the absence of TE. CSF specimens from 26 HIV-negative and immunocompetent patients were also tested. CSF specimens to be used for molecular analysis were stored frozen at −20°C and thawed on ice on the day of analysis, mixed with equal volume of lysis buffer (100 μL containing pK 200 μg mL^{-1} for 1 h at 56°C), and then heated (95°C, 10 min) and cooled at RT. DNA was extracted with phenol:chloroform:isoamyl alcohol (25:24:1) followed by precipitation with ethanol and resuspended in 50 μL of TE buffer. Using primers for the B1 gene, 75% of patients with first episode TE were positive, compared with 36.3% of those with relapse and 5.2% of AIDS-OLB. The SAG1 gene target was positive in 28.7% and 45.4% of patients with first episode TE or relapse, respectively, but not in the controls. Using the SAG4 and MAG1 gene targets, 72.7% of patients with TE relapse were detected, compared with 25% of patients with first episode TE and 5.2% with AIDS-OLB. None of the HIV-negative subjects was PCR positive, and these results demonstrate the usefulness of stage-specific detection (using the SAG4, MAG1, and SAG1 genes) for detecting relapse in AIDS positive TE patients when PCR targeting the B1 gene fails to detect *T. gondii* DNA [98].

15.4.6 PCR-BASED DETECTION OF BLOOD PARASITES USING SAMPLES OBTAINED FROM NONINVASIVE PROCEDURES

Animal models of human disease have demonstrated that PCR amplification of parasite DNA can assist the diagnosis of infection earlier than when using conventional diagnostic techniques (Section 15.1.1). In addition, the trend to avoid the use of invasive procedures for parasite detection wherever possible has resulted in alternative approaches to detecting blood and tissue parasites. The demonstration of parasite DNA in the urine of *S. mansoni* infected mice [3] (see above) before infection can be identified conventionally, offers further encouragement to investigate the usefulness of samples obtained by noninvasive procedures for detecting parasite-specific nucleic acids.

Urine was used to detect DNA from the blood parasites *P. falciparum* and *Wuchereria bancrofti* with encouraging results. *P. falciparum* DNA can be detected by PCR in urine and saliva [101] in cases of low parasitemia (708 μL^{-1} asexual parasites and 775 μL^{-1} sexual parasites) which were mostly asymptomatic (47 microscopy-positives and 4 microscopy-negatives following examination of thick blood film slides). The origin of the DNA detected in urine or saliva is unclear, but could originate from free molecular complexes released from lysed parasitized cells or traces of parasitized erythrocytes released into urine or saliva. DNA extractions from urine and saliva were compared using Chelex 100 and the Qiagen DNeasy kit. Whole urine or saliva (1 mL) was centrifuged (14,000 rpm, 3 min) and the sediment was resuspended and lysed in PBS/1% saponin solution by gentle tapping and vortexing, then left at RT for 20 min. Following lysis, samples were washed 1× in PBS, centrifuged (14,000 rpm, 2 min), and the supernatant discarded. 100 μL of 20% Chelex 100 was added to the pellet, which was boiled for 13 min and then centrifuged (14,000 rpm, 3 min). The supernatant was used in a nested PCR with *P. falciparum* MSP2 family-specific primers M2-FCR and M2FCF for MSP2 polymorphisms [82] and restriction fragment patterns of dihydrofolate reductase (DHFR) amino acid codon 59 [102]. The commercial kit was used according to the manufacturer's instructions on sediments of centrifuged samples of urine and saliva. Positive saliva or urine amplicons were the same as the corresponding types found in blood with the nested PCR *P. falciparum* MSP2 family-specific primers and the DHFR assay, however, variation was observed between patients. Of four microscopy negative samples, two were PCR positive. Amplicon yield was dependent on the DNA extraction method used, the parasite burden, and the primer sets used. The Qiagen DNeasy kit performed better than Chelex 100 for corresponding specimens; however, centrifugation after cell lysis and the subsequent wash in PBS may have contributed to DNA losses into the supernatant which was discarded. Amplicon yield in saliva was 1.6 times greater than in urine and shortest amplicons (PfDHFR primer set U 1–4) amplified better suggesting that DNA may be degraded in saliva or urine samples such that the chances of amplification are greater with short PCR fragments [101].

A sensitive and specific PCR based on a highly repetitive DNA sequence (188 bp; SspI repeat) was used to detect *Wuchereria bancrofti* DNA in blood and urine [103]. Five hundred microliters of plasma, thoroughly mixed with 500 μL PC (1:1) were centrifuged (5000 rev min^{-1}, 10 min) and the supernatant transferred to a fresh tube containing 100 μL of 8 M potassium acetate and 1000 μL of ice-cold absolute ethanol and then centrifuged (14,000 rev min^{-1}, 10 min). The pellet was washed with 70% ethanol, dried at RT in a vacuum centrifuge, resuspended in 25 μL of MilliQ autoclaved water, and used in the PCR. A 10 mL urine sample was precipitated by adding 20 mL of ice-cold absolute ethanol and 1 mL of 4 M sodium acetate, then centrifuged (5000 rev min^{-1}, 10 min). The supernatant was discarded and 1 mL of 4.5 M guanidine thiocyanate, 1.2% Triton X-100, 50 mM Tris–HCl (pH 6.4), 20 mM EDTA solution was added. DNA was purified by adding 40 μL of silica suspension for 10 min at RT and the centrifuged (5000 rev min^{-1}, 10 s) pellet was washed once with ice-cold absolute ethanol and dried by vacuum centrifugation. DNA was eluted by resuspending the pellet in 25 μL of MilliQ autoclaved water (42°C, 15 min) and following centrifugation (14,000 rev min^{-1}, 10 s) the supernatant was transferred to a fresh 0.5 mL microcentrifuge tube for use in the PCR. Samples were collected during the daytime from individuals in a *W. bancrofti* endemic area (Coque, Recife, Brazil). All microfilaraemic individuals were positive by PCR, irrespective of the sample used. PCR was also capable of detecting *W. bancrofti* DNA in amicrofilaraemic individuals: ~93% were positive by PCR when daytime blood samples were used and 59.7% when urine samples, collected at 0700 hours, were used. Thus, nocturnally periodic *W. bancrofti* infection can be detected in blood samples collected during the daytime, which is convenient for large-scale screening. In addition, noninvasive urine collection provided suitable samples for PCR, which is clearly advantageous for preliminary mass diagnosis.

15.4.7 DNA EXTRACTION FROM FORMALIN-FIXED FLUID SPECIMENS

Immersion in 10% buffered formalin is commonly used to preserve parasites in fecal specimens, for their transport from the field to the laboratory, for epidemiological studies, and for archiving. This preserves cysts, oocysts, and ova equally well [104]. Formalin fixation has a deleterious effect on DNA restricting its use in some PCR-based detection methods [105–108].

When axenically grown trophozoites of the human pathogen *E. histolytica* were suspended in either 1% or 10% formalin for 1, 4, and 7 days at 4°C, PCR amplification (amplicon size 1950 bp) was inhibited after 7 days in 10% formalin but not in 1% formalin. Exposure of naked DNA to the same treatment resulted in PCR amplification indicating that PCR failure was due to fracture of DNA during the extraction procedure (freezing in LN$_2$, followed by thawing at 37°C until the formalin-fixed trophozoites were lysed in 0.25% SDS, 0.1 M EDTA, pH 8). Trophozoites kept for 7 days in 10% formalin were difficult to break open. It was suggested that PCR failure occurred as a consequence of formalin-induced cross-linking of DNA and its associated proteins, which resulted in DNA chain breakage during commonly used methods for extraction [106].

Paglia and Visca [109] evaluated the QIAamp DNA Stool Mini Kit (Qiagen) for extracting DNA and detecting *E. histolytica/E. dispar* in formalin–ethyl acetate concentrated human fecal samples [8] stored at 4°C for a maximum of 90 days (samples were tested at 1, 7, 30, and 90 days). Prior to

DNA extraction, the residual volume of the concentrated fecal samples was centrifuged ($2650 \times g$, 5 min) and approximately 200 μL of sample was transferred to a 2 mL tube and washed twice, by centrifugation, with PBS (pH 7.2). Each sample was then resuspended in 1.2 mL of ASL buffer (provided with the kit) and extraction proceeded according to the manufacturer's instructions except for the sample incubation time in ASL, which was increased from 5 to 15 min at 70°C. A nested PCR assay for the SSU rRNA gene that amplified a 1076 bp fragment in both species in the first round of amplification and two different-sized fragments in the second amplification round (*E. histolytica* = 427 bp; *E. dispar* = 195 bp) was used. The absence of PCR inhibitors was confirmed by spiking DNA extracted from stools with unrelated human DNA, which was amplified using specific primers. The analysis of 30 formalin-fixed samples revealed cysts in 12/30 stools by microscopy and PCR yielding a 195 bp fragment characteristic of *E. dispar*. One formalin-fixed, microscopy positive sample (tested over a 6 weeks period following formalin-ethyl acetate concentration) was *E. histolytica* positive by PCR, with a similar amplicon yield over the 6 weeks period of testing [109].

Because of the deleterious effects of formalin on DNA, other preservatives for parasites, cysts, oocysts, and ova have been investigated. Suspension of *Cryptosporidium* oocysts in fecal specimens in 75% ethanol maintains oocyst morphology (as determined by morphometry) and DNA integrity. Complete amplification of the 2.2 kb *Cryptosporidium* thrombospondin-related adhesive protein (TRAP-C1) gene occurred with all 15 isolates tested, which had been stored at temperatures ranging from 22°C to 38°C for greater than 2 years. This procedure can be used in both temperate and tropical conditions when long-term storage of fecal samples is required [110].

15.4.8   DNA Extraction from Fresh or Frozen Biopsy Tissue

Skin biopsy tissues of suspicious cutaneous lesions are taken to assist in the laboratory diagnosis of cutaneous leishmaniasis (CL). Conventional CL diagnosis relies on microscopical examination of impression smears of skin lesions, taken from the boundaries between normal and affected skin, material taken from the edges of these lesions, and *in vitro* culture of infectious amastigotes which transform to promastigotes in *in vitro* culture. PCR-based diagnosis can be more sensitive and is more discriminatory when species identification is required [111–113]. Romero et al. [111] compared the sensitivity of PCR for diagnosing 35 untreated CL patients (caused by *Leishmania* (*Viannia*) *guyanensis* in the Brazilian Amazon) with *in vitro* culture and impression (imprint) smears. DNA extraction (RapidPrep genomic DNA isolation kits for cells and tissues [Pharmacia Biotech]) of biopsies incorporating the edges of recent active lesions that were previously stored at −70°C was performed according to the manufacturer's instructions. Extracted DNA was precipitated with isopropanol and eluted in 20 μL of TE. PCR was performed using primers that amplify the conserved region of the minicircle kinetoplast DNA (kDNA). PCR (100% sensitivity [95% CI from 90.0 to 100.0]) was as effective as microscopy of impression smears, but despite the high sensitivity of PCR, in this particular clinical setting of CL, the direct visualization of amastigotes on impression smears remains the method of choice, with PCR being reserved for patients with negative impression smear results.

Bensoussan et al. [112] analyzed 92 specimens from suspected cases of CL by *in vitro* culture, microscopy, and PCR. The authors compared the usefulness of needle aspiration samples for *in vitro* culture, impression smears from biopsy, and PCR-based detection using sterile Whatman 3 MM filter papers touched onto the cut edge of the lesion at the site of the biopsy. Air-dried filter papers were individually wrapped in aluminum foil and stored (4°C) prior to DNA extraction. A sample, cut from the filter paper, was incubated in 250 μL cell lysis buffer [114]. DNA extracted from lysates with PC was dissolved in 50 μL of TE buffer and analyzed by PCR. Either culture or microscopy alone detected 62.8% (49/78) or 74.4% (58/78) of positive samples, respectively, while culture and microscopy together improved overall sensitivity to 83.3% (65/78). Of three PCR assays tested, primers targeting kDNA (~10,000 minicircles per parasite) was most sensitive (98.7%, 77/78 of confirmed positive samples); however, six false positives were detected. The internal transcribed spacer 1 (ITS1) rRNA gene PCR (40–200 copies) had 91.0% sensitivity (71/78 positive,) and the spliced leader mini-exon (100 to 200 copies) PCR was least sensitive (53.8%, 42/78 positive). RFLP analysis of the ITS1 PCR product enabled identification of 74.6% of positive samples, which included strains of *L. major* (50.9%), *L. tropica* (47.2%), and the *L. braziliensis* complex (1.9%). Thus, PCR targeting kDNA is useful for diagnosing CL and an ITS1 PCR can be used reliably for diagnosing CL when rapid species identification is required.

15.4.9   DNA Extraction from Formalin-Fixed, Paraffin-Embedded Tissues

Formalin fixing and paraffin embedding of tissues or cells is a common method for preparing clinical samples for histological and immunohistochemical studies, and is a valuable resource as it represents a major component of archived material available to researchers. There is insufficient information available in the literature to recommend specific nucleic acid extraction and purification methods for parasite infected tissue, and the information presented below is meant to identify the best practice for extracting nucleic acid from formalin-fixed, paraffin-embedded tissues. The extraction of high-quality DNA from formalin-fixed, paraffin-embedded tissues is challenging and normally performed by processing sections of 5–10 μm thickness deposited on glass slides or in microcentrifuge tubes. Pretreatment to remove paraffin is often included and consists of one or two washes of sections in xylene either at RT or at 60°C followed by their rehydration in absolute ethanol (or gradual rehydration in increasing successive ethanol concentrations) followed by desiccation. The dried tissue is suspended in Tris–EDTA buffer, digested with pK, and the DNA is purified (Table 15.2). Paraffin removal is not always necessary. Boiling paraffin sections (single 5–10 μm sections) in 5% Chelex 100 solution was compared with pK digestion (200 μg mL⁻¹) for DNA extraction from normal tissues and archived biopsy specimens of colon cancer patients [23]. The chosen PCR amplified lengths of the p53 gene that varied from 80 to 214 bp equally well from both Chelex 100 and pK-digested DNA. Boiling in water degraded the DNA into 100–200 bp lengths, boiling in Chelex preserved the DNA from degradation and yielded fragments of 100–600 bp, however, boiling in pK (buffer 50 M KCl, 1.5 mMMgCl$_2$, 10 mM TRIS–HCl, 0.5% Tween 20, pH 9 at 25°C) produced DNA of increased molecular weight ranging from 100 to 10,000 bp with the majority being between 100 and 4,000 bp [23].

DNA from formalin-fixed, paraffin-embedded post-mortem archived material can only generate amplicons less than 90 bp by PCR due to the harsher conditions used for fixation. Bonin et al. [115] describe a simple treatment that enables the amplification of longer fragments by PCR. A pre-PCR restoration treatment by filling single strand breaks, followed by a vigorous denaturation step resulted in the amplification of 287 bp sequences of apoliprotein E and 291 bp of the prealbumin gene. A further explanation for PCR failures from formalin-fixed material is the blockage of polymerase elongation *in vitro* due to DNA cross-linking or an error prone trans-lesion synthesis across sites of damage, producing artifactual mutations during PCR [116]. This hypothesis was tested by sequencing PCR products (420–1007 bp) from fresh and fixed (1 to 7 days in formalin) paraffin-embedded human colon tissue. Compared to fresh tissue, formalin treatment generated less PCR product, and longer formalin contact (no PCR positives after 7 days) was associated with PCR failure which increased with larger PCR amplicons. In addition, significantly greater (three- to fourfold) mutations were observed with fixed specimens which suggests that formalin fixation produces random mutations that can be bypassed by *Taq* polymerase through error-prone trans-lesion synthesis [116].

The successes of DNA extraction from formalin-fixed, paraffin-embedded tissues and cells were shown to depend on the method of fixation and the age of the postfixed material. Buffered formalin (~37% w/v formaldehyde diluted x10 in phosphate buffer [3.7% formaldehyde, 29 mM NaH$_2$PO4, 45.8 mM Na$_2$HPO$_4$], known as 10% formalin) preserves DNA better than aqueous formalin. While fixing at RT for 24–48 h is standard, some laboratories may fix for longer periods of time or over weekends [107]. Fragmentation of DNA extracted from formalin-fixed, paraffin-embedded, oral inflammatory hyperplasia tissue samples was demonstrated on gel electrophoresis of genomic DNA extracted from archived material dating back 40 years, and the quality of DNA extracted from this material was best during the first decade of its storage [108]. The size of fragments depends on the conditions used to fix the specimen and the duration of contact with formalin, but sizes may vary from 300 to 400 bp for DNA and 200 nucleotides for RNA [107,108].

Coombs et al. [117] compared 10 protocols for DNA and RNA extraction from archived paraffin-embedded colonic tissue (1–30 years) using a combination of three methods for deparaffinization (xylene/ethanol, heating in a 650 W microwave oven for up to 45 s, and heating to 90°C for 10 min in a thermal cycler) and following digestion with pK (200 µg mL^{-1}, 3 h at 55°C), three purification methods were used for each sample (PC, heating [99°C, 10 min] in Tris–EDTA, heating [99°C, 10 min] in 5% Chelex 100 in Tris–EDTA, and a commercial kit [QIAmp DNA Mini Kit, Quiagen] following the manufacturer's instructions). The best method was using the thermal cycler and Chelex 100 for nucleic acid extraction which yielded PCR-amplifiable DNA from 61% of sections compared to 54% using microwave heating and Chelex 100, 15% using xylene, and 60% using the commercial kit. RNA was extracted effectively from 83.7% of sections using the thermal cycler and Chelex 100 method. RNA was also extracted efficiently from formalin-fixed, paraffin-embedded tissues using a denaturation solution that lyses tissue and rapidly denatures ribonucleases, based on guanidinium isothiocyanate (4 M guanidinium isothiocyanate, 0.25 M sodium citrate. 0.5% sarcosyl, 0.1 M 2-mercaptoethanol). Tissue sections of 1–6 µm were placed in a tube and 850 µL of the denaturing buffer and 250 µL of pK (20 mg mL^{-1}) solution in water, and incubated overnight at 55°C with vigorous agitation. The following day, samples were centrifuged (14,000 × g, 5 min, 4°C) and a white covering of paraffin formed on the surface of the solution. The digested material was transferred to a clean tube (avoiding the undigested sediment and paraffin) and the DNA extracted by PC and recovered by ethanol precipitation in the presence of glycogen. A detailed protocol for this technique is available [107].

DNA extraction-free methods for paraffin-embedded sections have the limitations of analyzing small areas (1–2 mm^2) of tissues, otherwise PCR inhibition is observed [118]. PCR interferents can be reduced using commercial kits. PCR amplification of small fragments of a *Leishmania*-specific genomic repetitive sequence (250 bp) [119] from paraffin-embedded skin biopsies of cutaneous lesions was used successfully to diagnose 19 German patients with cutaneous leishmaniasis who presented with unusual histopathologies and had been misdiagnosed. Total DNA was isolated (QIAamp DNA Mini Kit, Qiagen) from 15 (5 µm) sections. All were identified as *L. infantum* by sequencing [113]. While this substantiates the fact that PCR can be used with formalin-fixed tissues, both the amplification of small fragments of nucleic acid and the use of multicopy gene targets are the best options for successful outcomes.

Since formalin fixation of histological specimens causes DNA degradation and is inhibitory to PCR, Muller et al. [120] devised two quality-control-based PCRs to determine the likelihood of successful amplification using more specific primer sets. The target of the DNA substrate accessibility, quality control PCR for archived samples was the highly conserved α-actin gene sequence, common to human and animal *Leishmania* infections, using the α-ac1 and α-ac2 primers, which generate a 162 bp amplicon. A recombinant internal positive control was used to monitor possible sample-related inhibitory effects during PCR amplification. A total of 18 formalin-fixed samples from dogs with suspected or proven leishmaniasis were tested. Individual tissue sections were scraped and placed individually into a microcentrifuge tube containing 200 µL of digestion buffer (50 mM Tris–HCl, pH 8.5; 1 mM EDTA) for DNA extraction. The sample was incubated (95°C, 10 min) and then centrifuged (12,000 × g, 20 min). The paraffin ring that formed above the buffer was removed and the sample incubated (55°C, overnight) in pK solution (2 µL, 10 mg/mL). Following pK inactivation (95°C, 10 min) and centrifugation (12,000 × g, 10 min), the supernatant containing the extracted DNA was transferred to a clean tube and stored (−20°C) until used. A specific *Leishmania* PCR, amplifying a 250 bp of a repetitive DNA sequence [121] was used to determine the presence of all taxa of Old World *Leishmania* in these samples. Of the 18 samples, six were deemed PCR incompatible using the two quality-control-based PCRs, while 9 of the remaining 12 samples were PCR and immunocytochemistry positive. One sample that failed the DNA substrate accessibility, quality control PCR was positive in the *Leishmania* PCR.

Muller et al. [120] concluded that only if these two quality-control-based PCRs were included, could the specific PCR represent a reliable diagnostic tool that could supplement histological and immunohistochemical methods for diagnosing cutaneous leishmaniasis in formalin-fixed, paraffin-embedded skin biopsy material.

15.5 PARASITE CONCENTRATION AND DNA EXTRACTION FROM FOODS

The food-borne transmission of parasites is well recognized [122–128]. Mead et al. [129] reported that in the United States, an estimated 2.5 million (7%) food-borne illnesses were caused by parasitic diseases (300,000; 2,000,000; 225,000; and 52 for *C. parvum; G. lamblia* (= *G. duodenalis*), *T. gondii*, and *T. spiralis*, respectively), all with zoonotic implications. From 1993 to 1997, 19 food-borne outbreaks of parasitic origin occurred in the United States, with a total of 2325 cases reported [129]. Water contaminated with protozoan parasites is an important source of human infection not only following its direct consumption but also through using it in food processing or preparation. Water transports transmissible stages into drinking water supplies, recreational sites including fresh and marine waters and irrigation waters, which, in turn, can contaminate the food supply through agricultural and food industry practices from the farm to the fork. In addition to using parasite-contaminated water to irrigate crops, the food industry uses large volumes of water for its manufacturing and ancillary processes. Furthermore, consumer vogues such as consumption of raw vegetables and undercooking to retain natural taste and preserve heat-labile nutrients can increase the risk of food-borne transmission. Methods for concentrating protozoan parasites from water and extracting and purifying their DNA are presented in Section 15.6.

There are no standard methods for detecting the transmissive stages of protozoan parasites on foods, yet food-borne outbreaks of cryptosporidiosis, giardiasis, and cyclosporiasis have been documented both in developed and developing countries [123–127]. Published methods are primarily modifications of previously published methods utilizing steps and reagents devised for concentrating (oo)cysts from water [122–124] (Section 15.6). Similarly, methods for detecting ova as surface contaminants of food, such as salad vegetables, fruits, and herbs, are modifications of those used for feces and wastewater [7], and the addition of a detergent (e.g., Tween 20, Tween 80, Hyamine) discourages clumping and encourages the detachment of ova from the sample surface and other particulates. The transmission of (oo)cysts via food and water is enhanced by features such as zooanthroponotic and anthropozoonotic transmission, low infectious dose, monoxenous life cycle, the production of vast numbers of small-sized (oo)cysts, and their environmental and disinfection robustness.

Food-borne routes of *Cryptosporidium*, *Giardia*, and *Cyclospora cayetanensis* transmission include the consumption of raw (undercooked) vegetables, fruits, and cold drinks; the use of contaminated water used in food production; the consumption of oocyst- and cyst-contaminated shellfish, which filter and retain viable oocysts and cysts; and the contamination of foodstuffs by food handlers. A particular problem in investigating food-borne contamination with parasites is that foodstuff matrices are inherently variable, which poses problems for both conventional approaches to parasite concentration [130] and DNA extraction. Matrices implicated in food-borne transmission include nonalcoholic, pressed, apple cider; chicken salad; unpasteurized milk (food-borne cryptosporidiosis [125]) Christmas pudding; home-canned salmon; noodle salad; sandwiches; fruit salad; tripe soup; ice; raw sliced vegetables (food-borne giardiasis [126]) and raspberries; blackberries; blueberries; mesclun and mixed lettuce; basil; dill; chives; parsley; green onions; and snow peas (food-borne cyclosporiasis [127]). (Oo)cysts can also contaminate milk, juices, milk, other beverages, bottled waters, and potable quality water added without heating as a component of food [123]. Furthermore, oocysts and cysts can remain viable in marine and freshwater shellfish for protracted periods of time [125–127]. This diversity of matrices highlights the inherent problem of developing generic testing schedules as interferents in different matrices will influence recovery efficiency differently (e.g., Ref. [130]). As most methods for isolating (oo)cysts from foods are based on those used for water (Section 15.6; e.g., Refs. [131–133]) they have not been maximized for foods, even though the food matrix can often have a deleterious effect on recoveries [134–136] and PCR. The method developed by Cook et al. [134] addresses this issue specifically and is the only validated method available [135].

15.5.1 *CRYPTOSPORIDIUM* IN FOODS

Current methods use IMS to concentrate and separate (oo)cysts from the contaminating particulates and the inhibitory matrix. IMS has specific benefits not only for microscopic identification (e.g., vastly reducing particulates that can occlude (oo)cysts during microscopy) but also PCR, by capturing (oo)cysts in food samples and concentrating and processing them in a buffer free of PCR inhibitors, thus increasing the sensitivity of detection [12,37,137]. For IMS-based methods, the disruption of (oo)cysts prior to DNA extraction can be performed at one of two stages. If confirmation by microscopy is required, then (oo)cysts can be recovered from the semipermanent slide by removing the coverslip and scraping the sample from the well of the slide according to Nichols et al. [46]. If confirmation by microscopy is not required, then (oo)cysts can be disrupted either after their acid dissociation from IMS beads [138] or oocysts can be disrupted while still bound to IMS beads, potentially reducing oocysts losses [139]. Subsequent DNA extraction and purification can be performed using either commercial (e.g., High Pure PCR template preparation kit, Boehringer Mannheim; QIAamp DNA mini kit, Qiagen) kits [138,139] or in-house protocols [5,17,46]. The methods developed by Cook et al. [130,134] are microscopy based because amplifying naked DNA or DNA from morphologically unrecognizable life cycle forms is problematic in food and environmental samples, where the demonstration of an intact transmissive stage provides further confirmation. The authors have used this IMS/slide scraping/DNA isolation and extraction approach [46,134] successfully to detect as few as two *Cryptosporidium* sp. oocysts on salad products and herbs purchased at local markets.

Nichols et al. [5] described a nested PCR–RFLP method for detecting low densities of *Cryptosporidium* spp. oocysts in natural mineral waters and drinking waters, based on filtration, IMS, and oocyst filter entrapment, followed by direct extraction of DNA. DNA was released from polycarbonate filter-entrapped oocysts by disruption in lysis buffer using 15 freeze-thaw cycles, followed by pK digestion according to Nichols and Smith [20] (Protocol 15.1). Amplicons were readily detected from two to five intact oocysts, and the method consistently and routinely detected greater than or equal to five oocysts per sample.

An IMS pretreatment was used to concentrate *Cryptosporidium* oocysts from mussels prior to DNA extraction [138], which also reduces PCR interferents. Tissue was dissected from washed, trimmed, and dried mussels, which was chopped, then homogenized in a sterilized food blender. A portion of the homogenate (turbidity <2.5% ppv) was used for IMS, and IMS-purified oocyst suspensions were washed three times in double distilled water by centrifugation ($7000 \times g$, 10 min) and resuspended in 200 μL of lysis buffer (4 M urea, 200 mM Tris, 20 mM NaCl, 200 mM EDTA, pH 7.4), with 40 μL pK (2.0 mg mL^{-1}) for 1 h at 55°C. Following six cycles of freezing (LN$_2$, 2 min) and thawing (95°C, 5 min), the released DNA was purified using the High Pure PCR Template Preparation Kit (Boehringer Mannheim). PCR, using primers amplifying an 18S rRNA gene locus, detected *Cryptosporidium* DNA in 2 of 16 marine mussel samples used for human consumption (collected in Belfast Lough, Northern Ireland) [138].

Only three reports describe outbreaks associated with apple cider, one of which was associated with ozonated cider. The remaining contents of a jug of ozonated cider that a laboratory-confirmed case had partially drunk were concentrated by centrifugation, oocysts isolated by IMS and subjected to five freeze-thaw cycles, then incubated with 1 mg mL^{-1} pK (56°C, >1 h), and diluted with an equal volume of pure ethanol. Oocyst DNA was extracted by passing the oocyst–ethanol suspension through QIAamp DNA Mini isolate columns (Qiagen). The oocyst DNA was subtyped at the GP60 locus and yielded the same subtype of *C. parvum* (IIaA17G2R1) as did stools from this laboratory-confirmed case and from four other cases, all of whom drank this cider in the 2 weeks before their onset of illness.

15.5.2 CYCLOSPORA ON FOODS

No IMS method is available for concentrating *Cyclospora cayetanensis* oocysts from foods, and since berries (raspberries, blackberries, blueberries), leafy vegetables (mesclun and mixed lettuce), herbs (basil, dill, chives, parsley, green onions), and snow peas have been implicated in food-borne transmission, methods have been based on oocyst extraction from foodstuff eluates followed by oocyst disruption and DNA extraction. Foods, particularly soft berries, can hinder the detection of pathogens by PCR [140–142]. Acidity, from fruit extracts, and plant-derived polyphenolics and polysaccharides isolated during DNA extraction can inhibit PCR significantly [141,142], particularly when detecting small numbers of pathogens. Since the food-borne outbreaks of cyclosporiasis (caused by ingesting infectious *Cyclospora cayetanensis* oocysts) in the 1990s, much effort has been directed at developing sensitive PCR-based methods that can detect small numbers of oocysts in foods. As the foodstuffs implicated in food-borne outbreaks of cyclosporiasis are so numerous (see above; Ref. [127]), this variety brings with it matrix-specific problems with PCR interferents, which influence our ability to detect parasite nucleic acids reliably on such products.

Attempts to reduce such interferents, parallel those using other matrices, and include the adsorption of inhibitory substances from extracts with PVPP [34,141], binding DNA to a silica matrix in the presence of chaotropic reagents [34,143], template dilution [144], use of CTAB to eliminate polysaccharides during extraction [145], and extraction-free DNA preparation on FTA filter paper [56,146]. As for other matrices, the inclusion of these steps, alone or in combination, reduce not only PCR inhibition but also can reduce detection sensitivity.

In order to maximize *C. cayetanensis* oocyst isolation and DNA extraction from raspberries, Jinneman et al. [144] evaluated several approaches for PCR template preparation, including various washing and concentration steps, oocyst disruption protocols, resin matrix treatment, DNA precipitation, and/or the addition of nonfat dried milk solution to the PCR. The method they investigated consisted of washing the raspberry wash sediment (extract) in 1x PCR buffer I (Perkin-Elmer) by centrifuging (15,800 × g, 3 min) six times, adding a Chelex resin matrix (6% resin matrix, Instagene) to help reduce PCR inhibitors followed by six freeze-thaw cycles (LN$_2$, 2 min; 98°C, 2 min) to disrupt oocysts and release their DNA for PCR analysis. The resin matrix was removed by centrifugation (15,800 × g, 3 min). Ethanol precipitation of template DNA did not improve PCR sensitivity, but the addition of 2 μL (50 mg mL^{-1} solution) of nonfat dried milk solution to the template before the first round of the nested PCR reduced inhibition (representing a 400-fold increase in template volume above the maximum of 0.05 μL, which could be added previously to the PCR). PCR inhibition still occurred with these templates, and diluting the 50 μL template volume 1:1000 successfully overcame PCR inhibition, but reduced detection sensitivity as less template was included in the PCR. Approximately 19 *C. cayetanensis* oocysts per PCR were detected with this optimized template preparation method. The addition of 20 μL of raspberry wash sediment extract and nonfat dried milk solution did not inhibit the amplification of DNA from 25 *C. cayetanensis* oocysts in a 100 μL reaction PCR.

Steele et al. [147] evaluated the sensitivity of a PCR method for detecting *C. cayetanensis* oocysts seeded onto raspberries, basil, and mesclun lettuce. *C. cayetanensis* oocysts from human fecal samples were concentrated by water–ether extraction and sucrose density flotation, then washed in PBS and enumerated. Aliquots of 100 μL containing between 10 and 4000 partially purified oocysts were applied onto 100 g portions of raspberries, basil, and mesclun lettuce and air-dried at RT. Oocysts on basil and mesclun lettuce were extracted by washing for 2 min in a stomacher, whereas oocysts on raspberries were extracted by washing on a platform shaker (100 rpm, 20 min) in sealed stomacher bags. Oocysts were sedimented and washed in PBS by centrifugation (1800 × g, 5 min, RT), then disrupted in a volume of 2 mL in a beadbeater (500 rpm, 3 min) by mixing 500 μL of sediment with 1 mL of 600 μm glass beads in ASL buffer (QIAamp DNA Stool Mini Kit). After adding 100 μL of Instagene matrix (BioRad) to each sample, samples were subjected to three freeze-thaw (dry ice/ethanol bath; 100°C water bath; 2 min each) cycles, the lysed oocysts were centrifuged (12,000 × g, 3 min), and the supernatants extracted (QIAamp DNA Stool Mini Kit, Qiagen) to obtain template DNA. The assay detected 40 or less oocysts per 100 g of raspberries or basil, but only around 1000 per 100 g in mesclun lettuce and, although mesclun lettuce-specific PCR inhibitors were not detected, the authors emphasized the importance of testing PCR detection methods on different food types to determine how they affect PCR sensitivity.

In an attempt to standardize and maximize methods for detecting *C. cayetanensis* oocysts in foodstuffs, Orlandi and Lampel [56] used FTA Cards. As few as three oocysts were detected by PCR following their direct application onto FTA Cards. Eluates from berries seeded with *C. cayetanensis* oocysts were concentrated and applied onto FTA filters, washed in FTA purification buffer, dried (56°C) and 6 mm punched disks were used directly as template. FTA-PCR detected a DNA equivalent of 30 *C. cayetanensis* oocysts per 100 g samples of fresh raspberries [56]. Orlandi and Lampel recommended FTA-PCR for detecting various pathogens in foods, environmental samples, and clinical specimens.

Frazar and Orlandi [146] compared the FTA Concentrator-PS filter (Whatman) with DNA extraction using the MasterPure DNA kit (Epicentre) for detecting *C. parvum* oocysts in artificially contaminated foods (orange juice, apple cider, whole milk, strawberries, parsley, and lettuce) using varying numbers (5, 50, 500, 5000) of oocysts which were either seeded directly into 10 mL of liquid food or applied onto leaf surfaces and dried. Oocysts were eluted in BagPage filter bags containing NET buffer (100 mM Tris, pH 8.0, 150 mM NaCl, 1 mM EDTA) by gentle rocking (2 × 15 min) followed by centrifugation to reduce the volume to 10 mL for the IMS concentration stage. Following IMS, the acid dissociated oocysts were suspended in 10 mL of NET buffer and passed through the FTA Concentrator-PS filter under vacuum followed by two washes of the filter with diluted Tris-EDTA. Disks were dried in a heating block at 56°C and 3 × 6 mm disks were punched from the filter and used directly in the PCR. A greater variability occurred with the FTA Concentrator-PS filter concentrator compared to the DNA extraction kit when low concentrations of oocysts were seeded into the control buffer, and only 15% of low inoculum levels were detected. Oocysts were not detected in either the 50 or the 5 oocyst inocula. Although the FTA method was faster than using the DNA extraction kit, the pore size of the FTA paper (~25 μm) may have allowed oocysts (4.5–5.5 μm) to pass through the filters without lysing. The authors observed ~20% oocyst losses mainly during the initial filtration step in previous studies using this protocol. Sample turbidity and pH of the food matrices tested influenced the recovery of seeded oocysts by IMS, while PCR inhibitors influenced detection. Apple cider and orange juice produced low recovery rates [146].

Clearly, the sensitivity of PCR detection of *C. cayetanensis* on foods will be heavily influenced by the efficiency of the extraction procedure used to remove oocysts from foods, which will depend upon the food type analyzed. Published reports of PCR-based methods developed for detecting parasite DNA extracted from oocysts eluted from artificially contaminated foods indicate that they have similar sensitivities (30–50 *C. cayetanensis* oocysts per 100 g sample). Thus, the disparate oocyst and DNA extraction procedures used for the molecular detection of *C. cayetanensis* oocyst DNA appear to produce DNA template of similar qualities. It is clear from the examples above, that much effort has been focused on developing methods for extracting oocysts from artificially contaminated foods, some of which appear time consuming, and that if time is at a premium, FTA Cards or the FTA Concentrator-PS filter system should be considered. However, for *Cryptosporidium* detection, the observation that greater variability occurred with the FTA Concentrator-PS filter concentrator compared to the DNA extraction kit when low concentrations of oocysts were used, together with the fact that the pore size of the FTA paper may have allowed oocysts to pass through the filters without lysing them is of great concern.

The infectious doses for *Cryptosporidium* and *Giardia* in human volunteers is low (<10->1000) depending on the isolate [148–150]; and the current sensitivity of some PCR-based detection may not be sufficiently low to provide sufficient confidence for public health purposes [135]. This is likely to be the situation for *C. cayetanensis* PCR-based detection as well, as its infectious dose is also thought to be low (10–100 oocysts) [127,151]. Given the numerous *C. cayetanensis* outbreaks described [127], further emphasis should be placed on increasing the sensitivity of PCR-based detection methods, and importantly, greater focus should be placed on investigating the physico-chemical interactions between oocysts and the surfaces of food matrices, so that oocyst extraction from food surfaces can be maximized [134–136].

15.5.3 *Toxoplasma* in Foods

Aspinall et al. [145] investigated the frequency with which *T. gondii* DNA was present in 71 common meat products, and found *T. gondii* DNA (*SAG*2 locus) in 27 (pork, lamb, and beef products) samples. None carried the sulfonamide drug-resistant form of dihydropteroate synthase gene. To extract DNA, 1 g of each sample was cut into small pieces and ground thoroughly into a fine powder under LN_2. The powder was made up to 20 mL with sterile CTAB (Section 15.3.7) extraction buffer (100 mM Tris–HCl, pH 8.0, 1.4 M NaCl, 20 mM EDTA, 2% (w/v) CTAB) and incubated (60 min, 70°C) with frequent mixing by inversion. DNA was extracted using isopropanol precipitation of a chloroform extract and the pellet resuspended in 70% ethanol. The ethanol insoluble pellet was air-dried, then incubated (overnight, 4°C) in 1 mL of TE buffer and used in a nested PCR [152]. The authors confirmed that they had extracted DNA from the food samples by amplifying species-specific fragments of the mammalian mitochondrial cytochrome b genes [153].

15.6 EXTRACTION OF DNA FROM PARASITES FOUND IN THE ENVIRONMENT

Two areas of importance in environmental parasitology, which influence host development and demography, are water-borne parasites and invertebrate vectors of parasitic diseases. In recent years, the driving force behind method development has been the water-borne outbreaks of giardiasis and cryptosporidiosis. Many procedures have been published but, as yet, there is no universally accepted method. Although developed specifically for *Giardia* and *Cryptosporidium*, these methods should be useful for detecting the transmissive stages of other protozoan and helminth contaminants of water. As with foodstuffs, amplifying naked DNA or DNA from morphologically unrecognizable life cycle forms is problematic in water samples, where currently, the demonstration of an intact (and potentially infectious) viable transmissive stage is required. Methods for detecting parasite contamination of terrestrial environments, such as soils, are modifications or variants of those which are used for water, the focus of the modifications directed at addressing the removal of contaminating particulates in order to concentrate the target organisms. Here, the intention is to suspend the target organisms and to extract them into the liquid phase [7].

The identification of parasites in invertebrate life cycle stages is of paramount importance for epidemiological studies of major tropical diseases. Traditionally they are identified microscopically which, for arthropod vectors, requires extensive knowledge of the vector and parasite, the use of binomial identification keys, and dexterity in microdissection, none of which may be possible in field situations. Similarly, traditional methods for identifying freshwater snails which are intermediate hosts for human and livestock schistosomiasis, and whether they are infected, are difficult to undertake in field situations. Molecular detection offers practical advantages over traditional methods, and particularly for the snail intermediate hosts of human and livestock schistosomiasis, as they can determine whether snails are infected before the prepatent period and, hence, before they shed their infective stage (cercariae; Section 15.6.6).

Methods for detecting waterborne (oo)cysts consist of the following stages: (1) sampling; (2) elution, clarification, and concentration; and (3) identification. Such methods must be effective for a variety of matrices, including raw, potable, and wastewaters, in various countries and "standardized" methods, which are continually evolving, are available [7,12,154–159]. With the exceptions of feces and wastewater, the transmissive stages of gastrointestinal parasites tend to occur in low numbers in the environment; thus, methods appropriate for sampling large volumes of the suspected matrix are required. Two approaches to sampling have been promulgated by the U.K. and U.S. government regulators. In large volume water sampling, the sample is taken over a period of hours at a defined flow rate whereas, in small volume sampling, a volume of 10–20L is taken as a grab sample [156–159]. Protocols for sampling for protozoan parasites in water using small or large volume samples can be found at http://www.epa.gov/waterscience/methods/cryptsum.html and http://www.dwi.gov.uk/regs/crypto/legalindex.htm, Part 2). Methods for detecting helminth ova and larvae in the environment are traditionally those modified from methods used in clinical laboratories and have similar recovery efficiencies. Feces, soils, water and wastewater sludges, and foods are normally the matrices sampled [7].

In environmental (e.g., water) samples, humic or fulvic acids are the most commonly encountered PCR inhibitors [160,161]. IMS can reduce PCR inhibitors (e.g., clays, pH, humic and fulvic acids, polysaccharides and other organic compounds, salts and heavy metals, etc.) as well as other substances that co–purify with (oo)cysts and that are found in water concentrates [7]. As the methods above use IMS to capture (oo)cysts in crude samples and concentrate and process them in a buffer free of PCR inhibitors, they can increase the sensitivity of detection of parasite nucleic acids [12,22,162] (Section 15.5).

To determine and verify the presence of protozoan pathogens such as *Cryptosporidium*, *Giardia*, and *C. cayetanensis* in environmental samples for public health purposes, corroboration using as many different methods as possible increases the validity of the result. Obviating the need for a confirmatory method (microscopy) reduces available information (e.g., intactness and sporulation state for *C. cayetanensis*) and fails to identify whether PCR positivity might be the result of the presence of naked DNA in the sample.

As with food matrices, IMS purified oocysts can be separated from magnetizable beads by acid dissociation prior to DNA extraction [22,162]. Alternatively, dissociated oocysts can be placed on a membrane filter or microscope slide, and their morphology and morphometry determined by microscopy (following staining with FITC-C-mAb and the fluorogenic DNA intercalator, DAPI) prior to oocyst disruption and DNA extraction directly from the filter or from slide scrapings [5,46]. IMS-captured oocysts can be left attached onto beads during DNA extraction [163], which can increase the sensitivity of detection by reducing oocyst losses during acid dissociation; however, oocyst enumeration, oocyst morphometry, and assessment of oocyst integrity by microscopy is not possible using this method. Purified oocysts can be disrupted either by mechanical (glass bead beating) or physical (exposure to freeze-thawing cycles) injury to the oocyst wall to release sporozoites from which DNA is extracted.

15.6.1 Freeze-Thawing *Cryptosporidium* Oocysts Entrapped on Polycarbonate Membranes and Microscope Slides

As the standardized methods for detecting *Cryptosporidium* in water rely on IMS and confirmation by microscopy, Nichols et al. [5,46] and Nichols and Smith [20] extracted *Cryptosporidium* DNA from oocysts entrapped on either polycarbonate membranes [5] or on microscope slides following IMS [20,46] using a maximized freeze-thaw method (Protocol 15.1) in order to determine their species/genotypes/subgenotype at an 18S rRNA locus [22]. The method consistently detected small number of oocysts (≤2), validated with microscopy, and was more sensitive than when the Xiao et al. [164,165] 18S rRNA locus was used [46].

15.6.2   Disruption of *Cryptosporidium* Oocysts in Chelex **100**

Johnson et al. [22] mixed Chelex 100 solution (20% w/v in water) with oocyst seeded samples (1:5 ratio) and subjected the mixture to six freeze-thaw cycles (dry ice–ethanol bath for freezing; water bath at 98°C for thawing) for 1–2 min per cycle. After centrifugation, the supernatant was used in a direct PCR assay detecting the *Cryptosporidium* 18S rRNA gene, which is the most sensitive target known for this parasite (20 copies of the gene are present in one oocyst). While the sensitivity for purified oocysts was 1–10 oocysts per reaction following gel electrophoresis of the PCR product, some environmental samples required to be seeded with a minimum of 1000 oocysts before they were detected because of PCR interferents. IMS (Section 15.2.2) was used to concentrate oocysts from the inhibitory matrices in the water concentrates prior to freeze-thawing in Chelex 100 [22]. Similarly, Mahbubani et al. [166] detected *G. muris* cysts seeded into turbid river water concentrates by amplifying a 171 bp region of the giardin gene in a direct PCR, with a sensitivity of 3–30 cysts mL^{-1} using similar protocols (DNA release by freeze-thawing in Chelex 100 after separation of cysts from the matrix using IMS).

15.6.3   Disruption of *Cryptosporidium* Oocysts Chelex and **PVP360**

Guy et al. [167] designed a multiplex qPCR assay for the simultaneous detection of *Cryptosporidium* (targeting the COWP gene) and *Giardia* (β-giardin gene) in environmental (water and sewage) samples. The optimized DNA extraction procedure involved three freeze-thaw cycles (LN$_2$, 2 min followed by boiling) and sonication (three 20 s bursts) and the use of the tissue protocol for the DNeasy kit (Qiagen) (incubation in the proprietary ATL buffer with pK for 1 h) followed by separation on a silica gel column. Chelex 100, tested at a concentration of 5%–20%, had no effect on PCR inhibition in environmental samples; however, when 20% Chelex and 2% PVP 360 kDa were added during the extraction process, inhibition was removed. PVP 40 kDa was less effective at removing inhibitors from one of the environmental water samples tested as it probably contained more polyphenolics.

15.6.4   Proprietary DNA Extraction Kits Used
for *Cryptosporidium* Oocyst DNA Extraction

Standardized detection methods for *Cryptosporidium* oocysts in water samples rely on the isolation and concentration of oocysts from the inhibitory matrix by IMS prior to PCR, although there is little evidence to prove that IMS (or any other antibody-based purification step) can concentrate all *Cryptosporidium* species/genotypes/subgenotypes from water (or other matrices) equally as effectively. Jiang et al. [168] compared the effectiveness of six DNA extraction methods for extracting and purifying *Cryptosporidium* DNA. DNA extracted with the QIAamp DNA Mini Kit (Qiagen) following IMS concentration was compared with direct DNA extraction methods using the FastDNA SPIN kit for soil (Q.BIOgene), QIAamp DNA Stool Mini Kit (Qiagen), UltraClean soil kit (Mo Bio Labs), or QIAamp DNA Mini Kit (Qiagen) and PC using samples seeded with oocysts, DNA seeded samples, and environmental water samples. To control inhibition, nonacetylated bovine serum albumin (BSA), the T4 gene 32 protein, PVP, GeneReleaser (BioVentures), and ultra filtration (Microcon PCR reservoir; Amicon) were tested. PCR inhibitors occurred in DNA obtained using all direct extraction methods and the effect of PCR inhibitors could be relieved significantly following the addition of 400 ng mL^{-1} of BSA or 25 ng mL^{-1} of T4 gene 32 protein in the PCR reaction. When BSA was included in the PCR mixture, DNA extracted with the FastDNA SPIN kit for soil (Q.BIOgene), without IMS, resulted in PCR performance similar to that produced using the QIAamp DNA Mini Kit (Qiagen) following IMS.

15.6.5   Detection of Viable *Cryptosporidium parvum*
and *Giardia duodenalis* (Oo)cysts in Environmental Waters

Viable organisms can respond to external insults in their environment by producing increased amounts of messenger RNA (mRNA), and this has been used as a surrogate to determine *Cryptosporidium* and

Giardia (oo)cyst viability, with mRNAs of heat shock proteins being especially targeted. Separate reverse transcription-polymerase chain reactions (RT-PCR) to amplify a sequence of the mRNA for *Giardia* heat shock protein [169,170], (and a sequence of the mRNA for *C. parvum* heat-shock protein 70 (*hsp 70*) have been developed [171]. Stinear et al. [171] collected 20 L grab samples of environmental waters, expected to contain higher levels of PCR inhibitors than highly treated waters, mainly in the form of humic acids, fulvic acids, salts, and heavy metals, which were concentrated to 10 mL by calcium carbonate flocculation [172] and Percoll–sucrose density centrifugation. One milliliter volumes of seeded concentrate were centrifuged ($5000 \times g$, 3 min) to pellet oocysts, resuspended in 200 mL of InstaGene matrix (Bio-Rad), vortexed briefly to resuspend the pellet, and then incubated (45°C, 20 min) to induce production of *hsp70* mRNA. Then, 200 mL of lysis-binding buffer (100 mM Tris–HCl [pH 8.0], 500 mM LiCl, 10 mM EDTA [pH 8.0], 1% lithium dodecyl sulfate, 5 mM dithiothreitol) was added to each sample, which was subjected to five freeze-thaw cycles (LN_2, 1 min; 65°C, 1 min), centrifuged ($17,000 \times g$), and the supernatant transferred to a clean microcentrifuge tube containing the prepared oligo(dT)$_{25}$ beads (Dynal, Norway), essentially according to the protocol of the manufacturer. Hybridization was performed at 30°C with gentle mixing by rolling for 30 min. Beads were washed once with 200 mL of wash buffer (10 mM Tris–HCl [pH 7.5], 0.15 M LiCl, 1.0 mM EDTA) and once with 200 mL of 13 PCR buffer (10 mM Tris–HCl [pH 8.3], 50 mM KCl) using a magnetic particle concentrator. All traces of liquid were removed after the final wash, and the beads resuspended in 3 mL of pyrocarbonic acid diethyl ester (DEPC, diethyl pyrocarbonate)-treated water. To elute mRNA, the magnetic beads and hybridized mRNA were resuspended in elution buffer (2 mM EDTA) and incubated at 65°C for 2 min, then beads were pelleted in the magnetic concentrator, and the supernatant containing the mRNA was removed and used in RT-PCR. An RNA internal positive control was developed and included in each assay to safeguard against false negative results caused by inhibitory substances. Following RNA extraction and hybridization, the RT-PCR detected a single viable oocyst in all water types tested. Simultaneous detection of viable *Giardia* cysts and *C. parvum* oocysts from water and wastewater using RT-PCR was described by Kaucner and Stinear [170] using a similar mRNA extraction protocol to that described by Stinear et al. [171]. *Giardia* mRNA was detected in treated sewage effluent (0.8–1.0 NTU) and raw water (3.5–120 NTU), and *Giardia* and *Cryptosporidium* mRNA were detected in raw water concentrates. Sensitivity for both organisms was reported in the range of 1–2 viable organisms and, in a comparison with the immunofluorescence method for identifying (oo)cysts in water concentrates, the frequency of detection of viable *Giardia* cysts rose from 24% to 69% with RT-PCR. For *Cryptosporidium*, RT-PCR detected oocysts in one sample as compared with four samples by fluorescence microscopy. As the authors designed a *C. parvum*-specific *hsp70* assay, they suggested that the difference was due to the inability to distinguish oocysts of *C. parvum* from oocysts of other *Cryptosporidium* species with fluorescence microscopy.

15.6.6 EXTRACTION OF *CYCLOSPORA CAYETANENSIS* DNA FROM ENVIRONMENTAL WATER SAMPLES

Shields and Olson [173] developed a DNA extraction and purification protocol for extracting *C. cayetanensis* DNA from 10 L grab samples of environmental water samples flocculated according to Vesey et al. [172]. DNA in flocculated pellets was extracted according to da Silva et al. [57] with modifications. The amounts of digestion buffer (50 mM L^{-1} Tris–HCl, pH 8.5, 1 mM L^{-1} EDTA, 1% lauryl alcohol polyether [Laureth 12]) pK (final concentration 1 mg mL^{-1}) and silanized glass beads (Sigma G-9139, trimethylsilyl-silanized, 140–270 mesh) were adjusted dependent on the size and nature of the pellet. The 2:1 ratio of digestion buffer to packed pellet was increased if the pellet was difficult to resuspend and an equal weight to volume ratio (mg per mL) of silanized glass beads was used. Pellets were digested overnight, samples were centrifuged ($14,000 \times g$, 1 min), PC extracted, and the resultant aqueous layer was precipitated with ethanol–ammonium acetate, then washed with cold 70% ethanol. Samples were resuspended in 100 μL of sterile double distilled water. Chelex 100 was added at a concentration of 5%, and the samples boiled (2–3 min). After centrifugation, supernatants were further purified through a Pharmacia ion-exchange spin column (Amersham Pharmacia Biotech and Science) according to manufacturer's instructions. Samples were resuspended in 100 μL of sterile double distilled water and kept either at −20°C until PCR analysis or at −80°C for long-term storage. Five environmental water samples were PCR–RFLP positive for *C. cayetanensis*, but could not be confirmed by microscopy. Using *C. cayetanensis*-specific primers (CYCAO1, CYCAI2, and CYCAR1) none of the five environmental water samples could be confirmed as *C. cayetanensis*, but the authors calculated that the minimum detection limit was 0.75 oocyst (1.04 ng of genomic DNA as template), assuming 100% recovery of DNA from their seeded environmental sample.

15.6.7 SOIL

(Oo)cysts that contaminate land and soils can become potential reservoirs for (oo)cyst contamination of water following heavy rains or snowmelt through surface runoff or percolation through soils. Walker et al. (1998) seeded a *Cryptosporidium* oocyst-free Collamer silt loam with varying numbers (0–10,300 g^{-1}) of *C. parvum* oocysts and compared conventional detection by fluorescence microscopy (Section 15.6) with PCR. Oocysts were concentrated from Collamer silt loam using a combination of detergent (0.1% Tween 20) extraction, vortexing, centrifugation, gyratory shaking, sucrose flotation, and the final pellet, containing the oocysts, were resuspended to 1 mL. To 2 mL microcentrifuge tubes containing 2.5 g of sterilized 0.1 mm diameter zirconium beads (BioSpec Products) were added 300 μL of 100 mM $NaPO_4$ (pH 8.0), 300 μL of lysis buffer (100 mM NaCl, 500 mM Tris [pH 8.0], 10% sodium dodecyl sulphate), and 300 μL of phenol (equilibrated, pH 7.8), and the soil sample concentrate. The mixture was homogenized by bead mill, centrifuged, and both phenol and aqueous phases were collected. A second rinse of 300 μL of distilled water was collected, and the extracts were concentrated using butanol and SpinBind DNA extraction (FMC BioProducts) according to manufacturer's instructions. Over 30 trials, PCR amplification of an 18S rRNA gene locus proved more sensitive than the standardized fluorescence microscopy method, and as the percentage certainty of detection was higher with PCR, PCR could be used to screen soil samples prior to applying the standardized microscopy method for quantification.

15.6.8 INVERTEBRATE VECTORS OF DISEASE

Traditionally, invertebrate vectors of disease and the parasites they transmit are identified microscopically by their morphology, morphometry, and motility following dissection or culture. Extruded feces can also be examined and a few drops of rectal contents diluted in saline and placed between slide and coverslip can be used to demonstrate motile parasites by microscopy [174]. Furthermore, individual vectors can harbor more than one parasite species, not all of which may be infectious to humans, livestock, etc., necessitating the use of binomial identification keys, and parasite prevalence in such vectors can be low. Because of these limitations, microscopy-based methods are time consuming and require skilled analysts. PCR-based detection methods can be more sensitive than microscopy when the correct combination of sample preparation that minimizes coextraction of DNA with PCR inhibitors, and primer sensitivity and specificity are met [174]. There is no standardized method available for extracting DNA from arthropod vectors of blood-borne parasitic diseases, as each vector can contain specific, yet different, PCR interferents. Arthropod feces, larvae, dissected tissues, and intact organisms have been used for DNA extraction. Benefits of dissection include the reduction of PCR interferents and the concentration of parasites, particularly if they are localized to specific tissues (e.g., head, salivary glands, gut). Here we focus on DNA extraction from dissected tissues of arthropod vectors (e.g., Refs. [84,174–176]), intact vectors (e.g., Refs. [177–179]), and approaches to storage methods prior to DNA extraction [180].

The analysis of blood meal residues dried on filter paper has been used extensively in studies of the epidemiology of vector-borne diseases such as trypanosomiasis and provides information on host feeding behavior (e.g., Ref. [87]) and the infection status of vector populations (e.g., Ref. [175]).

Biting flies (Diptera, Tabanidae) and vampire bats (e.g., *Desmodus rotundus*) can transmit *T. evansi*, and PCR detection of *T. evansi* offers a sensitive and specific alternative to parasitological tests. Boid et al. [84] made gut smear preparations of newly emerged, colony-bred *Stomoxys calcitrans* artificially fed on blood infected with *T. evansi*, onto Whatman No.1 filter paper [181] which were left to dry (RT, 30 min), immersed in acetone (5 min), air-dried, and then stored at 4°C, in sealed plastic bags, until required. A disk (6 mm diameter) was punched from the center of the gut smear, placed into a 0.5 mL microcentrifuge tube, and a cold eluate made by incubating with 200 μL of sterile water (37°C, 30 min) in a thermocycler. The tube was then centrifuged and the supernatant stored for antibody analysis. For the hot eluate, a further 100 μL of sterile water was added to the filter paper disk in the original tube and incubated (100°C, 30 min) in the thermocycler. After brief high-speed centrifugation, the supernatant was removed and stored at 4°C. The hot eluate was used as DNA template for the PCR-based detection of *T. evansi* DNA [175] using the TB1 and TBR2 primers which produce a PCR product of ~170 bp, and *T. evansi* DNA was detected in hot eluates from smears from all flies fed on infected blood up to 1 h after feeding. The authors concluded that the elution used for the cold eluate removed PCR inhibitors and that sufficient *T. evansi* DNA remained bound to the filter paper that was released during the second elution process.

Tissue, dissected from the posterior end of the abdomen (avoiding the stomach, which has been reported to cause PCR inhibition), of individual nymphal or adult *Triatoma infestans* (the vector of *T. cruzii*) was used to diagnose *T. cruzii* infection in *T. infestans* [174]. Dissected tissue was placed in a tube and DNA extracted using the DNeasy kit (Qiagen) for animal tissues with 24 h lysis. The primers TCZ1 and TCZ2 [174] that amplify a 195 bp fragment of a repetitive element of the genome were used in a direct PCR. Pizarro et al. [174] found a high level of *T. cruzii* prevalence in 152 nymphal and adult vectors of *T. infestans* in Bolivia. PCR detection (81.16%) was more sensitive than microscopy (56.52%).

Michalsky et al. [179] assessed whether individual intact phlebotomine sand flies (a vector for cutaneous leishmaniasis) could be used for extracting *Leishmania* sp. DNA without dissection, using artificially seeded uninfected insect extracts and experimentally infected phlebotomine sand flies whose DNA was extracted 7 days postinfection. The whole insect DNA extraction method was evaluated against microscopy. Individual sand flies were macerated in a tube with 35 μL of Tris buffer (100 mM Tris–HCl, 100 mM NaCl, 25 mM EDTA, 0.5% SDS pH 8.0), digested overnight at 37°C with pK (1.25 μL of a 10 mg mL^{-1} solution), and DNA was extracted with PC. The DNA pellet was resuspended in 20 μL of TE. PCR was performed with *Leishmania* complex-specific primer sets (*L. braziliensis* [182] and *L. mexicana* [183]) and genus-specific primers [184]. In simulated infections (containing known numbers of promastigotes added to each sand fly), PCR could detect 10 parasites per sand fly. The contents of the digestive tract of the *Lutzomyia* spp. sand fly including ingested mouse blood did not affect the PCR and similar rates of infection were detected by PCR (87%, $n = 30$) and by microscopy (70%, $n = 30$). The percentages of PCR-positive samples following experimental infection were similar to the rates of infection determined by insect dissection. Satisfactory results were obtained using intact sand flies, suggesting that there was no need to dissect or pool insects in field surveys [179].

The minicircle kDNA, which contains ~10,000 minicircles per *Leishmania* genome and whose sequence is known for the majority of species, can make an ideal PCR target. A seminested PCR increased the sensitivity of PCR from >5 to 0.25 parasites per reaction and could detect eight species of *Leishmania* [177]. Individual dissected sand flies bodies (heads and last abdominal segment were kept for morphological identification) were homogenized, placed in 150 μL of extraction buffer (25 mM EDTA, 25 mM NaCl, 1% SDS), and incubated (65°C, 30 min). Following the addition of 100 μL of 3M potassium acetate (pH 7.2), incubation on ice (30 min), and centrifugation (13,000 × g), the supernatant was precipitated with 600 μL of ethanol. The DNA pellet was resuspended in TE and 5 μL used for PCR. No PCR inhibition was observed with this preparation method; however, 6.7% of females ($n = 522$) were PCR positive without a blood meal and 4.8% of females ($n = 123$) that contained blood were positive by PCR [177]. Evidently, the detection of *Leishmania* DNA in an individual sand fly does not imply that it is a vector, as the PCR assay detects both *Leishmania* amastigote and promastigote DNA, and amastigotes (the stage in macrophages) in a blood meal have to develop into promastigotes, which are the infective stage, transmitted in the bite of the sand fly vector. This assay is typical of those that can be used to study sand fly populations.

Vasuki et al. [178] developed a simple DNA preparation method for detecting the filarial parasite, *Brugia malayi*, in vector mosquitoes which consisted of drying and crushing the mosquitoes to a powder, which was homogenized in 100 µL of TE buffer, vortexed, boiled (10 min), centrifuged (10,000×g, 10 min), and the supernatant used for PCR. The method detected one microfilaria in pools of 25 mosquitoes [178]. Modifications to this method are required to obtain the best results when detecting *W. bancrofti* in *Culex quinquefasciatus*. The pool of mosquitoes was initially dried (95°C, 3 h), crushed into powder using a micro pestle, and homogenized in 30 µL of TE buffer. The pestle was washed in a further 30 µL of TE buffer and the sample vortexed, boiled (10 min) and centrifuged (10,000×g, 10 min). The supernatant (4 µL) was used for PCR with primers NV1 and NV2 [185] that amplify a 188 bp of a segment of a gene containing 300 copies in *W. bancrofti* genome. Using field trapped insects, the PCR-based method gave similar results to the conventional method (3.35% and 3.01% infections, respectively) indicating that it could prove useful for xeno-monitoring. The DNA extracted was suitable for PCR detection of *W. bancrofti* infection from pools of 10–30 mosquitoes, and its success depended on the species of mosquitoes used (some species contain more PCR inhibitors) and on the number of copies of the amplified gene [186].

Archiving arthropod samples collected during field investigations for the subsequent determination of their parasite vector status by microscopy is a standard procedure, but certain storage conditions and preservatives can reduce the sensitivity of detection by PCR. Larval and adult black flies (Diptera, Simuliidae; vectors of *Onchocerca volvulus*) preserved in ethanol, card point mounting, and sun drying provided good material for DNA extraction, but DNA extracted from specimens preserved in Carnoy's solution (ethanol: acetic acid, 3:1) yielded degraded DNA. Other parasitic nematodes and, to a lesser extent, gut contents resulted in extra products when amplified with randomly amplified polymorphic DNA primers, and sufficient DNA could be extracted from the head of one larva which resulted in the coextraction of fewer PCR interferents found in other tissues such as the digestive tract, and more positive PCRs [178].

The influence of different methods of storing phlebotomine samples on the amplification of *Leishmania* DNA was investigated by Cabrera et al. [180] Females of *Lutzomyia longipalpis* experimentally infected with *Leishmania chagasi* (=*L. infantum*) were preserved in 100% ethanol, 70% ethanol, and TE and subsamples were stored at −80°C, −20°C, and RT. Infection rates were determined by microscopy on dissected tissue. DNA was extracted with Chelex 100 and the kinetoplastic minicircle DNA primers OL1 and OL2 were used to amplify a *Leishmania* specific ~120 bp product. All storage methods were effective and the preservation of phlebotomine sand fly samples in 70% ethanol at RT was reported to be the most cost-effective way of yielding amplifiable DNA.

These results highlight the advantage of PCR over microscopy as a routine screening method for detecting infected flies in endemic foci of onchocerciasis and visceral leishmaniasis, and should simplify further vector incrimination studies.

15.6.9 DETECTION OF PARASITE LIFE CYCLE STAGES IN INTERMEDIATE AQUATIC HOSTS

Freshwater snails belonging to the genera *Biomphalaria* and *Bulinus* are intermediate hosts for trematode parasites of the genus *Schistosoma*. There are five species of *Schistosoma* that infect humans with similar life cycles. *Schistosoma* ova are shed by adult worms and released in the feces or urine of infected human hosts and hatch in freshwater to release the miracidium life cycle stage. The miracidium penetrates the snail and develops into the cercariae, which are released in water after a prepatent period of 3–4 weeks after infection. The presence of cercariae in snails is usually determined by squeezing snails between two glass slides (squash preparation) or by exposing snails to a source of light in order to release cercariae (not possible with dead snails or in the prepatent period). The detection of *Schistosoma* DNA in snails by PCR enables the detection of early infection and facilitates large-scale surveys for detecting the prevalence of infected snails in suspected sites.

Vidigal et al. [187] detected *Schistosoma* DNA in snails by extracting total DNA from the foot of *B. glabrata* snails. The tissue was mechanically disrupted in buffer (50 mM Tris–HCl pH 8.0, 100 mM NaCl, 50 mM EDTA, 0.5% SDS) incubated with 50 μg mL^{-1} pK overnight at 37°C followed by PC extraction and ethanol precipitation. DNA was resuspended in TE buffer. Hamburger et al. [188] identified infected snails at very early prepatency by nested PCR amplification (ladder of 42–121 bp size amplicons) of a repetitive gene sequence in *S. mansoni* [189] from DNA extracted from whole *B. glabrata* snails. Individual snails (shell and soft body) were placed in a plastic tube containing 300 μL of 1 M NaOH and 1% Triton X-100 and triturated with a wooden spatula. The mixtures were left at RT for 3 days (or heated at 65°C, 1 h), then neutralized with HCl, boiled (5 min), cooled on ice, and centrifuged (10,000 rpm, 15 min). The DNA was precipitated from the supernatant by adding 2.5 volumes of ethanol (without the addition of salts), cooling at −70°C for 30 min, and centrifuged. The sediment was washed three times with 70% ethanol and dissolved in 100 μL of TE buffer. *Schistosoma* DNA from snails experimentally infected with a single miracidium were detected in 80% of samples (16/20 snails tested) after 1 day, 3 days, and 1 week postinfection. This early detection by PCR contrasts with the cercarial shedding that occurred between 6 and 8 weeks postinfection. When extracted DNA from a single, 3 day infected snail, infected with a single miracidium was mixed with DNA from uninfected snails at different ratios, PCR amplification of *Schistosoma* DNA occurred at ratios up to 1:40. Doubling the amount of primers enabled the detection of a single, infected snail in pooled material from 100 snails, indicating the feasibility of rapid mass screening of prepatent infections in infected snails.

PCR proved to be a useful adjunct to conventional methods for detecting *S. mansoni* infected snails (see above). Jannotti-Passos et al. [190] extracted DNA from *Biomphalaria* spp. snails artificially infected with *S. mansoni* miracidia using a commercial DNA extraction kit (Wizard genomic DNA purification kit [Promega]). *S. mansoni* DNA profile was detected in both species after 7 day exposure to miracidia, using a low stringency PCR (LS-PCR) that amplifies adjacent tandem minisatellite units from *S. mansoni* mtDNA. Infection rates of 15% (*B. straminea*) and 50% (*B. tenogophila*) were detected by LS-PCR, whereas exposing snails to a source of light 42 days after miracidial exposure identified infection rates of 20.0% and 45.0%, respectively. When LS-PCR was used on snails which did not shed cercariae until 42 days following miracidial exposure, infection rates increased to 55.0% and 67.6% for *B. straminea* and *B. tenogophila*, respectively. Jannotti-Passos et al. [190] concluded that DNA extraction using the Wizard genomic DNA purification kit and LS-PCR enabled the early detection of schistosomiasis transmission focuses, in endemic areas, prior to the commencement of cercarial shedding.

15.7 FUTURE DEVELOPMENT TRENDS

Molecular biology has provided insights into the taxonomy and epidemiology of parasitic infections and diseases that were immeasurable using conventional diagnostic methods. In the diagnostic setting, molecular tools appropriate for species, genotype and sub-genotype analysis have unearthed previously unrecognized differences in disease, symptomatology, zoonotic potential, risk factors, and environmental occurrence and distribution, mainly because of their ability to differentiate between morphologically similar organisms. In this chapter we identify that PCR-based diagnosis can be more sensitive than conventional diagnostic methods and is more discriminatory when species or subgenotype identification is required, can avoid the use of more invasive diagnostic procedures and detect low parasite abundance, as long as nucleic acid extraction and purification is maximized.

Parasite abundance differs in the differing matrices identified. Where parasite abundance is high (e.g., in some clinical and environmental samples depending on the parasite), there is less of a requirement for developing specific nucleic acid extraction methods, which can be time consuming, and many investigators have used either combinations of in-house approaches and proprietary kits or one or the other. Conversely, where parasite abundance is low (e.g., in some clinical and most food and environmental samples), there is a fundamental requirement to maximize parasite recovery from its matrix. In the absence of selective enrichment by *in vitro* culture for the majority of parasites highlighted in the chapter, enrichment by concentration is the mainstay of clinical and environmental parasitology diagnostics, as it increases parasite abundance and hence parasite nucleic acid to detectable levels.

Current methods to prepare DNA templates vary, can be inefficient and labor intensive yielding inconsistent results. Freeze-thawing, bead disruption, and sonication are frequently employed, as is DNA binding in the presence of chaotropic agents, but additional nucleic acid purification steps are frequently necessary. Proprietary kits have also been used to good effect. With lower abundances of parasites, however, these methods can result in significant losses and yield variable and inconsistent results. This chapter highlights the variety of methods, and combinations thereof, that have been used to extract parasite nucleic acids from clinical, food, environmental, and invertebrate life cycle stages samples for direct molecular applications, primarily for detection. This indicates that there is no consensus on whether one, or a combination of approaches, either in-house or commercial, maximizes nucleic acid recovery from individual matrices. Commercial kits and the use of filter paper matrices offer a certain degree of quality assurance, which are requirements with clinical, food, and environmental samples, but many have not been tested exhaustively with parasites. While it is obvious that researchers have undertaken comparisons of methods, the parasites, parasite isolates, and matrices that they have used differ. In addition, the use of different PCR reactions, using different primers and conditions, further complicates interlaboratory comparisons. Clearly, to progress, standardized, validated methods, based on round robin trials and using agreed parasite isolates with known biophysical and biochemical characteristics in defined clinical, food, and environmental matrices should identify best options.

DNA extraction is at the centre of efficient PCR amplification and the detection of small numbers of parasites by molecular methods. Standardized, maximized methods for DNA extraction are essential both for detecting small numbers of parasites and for evaluating the sensitivity of detection using different primers. Disruption of these robust transmissive stages is a prerequisite for the release of nuclei and effective DNA extraction, while the liberation of DNA from bound protein, is essential both for efficient primer annealing and successful PCR amplification. DNA extraction and purification steps are often performed separately, but methods are available that combine these steps, which provide economy of time. As the sensitivity of detection is dependent on the abundance of the target template, coupled with maximizing DNA extraction, both the use of multicopy gene targets and the amplification of smaller fragments of nucleic acid increase sensitivity. This is particularly the case for parasites in low abundance and those in formalinized matrices. Whenever possible, a search of the parasite genome for multicopy genes for diagnosis is advisable so that the requirement for nested PCR assays can be minimized and the sensitivity of PCR can be increased. Some examples where this has been achieved include *Leishmania*, *Toxoplasma*, *Cryptosporidium*, and filaria.

Whereas current DNA template preparation and PCR detection methods using partially purified or purified parasite isolates may yield satisfactory results, detection sensitivities can be greatly affected by substances derived from the sample matrix and its processing which can result in a high percentage of false-negative results. As the major contributor of PCR interferents is the matrix, its inherent variability in clinical, food, and environmental samples may prevent the development of generic approaches to DNA extraction and purification, and hence the use of commercial kits for specific or individual parasite applications. Although parasites are a major cause of morbidity and mortality, worldwide, many occur in developing countries where the costs of current molecular methods are deemed prohibitive, and as such these potential customers are peripheral to the commercial field of view of most large proprietary kit manufacturers. This has led to the in-house development of DNA extraction and purification methods. While many of the PCR interferents have been identified, there are many others which have not or will not, given the breadth of matrices that parasites are sought in.

Some benefits of storing parasites and their nucleic acid on "alternative" matrices such as air-dried microscope slides, filter paper, and in ethanol have been identified. These offer distinct advantages over the more traditional methods of formalin fixation and freezing in terms of sensitivity, cost, and time, particularly as DNA extraction and purification methods become more effective. They have already enhanced the ease with which we can undertake field trials. The search for high specificity and sensitivity primers should complement improved DNA extraction and purification methods and tip the balance toward the use of alternative storage matrices for clinical samples, particularly for feces and blood spotted onto filter paper instead of using whole feces, blood, or buffy coats, since the practicality and safety aspects of this application outweigh the possible loss of some sensitivity; however, PCR reactions have to be optimized. While their acceptance for blood parasitology diagnostics and research clearly indicates their usefulness, it also highlights the need for reliable, sensitive, and cost-effective DNA extraction methods. A possible limitation of the alternative storage methods identified is that, as they often utilize less initial material, they can be compromised when used to detect parasite nucleic acid present in low abundance in clinical (feces, blood), food, and environmental samples. As identified, this can result in the uneven distribution of parasites and nucleic acid on filter paper disks. Increasing the number of disks analyzed is one option as long as the PCR interferents in the increased number of disks does not influence the outcome. A further option is to identify primers that amplify multicopy genes and/or diagnostically important, shorter nucleic acid sequences, as stated above. Furthermore, the report that the pore size of FTA paper ($\sim 25\,\mu m$) may have allowed *Cryptosporidium* oocysts (4.5–$5.5\,\mu m$) to pass through the filters without lysing is a cause for concern when attempting to identify standardized, validated methods for extracting and purifying nucleic acids from parasites in clinical, food, and environmental matrices, particularly when $\sim 20\%$ oocyst losses, mainly during the initial filtration step, can occur.

The sensitivity of validated DNA extraction and purification methods coupled with optimized detection methods also offers the possibility of detecting minute quantities of parasite nucleic acids as a component of bodily fluids in which parasites do not reside. Undoubtedly, PCR-based detection of blood and other tissue dwelling parasites using samples obtained from noninvasive procedures is a major development in clinical diagnostics and deserves further intense investigation. Again, maximizing nucleic acid extraction and purification from bodily secretions is a prerequisite to identifying specific benefits such as carrier status, drug intervention therapies, and posttreatment follow up using clinically relevant PCR-based approaches.

Previously, our capability to determine benefits from using molecular tools for detection depended on our ability to compare molecular outcomes with outcomes using conventional detection methods, particularly morphologically and morphometrically identifiable parasite stages. It is clear that parasite nucleic acid can be detected in samples that are negative by conventional methods, and here detecting naked DNA or DNA present in unrecognizable morphological forms is a valuable asset to early diagnosis (e.g., cryptosporidiosis, *S. mansoni* DNA in snails). In order to maximize detecting nucleic acid by PCR in the breadth of matrices identified herein, we require more effective methods both for neutralizing inhibitory effects and extracting nucleic acids, which, when allied with better internal (e.g., PCR internal controls) and external quality assurance should enable us to determine the real costs of morbidity and mortality caused by parasites. Certainly, this represents one of the most interesting challenges to diagnostic PCR today.

Huw Vaughan Smith, PhD, is a consultant clinical parasitologist and the director of the Scottish Parasite Diagnostic Laboratory and the Drinking Water Inspectorate *Cryptosporidium* External Quality Assurance Laboratory, Stobhill Hospital, Glasgow, Scotland. He received his BSc (special honours, applied zoology) and PhD (immunology of nematode parasites) from the University of Wales. He has researched widely in the immunobiology of nematode parasites and the epidemiology of protozoan parasites. He has a major interest in zoonoses. He has led numerous research projects that focused on the development of innovative methods for the detection, concentration, species identification, and determination of the viability of protozoan parasites, in particular, *Cryptosporidium*,

Giardia, Cyclospora, Microsporidia, and *Blastocystis*. His research outcome is documented in over 250 publications in scientific journals and books. He has over 30 years experience of teaching at both undergraduate and postgraduate levels at various British universities and overseas. He is a member of the College of Experts, UK Medical Research Council, four times member of the UK Expert Group on *Cryptosporidium* and water, and has served as a consultant for WHO, PAHO, and IAEA; he is also the lead scientist for the WHO center for drinking water quality; external referee for the UK Department for the Environment, Food and Rural Affairs (foodborne zoonoses research programme); adviser on protozoan parasites and water for the Drinking Water Inspectorate; specialist adviser for Health Protection Scotland; and consultant for various biotechnology companies. He is an honorary professor at the University of Glasgow and holds adjunct faculty positions at other UK and overseas universities.

Rosely Angela Bergamin Nichols received a biological sciences degree from the University of São Paulo, São Paulo, Brazil. During her subsequent employment at the Department of Immunology at the Adolf Lutz Institute of Public Health, São Paulo, Brazil, she acquired a keen interest in parasitology while developing immunodiagnostic methods for the detection of *Toxoplasma gondii* infections. In the United Kingdom since 1975, she has worked at the Beatson Institute for Cancer Research and at Stobhill Hospital, Glasgow, Scotland. She received her PhD in environmental health at the University of Strathclyde, Glasgow, Scotland. During her thesis research, she developed methods for the recovery and identification of small numbers of *Cryptosporidium* spp. oocysts from water focusing on DNA extraction and polymerase chain reaction (PCR)-based methods for species identification. For the past 12 years, she has been a clinical scientist at the Scottish Parasite Diagnostic Laboratory, Stobhill Hospital, Glasgow, Scotland. Her main interests are the development of methods for the molecular detection of protozoan parasites and their application to water, food, and feces for epidemiological studies. She has supervised students at graduate and postgraduate levels and she is an honorary research fellow at the Division of Environmental Health, University of Strathclyde.

REFERENCES

1. Kirchhoff, L.V. et al., Comparison of PCR and microscopic methods for detecting *Trypanosoma cruzi*, *J. Clin. Microbiol.*, 34: 1171, 1996.
2. Weiss, L.M. et al., Sensitive and specific detection of *Toxoplasma* DNA in an experimental murine model: Use of *Toxoplasma gondii*-specific cDNA and the polymerase chain reaction, *J. Infect. Dis.*, 163: 180, 1991.
3. Sandoval, N. et al., A new PCR-based approach for the specific amplification of DNA from different *Schistosoma* species applicable to human urine samples, *Parasitology*, 133: 581, 2006.
4. Smith, H.V., Diagnostics, in *Cryptosporidium and Cryptosporidiosis*, 2nd edn., Fayer, R. and Xiao, L., (Eds.), Taylor & Francis, Boca Raton, Florida, 2007, pp. 173–208.
5. Nichols, R.A.B., Campbell, B.M., and Smith, H.V., Identification of *Cryptosporidium* spp. oocysts in United Kingdom noncarbonated natural mineral waters and drinking waters by using a modified nested PCR-restriction fragment length polymorphism assay, *Appl. Environ. Microbiol.*, 69: 4183, 2003.
6. Homan, W. et al., Characterization of *Cryptosporidium parvum*_in human and animal feces by single-tube nested polymerase chain reaction and restriction analysis, *Parasitol. Res.*, 85: 707, 1999.
7. Smith, H.V. Detection of parasites in the environment, in *Infectious Diseases Diagnosis: Current Status and Future Trends*, Smith, H.V. and Stimson, W.H., (Eds.); *Parasitology*, 117: S113, 1999 (Chappel, L.H., co-ordinating editor).
8. Ash, R.L. and Orihel, T.C., Collection and preservation of feces, in *Parasites: A Guide to Laboratory Procedures and Identification*, American Society of Clinical Pathologists, (Eds.), ASCP Press, Chicago, 1987, p. 5.
9. Allen, A.V.H. and Ridley, D.S., Further observations on the formol ether concentration technique for faecal parasites, *J. Clin. Pathol.*, 23: 545, 1970.
10. Smith, H. V., Intestinal protozoa, in *Medical Parasitology: A Practical Approach*, Hawkey, P.M. and Gillespie, S.H., (Eds.), IRL Press, Oxford University Press, Oxford, UK, 1992, p. 79.
11. Campbell, A.T. and Smith, H.V., Immunomagnetic separation of *Cryptosporidium* oocysts from water samples: Round robin comparison of techniques, *Water Sci. Technol.*, 35: 397, 1997.

12. Smith, H.V., Detection of *Cryptosporidium* and *Giardia* in water, in *Molecular Approaches to Environmental Microbiology*, Pickup, R.W. and Saunders, J.R., (Eds.), Ellis-Horwood, Hemel Hempstead, Hertfordshire, UK, 1996, p. 195.

13. Balinger, J., Mechanisms of parasitological concentration in coprology and their practical consequences, *Am. J. Med. Technol.*, 41: 65, 1979.

14. Bouhoum, K. and Schwartzbrod, J., Quantification of helminth eggs in wastewater, *Int. J. Hyg. Environ. Health*, 188: 322, 1989.

15. Ayers, R.M. and Mara, D.D., *Analysis of Wastewater for Use in Agriculture: A Laboratory Manual of Parasitological and Bacteriological Techniques*, WHO, Geneva, 1996.

16. Bukhari, Z. and Smith, H.V., Effect of three concentration techniques on viability of *Cryptosporidium parvum* oocysts recovered from bovine faeces. *J. Clin. Microbiol.*, 33: 2592, 1995.

17. Nichols, R.A.B., Moore, J.E., and Smith, H.V., A rapid method for extracting oocyst DNA from *Cryptosporidium*-positive human faeces for outbreak investigations, *J. Microbiol. Methods*, 65: 512, 2006a.

18. Grimason, et al., Application of DAPI and immunofluorescence for enhanced detection of *Cryptosporidium* spp. oocysts in water samples, *Water Res.,* 28: 733, 1994.

19. Smith, H.V., Campbell, B.M., Paton, C.A., and Nichols, R.A.B., Significance of enhanced morphological detection of *Cryptosporidium* sp. oocysts in water concentrates using DAPI and immunofluorescence microscopy, *Appl. Environ. Microbiol.*, 68: 5198, 2002.

20. Nichols, R.A.B. and Smith, H.V., Optimisation of DNA extraction and molecular detection of *Cryptosporidium parvum* oocysts in natural mineral water sources, *J. Food Prot.*, 67: 524, 2004.

21. Mahabubani, M.H. and Bej, A.K., Applications of polymerase chain reaction methodology in clinical diagnostics, in *PCR Technology: Current Innovations*, 1st edn., Griffin, H.G. and Griffin, A.M., (Eds.), CRC Press Inc., Boca Raton, Florida, 1994, Chap 31.

22. Johnson, D. et al., Development of a PCR protocol for sensitive detection of *Cryptosporidium* in water samples, *Appl. Environ. Microbiol.*, 61: 3849, 1995.

23. Sepp, R. et al., Rapid techniques for DNA extraction from routinely processed archival tissue for use in PCR, *J. Clin. Pathol.*, 47: 318, 1994.

24. Boom, R. et al., Rapid and simple method for purification of nucleic acids, *J. Clin. Microbiol.*, 28: 495, 1990.

25. Sean, L.H. and Burgoygne, L., DNA database of FTA paper: Biological assault and techniques for measuring photogenic damage, *Prog. Forensic Genet.*, 8: 74, 2000.

26. Anonymous, Room temperature sample collection, storage and purification of nucleic acids, Technology and Services, a report by Whatman Int. Ltd., 2003, 1.

27. Del Rio, S.A., Marino, M.A., and Belgrader, P., Reusing the same blood-stained punch for sequential DNA amplification and typing, *Biotechniques*, 20: 970, 1996.

28. Clark, C.G., DNA purification from polysaccharide-rich cells, in *Protocols in Protozoology*, Vol. 1, Lee, J.J. and Soldo, A.T., (Eds.), Allen Press, Lawrence, Kansas, 1992, D-3.1–D-3.2.

29. Ali, I.K.M., Zaki, M., and Clark, C.G., Use of PCR amplification of tRNA gene-linked short tandem repeats for genotyping *Entamoeba histolytica*, *J. Clin. Microbiol.*, 43: 5842, 2005.

30. Ebeling, W. et al., Proteinase K from *Tritirachium album Limber*, *Eur. J. Biochem.*, 47: 91, 1974.

31. Hilz, H., Wiegers, U., and Adamietz, P., Stimulation of proteinase K action by denaturing agents: application to the isolation of nucleic acids and the degradation of 'masked' proteins, *Eur. J. Biochem.*, 56: 103, 1975.

32. Bialek, R. et al., Comparison of fluorescence, antigen and PCR assays to detect *Cryptosporidium parvum* in faecal specimens, *Diagn. Microbiol. Infect. Dis.*, 43: 283, 2002.

33. Ward, L.A. and Wang, Y., Rapid methods to isolate *Cryptosporidium* DNA from frozen feces for PCR, *Diagn. Microbiol. Infect. Dis.*, 41: 37, 2001.

34. da Silva, A.J. et al., Fast and reliable extraction of protozoan parasite DNA from fecal specimens, *Mol. Diagn.*, 4: 57, 1999.

35. McLauchlin, J. et al., Genetic characterization of *Cryptosporidium* strains from 218 patients with diarrhea diagnosed as having sporadic cryptosporidiosis, *J. Clin. Microbiol.*, 37: 3153, 1999.

36. Xiao, L. et al., Identification of species and sources of *Cryptosporidium* oocysts in storm waters with a small-subunit rRNA-based diagnostic and genotyping tool, *Appl. Environ. Microbiol.*, 66: 5492, 2000.

37. Di Pinto, A. and Tantillo, M.G., Direct detection of *Cryptosporidium parvum* oocysts by immunomagnetic separation – polymerase chain reaction in raw milk, *J. Food Prot.*, 65: 1345, 2002.

38. Deng, M.Q. and Cliver, D.O., Comparative detection of *Cryptosporidium parvum* oocysts from apple juice, *Int. J. Food Microbiol.*, 54: 155, 2000.

39. Lawson, A.J. et al., Polymerase chain reaction detection and speciation of *Campylobacter upsaliensis* and *C. helveticus* in human faeces and comparison with culture techniques, *J. Appl. Microbiol.*, 83: 375, 1997.
40. Nichols, R.A.B., Development of methods for the concentration, recovery and molecular identification of small numbers of *Cryptosporidium* spp. oocysts in natural mineral waters and its application for drinking waters, PhD thesis, Department of Civil Engineering, Division of Environmental Health, Strathclyde University, Glasgow, 2002.
41. Fedorko, D.P., Nelson, N.A., and Cartwright, C.P., Identification of microsporidia in stool specimens by using PCR and restriction endonucleases, *J. Clin. Microbiol.*, 33: 1739, 1995.
42. Sluter, S.D., Tzipori, S., and Widmer, G., Parameters affecting polymerase chain reaction detection of waterborne *Cryptosporidium parvum* oocysts, *Appl. Microbiol. Biotechnol.*, 48: 325, 1997.
43. Leng, X., Mosier, D.A., and Oberst, R.D., Differentiation of *Cryptosporidium parvum, C. muris, and C. baileyi* by PCR-RFLP analysis of the 18S rRNA gene, *Vet. Parasitol.*, 62: 1, 1996.
44. Champliaud, D. et al., Failure to differentiate *Cryptosporidium parvum* from *C. meleagridis* based on PCR amplification of eight DNA sequences, *Appl. Environ. Microbiol.*, 64: 1454, 1998.
45. Chung, E. et al., Detection of *Cryptosporidium parvum* oocysts in municipal water samples by the polymerase chain reaction, *J. Microbiol. Methods*, 33: 171, 1998.
46. Nichols, R.A.B., Campbell, B., and Smith, H.V., Molecular fingerprinting of *Cryptosporidium* species oocysts isolated during water monitoring, *Appl. Environ. Microbiol.*, 72: 5428, 2006b.
47. Belosevic, M. et al., Nucleic acid stains as indicators of *Cryptosporidium parvum* oocyst viability, *Int. J. Parasitol.*, 27: 787, 1997.
48. Morgan, U.M. et al., Comparison of PCR and microscopy for detection of *Cryptosporidium parvum* in human fecal specimens: clinical trial, *J. Clin. Microbiol.*, 36: 995, 1998.
49. Verweij, J.J., Polderman, A.M., and Clark, G., Genetic variation among human isolates of uninucleated cyst-producing *Entamoeba* species, *J. Clin. Microbiol.*, 39: 1644, 2001.
50. Verweij, J.J. et al., Real-time PCR for the detection of *Giardia lamblia, Mol. Cell. Probes*, 17: 223, 2003a.
51. Verweij, J.J. et al., Prevalence of *Entamoeba histolytica* and *Entamoeba dispar* in northern Ghana, *Trop. Med. Int. Health*, 8: 1153, 2003b.
52. Verweij, J.J. et al., Simultaneous detection and quantification of *Ancylostoma duodenale, Necator americanus*, and *Oesophagostomum bifurcum* in fecal samples using multiplex real-time PCR, *Am. J. Trop. Med. Hyg.*, 77: 685, 2007.
53. Nichols, R.A.B. et al., A sensitive, semi-quantitative direct PCR-RFLP assay for simultaneous detection of five *Cryptosporidium* species in treated drinking waters and mineral waters, *Water Sci. Technol.: Water Supply*, 2: 1, 2002.
54. Nichols, R.A.B. and Smith, H.V., unpublished data, 2005.
55. Koonjul, P.K. et al., Inclusion of polyvinylpyrrolidone in the polymerase chain reaction reverses the inhibitory effects of polyphenolic contamination of RNA, *Nucleic Acids Res.*, 27: 915, 1998.
56. Orlandi, P.A. and Lampel, K.A., Extraction-free, filter-based template preparation for rapid and sensitive PCR detection of pathogenic parasitic protozoa, *J. Clin. Microbiol.*, 38: 2271, 2000.
57. da Silva, A.J. et al., Detection of *Septata intestinalis* (microsporidia) using polymerase chain reaction primers targeting the small subunit ribosomal RNA coding region, *Mol. Diagn.*, 2: 47, 1997.
58. Subrungruang, I. et al., Evaluation of DNA extraction and PCR methods for detection of E*nterocytozoon bieneusi* in stool specimens, *J. Clin. Microbiol.*, 42: 3490, 2004.
59. Katzwinkel-Wladarsch et al., Direct amplification and species determination of microsporidian DNA from stool samples, *Trop. Med. Int. Health*, 1: 373, 1996.
60. da Silva, A.J. et al., Sensitive PCR diagnosis of infections by *Enterocytozoon bieneusi* (microsporidia) using primers based on the region coding for small-subunit rRNA, *J. Clin. Microbiol.*, 34: 986, 1996.
61. Schuitema, A.R.J. et al., Application of the polymerase chain reaction for the diagnosis of microsporidiosis, *AIDS*, 7: S62, 1993.
62. Velasquez, J.N. et al., Detection of the microsporidian parasite *Enterocytozoon bieneusi* in specimens from patients with AIDS by PCR, *J. Clin. Microbiol.*, 34: 3230, 1996.
63. Manfield, K.G. et al., Identification of an *Enterocytozoon bieneusi*-like microsporidian parasite in simian-immunodeficiency-virus-inoculated macaques with hepatobiliary disease, *Am. J. Pathol.*, 150: 1395, 1997.
64. Carnevale, S. et al., Diagnosis of *Enterocytozoon bieneusi* by PCR in stool samples eluted from filter paper disks, *Clin. Diagn. Lab. Immunol.*, 7: 504, 2000.

65. Dowd, S.E. et al., PCR amplification and species determination of microsporidia in formalin-fixed feces after immunomagnetic separation, *Appl. Environ. Microbiol.*, 64: 333, 1998.

66. Nantavisai, K. et al., Evaluation of the sensitivities of DNA extraction and PCR methods for detection of *Giardia duodenalis* in stool specimens, *J. Clin. Microbiol.*, 45: 581, 2007.

67. Hopkins, R.M. et al., Ribosomal RNA sequencing reveals differences between the genotypes of *Giardia* isolates recovered from humans and dogs living in the same locality, *J. Parasitol.*, 83: 44, 1997.

68. Casemore, D.P., Laboratory methods for diagnosing cryptosporidiosis, *ACP Broadsheet*, 128: 445, 1991.

69. Amar, C., Pedraza-Diaz, S., and McLauchlin, J., Extraction and genotyping of *Cryptosporidium parvum* DNA from fecal smears on glass slides stained conventionally for direct microscope examination, *J. Clin. Microbiol.*, 39: 401, 2001.

70. Morse, T.D. et al., Incidence of cryptosporidiosis species in paediatric patients in Malawi, *Epidemiol. Infect.*, 135: 1307, 2007.

71. Goldenberger, D. et al., A simple "universal" DNA extraction procedure using SDS and Proteinase K is compatible with direct PCR amplification, *PCR Methods Appl.*, 4: 368, 1995.

72. Ali, I.K.M., Zaki, M., and Clark, C.G., Use of PCR amplification of tRNA gene-linked short tandem repeats for genotyping *Entamoeba histolytica*, *J. Clin. Microbiol.*, 43: 5842, 2005.

73. Singh, B. et al., Detection of malaria in Malaysia by nested polymerase chain reaction amplification of dried blood spots on filter papers, *Trans. R. Soc. Trop. Med. Hyg.*, 5: 519, 1996.

74. Cox-Singh, J. et al., Simple blood-spot sampling with nested polymerase chain reaction detection for epidemiology studies on *Brugia malayi*, *Int. J. Parasitol.*, 29: 717, 1999.

75. Shigidi, M.M.T. et al., Parasite diversity in adult patients with cerebral malaria: A hospital-based, case-control study, *Am. J. Trop. Med. Hyg.*, 71: 754, 2004.

76. Bereczky, S. et al., Rapid DNA extraction from archive blood spots on filter paper for genotyping of *Plasmodium falciparum*, *Am. J. Trop. Med. Hyg.*, 72: 249, 2005.

77. Alhassan, A. et al., Comparison of polymerase chain reaction methods for the detection of *Theileria equi* infection using whole blood compared with pre-extracted DNA samples as PCR templates, *Trop. Anim. Health Prod.*, 39: 369, 2007.

78. Gonzales, J.L., Loza, A., and Chacon, E., Sensitivity of different *Trypanosoma vivax* specific primers for the diagnosis of livestock trypanosomosis using different DNA extraction methods, *Vet. Parasitol.*, 136: 119, 2006.

79. Makowski, G.S., Davis, E.L., and Hopfer, S.M., The effect of storage on Guthrie cards: Implications for deoxyribonucleic acid amplification, *Ann. Clin. Lab. Sci.*, 26: 458, 1996.

80. Gil, J.P. et al., Detection of atovaquone and Malarone resistance conferring mutations in *Plasmodium falciparum* cytochrome b gene (cytb), *Mol. Cell. Probes*, 17: 85, 2003.

81. Plowe, C.V. et al., Pyrimethamine and proguanil resistance-conferring mutations in *Plasmodium falciparum* dihydrofolate reductase: Polymerase chain reaction methods for surveillance in Africa, *Am. J. Trop. Med. Hyg.*, 52: 565, 1995.

82. Snounou, G. et al., Biased distribution of msp1 and msp2 allelic variants in *Plasmodium falciparum* populations in Thailand, *Trans. R. Soc. Trop. Med. Hyg.*, 93: 369, 1999.

83. de Almeida, P.P., et al., Diagnostic evaluation of PCR on dried blood samples from goats experimentally infected with *Trypanosoma brucei*, *Acta Trop.*, 70: 269, 1998.

84. Boid, R., Jones, T.W., and Munro, A., A simple procedure for the extraction of trypanosome DNA and host protein from dried blood meal residues of haematophagous diptera, *Vet. Parasitol.*, 85: 313, 1999.

85. Masaque, R.A. et al., Sensitive and specific detection of *Trypanosoma vivax* using the polymerase chain reaction, *Exp. Parasitol.*, 85: 193, 1997.

86. Masiga, D.K. et al., A high prevalence of mixed trypanosome infections in tsetse flies in Sifra, Cote d'Ivoire, detected by DNA amplification, *Parasitology*, 112: 75, 1996.

87. Clausen, P.H. et al., Host preferences of tsetse (Diptera: *Glossinidae*) based on bloodmeal identifications, *Med. Vet. Entomol.*, 12: 169, 1998.

88. Davila, A.M.R. et al., unpublished data, 2006.

89. Reithinger, R. et al., Use of PCR to detect *Leishmania (Viannia)* spp. in dog blood and bone marrow, *J. Clin. Microbiol.*, 38: 748, 2000.

90. Lachaud, L. et al., Comparison of various sample preparation methods for PCR diagnosis of visceral leishmaniasis using peripheral blood, *J. Clin. Microbiol.*, 39: 613, 2001.

91. Kazemi, B. et al., Diagnosis of *Trichomonas vaginalis* infection by urine PCR analysis compared to wet Mount microscopic screening, *J. Med. Sci.*, 4: 206, 2004.

92. Mayta, H. et al., 18S ribosomal DNA–based PCR for diagnosis of *Trichomonas vaginalis, J. Clin. Microbiol.*, 38: 2683, 2000.

93. Leibovitz, E. et al., Comparison of PCR and standard cytological staining for detection of *Pneumocystis carinii* from respiratory specimens from patients with or at high risk for infection by human immuno-deficiency virus, *J. Clin. Microbiol.*, 33: 3004, 1995.

94. Nuchprayoon, S. et al., Flinders technology associates (FTA) filter paper-based DNA extraction with polymerase chain reaction (PCR) for detection of Pneumocystis jirovecii from respiratory specimens of immunocompromised patients, *J. Clin. Lab. Anal.*, 21: 382, 2007.

95. Parmley, S.F., Goebel, F.D., and Remington, J.S., Detection of *Toxoplasma gondii* in cerebrospinal fluid from AIDS patients by polymerase chain reaction, *J. Clin. Microbiol.*, 30: 3000, 1992.

96. Cingolani, A. et al., PCR detection of *Toxoplasma gondii* DNA in CSF for the differential diagnosis of AIDS-related focal brain lesions, *J. Med. Microbiol.*, 45: 472, 1996.

97. Julander, I. et al., Polymerase chain reaction for diagnosis of cerebral toxoplasmosis in cerebrospinal fluid in HIV-positive patients, *Scand. J. Infect. Dis.*, 33: 538, 2001.

98. Contini, C. et al., The role of stage-specific oligonucleotide primers in providing effective laboratory support for the molecular diagnosis of reactivated *Toxoplasma gondii* encephalitis in patients with AIDS, *J. Med. Microbiol.*, 51: 879, 2002.

99. Goto, M. et al., Detection of *Toxoplasma gondii* by polymerase chain reaction in cerebrospinal fluid from human immunodeficiency virus-1-infected Japanese patients with focal neurological signs, *J. Int. Med. Res.*, 32: 665, 2004.

100. Vidal, J.E. et al., PCR assay using cerebrospinal fluid for diagnosis of cerebral toxoplasmosis in Brazilian AIDS patients, *J. Clin. Microbiol.*, 42: 4765, 2004.

101. Mharakurwa, S. et al., PCR detection of *Plasmodium falciparum* in human urine and saliva samples, *Malaria J.*, 5: 103, 2006.

102. Duraisingh, M.T., Curtis, J., and Warhurst, D.C., *Plasmodium falciparum*: Detection of polymorphisms in the dihydrofolate reductase and dihydropteroate synthetase genes by PCR and restriction digestion, *Exp. Parasitol.*, 89: 1, 1998.

103. Lucena, W.A. et al., Diagnosis of *Wuchereria bancrofti* infection by the polymerase chain reaction using urine and day blood samples from microfilaraemic patients, *Trans. R. Soc. Trop. Med. Hyg.*, 92: 290, 1998.

104. Ash, L.R. and Orihel, T.C. Formalin-ethyl acetate or formalin-ether sedimentation technique for fresh material, in *Parasites: A Guide to Laboratory Procedures and Identification*, 3rd volume, ASCP Press, Chicago, 1991, p. 23.

105. Dowd, S.E. et al., PCR amplification and species determination of microsporidia in formalin-fixed feces after immunomagnetic separation, *Appl. Environ. Microbiol.*, 64: 333, 1998.

106. Ramos, F. et al., The effect of formalin fixation on the polymerase chain reaction characterization of *Entamoeba histolytica, Trans. R. Soc. Trop. Med. Hyg.*, 93: 335, 1999.

107. Lehmann, U. and Kreipe, H., Real-time PCR analysis of DNA and RNA extracted from formalin-fixed and paraffin-embedded biopsies, *Methods*, 25: 409, 2001.

108. Libório, T.N. et al., Evaluation of the genomic DNA extracted from formalin-fixed, paraffin-embedded oral samples archived for the past 40-years, *J. Bras. Patol. Med. Lab.*, 41: 405, 2005.

109. Paglia, M.G. and Visca, P., An improved PCR-based method for detection and differentiation of *Entamoeba histolytica* and *Entamoeba dispar* in formalin-fixed stools, *Acta Trop.*, 92: 273, 2004.

110. Jongwutiwes, S. et al., Simple method for long-term copro-preservation of *Cryptosporidium* oocysts for morphometric and molecular analysis, *Trop. Med. Int. Health*, 7: 257, 2002.

111. Romero, G.A.S., et al., Sensitivity of the polymerase chain reaction for the diagnosis of cutaneous leishmaniasis due to *Leishmania (Viannia) guyanensis, Acta Trop.*, 79: 225, 2001.

112. Bensoussan, E., et al., Comparison of PCR assays for diagnosis of cutaneous leishmaniasis, *J. Clin. Microbiol.*, 44: 1435, 2006.

113. Böer, A. et al., Unusual histopathological features of cutaneous leishmaniasis by polymerase chain reaction specific for *Leishmania* on paraffin-embedded skin biopsies, *Br. J. Dermatol.*, 155: 815, 2006.

114. Schonian, G.A. et al., PCR diagnosis and characterization of *Leishmania* in local and imported clinical samples, *Diagn. Microbiol. Infect. Dis.*, 47: 349, 2003.

115. Bonin, S. et al., PCR analysis in archival postmortem tissues, *J. Clin. Pathol.: Mol. Pathol.*, 56: 184, 2003.

116. Quach, N., Goodman, M.F., and Shibata, D., *In vitro* mutation after formalin fixation and error prone translesion synthesis during PCR, *BMC Clin. Pathol.*, 4: 1, 2004.

117. Coombs, N.J., Gough, A.C., and Primrose, J.N., Optimisation of DNA and RNA extraction from archival formalin-fixed tissue, *Nucleic Acids Res.*, 27: e12, 1999.
118. Cawkwell, L. and Quirke, P., Direct multiplex amplification of DNA from a formalin fixed, paraffin wax embedded tissue section, *J. Clin. Pathol.: Mol. Pathol.*, 53: 51, 2000.
119. Minodier, P. et al., Rapid identification of causative species in patients with Old World leishmaniasis, *J. Clin. Microbiol.*, 35: 2551, 1997.
120. Muller, N. et al., PCR-based detection of canine *Leishmania* infections in formalin-fixed and paraffin embedded skin biopsies: Elaboration of a protocol for quality assessment of the diagnostic amplification reaction, *Vet. Parasitol.*, 114: 223, 2003.
121. Piarroux, R. et al., Phylogenetic relationships between Old World Leishmania strains revealed by analysis of a repetitive DNA sequence, *Mol. Biochem. Parasitol.*, 73: 249, 1995.
122. Slifco, T.R., Smith, H.V., and Rose, J.B., Emerging parasite zoonoses associated with food and water, *Int. J. Parasitol.*, 30: 1379, 2000.
123. Nichols, R.A.B. and Smith, H.V., *Cryptosporidium, Giardia* and *Cyclospora* as foodborne pathogens, in *Foodborne Pathogens: Hazards, Risk and Control*, Part III, *Non-Bacterial and Emerging Foodborne Pathogens*, Blackburn, C. and McClure, P., (Eds.), Woodhead Publishing Limited, Cambridge, 2002, p. 453.
124. Smith, H.V. et al., *Cryptosporidium* and *Giardia* as foodborne zoonoses, *Vet. Parasitol.*, 149: 29, 2007.
125. Smith, H.V. and Nichols, R.A.B., *Cryptosporidium*, in *Foodborne Diseases*, Simjee S., (Ed.), Humana Press Totowa, New Jersey, 2007, p. 303.
126. Smith, H.V. and Paget, T., *Giardia*, in *Foodborne Diseases*, Simjee S., (Ed.), Humana Press Totowa, New Jersey, 2007, p. 303.
127. Smith, H.V., *Cyclospora*, in *Foodborne Diseases*, Simjee S., (Ed.), Humana Press, Totowa, New Jersey, 2007, 277.
128. Gamble, H.R. and Murrell, K.D., Diagnosis of parasites in food, in *Infectious Diseases Diagnosis: Current Status and Future Trends*, Smith, H.V. and Stimson, W.H., (Eds.), Chappel; Parasitology, 117: S97, 1999 (L.H., co-ordinating editor).
129. Mead, P.S. et al., Food-related illness and death in the United States, *Emerg. Infect. Dis.*, 5: 607, 1999.
130. Cook, N. et al., Development of a method to detect *Giardia duodenalis* cysts on lettuce and the simultaneous analysis of salad products for the presence of *Giardia* spp. cysts and *Cryptosporidium* spp. oocysts, *Appl. Environ. Microbiol.*, 73: 7388, 2007.
131. Robertson, L.J. and Gjerde, B., Isolation and enumeration of *Giardia* cysts, *Cryptosporidium* oocysts, and Ascaris eggs from fruits and vegetables, *J. Food Prot.*, 63: 775, 2000.
132. Robertson, L.J. and Gjerde, B., Factors affecting recovery efficiency in isolation of *Cryptosporidium* oocysts and *Giardia* cysts from vegetables for standard method development, *J. Food Prot.*, 64: 1799, 2001a.
133. Robertson, L.J. and Gjerde, B., Occurrence of parasites on fruits and vegetables in Norway, *J. Food Prot.*, 64: 1793, 2001b.
134. Cook, N., et al., Towards standard methods for the detection of *Cryptosporidium parvum* on lettuce and raspberries. Part 1: Development and optimization of methods, *Int. J. Food Microbiol.*, 109: 215, 2006.
135. Cook, N., et al., Towards standard methods for the detection of *Cryptosporidium parvum* on lettuce and raspberries. Part 2. Validation, *Int. J. Food Microbiol.*, 109: 222, 2006.
136. Smith, H.V. and Cook, N., Lessons learnt in development and application of detection methods for zoonotic foodborne protozoa on lettuce and fresh fruit, in *Global Issues in Food Science and Technology*, Mortimer, A., Colonna, P., Lineback, D., Spiess, W., Buckle, K., and Barbosa-Cánovas, G.V., (Eds.), Elsevier, The Netherlands, 2008, in press.
137. Deng, M.Q. and Cliver, D.O., *Cryptosporidium parvum* studies with dairy products, *Int. J. Food Microbiol.*, 46: 113, 1999.
138. Lowery, C.J. et al., PCR–IMS detection and molecular typing of *Cryptosporidium parvum* recovered from a recreational river source and an associated mussel (*Mytilus edulis*) bed in Northern Ireland, *Epidemiol. Infect.*, 127: 545, 2001.
139. Xiao, L., Alderisio, K.A., and Jiang, J., Detection of *Cryptosporidium* oocysts in water: Effect of the number of samples and analytic replicates on test results, *Appl. Environ. Microbiol.*, 72: 5942, 2006.
140. De Boer, S.H. et al., Attenuation of PCR inhibition in the presence of plant compounds by addition of BLOTTO, *Nucleic Acids Res.*, 23: 2567, 1995.
141. Jones, C.S. et al., The isolation of RNA from raspberry (*Rubus idaeus*) fruit, *Mol. Biotech.*, 8: 219, 1997.

142. Manning, K., Isolation of nucleic acids from plants by differential solvent precipitation, *Anal. Biochem.*, 195: 45, 1995.
143. Lorenz, H. et al., PCR detection of *Coxiella burnetti* from different clinical specimens, especially bovine milk, on the basis of DNA preparation with a silica matrix, *Appl. Environ. Microbiol.*, 64: 4234, 1998.
144. Jinneman, K.C. et al., Template preparation for PCR and RFLP of amplification products for the detection and identification of *Cyclospora* sp. and *Eimeria* spp. oocysts directly from raspberries, *J. Food Prot.*, 61: 1497, 1998.
145. Aspinall, T.V. et al., Prevalence of *Toxoplasma gondii* in commercial meat products as monitored by polymerase chain reaction—food for thought?, *Int. J. Parasitol.*, 32: 1193, 2002.
146. Frazar, C.D. and Orlandi, P.A., Evaluation of two DNA template preparation methods for post-immunomagnetic separation detection of *Cryptosporidium parvum* in foods and beverages by PCR, *Appl. Environ. Microbiol.*, 73: 7474, 2007.
147. Steele, M., Unger, S., and Odumeru, J., Sensitivity of PCR detection of *Cyclospora cayetanensis* in raspberries, basil, and mesclun lettuce, *J. Microbiol. Methods*, 54: 277, 2003.
148. Okhuysen, P.C. et al., Virulence of three distinct *Cryptosporidium parvum* isolates for healthy adults, *J. Infect. Dis.*, 180: 1275, 1999.
149. Rendtorff, R.C., The experimental transmission of human intestinal protozoan parasites. II. Giardia lamblia cysts given in capsules, *Am. J. Hyg.*, 59: 209, 1954.
150. Rendtorff, R.C., The experimental transmission of *Giardia lamblia* among volunteer subjects, in *Waterborne Transmission of Giardiasis*, Jakubowski, W. and Hoff, J.C., (Eds.), U.S. Environmental Protection Agency, Office of Research and Development, Environmental Research Centre, Cincinnati, Ohio, EPA-600/9-79-001, 1979, p. 64.
151. Adams, A.M., Jinneman, K.C., and Ortega, Y.R., *Cyclospora*, in *Encyclopaedia of Food Microbiology*, Volume 1, Robinson R., Batt, C., and Patel, P., (Eds.), Academic Press, London and New York, 1999, p. 502.
152. Howe, D.K. et al., Determination of genotypes of *Toxoplasma gondii* strains isolated from patients with toxoplasmosis, *J. Clin. Microbiol.*, 35: 1411, 1997.
153. Matsunaga, T. et al., A quick and simple method for the identification of meat species and meat products by PCR assay, *Meat Sci.*, 51: 143, 1999.
154. Anonymous, isolation and identification of *Giardia* cysts, *Cryptosporidium* oocysts and free living amoebae in water, Department of the Environment, Standing Committee of Analysis, London, HMSO, 1990, 30 pp.
155. Anonymous. Method 1623, *Cryptosporidium* in water by filtration/IMS/FA. United States Environmental Protection Agency, Office of Water, Washington DC. Consumer confidence reports final rule, Federal Register, 1998, 63: 160.
156. Anonymous, Water Supply (Water Quality) (Amendment) Regulations 1999, SI No 1524. 1999a.
157. Anonymous, Isolation and identification of *Cryptosporidium* oocysts and *Giardia* cysts in waters 1999. Methods for the Examination of Waters and Associated Materials, London, HMSO, 1999b, 44 pp.
158. Anonymous, DWI Information letter 2005. The water supply (water quality) (amendment) regulations 2000, SI No. 3184 England and 2001, SI No. 3911 (W.323) Wales: *Cryptosporidium* in water supplies: Laboratory and analytical procedures. Part 2, June 2005a.Protocol containing standard operating protocols (SOPs) for the monitoring of *Cryptosporidium* oocysts in water supplies. UK Drinking Water Inspectorate, Available at www.dwi.detr.gov.uk.
159. Anonymous, Method 1623: *Cryptosporidium* and *Giardia* in water by filtration/IMS/FA; December 2005b. Office of Water (4607) U.S. Environmental Protection Agency 815-R-05-002, Available at http://www.epa.gov/microbes/1623de05.pdf.
160. Tebbe, C.C. and Vahjen, W., Interference of humic acids and DNA extracted directly from soil in detection and transformation of recombinant DNA from bacteria and a yeast, *Appl. Environ. Microbiol.*, 59: 2657, 1993.
161. Sluter, S.D., Tzipori, S., and Widmer, G., Parameters affecting polymerase chain reaction detection of waterborne *Cryptosporidium parvum* oocysts, *Appl. Microbiol. Biotechnol.*, 48: 325, 1997.
162. Lowery, C.J. et al., Detection and speciation of *Cryptosporidium* spp. in environmental water samples by immunomagnetic separation, PCR and endonuclease restriction, *J. Med. Microbiol.*, 49: 779, 2000.
163. Zhou, L. et al., Molecular surveillance of *Cryptosporidium* spp. in raw wastewater in Milwaukee: implications for understanding outbreak occurrence and transmission dynamics, *J. Clin. Microbiol.*, 41: 5254, 2003.

164. Xiao, L. et al., Phylogenetic analysis of *Cryptosporidium* parasites based on the small-subunit rRNA gene locus, *Appl. Environ. Microbiol.*, 65: 1578, 1999.

165. Xiao, L. et al., Molecular characterisation of *Cryptosporidium* oocysts in samples of raw surface water and wastewater, *Appl. Environ. Microbiol.*, 67: 1091, 201.

166. Mahbubani, M.H. et al., Detection of *Giardia* in environmental waters by immuno-PCR amplification methods, *Curr. Microbiol.*, 36: 107, 1998.

167. Guy, R.A. et al., Real-time PCR for quantification of *Giardia* and *Cryptosporidium* in environmental water samples and sewage, *Appl. Environ. Microbiol.*, 69: 5178, 2003.

168. Jiang, J. et al., Development of procedures for direct extraction of *Cryptosporidium* DNA from water concentrates and for relief of PCR inhibitors, *Appl. Environ. Microbiol.*, 71: 1135, 2005.

169. Abbaszadegan, M. et al., Detection of viable *Giardia* cysts by amplification of heat shock-induced mRNA, *Appl. Environ. Microbiol.*, 63: 324, 1997.

170. Kaucner, C. and Stinear, T., Sensitive and rapid detection of viable *Giardia* cysts and *Cryptosporidium parvum* oocysts in large-volume water samples with wound fiberglass cartridge filters and reverse transcription-PCR, *Appl. Environ. Microbiol.*, 64: 1743, 1998.

171. Stinear, T. et al., Detection of a single viable *Cryptosporidium parvum* oocyst in environmental water concentrates by reverse transcription-PCR, *Appl. Environ. Microbiol.*, 62: 3385, 1996.

172. Vesey, G. et al., A new method for the concentration of *Cryptosporidium* oocysts from water, *J. Appl. Bacteriol.*, 75: 82, 1993.

173. Shields, J.M. and Olson, B.H., PCR-restriction fragment length polymorphism method for detection of *Cyclospora cayetanensis* in environmental waters without microscopic confirmation, *Appl. Environ. Microbiol.*, 69: 4662, 2003.

174. Pizarro, J.C. et al., PCR reveals significantly higher rates of *Trypanosoma cruzi* infection than microscopy in the Chagas vector, *Triatoma infestans*: High rates found in Chuquisaca, Bolivia, *BMC Infect. Dis.*, 7: 66, 2007.

175. Masiga, D.K. et al., Sensitive detection of Trypanosomes in tsetse flies by DNA amplification, *Int. J. Parasitol.*, 85: 313, 1992.

176. Koch, D.A. et al., Effects of preservation methods, parasites, and gut contents of black flies (*Diptera: Simuliidae*) on polymerase chain reaction products, *J. Med. Entomol.*, 35: 314, 1998.

177. Aransay, A.M., Scoulica, E., and Tselentis, Y., Detection and identification of *Leishmania* DNA within naturally infected sand flies by semi nested PCR on minicircle kinetoplastic DNA, *Appl. Environ. Microbiol.*, 66: 1933, 2000.

178. Vasuki, V., Patra, K.P., and Hoti, S.L., A rapid and simplified method of DNA extraction for the detection of Brugia malayi infection in mosquitoes by PCR assay, *Acta Trop.*, 79: 245, 2001.

179. Michalsky, E.M. et al., Assessment of PCR in the detection of *Leishmania* spp. in experimentally infected individual phlebotomine sandflies (*Diptera: Psychodidae: Phlebotominae*), *Rev. Inst. Med. Trop. S. Paulo*, 44: 255, 2002.

180. Cabrera, O.L. et al., Definition of appropriate temperature and storage conditions in the detection of *Leishmania* DNA with PCR in phlebotomine flies, *Biomedica*, 22: 296, 2002.

181. F.A.O., Training manual for tsetse control personnel, in *Tsetse Biology, Systematics and Distribution Techniques*, Pollock, J.N. (Ed.), Food and Agriculture Organisation of the United Nations (FAO), Rome, 1982.

182. de Bruijn, M.H.L. and Barker, D.C., Diagnosis of New World leishmaniasis: specific detection of species of the *Leishmania braziliensis* complex by amplification of kinetoplast DNA, *Acta Trop. (Basel)*, 52: 45, 1992.

183. Eresh, S., McCallum, S.M., and Barker, D.C., Identification and diagnosis of *Leishmania mexicana* complex isolates by polymerase chain reaction, *Parasitology*, 109: 423, 1994.

184. Degrave, W. et al., Use of molecular probes and PCR for detection and typing of *Leishmania*: A mini review, *Mem. Inst. Oswaldo Cruz*, 89: 463, 1994.

185. Zhong, K.J.Y. et al., Comparison of IsoCode STIX and FTA gene guard collection matrices as whole-blood storage and processing devices for diagnosis of Malaria by PCR, *J. Clin. Microbiol.*, 39: 1195, 1996.

186. Vasuki, V., et al., A simple and rapid DNA extraction method for the detection of Wuchereria bancrofti infection in the vector mosquito, Culex quinquefasciatus by Ssp I PCR assay, *Acta Trop.*, 86: 109, 2003.

187. Vidigal, T.H.D.A. et al., *Biomphalaria glabrata*: Extensive genetic variation in Brazilian isolates revealed by random amplified polymorphic DNA analysis, *Exp. Parasitol.*, 79: 187, 1994.

188. Hamburger, J. et al., A polymerase chain reaction assay for detecting snails infected with bilharzia parasites (*Schistosoma mansoni*) from very early prepatency, *Am. J. Trop. Med. Hyg.*, 59: 872, 1998.
189. Hamburger, J. et al., Highly repeated short DNA sequences in the genome of *Schistosoma mansoni* recognises by a species specific probe, *Mol. Biochem. Parasitol.*, 44: 73, 1991.
190. Jannotti-Passos, L.K. and de Souza, C.P., Susceptibility of *Biomphalaria tenagophila* and *Biomphalaria straminea* to *Schistosoma mansoni* infection detected by low stringency polymerase chain reaction, *Rev. Inst. Med. Trop. Sao Paulo*, 42: 291, 2000.

Part V

Purification of Nucleic Acids from Mammals

16 Isolation of Nucleic Acids from Body Fluids

Karen Page and Jacqueline Amanda Shaw

CONTENTS

16.1 INTRODUCTION

Recent decades have been marked by a significant advancement of technology. The development of sensitive molecular biology techniques to detect even the smallest amounts of nucleic acids has been welcomed in many areas of research. Increasingly, attention is being drawn to the presence of nucleic acids circulating in blood and their potential diagnostic and prognostic value for a number of diseases. In addition to blood, nucleic acids have also been found in other body fluids including urine, bronchial lavage, bone marrow aspirates, saliva, and sputum. This has had a huge impact on the field of forensic science, as DNA can readily be isolated from such fluids, providing crime scene investigations with a wide range of biological evidence.

16.1.1   CELL-FREE DNA IN PLASMA AND SERUM

The presence of circulating nucleic acids in blood plasma was first described by Mandel and Métais in 1947 [1], who demonstrated that extracellular nucleic acids could be detected in the peripheral blood plasma of both sick and healthy individuals, using a perchloric acid precipitation methodology. Due to technical limitations at that time, other developments in this field were not made until the 1960s, when high levels of DNA (up to 60 µg/mL) were reported in the serum of patients with systemic lupus erythematosus [2]. The potential diagnostic implications of Mandel and Métais' discovery was realized when in 1977, Leon et al. reported that cancer patients showed increased levels of DNA in serum compared to healthy controls, with greater amounts of DNA found in the serum of patients with metastases, compared to localized disease. They showed levels of free DNA decreased by 90% after radiotherapy for lymphomas, ovarian, lung, endometrial, and cervical cancer, while high or increasing concentrations of DNA were associated with a lack of response to treatment [3]. However, again, owing to limitations in the available technology, the precise cellular origin of this extracellular DNA in the circulation of cancer patients could not be determined. Subsequent studies showed that increased levels of cell-free DNA are not only found in patients with cancer, but also in patients with rheumatoid arthritis, ulcerative colitis, pancreatitis, and other inflammatory conditions [4–6].

In 1989, Stroun and coworkers established that plasma DNA has malignant properties. They reported DNA strand instability in both circulating DNA and malignant tissue from cancer patients [7]. This study was followed closely by work in the early 1990s, in which the importance of circulating nucleic acids was brought to the fore by two different research groups, providing conclusive evidence that at least a proportion of this DNA does originate from tumor cells. Sorenson et al. reported the presence of tumor-specific mutations of K-*RAS* in the plasma/serum of pancreatic cancer patients [8], while later that year, Philippe Anker's research group demonstrated N-*ras* mutations in patients with myelodysplastic syndrome [9]. These reports were followed by two publications in 1996, which were the first to show that circulating DNA exhibits tumor-specific loss of heterozygosity (LOH). LOH was detected using the polymerase chain reaction (PCR) in plasma and serum of patients with advanced small cell lung cancer [10] and head and neck cancer [11], respectively. Investigations into aberrant methylation of cell-free DNA in cancer patients then followed. Methylation of the tumor suppressor gene, p16, was identified in the plasma/serum of both nonsmall cell lung and liver cancer patients [12,13].

Since these first studies, it has been reported that mutant cell-free circulating nucleic acids, derived from plasma and serum are found in many different malignant diseases. Hence, mutant plasma and serum DNA have been found in nasopharyngeal [14], esophageal [15,16], skin [17,18], colorectal [19,20], breast [21,22], lung [23,24], kidney [25], liver [13], bladder [26], pancreatic [27], ovarian [28], cervical [29], and prostate cancer [30], in addition to hematological malignancies, including lymphoma [31].

16.1.2   ORIGIN OF CIRCULATING NUCLEIC ACIDS

In healthy individuals, it can be assumed that circulating cell-free DNA originates from lymphocytes or other nucleated cells; however, its origins in malignancies are less clear [32]. Molecular studies indicate that a certain amount, but not all, of the circulating cell-free DNA originates from degenerating tumor cells, as reports commonly include cases where mutated genes can be detected in tumor tissue, but not circulating DNA [33]. The biological mechanisms by which circulating DNA enters the bloodstream and escapes degradation by deoxyribonucleases (DNases) are not fully understood; however, a number of mechanisms have been proposed. One possible mechanism is via cell necrosis, as higher amounts of DNA have been found in the plasma of patients with large or advanced/metastatic tumors [6,7,11]. It was noted, however, that after radiotherapy, presumed to induce cell death/necrosis, there was initially a decrease, opposed to an increase, in amounts of circulating DNA in 40% of patients [3], although it is possible that cell death caused by radiotherapy reduced the amount of DNA released. Apoptosis is another possible mechanism by which cell-free circulating DNA can enter the bloodstream, proposed by the fact that in some patient samples, apoptotic cell death can be clearly characterized by a typical pattern of DNA fragmentation. This is known as an apoptotic ladder, which results from internucleosomal cleavage of genomic DNA [34]. However, many proliferating cancer cells lose their ability to induce apoptosis, therefore, other mechanisms may be more important. Another possible mechanism is the spontaneous and active release of DNA by proliferating cancer cells, as activated lymphocytes have been shown to release DNA *in vitro* [35]. This may explain the presence of very low concentrations of cell-free DNA in some cancer patients, where the cancer may have been quiescent at the time of sample collection [32]. A final mechanism is the release of intact cells into the bloodstream and their subsequent lysis [36]. These cells can be detected by techniques such as immunocytology [37] and PCR-based methods [38,39] using specific antibodies and gene targets. It must be noted, however, that the number of circulating tumor cells does not correlate with the total amount of circulating cell-free DNA. Chen et al. calculated there would need to be 1,000–10,000 cancer cells/mL of plasma to give rise directly to the amounts of cell-free DNA detected [40].

Tumors are known to exhibit high-cellular turnover through normal programmed cell death or apoptosis. However, it is also common to detect very large DNA fragments, which suggests cell death by necrosis. Jahr et al. [34] reported that the source of the DNA in blood plasma of cancer patients are cells that disintegrate by apoptosis or necrosis in expanding tumor tissue, and that it is possible to distinguish between the two modes of cell death. They concluded that several complex processes determine the fate of the DNA released from degenerating tumor cells and the differing efficiencies of these processes may explain the wide range in the amounts and composition of cell-free circulating DNA present in the blood of cancer patients. Recently, Diehl et al. [36] have shown that in colorectal cancer, most of the circulating DNA fragments that contained adenomatous polyposis coli mutations were relatively small in size, opposed to larger fragments, which were predominantly wild type. They proposed that mutant DNA fragments may arise from necrotic cells, which have been engulfed by macrophages, and release partially digested DNA.

In addition to these mechanisms, it is important to mention the role of DNase enzymes. Minimal levels of plasma DNA (approximately, 10 ng/mL) are found in healthy people, mainly due to efficient activity of enzymes DNase I and II, whereas low activity of these enzymes is often seen in patients with malignant disease. DNase inhibitors have been detected in tumors [41], thus elevated levels of DNA can be seen in cancer patients, although they have also been reported in healthy cells such as thrombocytes [41]. The stability of plasma DNA and its clearance from the circulation have not been extensively studied. Nonetheless, a study showing the rapid clearance of fetal DNA from the maternal circulation gives us some insight into how this may work in cancer, where the authors suggest the rapid kinetics of circulating DNA will potentially make plasma-based molecular diagnostics very useful for monitoring dynamic changes in cancer patients [42].

Silva and coworkers investigated the persistence of tumor DNA in the plasma of breast cancer patients following mastectomy [43]. They reported that patients with detectable tumor DNA in plasma, postmastectomy, had characteristics of poor prognosis, such as vascular invasion. They suggested that the persistence of plasma DNA after mastectomy may identify a group of patients whose disease has more aggressive features, and that this could also have facilitated micrometastatic dissemination, the possible origin of the circulating DNA detected at this time.

16.1.3 DNA CONCENTRATION AND INTEGRITY

Using a quantitative real-time PCR approach, we and others [44,45] have shown a higher DNA integrity index in patients with malignant disease, compared to healthy controls. Data from our research group with regard to cell-free plasma DNA are shown in Figure 16.1. Using quantitative real-time PCR of the glyceraldehyde 3-phosphate dehydrogenase (GAPDH) gene, we found the mean plasma DNA concentration was significantly higher in late stage (metastatic) breast cancer

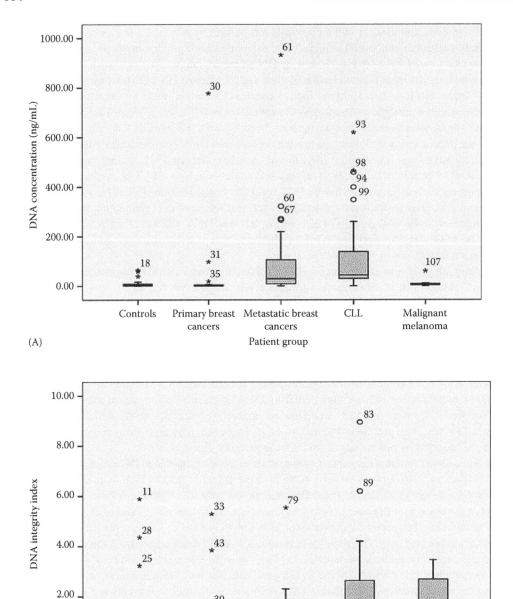

FIGURE 16.1 Measurement of cell-free plasma DNA concentration (A) and DNA integrity index (B) in patients from different sample groups.

and chronic lymphocytic leukemia (CLL) patients, but not those with early stage breast cancer, compared to healthy controls. We also found mean plasma DNA integrity to be significantly higher in metastatic breast cancer, CLL, and malignant melanoma patients, when compared to healthy controls [46]. Similar findings have also been reported for cell-free DNA in serum. Using quantitative real-time PCR analysis of ALU repeats (short stretch of DNA originally characterized by the action of the restriction endonuclease), the mean serum DNA integrity was shown to be significantly higher in late stage (II–IV), but not in early stage (0 and I) breast cancers, when compared to healthy female controls [47]. The same group also observed similar findings in colorectal cancer, with increased serum DNA integrity seen in Stage I/II and Stage III/IV cancers, when compared to healthy controls [48].

16.1.4 Nucleic Acids from Other Body Fluids

Cell-free DNA is also found in other body fluids. Cell-free DNA has been detected in urine [49,50], bone marrow aspirates [51], bronchial lavage [52,53], and sputum [54].

In urine, tumor-specific DNA alterations have been used as a tool for detection of primary [55] and recurrent [56] bladder cancer, with Su et al. detecting K-ras mutations in urine sediment [57]. More recently, Miyake et al. reported detection of FGFR3 mutations in DNA isolated from urine [58]; however, it has been shown that DNA has a higher mutation detection rate in the urine supernatant, rather than sediment [59]. Quantification of cell-free fetal DNA from maternal urine was performed by Majer et al., and very low concentrations of DNA were detected, so this approach was inappropriate for prenatal diagnosis [60].

Taback et al. reported LOH in the plasma obtained from bone marrow aspirates from 48 breast cancer patients, using 8 polymorphic markers and demonstrated novel findings of tumor-related genetic markers [51]. Recently, Schwarzenbach et al. demonstrated, for the first time, the presence of cell-free tumor-specific DNA in blood and bone marrow aspirates of prostate cancer patients, and they suggested a possible relationship to bone marrow micrometastasis [61].

A study by Carstensen and coworkers showed, for the first time, that DNA could be isolated from cell-free bronchial lavage supernatants, a sample that is usually discarded after cell harvest. In a cohort of 30 lung cancer patients, they examined isolated cell-free DNA for microsatellite alterations. They found that intact DNA could be isolated from all cell-free bronchial lavage supernatants, and that tumor-associated changes were detected in the DNA of 47% of patients [52].

Belinsky et al. compared alterations in DNA isolated from sputum and serum, by detecting and evaluating levels of methylation across a panel of 8 genes in the primary tumor biopsies, serum, and sputum obtained from 72 patients with Stage III lung cancer. The prevalence of methylation in the 8 genes in sputum (21%–43%) approximated to that seen in tumors, but was 0.7–4.3-fold greater than detected in serum. Their study demonstrates that sputum can be used effectively as a surrogate for tumor tissue, to predict the methylation status of advanced lung cancer where biopsy is not feasible [54]. This study was in agreement with Wang et al. who detected levels of LOH, microsatellite instability, and DNA methylation in sputum samples and reported that these biomarkers may be used as a sensitive and reliable molecular diagnostic method for lung cancer [62].

16.1.5 Circulating Cell-Free RNA

In addition to the recent achievements in the study of circulating cell-free DNA, there is also growing interest in the presence of circulating RNA. Similar to DNA, studies on circulating RNA have focused on cancer patients, with elevated levels seen in patients with malignant melanoma [63], breast and thyroid cancer [64], and hepatocellular carcinoma [65]. However, circulating RNA has also been found in the plasma and serum of healthy individuals [66]. This is perhaps surprising as RNase is known to be present in the blood, and all free RNA in the blood might be expected to be rapidly degraded or destroyed. Kopreski et al. demonstrated the presence of tyrosine kinase mRNA in patients with malignant melanoma, and found no detectable mRNA in healthy controls [63]. Both this group and Lo et al. [67] suggested there were extracellular RNA molecules present in the circulation of cancer patients that are possibly protected from RNase. It was suggested that RNA could be contained in apoptotic bodies or bound to protein/phospholipids and therefore protected from nuclease degradation. It has also been suggested that a large proportion of plasma RNA is associated with such filterable particulate matter [68]. This study is also in agreement that the RNA is surprisingly stable against RNase degradation [69]. It was later demonstrated that fetal RNA is present in maternal plasma. Other developments in this area demonstrated that placental-derived mRNA such as corticotrophin-releasing hormone and human placental lactogen were not only detectable in maternal plasma, but also their protein product levels correlated with mRNA expression [70]. These targets may be utilized in noninvasive prenatal diagnosis, as gender- and polymorphism-independent fetal markers. These findings have prompted further investigation in linking circulating RNA, like DNA, to malignant diseases.

Several groups have demonstrated different types of mRNA in the plasma/serum of cancer patients, but these studies do not directly relate the presence of tumor-specific mRNA in plasma to tumor characteristics. However, an association has been shown between the presence of circulating cytokeratin 19 and mammaglobin mRNA in patients with breast cancer and poor prognostic features [71]. In addition, telomerase mRNA has been detected in breast cancer [72], and the presence of epithelial mRNA in plasma significantly correlated with tumor size and proliferative index [21]. Recently, a report demonstrated reduced plasma RNA integrity in nasopharyngeal cancer patients, by analysis of 5' and 3' transcript fragments of the GAPDH gene [73]. The 3' to 5' GAPDH ratio was significantly lower in the plasma of untreated cancer patients, compared to healthy controls, with 74% showing a significant increase following radiotherapy.

There have been also a small number of reports analyzing extracellular mRNA profiling in the area of forensic science, and these studies suggest this approach may have potential use in this area. Multiplex PCR studies were carried out by Juusola and Ballantyne in 2005 [74], and more recently, they have reported the development of a sensitive and robust multiplex quantitative reverse transcriptase-PCR (RT-PCR) assay to identify specific genes from blood, saliva, semen, and menstrual blood [75]. Their multiplex assay could detect up to two body fluids/tissue-specific genes and one housekeeper or control gene simultaneously, and by using a quantitative PCR approach, they could establish the tissue specificity of a gene product, particularly when relative abundance of a number of different mRNA molecules demonstrates a unique or restricted pattern of expression. To date, there have been fewer published studies on circulating RNA than on DNA, and this may be due in part to problems with storage and freeze–thawing of samples and reproducibility of RNA analyses [76], in addition to demanding and more variable extraction methods.

16.1.6 Circulating DNA: A Potential Biomarker for Cancer?

Circulating nucleic acids present in body fluids provide us with a potential and attractive target for molecular diagnosis of diseases such as cancer, as they can be detected by minimally invasive measures. There are few studies to date with long patient follow-up times, however, the presence of cell-free tumor-derived DNA has been shown to be an indicator of poor prognosis and shorter survival [21]. More recently, the presence of cell-free tumor DNA at diagnosis has been found to be a reliable predictor of overall survival [77]. The development of cancer involves a complex and progressive accumulation of molecular genetic changes, such as mutation, LOH, and hypermethylation. The invasive nature of biopsy sampling makes mass cancer screening and regular follow-up impractical; therefore, the development of noninvasive tumor biomarkers is highly attractive. The potential benefit of using plasma DNA in cancer clinical medicine is twofold. First, there is the possibility of a plasma DNA test for cancer diagnosis. This is challenging, as a reliable, plasma DNA-based assay would require a number of tumor biomarkers, however it would require only a simple blood sample. Second, the association between mutations in plasma DNA and tumor may be valuable in the use of plasma DNA in the follow-up of patients, assessing their responses to therapy, and potentially to identify disease recurrence. Tumor heterogeneity may mean a combination of biomarkers is required; therefore, multiple markers or techniques will be required. It will be necessary, then also, to tailor each disease with a panel of associated genes to increase the reliability and sensitivity of testing.

In addition to the area of cancer, cell-free DNA has also proved to be a valuable resource in the fields of organ transplant and posttrauma monitoring, where circulating DNA concentrations have been correlated with both severity of injury in trauma patients, as reported by Lo et al. [78], and severity of stroke [79]. It also plays a role in noninvasive prenatal diagnosis [80], an essential part of modern obstetrics, since the discovery of cell-free fetal DNA in maternal plasma in 1997 [81]. Compared with circulating fetal cells, gained via amniocentesis for example, cell-free fetal DNA in maternal plasma appears to offer an advantage in that the concentration in the second trimester is some 970-fold higher, offering an easily accessible source for analysis [82]. In the last decade, this noninvasive source of fetal DNA has been used in a number of areas including prenatal diagnosis of sex-linked disorders [83], fetal RhD status [84], and β-thalassemia [85]. In addition, abnormal concentrations of this DNA have also been described in pregnancy-associated disorders including preeclampsia [86] and trisomy 21 [87]. Recent research in this field is targeting epigenetic alterations, such as DNA methylation, for use as a potential marker for prenatal diagnosis.

The increasing number of publications in the area of circulating nucleic acids is the evidence of this growing research field. However, direct comparison of data is not always possible due to differences in parameters analyzed and a lack of standardized methods. There have been considerable advances made toward applications of circulating nucleic acids, with early methods such as radioimmunoassay only detecting nanogram quantities of DNA [88], whereas with PCR, detection is possible to picogram levels. Methods used for isolation of nucleic acids are numerous, and even when using commercially available kits, there is no accepted gold standard method thus far. More importantly, consideration also needs to be given to the very first steps in the collection of the body fluid in question, for example blood, where extreme care needs to be taken when separating plasma from the buffy coat (containing lymphocytes, and therefore cellular DNA), so as to ensure the sample is free from contaminating cellular DNA. Collection and storage, isolation of nucleic acids, and their quantification represent the most critical aspects to cell-free determination. The following section attempts to deal with methodology issues and provide a comparison of cell-free nucleic acid isolation methods.

16.2 PURIFICATION PRINCIPLES

16.2.1 ISOLATION OF NUCLEIC ACIDS FROM WHOLE BLOOD

Blood (whether for plasma or serum) is the most commonly used bodily fluid for isolation of cell-free nucleic acids. DNA yields are usually higher in serum, opposed to plasma [89,90], and this may be reflected by an *in vitro* process that occurs during clotting or during the release of DNA from destroyed white blood cells [89,91]. Lee et al. observed that most cell-free DNA in serum is generated by the lysis of leukocytes by performing blood spiking experiments, and concluded that serum is not suited to monitor the concentration of cell-free DNA [89]. However, while Umetani et al. are in agreement with regard to increased levels of cell-free DNA in serum opposed to plasma, they also suggest an unequal distribution of DNA during separation from whole blood, and the use of serum would provide increased sensitivity. Also, if DNA is lost during purification from plasma, but not serum, using the latter may be more efficient [92]. Therefore, measurement of cell-free DNA concentration may provide erroneous results if serum is used; however, serum may provide a better specimen as a biomarker, as advocated by Umetani et al. [92].

Methodological differences are present throughout the isolation process, but are evident even at the preanalytical stage, for example, delays in blood processing and storage temperatures can influence DNA yields. Jung et al. found no changes in DNA concentrations from plasma samples stored at room temperature or 4°C for up to 24 h; however, this time delay had a substantial impact on serum DNA levels [93]. Extensive analysis of the effects of prolonged storage of plasma/serum that affects nucleic acid yield is not published, but Kopreski et al. were able to detect intact mRNA after 2 years of storage at −70°C [63], and our group has successfully isolated DNA from plasma samples after more than 5 years of storage at −70°C. Different coagulants (ethylenediaminetetraacetic acid [EDTA]), heparin, and citrate) can be used for plasma, this appears to make no difference to DNA quality and yield; however, if there is a delay in blood processing, EDTA is superior to citrate and heparin [94], while for serum, different time periods have been allowed for clotting. After collection, we, and others, have reported that the centrifugation steps taken to separate the plasma from the buffy coat and erythrocytes are also important. We demonstrated that a third centrifugation step was necessary to eliminate cellular contamination in healthy controls [44], in agreement with Chiu et al. [95].

Further variation can also be introduced at the level of isolation of DNA. Many methods and commercial kits are currently available for isolation of cell-free DNA from serum or plasma, ranging from column-based methods to magnetic bead technology, which can provide a large variation in both DNA quantity and yield. These variations may lead to difficulties when comparing results between reports from different groups, due to experimental inconsistencies. Quantification of isolated DNA is commonly performed by laboratories, and whether using methods utilizing PCR, fluorescent dyes, or standard spectrophotometry, it is vital that the initial starting fluid is carefully prepared, so as to avoid contamination with other cellular DNA.

16.2.2 ISOLATION OF NUCLEIC ACIDS FROM OTHER BODILY FLUIDS

Nucleic acids are also routinely isolated from urine, sputum, and bronchial lavage, using a variety of methods. The initial processing of the particular type of sample may vary, with the isolation of DNA occurring either by a standard procedure such as proteinase K digestion and phenol/chloroform extraction, or using a kit-based method. Schmidt et al. were the first to demonstrate that it is possible to isolate intact DNA and RNA from cell-free bronchial lavage supernatants [96]. They compared isolation of DNA using (1) a modified salting-out protocol, first described by Miller et al. in 1988 [97], and (2) the QIAamp DNA blood mini kit, from plasma and cell-free bronchial lavage supernatants [98]. They reported, first, that higher DNA yields were present in bronchial lavage samples, compared to those from plasma, and second, the increased yield was consistent when using the modified salt protocol, showing an eightfold elevation in levels of cell-free DNA. The same group also isolated RNA from serum and bronchial lavage samples, with increased RNA yields in the cell-free bronchial lavage supernatant [53]. As for urine, it is useful to mix the sample with 0.5 mol/L EDTA pH 8.0, to give a final concentration of 10 mmol/L EDTA prior to isolation of nucleic acids. This inhibits any possible nuclease activity in urine, and the sample is then stored at −70°C [99]. Isolation of cell-free DNA is performed using similar methods as for plasma and serum, such as using column-based or magnetic bead technology to bind DNA.

In a study based on lung cancer, DNA was isolated from sputum by protease digestion, followed by phenol/chloroform extraction and ethanol precipitation, a routinely used method. The fact that cell-free DNA can be successfully isolated from easily accessible plasma, serum, and other bodily fluids, such as urine and bronchial lavage, makes them an extremely valuable resource for research. It should be emphasized that the preanalytical stages are just as important as the DNA isolation itself. Methods for isolating DNA for forensic casework have been modified in recent years, with groups initially favoring the Chelex method (BioRad, Hercules, California), where cells are gently removed from the sample material, lysed in an alkaline Chelex bead suspension, and the DNA is freed in the presence of chelators, ready for PCR amplification of short tandem repeat loci. These are highly informative genetic markers, which are a powerful system of human identification. However, Greenspoon et al. identified problems with DNA isolated using this method, in that storage of DNA for more than 1 year at −20°C resulted in signal loss at a locus was observed in 30% of samples, compared to no loss when the DNA was freshly isolated [100]. By using QIAamp spin columns (Qiagen, West Sussex, United Kingdom), they eliminated this problem.

Historically, methods for isolation and quantification of DNA lacked both specificity and sensitivity, were time-consuming, and also associated with exposure to potentially toxic chemicals and radioactive isotopes [88]. In 1989, a method for the extraction of DNA from whole blood that used sodium perchlorate and chloroform instead of phenol was reported by Johns et al. [101], noting significant time savings in addition to fewer hazards to the user. The following year, Boom et al. described the use of a silica-based method of DNA isolation [102]. This method used a chaotropic agent, guanidinium thiocyanate, to lyse cells and inactivate nucleases, whilst simultaneously facilitating the binding of the freed nucleic acids to silica particles. This method was extremely sensitive due to the strong binding affinity of silica particles for nucleic acids in the presence of chaotropic agents. The method was designed to be more rapid and involves less tube transfer stages, to reduce the risk of sample contamination or DNA loss than an organic method; however, care must be taken while handling agents such as guanidinium thiocyanate. Jen et al. reported that, in their experience, SDS/proteinase K digestion of DNA from plasma or serum followed by a phenol/chloroform extraction still provided the best quality and yield of DNA; however, they also agreed the method had its disadvantages [103]. In recent years, the silica–DNA binding affinity method first described by Boom et al. has been utilized in the development of many commercial kits such as the DNAIQ system produced by the Promega Corporation (Madison, Wisconsin).

16.2.3 ISOLATION OF NUCLEIC ACIDS FROM PLASMA AND SERUM

Methods for isolating nucleic acids from plasma and serum have increased in number in recent years, moving on from initial SDS/proteinase K digestion followed by phenol/chloroform methods to commercially available column-based kits, these being the most common method used at present. Such kits are manufactured by different companies, each with slight modifications and variations on the protocol to be followed. More recently, automated large scale DNA extraction systems, such as the Maxwell 16 system from Promega and the X-tractor gene from Corbett Life Sciences, Sydney, Australia, have been introduced, which claim to offer automation and purification that save time and labor by eliminating reagent preparation, pipetting, and centrifugation steps [104].

When isolating extracellular nucleic acids from different samples, the main problem to overcome is the low quantity and poor quality of DNA gained, irrespective of the isolation method used. Therefore, when deciding on a method for isolating DNA, there are a number of other factors to consider. First, what is the source of the genomic DNA? Commercially available kits are able to extract DNA from a wide variety of samples, such as blood, plasma, serum, and urine, with just minor modification to the protocol. Some kits can be used for a number of different starting materials, therefore reducing the need to purchase several kits. Other kits are designed for a particular tissue type, and in some cases species-specific kits are available, should these be relevant. These types of kit would be particularly important in the area of forensic science, for example, when hair, bone, or degraded samples need to be analyzed. Second, consideration should be given to the sample volume and the amount of DNA to be isolated. Kits can be restricted to a certain sample size or volume because of the limits of the columns or reagents used. It may prove worthwhile having a closer look at individual protocols, as there are a number of different ways to lyse cells (e.g., sonication or detergent-based cell lysis), remove proteins and RNA (e.g., enzymatically), and the total processing time can vary. Third, consideration of how the DNA is to be isolated needs to be taken into account. Some kits use organic solvents and others may use an alcohol precipitation or wash steps. These steps may potentially affect downstream applications of the DNA, such as PCR, if not removed completely. Last, thought should be given to the throughput of samples to be processed and the frequency of kit used. Some kits may be more suitable for managing multiple samples at once and high-throughput-specific kits may also be available.

16.3 CURRENT STATUS OF TECHNOLOGY

Over the years a variety of methods have been applied to the isolation of DNA and RNA from blood and tissues. These range from standard laboratory-based protocols, followed by the use of magnetic bead technology, to the present, predominantly kit-based column methods as the method of choice when extracting nucleic acids from body fluids.

16.3.1 IN-HOUSE METHODS FOR EXTRACTION OF DNA
FROM LYMPHOCYTES AND PLASMA

The standard in-house protocol for extraction of DNA from plasma and lymphocytes is based on the
use of phenol/chloroform. Samples from −80°C storage are thawed and their volume is increased to
5 (lymphocytes) and 1 mL (plasma) with sterile 1x phosphate buffered saline (PBS). For lymphocytes, the samples are centrifuged to remove the supernatant. The pellet is resuspended in in-house
rapid extraction buffer (1 M KCl, 1 M Tris pH 8.3, 1 M MgCl$_2$, Tween 20, and Tergitol) containing
proteinase K (10 mg/mL), and incubated overnight at 58°C. The samples are heated to 99°C, centrifuged, and the supernatant is transferred to a clean tube. The plasma samples are also heated,
centrifuged, and the supernatant reserved for later manipulation. Following these initial preparation
steps, a standard phenol/chloroform extraction is then performed. The purified DNA is air-dried and
resuspended in 1x Tris/EDTA (TE) buffer.

We have also isolated cell-free DNA from blood plasma samples using magnetic beads, via a kit
utilizing CST (ChargeSwitch technology) magnetic bead technology. Using this method, up to
100 μL whole blood or plasma is lysed using a buffer containing proteinase K, followed by incubation at room temperature for 5 min. Purification buffer (containing magnetic beads) is added to each
sample and separation performed using a magnetic rack. The magnetic beads are washed twice and
the DNA eluate is transferred to a fresh tube.

More recently, the most widely reported kit-based method for isolating DNA from plasma or
serum is the QIAamp DNA blood mini kit, manufactured by Qiagen. This is evident from the literature, with many groups using this as their method of choice [34,43,45,77,105,106]. The QIAamp
DNA blood mini kit is a column-based method where the starting fluid (plasma, serum, or buffy
coat) is lysed using a mixture of buffer and protease, the samples are manually passed through a
column utilizing silica technology and centrifugation to bind the DNA, followed by two washes, and
the DNA is finally eluted into a small volume, typically between 100 and 200 μL. Depending on the
downstream application, the volume can be reduced to 50 μL. Modifications have been made to the
manufacturer's protocol by some research groups when isolating DNA from plasma, including ours,
with differences in initial and final elution volumes. We use a starting volume of 500 μL of plasma
and 200 μL of lymphocytes with this method. The DNA from plasma and lymphocytes is eluted in
a volume of 70 and 100 μL, respectively, as we find this provides isolated cell-free DNA at ideal
concentrations for downstream applications such as quantitative real-time PCR.

16.3.2 SMALL-SCALE DNA EXTRACTIONS

The column-based methods are the basis for a number of commercially available kits, with a growing
number of companies now manufacturing such kits for the isolation of genomic DNA from bodily
fluids. A comparison of some of the currently available small-scale kits is summarized in Table 16.1.

The most commonly used kits are column-based, as they do not require the use of potentially
harmful organic solvents, such as phenol and chloroform, and do not require ethanol precipitation
of DNA. These methods also serve to efficiently remove cellular debris and inhibitors, which may
potentially affect downstream applications such as PCR. Many companies also now offer mini-,
midi-, and maxi-scale editions of the same kits, which can be used for materials in a range of starting
volumes, from 100 μL up to 10 mL. As previously mentioned, the same kit can also often be used
with a variety of starting materials, for example, the Flowgen genomic DNA extraction kit (Flowgen,
Nottingham, UK) (Table 16.1) can be used for whole blood, serum, buffy coat, and other fluids.
Qiagen has expanded their kit design and offered a wide range of kits, for whole blood, their
FlexiGene DNA system enables isolation of genomic DNA from variable amounts of whole blood
and buffy coat. The purification step is performed in a single tube that not only reduces cost, but also
simplifies handling. For isolation of genomic DNA from other body fluids, including bone marrow
aspirates and saliva, their Generation Capture Column kit can be used.

TABLE 16.1

Comparison between Manufacturer's Small-Scale DNA Isolation Methods

Manufacturer's Kit	Method	Recommended for	Starting Volume (µL)	Time (min)	Reported Total Yield (µg)
Chemicon Non-organic DNA kit	Noncolumn	Whole blood and body fluids	Not stated	240	Up to 150
Invitrogen Easy DNA kit	Noncolumn	Whole blood, single hair, and tissue	Up to 350	<90	2–5
Flowgen Genomic DNA extraction kit column	Glass/fiber matrix	Whole blood and body fluids	Up to 300	40–60	6 from whole blood
					50 from buffy coat
Invitrogen Gene-Catcher kit	Magnetic beads	Whole blood	300–1000	Not stated	30
FavorPrep Blood genomic DNA kit	Column	Whole blood, plasma, serum, and body fluids	Up to 200	60	3–6 from whole blood
					30–60 from buffy coat
Promega Wizard genomic DNA kit	Column	Whole blood	Up to 300	60	5–15
Qiagen DNA blood mini kit	Column	Whole blood and body fluids	Up to 200	20–40	4–12
Qiagen Generation capture column kit	Column	Whole blood, plasma, buffy coat, and body fluids	200	20–40	3–8
Sigma GenElute blood genomic kit	Column	Whole blood	Up to 500	<40	Up to 10
Spin Clean genomic DNA kit	Column	Whole blood	Not stated	Not stated	2–5

The recovery of nucleic acids from bodily fluids can vary between methods, with the standard phenol/chloroform extraction method providing high yields of DNA from serum and plasma [103]. Companies are keen to advertise the fact that their kits will provide the user with high yields of excellent quality DNA for downstream applications, and comparisons have been made between different extraction methods. de Kok et al. evaluated the ABI 7700 sequence detection system (Applied Biosystems, Foster City, California) for standardization of four isolation methods: the PureGene DNA isolation kit, with or without further purification (Gentra Systems, Plymouth, Minnesota); a standard phenol/chloroform method; and the QIAamp blood kit [107]. They used a known amount of DNA marker copies, close to the detection limit of the PCR, isolated the DNA using the four different methods, and used the ABI sequence detection system to determine isolation efficiency and repeatability of each method. They reported significant differences between the four methods. The PureGene method had the highest isolation efficiency but relatively poor precision (mean relative variance); the use of a QiaQuick column following the PureGene method markedly decreased its isolation efficiency and the QiaAmp method showed poor efficiency when only a few copies of marker DNA were present. A similar study was carried out by Lee et al., in which they evaluated five protocols for DNA isolation from plasma [89]. Ten HIV-1-positive plasma samples were processed in parallel, with the GlasPac/GS and the HIV monitor assay yielding the highest recovery of genomic DNA, more so than GeneClean kit (Q-bioGene, Cambridge, United Kingdom), Dynal beads (Invitrogen, Paisley, United Kingdom), and QIAamp Blood Kit (Qiagen).

TABLE 16.2

Comparison between Manufacturer's Large-Scale DNA/RNA Isolation Methods

Company/Kit	Number of Samples	Sample	Time (min)	Used to Purify	Notes
Applied Biosystems 6100 Nucleic Acid PrepStation	Up to 96	Whole blood and plasma; up to 750 μL	30	DNA and RNA	High-throughput integral vacuum system
Autogen Quick Gene 810 nucleic acid system	1–8	Whole blood and tissue; up to 200 μL	6	DNA and RNA	DNA/RNA isolation via a 80 μm membrane/film. No need for centrifugation
Corbett X-tractor membrane gene system	8–96	Whole blood, urine, and buffy coat; up to 200 μL	60	DNA and RNA	8–96 high-throughput silica bind/elute protocols with vacuum processing
Enzyme 12 GC system	1–12	Various fluids	30–60	DNA and RNA	Small-scale isolation instrument
Invitrogen IPreppurification instrument	12	Whole blood; up to 350 μL	30	DNA	Utilizing CST
Promega Maxwell purification system	1–16	Whole blood and buffy coat; up to 500 μL	30	DNA, RNA, and protein	Binding via paramagnetic particles, washing, and eluting for direct use in downstream applications
Qiagen QIAcube	12	Whole blood, serum, plasma, and buffy coat	30–60	DNA, RNA, and protein	Allows processing of most Qiagen spin column kits

16.3.3 LARGE-SCALE DNA EXTRACTIONS

As mentioned, more recently, there has been increasing interest in automated DNA isolation methods, and a number of different robotic systems are available (Table 16.2), for both large-and small-scale extractions. These systems also vary, like manual kit-based methods, in the way the DNA is extracted. Automated systems currently available include the column-based Promega Maxwell 16 purification system; the QuickGene-810 from Autogen (Autogen, MA, USA), which employs an ultrathin membrane film for efficient capture of nucleic acids; the Corbett Life Sciences X-tractor gene, an automated nucleic acid extraction system based on silica-column technology; and the iPrep purification instrument from Invitrogen, which utilizes CST, a unique, ionizable nucleic-acid-binding ligand whose charge can be switched based on the pH of the surrounding medium. Other systems based on similar technology to those mentioned above are the Applied Biosystems PRISM 6100 Nucleic Acid PrepStation; the Qiagen Qiacube, which can be used for most bodily fluids; or the Qiagen BioRobot M48 for larger scale automated DNA extraction.

Automated systems, such as Personal Automation technology from Promega, incorporating the Maxwell 16 instrument, can potentially improve laboratory productivity while maintaining high-quality, reproducible results. Traditionally, automated systems have been large, expensive, and complicated, with the emphasis solely on high-throughput use. In contrast, the trend is now moving toward more integrated solutions, combining compact, low-cost instruments with optimized reagents and methods. This has the potential to maximize flexibility, productivity, and reliability for an individual, to reduce the time and labor spent on sample preparation, in addition to reducing inter-run variation. Instruments can be used for the isolation of DNA, RNA, and protein, and are also moving toward working with low-elution volumes, thus expanding their use in the area of forensic science.

16.3.4 ISOLATION OF RNA FROM NUCLEIC ACIDS

Isolation of RNA from bodily fluids can be performed using similar commercially available kits, with methods for isolating RNA from plasma and serum also coming under scrutiny. El-Hefnawy et al. performed a comparison between nine different RNA isolation protocols [76]. The methods in the study included RNeasy mini kit and QIAamp viral RNA mini kit (Qiagen), SV total RNA isolation system (Promega), TriBD reagent (Sigma-Aldrich, Poole, United Kingdom), and a modified guanidinium isothiocyanate (GIT)/phenol extraction, based on the original method of Chomczynski and Sacchi [108]. As most plasma RNA is probably present as short fragments, the efficiency of the different methods was tested using a short synthetic bacterial β-galactosidase transcript, in addition to endogenous plasma 18S rRNA. The methods that gave the best recovery were the precipitation-based RNA isolation methods: GIT/phenol extraction and TriBD reagent. In comparison, the recovery of short RNA transcripts by nonspecific binding of RNA to resins/columns or magnetic beads was much lower. While precipitation-based methods of RNA extraction provide the best yields, they are not practical for scaling up reactions, as difficulties are encountered when precipitating low concentrations from excessively large-aqueous volumes.

A procedure based on the binding of nucleic acids with glass-milk in the presence of chaotropic salts has been adapted for efficient isolation of 100–10,000 bp DNA fragments and 50–10,000 bp RNA fragments by Tamkovich et al., providing 90% and 85% efficacy of isolation of 100 bp DNA and RNA, respectively [109]. They reported that the extracted nucleic acids are free from contamination, enzymes, and fluorochromes, and that it is a simple, rapid, and cost-effective method. A recent study by Cerkovnik et al. reported the introduction of a new step in the RNA isolation procedure, in that the plasma was concentrated by evaporation prior to isolation of RNA [110]. The effectiveness of the new isolation protocol and its influence on RNA integrity were evaluated by direct determination of RT-PCR transcripts for porphobilinogen deaminase and GAPDH genes. They observed that plasma RNA was most efficiently isolated from large volumes of samples after the introduction of an evaporation step and by using TRIzol LS reagent. Additionally, they also reported that a single freeze–thaw process had no significant effect on RNA integrity and quantity of plasma RNA. Large-scale automated processes are also available for RNA isolation, including the QuickGene 810 (Fuji-Film, United Kingdom), QIAcube (Qiagen), Applied Biosystems PRISM 6100 Nucleic Acid PrepStation, and the Promega Maxwell 16 purification system.

16.3.5 QUANTITATIVE ANALYSIS OF NUCLEIC ACIDS

Studies show elevated levels of nucleic acids in patients with malignant disease (in addition to other inflammatory conditions), compared to healthy controls, irrespective of the use of serum or plasma, or other bodily fluids such as urine. A number of different methods have been employed by researchers to assess the amount of DNA or RNA present in body fluids, and a comparison of methods has been summarized in Table 16.3. For DNA, these include radioimmunoassay [6,88], fluorescent nucleic acid stains such as PicoGreen dsDNA quantitation reagent [91,111], and Hoechst 33258 [112], the DNA dipstick test (Invitrogen) [113], competitive PCR (34), and spectrophotometry [114]. Early methods of nucleic acid quantitation, such as radioimmunoassay, were less sensitive and involved the use of hazardous materials. Colorimetric assays were also used, in which reagents such as diphenylamine were added to the patients' plasma or serum, and a color change was produced, which could be correlated with DNA concentration [115]. The method of quantitation varies between research groups, but whatever the method chosen, it must be deemed to be sensitive, robust, and safe enough to use.

Most recently, many groups have developed quantitative real-time PCR [15,91,105,116] to accurately quantify picogram amounts of circulating nucleic acids. DNA isolated from body fluids is measured by analyzing the amounts of a given housekeeping gene. For example, plasma DNA concentration has been measured by real-time PCR analysis of the β-globin [105,117], β-actin [15],

TABLE 16.3

Comparison of DNA Concentration Determined by Different Quantitation Methods

Method	Sample	DNA Concentration Reported (ng/mL)	Reference
Radioimmunoassay	Serum	Benign 118 ± 14; malignant 412 ± 63	[6]
	Serum	Range 25–1000	[87]
PicoGreen dsDNA reagent	Plasma	Range 17–256	[110]
Hoechst 33258 dye	Plasma	Controls 0–66; malignant 0–1054	[111]
DNA dipstick test (Invitrogen)	Plasma	Range 5–1000	[112]
Spectrophotometry	Plasma	Controls 0–49; malignant 21–195	[113]
Quantitative real-time PCR			
β-Actin	Plasma	Controls 7–14; benign 6–288; malignant 46–4738	[15]
hTERT	Plasma	Controls 8.4–27; malignant 1–3000	[115]
β-Globin	Plasma	Controls 3–73; benign 5–96; malignant 9–566	[116]
GAPDH	Plasma	Controls 1–97; benign 9–91; malignant 4–136	[117]

hTERT [116], and GAPDH genes [118]. Measurement of DNA concentration by a real-time PCR assay is based on the construction of a standard curve using serial dilutions, usually in triplicate, of a genomic DNA standard. After completion of the PCR assay, the concentrations of unknown samples are extrapolated from the standard curve using manufacturer's software, and can be represented in different ways. The DNA concentration can be either expressed as nanogram per milliliters (taking into account the relevant dilution factors) or as genome equivalents (GEs) per milliliter of plasma. Lo et al. described an equation, used by a number of groups, expressing concentration of DNA in copies per milliliter, based on using 6.6 pg of DNA per cell (GE) as a conversion factor [82].

It is therefore easy to quantify cell-free DNA to discriminate between very advanced stage disease and early stage cancers/healthy controls. However, identifying differences between early stage cancer and healthy controls in terms of DNA concentration is more difficult as the ranges between the groups overlap, and the use of this as a sole breast cancer marker is unlikely.

Methods for quantifying RNA from body fluids are similar to those used for DNA, with quantitative RT-PCR in current use [53]. Lledó et al. demonstrated clear differentiation between healthy controls and colorectal cancer patients and demonstrated that hTERT mRNA could be detected and quantified in plasma [119].

16.4 METHODS

General reagents:

Depending on the method of choice, a variety of reagents are required for isolation of nucleic acids from body fluids. However, if a kit-based method is used, most of the reagents required are included within the kit itself.

As previously mentioned, the QIAamp DNA blood mini kit is the most commonly used kit for extraction of DNA from body fluids. With this kit, the user will only need to provide absolute ethanol and sterile PBS (if sample volume needs to be increased), as all other reagents are included. Other kits, such as those that utilize magnetic bead technology, require no additional reagents. Reagents required for a nonkit method, such as a routine, in-house phenol/chloroform, extraction require the user to provide phenol/chloroform/isoamyl alcohol (IAA) (24:24:1), chloroform/IAA (24:1), sodium chloride, absolute ethanol, 70% ethanol, and 1x TE buffer.

Similarly, kit-based methods that isolate RNA from body fluids usually contain most, if not all, of the required reagents. However, if the user follows in-house protocols, such as described below, the user will need to provide guanidinium thiocyanate, sodium acetate, water-saturated phenol, isopropanol, and diethylpyrocarbonate (DEPC)-treated water/deionized formamide.

Supplies and equipment:

There is no special equipment necessary for the isolation of nucleic acids from body fluids, and most, if not all, of the items will already be in use within the laboratory. Items required are disposable gloves; 10, 20, 200, and 1000 μL pipettes and tips; 0.5 and 1.5 mL tubes; water bath or incubator, microfuge or vacuum manifold; vortexer, pH meter, and a spectrophotometer. These items are needed when following either a commercial kit-based or an in-house protocol.

16.4.1   PURIFICATION OF DNA FROM BODY FLUIDS

16.4.1.1   Phenol/Chloroform Method

This in-house protocol is based on a standard phenol/chloroform extraction and can be used to isolate DNA from body fluids.

Procedure:

1. Respective samples from −80°C storage are thawed and their volume increased with sterile 1x PBS to 5 and 1 mL for lymphocytes and plasma, respectively.
2. For lymphocytes, the samples are centrifuged at $850 \times g$ for 5 min and the supernatant removed. The pellet is resuspended in our in-house rapid extraction buffer (1 M KCl, 1 M Tris pH 8.3, 1 M $MgCl_2$, Tween 20, and Tergitol) containing proteinase K (10 mg/mL), and incubated overnight at 58°C.
3. The lymphocyte samples are heated to 99°C, centrifuged, and the supernatant is transferred to a clean tube. The plasma samples are also heated to 99°C, centrifuged, and the supernatant is reserved for later manipulation.
4. An equal volume of phenol/chloroform/IAA (25:24:1) is added, the sample is mixed and centrifuged at top speed ($20,000 \times g$) for 2 min.
5. The top aqueous layer is removed, to which an equal volume of chloroform/IAA is added, mixed, and the sample is centrifuged at top speed for 2 min.
6. The top aqueous layer is removed and the DNA is precipitated by adding 1/10 of the total volume of sodium chloride (1 M) and 3 times the volume of ice-cold absolute ethanol.
7. Samples are incubated at −20°C for 30 min, after which they are centrifuged at top speed, washed with ethanol (70%), recentrifuged ,and the supernatant is removed.
8. The pellet is air-dried and resuspended in a suitable volume of 1x TE buffer. Lymphocytes and plasma are resuspended in 500 and 100 μL, respectively.

16.4.1.2   QiaAmp DNA Blood Mini Kit

In our laboratory, we routinely use the Qiagen QiaAmp DNA blood mini kit for the isolation of DNA from plasma and lymphocytes, with slight modifications to the manufacturer's protocol.

Procedure:

1. Buffer AL (200 μL) and protease (40 μL) are added to plasma (500 μL) or lymphocytes (200 μL), the sample is mixed thoroughly by pulse-vortexing for 15 s, and incubated at 56°C for 15 min.

2. Ethanol (96%–100%, 200 μL) is added to the sample, mixed thoroughly by pulse-vortexing, and applied carefully to a Qiagen column.
3. The column is microfuged at $6000 \times g$ for 1 min, after which buffer AW1 (500 μL) is carefully added and centrifuged for a further min. The column is placed in a clean tube, and the filtrate is discarded.
4. Buffer AW2 (500 μL) is carefully added to column and centrifuged at top speed ($20,000 \times g$) for 3 min, after which the column is placed in a clean tube and the filtrate discarded.
5. The column is centrifuged at top speed for a further min to eliminate any residual buffer and to dry the column.
6. Buffer AE (plasma 70 μL and lymphocytes 100 μL) is applied to the center of the column and incubated at room temperature for 5 min (to increase DNA yield), after which the column is centrifuged at $6000 \times g$ for 1 min.
7. The DNA eluate is stored at 4°C.

16.4.2 PURIFICATION OF RNA FROM BODY FLUIDS

16.4.2.1 Guanidinium Isothiocyanate/Phenol Extraction

This method, originally reported by Chomczynski and Sacchi [108], uses a GIT/phenol extraction to isolate RNA from body fluids [120].

Procedure:

1. To a sample (1 mL) in a 4 mL polypropylene tube, sequentially add the following: 2 M sodium acetate (pH 4.0, 100 μL), water-saturated phenol (1 mL), and chloroform/IAA(49:1, 200 μL). After addition of each reagent, mix thoroughly by inversion and shake vigorously by hand for 10 s.
2. Cool the samples on ice for 15 min, after which centrifuge at $10,000 \times g$ for 20 min at 4°C.
3. Transfer the upper aqueous phase, which contains RNA, carefully to a clean tube.
4. Add isopropanol (1 mL) to the aqueous phase to precipitate the RNA and incubate the samples for ≥1 h at −20°C.
5. Centrifuge at $10,000 \times g$ for 20 min at 4°C and discard the supernatant. The RNA precipitate should form a gel-like pellet. Dissolve the RNA pellet in denaturing solution (300 μL), transfer to a 1.5 mL microcentrifuge tube, and add isopropanol (300 μL).
6. Incubate the samples for ≥30 min at −20°C, and centrifuge at $10,000 \times g$ for 10 min at 4°C and discard the supernatant.
7. Resuspend the RNA pellet with 75% ethanol (500 μL–1 mL) and vortex for a few seconds.

Incubate samples for 10–15 min at room temperature, to dissolve possible residual traces of guanidinium.

8. Centrifuge at $10,000 \times g$ for 5 min at 4°C, discard the supernatant, and air-dry the RNA pellet for 5–10 min at room temperature.
9. Dissolve the RNA pellet in either DEPC-treated water, 0.5% SDS, or freshly deionized formamide (100–200 μL) and incubate the RNA for 10–15 min at 60°C.

The choice of RNA solvent depends on both storage and the nature of the subsequent RNA application. For example, if using RT-PCR, RNA must be dissolved in DEPC-treated water, as both SDS and formamide can interfere with any subsequent enzymatic reactions.

16.4.2.2 Qiagen MinElute Virus Kit

Isolation of RNA from body fluids can also be performed using commercially available kits, and Qiagen is the company of choice for a number of research groups. The Qiagen MinElute virus kit has been utilized to isolate cell-free RNA from bronchial lavage and serum samples [52,97].

Procedure:

1. Protease (25 μL) is added to a 1.5 mL microcentrifuge tube containing plasma, serum, or other body fluid (200 μL). (If the sample volume is less than 200 μL, add the appropriate volume of 0.9% sodium chloride solution to bring the volume of protease and sample up to a total of 225 μL.)
2. Buffer AL (200 μL, containing 28 μg/mL of carrier RNA) is added, mixed thoroughly by pulse-vortexing for 15 s, and incubated at 56°C for 15 min.
3. Ethanol (96%–100%, 250 μL) is added to the sample, mixed thoroughly by pulse-vortexing, and the lysate/ethanol mix is incubated for 5 min at room temperature (15°C–25°C).
4. The lysate is applied to the QIAamp MinElute column without wetting the rim and centrifuged at $6000 \times g$ for 1 min.
5. The QIAamp MinElute column is placed in a clean tube, and the filtrate is discarded. (If the lysate has not completely passed through the column after centrifugation, spin again at higher speed until the QIAamp MinElute column is empty.)
6. Buffer AW1 (500 μL) is added to the column without wetting the rim, and centrifuged at $6000 \times g$ for 1 min. The QIAamp MinElute column is placed in a clean tube and the filtrate is discarded.
7. Step 6 is repeated with Buffer AW2.
8. Ethanol (96%–100%, 500 μL) is added to the column without wetting the rim, and centrifuged at $6000 \times g$ for 1 min, after which the filtrate is discarded.
9. The QIAamp MinElute column is placed in a clean tube and centrifuged at full speed ($20,000 \times g$) for 3 min.
10. The QIAamp MinElute column is placed into a new tube and incubated at 56°C for 3 min in order to completely dry the membrane.
11. The QIAamp MinElute column was placed in a clean 1.5 mL microcentrifuge tube, the filtrate discarded, and buffer AVE (20–150 μL) or RNase-free water is applied to the center of the membrane. It is incubated at room temperature for 1 min, followed by centrifugation at full speed for 1 min. (The elution buffer must be equilibrated to room temperature. If elution is carried out in small volumes [<50 μL], the elution buffer must be dispensed onto the center of the membrane for complete elution of bound RNA and DNA.)

16.5 FUTURE DEVELOPMENT TRENDS

Nucleic acids can be detected in many bodily fluids, including serum, plasma, saliva, urine, and bronchial lavage. Additional research is needed to fully characterize the nature and the structure of this extracellular DNA. Methods such as whole genome scanning may help to determine whether the DNA released into the bloodstream is representative of the global content of the genome, or whether particular types of DNA fragments or sequences are released at a high rate [121]. Microarray analysis of tumors is a recently developed method, which can provide large amounts of information [122]. This type of analysis was used by Sung et al. who demonstrated genome-wide expression analysis using microarrays, which identified complex signaling pathways modulated by hypoxia in nasopharyngeal carcinoma [123], and it has also been used to profile augmented expression of genes from bone marrow mononuclear cells involved in rheumatoid arthritis [124]. Tsui and coworkers also showed microarray-based identification of placental mRNA in maternal plasma [125], while another study that reported RNA isolation from a cell-free saliva supernatant was used in the detection of oral cancer [126]. Li et al. also demonstrated informative mRNA exists in cell-free saliva, and proposed a novel clinical approach to salivary diagnostics. Developments in detection of cancer will no doubt enhance our knowledge in a wide range of diseases.

There is a clear need for standardization of techniques used for isolation of nucleic acids, as methods for isolating and quantifying nucleic acids from body fluids are crucial to all studies, as accurate and reproducible data are essential. If method standardization is going to be successful, then it needs to start at the very beginning of the study, at the preanalytical level. Ideally, research groups need to adhere to very similar, if not the same methods, for collection, processing, and storage of bodily fluids. This should then be followed by a standard method of DNA isolation specific for that particular starting material, either manually or by automated means. Recent methods, such as quantitative PCR, are an accurate method for analyzing levels of isolated cell-free nucleic acids, and if this again is standardized, then it will be easier to compare results between research groups. In turn, this may serve to facilitate the development of circulating nucleic acids in prognosis and diagnosis of disease.

Future applications of circulating nucleic acids may rely upon such data, which can only be gained by optimizing procedures. It is clear there are discrepancies between studies relating to quantitation of nucleic acids, and this can hinder the estimation of their potential in a prognostic setting. For example, in a study by Sozzi et al., the mean DNA concentration in lung cancer patients was 8-fold higher than controls, but was 13-fold lower than they reported in a previous study [116,127]. Different methods of quantifying DNA accounted for this difference, with the original study using spectrophotometry, thus detecting all DNA fragments, whereas the second study only measured amplified DNA.

Large disparities among quantitative and qualitative reports at present may be due to lack of such standards to evaluate nucleic acids, using different methods of quantification and differences in sample processing techniques. The findings from the ever-increasing number of quantitative and qualitative studies based on isolation of nucleic acids from body fluids suggest it may be possible to develop simple, rapid, and noninvasive blood test that could improve diagnosis and monitoring of a number of malignancies, in addition to a role in prenatal obstetrics.

The advent of both large- and small-scale automated DNA extraction methods available from a number of companies will potentially revolutionize isolation of DNA from body fluids, in that time-consuming, manual methods will decrease in popularity. Automated technology will enable the researcher to manage their time more effectively, and will produce reliable and reproducible results, thereby increasing sample preparation efficiency without sacrificing quality. Commercial companies will strive to improve methods and reagents to such an extent that the techniques will be able to become more standardized between laboratories. It is therefore hoped that easy, cheap, and faster techniques will make the isolation and quantitation of nucleic acids a routine laboratory practice.

Karen Page is a postdoctoral research associate in the Department of Cancer Studies and Molecular Medicine at the University of Leicester, Leicester, United Kingdom. She received her PhD from the University of Leicester in 2000 entitled "Mechanisms of action of the chemopreventive agent genistein in breast cancer" and followed this by research into thrombophilia and pregnancy, utilizing DNA biomarkers in plasma and serum. Since 2002, she has been undertaking research into cell-free circulating DNA in breast cancer, together with Dr. Jacqueline A. Shaw. Her research focus lies in the detection of tumor-specific alterations in the plasma of breast cancer patients, including loss of heterozygosity and DNA methylation; biomarkers that can be correlated with those seen in the primary tumor and may be used in monitoring breast disease. During her PhD and postdoctoral research, she has presented her work both orally and using posters at both the national and international level. She has extensive experience in isolation of nucleic acids from a number of body fluids, including blood plasma, serum, and bone marrow samples, using a variety of in-house and kit-based techniques, and has published methods for reducing cellular contamination in isolated blood plasma DNA.

Jacqueline Amanda Shaw is a senior lecturer and postgraduate tutor in the Department of Cancer Studies and Molecular Medicine at the University of Leicester, Leicester, United Kingdom. She received her PhD from Imperial College London, London, United Kingdom in 1991 for "A molecular genetic study of Freidreich's ataxia." Having completed a postdoctoral position in the ataxia research group at Imperial College, she moved to a lectureship and later to a senior lectureship in the Department of Pathology at Leicester. This department became part of the Department of Cancer Studies and Molecular Medicine following reorganization of the Leicester Medical School, Leicester in 2003. She has received prizes from the Pathological Society of Great Britain for plenary oral presentations and has given several invited presentations. Her major research is in investigating the clinical utility of cell-free plasma DNA as a surrogate biomarker for breast cancer diagnosis and disease monitoring (in collaboration with Professor R.C. Coombes, Imperial College). Other research interests included molecular analysis of sporadic breast cancers in young women using cDNA microarray, myoepithelial cells as a key player in tumor suppression, and studies of the matrix protein Tenascin-C in normal, noninvasive, and invasive breast cancer. In the last 2 years, she has received grant/research funding from Cancer Research (United Kingdom), the Breast Cancer Campaign, Medisearch (the Leicestershire Medical Research Foundation), and from the Leicester Breast Care Appeal.

REFERENCES

1. Mandel, P. and Métais, P., Les acides nucleiques du plasma sanguin l'homme, *CR Acad. Sci. Paris*, 142: 241, 1947.
2. Tan, E.M. et al., Deoxyribonucleic acid (DNA) and antibodies to DNA in the serum of patients with systemic lupus erythematosus, *J. Clin. Invest.*, 45: 1732, 1966.
3. Leon, S.A. et al., Free DNA in the serum of cancer patients and the effect of therapy, *Cancer Res.*, 37: 646, 1977.
4. Koffler, D. et al., The occurrence of single-stranded DNA in the serum of patients with systemic lupus erythematosus and other diseases, *J. Clin. Invest.*, 52: 198, 1973.
5. Leon, S.A. et al., DNA in synovial fluid and the circulation of patients with arthritis, *Arthritis Rheum.*, 24: 1142, 1981.
6. Shapiro, B. et al., Determination of circulating DNA levels in patients with benign or malignant gastrointestinal disease, *Cancer*, 51: 2116, 1983.
7. Stroun, M. et al., Neoplastic characteristics of the DNA found in the plasma of cancer patients, *Oncology*, 46: 318, 1989.
8. Sorenson, G.D. et al., Soluble normal and mutated DNA sequences from single-copy genes in human blood, *Cancer Epidemiol. Biomark. Prev.*, 3: 67, 1994.
9. Vasioukhin, V. et al., Point mutations of the N-ras gene in the blood plasma DNA of patients with myelodysplastic syndrome or acute myelogenous leukemia, *Br. J. Haematol.*, 86: 774, 1994.
10. Chen, X.Q. et al., Microsatellite alterations in plasma DNA of small cell lung cancer patients, *Nat. Med.*, 2: 1033, 1996.
11. Nawroz, H. et al., Microsatellite alterations in serum DNA of head and neck cancer patients, *Nat. Med.*, 2: 1035, 1996.
12. Esteller, M. et al., Detection of aberrant promoter hypermethylation of tumor suppressor genes in serum DNA from non-small cell lung cancer patients, *Cancer Res.*, 59: 67, 1999.
13. Wong, I.H. et al., Detection of aberrant p16 methylation in the plasma and serum of liver cancer patients, *Cancer Res.*, 59: 71, 1999.
14. Lo, Y.M. et al., Quantitative analysis of cell-free Epstein-Barr virus DNA in plasma of patients with nasopharyngeal carcinoma, *Cancer Res.*, 59: 1188, 1999.
15. Herrera, L.J. et al., Quantitative analysis of circulating plasma DNA as a tumor marker in thoracic malignancies, *Clin. Chem.*, 51: 113, 2005.
16. Sai, S. et al., Quantification of plasma cell-free DNA in patients with gastric cancer, *Anticancer Res.*, 27: 2747, 2007.
17. Taback, B. et al., Prognostic significance of circulating microsatellite markers in the plasma of melanoma patients, *Cancer Res.*, 61: 5723, 2001.
18. Vdovichenko, K.K., Markova, S.I., and Belokhvostov, A.S., Mutant form of BRAF gene in blood plasma of cancer patients, *Ann. N. Y. Acad. Sci.*, 1022: 228, 2004.

19. de Kok, J.B. et al., Detection of tumor DNA in serum of colorectal cancer patients, *Scand. J. Clin. Lab. Invest.*, 57: 601, 1997.

20. Kopreski, M.S. et al., Detection of mutant K-ras DNA in plasma or serum of patients with colorectal cancer, *Br. J. Cancer*, 76: 1293, 1997.

21. Silva, J.M. et al., Presence of tumor DNA in plasma of breast cancer patients: Clinicopathological correlations, *Cancer Res.*, 59: 3251, 1999.

22. Shaw, J.A. et al., Microsatellite alterations plasma DNA of primary breast cancer patients, *Clin. Cancer Res.*, 6: 1119, 2000.

23. Silva, J.M. et al., TP53 gene mutations in plasma DNA of cancer patients, *Genes Chromosom. Cancer*, 24: 160, 1999.

24. Bruhn, N. et al., Detection of microsatellite alterations in the DNA isolated from tumor cells and from plasma DNA of patients with lung cancer, *Ann. N. Y. Acad. Sci.*, 906: 72, 2000.

25. Goessl, C. et al., Microsatellite analysis of plasma DNA from patients with clear cell renal carcinoma, *Cancer Res.*, 58: 4728, 1998.

26. Utting, M. et al., Microsatellite analysis of free tumor DNA in urine, serum, and plasma of patients: A minimally invasive method for the detection of bladder cancer, *Clin. Cancer Res.*, 8: 35, 2002.

27. Mulcahy, H.E. et al., A prospective study of K-ras mutations in the plasma of pancreatic cancer patients, *Clin. Cancer Res.*, 4: 271, 1998.

28. Chang, H.W. et al., Assessment of plasma DNA levels, allelic imbalance, and CA 125 as diagnostic tests for cancer, *J. Natl. Cancer Inst.*, 94: 1697, 2002.

29. Kersemaekers, A.M. et al., Loss of heterozygosity for defined regions on chromosomes 3, 11 and 17 in carcinomas of the uterine cervix, *Br. J. Cancer*, 77: 192, 1998.

30. Goessl, C. et al., Fluorescent methylation-specific polymerase chain reaction for DNA-based detection of prostate cancer in bodily fluids, *Cancer Res.*, 60: 5941, 2000.

31. Frickhofen, N. et al., Rearranged Ig heavy chain DNA is detectable in cell-free blood samples of patients with B-cell neoplasia, *Blood*, 90: 4953, 1997.

32. Pathak, A.K., et al., Circulating cell-free DNA in plasma/serum of lung cancer patients as a potential screening and prognostic tool, *Clin. Chem.*, 52: 1833, 2006.

33. Anker, P. et al., Detection of circulating tumor DNA in the blood (plasma/serum) of cancer patients, *Cancer Metastasis Rev.*, 18: 65, 1999.

34. Jahr, S. et al., DNA fragments in the blood plasma of cancer patients: Quantitations and evidence for their origin from apoptotic and necrotic cells, *Cancer Res.*, 61: 1659, 2001.

35. Anker, P., Stroun, M., and Maurice, P.A., Spontaneous release of DNA by human blood lymphocytes as shown in an in vitro system, *Cancer Res.*, 35: 2375, 1975.

36. Diehl, F. et al., Detection and quantification of mutations in the plasma of patients with colorectal tumors, *Proc. Natl. Acad. Sci. U.S.A.*, 102: 16368, 2005.

37. Moss, T.J. and Sanders, D.G., Detection of neuroblastoma cells in blood, *J. Clin. Oncol.*, 8: 736, 1990.

38. Foss, A.J. et al., The detection of melanoma cells in peripheral blood by reverse transcription-polymerase chain reaction, *Br. J. Cancer*, 72: 155, 1995.

39. Mattano, L.A., Jr., Moss, T.J., and Emerson, S.G., Sensitive detection of rare circulating neuroblastoma cells by the reverse transcriptase-polymerase chain reaction, *Cancer Res.*, 52: 4701, 1992.

40. Chen, X. et al., Detecting tumor-related alterations in plasma or serum DNA of patients diagnosed with breast cancer, *Clin. Cancer Res.*, 5: 2297, 1999.

41. Frost, P.G. and Lachmann, P.J., The relationship of deoxyribonuclease inhibitor levels in human sera to the occurrence of antinuclear antibodies, *Clin. Exp. Immunol.*, 3: 447, 1968.

42. Lo, Y.M. et al., Rapid clearance of fetal DNA from maternal plasma, *Am. J. Hum. Genet.*, 64: 218, 1999.

43. Silva, J.M. et al., Persistence of tumor DNA in plasma of breast cancer patients after mastectomy, *Ann. Surg. Oncol.*, 9: 71, 2002.

44. Page, K. et al., The importance of careful blood processing in isolation of cell-free DNA, *Ann. N. Y. Acad. Sci.*, 1075: 313, 2006.

45. Wang, B.G. et al., Increased plasma DNA integrity in cancer patients, *Cancer Res.*, 63: 3966, 2003.

46. Page, K., unpublished data, 2007.

47. Umetani, N. et al., Prediction of breast tumor progression by integrity of free circulating DNA in serum, *J. Clin. Oncol.*, 24: 4270, 2006.

48. Umetani, N. et al., Increased integrity of free circulating DNA in sera of patients with colorectal or periampullary cancer: Direct quantitative PCR for ALU repeats, *Clin. Chem.*, 52: 1062, 2006.

49. von Knobloch, R. et al., Serum DNA and urine DNA alterations of urinary transitional cell bladder carcinoma detected by fluorescent microsatellite analysis, *Int. J. Cancer*, 94: 67, 2001.

50. Berger, A.P. et al., Microsatellite alterations in human bladder cancer: Detection of tumor cells in urine sediment and tumor tissue, *Eur. Urol.*, 41: 532, 2002.

51. Taback, B. et al., Detection of tumor-specific genetic alterations in bone marrow from early-stage breast cancer patients, *Cancer Res.*, 63: 1884, 2003.

52. Carstensen, T. et al., Detection of cell-free DNA in bronchial lavage fluid supernatants of patients with lung cancer, *Ann. N. Y. Acad. Sci.*, 1022: 202, 2004.

53. Schmidt, B. et al., Quantification of free RNA in serum and bronchial lavage: A new diagnostic tool in lung cancer detection?, *Lung Cancer*, 48: 145, 2005.

54. Belinsky, S.A. et al., Gene promoter methylation in plasma and sputum increases with lung cancer risk, *Clin. Cancer Res.*, 11: 6505, 2005.

55. Mao, L. et al., Molecular detection of primary bladder cancer by microsatellite analysis, *Science*, 271: 659, 1996.

56. Steiner, G. et al., Detection of bladder cancer recurrence by microsatellite analysis of urine, *Nat. Med.*, 3: 621, 1997.

57. Su, Y.H. et al., Detection of a K-ras mutation in urine of patients with colorectal cancer, *Cancer Biomark.*, 1: 177, 2005.

58. Miyake, M. et al., Sensitive detection of FGFR3 mutations in bladder cancer and urine sediments by peptide nucleic acid-mediated real-time PCR clamping, *Biochem. Biophys. Res. Commun.*, 362: 865, 2007.

59. Szarvas, T. et al., Deletion analysis of tumor and urinary DNA to detect bladder cancer: Urine supernatant versus urine sediment, *Oncol. Rep.*, 18: 405, 2007.

60. Majer, S. et al., Maternal urine for prenatal diagnosis—An analysis of cell-free fetal DNA in maternal urine and plasma in the third trimester, *Prenat. Diagn.*, 27: 1219, 2007.

61. Schwarzenbach, H. et al., Detection of tumor-specific DNA in blood and bone marrow plasma from patients with prostate cancer, *Int. J. Cancer.*, 120: 1465, 2007.

62. Wang, Y.C. et al., Molecular diagnostic markers for lung cancer in sputum and plasma, *Ann. N. Y. Acad. Sci.*, 1075: 179, 2006.

63. Kopreski, M.S. et al., Detection of tumor messenger RNA in the serum of patients with malignant melanoma, *Clin. Cancer Res.*, 5: 1961, 1999.

64. Novakovic, S. et al., Detection of telomerase RNA in the plasma of patients with breast cancer, malignant melanoma or thyroid cancer, *Oncol. Rep.*, 11: 245, 2004.

65. Miura, N. et al., Sensitive detection of human telomerase reverse transcriptase mRNA in the serum of patients with hepatocellular carcinoma, *Oncology*, 64: 430, 2003.

66. Guin, L.W. et al., Electrophoretic characterization of plasma RNA, *Biochem. Med.*, 13: 224, 1975.

67. Lo, K.W. et al., Analysis of cell-free Epstein-Barr virus associated RNA in the plasma of patients with nasopharyngeal carcinoma, *Clin. Chem.*, 45: 1292, 1999.

68. Ng, E.K. et al., The concentration of circulating corticotrophin-releasing hormone mRNA in maternal plasma is increased in pre-eclampsia, *Clin. Chem.*, 49: 727, 2003.

69. Tsui, N.B., Ng, E.K., and Lo, Y.M., Stability of endogenous and added RNA in blood specimens, serum, and plasma, *Clin. Chem.*, 48: 1647, 2002.

70. Tong, Y.K. and Lo, Y.M., Diagnostic developments involving cell-free (circulating) nucleic acids, *Clin. Chim. Acta*, 363: 187, 2006.

71. Silva, J.M. et al., Detection of epithelial messenger RNA in the plasma of breast cancer patients is associated with poor prognosis tumor characteristics, *Clin. Cancer Res.*, 7: 2821, 2001.

72. Chen, X.Q. et al., Telomerase RNA as a detection marker in the serum of breast cancer patients, *Clin. Cancer Res.*, 6: 3823, 2000.

73. Wong, B.C. et al., Reduced plasma RNA integrity in nasopharyngeal carcinoma patients, *Clin. Cancer Res.*, 12: 2512, 2006.

74. Juusola, J. and Ballantyne, J., Multiplex mRNA profiling for the identification of body fluids, *Forensic Sci. Int.*, 152: 1, 2005.

75. Juusola, J. and Ballantyne, J., mRNA profiling for body fluid identification by multiplex quantitative RT-PCR, *J. Forensic Sci.*, 52: 1252, 2007.

76. El-Hefnawy, T. et al., Characterization of amplifiable, circulating RNA in plasma and its potential as a tool for cancer diagnostics, *Clin. Chem.*, 50: 564, 2004.

77. Garcia, J.M. et al., Extracellular tumor DNA in plasma and overall survival in breast cancer patients, *Genes Chromosom. Cancer*, 45: 692, 2006.

78. Lo, Y.M. et al., Plasma DNA as a prognostic marker in trauma patients, *Clin. Chem.*, 46: 319, 2000.

79. Rainer, T.H. et al., Prognostic use of circulating plasma nucleic acid concentrations in patients with acute stroke, *Clin. Chem.*, 49: 562, 2003.

80. Tsang, J.C. and Lo, Y.M., Circulating nucleic acids in plasma/serum, *Pathology*, 39: 197, 2007.

81. Lo, Y.M. et al., Presence of fetal DNA in maternal plasma and serum, *Lancet*, 350: 485, 1997.

82. Lo, Y.M. et al., Quantitative analysis of fetal DNA in maternal plasma and serum: Implications for non-invasive prenatal diagnosis, *Am. J. Hum. Genet.*, 62: 768, 1998.

83. Costa, J.M., Benachi, A., and Gautier, E., New strategy for prenatal diagnosis of X-linked disorders, *N. Engl. J. Med.*, 346: 1502, 2002.

84. Lo, Y.M. et al., Prenatal diagnosis of fetal RhD status by molecular analysis of maternal plasma, *N. Engl. J. Med.*, 339: 1734, 1998.

85. Chiu, R.W. et al., Prenatal exclusion of beta thalassemia major by examination of maternal plasma, *Lancet*, 360: 998, 2002.

86. Lo, Y.M. et al., Quantitative abnormalities of fetal DNA in maternal serum in pre-eclampsia, *Clin. Chem.*, 45: 184, 1999.

87. Lo, Y.M. et al., Increased fetal DNA concentrations in the plasma of pregnant women carrying fetuses with trisomy 21, *Clin. Chem.*, 45: 1747, 1999.

88. Leon, S.A. et al., Radioimmunoassay for nanogram quantities of DNA, *J. Immunol. Methods*, 9: 157, 1975.

89. Lee, T.H. et al., Quantitation of genomic DNA in plasma and serum samples: Higher concentrations of genomic DNA found in serum than in plasma, *Transfusion*, 41: 276, 2001.

90. Taback, B., O'Day, S.J., and Hoon, D.S., Quantification of circulating DNA in the plasma and serum of cancer patients, *Ann. N. Y. Acad. Sci.*, 1022: 17, 2004.

91. Thijssen, M.A. et al., Difference between free circulating plasma and serum DNA in patients with colorectal liver metastases, *Anticancer Res.*, 22: 421, 2002.

92. Umetani, N., Hiramatsu, S., and Hoon, D.S., Higher amount of free circulating DNA in serum than in plasma is not mainly caused by contaminated extraneous DNA during separation, *Ann. N. Y. Acad. Sci.*, 1075: 299, 2006.

93. Jung, M. et al., Changes in concentration of DNA in serum and plasma during storage of blood samples, *Clin. Chem.*, 49: 1028, 2003.

94. Lam, N.Y. et al., EDTA is a better anticoagulant than heparin or citrate for delayed blood processing for plasma DNA analysis, *Clin. Chem.*, 50: 256, 2004.

95. Chiu, R.W., et al., Effects of blood-processing protocols on fetal and total DNA quantification in maternal plasma, *Clin. Chem.*, 47: 1607, 2001.

96. Schmidt, B. et al., Detection of cell-free nucleic acids in bronchial lavage fluid supernatants from patients with lung cancer, *Eur. J. Cancer*, 40: 452, 2004.

97. Miller, S.A., Dykes, D.D., and Polesky, H.F., A simple salting out procedure for extracting DNA from human nucleated cells, *Nucleic Acids Res.*, 16: 1215, 1988.

98. Schmidt, B. et al., Improved method for isolating cell-free DNA, *Clin. Chem.*, 51: 1561, 2005.

99. Su, Y.H. et al., Human urine contains small, 150 to 250 nucleotide-sized, soluble DNA derived from the circulation and may be useful in the detection of colorectal cancer, *J. Mol. Diagn.*, 6: 101, 2004.

100. Greenspoon, S.A. et al., QIAamp spin columns as a method of DNA isolation for forensic casework, *J. Forensic Sci.*, 43: 1024, 1998.

101. Johns, M.B., Jr. and Paulus-Thomas, J.E., Purification of human genomic DNA from whole blood using sodium perchlorate in place of phenol, *Anal. Biochem.*, 180: 276, 1989.

102. Boom, R. et al., Rapid and simple method for purification of nucleic acids, *J. Clin. Microbiol.*, 28: 495, 1990.

103. Jen, J., Wu, L., and Sidransky, D., An overview on the isolation and analysis of circulating tumor DNA in plasma and serum, *Ann. N. Y. Acad. Sci.*, 906: 8, 2000.

104. Kephart, D.K.S., Grunst, T., and Shenoi, H., Introducing the Maxwell 16 instrument: A simple, robust and flexible tool for DNA purification, *Promega Notes*, 20, 2006.

105. Allen, D. et al., Role of cell-free plasma DNA as a diagnostic marker for prostate cancer, *Ann. N. Y. Acad. Sci.*, 102: 276, 2004.

106. Gal, S. et al., Quantitation of circulating DNA in the serum of breast cancer patients by real-time PCR, *Br. J. Cancer*, 90: 1211, 2004.

107. de Kok, J.B. et al., Use of real-time quantitative PCR to compare DNA isolation methods, *Clin. Chem.*, 44: 2201, 1998.

108. Chomczynski, P. and Sacchi, N., Single-step method of RNA isolation by acid guanidinium thiocyanate-phenol-chloroform extraction, *Anal. Biochem.*, 162: 156, 1987.
109. Tamkovich, S.N. et al., Simple and rapid procedure suitable for quantitative isolation of low and high molecular weight extracellular nucleic acids, *Nucleos. Nucleot. Nucleic Acids*, 23: 873, 2004.
110. Cerkovnik, P. et al., Optimization of an RNA isolation procedure from plasma samples, *Int. J. Mol. Med.*, 20: 293, 2007.
111. Xie, G.S. et al., Quantification of plasma DNA as a screening tool for lung cancer, *Chin. Med. J. (Engl)*, 117: 1485, 2004.
112. Tamkovich, S.N. et al., Circulating DNA and DNase activity in human blood, *Ann. N. Y. Acad. Sci.*, 1075: 191, 2006.
113. Bearzatto, A. et al., p16 (INK4A) hypermethylation detected by fluorescent methylation-specific PCR in plasmas from non-small cell lung cancer, *Clin. Cancer Res.*, 8: 3782, 2002.
114. Silva, J.M. et al., Tumor DNA in plasma at diagnosis of breast cancer patients is a valuable predictor of disease-free survival, *Clin. Cancer Res.*, 8: 3761, 2002.
115. Kamm, R.C. and Smith, A.G., Nucleic acid concentrations in normal human plasma, *Clin. Chem.*, 18: 519, 1972.
116. Sozzi, G. et al., Quantification of free circulating DNA as a diagnostic marker in lung cancer, *J. Clin. Oncol.*, 21: 3902, 2003.
117. Huang, Z.H., Li, L.H., and Hua, D., Quantitative analysis of plasma circulating DNA at diagnosis and during follow-up of breast cancer patients, *Cancer Lett.*, 243: 64, 2006.
118. Zhong, X.Y. et al., Elevated level of cell-free plasma DNA is associated with breast cancer, *Arch. Gynecol. Obstet.*, 276: 327, 2007.
119. Lledó, S.M. et al., Real time quantification in plasma of human telomerase reverse transcriptase (hTERT) mRNA in patients with colorectal cancer, *Colorectal Dis.*, 6: 236, 2004.
120. Chomczynski, P. and Sacchi, N., The single-step method of RNA isolation by acid guanidinium thiocyanate-phenol-chloroform extraction: Twenty-something years on, *Nat. Prot.*, 1: 581, 2006.
121. Gormally, E. et al., Circulating free DNA in plasma or serum as biomarker of carcinogenesis: Practical aspects and biological significance, *Mutat. Res.*, 635: 105, 2007.
122. Quackenbush, J., Microarray analysis and tumor classification, *N. Engl. J. Med.*, 354: 2463, 2006.
123. Sung, F.L. et al., Genome-wide expression analysis using microarray identified complex signaling pathways modulated by hypoxia in nasopharyngeal carcinoma, *Cancer Lett.*, 253: 74, 2007.
124. Nakamura, N. et al., Isolation and expression profiling of genes upregulated in bone marrow-derived mononuclear cells of rheumatoid arthritis patients, *DNA Res.*, 13: 169, 2006.
125. Tsui, N.B. et al., Systematic micro-array based identification of placental mRNA in maternal plasma: Towards non-invasive prenatal gene expression profiling, *J. Med. Genet.*, 41: 461, 2004.
126. Li, Y. et al., Salivary transcriptome diagnostics for oral cancer detection, *Clin. Cancer Res.*, 10: 8442, 2004.
127. Sozzi, G. et al., Analysis of circulating tumor DNA in plasma at diagnosis and during follow-up of lung cancer patients, *Cancer Res.*, 61: 4675, 2001.

17 Isolation of Nucleic Acids from Soft Tissues

Sergey Kovalenko

CONTENTS

17.1 INTRODUCTION

In biological term, tissue is a collection of interconnected cells that perform a similar function within an organism. In medicine, soft tissues include muscles, tendons, ligaments, fascia (muscle sheath), nerves, fibrous tissues, fat, blood vessels, and synovial tissues, and they connect, support, or surround other structures and organs of the body. Chief among them, epithelial tissues cover the surfaces such as surface of the skin and inner lining of digestive tract that serve for protection, secretion, and absorption; connective tissue holds everything together; muscle tissues, in the forms of visceral or smooth muscle (located in the inner linings of organs), skeletal muscle (attached to bone in order for mobility to take place), and cardiac muscle (found in the heart) produce force and cause motion, either locomotion or movement within internal organs; and nervous tissues make up the brain, spinal cord, and peripheral nervous system.

Given that soft tissues constitute essential components and play vital roles in the normal functions of human body, disorders in soft tissues (e.g., muscle, fat, fibrous tissues, heart, and blood vessels) often have severe consequences in human health and well-being. Indeed, like other parts in human body, soft tissues are liable to suffer from a range of microbial infections (e.g., *Streptococcus pyogenes*, *Staphylococcus aureus*, *Haemophilus influenzae*, *Clostridium perfringens*, and *Mycobacterium* species). In addition, the performance of soft tissues can also be undermined by underlying genetic mutations (e.g., Ehlers–Danlos syndrome, Marfan syndrome, and osteogenesis imperfecta) and cancers (e.g., sarcomas).

Among the more severe soft tissue infections, necrotizing soft tissue infections involving skin, subcutaneous fat, the muscle sheath (fascia), and the muscle can lead to abscess, gangrene, tissue death, systematic disease, and death. Soft tissue sarcoma is a malignant tumor that can begin in any types of soft tissues that connect, support, or surround organs and other body structures, but about half of them are found in the arms, legs, hands, or feet, and another 40% occur in the trunk, which includes the chest, back, hips, shoulder, and abdomen. One of the most common soft tissue sarcomas is rhabdomyosarcoma, a tumor of skeletal muscles; and another common soft tissue tumor is desmoid tumor (or aggressive fibromatosis), a tumor affecting the fibrous tissues that make up tendons and ligaments. Other less notable soft tissue sarcomas include fibrosarcoma (also affecting tendons and ligaments), leiomyosarcoma (affecting involuntary muscle tissue found in abdomen, bowels, uterus, and blood vessels), liposarcoma (affecting fat tissue, often in the abdominal cavity), malignant fibrous histiocytoma (affecting fibrous tissue in the legs), peripheral nerve sheath tumor (affecting the cells that surround nerves), and synovial cell sarcoma (affecting cells around joints).

To accurately determine the types of soft tissue disorders and assess whether a soft tissue tumor is benign or malignant, a biopsy is often obtained via needle biopsy or with surgical biopsy. Traditionally, the resulting biopsy is examined under microscope utilizing a variety of special stains and immunohistochemistry to ascertain the nature of the originating cell. Due to the fact that conventional microscopic procedures usually lack desired specificity and sensitivity, a range of molecular biological techniques targeting nucleic acids from soft tissues have been applied in recent decades to the diagnosis of soft tissue disorders, especially sarcomas. This is possible as many soft tissue sarcomas demonstrate unique genetic translocations and other molecular defects. The molecular techniques not only provide a useful diagnostic tool, but also shed light on the pathogenesis of soft tissue tumors.

For efficient extraction, isolation, and purification of nucleic acids from soft tissues, a large number of methods have been developed [1–5]. The selection of a nucleic acid extraction methodology is generally based on pragmatic considerations that include extraction sample cost, time commitment, technical complexity of the methodology, quantity of the starting biological sample, and suitability and adaptability of the methodology for high-throughput automation. The purity requirements vary tremendously depending on the downstream applications. As a result, rarely is one purification procedure optimized for every need. Although contemporary extraction methodologies such as the modified-single step method [6] still employ chaotropic agents, alternative nucleic acid extraction technologies now include a variety of different biochemical approaches such as adsorption to silica- or glass-based fiber spin columns, or affinity binding to insoluble cellulose matrix, latex beads, and paramagnetic particles. The purpose of this chapter is to provide a quick overview of current most advanced methods for nucleic acid isolation and purification from soft tissues. We also present here the extraction methods currently successfully utilized in our laboratory for both small- and large-scale nucleic acid analyses.

17.2 NUCLEIC ACID EXTRACTION FROM SOFT TISSUES: PRINCIPLES AND APPLICATION

There are few basic steps in nucleic acid extraction, isolation, and purification, the details of which may vary depending on the type of sample and any substances that may interfere with the extraction and subsequent analysis. These steps involve the following:

1. Tissue handling: Immediate freezing or fixing of tissue specimens after surgery is a standard procedure for the procurement of surgical specimens, and avoiding RNA degradation is a major challenge in this process.

2. Tissue preservation before nucleic acid isolation: Storage of tissue at +4°C, −20°C, or −80°C in liquid nitrogen, in RNA-later solution, and in fixatives.

3. Breaking open cells: Mechanically by grinding, sonication, and pressure shearing; and nonmechanically by enzymatic lysis, osmotic lysis, freezing/thawing, detergent-based lysis, and electroporation.

4. Removing solids and debris by centrifugation, filtration, membrane separation, and precipitation.

5. Nucleic acid purification methods such as solvent extraction and precipitation, gel electrophoresis, and chromatography (e.g., size exclusion chromatography [SEC], ion exchange chromatography, solid-phase extraction [SPE], solid-phase reversible immobilization [SPRI], and affinity purification).

6. Isolation of purified nucleic acid by washing, elution, precipitation, and centrifugation.

7. Preservation and storage of purified nucleic acid.

17.2.1 NUCLEIC ACID PRESERVATION

It is important always to remember that nucleic acids require to be protected before, during, and after isolation. RNA is subject to degradation by ubiquitous ribonucleases (RNases) and is also subject to rapid hydrolysis especially at higher temperature, and in the presence of divalent cations, it requires more protection than DNA [7]. RNA*later* is a tissue storage solution that stabilizes and protects cellular RNA in fresh tissues (Applied Biosystems/Ambion, Foster City, USA). The new RNA*later*-ICE is designed specifically for use with frozen tissues. Tissue harvested and immediately submerged in RNA*later* can be stored indefinitely. In doing so, the quality and quantity of the RNA will be maintained, eliminating the need to immediately process or snap-freeze samples. RNA*later* is beneficial for preserving RNA integrity in whole renal cortex during storage and processing, but is not suitable for implementation in routine diagnostic histological staining combined with RNA expression studies in dissected biopsy material [8]. Recently, it has been shown that RNA degradation is a minor problem during handling of fresh tissue before processing. Data indicate that nonfixed tissue specimens may be transported on ice for hours without any major influence on RNA quality and expression of the selected genes [9]. The issue of DNA and RNA stability has been studied mostly with respect to various fixatives [10–12].

HOPE (HEPES-glutamic acid buffer mediated organic solvent protection effect) fixative preserves DNA and RNA suitable for polymerase chain reaction (PCR) and reverse transcription (RT)-PCR; and reversible cross-linker dithio-bis(succinimidyl propionate) is useful for immunostaining, microdissection, and expression profiling [13]. The potential value of a new universal molecular fixative for preservation of macromolecules in paraffin-embedded tissue has been tested that can preserve morphology and macromolecules in paraffin-embedded tissue [14]. These new generation fixatives allow to preserve the tissue architecture and also to achieve the extraction of high molecular weight DNA and RNA of >20 kb from paraffin-embedded tissues.

Purified DNA is stable in the TE buffer if stored at +4°C, and prolonged storage at −20°C can cause DNA shearing. The best option to keep DNA intact is to store DNA in ethanol solution and precipitate DNA when it is needed for analysis. Purified RNA is best stored in aliquots (in ethanol or isopropanol) at −80°C, and ethanol is removed by centrifugation, and RNA is resuspended in appropriate RNase-free buffer before use.

17.2.2 TISSUE HOMOGENIZATION METHODS

Complete homogenization of the tissue is a critical step in isolating high-quality nucleic acid from tissue. Incomplete homogenization may result in lower yields and decreased purity of the isolated nucleic acid. The ability to isolate nucleic acid can vary greatly between tissue types due to many factors including differences in endogenous levels of RNases and the fibrous or lipid-rich nature of certain tissues. The efficiency of extraction is greatly improved if the tissue is reduced to powder before homogenization in lysis buffer.

Several protocols for the homogenization of various tissues are commercially available. Selection of the best method for a particular range of uses should be determined experimentally. The following are a few key points that help to maintain nucleic acid quality and maximize its overall yield from soft tissues:

- Keep tissue frozen on dry ice as much as possible to minimize degradation.
- Keep plastics, forceps, and cutting tools on dry ice to minimize degradation.
- Weigh tube without tissue and reweigh with tissue to make sure that the appropriate amounts of lysis buffer and proteinase K are added to the sample.
- We recommend doing the homogenization in a 15 or 50 mL tube. In a smaller tube, foaming can create a problem.
- If you have enough starting material, make up an extra sample to compensate for dead volume and for the foaming that may occur during and post-homogenization. For example, if processing two tissues samples (20 mg) homogenize enough for three (30 mg).
- If using a roto-stator homogenizer, homogenize using an up-and-down motion in the tube. Avoid heating up the sample because this can degrade the RNA.
- For RNA extraction: Tissue lysis buffer should contain RNase inhibitors. RNA will begin to degrade when thawed, if it is not in contact with lysis buffer. Therefore, place the frozen tissue directly into the lysis buffer before it begins to thaw. Homogenize as quickly as possible to stabilize the entire piece of tissue.

The homogenization protocols and equipment outlined below have been used successfully in our laboratory.

17.2.2.1   Low-Throughput Homogenization

For simple and rapid homogenization of cell and tissue lysates, QIAshredder homogenizer has the advantages of replacing syringe-and-needle homogenization, reducing loss of sample material, eliminating cross-contamination between samples, and filtering out insoluble debris and reduces viscosity.

Using QIAshredder homogenizer is a fast and efficient way to homogenize cell and tissue lysates without cross-contamination of the samples. The QIAshredder homogenizer consists of a unique biopolymer shredding system in a microcentrifuge spin-column format. The lysate is loaded onto the QIAshredder homogenizer placed in a collection tube and centrifuged; the homogenized lysate is then collected. In general, similar yields and quality of RNA are obtained as with rotor–stator.

Other recommended equipment for tissue homogenization include the following:

1. For tube format: Tissue dispersing device IKA Ultra Turrax (http://www.ika.de/ika/home.html) using a 5 mm dispersing element or Brinkman polytron homogenizer
2. For 96 well plate format: Plate vortexer (Troemner VX2400 multitube vortexer) with two 3.2 mm stainless steel beads (http://www.biospec.com/ part number 11079132ss) per well, 2 mL deep well plate (http://www.abgene.com/ Product No. AB-0661), and plate seal (http://www.abgene.com/ Product No. AB-0580)

17.2.2.2   Protease Tissue Digestion (Liquid Homogenization)

Instead of using mechanical tissue disruption, the MELT total nucleic acid isolation system (Applied Biosystems/Ambion, Cat. No. AM1983) employs a cocktail of proteases, an optimized buffer, and the Ambion Vortex Adapter-60 (Cat. No. AM10014) to liquefy tissue samples in ~10 min in a closed-container system while maintaining RNA integrity and maximizing RNA yield. Unlike chaotropes such as guanidine, MELT reagents irreversibly destroy RNases, and thus MELT lysates can be safely stored for up to 1 week at ambient temperatures or at −20°C or −80°C for up to 1 week without compromising RNA quality or yield. This creates opportunities for convenient storage and shipment of MELT lysates. The MELT reagent is designed for use with most animal tissues that do not contain comparatively high levels of RNases and that are not extremely hard or fibrous. It is not compatible with adipose tissue or tissues that have been stored in Ambion RNA*later* or RNA*later*-ICE.

17.2.3   NUCLEIC ACID PURIFICATION METHODS

17.2.3.1   Solvent Extraction and Precipitation

The classical liquid method for nucleic acid preparation is phenol/chloroform extraction followed by ethanol or isopropanol precipitation. The extraction is carried out with consequent mixing phenol and then phenol/chloroform/isoamyl alcohol (IAA) mix to the sample containing DNA, inverting the tube several times in order to mix the phases, and final separation of the phases by centrifugation. Phenol/chloroform extraction removes other macromolecules such as proteins and lipids. An alternative to phenol-based methods for DNA extraction is the use of guanidinium salts and detergents for homogenization of tissues followed by alcohol precipitation of DNA [15]. A commercially available reagent (DNazol, TRIZOL) available from Invitrogen (Carlsbad, California, USA) uses a proprietary formulation of guanidine/detergent lysis solution for nucleic acid isolation from tissue [6]. Alternative and more advanced liquid methods avoiding application of phenol and chloroform are available from QIAGEN, Hilden, Germany, (Gentra Puregene tissue kit) and from Promega (Wizard tissue DNA kit). These methods use a specially formulated protein precipitation solution in order to eliminate proteins from DNA solution before isopropanol-precipitation step.

17.2.3.2   Membrane Filtration

Basic features of this purification method are (1) sample is run through microchannel; (2) DNA binds to the channel, all other molecules remain in buffer solution; (3) channel is washed of impurities; (4) an elution buffer removes DNA from channel walls; and (5) DNA is collected at the end of the channel.

This method exploits a unique characteristic of DNA that shows affinity to silica under condition created by high concentration of chaotropic reagents such as guanidine HCl. A chaotrope denatures biomolecules by disrupting the shell of hydration around them. This allows a positively charged ion to form a salt bridge between the negatively charged silica and the negatively charged DNA backbone in high salt concentration [16]. The DNA can then be washed with high salt and EtOH, and ultimately eluted with low salt. After the DNA is adsorbed to the silica surface, all other molecules pass through the column. Most likely, these molecules are sent to a waste section on the chip, which can then be closed off using a gated channel or a pressure or voltage-controlled chamber. Once these are washed from the DNA, the DNA is washed to remove any excess waste particles from the sample and then eluted from the channel using an elution buffer (of low ionic strength).

17.2.3.3   Chromatography

17.2.3.3.1   Size Exclusion Chromatography (Gel Filtration)
The underlying principle of SEC is that particles of different sizes will go (filter) through a stationary phase at different rates [17]. This results in the separation of a solution of particles based on size. Provided that all the particles are loaded simultaneously, particles of the same size should elute together. This is usually achieved with the use of a column, which consists of a hollow tube tightly packed with extremely small porous polymer beads designed to have pores of different sizes. These pores may be depressions on the surface or channels through the bead. As the solution travels down

the column, some particles enter into the pores. Larger particles cannot enter into as many pores. The larger the particles, the less overall volume to traverse over the length of the column, lead to the faster elution.

17.2.3.3.2 Ion Exchange Chromatography

Ion exchange chromatography retains analyte molecules based on coulombic (ionic) interactions. The stationary phase surface displays ionic functional groups that interact with analyte ions of opposite charge [18]. This type of chromatography is further subdivided into cation exchange chromatography and anion exchange chromatography, with the former retaining positively charged cations because the stationary phase displays a negatively charged functional group such as a phosphoric acid, and the latter retaining anions uses positively charged functional group such as a quaternary ammonium cation.

17.2.3.3.3 Solid-Phase Extraction

The separation ability of SPE is based on the preferential affinity of desired or undesired solutes in a liquid, mobile phase for a solid, stationary phase through which the sample is passed. Impurities in the sample are either washed away while the analyte of interest is retained on the stationary phase, or vice versa. Analytes that are retained on the stationary phase can then be eluted from the SPE cartridge with the appropriate solvent.

A typical SPE involves four basic steps. First, the cartridge is equilibrated with a nonpolar solvent, which wets the surface and penetrates the bonded phase. Then water, or buffer of the same composition as the sample, is typically washed through the column to wet the silica surface. The sample is then added to the cartridge. As the sample passes through the stationary phase, the analytes in the sample will interact and retain on the sorbent while the solvent, salts, and other impurities pass through the cartridge. After the sample is loaded, the cartridge is washed with buffer or solvent to remove further impurities. Then, the analyte is eluted with a nonpolar solvent or a buffer of the appropriate pH.

17.2.3.3.4 Solid-Phase Reversible Immobilization

This is a popular magnetic bead technology that was first debuted in 1995 and has been adopted by many of the largest genome research facilities around the world and utilized extensively during human genome project [19,20]. Efficient DNA and RNA isolation from the cell lysate solution relies on the binding of nucleic acid to the surface of paramagnetic beads coated with sorbent material such as silica or carboxylate coatings. Silica-based purification does not support a size selectivity such as carboxy-based beads. The key feature of reverse immobilization is that unlike Si-OH groups of silica beads, carboxy groups have pK_a of 4.7, so they are negatively charged at neutral pH [21]. This is critical for complete elution so that the adsorbed nucleic acid can be released from solid support. The carboxy-based method is capable of capturing two to three times more product over silica counterpart. In addition to quickly separating magnetically bound nucleic acid, this method does not create the shear forces generated by spin centrifugation that may lead to nucleic acid degradation.

Since its inception in 1995 [19], SPRI technology has demonstrated great utility and is widely used to provide efficient and cost-effective nucleic acid purification from different sources. Agencourt Bioscience Corporation (Beckman Coulter, Fullerton, USA) offers SPRI-based nucleic acid isolation systems to purify DNA and RNA from samples such as cultured cells, whole blood, human cheek cells, plant material, viral particles, fresh, frozen, or formalin-fixed, paraffin-embedded (FFPE) tissue. The success of SPRI is attributed not only to the low cost per preparation and great adaptability to automation, but also to the facts that the system has demonstrated a high reproducibility, higher yield, greater purity, and flexibility. The SPRI technology allows for easy automation of nucleic acid purification with superior performance. While magnetic beads protocols are available for many nucleic acid isolation procedures, the SPRI technology allows a single-core nucleic acid engine to be built, and the infrastructure and troubleshooting overhead required in any given molecular laboratory can

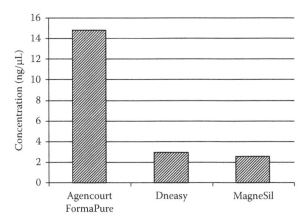

FIGURE 17.1   Average DNA concentration from processing one 10 μ rat liver FFPE sample using Agencourt FormaPure, DNeasy tissue kit, and MagneSil Genomic fixed tissue system.

be simplified by centralizing the technology platform for all nucleic acid isolation. SPRI chemistry has many advantages over other solid-support purification methods:

1. *Superior recovery*:   SPRI-based Agencourt FormaPure kit yields significantly more nucleic acid than competitive approaches. The experiment listed below was performed on rat liver tissue fixed at Agencourt Bioscience to ensure consistent starting material. When compared to currently available technologies, Agencourt FormaPure produced nucleic acid yields in excess of five times that of competitive techniques (Figure 17.1).

The spectrophotometer readings were confirmed with quantitative real-time PCR method. A real-time PCR reaction volume of 2 μL was used in the amplification of the rat β-actin gene. Ct values obtained in qPCR reactions indicate that samples purified by the Agencourt FormaPure system also produce more amplifiable DNA than competitor kits (Figure 17.2).

High-quality DNA can be obtained using SPRI from fresh and frozen tissue. The DNA is sufficiently clean to be utilized in downstream PCR amplification, and can be stored at −20°C for extended periods without losing activity in PCR. In addition, the SPRI DNA isolation procedure returns a higher yield of DNA compared to competing DNA isolation procedures (Figure 17.3).

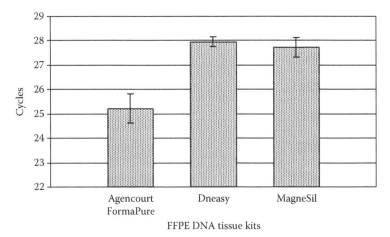

FIGURE 17.2   Comparison of Ct values for nucleic acid purified using the Agencourt FormaPure system, DNeasy tissue kit, and MagneSil Genomic fixed tissue system.

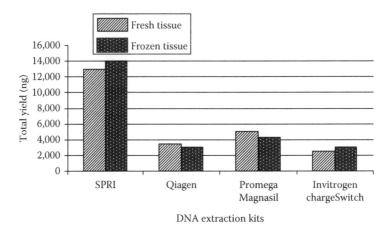

FIGURE 17.3 Comparison of DNA yields using SPRI and competitor kits.

The large binding capacity of the SPRI beads (up to 4 μg of DNA per 1 μg of beads) appears to be due to a bimodal binding mechanism and possibly to a different charge switching of the functional groups. It has been reproducibly shown that steadily increasing the amount of DNA in a fixed volume reveals that SPRI recovers 100% of the product up to a certain level of input, then recovers 70% of the product thereafter, up to the highest amount tested. In addition, solution-suspended solid phase particles like SPRI beads are free to move throughout the solution and greatly accelerate the binding time. Fixed supports also lose sample volume due to fluid retention in the filter.

2. *Consistent RNA purity*:   The Agencourt FormaPure system provides consistent RNA yield and purity. Figure 17.4 shows data for RNA recovered from five FFPE rat lung biological replicates, which were purified using Agencourt FormaPure. Consistent ratios ranging from 1.79 to 1.83 for $A260/A280$ and 1.06 to 1.21 for $A260/A230$ are observed. Agilent bioanalyzer traces from the same samples showed a consistent RNA purity and fragment recovery. When compared to RNA purification methods such as Trizol, silica columns, silica magnetic beads, or charge switch methods, SPRI stands out in terms of RNA quality or quantity resulted.

3. *Recovery of fragmented nucleic acid*:   The fixation of tissue samples in formaldehyde can cause extensive cross-linking of tissue components [22–24]. It reacts primarily with proteins, creating a tightly locked three-dimensional network of proteins, which is also linked to other macromolecules.

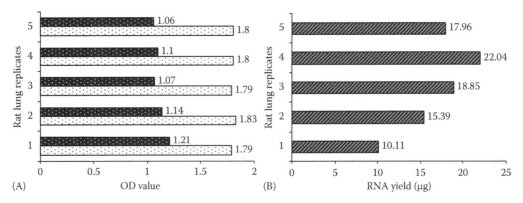

FIGURE 17.4 Consistent RNA yield and purity: (A) OD values ($A260/A230$, black horizontal bar; $A260/A280$, white horizontal bar) and (B) RNA yield.

To date, formaldehyde as a 10% neutral buffered formalin is the most widely used universal fixative because it preserves a wide range of tissues and tissue components. However, attempts to extract usable DNA from formalin-fixed tissues for molecular biological studies have been variably successful [25–27]. Even short-term treatment of sections with formalin has been shown to significantly reduce the DNA solubility [28]. Although considerable evidence suggests that formaldehyde induces DNA degradation, few studies have reported yield of high molecular weight DNA from formalin-fixed tissues [29].

It was demonstrated that DNA isolated using the Agencourt FormaPure from a variety of FFPE sample types and ages was successfully amplified by PCR (300 bp region of the glyceraldehyde 3-phosphate dehydrogenase [GAPDH] gene). Silica-based kits appear to capture less material from equal percentage regardless of size and fail to capture any small products. The SPRI-based kit was able to recover the fragments from 25 up to 500 bases in length.

Extraction of RNA from FFPE tissue would be of vital importance as pathology departments all over the world possess huge archives of various illnesses, which could be analyzed. The RNA in tissue blocks degrades fast during standard storage. The chemistry behind this is unclear. Quality of nucleic acid will vary depending on a variety of factors, including how the formalin-fixation process was carried out, age of sample, storage conditions, etc. Real-time PCR techniques require short RNA segments to be amplified for RNA quantitation. This approach seemed to be particularly suitable for quantitative determination of gene transcript levels even in tissue extract containing fragmented RNA, so with limited amplicon sizes, the paraffin tissue could offer a reliable source of information about gene expression.

For SPRI technology, similar results to DNA have also been seen with RNA capture. SPRI carboxy-based beads have proved to capture more RNA and DNA material than their competitive silica-diethylaminoethyl cellulose (DEAE) columns (Figure 17.5).

It is difficult to ascertain whether this is a result of the silica functional group or the design of the solid support, because there are no silica magnetic beads that have the same functional group density and geometry as commercially available carboxylated beads.

The SPRI system (Agencourt RNAdvance tissue) has been shown to work well with a wide variety of tissue samples. Figure 17.6 shows the yield of RNA obtained from extraction of five different tissue types including fibrous and fatty tissues. The RNA RIN scores obtained with Agilent's 2100 bioanalyzer were 9.8, 9.6, 9.8, and 9.4 for liver, lung, thymus, and spleen, respectively.

Other advantages of the SPRI technology include isolation of high-quality nucleic acids, scalability, no centrifugation, low cost of preparation and equipment, ease of buffer exchanges, fast processing, and high reproducibility. The main limitation of magnetic particles method, as solution suspension

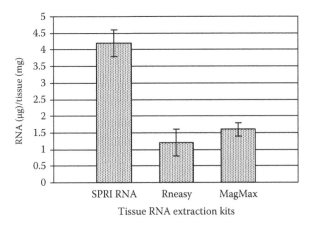

FIGURE 17.5 The Agencourt RNAdvance tissue and competitor kits were compared by extracting 10 mg of rat liver tissue in a multiwell format, $N = 24$ for each kit.

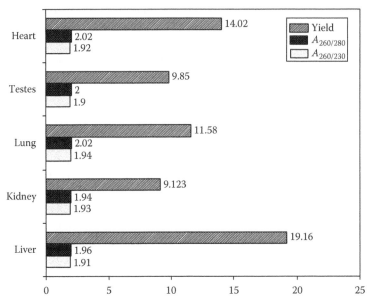

FIGURE 17.6 RNA extraction from several different rat tissues using the Agencourt RNAdvance tissue system on a Biomek NX span. Yields are reported as microgram per milligram of tissue. $N = 24$ for liver tissue and $N = 8$ for all other tissues.

solid phase system, is particle agglutination. The predominant cause of clumps is protein adsorption to the beads. Agglutination often occurs during purifying tissue preparation and the proteins vastly outnumber the beads, therefore the proteins begin cross-binding to multiple beads and coalescing beads in solution. Thus to overcome this problem, either the percentage of solid mass should be increased, or the larger volume of surfactant in solution (Tween 20 or sodium dodecyl sulfate [SDS]) should be utilized.

17.2.3.3.5 Affinity Purification

The process itself can be thought of as an entrapment, with the target molecule becoming trapped on a solid or stationary phase or medium. The other molecules in solution will not become trapped as they do not possess this property. The solid medium can then be removed from the mixture, washed, and the target molecule released from the entrapment in a process known as elution. Simply, DNA or RNA molecule can be separated from a solution using a complementary probe. For instance, the strong affinity between biotin and streptavidin has been used for the purification of nucleic acids [30–32]. Triple-helix affinity capture has proven to be a useful approach for DNA targeting [33]. The method is based on the specific binding of pyrimidine oligonucleotides to the purine strand in duplex DNA, forming a local triple helical structure.

17.2.4 Isolation of Total RNA from Difficult Tissues

The success of isolation of a good quality RNA depends not only on particular isolation method and reagents, but also on how the tissue is handled (storage condition and the time from dissection) and how rapidly the tissue is homogenized for RNA isolation [9].

The methods of RNA extraction depend on the tissue type and type of RNA to be extracted. Special attention should be paid to the type of tissue used for RNA isolation. Certain tissues, such as pancreas and spleen, are particularly abundant in RNases that rapidly degrade RNA. It is possible to curtail endogenous tissue RNase activity by rapid disruption using tissue homogenizer in the presence of a strong chaotropic agent (a biologically disruptive agent) such as the guanidinium salts, phenol, and a detergent (SDS).

The often exacting process of isolating intact total RNA from tissue becomes even more difficult when processing certain problematic tissues. Fibrous tissues and tissues rich in protein, DNA, and nucleases present distinct challenges for total RNA isolation. Some of the demanding tissues requiring more manipulation and fine-tuning during the RNA isolation procedure are heart, brain, thymus, and spleen. For fibrous tissues such as rat and mouse heart and skeletal muscle, the most difficult step in the isolation process can be complete disruption of all the cells when preparing tissue homogenates. Preparation for homogenization should be carried out on dry ice, under liquid nitrogen. Pulverizing the tissue into a powder while keeping the tissue completely frozen is the key to isolating intact total RNA. Brain tissue is rich in lipids, which can complicate the RNA extraction process, making it difficult to get a clean separation of RNA. To cure this procedure, remix the aqueous and organic phases, add more lysis solution, effectively diluting the protein and lipids, and reextract with phenol/chloroform/IAA. Rat spleen and thymus are high in nucleases and nucleic acids. Multiple phenol/chloroform/IAA extractions can be performed to ensure the partitioning of DNA into the organic phase during the acid–phenol extractions of the RNA isolation procedure. The high DNA and RNA content of these tissues cause the homogenates to be unusually viscous. To avoid high viscosity of lysates, add more lysis solution or re-extract with phenol/chloroform/IAA.

17.2.5   RAPID NUCLEIC ACID EXTRACTION PROTOCOLS FROM FRESH, FROZEN, AND FFPE TISSUES

Recently, MultiTarget Pharmaceuticals (Salt Lake City, Utah, USA) has developed nontoxic aqueous solutions, AquaGenomic and AquaRNA, for genomic DNA and total RNA isolation and purification. AquaGenomic is a one-solution genomic DNA isolation solution. This single solution achieves cell lysis, debris removal, and DNA extraction in one-step.

AquaGenomic tissue protocol does not require the usual 4–16 h of proteinase K digestion to extract genomic DNA from solid tissues. DNA can be isolated from tissues in as short as 15 min. About 10–20 µg of genomic DNA can be isolated from 10 mg of tissues. AquaGenomic solution contains detergents. Personal protections, such as rubber gloves, chemical safety goggles, and laboratory coat, should be worn when handling AquaGenomic solution.

AquaRNA is a multifunctional aqueous solution for total RNA isolation and purification. This single solution functions to lyse cells, inactivate and remove RNases, extract DNA and RNA, and precipitate cell debris. The protocol is simple, fast, and scalable. Approximately, 5–10 µg RNA can be isolated from 20 mg of animal tissues, using 0.5 mL of AquaRNA solution. You may scale up or down the starting material by using 0.5 mL AquaRNA solution for every 50 mg of tissue. The overriding advantages of AquaGenomic and AquaRNA are their simplicity and low cost. Both kits are less than 50% of the costs of other DNA isolation kits on a perminiprep basis. The advantage of AquaRNA is its effective inactivation and removal of endogenous RNases from your RNA samples. AquaRNA isolated RNA is so much more stable, it would provide you consistent RNA analysis results time after time, and would likely erase your fear of working with RNA forever.

Another simple protocol for rapid DNA extraction from tissue is EDNA. The Easy DNA (EDNA) high-speed extraction tissue kit (Fisher Biotech, Perth, Australia) has been developed to rapidly produce denatured DNA suitable for PCR and related processes. It has been designed for extraction of DNA using manual or high-throughput robotic systems and does not require centrifugation. The resulting DNA is not suitable for processes that require double-stranded DNA (e.g., restriction digestion). The kit has been validated for extraction of PCR-ready DNA from many types of tissues including muscle, shellfish muscle, feather tips, hair roots, kidney, lung, heart, buccal cells, tissue culture, and blood. In our laboratory, we have successfully utilized this system for a number of applications such as PCR, microsatellite instability (MSI), and sequencing of DNA isolated from FFPE tumour tissue. The use of 1 µL of extract as template in a 10 µL PCR reaction is suggested as a starting point. Some PCRs may require optimization of extract volume used.

17.2.6 Comparison of Commercial DNA/RNA Purification Protocols

Since phenol/chloroform purification process is time-consuming and exposes laboratory personnel to toxic reagents, commercial kits that do not involve such hazardous chemicals are widely utilized for DNA and RNA extraction. Over 30 companies [34] are marketing a spectrum of nucleic acid extraction kits based on variety of technologies that include classic and modified aquatic methods, adsorption to silica- or glass-based fiber, spin columns or affinity binding to insoluble cellulose matrix, latex beads, and paramagnetic particles. Table 17.1 presents a quick overview of most popular kits currently available in the market. However, when isolating nucleic acids from hundreds of samples, the best choice is to test which kit best suits particular needs. (Tip: Try the trial kits most suppliers give you for free or startup kit for reduced price.)

17.3 REAGENTS AND EQUIPMENT

1. Cell lysis solution (10 mM Tris-HCl pH 8.0, 25 mM EDTA, and 0.5% SDS)
2. Protein precipitation solution (5 M ammonium acetate, Mr = 77.09 is Formula Weight: 77.09)
3. Proteinase K (Merck, Cat. No. 1.24568)
4. RNase A solution (QIAGEN, Cat. No. 158922)
5. Glycogen solution (QIAGEN, Cat. No. 158930)
6. Agencourt FormaPure 96 prep kit (Beckman Coulter, Cat. No. A33342)
7. Agencourt Supermagnet magnetic plate (Agencourt, Cat. No. 000322) or SPRIStand for 1.7 mL tubes (Agencourt, Cat. No. 001139; http://www.agencourt.com)
8. 70°C–72°C and 55°C water bath (Optional: 37°C water bath for DNase incubation)
9. For tube format only, 1.7 mL microcentrifuge tubes (ABGene, UK, Cat. No. T6050G)
10. Microtome for tissue sectioning
11. 100% isopropanol, ultrapure (American Bioanalytical, USA, Cat. No. AB-07015; http://www.americanbio.com/)
12. 90% isopropanol made from ultrapure isopropanol and nuclease-free water
13. Fresh 70% ethanol made with nuclease free water (Note: 70% ethanol is hygroscopic. Fresh 70% ethanol should be prepared for optimal results) (American Bioanalytical, Cat. No. AB-00138; http://www.americanbio.com/)
14. Reagent grade water, nuclease-free (Ambion, Cat. No. 9932; http://www.ambion.com)

17.4 METHODS

17.4.1 Isolation of DNA from Fresh/Frozen/Fixed Soft Tissue (Small Scale)

This is a very simple, robust, small scale, and low-cost DNA isolation procedure for all types of soft tissues. Cells are lysed with an anionic detergent in the presence of a DNA stabilizer. The DNA stabilizer limits the activity of intracellular DNases and also DNases found elsewhere in the environment. RNA is then removed by treatment with an RNA digesting enzyme. Other contaminants, such as proteins, are removed by salt precipitation. Finally, the genomic DNA is recovered by precipitation with alcohol and dissolved in a buffered solution containing a DNA stabilizer. Purified DNA typically has an $A260/A280$ ratio between 1.7 and 1.9, and is up to 200 kb in size. This protocol is for purification of genomic DNA from 5 to 10 mg fresh or frozen solid tissue. Range of expected yield from 1 mg of tissue is 0.5–10 μg of DNA.

Procedures

For tissue in fixative: Briefly blot excess fixative from tissue on clean absorbent paper. Dispense 300 μL cell lysis solution into a 1.5 mL microcentrifuge tube, and add 5–10 mg tissue (3 mm × 3 mm pieces). Incubate for 15 min at 65°C to soften the tissue. Homogenize using 30–50 strokes with a microcentrifuge tube pestle.

Add 1.5 μL of proteinase K, mix by inverting 25 times, and incubate at 55°C for 3 h. Samples can be incubated at 55°C over night for maximum homogenization. Invert tube periodically during the incubation.

If tissue is not completely digested after an overnight incubation, add an additional 1.5 μL proteinase K and continue incubation at 55°C for 3 h. Samples can be incubated overnight for maximum homogenization. Invert tube periodically during the incubation. Continue to Step 13.

For fresh/frozen tissue: Cool mortar, pestle, and homogenizer in −70°C freezer for 1–2 h or overnight. Remove the tissue sample from −70°C freezer. Cut it into small pieces (0.3–0.5 cm²) using scalpel blade. Refreeze the tissue for processing by tipping a small amount of liquid nitrogen into the specimen container. Transfer tissue to base of cold homogenizer using scalpel blade. Place top of metal homogenizer on top of tissue pieces. With a hammer hit the top of the metal pestle three to five times until tissue is shattered to the consistency of powder.

1. Transfer the powdered tissue to the cold ceramic mortar and pestle (using scalpel blade) and grind it to a very fine powder. Add liquid nitrogen to mortar if tissue starts to thaw.
2. Dispense 300 μL cell lysis solution into a 1.5 mL grinder tube on ice, and add the ground tissue from the previous step.
3. Heat at 65°C for 15 min to 1 h. If maximum yield is required, add 1.5 μL of proteinase K (20 mg/mL), mix by inverting 25 times, and incubate at 55°C for 3 h or until tissue has completely lysed. Invert tube periodically during the incubation. The sample can be incubated at 55°C overnight for maximum yields.
4. Add 1.5 μL of RNase A solution, and mix the sample by inverting 25 times. Incubate at 37°C for 15–60 min.
5. Incubate for 1 min on ice to quickly cool the sample.
6. Add 100 μL of protein precipitation solution, and vortex vigorously for 20 s at high speed.
7. Centrifuge for 3 min at 13,000–16,000 × g. The precipitated proteins should form a tight pellet. If the protein pellet is not tight, incubate on ice for 5 min and repeat the centrifugation.
8. Pipet 300 μL of isopropanol into a clean 1.5 mL microcentrifuge tube and add the supernatant from the previous step by pouring carefully. Be sure the protein pellet is not dislodged during pouring. Note: If the DNA yield is expected to be low (<1 μg), add 0.5 μL glycogen solution.
9. Mix by inverting gently 50 times and keep for 15–30 min at RT.
10. Centrifuge for 1 min at 13,000–16,000 × g
11. Carefully discard the supernatant, and drain the tube by inverting on a clean piece of absorbent paper, taking care that the pellet remains in the tube.
12. Add 300 μL of 70% ethanol and invert several times to wash the DNA pellet.
13. Centrifuge for 1 min at 13,000–16,000 × g.
14. Carefully discard the supernatant. Drain the tube on a clean piece of absorbent paper, taking care that the pellet remains in the tube. Allow to air-dry for up to 15 min. The pellet might be soft and easily dislodged.
15. Add 50–100 μL of TE buffer and vortex for 5 s at medium speed to mix.
16. Incubate at 65°C for 1 h to dissolve the DNA.

Samples can be incubated at room temperature overnight with gentle shaking. Ensure tube cap is tightly closed to avoid leakage. DNA in TE buffer can be stored at room temperature for up to 1 year.

17.4.2 ISOLATION OF TOTAL RNA AND DNA ISOLATION FROM FRESH/FROZEN/FIXED SOFT TISSUE (SMALL AND LARGE SCALES)

The Agencourt FormaPure nucleic acid purification kit utilizes the patented Agencourt SPRI paramagnetic bead-based technology to isolate nucleic acids (both DNA and RNA) from a maximum

TABLE 17.1
Comparison of the Commercial Kits for Nucleic Acids Purification

Vendor	QIAGEN	QIAGEN	QIAGEN	QIAGEN	QIAGEN	QIAGEN	Mo Bio Laboratories	MultiTarget Pharmaceuticals	MultiTarget Pharmaceuticals	BeckMan Coulter	BeckMan Coulter
System	Gentra Puregene tissue kit	QIAamp DNA mini kit	QIAamp DNA FFPE tissue kit	RNeasy protect midi kit	EZ1 RNA tissue mini kit	Oligotex mRNA kit	UltraClean tissue DNA kit	AquaRNA	AquaGenomics	Agencourt RNAdvance	Agencourt FormaPure
Description	For purification of archive-quality DNA from tissues	For purification of genomic, mitochondrial, bacterial, parasite, or viral DNA	For purification of genomic DNA from FFPE tissues	For RNA*later* stabilization and RNeasy purification of up to 1 mg total RNA from animal tissues	For automated purification of high-quality total RNA from tissue samples up to 10mg using the BioRobot EZ1 workstation	For maxiprep purification of poly(A) + mRNA from total RNA and cleanup of *in vitro* transcripts	For purification of genomic, mitochondrial, bacterial, parasite, or viral DNA	A multifunctional aqueous solution for total RNA isolation and purification	A multifunctional aqueous solution for total DNA isolation and purification	Total RNA extraction from tissue	Total RNA and genomic DNA extraction from tissue
Applications	PCR, restriction digest, Southern analysis, and SNP analysis	PCR and Southern blotting	Real-time PCR, STR analysis, and LMD-PCR	PCR, qPCR, real-time RT-PCR, microarray	PCR, qPCR, real-time RT-PCR, and microarray	PCR, qPCR, real-time RT-PCR, and microarray	PCR and RFLP	PCR, qPCR, and real-time RT-PCR	PCR and RFLP	qRT-PCR and microarray analysis	PCR, Southern blotting, qRT-PCR1, and microarray analysis
Format	Aqueous solution	Spin column	Spin column	Spin column	Spin column	Spin column	Spin column	Aqueous solution	Aqueous solution	Magnetic beads	Magnetic beads
Main sample type	Tissue samples: fresh, frozen, and fixed and paraffin-embedded tissues	Whole blood, fresh and frozen tissues, and cells	FFPE tissue	Fresh or frozen tissue (liver, kidney, and spleen)	Fresh and frozen tissues	Fresh and frozen tissues	Fresh or frozen all types of tissue	Fresh or frozen and fixed tissues	Fresh or frozen and fixed tissues	Fresh or frozen tissue	Fresh or frozen and fixed tissues
Processing	Manual (centrifugation)	Manual (centrifugation or vacuum)	Manual (centrifugation or vacuum)	Manual (centrifugation)	Manual (centrifugation or vacuum)	Manual (centrifugation or vacuum)	Manual (centrifugation or vacuum)	Manual (centrifugation)	Manual (centrifugation)	Manual (small scale)	Manual (small scale)
Purification of total RNA, miRNA, poly(A) + mRNA, DNA, or protein	Genomic DNA	Genomic DNA, mitochondrial DNA, bacterial DNA, parasite DNA, and viral DNA	Genomic DNA and mitochondrial DNA	Total RNA	Total RNA	mRNA	Genomic DNA	Total RNA, microRNA, and siRNA	Genomic DNA and mitochondrial DNA	Total RNA	Genomic DNA and total RNA
Sample amount	5–100 mg	25 mg	Up to 8 sections, 20–250 mg each with a thickness of up to 10 μm and a surface area of up to 250 mm²		Up to 10 mg	Up to 10 mg tissue or 1–3 μg tRNA	1–25 mg of tissue or 5 million cultured cells	50 mg tissues	10–20 mg tissues	Up to 100 mg	Maximum 10 mg

	Modified salting-out precipitation method	Silica technology	Silica technology	Silica technology	Silica column	Silica column	Spin filters	Selective extraction	Selective extraction	SPRI	SPRI
Technology	Modified salting-out precipitation method	Silica technology	Silica technology	Silica technology	Silica column	Silica column	Spin filters	Selective extraction	Selective extraction	SPRI	SPRI
Time per run or per preparation	25–60 min	20 min	24 h	60 min	50–60 min	30 min	30 min–1 h	30 min	15 min	One tube processing for 30 min or 96-well plate processed in about 4 h	One tube processing for 60 min or 96-well plate processed in about 4 h
Yield	Varies	4–12 µg	Varies	100–800 µg	30 µg total RNA	N/A	<4 µg	5–20 µg from 50 mg tissues	5–10 µg from 10–20 mg tissues	4–6 mg from 10 mg of soft tissue	From 10 µm FFPE tissue section—10–20 gµ total RNA and 400 ng of DNA
Technical complexity of the methodology	Precipitation, centrifugation, and no organic hazardous chemicals	Spin column containing a silica membrane	Spin column containing a silica membrane	Spin column containing a silica membrane	Flexible purification of 1–6 samples per run; and credit card ease-of-use for protocol and worktable setup	No oligo-dT cellulose or ethanol precipitation; and flexibility for use with widely varying amounts of starting RNA	Spin column containing a silica membrane	Single solution acts to lyse cells, inactivate and remove RNases, extract DNA and RNA, and precipitate cell debris	Single solution to lyse the cells, extract the DNA, and precipitate the cell debris, and then add isopropanol to precipitate the purified genomic DNA	Based on paramagnetic bead-based technology; no centrifugation or vacuum filtration required; And no organic extraction steps	Based on paramagnetic bead-based technology; no centrifugation or vacuum filtration required; and no organic extraction steps
Adaptability of the methodology for high-throughput automation	No	Yes	Yes	Yes	Yes	Yes	Yes	No	No	Yes	Yes
Quality of nucleic acid	A260/A280 is 1.7–1.8 (50–200 kb in length)	A260/A280 is 1.8–1.9 (50–200 kb in length)	PCR ready	PCR fragments up to 300bp	PCR fragments up to 300bp	PCR fragments up to 300bp	A260/A280 is 1.7–1.9 (up to 50 kb in length)	No data	A260/A280 of the isolated genomic DNA is 1.6–1.8. The genomic DNA is free from enzyme inhibitors	RNA—A260/A280 is 1.79–1.84 and RIN score is 9.1–9.8	RNA—A260/A280 is 1.79–1.84; DNA—A260/A280 is 1.7–1.9 (up to 200kb in length for fresh/frozen tissue); and 300–400bp amplicons from DNA FFPE tissue
Cost per preparation ($)	1.45	5.30–6.00	7.90	24.50	10.80	49.50–186.00	2.10	0.66	0.66	3.47–6.64	6.59–8.24

(continued)

TABLE 17.1 (continued)
Comparison of the Commercial Kits for Nucleic Acids Purification

Vendor	Promega	Promega	Promega	Promega	Promega	Ambion	Ambion	Ambion	ROCHE	ROCHE	Corbett Robotics
System	Wizard SV genomic DNA	Wizard genomic DNA	SV total RNA isolation system	PureYield RNA midiprep	PolyATtract System 1000	MELT total nucleic acid isolation system	RNAqueous	RiboPure	High Pure FFPE RNA micro kit	High Pure RNA paraffin kit	Corbett X-tractor gene
Description	Genomic DNA extraction from tissue	Genomic DNA extraction from tissue	Total RNA from tissues	Total RNA from variety of tissues	Isolates mRNA directly from crude cell or tissue lysates	Total RNA or genomic DNA	Total RNA from variety of tissues	Total RNA from variety of tissues	Isolation of total RNA from FFPE tissue	Isolation of total RNA from fresh, frozen, and FFPE tissues	Generic protocols for the purification of tissue or liquid samples
Applications	PCR and Southern blotting	PCR and digestion with restriction endonucleases and membrane hybridizations (e.g., Southern and dot/slot blots)	All routine molecular biology applications, including RT-PCR and Northern blotting	All routine molecular biology applications, including RT-PCR and Northern blotting	All molecular biology applications, including *in vitro* translation, cDNA synthesis, PCR analysis, RNase protection assays (RPAs), primer extension, and Northern blots	qRT-PCR, Northern blot analysis, RPA, and RNA amplification and microarray analysis	Most common applications, including Northern blots, RT-PCR, nuclease protection assays, array probe labeling, and *in vitro* translation	cDNA synthesis, real-time and end-point RT-PCR, microarray analysis, Northern blots, and RPAs	Isolate total RNA for direct use in RT-PCR, differential display RT-PCR, and cDNA synthesis/primer extension	RT-PCR	PCR, RT-PCR, Southern blotting, SNP, and marker analysis
Format	Spin or vacuum formats of silica column	Aqueous solution	Spin or vacuum formats of silica column	Spin or vacuum formats of silica column	Magnetic beads	Magnetic beads	Spin or vacuum formats column	Spin or vacuum formats column	Spin column	Spin column	Automated 8–96 samples
Main sample type	Any type of fresh or frozen tissue	Any type of fresh or frozen tissue	Any type of fresh or frozen tissue	Any type of fresh or frozen tissue	Any type of fresh or frozen tissue	Fresh or frozen tissue	Fresh or frozen tissue	Fresh or frozen tissue	FFPE tissue sections	Fresh/frozen or FFPE tissue	Fresh or frozen tissue
Processing	Manual (small scale)	Manual	Manual (small scale)	Manual (small scale)	Manual (small scale)	Manual (small scale)	Manual (small scale)	Manual (small scale)	Manual	Manual	Automated
Purification of total RNA, miRNA, poly(A) + mRNA, DNA, or protein	Genomic DNA	Genomic DNA	Total RNA	Total RNA	mRNA	Total RNA	Total RNA	Total RNA	Total RNA	Total RNA	Genomic DNA, mitochondrial DNA, bacterial DNA, parasite DNA, and viral DNA
Sample amount	Up to 20 mg	Scalable	Up to 60 mg	Up to 300 mg	5 mg–2 g	Up to 10 mg	75 mg	1–100 mg	1–10 μm FFPE tissue sections	5–10 μm FFPE sections, 10–30 mg fresh/frozen solid tissue, and 3–5 μm fresh/frozen tissue sections	Varies for sample quality, type, and source

	Silica column	Modified salting-out precipitation method	Silica column	Silica column	MagneSphere magnetic separation	MagMAX magnetic bead	Glass-fiber filter membrane	Glass-fiber filter membrane	Silica column	Silica column	Glass-fiber membrane
Technology	Silica column	Modified salting-out precipitation method	Silica column	Silica column	MagneSphere magnetic separation	MagMAX magnetic bead	Glass-fiber filter membrane	Glass-fiber filter membrane	Silica column	Silica column	Glass-fiber membrane
Time per run or per preparation	20 min	60 min	20 min	20–30 min	45 min	20 min	20 min	30 min	60 min	2 h	40 min for 8 samples; and 90 min for 96 samples
Yield	20–30 μg	Varied	10 μg	Up to 1 mg	N/A	4–12 μg total RNA from 3–5 mg, and 1 μg of DNA/mg of tissue	20–50 μg	100–500 μg/100 mg	1.5–3.5 μg/5 μm	0.3–1.5 μg/5 μm section, and 2–6 μg/20 mg fresh/frozen solid tissue	Up to 10 μg (sample dependent)
Technical complexity of the methodology	Spin silica column containing a silica membrane	Precipitation, centrifugation, no organic hazardous chemicals	Microcentrifugation (spin) or vacuum	Microcentrifugation (spin) or vacuum	Based on paramagnetic bead-based technology; and no centrifugation or vacuum filtration required	Based on paramagnetic bead-based technology; and no centrifugation or vacuum filtration required for RNA prep	Microcentrifugation (spin) or vacuum	Robust lysis/denaturant and TRI reagent, with glass-fiber filter purification	Spin column containing a silica membrane	Spin column containing a silica membrane	SBS 96 well plate with a glass-fiber membrane
Adaptability of the methodology for high-throughput automation	No	Yes	Yes	Yes	Yes	Yes	Yes	No	No	Yes	Yes
Quality of nucleic acid	A260/A280 is 1.7–1.8 (50–200kb in length)	A260/A280 is 1.8–1.9 (50–200kb in length)	PCR ready	PCR fragments up to 300bp	PCR fragments up to 300bp	PCR fragments up to 300bp	A260/A280 is 1.7–1.9 (up to 50kb in length)	No data	A260/A280 of the isolated genomic DNA is 1.6–1.8. The genomic DNA is free from enzyme inhibitors	RNA—A260/A280 is 1.79–1.84 and RIN score is 9.1–9.8	RNA—A260/A280 is 1.79–1.84; DNA—A260/A280 is 1.7–1.9 (up to 200kb in length for fresh/frozen tissue); and 300–400bp amplicons from DNA FFPE tissue
Cost per preparation ($)	1.45	5.30–6.00	7.90	24.50	10.80	49.50–186.00	2.10	0.66	0.66	3.47–6.64	6.59–8.24

input of 10 mg of FFPE tissue. We have adapted this protocol to work successfully with fresh, frozen tissue as well as with FFPE material. The protocol can be performed in both 96 well plates (manually and fully automated) and in 1.5 mL tubes. Nucleic acid extraction begins with the addition of a reagent that melts paraffin and de-crosslinks nucleic acids. Proteinase K is then added to complete tissue digestion and inactivate nucleases. Next, binding buffer is added to facilitate immobilization of the nucleic acids to the surface of paramagnetic beads. The contaminants are rinsed away using a simple washing procedure. The Agencourt FormaPure procedure does not require vacuum filtration or centrifugation. The FormaPure system allows researchers to process a few to hundreds of samples per day.

Procedure

Tube format: Up to 10 mg of tissue may be homogenized in 400 μL of lysis buffer/proteinase K in a 1.7 μL microcentrifuge tube. Larger amounts of tissue may be homogenized simply by using a 15 or 50 mL conical tube and scaling up the volume of the homogenization. Use an additional 400 μL of lysis/proteinase K for every additional 10 mg of tissue. For volumes above 2 mL total, a larger tissue dispersing element should be used (8–10 mm) to ensure complete homogenization. Foaming of the sample during lysis can be minimized by keeping the dispersing element in a fixed location in the center of the tube and as close to the bottom as possible. Fresh and frozen tissue should be pulverized to powder as described in the protocol above.

- Soft tissue, tube format: Samples should be homogenized at the highest speed setting for about 2 min.
- Fibrous tissue, tube format: Samples should be homogenized at the highest speed setting for about 5 min. In addition, the 37°C lysis incubation can be extended to 45 min for particularly tough tissue such as vascular smooth muscle.
- Lipid rich tissues, tube format: Samples should be homogenized at the highest speed setting for about 30–90 s. Special care should be taken to avoid excessive foaming of lipid-rich tissues during homogenization.

96 well plate format: Homogenization of up to 10 mg of tissue in 400 μL of lysis buffer/proteinase K may be accomplished using either a tissue dispersing device or metal beads and agitation (bead beating). The bead-beating method provides a higher throughput solution for 96 well plate isolations. Add 400 μL of lysis buffer per well and up to 10 mg of tissue. Seal the plate with a plastic plate seal and shake vigorously (2400 rpm) in a plate vortexer for homogenization. Incubate the deep well plate at 37°C without removing the metal beads. Upon completion of the incubation, the lysate should be transferred to a 1.2 mL plate for processing.

- Soft tissue, plate format: Samples should be homogenized for about 10 min at 2400 rpm.
- Fibrous tissue, plate format: Vortex for about 20–25 min at 2400 rpm. (Note: Some fibrous tissue may not be completely dispersed using metal beads—specific homogenization requirements for some fibrous tissues may need to be determined experimentally.)
- Lipid rich tissues, plate format: Vortex for about 10 min at 2400 rpm.

1. For fresh/frozen tissue, use liquid nitrogen method of homogenization as described above.
2. For fixed tissue, cut 10 mg of fixed tissue into sections of 10 μm each. Transfer FFPE tissue sections (up to 5 sections) into the plate/tube. For tissue sections attached to glass slides, wet the section with 20 μL of lysis buffer prior to scraping them off with a clean single-edge razor blade. This allows the tissue sections to be more easily transferred to the plate/tube. Push the tissue section into the plate/tube with a pipette tip. Optimal amount of starting material needs to be scaled according to the size of tissue from 1-5 of 10 μm slices.
3. Seal and incubate the plate/tube at 70–72°C in a water bath for 60 min.

Note: Prolonged incubation at 70–72°C may cause damage to the RNA. When using this plate in conjunction with a water bath, make sure the plate does not tip over and the seal does not get wet. Should the seal get wet or condensation form on it, spin the liquid down and very carefully remove the seal.

4. Following the 70–72°C incubation, pipette 20 μL of proteinase K (40 mg/mL to each well/tube), pipette mix twice with a volume of 200 μL.
5. Seal and incubate the plate/tube in a water bath at 55°C for 60 min.
6. Cool the plate on ice for 2 min.
7. Transfer the lysate to a new 1.2 mL plate/1.7 mL tube for nucleic acid extraction.
8. Add 150 μL of bind I buffer and 320 μL of bind II buffer to each well/tube. Pipette mix 5 times with a volume of 600 μL. Seal plate with a plate seal and incubate in 55°C water bath for 5 min.
9. Move the plate onto the Agencourt supermagnet (or SPRIStand for tubes) and separate for 5 min. Wait for the solution to clear before proceeding to the next step.
10. Slowly aspirate the cleared solution from the plate/tube and discard. This step must be performed while the plate/tube is situated on the magnet. Do not disturb the separated magnetic beads. If beads are drawn out, leave a few microliters of supernatant behind.

For RNA only extraction: Go to Step 13.
It is not necessary to perform the wash step if you plan to do the DNase treatment.
For total nucleic acid (RNA and DNA) extraction:

11. Remove the plate/tube from the magnet and add 300 μL of wash buffer. Pipette mix 5 times and incubate for 1 min.
12. Return plate to the magnet and separate for 1 min. Wait for the solution to clear before proceeding to the next step.
13. Repeat Step 10.
14. Remove the plate/tube from the magnet and add 750 μL of 70% ethanol. Pipette mix 5 times with a volume of 500 μL to resuspend the beads.
15. Repeat Step 12.
16. Repeat Step 10.

For total nucleic acid extraction: Go to Step 22.
For RNA only extraction:

17. Add 100 μL of DNase solution with the plate off the magnet.
18. Pipette mix 5 times to resuspend the beads in the DNase solution.
19. Seal and incubate plate/tube in a 37°C water bath for 15 min to facilitate digestion of DNA.
20. Do not remove the DNase solution. Add 550 μL of wash buffer and pipette mix 5 times. Incubate at room temperature for 5 min.
21. Place plate/tube onto the magnet and separate for 10 min. Wait for the solution to clear before proceeding to the next step.
22. Repeat Step 10.
23. Remove the plate/tube from the magnet and add 750 μL of 70% ethanol. Pipette mix 5 times.
24. Place plate/tube onto the magnet and separate for 5 min. Wait for the solution to clear before proceeding to the next step.
25. Slowly aspirate the cleared solution from the plate/tube and discard. This step must be performed while the plate/tube is situated on the magnet. Do not disturb the separated magnetic beads. If beads are drawn out, leave a few microliters of supernatant behind.

26. Remove the plate/tube from the magnet and add 500 μL of 90% isopropanol. Pipette mix 5 times with a volume of 400 μL to resuspend the beads.
27. Seal and incubate the plate/tube in a 70°C water bath for 3 min.
28. Repeat Step 12.
29. Aspirate the cleared solution from the plate/tube and discard. This step must be performed while the plate is situated on the magnet. Do not disturb the separated magnetic beads.
30. Repeat Steps 31–35 for a total of two isopropanol washes.
31. Remove the plate from the magnet and add 750 μL of 70% ethanol. Pipette mix 5 times with a volume of 500 μL.
32. Repeat Step 12.
33. Repeat Step 10.
34. Let the plate/tube air-dry for 10 min.
35. The plate/tube should air-dry until the last visible traces of ethanol evaporate. Overdrying the sample may result in a lower recovery.
36. Remove the plate/tube from the magnet and add 80 μL of nuclease-free water. Resuspend the beads by pipette mixing 5 times. Smaller or larger elution volumes can be used for more or less concentrated product; however, the minimum elution volume should be 40 μL to ensure complete elution. Optimal elution volumes need to be experimentally determined, higher yielding samples require larger elution volumes due to potential bead carryover during the final transfer.
37. Incubate the plate/tube at 65°C–70°C for 30 s.
38. Place the plate/tube on the magnet for 1 min and transfer eluted nucleic acid to a suitable 96 well storage plate or a fresh tube. Wait for the solution to clear before transferring sample.

17.5 FUTURE DEVELOPMENT TRENDS

The current generation of automated extractors all use some form of chemical lysis, which limits the initial volume that can be extracted. However, next generation extractors are moving away from small volumes measured in microliters to relatively large starting volumes of 5–10 mL, using alternatives to liquid lysis. There is also a trend away from the movement of liquid to moving nucleic acid captured in beads. Looking into the future, further advances in the areas of automated extraction, amplification, and detection can be foreseen. The most recent introduction of semi- and fully automated methods for DNA and RNA extraction such as Maxwell from Promega (http://www.promega.com/maxwell16/default.htm), QIAcube from QIAGEN (www1.qiagen.com/myqiacube/), iPrep from Invitrogen (www.invitrogen.com/iprep), and GeneXpert system from Cepheid, Sunnyvale, USA (www.cepheid.com/genexpert) could be considered as the most likely choices for small to medium size laboratories. These automated microfluidic systems are fulfilling the requirement for molecular pathology laboratories to consolidate and simplify workflow while improving efficiency and maintaining quality of results. Among the systems mentioned above, the GeneXpert self-contained automated molecular analysis system warrants a more detailed description. This system utilizes proprietary homogeneous magnetic bead extraction technology on an automated platform, combined with a thermal cycler and software to achieve highly sensitive detection and quantitation through real-time PCR. It was reported [35,36] that the GeneXpert assays successfully detected the presence of metastatic melanoma, breast cancer, and lung cancer in lymph nodes and also differentiated between metastatic melanoma, lung adenocarcinoma, and healthy lung. In addition, the system demonstrated a high specificity and sensitivity in the detection of Philadelphia chromosome (BCR-ABL) transcripts in patients with chronic myeloid leukemia [37,38]. The GeneXpert can perform RNA isolation, reverse transcription, and quantitative PCR in 35 min and could therefore be used for intraoperative testing when applicable. By providing faster turnaround times, requiring less hands-on time, fewer technical skills, and offering the possibility of more convenient cross-laboratory standardization, this assay system could provide benefits over currently used home-brew or commercial methods. Importantly, the ability to perform rapid molecular assays outside a specialized laboratory setting could lead to a major shift in the approach to patient care and management [39].

Sergey Kovalenko graduated from the Stavropol Agriculture Academy (Stavropol, Russia) in 1984 and completed his PhD at the All-Russian Research Institute of Animal Genetics & Breeding (St. Petersburg) in Russia. After spending 3 years as a postdoctoral fellow in Nagoya University (Nagoya, Japan) investigating the effect of different mutagens on the formation of somatic mutations in postmitotic tissues, he is now a scientist in charge of molecular pathology at the Peter MacCullum Cancer Centre (Melbourne, Australia). His main scientific interest is aging and cancer research, and he has published over 40 scientific papers related to his research and diagnostics.

REFERENCES

1. Delgado, R., Mikuz, G., and Hofstadter, F., DNA-Feulgen-cytophotometric analysis of single cells isolated from paraffin embedded tissue, *Pathol. Res. Pract.*, 179: 92, 1984.
2. Takahashi, H. et al., Cancer diagnosis marker extraction for soft tissue sarcomas based on gene expression profiling data by using projective adaptive resonance theory (PART) filtering method, *BMC Bioinform.*, 7: 399, 2006.
3. Eilber, F.R. et al., Progress in the recognition and treatment of soft tissue sarcomas, *Cancer*, 65: 660, 1990.
4. Randall, R.L. et al., Transit tumor retrieval preserves RNA fidelity and obviates snap-freezing, *Clin. Orthop. Relat.* Res., 438: 149, 2005.
5. Chomczynski, P., A reagent for the single-step simultaneous isolation of RNA, DNA and proteins from cell and tissue samples, *Biotechniques*, 15: 532, 1993.
6. Chomczynski, P. et al., DNAzol: A reagent for the rapid isolation of genomic DNA, *Biotechniques*, 22: 550, 1997.
7. Nabel, G.J. et al., Direct gene transfer with DNA-liposome complexes in melanoma: Expression, biologic activity, and lack of toxicity in humans, *Proc. Natl. Acad. Sci. U. S. A.*, 90: 11307, 1993.
8. Roos-van Groningen, M.C. et al., Improvement of extraction and processing of RNA from renal biopsies, *Kidney Int.*, 65: 97, 2004.
9. Micke, P. et al., Biobanking of fresh frozen tissue: RNA is stable in nonfixed surgical specimens, *Lab. Invest.*, 86: 202, 2006.
10. Wiedorn, K.H. et al., HOPE—A new fixing technique enables preservation and extraction of high molecular weight DNA and RNA of >20 kb from paraffin-embedded tissues. Hepes-Glutamic acid buffer mediated organic solvent protection effect, *Pathol. Res. Pract.*, 198: 735, 2002.
11. Nassiri, M. et al., Preservation of skin DNA for oligonucleotide array CGH studies: A feasibility study, *Arch. Dermatol. Res.*, 299: 353, 2007.
12. Vincek, V. et al., Preservation of tissue RNA in normal saline, *Lab. Invest.*, 83: 137, 2003.
13. Xiang, C.C. et al., Using DSP, a reversible cross-linker, to fix tissue sections for immunostaining, microdissection and expression profiling, *Nucleic Acids Res.*, 32: e185, 2004.
14. Vincek, V. et al., A tissue fixative that protects macromolecules (DNA, RNA, and protein) and histomorphology in clinical samples, *Lab. Invest.*, 83: 1427, 2003.
15. Cox, R.A., The use of guanidinium chloride in the isolation of nucleic acids: In Grossman LaM, E., ed., *Methods in Enzymology*, Vol. 12, New York: Academic Press, 1968, pp. 120–129.
16. Cady, N.C., Stelick, S., and Batt, C.A., Nucleic acid purification using microfabricated silicon structures, *Biosens. Bioelectron.*, 19: 59, 2003.
17. Lathe, G.H. and Ruthven, C.R., The separation of substances and estimation of their relative molecular sizes by the use of columns of starch in water, *Biochem. J.*, 62: 665, 1956.
18. Haddad, P.R. and Jackson, P.E., *Ion Chromatography*, Vol. 46, Elsevier Science, 1990.
19. DeAngelis, M.M., Wang, D.G., and Hawkins, T.L., Solid-phase reversible immobilization for the isolation of PCR products, *Nucleic Acids Res.*, 23: 4742, 1995.
20. Hawkins, T.L. et al., A magnetic attraction to high-throughput genomics, *Science*, 276: 1887, 1997.
21. Elkin, C.J. et al., High-throughput plasmid purification for capillary sequencing, *Genome Res.*, 11: 1269, 2001.

22. Paska, C. et al., Effect of formalin, acetone, and RNAlater fixatives on tissue preservation and different size amplicons by real-time PCR from paraffin-embedded tissue, *Diagn. Mol. Pathol.*, 13: 234, 2004.

23. Jewell, S.D. et al., Analysis of the molecular quality of human tissues: An experience from the Cooperative Human Tissue Network, *Am. J. Clin. Pathol.*, 118: 733, 2002.

24. Srinivasan, M., Sedmak, D., and Jewell, S., Effect of fixatives and tissue processing on the content and integrity of nucleic acids, *Am. J. Pathol.*, 161: 1961, 2002.

25. Bramwell, N.H. and Burns, B.F., The effects of fixative type and fixation time on the quantity and quality of extractable DNA for hybridization studies on lymphoid tissue, *Exp. Hematol.*, 16: 730, 1988.

26. Serth, J. et al., Quantitation of DNA extracted after micropreparation of cells from frozen and formalin-fixed tissue sections, *Am. J. Pathol.*, 156: 1189, 2000.

27. Yagi, N. et al., The role of DNase and EDTA on DNA degradation in formaldehyde fixed tissues, *Biotech. Histochem.*, 71: 123, 1996.

28. Douglas, M.P. and Rogers, S.O., DNA damage caused by common cytological fixatives, *Mutat. Res.*, 401: 77, 1998.

29. Diaz-Cano, S.J. and Brady, S.P., DNA extraction from formalin-fixed, paraffin-embedded tissues: Protein digestion as a limiting step for retrieval of high-quality DNA, *Diagn. Mol. Pathol.*, 6: 342, 1997.

30. Tong, X. and Smith, L.M., Solid phase purification in automated DNA sequencing, *DNA Seq.*, 4: 151, 1993.

31. Olsvik, O. et al., Magnetic separation techniques in diagnostic microbiology, *Clin. Microbiol. Rev.*, 7: 43, 1994.

32. Thorp, B.H. et al., Type II collagen expression in small, biopsy-sized samples of cartilage using a new method of RNA extraction, *Clin. Exp. Rheumatol.*, 12: 169, 1994.

33. Johnson, A.F. et al., Purification of single-stranded M13 DNA by cooperative triple-helix-mediated affinity capture, *Anal. Biochem.*, 234: 83, 1996.

34. Chua, J., A buyer's guide to DNA and RNA prep kits, *Scientist*, 18: 38, 2004.

35. Raja, S. et al., Technology for automated, rapid, and quantitative PCR or reverse transcription-PCR clinical testing, *Clin. Chem.*, 51: 882, 2005.

36. Hughes, S.J. et al., A rapid, fully automated, molecular-based assay accurately analyzes sentinel lymph nodes for the presence of metastatic breast cancer, *Ann. Surg.*, 243: 389, 2006.

37. Jobbagy, Z. et al., Evaluation of the Cepheid GeneXpert BCR-ABL assay, *J. Mol. Diagn.*, 9: 220, 2007.

38. Winn-Deen, E.S. et al., Development of an integrated assay for detection of BCR-ABL RNA, *Clin. Chem.*, 53: 1593, 2007.

39. McMillan, W.A., Real-time point-of-care molecular detection of infectious disease agents, *Am. Clin. Lab.*, 21: 29, 2002.

18 Isolation of Nucleic Acids from Hard Tissues

Judith Weidenhofer and Jennifer A. Byrne

CONTENTS

18.1 INTRODUCTION

The extraction of nucleic acids from hard tissues has always been considered challenging. Methods need to meet the requirements of disrupting tissue such that nucleic acids are released, without inducing or exacerbating nucleic acid degradation, or copurifying inhibitors of subsequent enzyme-catalyzed reactions [1]. In addition, tissues such as bone harbor comparatively low levels of endogenous nucleic acid due to their low cellularity, and may include greater amounts of exogenous microbial or fungal DNA [2]. There are however fundamental differences in the types of applications where DNA or RNA would be extracted from these tissues, which produce challenges specific to each type of extraction. Where RNA is being sought from hard tissues, this is because the study is focusing upon the biology of the tissues in question, and in such studies the tissues will have been deliberately chosen. This choice implies some degree of control over sample conditions, which greatly facilitates the extraction of labile RNA molecules. In contrast, where DNA is extracted from hard tissues, this is rarely an issue of choice, but instead reflects that more accessible tissues (Chapters 16 and 17) are not or no longer available. This means that the hard tissue samples commonly used for DNA extraction present additional challenges, through being aged, environmentally exposed or contaminated, available in limited quantity, or through being available in large numbers, with none of these parameters necessarily reflecting the choice of the investigators.

This chapter reviews methods and their application for extracting both DNA and RNA from hard tissues, focusing predominantly on bone. While a vast number of studies have carried out DNA extractions from bones and teeth, this chapter focuses upon articles that have developed protocols or analyzed parameters affecting DNA or RNA extraction from these tissues. Readers should also note that major applications of DNA extractions from hard tissues include the analysis of ancient and forensic samples. As these are reviewed elsewhere (Chapters 19 and 20, respectively), this chapter does not review methods that are highly specific to these applications in detail. Due, however, to the predominance of these applications in the literature, all such studies cannot be excluded. This chapter therefore discusses sample handling issues and methods that have been developed for either forensics or ancient DNA studies, where these may have broader relevance.

18.2   KEY ISSUES CONCERNING DNA ISOLATION FROM HARD TISSUES

18.2.1   GENERAL CONSIDERATIONS

As mentioned previously, where DNA is being extracted from bone or teeth, this usually reflects the fact that no alternative or better source of DNA is available. Despite difficulties in extracting nucleic acids from hard samples, immense scientific and societal benefits have arisen from the development and application of techniques for extracting DNA from bones and teeth. The longevity of DNA in bones and teeth means that these tissues have provided templates for evolutionary studies of both living and extinct species, and DNA extracted from ancient bone and tooth specimens almost entirely underpins the field of ancient DNA [3]. DNA extraction from museum specimens has also made analysis of living species more accessible, with the availability of museum specimens reducing field trip costs, and disturbances to wild populations [4,5]. Clinical repositories of human bone can provide a source of genomic DNA for genetic studies, thereby expanding patient cohorts for gene association or mutation studies [6]. As bone samples will almost always harbor microbial and/or fungal DNA [1], human skeletal specimens can be used to confirm past clinical diagnoses, investigate causes of human disease throughout history, and study pathogen evolution [7–9]. New applications for DNA extraction from tissues such as bone continue to emerge, with DNA extraction coupled with species-specific polymerase chain reaction (PCR) primers being applied to detect bone fragments in animal feed [10,11]. This has been proposed as a technique to control transmission of bovine spongiform encephalopathy, a disease with major public health and economic consequences [10,11].

At a societal level, DNA extraction from skeletal remains has also provided immense, less tangible benefits. Bones and teeth are often the only source of DNA for identifying crime victims, deceased persons whose identities are in dispute, including those of historical interest, with the longevity of DNA within bones and teeth permitting identification long after time of death [12,13]. The ability to extract and analyze DNA from skeletal remains also allows victim identification where this must be delayed, due to requirements of criminal investigations, or lack of local infrastructure [14]. Similarly, bones and teeth may represent the only source of DNA for victim identification following disasters and mass fatalities through terrorist attacks, armed conflict, or civil unrest. The existence of missing persons is one of the most significant barriers to individual and societal healing or reconciliation following the latter events [14]. DNA extraction from skeletal or dental remains played a major role in allowing accurate and timely victim identification following the World Trade Center attacks [15,16], and as a result of genocide in the former Yugoslavia [17]. Victim identification also brings psychological benefits to individuals and societies, by helping bring perpetrators and supporting organizations to justice [17]. As DNA typing of skeletal remains can overcome deliberate attempts to conceal victim identity, it has recently been proposed that this may help deter future acts of violence and genocide [17].

The results of nucleic acid extraction profoundly influence those of subsequent molecular analyses. In few applications is this more apparent than those extracting DNA from bones and teeth. Sample parameters, both environmental and biological, play a major role in influencing DNA extraction, but as has previously been explained, these parameters may not be within the investigator's control. However, investigator choices in terms of sample preparation and cleaning, tissue harvesting and disruption, and the DNA extraction technique employed have all been shown to influence the quantity and/or quality of DNA extracted from hard samples. The major decision points affecting sample handling prior to DNA extraction have been summarized in Figure 18.1. The purpose of Sections 18.2.3 through 18.2.6 is to outline research underpinning choices at each stage of the extraction process, to allow investigators to make informed choices among the many techniques or approaches described. This will also highlight existing uncertainties, which could form the subject of future investigation.

18.2.2 Sample Parameters

Bones and teeth include a calcified mineral matrix to which DNA can adsorb and thereby be protected from degradation for periods up to 100,000 years, and possibly beyond [18]. The major mineral component of bone is hydroxyapatite, and DNA's capacity for hydroxyapatite binding has indeed been exploited in molecular techniques such as hydroxyapatite chromatography [19]. DNA extractions from bones and teeth therefore need to balance the need to both release DNA from tissue components and separate DNA from inhibitors, with the desire to simultaneously minimize DNA degradation [20].

18.2.2.1 Environmental Factors

The survival of endogenous nucleic acids within bone depends upon a myriad of factors. Microbial and fungal attack can cause rapid bone degradation [21], but specimens available for DNA extraction have, almost by definition, resisted this to a large degree. Otherwise, DNA degradation in bone occurs slowly under favorable storage conditions, through oxidation, the direct and indirect effects of background radiation, hydrolysis, and other processes [18]. In studies of recent samples, DNA yields from fox teeth were found to decline in a nonlinear fashion over approximately 30 years [22], and similarly, PCR success rates using human bone DNA templates declined according to a 7 year difference in sample age [23]. Specimen age may more indirectly influence the quantity of DNA that can be extracted, as older or ancient samples may be available in limiting amounts. Furthermore, benefits predicted through DNA extraction need to be compared with those to be obtained by preserving specimen appearance, which could be of independent artistic, cultural, or scientific value [3,4,24].

In addition to sample age, ambient storage temperature is also critical, with burial serving to stabilize temperature fluctuations around a mean [21]. The influence of ambient temperature upon DNA survival has led to the concept of a specimen's "thermal age" being more significant than its temporal age [21]. Other environmental factors have important effects, such as ambient humidity, and exposure to fire, water, soil, and other chemicals, through accelerating DNA degradation or introducing subsequent PCR inhibitors [1]. While environmental factors are accepted as critical in determining whether DNA can be extracted from ancient samples [18], even recent samples may have suffered environmental extremes such that DNA extraction is very challenging [15].

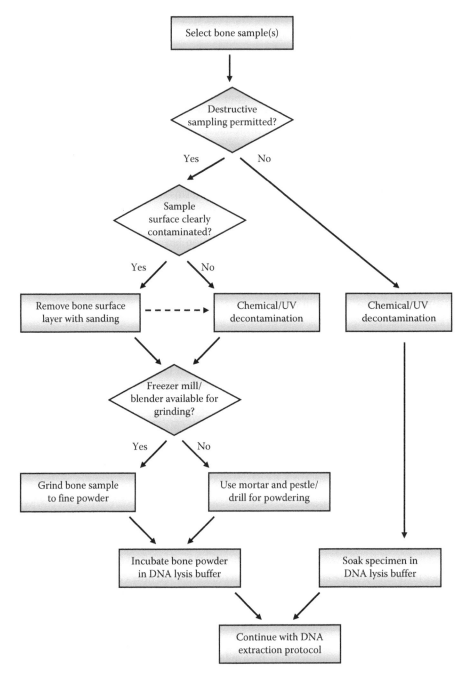

FIGURE 18.1 Decision tree summarizing the major considerations when processing hard samples for subsequent DNA extraction. The dotted arrow indicates the option of carrying out chemical decontamination in addition to surface removal.

18.2.2.2 Biological Factors

Biological factors may also impact upon the quantity and/or quality of DNA that can be extracted from hard tissues. Different bones are likely to contain different concentrations of endogenous DNA, with spongier bones of greater cellularity likely to contain higher DNA concentrations [23]. However, as bone density or mineralization is one of the most important factors determining the survival of DNA in bone [25], less compact bones are also less likely to show good DNA preservation over time [23]. Gonzalez-Andrade and Sanchez [26] reported improved DNA quality extracted from long versus other human bones, and in a larger study, improved PCR success rates were reported for human bones of the lower versus upper body [23]. Mid-shaft sections of human femur and intact teeth samples gave the greatest rates of nuclear short tandem repeat (STR) typing success, and were also the most numerous samples in the cohort, possibly reflecting their density and capacity to resist degradation [23]. Bone density can also be influenced by gender, with women showing lower bone density values than men, and by age, with bones from older individuals showing relative loss of bone mineralization [23].

Knowledge of parameters influencing DNA concentration and/or survival in bone and teeth should be applied to sample selection, where multiple samples are available. Ideally, bone or tooth specimens should be as complete and well preserved as possible [3], with large bones being preferred [3,23]. However, where forensic samples are collected by field teams, sample choice may reflect convenience, as opposed to current knowledge of critical sample parameters [23]. It has been suggested that in these cases, guideline formulation and field team training could do much to improve DNA extraction results from skeletal samples, irrespective of other technical developments [23].

18.2.3 Sample Preparation and Cleaning

Most bone or tooth samples require some form of cleaning or decontamination prior to DNA extraction. This serves to both remove external contaminants that will act as subsequent PCR inhibitors, and exogenous contaminating DNA. Extractions from bone produce a mix of endogenous nuclear or mitochondrial DNA, and DNA from other sources. Microbial DNA may exceed endogenous DNA in bone [20], and while this cannot be avoided, use of species-specific PCR primers can usually distinguish microbial from mammalian DNA. Microbial DNA may even act as a carrier when extracting low levels of endogenous DNA from bones and teeth.

Sample contamination with DNA from the same species represents a much greater problem, particularly where endogenous DNA is available in limited quantity and quality [6,18]. Human specimens are particularly sensitive to contamination, as every individual handling these, from collection to final analysis, represents a source of contaminating DNA [3]. Furthermore, as contaminating DNA is usually introduced comparatively recently during the life of a sample, this can show preferential PCR amplification, thereby leading to erroneous and misleading conclusions [3,27]. It should be noted that the methods outlined below do not control DNA contamination occurring during DNA extraction or analysis. Such methods have been described elsewhere [1], with particularly rigorous measures taken in institutes devoted to the study of ancient DNA [3,12,27]. In addition, all the methods described below are likely to be incompletely effective, but should at least reduce the proportion of contaminating DNA present [20].

A major decision point when planning DNA extractions from hard tissues is whether or not the sample can be physically modified during the course of DNA extraction (Figure 18.1). There are applications where this is undesirable, such as those using museum specimens, which may have value beyond acting as sources of DNA [24]. Considering the various applications for DNA extracted from bones and teeth reveals a paradox, namely that recent specimens (containing more abundant, less degraded DNA) are more likely to be available for unrestricted sampling, whereas old or ancient specimens (containing lesser quantities of more degraded DNA) are less likely to be available for unrestricted destructive sampling (Figure 18.2). In some cases, specimens may be unlikely to yield sufficient DNA for analyses without destructive sampling, in which cases DNA extraction should not be attempted.

Where bone samples are available for destructive sampling and show obvious external contamination, the surface layers of bone are typically cleaned with a single use grinding tool and samples are then cut from the cleaned bone (Figure 18.1) [20,28]. Where large numbers of samples need to be processed, and small amounts of sample required for analyses, a variable speed drill can be used

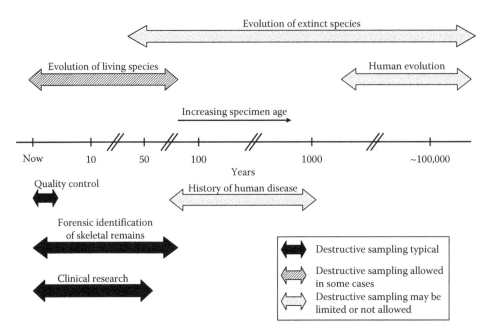

FIGURE 18.2 Time-line showing approximate age ranges (broad arrows) for bone and tooth specimens used for the indicated applications requiring DNA extraction (shown above each arrow). Solid arrows indicate applications using specimens usually available in nonlimiting quantities, which may be destroyed for analysis. Hatched arrows indicate applications using both specimens available in nonlimiting and limited quantities, which may or may not be destroyed for analysis. Shaded arrows indicate specimens frequently available in limited quantities, for which limited or no destructive sampling may be permitted. This figure highlights a paradox inherent to the extraction of DNA from hard specimens. Recent specimens (containing more abundant, less-degraded DNA) are more likely to be available for destructive sampling, whereas old or ancient specimens (containing lesser quantities of more degraded DNA) are less likely to be available for unrestricted destructive sampling.

to both remove the external bone layer, and then sample an internal portion of the bone [15]. When using an electric drill for sampling, drilling speed needs to be controlled to minimize heat production [24]. Where only limited sampling of specimens is permitted, narrow drill bits can be used to drill holes where these preserve the specimen's external appearance, such as within the root cavity of teeth [24]. Sampling from tooth pulp may also be desirable as this should contain the highest concentrations of DNA [1], and may also be relatively protected from environmental DNA contamination [7].

Many investigators follow physical sample preparation with chemical cleaning, and this represents an alternative method of surface decontamination where destructive sampling is not permitted (Figure 18.1). Bone samples can be cleaned through extensive washing in mild detergent or bleach, sterile distilled water, and finally ethanol before air or oven drying [13,28]. The use of bleach as a method for decontamination was formally analyzed by Kemp and Smith [29] and reported to be an effective and inexpensive method for the removal of exogenous DNA introduced by manual handling of ancient bones. Furthermore, bleach treatment did not appear to degrade endogenous DNA, which may be protected through its association with the bone matrix [29]. These authors recommend immersing ancient bones and teeth in 6.0% sodium hypochlorite for 15 min prior to DNA extraction [29]. However, Gilbert et al. [7] reported that comparable bleach treatment of medieval teeth did not remove exogenous contaminating DNA. In this study, the most effective measure to prevent DNA contamination was to encase teeth in silicone to minimize handling, and then carry out sampling with a dental drill [7]. Samples can also be subjected to UV irradiation as an additional measure to destroy contaminating surface DNA [7].

18.2.4 TISSUE DISRUPTION

Effective tissue disruption is important for DNA extraction from hard tissues, in order to maximize exposure of tissues to extracting agents [30]. When applied to ancient specimens, Rohland and Hofreiter [30] reported that samples ground in a Spex freezer mill produced higher DNA yields than those more coarsely ground with a mortar and pestle. This also implies that grinding produced by freezer mills does not produce excessive DNA degradation. However, while freezer mills have been reported to grind bone more finely than a Waring blender, the choice of instrument did not seem to significantly affect DNA yields from human bone samples up to 100 years postmortem [2]. These studies indicate that commercially available tissue homogenizers should be used to disrupt hard tissues where these are available (Figure 18.1), and examples of homogenizers that have been used for DNA extraction from bones and teeth are listed in Table 18.1. Where a freezer mill or blender is not available, other methods can be chosen (Figure 18.1), such as a mortar and pestle, which may be combined with sieving [25], or scraping bones with sterile surgical blades [11]. Small dried bones can also be placed inside folded weighing paper and crushed with needle nose pliers [31]. The use of drills for sampling can serve to simultaneously render bone or tooth material into powder [7,15,24].

18.2.5 DNA EXTRACTION

Methods for nucleic acid extraction can be broadly defined as either affinity or nonaffinity methods (Figure 18.3) [1]. The former involve the specific binding of DNA or RNA to a matrix with chemical affinity for nucleic acids. In this way, nucleic acid is actively bound and impurities are more passively excluded through their comparative failure to bind the matrix. Nonaffinity methods instead actively remove impurities, leaving the nucleic acid in solution, from where it can be concentrated. As will be outlined below, both methods have been employed in DNA extractions from bone and teeth. Regardless of the approach chosen, DNA extraction methods for hard tissues are evolving such that these contain fewer steps, as it is increasingly recognized that each extraction step involves loss of DNA, while providing opportunities for DNA contamination [2,3].

18.2.5.1 DNA Lysis

The first phase of DNA extraction from hard tissues involves dissolving as much material as possible, and dissociating DNA from surrounding proteins and tissue [1]. DNA lysis buffers therefore contain (1) ethylenediaminetetraacetic acid (EDTA), which releases DNA from calcium-containing proteins in bone and tooth matrix, and chelates DNase metal cofactors, (2) proteinase K for broad-spectrum protein digestion, and (3) some form of detergent, to lyze remaining cell membranes [30].

Previous bone and tooth extraction protocols have involved decalcifying material in EDTA-containing buffers, and then either extracting DNA from the supernatant, discarding any precipitate, or extracting the precipitate, discarding any washes [2]. For maximal yields, however, this step should produce complete dissolution of material to be extracted [2]. Complete dissolution requires high EDTA concentrations (0.45–0.5 M EDTA), and a low ratio of bone powder to lysis buffer [2,20]. Interestingly, improved STR profiles were also obtained from DNA extracted by complete dissolution [2]. This could be not only due to improved yields but possibly also through recovery of higher quality templates from dense bone matrix [2,25].

In a study comparing the effectiveness of different lysis conditions, Prado et al. [11] reported superior DNA yields from bone meal when samples were incubated in the presence of both 0.5 M EDTA and proteinase K. However, the addition of 0.05% SDS did not present any advantage [11]. Similar findings were reported by Rohland and Hofreiter [30], who found that only the combination of EDTA at a minimum concentration of 0.2 M, and proteinase K enhanced DNA yields, whereas detergents and other additives generally had inhibitory effects [30]. This could reflect the fact in many skeletal samples, cell membranes are no longer present, and so including detergents in extraction

TABLE 18.1

Commercially Available Tissue Disruptors Used for DNA or RNA Extraction from Hard Tissues

Nucleic Acid	Tissue	Instrument	Comments	Subsequent DNA or RNA Extraction Method	Reference
RNA	Frozen rat tibiae	SPEX CertiPrep Freezer/Mill	Samples chilled in liquid nitrogen, then pulverized with magnetically driven impactor. Each sample placed in separate grinding vial, immersed in liquid nitrogen bath inside mill. No sample cross-contamination, low temperature maintained during grinding.	Single-step method [42]	[46]
RNA	Frozen rat femora	SPEX CertiPrep Freezer/Mill		Single-step method [42]	[47]
DNA	Ancient cave bear bones and teeth	SPEX CertiPrep Freezer/Mill		Numerous	[30]
DNA	Human bones, 5–100 years postmortem	SPEX CertiPrep 6750 freezer/mill		Phenol/chloroform	[2]
RNA	Frozen human articular cartilage	SPEX CertiPrep 6800 freezer/mill		Qiagen RNeasy midi kit	[37]
RNA	Frozen human chondrosarcoma sections	Ultra Turrax Homogenizer	Ultra Turrax homogenizers have been widely used for sample disruption for subsequent total RNA extractions. In this study, sections were homogenized in Trizol reagent	Trizol reagent followed by Qiagen RNeasy mini column	[44]
DNA	Human bone, 5–100 years post mortem	Waring blender (Waring MC2 blender cup)		Phenol/chloroform	[2]
DNA	Bone fragments, teeth, Second World War period	Waring blender		Phenol/chloroform and silica extraction	[13]
RNA	Frozen human bone biopsies	Mikro Dismembrator II	Tissue disruption achieved through shaking with grinding balls. Can be used with liquid nitrogen	RNA STAT-60 kit	[39]

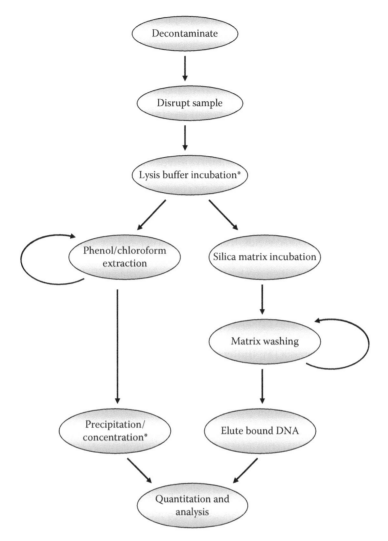

FIGURE 18.3 Outline of the essential steps of DNA extraction from bone/tooth when performed using either the phenol/chloroform or silica methods. Asterisks indicate steps that may be performed overnight.

buffers is unnecessary [30]. Incubating samples at 60°C gave superior DNA yields compared with incubations at 100°C [11], presumably because the higher temperature produced some DNA degradation. However, Rohland and Hofreiter [30] reported that varying incubation temperatures from room temperature to 56°C did not appreciably alter DNA yields.

18.2.5.2 DNA Purification and Concentration

Phenol/chloroform extraction is considered by some authors as the preferred method of DNA purification from hard tissues [3], and has been used in many applications [2,13,25,29]. However, disadvantages include the use of toxic organic solvents, increased sample loss where successive phenol/ chloroform extractions must be performed, and the fact that nonaffinity methods may less effectively remove nonprotein PCR inhibitors [1,3]. Direct comparison of a phenol/chloroform extraction technique with a salting-out technique using saturated sodium acetate showed that the latter method more frequently produced amplifiable DNA than the phenol/chloroform method [32]. Phenol/chloroform extraction is typically followed by ethanol precipitation, which may less effectively precipitate the degraded DNA frequently extracted from hard samples than contaminating DNA [3]. Centricon columns have also been widely used to further concentrate DNA samples from bone after purification [2,13,28]. However, as DNA concentration steps can serve to introduce contaminating DNA [3], they should only be performed when required.

While the best extraction method may indeed be sample dependent [3], recent studies are increasingly preferring variations of the most common affinity method, namely the glass milk or silica method [33]. This relies on glass particles bound to positively charged guanidinium thiocyanate, which retain negatively charged DNA [33]. Commercial DNA extraction kits overwhelmingly rely upon the silica method (Section 18.2.6), which has no doubt contributed to its increased uptake. Advantages of this method include the possibility of repeatedly washing the DNA-bound matrix to dilute PCR inhibitors (Figure 18.3). The fact that DNA may be eluted without further concentration (Figure 18.3) also contributes to the method's convenience. This may also improve yields, while avoiding additional sample manipulation which could introduce DNA contamination. Disadvantages include the possibility that contaminants may hinder DNA binding to the matrix, which is supported by findings that DNA yields from bone powder reduce as a function of excess input material [15]. DNA complexed with protein may also not bind with high efficiency [3].

A further advantage of the silica method is that it can be conveniently used to extract DNA from specimens where destructive sampling is not permitted. Rohland et al. [4] soaked intact bone and tooth specimens in guanidinium thiocyanate-containing buffer for up to 7 days, and then extracted DNA using the silica method. This produced mitochondrial DNA samples that could be successfully PCR amplified in most cases [4]. A subsequent application of this method obtained amplifiable nuclear DNA from Tenrec crania [5]. Buffer treatment did not obviously affect sample appearance, even after repeated extractions [4].

18.2.6   COMMERCIAL DNA EXTRACTION KITS

Numerous commercial DNA extraction kits have been reported to be suitable for extracting DNA from hard tissues. These are typically generic kits based on the silica extraction method, which can be used with tissues such as bone and teeth, often with substantial modification of manufacturer's protocols [28]. This section summarizes the key features of DNA extraction kits whose application to hard tissues has been described in the literature (Table 18.2), focusing upon studies where kit performance was compared with other DNA extraction methods.

Where comparisons have been made, these have frequently shown that kits provide superior results or convenience over laboratory methods. Staiti et al. [34] reported that use of the Promega DNA IQ system to extract nuclear DNA from forensic samples was faster and safer than conventional phenol/chloroform extraction. Davoren et al. [28] compared the use of the QIAamp DNA blood maxi kit with standard phenol/chloroform extraction for obtaining DNA from human femur samples exhumed from mass graves. This study reported higher DNA yields per gram of bone using the silica method, and lower levels of inhibitors and improved STR profiling results [28]. Similarly, Lahiff et al. [10] compared the use of a QIAamp blood kit with a laboratory silica-based method [35] for extracting DNA from meat and bone meal. Use of the kit yielded higher concentrations of DNA, as confirmed by gel electrophoresis and PCR results [10]. Two commercial kits, one from Biotools and a DNeasy Plant mini kit from Qiagen, were compared for DNA extractions of meat and bone meal [11]. This revealed similar performance in terms of PCR results, but the Qiagen kit was preferred for convenience and speed [11]. Availability of DNA extraction kits in 96 well format, such as the QIAmp 96 DNA blood kit, also vastly improves the handling of large numbers of bone samples where required [15]. These favorable comparisons, combined with a growing range of extraction kits, including those for specialized applications (Table 18.2), indicate that use of kits to extract DNA from hard samples is likely to further increase. The only formal comparison of DNA extraction methods for ancient samples found that a number of commercial kits were not superior to a laboratory silica method for extracting DNA from ancient cave bear bones and teeth [30]. This study

TABLE 18.2

Examples of Commercial Kits Used for DNA Extraction from Hard Tissues

Kit or Reagent Name	Basis for DNA Purification	Tissue Types Employed	Sample Mass/ Extraction	Reference
Qiagen QIAamp 96 DNA blood kit	Protease lysis, then silica-gel membrane affinity purification in 96 well plate format	Human skeletal remains from World Trade Center attack, 2001	25–50 mg bone powder	[15]
Qiagen QIAamp DNA blood maxi kit	Protease lysis, silica-gel membrane affinity purification using spin columns	Human femur samples from victims of 1992–1995 Balkan conflicts	5.6–9.8 g bone powder	[28]
Qiagen QIAamp DNA blood kit	As above	Industrial meat and bone meal samples	500 mg meat and bone meal	[10]
Qiagen QIAamp DNA blood kit	Lysis in proprietary buffer, debris removed by filtration unit, affinity purification using spin columns	Industrial meat and bone meal samples	100 mg meat and bone meal	[11]
Qiagen DNeasy tissue kit	Proteinase K lysis in proprietary buffer, affinity purification on silica columns	Ribs and wing bones from museum bat specimens	2–9 mm of bone (<10 mg)	[31]
Promega DNA IQ system	Proteinase K lysis, affinity purification using paramagnetic resin	Forensic human bone samples, including burnt bone and bone found in seawater	Not described	[34]
bioMérieux Nucli SENS isolation kit	Lysis in proprietary buffer, silica-gel membrane affinity purification	Vertebrae and ribs from skeleton of Iron Age	60–200 mg bone powder	[8]
GENECLEAN kit for ancient DNA	Lysis in proprietary buffer, silica-gel membrane affinity purification using spin columns	Medieval human teeth	Not described	[7]
Invisorb forensic kit I	Lysis in proprietary buffer (no proteinase K), affinity purification using Invisorb nanoparticles	Human femoral bone powder	500 mg	[57]

compared kit performance to a laboratory method that had been extensively optimized, which could account for these less favorable comparisons. Furthermore, as studies preferring kits have compared extraction performance using recent samples [10,28], whether to best use a kit or laboratory protocol may differ according to sample type and operator.

18.2.7 FUTURE DEVELOPMENT TRENDS

Recent developments in sample handling, DNA extraction procedures, and information transfer mean that forensic testing of skeletal remains is no longer used in rare cases, but is now available for samples in conditions and numbers unimaginable only a decade ago [17]. These developments have permitted large-scale victim identification in the former Yugoslavia [17], and following the attacks on the World Trade Centre on September 11, 2001 [15]. The World Trade Centre collapse provided an unprecedented challenge in terms of victim identification, requiring the extraction of DNA from over 12,000 fragmented skeletal samples exposed to a unique set of damaging conditions from some 2700 individuals [15]. The fact that a greater percentage of skeletal than soft tissue remains permitted victim identification highlights the importance of further developments in methods for extracting DNA from hard tissues [15]. These will similarly continue to be responsive to the public, legal, and technical demands resulting from major disasters with loss of human life.

The analysis of large sample numbers additionally provides scientific benefits, through highlighting sample parameters or DNA extraction procedures, which can contribute to more successful techniques [28]. Due to the generally improved performance of commercial DNA extraction kits [10,28], or optimized laboratory protocols using the silica method [20,30], DNA extractions from hard tissues are likely to increasingly use commercial kits, where costs permit. However, as few studies have performed comparative analyses of DNA extraction kit performance [10,28,30], there is clearly scope for further testing and comparison, particularly for specimens of different types and ages [30]. Researchers may also test technical improvements made in the fields of forensics or ancient DNA in their chosen applications.

Improved understanding of the factors influencing the survival of DNA in ancient bones and teeth is also of vital importance, given that DNA extraction frequently involves the destruction of specimens, and consumes researcher time, resources, and funding [21]. In many cases, specimen destruction can be unrestricted, but this is more frequently true for recent specimens, which are more likely to contain more abundant, less degraded DNA. In contrast, old or ancient specimens, containing less, more degraded DNA, may have additional value, and may not be available for unrestricted sampling. This can lead to situations where the sample quantity required to yield sufficient DNA for analysis exceeds that is available, in which case sample destruction cannot be justified [8]. In some cases, delaying extraction of particular samples has allowed developments in DNA extraction techniques which have permitted their successful use [36].

18.3 KEY ISSUES CONCERNING RNA ISOLATION FROM HARD TISSUES

18.3.1 GENERAL CONSIDERATIONS

Hard tissues such as bone perform many vital physiological functions. These include providing a framework for body shape and movement, protecting internal organs, providing mineral stores, and assisting in overall pH homeostasis. Common diseases affecting bone or cartilage such as osteoporosis and arthritis are of immense clinical and socioeconomic importance, and as many such conditions are also associated with aging, their incidence and impact are likely to further increase in future.

Due to their highly specialized structures, it is obvious that the molecular correlates of hard tissue physiology and pathology will not always be revealed through studying more accessible tissues. Many studies interested in bone and tooth biology have chosen to isolate cells from primary tissue and grow these in culture, or to use established cell lines whose phenotypes would be expected to mimic those of cells *in vivo*, and these will not be reviewed here. However, cells from *in vitro* cultures are unlikely to faithfully mimic all *in vivo* phenotypes, and the study of RNA from primary material cannot always be replaced [37]. This is becoming more apparent through gene profiling experiments, where large numbers of transcripts can be monitored in a single experiment. While still of comparatively limited number, total RNA extractions have been carried out to achieve diverse aims, such as examining changes in gene expression in response to physical bone loading [38], detecting pathogenic viruses in human clinical bone samples [39], and detecting gene fusion transcripts in bone tumors [40].

RNA isolation is usually more technically challenging than DNA isolation from the same tissue, due to the relative stability of each molecule type and the susceptibility of RNA to attack by ribonucleases, which are both ubiquitous and difficult to denature [41]. Common to DNA extractions from hard tissues, total RNA extractions require RNA isolation from materials of low cellularity and high extracellular matrix content [30], and effective disruption of hard matrices without simultaneously producing RNA degradation. However as previously mentioned, in contrast to applications where DNA is extracted from hard tissues, RNA extractions usually reflect some degree of experimental control over sample conditions, particularly where material has been collected from laboratory animals.

As for DNA extractions, RNA extraction methods can essentially be divided into affinity and nonaffinity methods. For extractions from bone and similar tissues, the preferred method has overwhelmingly represented variations of the nonaffinity single-step method of Chomczynski and Sacchi [42]. This method involves protein denaturation by guanidinium thiocyanate, and subsequent phenol/chloroform extraction. By not requiring ultracentrifugation, and involving limited sample handling, this protocol provided improvements in terms of both investigator convenience, and the number of samples that could be processed simultaneously, while also reducing RNA degradation [42]. The one-step method also underlies commercial RNA extraction reagents such as Trizol (to be discussed further in Section 18.3.5). It is possible that nonaffinity RNA extraction methods outperform affinity methods for hard tissues, because extracellular contaminants such as remaining tissue fragments may sterically hinder RNA binding to the matrix. However, this explanation is difficult to reconcile with the fact that affinity methods perform well when DNA is extracted from similar tissues. This differential success may reflect subtle differences in RNA versus DNA binding to silica matrices. Successful RNA extraction from hard tissues may also require the active removal of RNases through phenol extraction.

18.3.2 Sample Parameters

In general, the impact of sample storage and handling conditions on subsequent RNA extractions from hard tissues has not been well explored. Where RNA is extracted from clinical samples, investigators usually do not have control over all sample parameters, with sampling delays and prolonged storage at $-70°C$ often being unavoidable. Bone samples may be subjected to frozen storage, but the impact of this on subsequent RNA extraction results is unknown. When analyzing gene expression using total RNA from frozen bone marrow, Barbaric et al. [43] found that median freezer storage times differed significantly according to gene expression status, and that freezer storage could differentially affect particular gene transcripts. While not extended to the tissue types under discussion, this result may indicate that transcripts within bone samples may also differentially degrade during frozen storage. Formalin-fixed bone samples can be subject to decalcification prior to paraffin embedding, and this can lead to RNA fragmentation [40]. Samples decalcified with nitric acid produced RNA of poor quality, whereas untreated samples or those decalcified with formic acid showed improved RNA integrity [40]. As with RNA extractions in general, precautions need to be taken to avoid exogenous RNase contamination, and these are described elsewhere [41].

18.3.3 Tissue Disruption

For both DNA and RNA isolation from tissue, it is important that samples be disrupted effectively. In the particular case of RNA extraction from frozen tissue, it is essential that samples are disrupted without thawing and excessive heat generation [37,44,45], as both will contribute to RNA degradation. A number of methods have been employed to disrupt bone or cartilage for subsequent RNA extraction, and these are summarized in Table 18.1. Many studies use commercial tissue homogenizers that can be used with liquid nitrogen to maintain chilling [37,39,44,46,47]. When using a freezer mill, cooling periods between grinding cycles may be required, to prevent samples thawing [37]. Samples have also been ground [38] or otherwise homogenized [44,48] in Trizol, which is then used for RNA extraction. Heinrichs et al. [49] extracted total RNA from rabbit growth plates, by submerging distal femora and proximal tibiae which had been fractured at the growth plates in guanidinium thiocyanate buffer. Growth plate cartilage was then physically scraped from the bone, and snap frozen [49].

An alternate approach that facilitates subsequent tissue disruption is to extract total RNA from tissue sections [44,48,50], which may in addition have been microdissected [51,52]. The use of smaller tissue fragments which may have already been subject to dissection requires less homogenization. In addition, microdissection allows the analysis of separate tissue components for more refined comparisons and analyses.

18.3.4 RNA Extraction

As previously mentioned, methods for total RNA extraction from bone and cartilage have overwhelmingly represented variations of the single-step method of Chomczynski and Sacchi [42]. An early application of this method was described for frozen rat tibiae [46] and femora [47]. Another early study [49] extracted total RNA from rabbit growth plates using guanidinium thiocyanate buffer. Total RNA was extracted with phenol/chloroform and precipitated twice, once with isopropanol and once with lithium chloride. This method was subsequently scaled down and used to extract total RNA from microdissected rat tibial epiphyses, which had been first embedded in Tissue-Tek O.C.T. and stored frozen at −80°C [52]. Total RNA was then used for expression microarray analyses and RT-PCR [52].

18.3.5 Commercial RNA Extraction Reagents and Kits

Commercial reagents and kits have featured prominently in RNA extractions from hard tissues (Table 18.3). This reflects reliance upon the use of the single-step method, and the fact that an early commercial formulation of single-step reagents gave superior RNA extraction results, when compared with in-house solutions [53]. This finding has since been replicated in studies extracting total

TABLE 18.3
Examples of Commercial Kits Used for Total RNA Extraction from Hard Tissues

Kit or Reagent Name	Commercial Supplier	Chemical Composition	Tissue Types Employed	Reference
Trizol reagent	Invitrogen	Proprietary formulation of one-step extraction buffer with phenol	Archival celloidin-embedded guinea pig temporal bones	[48]
			Equine articular cartilage	[50]
RNeasy mini column[a]	Qiagen	Silica gel-membrane affinity purification using spin columns	Frozen human chondrosarcoma sections	[44]
RNeasy mini kit			Frozen equine articular cartilage	[50]
RNeasy[a] lipid tissue kit			Frozen mouse tibiae	[38]
DNeasy tissue kit			Equine articular cartilage	[50]
RNeasy midi kit with on-column DNA digestion			Frozen human articular cartilage	[37]
RNA STAT-60 kit	Tel-test	Proprietary formulation of one-step extraction buffer with phenol	Frozen human bone biopsies	[39]
Paradise RNA extraction and isolation system	Arcturus/molecular devices	Proprietary formulation of proteinase K solution, MiraCol affinity columns	Microdissected formalin-fixed, paraffin-embedded archival human temporal bone sections	[51]

[a] Trizol reagent was employed prior to use of RNeasy reagents.

RNA from bone. A commercial formulation of guanidinium thiocyanate/phenol reagent, in the form of the RNA Stat-60 kit, was reported to be superior to the classical guanidinium thiocyanate/phenol method [42] when extracting RNA from human clinical bone samples [39]. The RNA Stat-60 method was noted to extract increased amounts of RNA, as assessed by denaturing gel electrophoresis, and RNA extracted with this method also gave superior RT-PCR results [39]. This study is notable for having compared two similar guanidinium thiocyanate/phenol methods of RNA extraction, and finding that the method chosen could nonetheless affect whether an RT-PCR product could be obtained from extracted RNA. This suggests that ability to RT-PCR amplify low abundance species is highly dependent upon the quality of the starting RNA, a finding supported by a later study also examining RNA extracted from frozen bone marrow samples [43].

Hall et al. [48] also compared total RNA extraction from archival guinea pig temporal bones using Trizol and another phenol/chloroform extraction technique, which first involved sample digestion in proteinase K. The Trizol method was preferred as this involved fewer steps, and was associated with more frequent RT-PCR success [48]. Another study directly compared several methods to extract total RNA from equine articular cartilage samples [50]. These methods involved storing small tissue samples in Trizol for subsequent Trizol extraction, or use of either the Qiagen RNeasy mini kit or DNeasy tissue kit for RNA extraction from snap frozen or paraffin-embedded tissue, respectively [50]. The authors reported that use of Trizol for sample storage and extraction was the superior method, as judged by comparing PCR cycle threshold values. In addition, the cost per sample was less when Trizol was used [50]. A further appealing factor of commercial one-step reagents is that many allow the simultaneous isolation of RNA, DNA, and protein from what may be limited samples [41].

A number of studies have also reported successful RNA extraction from hard tissues through combining the use of Trizol with subsequent affinity purification (Table 18.3). In a study extracting total RNA from frozen chondrocytoma tissue sections [44], samples were first homogenized in Trizol, and then extracted with chloroform, and phenol/chloroform for a second time. After isopropanol precipitation, RNA pellets were dissolved in water and further purified using an RNeasy mini column. The investigators attributed the success of their method to the double extraction performed, as Trizol extraction alone produced degraded RNA contaminated with DNA, whereas column purification alone gave low yields [44]. Fitzgerald et al. [54] similarly combined Trizol lysis with subsequent Qiagen RNeasy mini column purification for total RNA extraction from calf cartilage explants. A Qiagen lipid extraction kit was also used with modification to extract total RNA from mouse bone subjected to mechanical loading [38]. However when isolating total RNA from human articular cartilage, McKenna et al. [37] reported the successful use of Qiagen RLT lysis buffer, which does not appear to contain phenol, followed by Qiagen RNeasy midi columns and on-column DNase digestion.

The increasing use of tissue microdissection to obtain nucleic acids from morphologically identified pure cell populations is also driving the development of kits specifically for the extraction of RNA from paraffin-embedded samples. One such kit is the Paradise RNA extraction and isolation system, which was used by Pagedar et al. [51] to purify total RNA from microdissected formalin-fixed and paraffin-embedded archival human temporal bone sections. This was subsequently amplified and used for RT-PCR.

18.3.6 FUTURE DEVELOPMENT TRENDS

The comparative success or failure of RNA extraction can critically determine results obtained through subsequent molecular analyses [43]. The fact that use of two similar guanidinium thiocyanate/phenol methods of RNA extraction was associated with differential RT-PCR success highlights the degree to which even subtle extraction protocol changes may impact upon subsequent molecular analyses [39]. Given the dominance of the single-step method for extracting RNA from hard tissues, it seems likely that future developments will arise from continued comparison of these applications, which has to date taken place in a limited number of studies [39,48,50]. The use of affinity purification methods is likely to remain as a complement to use of reagents such as Trizol.

Continued modification of RNA extraction methods to suit particular archival tissues will also be required for full exploitation of clinical tissue banks [45,48]. Here, we will also require a much greater understanding of the determinants of RNA survival in bone stored in different conditions, given that existing tissue banks may house specimens collected over many years. As differences in RNA integrity may significantly compromise comparative gene expression analyses [55], it may not be valid to compare gene expression findings between clinical samples with significantly different ages and storage periods [43], even though this may be required to achieve statistically meaningful cohorts.

18.4 METHODS

Generic reagents and equipment

Grinding implement (e.g.. Mixermill/freezer mill), liquid nitrogen, microfuge tubes (1.5 and 2 mL), microfuge, micropipettors (10 μL, 100 μL, and 1 mL), RNase/DNase-free tips (preferably filter), 100% ethanol, 70–80% ethanol, gloves.

18.4.1 EXTRACTION OF DNA FROM BONE (ROHLAND AND HOFREITER, 2007) [20]

Specific reagents

Sodium hypochlorite, DNase-free water, EDTA disodium salt hydrate, proteinase K, GuSCN, Tris, sodium chloride, silicon dioxide, TE buffer, HCl, extraction solution (0.45 M EDTA, 0.25 mg/mL proteinase K pH 8.0, prepare fresh), binding buffer (5 M GuSCN, 25 mM NaCl, 50 mM Tris prepare fresh or use within 3 weeks if stored in the dark at room temperature), washing buffer (50% v/v ethanol, 125 mM NaCl, 1 mM EDTA pH 8.0), silica suspension (mix 4.8 g of silicon dioxide in 40 mL of DNase-free water and allow to settle for 1 h, transfer 39 mL to a new tube and allow to settle for 4 h, discard 35 mL of the supernatant and add 48 μL of 30% w/v HCl to the pellet, store aliquots in the dark at room temperature for up to 1 month), table top-centrifuge suitable for 15 mL and 50 mL tubes, graduated pipettes, rotary wheel, pH indicator strips.

Procedure

Ensure bone is clean and necessary reagents are prepared:

1. Excise bone fragment to be extracted, preferably choose a compact section of bone.
2. Grind the bone fragment, the finer the powder the greater the DNA yield (see Section 18.2.4 and Table 18.1 for tissue disruption and grinding mechanism choices).
3. Transfer less than 500 mg of sample powder to a 15 mL tube.
4. Add 10 mL of extraction solution to the tube, seal the capped tube with parafilm, and incubate with gentle agitation such as a slow rotation overnight in the dark.
5. If the powder was only coarsely ground include an additional 1–3 h agitation at 56°C to aid DNA release (due to the high temperature and risk of DNA degradation this step is not advised if it is not necessary to obtain a high yield).
6. Centrifuge the samples at 5000 × g for 2 min.
7. Transfer the supernatant to 40 mL of binding solution in a 50 mL tube (the pellet may be retained for a second round of extraction if required).
8. Add 100 μL of silica suspension (vortex before use) and adjust the pH to 4.0 by testing 25 μL on a pH strip. Initially add 200 μL of 30% w/v HCl (do not add too much HCl, it is better to have a pH closer to 4.5 than 3.5).
9. Agitate the tubes for 3 h in the dark to allow binding of the DNA to the silica.
10. Centrifuge at 5000 × g for 2 min, retain the supernatant at 4°C, for subsequent DNA binding if necessary (by adding additional silica and repeating).
11. Add 1 mL of binding buffer to the pellet and resuspend the silica by pipetting, then transfer to a 2 mL tube for convenience.

12. Centrifuge at $16000 \times g$ for 15 s, remove the supernatant (ensure all supernatant is removed as the high salt concentration can inhibit the release of DNA from the silica).
13. Add 1 mL of washing buffer, mix by pipetting, and centrifuge at $16000 \times g$ for 15 s, discard the supernatant.
14. Repeat Step 13 and then centrifuge for an additional 15 s and remove the supernatant.
15. Dry the silica by opening the lids and leaving at room temperature for 15 min.
16. Add 50 μL of TE buffer and resuspend the silica by pipetting, incubate for 10 min at room temperature with occasional gentle shaking.
17. Centrifuge at $16000 \times g$ for 2 min.
18. Transfer the supernatant to a fresh tube and store for subsequent applications. Further DNA can be eluted from the silica by repeating Steps 16 and 17; however, this will have lower DNA concentration.

18.4.2 EXTRACTION OF RNA FROM BONE

18.4.2.1 Trizol Method

Specific reagents

Trizol reagent (Invitrogen Cat. No. 15596-026), chloroform, and isopropanol.

Procedure

1. Excise bone section to be extracted (minimum 50 mg and maximum 100 mg per mL of Trizol, process additional samples separately and resuspend pellets in the same water at the end) grind in a freezer mill or alternative grinder, ensuring the sample remains as cold as possible once the grinding has begun.
2. Homogenize the powder in 1 mL of Trizol in a 1.5 mL microfuge tube by passing through a blunt 20 gauge needle with a RNase-free syringe five times, allow to stand at room temperature 5 min.
3. Add 200 μL of chloroform and mix by vigorously shaking for 15 s, allow to stand for 2–3 min.
4. Centrifuge at $12000 \times g$ for 15 min at 4°C, carefully remove the top layer to a new tube (the middle DNA layer could also be collected if DNA was required as well).
5. Add 500 μL of isopropanol and mix by inversion, allow to stand for 10 min.
6. Centrifuge at $12000 \times g$ for 10 min at 4°C, discard supernatant.
7. Wash the pellet with 1 mL of 70%–80% ethanol, centrifuge at $7500 \times g$ for 5 min at 4°C.
8. Discard the supernatant and air-dry the pellet (do not allow to over dry as it will be hard to resuspend).
9. Resuspend the pellet in an appropriate amount of RNase-free water, 0.5% SDS, or TE buffer as preferred, 10 μL is often a good place to start and additional solution can be added if needed to ensure the RNA is completely dissolved, samples can be heated to 55°C–65°C to aid dissolving for 10 min.

18.4.2.2 RNeasy Mini Kit

The RNeasy mini kit (an affinity method) can also be used for additional purification following isolation with Trizol. Note that the maximum quantity of starting material is 30 mg, due to the binding capacity of the column. Only molecules greater than 200 nucleotides will be retained and eluted, therefore this method is not suitable for highly degraded samples.

Specific reagents

RNeasy mini kit (Qiagen, Cat. No. 74104), and 14.3 M β-mercaptoethanol.

Procedure

1. Excise bone section to be extracted (not more than 30 mg) and grind in a mixer mill or alternative grinding tool (see Table 18.1).
2. Homogenize ground bone in 600 µL of buffer RLT with β-mercaptoethanol (use 350 µL if <20 mg of bone).
3. Centrifuge at full speed for 3 min, and transfer the supernatant to a clean microfuge tube.
4. Add 1 volume of 70% ethanol to the supernatant and mix by pipetting.
5. Immediately transfer less than 700 µL of the sample (include any precipitate that may have formed) to a spin column in a 2 mL collection tube. Centrifuge at 8000×g for 15 s and discard the flow through. If the sample volume is greater than 700 µL then continue adding 700 µL volumes to the column and centrifuging until all the sample is processed.
6. An on-column DNase digestion can be performed at this stage if desired, in this case skip Step 7.
7. Add 700 µL of buffer RW1 to the column and centrifuge at 8000×g for 15 s to wash the column. Discard the flow through and be careful not to allow the spin column to contact the flow-through (alternatively DNA can be extracted from this flow-through if desired).
8. Wash the column with 500 µL of buffer RPE by centrifuging at 8000×g for 15 s and discarding the flow-through.
9. Repeat Step 8 centrifuging for 2 min at this stage.
10. To ensure complete removal of the ethanol in buffer RPE, it is recommended to centrifuge the spin column for 1 min at full speed in a new collection tube.
11. Place the spin column in a clean 1.5 mL microfuge tube, add 30–50 µL of RNase-free water and centrifuge at 8000 × g for 1 min to elute the RNA. If a high yield of RNA is expected (<30 µg), then add an additional 30–50 µL of water (or if a high concentration is required, the eluate can be respun through the column instead of using additional water) and repeat the spin to elute more RNA off the column.

18.5 CONCLUSION

Perceived technical feasibility is one of many factors influencing the biological systems chosen for study. Whereas sample availability (and therefore necessity) has driven researchers to develop methods for extracting DNA from hard tissues, perceived technical difficulties surrounding RNA extraction have undoubtedly reduced the number of studies examining gene expression in tissues such as bone. A crude measure of this can be made through PubMed searches. A search conducted in January 2008 of the terms "gene expression blood" produced 66,027 entries, whereas a PubMed search of "gene expression bone" identified 17,800 entries, of which most (at least 7912 entries) are likely to reflect analyses of bone marrow, as opposed to bone. This could be considered a manifestation of the Matthew effect, which refers to the unequal distribution of reward and recognition to scientists based upon prior achievements [56]. Similarly, fields that are strongly represented in the literature are likely to continue to grow (in at least the short term), through being perceived to be important by both scientists and supporting funding bodies, whereas small fields may be perceived as less important, and less worthy of future study. Clearly, without active technical development to counteract this form of bias, we will ultimately know a great deal about the molecular biology of certain cells and tissues, and very little about others. At present, our understanding of the *in vivo* molecular biology of hard tissues poorly reflects their physiological significance [37], or the impact of diseases affecting these tissues on patients and supporting health systems.

This chapter serves to highlight that the research community is currently well served by methods to extract both DNA and RNA from hard tissues. This may be less well recognized by the community aiming to extract RNA, than the larger communities extracting DNA for an essentially nonoverlapping set of applications. Overall, silica gel-based methods are emerging as preferred methods for extracting DNA from hard tissues, whereas variations of the one-step method using phenol are preferred for extracting total RNA. Future efforts to directly compare different extraction methods, and to compare the performance of commercial kits and reagents would represent valuable services to the research community, and should be viewed as such by editors and reviewers of relevant journals. Application of techniques to novel tissue and sample types will also continue to unlock existing tissue repositories and their supporting clinical data to molecular analysis. A better understanding of sample parameters and their influence on nucleic acid integrity and yields will also be important for the appropriate selection of samples for analysis.

ACKNOWLEDGMENTS

The authors are grateful to Drs. Robert J. Sokol, Nadin Rohland, Nicole von Wurmb-Schwark, and Cliff Megerian for providing reprints for this chapter. Relevant work in the Molecular Oncology Laboratory has been supported by grants from the Children's Hospital Fund, Perpetual Trustees (the Derham Green fund and the Margaret Augusta Farrell fund), and the Cancer Institute NSW. We also gratefully acknowledge financial support from the Parramatta Leagues Club and the Oncology Children's Foundation. The authors declare no competing financial interests.

Judith Weidenhofer started her career with a BAppSc (honors) (medical and biotechnology), investigating mutations in colon cancer at Charles Sturt University in Wagga Wagga, New South Wales, Australia in 2001. Following this, she commenced her PhD at the University of Newcastle, Newcastle, New South Wales, Australia, investigating gene and protein expression profiles in schizophrenia using postmortem brain tissue. This study required the optimization of RNA extraction techniques, to ensure high-quality RNA for microarray analyses from tissue that had been collected with less-than-ideal postmortem delays. From 2005–2008, Judith undertook postdoctoral training in the molecular oncology laboratory at the Children's Hospital at Westmead, generating mouse models to investigate the role of novel genes in cancer. Judith is currently a lecturer in the School of Biomedical Sciences at the University of Newcastle where her research interest is focused on the molecular mechanisms of cancer.

Jennifer A. Byrne began her career with a BSc honors at the University of Queensland, Brisbane, Australia in 1988. As an undergraduate neuroscience student, she investigated functional plasticity in somatosensory cerebral cortex, before taking a molecular turn for her PhD, which focused on determining the structure of the *RRM1* gene on human chromosome 11p, and using this and other loci for the molecular analysis of embryonal tumors of childhood. During her postdoctoral studies at the IGBMC institute in Strasbourg, France, from 1993 to 1996, she developed a differential screening method to identify genes overexpressed in human breast cancer, leading to the characterization of three human genes forming a novel gene and protein family. Returning to Australia in early 1996, she continued her work at the Children's Medical Research Institute, Sydney, New South Wales, Australia, before moving to the adjacent Children's Hospital at Westmead, New South Wales, Australia. She is now head of the molecular oncology laboratory within the hospital's Oncology Research Unit, and associate professor within the Discipline of Paediatrics and Child Health at the University of Sydney, Sydney, New South Wales, Australia. Her current research interests include determining the functions of amplification target genes which are overexpressed in multiple cancers, and the clinical significance of target gene overexpression. She established the hospital's tumor bank in 1997, and she also has a keen interest in the development and analysis of nucleic acid extraction methods, and their application to clinical samples.

REFERENCES

1. Cattaneo, C., Gelsthorpe, K., and Sokol, R.J., DNA extraction methods in forensic analysis, In *Encyclopedia of Analytical Chemistry*, Meyers, R.A., (Ed.), John Wiley & Sons, New York, 2000.
2. Loreille, O.M. et al., High efficiency DNA extraction from bone by total demineralization, *Forensic Sci. Int. Genet.*, 1: 191, 2007.
3. Mulligan, C.J., Isolation and analysis of DNA from archaeological, clinical, and natural history specimens, *Methods Enzymol.*, 395: 87, 2005.
4. Rohland, N., Seidel, H., and Hofreiter, M., Nondestructive DNA extraction method for mitochondrial DNA analyses of museum specimens, *Biotechniques*, 36: 814, 2004.
5. Asher, R.J. and Hofreiter, M., Tenrec phylogeny and the noninvasive extraction of nuclear DNA, *Syst. Biol.*, 55: 181, 2006.
6. McKenna, M.J. et al., Deoxyribonucleic acid contamination in archival human temporal bones: A potentially significant problem, *Otol. Neurotol.*, 23: 789, 2002.
7. Gilbert, M.T.P. et al., Absence of Yersinia pestis-specific DNA in human teeth from five European excavations of putative plague victims, *Microbiology*, 150: 341, 2004.
8. Taylor, G.M., Young, D.B., and Mays, S.A., Genotypic analysis of the earliest known prehistoric case of tuberculosis in Britain, *J. Clin. Microbiol.*, 43: 2236, 2005.
9. Barnes, I. and Thomas, M.G., Evaluating bacterial pathogen DNA preservation in museum osteological collections, *Proc. Biol. Sci.*, 273: 645, 2006.
10. Lahiff, S. et al., Species-specific PCR for the identification of ovine, porcine and chicken species in meat and bone meal (MBM), *Mol. Cell. Probes*, 15: 27, 2001.
11. Prado, M. et al., Comparison of extraction methods for the recovery, amplification and species-specific analysis of DNA from bone and bone meals, *Electrophoresis*, 23: 1005, 2002.
12. Gill, P. et al., Identification of the remains of the Romanov family by DNA analysis, *Nat. Genet.*, 6: 130, 1994.
13. Marjanović, D. et al., DNA identification of skeletal remains from the World War II mass graves uncovered in Slovenia, *Croat. Med. J.*, 48: 513, 2007.
14. Keough, M.E., Simmons, T., and Samuels, M., Missing persons in post-conflict settings: best practices for integrating psychosocial and scientific approaches, *J. R. Soc. Health*, 124: 271, 2004.
15. Holland, M.M. et al., Development of a quality, high throughput DNA analysis procedure for skeletal samples to assist with the identification of victims from the World Trade Center attacks, *Croat. Med. J.*, 44: 264, 2003.
16. Biesecker, L.G. et al., DNA identifications after the 9/11 World Trade Center attack, *Science*, 310: 1122, 2005.
17. Huffine, E., Crews, J., and Davoren, J., Developing role of forensics in deterring violence and genocide, *Croat. Med. J.*, 48: 431, 2007.
18. Hofreiter, M. et al., Ancient DNA, *Nat. Rev. Genet.*, 2: 353, 2001.
19. Byrne, J.A. et al., A screening method to identify genes commonly overexpressed in carcinomas and the identification of a novel complementary DNA sequence, *Cancer Res.*, 55: 2896, 1995.
20. Rohland, N. and Hofreiter, M., Ancient DNA extraction from bones and teeth, *Nat. Protoc.*, 2: 1756, 2007.
21. Smith, C.I. et al., The thermal history of human fossils and the likelihood of successful DNA amplification, *J. Hum. Evol.*, 45: 203, 2003.
22. Wandeler, P. et al., Patterns of nuclear DNA degeneration over time—a case study in historic teeth samples, *Mol. Ecol.*, 12: 1087, 2003.
23. Miloš, A. et al., Success rates of nuclear short tandem repeat typing from different skeletal elements, *Croat. Med. J.*, 48: 486, 2007.
24. Pichler, F.B., Dalebout, M.L., and Baker, C.S., Nondestructive DNA extraction from sperm whale teeth and scrimshaw, *Mol. Ecol. Notes*, 1: 106, 2001.
25. Salamon, M. et al., Relatively well preserved DNA is present in the crystal aggregates of fossil bones, *Proc. Natl. Acad. Sci. U.S.A.* 102: 13783, 2005.
26. González-Andrade, F. and Sánchez, D., DNA typing from skeletal remains following an explosion in a military fort- first experience in Ecuador (South-America), *Leg. Med.*, 7: 314, 2005.
27. Cooper, A. and Poinar, H.N., Ancient DNA: Do it right or not at all, *Science*, 289: 1139, 2000.
28. Davoren, J. et al., Highly effective DNA extraction method for nuclear short tandem repeat testing of skeletal remains from mass graves, *Croat. Med. J.*, 48: 478, 2007.

29. Kemp, B.M. and Smith, D.G., Use of bleach to eliminate contaminating DNA from the surface of bones and teeth, *Forensic Sci. Int.*, 154: 53, 2005.
30. Rohland, N. and Hofreiter, M., Comparison and optimization of ancient DNA extraction, *Biotechniques*, 42: 343, 2007.
31. Iudica, C.A., Whitten, W.M., and Williams, N.H., Small bones from dried mammal museum specimens as a reliable source of DNA, *Biotechniques*, 30: 732, 2001.
32. Cattaneo, C. et al., A simple method for extracting DNA from old skeletal material, *Forensic Sci. Int.*, 74: 167, 1995.
33. Boom, R. et al., Rapid and simple method for purification of nucleic acids, *J. Clin. Microbiol.*, 28: 495, 1990.
34. Staiti, N., Di Martino, D., and Saravo, L., A novel approach in personal identification from tissue samples undergone different processes through STR typing, *Forensic Sci. Int.*, 146: S171, 2004.
35. Boom, R. et al., Improved silica-guanidiniumthiocyanate DNA isolation procedure based on selective binding of bovine alpha-casein to silica particles, *J. Clin. Microbiol.*, 37: 615, 1999.
36. Lindahl, T., Facts and artifacts of ancient DNA, *Cell*, 90: 1, 1997.
37. McKenna, L.A. et al., Effective isolation of high-quality total RNA from human adult articular cartilage, *Anal. Biochem.*, 286: 80, 2000.
38. Kesavan, C. et al., Mechanical loading-induced gene expression and BMD changes are different in two inbred mouse strains, *J. Appl. Physiol.*, 99: 1951, 2005.
39. Hoyland, J.A. et al., A comparison of in situ hybridisation, reverse transcriptase-polymerase chain reaction (RT-PCR) and in situ-RT-PCR for the detection of canine distemper virus RNA in Paget's disease, *J. Virol. Methods*, 109: 253, 2003.
40. Mangham, D.C. et al., Ewing's sarcoma of bone: the detection of specific transcripts in a large, consecutive series of formalin-fixed, decalcified, paraffin-embedded tissue samples using the reverse transcriptase-polymerase chain reaction, *Histopathology*, 48: 363, 2006.
41. Kingston, R.E., Preparation and analysis of RNA, In *Current Protocols in Molecular Biology*, Ausubel, F.M. et al., (Eds.), John Wiley and Sons, New York, 2002, Chapter 4.
42. Chomczynski, P. and Sacchi, N., Single-step method of RNA isolation by acid guanidinium thiocyanate-phenol-chloroform extraction, *Anal. Biochem.*, 162: 156, 1987.
43. Barbaric, D. et al., Expression of tumor protein D52-like genes in childhood leukemia at diagnosis: Clinical and sample considerations, *Leuk. Res.*, 30: 1355, 2006.
44. Baelde, H.J. et al., High quality RNA isolation from tumours with low cellularity and high extracellular matrix component for cDNA microarrays: Application to chondrosarcoma, *J. Clin. Pathol.*, 54: 778, 2001.
45. Barbaric, D., Dalla-Pozza, L., and Byrne, J.A., A reliable method for total RNA extraction from frozen human bone marrow samples taken at diagnosis of acute leukaemia, *J. Clin. Pathol.*, 55: 865, 2002.
46. Westerlind, K.C. et al., Estrogen does not increase bone formation in growing rats, *Endocrinology*, 133: 2924, 1993.
47. Westerlind, K.C. et al., The effect of long-term ovarian hormone deficiency on transforming growth factor-beta and bone matrix protein mRNA expression in rat femora, *Biochem. Biophys. Res. Commun.*, 200: 283, 1994.
48. Hall, K.L. et al., Optimization of ribonucleic acid detection from archival Guinea pig temporal bone specimens, *Otol. Neurotol.*, 28: 116, 2007.
49. Heinrichs, C. et al., Dexamethasone increases growth hormone receptor messenger ribonucleic acid levels in liver and growth plate, *Endocrinology*, 135: 1113, 1994.
50. Hume, A. and Frisbie, D., Evaluation of methods for RNA isolation and cDNA synthesis using small articular cartilage samples, In *Summaries: Focus 2, Early diagnosis of bone and joint disease*, 2002–2003 Colorado State University Orthopaedic Research Laboratory and Bioengineering Laboratory Report, Colorado State University, 2003, 67.
51. Pagedar, N.A. et al., Gene expression analysis of distinct populations of cells isolated from mouse and human inner ear FFPE tissue using laser capture microdissection—a technical report based on preliminary findings, *Brain Res.*, 1091: 289, 2006.
52. Nilsson, O. et al., Gradients in bone morphogenetic protein-related gene expression across the growth plate, *J. Endocrinol.*, 193: 75, 2007.
53. Chomczynski, P. and Mackey, K., Substitution of chloroform by bromo-chloropropane in the single-step method of RNA isolation, *Anal. Biochem.*, 225: 163, 1995.

54. Fitzgerald, J.B., Jin, M., and Grodzinsky, A.J., Shear and compression differentially regulate clusters of functionally related temporal transcription patterns in cartilage tissue, *J. Biol. Chem.*, 281: 24095, 2006.

55. Imbeaud, S. et al., Towards standardization of RNA quality assessment using user-independent classifiers of microcapillary electrophoresis traces, *Nucleic Acids Res.*, 33: e56, 2005.

56. Merton, R.K., The Matthew effect in science, *Science*, 159: 56, 1968.

57. von Wurmb-Schwark, N., et al., Extraction and amplification of nuclear and mitochondrial DNA from ancient and artificially aged bones, *Leg. Med.*, 5: S169, 2003.

19 Isolation of DNA from Ancient Samples

*Paula F. Campos, Eske Willerslev,
and M. Thomas P. Gilbert*

CONTENTS

19.1   INTRODUCTION

Understanding the key features that differentiate ancient DNA (aDNA) from other DNA sources underlies successful aDNA extractions. Intrinsic problems such as low template quantity, poor template quality, and the presence of polymerase chain reaction (PCR) inhibitors need to be taken into account. The postmortem instability of nucleic acids is central to the methodological problems inherent in aDNA research. In metabolically active tissues, damage to the DNA molecules is rapidly and efficiently repaired via a host of repair pathways [1]; however, after cell death, DNA is quickly altered and degraded. The degradation of endogenous DNA (i.e., that belonging to a sample of interest) starts shortly after the death of the sample, and consequently most ancient specimens do not contain any amplifiable endogenous DNA, while those that do, possess only fragments in the 100–500 bp size range [2–4].

19.1.1   DNA Degradation and Its Implications

Several processes contribute to DNA degradation. These biochemical modifications act both via the cross-linking and fragmentation of the molecule's chemical backbone, and the alteration of individual nucleotide bases. The end result of most of these processes is the same—the length of intact DNA molecules available for PCR amplification and sequencing analysis rapidly decreases. Although several factors such as temperature, proximity to free water, environmental salt content, and exposure to radiation affect the rate of this decay [1], of all these factors, temperature plays perhaps the central role in the longevity of the aDNA molecules. In brief, the exponentially linked relationship between temperature and degradation ensures that rates of DNA degradation rapidly increase with temperature, and thus low, relatively constant, temperatures provide the most optimal conditions for DNA preservation. A further implication of this temperature–degradation relationship is that for any given sample, cold preserved samples are more likely to provide useable genetic material than those that have been preserved at warmer temperatures. In addition to low temperatures, rapid desiccation and high salt concentrations may also prolong DNA survival [1]; however, despite this, the DNA content of all dead biological tissues will decrease with time to levels where the remaining fragments are too short for any meaningful information to be recovered.

An additional implication of the nature of DNA degradation is that in ancient or otherwise degraded samples, the total amount of PCR amplifiable DNA fragments increases rapidly as template size is decreased [5]. Although ultimately a natural limit exists beyond which no DNA sequence information can be recovered, with the increased use of sequencing-by-synthesis and related platforms (e.g., GS-FLX [Roche Diagnostics, Basel, Switzerland], Solexa [Illumina Inc, San Diego, California], and SOLiD [Applied Biosystems, Foster City, California]), the size of template molecules that can be used to generate meaningful data is decreasing. This is predominantly because, while conventional PCR approaches require sufficient DNA template survival to allow the placement of sequence-specific primers on single undamaged DNA fragments, which prevents amplification if the surviving fragments are below this size, the sequencing-by-synthesis platforms bypass this requirement through the ligation of DNA template molecules to predetermined adaptor sequences, which in turn are used as the sites for PCR primer binding. Indeed, in such cases, the sole theoretical requirement is that some DNA survives, so it can be amplified by emulsion PCR. In practice, however, and in contrast to PCR-based approaches that can in theory be successfully performed as long as a single initially amplifiable template molecule is present in the DNA extract, it is worth noting that in their current form, sequencing-by-synthesis analyses require a large amount of initial DNA (e.g., 1–3 μg recommended for the FLX by its manufacturer) from which to construct the initial DNA libraries.

19.1.2 DNA Miscoding Lesions

In addition to simply reducing the absolute size and quantity of template molecules available for analysis, DNA degradation plays a further role that can have serious effects on the data generated. The earliest studies that investigated the qualities of DNA recovered from ancient remains [2,6] demonstrated that a small, but apparently common, number of DNA damage processes can impair DNA molecules in such a way that the sequence is modified, yet still PCR amplifiable and thus sequenceable. Although these reactions do not block PCR, they lead to sequence modification, with the end result that the generated sequence differs from the undamaged original template. With the advent of sequencing-by-synthesis analyses, it has been conclusively demonstrated that these differences originate predominantly from the hydrolytic deamination of cytosine to uracil or its analogues, although a small number of other modifications may play a role [7]. To complicate matters further, all DNA polymerases have innate error rates, leading to a small number of nucleotide misincorporations during replication of the DNA molecule. Enzyme and template error are rarely an issue in conventional PCR-based analyses. Errors are essentially randomly distributed along the template molecules, and the number of DNA templates at the start of a PCR reaction is normally sufficiently high. Therefore, the final sequence produced from the resultant amplicons will be generated from predominantly unmodified templates, and will not show the errors.

19.1.3 Contamination Challenges Facing aDNA

Besides degradation, another major challenge to almost all aDNA studies is contamination. In general, as the old or degraded samples used in aDNA studies contain very low concentrations of fragmented endogenous DNA, modern contaminant molecules (if present) will be preferentially amplified during subsequent PCR reactions, leading to the generation of misleading results. This is a particular problem in aDNA extracts, where PCRs are often used with a very high number of cycles (that can detect as little as 10 copies of modern DNA template per reaction) [8], and where, almost without exception, the ancient samples (and thus extracts) contain a large amount of nonendogenous DNA. The sources and types of contamination vary, but can be divided into two broad classes. The first, and probably dominant, is sources of bacterial or other environmental organism DNA derived either from the environment that the samples have been preserved in, or even from initial sample putrefaction [9]. The second potential source is DNA derived from human (or other) handling once excavated/sampled, or even reagents or conservation treatments that may contain DNA [10–12]. This second source is especially important in aDNA studies where DNA from modern equivalents of the target species is likely to be present in the environment (e.g., humans, bacteria, and domestic animals), or if the so-called universal primers are used that may amplify DNA from a wide range of taxa. The key implication of contamination is that it is extremely important to undertake all aDNA extractions and additional pre-PCR manipulation in dedicated controlled facilities. In such laboratories, background levels of contamination can be limited by, for example, regular cleaning of the work surfaces with dilute bleach (e.g., 10% commercial strength), HCl (0.1 M) or commercial DNA destroying solutions (e.g., DNAaway, MBP, San Diego, California), frequent irradiation of the work surfaces with ultraviolet light ($\lambda = 254$ nm), the use of positive air pressure, and so on. In addition to the environment itself, a further challenge comes from DNA present in the reagents used, for example, a recent study has demonstrated cow (*Bos taurus*), pig (*Sus scrofa*), and chicken (*Gallus gallus*) DNA in PCR reagents [12]. Therefore, those used must be of the highest purity available, and sometimes may even require additional purification (e.g., through autoclaving, irradiation, or ultrafiltration).

In light of the problems of DNA damage and contamination, the choice of extraction method and strategy depends heavily on the subsequent analysis planned for the study. Where conventional PCR is to be employed, the key challenges are template molecule length and contaminant sequences that closely match the PCR target. In particular, if the DNA molecules are overall degraded to an extent where they are shorter than the desired amplicon, little or no amplification will result. With regard to contamination, if even a small number of contaminant molecules exist that match the PCR target, they will be coamplified during subsequent PCR, leading to the generation of erroneous sequence. As referred to earlier, the use of sequencing-by-synthesis platforms, with their lack of predetermined target-specific primer-based amplification, makes contamination an altogether different problem. In essence, as this process results in the random emulsion PCR amplification and sequencing of molecules from the DNA extract, the presence of contaminant sequences among the resulting sequence data simply reflects the contaminant molecules' frequencies in the extract. Furthermore, the sequence itself of the contaminant becomes irrelevant in this context. The effect of this contamination is extensive and can be readily observed in the data of the initial neanderthal and mammoth paleogenomic publications. For example, the study of Poinar et al. [13] used a mammoth bone recovered frozen from Siberia and kept frozen since then. Nonetheless, nearly 55% of the sequences generated in the initial 454-GS20 analysis of the library were likely of nonmammoth origin (Figure 19.1). From poorer quality samples, like the 38,000 year old neanderthal bone from Green et al. [14] study, only 6.2% of the 254,933 sequences generated aligned with primate DNA (Figure 19.1c). Moreover, an unknown, although likely high, proportion derive from anatomically modern human contaminant DNA (Figure 19.1c) [15].

19.2   PRINCIPLES OF DNA ISOLATION FROM ANCIENT SAMPLES

Ancient DNA extraction methods are generally based on similar principles to modern DNA extractions. The process involves several steps: digestion of the structural biomolecules in the tissue to release DNA into solution, separation of the DNA from other molecules, and often an additional concentration of the nucleic acids. The digestion process itself normally facilitates the liberation of cellular DNA through the breakage of cell walls or membranes to release the cellular constituents into the extraction buffer, and the digestion of proteins or other molecules that may be complexed with the DNA. A wide variety of ingredients have been used in different studies and are thus available for use.

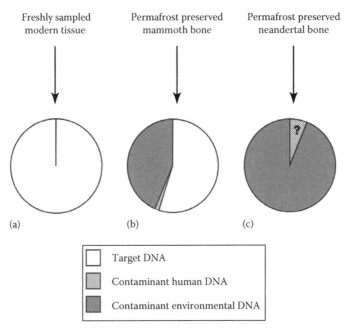

FIGURE 19.1   Observed, estimated, and hypothetical sequence distributions from different tissues (for details refer to Section 19.1.3).

To ensure maximum digestion efficacy, the digestion buffers can be tailored to the specific tissue type from which DNA has to be extracted. Below, we briefly discuss the key features of some of the common ingredients in addition to the adoption of several crucial prepreparation steps.

19.2.1 Prepreparation Steps

Prior to DNA extraction from ancient samples, a number of prepreparation steps are often advocated. These steps have two predominant functions: the first is simply to enhance the purity of extracted nucleic acids and the digestion mix (so as to ensure optimal digestion), the second to remove contaminant DNA sequences. In particular, this has often referred in the aDNA context to simply cleaning the external surfaces of specimens to remove common forms of impurities and contaminants that can be expected bearing in mind the burial origin of many samples (underground!) and handling history since. Most commonly this is applied to bone and tooth samples, and may involve either washing of the specimens in detergents or dilute bleach (10% commercial strength) or HCl (0.1 M) solutions, or the removal of external surfaces using sandpaper, shot-blasting, sand-blasting, or other methods. However, it should be stressed that the above methods are not without problems, and may in some situations lead to extra complications. In particular, we refer to the complications of further contaminating vulnerable samples. For example, ancient, and thus degraded, bone and tooth (probably the most common aDNA source tissue) are extremely porous materials. When fresh, human compact bone is estimated to be ~8% air, and when degraded, this level can be significantly higher [16]. The implication of this porosity is that contaminant molecules (both dirt but more importantly contaminant DNA sources, such as DNA derived from environmental microorganisms, conservational preparations, or human handling) can penetrate deep into the material, to the extent that external cleansing methods fail to remove them [9,17]. As such, several recent studies have demonstrated that the optimal method for the cleansing of bone and tooth material is an initial external clean as described above, followed by a subsequent incubation of the material in dilute bleach postgrinding [18,19]. The logic underlying these methods is that grinding of the material prior to incubation disrupts the bone/tooth to an extent where contaminant molecules are exposed to the bleach, while true endogenous DNA is protected in osteocytes or crystal aggregates that form during bone digenesis.

For less porous materials, in particular keratinous sources such as hair shaft, hoof, nail, or feather, cleansing is a simpler matter and can be achieved by short incubations (e.g., 10–30 s) in dilute bleach solution [20,21].

19.2.2 Digestion of Structural Molecules

19.2.2.1 Tris–HCl and EDTA

The stability of DNA molecules is greatest at approximately neutral pHs (e.g., pH 8). Therefore most digestion solutions incorporate Tris–HCl to achieve this. In addition, ethylenediaminetetraacetic acid (EDTA) is almost always included, due to its properties in chelating (sequestering) di- and trivalent metal ions (e.g., Fe^{2+}, Mg^{2+}, and Ca^{2+}). The benefits of EDTA are twofold. Firstly, magnesium ions are a cofactor of most nucleases, thus their sequestering by the EDTA helps protect the DNA molecules from nuclease degradation during the digestion process itself. Secondly, where bone or tooth is the source of DNA, the EDTA plays the role of demineralizing the bone (which is ~70% hydroxyapatite, a calcium-rich mineral) and thus releases entrapped DNA.

19.2.2.2 Surfactants, Reducing Agents, and Antioxidant Proteases

Many digestion mixes also include surfactants (e.g., sodium dodecyl sulfate [SDS] or Triton-X), reducing agents (e.g., dithiothreitol [DTT] and β-mercaptoethanol), and proteases (e.g., proteinase K).

The function of anionic surfactants such as SDS, and Triton-X is predominantly to disrupt the cell, mitochondrial, and nuclear membranes, by breaking up the lipids that they are based on, and to denature proteins (which aids their subsequent digestion by proteases, most commonly proteinase K). However, in some of the old specimens used for aDNA studies, where even small molecules like DNA are fragmented into small pieces, debate exists as to whether cell membranes are present at all. Furthermore, the proteins may already be partially degraded. As such, Rohland and Hofreiter [22] report that when used for the DNA extraction of ancient bone, such surfactants do not improve DNA yields (although the authors also report they have no negative impacts on the results). We would caution, however, that this is likely to be sample dependent, and will vary with preservation. Moreover, SDS appears to be extremely beneficial for the recovery of DNA from keratinous tissues such as hair and nail [20], due to its role in unraveling the keratin tertiary structure [23].

The role of reducing agents, such as DTT or β-mercaptoethanol, is to reduce disulphide bonds and to act as an antioxidant by scavenging hydroxyl radicals in the digestion solution. Antioxidants are commonly used to address problems related to phenolics that may cause PCR inhibition post extraction. Examples include the use of β-mercaptoethanol, ascorbic acid, bovine serum albumin (BSA), sodium azide, and polyvinylpyrrolidone (PVP) [24]. PVP, a water soluble antioxidant, is used especially in plant DNA extractions for its extremely good adsorption to polyphenols, which if present can inhibit downstream applications such as PCR.

19.2.2.3 N-Phenacylthiazolium Bromide

N-phenacylthiazolium bromide (PTB) is a common, although controversial, component of many aDNA extraction buffers. Originally postulated to be useful in the reversal of cross-links caused by Maillard reactions, and in particular the liberation of DNA from DNA–protein complexes [25], its usefulness remains debated. While Rohland and Hofreiter [22] have recently argued that, for bone samples at least, it offers no positive effect on final DNA yields, in other situations (e.g., coprolites) it appears to significantly improve DNA yields [25]. This discrepancy is most likely due to tissue specific differences in DNA survival.

19.2.3 DNA Purification

Following tissue digestion, a large number of different biomolecules will be free in the solution, including RNA (if any survives), proteins, polysaccharides, tannins pigments and other kind of inhibitors that interfere with subsequent analyses. As such, it is the norm to subsequently employ one of a variety of different methods to recover pure DNA.

19.2.3.1 Silica-Based DNA Purification

Among the most common purification methods are those that employ the DNA-binding properties of silica [26]. Key to the efficacy of this method is the observation that nucleic acids readily bind to silica when at low pH and in the presence of chaotropic salts. Although the chemistry behind this process is not fully understood, it likely results from the dehydration of both DNA and silica surfaces and hydrophobic forces acting under high salt concentrations [27]. As polysaccharides, lipids, amino acids, protein fragments, and other nonnucleic acid biomolecules do not bind to the silica matrix, they can be removed during subsequent wash steps. The most common chaotropic salt used is guanidinium thiocyanate (GuSCN) but sodium iodine (NaI) can also be used. The use of nonchaotropic salts has been tested in a recent study and although the use of NaCl gave better DNA yields, it also co–purified a large amount of PCR inhibitors [22]. Silica-based purification methods have a number of advantages over current alternatives, in particular as they are both faster and easier to perform than organic-based extraction methods such as those involving phenol and chloroform. Despite its ease of use, some precautions need to be taken to ensure success. For example, when solutions containing loose silica are used (as opposed to solid filters), the addition of too much ($>100\,\mu g$) silica can result in a solid pellet during centrifugation that is hard to disrupt during the final stages of the extraction. Furthermore, the total amount of DNA recovered during the extraction depends on not only how much is initially bound to the silica but also how well it can be eluted off the silica in the last stage of the extraction. However, the greatest disadvantage of this purification method is that if silicon dioxide solution is used, any particulate carryover into the purified DNA can interfere with downstream PCR, since this compound is itself a powerful PCR inhibitor.

A number of commercially available DNA extraction kits are based on the silica method, including the commonly used DNeasy tissue kit or QIAquick DNA purification kit (Qiagen, Valencia, California). In these and many other related commercial kits, the silica suspension solution of the original protocol [26] is replaced by a silica-gel-membrane technology over which the digestion mixture is washed through centrifugation. As in the previously described method, nucleic acids bind to the silica-gel-membrane and after a wash step, pure nucleic acids are eluted under low- or no-salt conditions in small volumes, ready for immediate use. The great advantage of using these kits is avoiding the problem of silica carryover, which can interfere with downstream applications.

19.2.3.2 Phenol: Chloroform Extraction and Precipitation Methods

The predominantly used alternatives to silica purification methods are those that incorporate the organic solvents phenol and chloroform. Phenol–chloroform extraction is a liquid–liquid technique used mainly for purifying DNA contaminated by histones, other proteins, and their degraded derivatives. Using this method, most proteins are removed from the extract through first denaturation, then precipitation at the interface between the separate organic and aqueous layers that form when equal volumes of a phenol:chloroform (or phenol then chloroform) mixture and the aqueous DNA extract sample are mixed (the organic and aqueous fractions form a biphasic mixture). The proteins will partition into the organic phase while the DNA (as well as other contaminants such as salts, sugars) remain in the aqueous phase.

This procedure is usually repeated at least once, and depending on the downstream applications and the need of purified DNA is often followed by isopropanol or ethanol plus salt precipitation of the nucleic acids. Isopropanol precipitation is often preferred over ethanol precipitation not only because less alcohol is necessary but also because, as it can be performed at room temperature, it minimizes co–precipitation of salt and minimizes the risk that co–precipitated salt will interfere with any downstream application. This is usually followed by a wash with 70% room temperature ethanol. This removes residual salts and replaces the isopropanol with the more volatile ethanol-making DNA easier to dissolve. After the pellet is dried, it is resuspended with 1x TE pH 8, as DNA is not easily dissolved in acidic buffers. Both phenol and chloroform are hazardous chemicals and the extraction is laborious, so in recent years alternative ways to isolate DNA have been used. However, it is still extremely useful in DNA extractions from hair and other keratinous tissues [20].

19.3 CURRENT TECHNIQUES FOR DNA ISOLATION FROM ANCIENT SAMPLES

Although a number of dedicated commercial aDNA extraction kits have been produced over recent years (e.g., Geneclean for aDNA [Bio101, QBiogene, Irvine, California]), and although the variability in DNA content and quality of many aDNA sources are such that some samples are so close in quality to "fresh" material as to facilitate the use of commercial "modern" DNA extraction kits for DNA extraction (e.g., Qiagen's DNeasy tissue extraction kit), we focus predominantly on non-kit-based methods. In particular, we argue that the use of kits limits the extent to which extraction methods can be tailored to particular aDNA sources and the problem of limited/degraded nucleic acids in the specimens. In essence, this arises solely due to the lack of available information as to exactly what is in the commercial digestion buffers, thus rendering it difficult to modify them. Thus we focus on nonkit-based protocols that we have found to be effective in the aDNA context.

19.3.1 BONE AND TOOTH

Bones and teeth are the most used tissues in ancient DNA studies, as they are the long-lasting physical evidence of human or animal presence at an archaeological site. The reason why these are very suitable for ancient DNA studies has been a subject of some debate. It has been argued by some that DNA in bone and teeth undergoes a retarded rate of decomposition, because of its hypothetical adsorption to hydroxyapatite [1]. Their low water content [28], the mummification of individual cells [29], and the physical exclusion of microbes and other external contaminants [28] also seem to be important features. The recent awareness that handling may be a source of contamination has led researchers to use teeth for DNA studies. Also, several studies have reported better DNA yields in teeth than in bone [30,31].

In order to extract DNA from bone or teeth, the first step is to obtain a homogenized powder from the sample (unless a nondestructive extraction method is to be attempted, see Section 19.5.2.5). This can be done in several different ways. The most common procedure is to cut a section of hard, compact bone (Figure 19.2a) and then powder it using specialized equipment such as a mikrodismembrator (Sartorius Stedim Biotech S.A., Aubagne, France) or a freezer mill. However, with the recent shift to using smaller amounts of tissue (as little as 0.01 g), many now prefer simply obtaining the powder through direct drilling into the sample using suitable drill bits. Not only is the use of small drill bits the fastest way of getting bone powder but also it conveys the additional benefit of minimizing bone destruction as only discrete holes are left (Figure 19.2b). However, if other analyses like isotope levels or radiocarbon dating by accelerator mass spectrometry are necessary, more bone powder should be drilled. Drilling is a similarly useful technique for enabling the recovery of powder from tooth dentine. Due to the hardness of the enamel crown of most teeth, it is recommended that in such samples drilling be performed from, and into, the un-enameled root.

EDTA is the base of most digestion buffers that are subsequently applied to the powdered bone or tooth. In many previous studies, DNA purification commenced with a 24 h incubation of the powder in 0.5 M EDTA (pH 8.0) at room temperature to demineralize the bone/tooth, following which the EDTA was often thrown away prior to a second round of enzymatic digestion on the remaining tissue. However recent unpublished data by our group indicates that in many

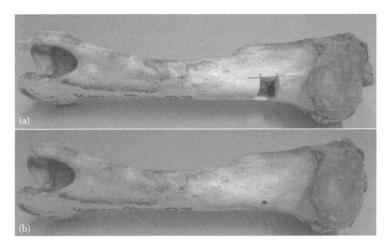

FIGURE 19.2   Musk ox bone sampled for ancient DNA extraction (a) using a cutting disk and (b) using a drill bit.

aDNA sources significant amounts of DNA are present in this solution—in some cases up to 90% of the total DNA present in the sample can be lost in this way! As such, alternative methods are preferable.

One recent study has shown that a direct incubation of the powder in 0.45 M EDTA and 0.25 mg/mL proteinase K solution (pH 8.0) is an efficient digestion buffer, if a large extraction (using at least 0.5 g of bone powder) is possible [22]. Subsequently, DNA can be purified with the addition of concentrated guanidinium-salts and a silica-suspension. While this method has the advantage of being specific to DNA and less likely to purify PCR inhibitors, as the silica particles in the suspension are a powerful PCR inhibitor [32], care has to be taken to ensure that the extract is free of residual silica particles. Furthermore, this method requires the use of a large amount of EDTA. Thus an alternative is to remove the EDTA–proteinase K solution from any residual solids following incubation (and centrifugation), then concentration of this solution using a spin-filter such as a Centricon microconcentrator (30 kD cutoff) (Millipore, Billerica, Massachusetts). Such filters contain an anisotropic membrane that retains macrosolutes (including DNA) while allowing low-molecular-weight solutes to pass through. Therefore, the volume of the solute is reduced, and the concentration of retained DNA increases. It is worth bearing in mind that one drawback is that potential PCR inhibitors, which fall above the molecular weight cutoff of the Centricon, are also concentrated. Following concentration, DNA can be purified using solid-state silica filters, such as the QIAquick purification kit (Qiagen) [32]. Lastly, and if required, DNA can be recovered from any remaining demineralized solids using a silica-spin based extraction kit such as the DNeasy tissue kit (Qiagen) (Figures 19.3a and 19.3b).

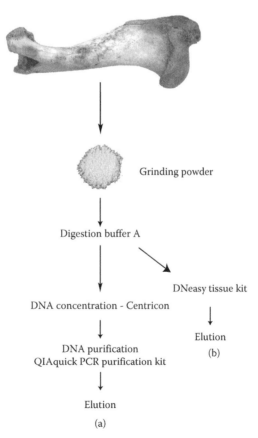

FIGURE 19.3 Scheme of DNA extraction from ancient bone.

19.3.2 Keratinous and Chitinous Tissues

In recent years a number of aDNA studies have used keratinous tissues (e.g., hair, nails, or horn sheath) and chitinous tissues (e.g., insect cuticles) as a source of aDNA [20,21,33]. Where available, such materials appear to have several advantages over bone and tooth, including (for hair at least) the ease with which they can be decontaminated [20,21] and lower than expected levels of hydrolytic DNA damage [20]. Key to the extraction of DNA from keratinous tissues is the breaking down of the keratin in order to liberate the DNA. To do this, special digestion buffers containing enhanced levels of detergents (e.g., SDS, DTT, or Cleland's reagent) and proteinase K are normally used (buffer B). The DNA is then purified from the solution using a phenol:chloroform extraction [34] followed by isopropanol purification (Figure 19.4). Although the biochemical mechanisms underlying the method have not been fully elucidated, the above also works well on chitin.

19.3.3 Ethanol/Dried Tissues

Ethanol- and dry-preserved tissues are often used as sources of ancient DNA, in particular with regard to samples from natural mummies, or from historic spirit collections. The recovery of DNA from both sources is not normally problematic (assuming DNA survives in the specimen) and can be achieved using conventional DNA extraction kits, or buffers similar to those used on bone, tooth,

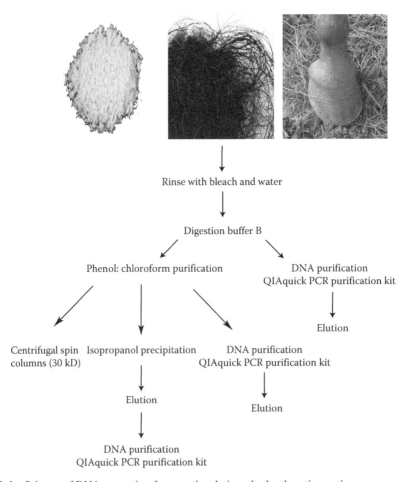

FIGURE 19.4 Scheme of DNA extraction from ancient hair and other keratinous tissues.

or keratinous materials. However, it is worth stressing that it is very important that all the residual ethanol present in relevant samples is removed prior to digestion, for example, through incubation for a few min (5–10, although dependent on the sample) at 55°C. This is vital as ethanol is a strong inhibitor of downstream digestion (among other things it may denature the proteinase K).

19.3.4 PLANT MATERIALS

Although most previous aDNA research has focused on human and other animals, in recent years there has been an increasing interest in the application of aDNA techniques to plants [35]. Examples where ancient plant DNA has been successfully extracted include specimens from herbarium collections [36], fruit stones [37], pollen [38], wood [39], and sediments [40]. The main difficulties that aDNA plant research encounters are identical to the ones facing modern samples. Plant tissues possess a myriad of substances that inhibit the success of both PCR and post-PCR reactions. During the DNA extraction, these contaminants often co–precipitate or co–purify with the DNA. To minimize these undesirable effects, substances such as PVP and cetyltrimethylammonium bromide (CTAB) can be used during the digestion step. PVP is a water-soluble polymer widely used in plant DNA extractions due to its exceptionally good adsorption to polyphenols during DNA purification. Polyphenols are common in many plant tissues and can deactivate proteins if not removed and therefore inhibit many downstream reactions like PCR. CTAB is also routinely used because of its ability to eliminate enzyme-inhibitory polysaccharides. 2-Mercaptoethanol is also frequently included in DNA extraction buffers because it prevents oxidation of polyphenols present in the plant extract. As the methods used to extract ancient plant DNA rarely differ from those used to extract modern plant DNA, readers should refer to Chapter 21 on Plant DNA extraction for methods that are also useful for ancient DNA.

19.3.5 FORMALIN-FIXED TISSUES

Although not normally "ancient," DNA recovered from formalin-fixed materials (or other, such as Bouin's solution) shares many characteristics with aDNA. In particular, it is usually heavily degraded, both by cross-linking to proteins [41], and when unbuffered fixatives are used, fragmented [42]. Therefore, such samples are normally characterized by relatively low levels of amplifiable DNA template postextraction. A number of alternative techniques have been published over recent years for dealing with such materials, and a recent cross-comparison of the methods has indicated that the method of choice is heavily dependent on the desired downstream applications. Although the topic is sufficiently complicated that we recommend consultation of the primary literature on this matter to ensure optimal results [42], some initial details of interest follow. Firstly, although many fixed materials are often stored embedded in paraffin or other similar waxes, experiments have demonstrated that in contrast to most published methodologies it is often not necessary to remove this (e.g., through xylene washes) prior to DNA extraction [42]. Secondly, several studies have demonstrated that the DNA–protein cross-linking derived from interaction with the formaldehyde present in the fixatives can, to some extent, be reversed. Most effective in this sense seems to be a brief incubation at high temperature in alkali (e.g., 20 min in 0.1 M NaOH at 100–120°C) [43,44]. While such treatment has the negative effect of significantly reducing the total level of DNA in the extract, it significantly increases the level of amplifiable molecules available, a benefit if PCR is a downstream goal.

19.3.6 SEDIMENTS, ICE, AND COPROLITES

Ice and sediments (whether frozen or from temperate sources, often spanning back as far as hundreds of thousand of years old) have started to play an important role in aDNA studies. Although the types of DNA that have so far been recovered from the sediments vary widely (including vertebrate, insect, plant fungal, and bacterial [40,45–47]), it seems that different genetic sources do not require specialized extraction methods. This is similarly the case for the recovery of DNA from coprolites (dried feces), which to date have been used to investigate a number of questions, including the genetics of the source species [48–50] and their diets [25,50,51]. In general, extracting DNA from the above materials is often problematic. Firstly, DNA may need to be liberated from tough protective materials (e.g., some bacterial cell walls and the outer coat of pollen and some parasite eggs). Secondly, the direct lysis of cells within the soil matrix, often results in the coextraction of other soil components, such as humic and fulvic acids. Humic substances are a major component of soil organic matter and though their chemical compositions are highly complex, these compounds, if co–purified with DNA, will inhibit downstream applications like endonuclease restrictions and PCR. Thirdly, as mentioned above for coprolites at least, significant amounts of protein-DNA cross-linkage may have occurred. In our experience, we have found that the published protocol originally used by Willerslev et al. [45] for use on permafrost sediments works well on both sediment and coprolite material. Specifically, this protocol involves an initial step where the material is solubilized in a buffer then subjected to vigorous shaking in the presence of glass beads using a FastPrep platform (QBiogene, Irvine, California). This step acts to physically break open the protective layers that might otherwise hinder DNA extraction. Subsequently, the DNA can be extracted from the source material using a tailored digestion buffer that includes PTB to help reverse the protein–DNA cross-linkages. Lastly, the DNA is purified from the waste biomolecules and other inhibitors that may be present using a combination of organic solvents and silica filters.

19.3.7 "NONDESTRUCTIVE" EXTRACTION TECHNIQUES

Museum specimens have provided the material for a large proportion of ancient DNA studies conducted in the last few decades. They are a convenient source of species-wide sampling from many mammalian, bird, or reptile species that otherwise requires extensive fieldwork to get. Moreover, museum specimens are the only source for samples of nowadays-extinct species, so they are very important in population genetic studies over time [52]. However, a major drawback is that the specimens investigated are usually damaged on some observable way or other, as "standard" DNA extraction methods involve at least partial specimen destruction. Parts of skin, bone, tooth or hair have to be removed for the DNA extraction. This is obviously undesirable when dealing with rare species or otherwise important specimens, such as type or voucher specimens. In realization of these problems, several recent studies have pioneered the use of so-called "nondestructive" methods for the recovery of DNA— although it is worth highlighting that nondestructive is a relative term in this sense, as sampling and recovery of DNA is of course always destructive in some way.

We therefore here refer to nondestructive in so far as methods that confer no visible external morphological damage to the samples used. To date, such methods have been used in two contexts, firstly on mammalian teeth [53] and secondly on dried arthropod samples [54,55]. In essence the key to both methods is the leaching of DNA from within the sample through the use of appropriate buffers. In the first case, the authors achieve this through immersion of the teeth in a guanidinium–thiocyanate (GuSCN) based buffer for 2 or 7 days at 40°C in the dark, followed by a DNA purification by binding to silica, using a 50 µL silica suspension and washing of the silica pellet with binding and washing buffer as described in Rohland and Hofreiter [22]. The data indicates that GuSCN is very effective in leaching DNA from the hydroxyapatite matrix, probably because GuSCN is a strong protein-denaturing agent that also breaks certain chemical cross-links. For arthropods, the complete specimen is immersed in either a GuSCN [55] or detergent-rich digestion buffer [54] (high levels of detergent reduce the surface tension of the buffer, enabling enhanced penetration into the insect via spiracles, the anus and other natural holes) prior to incubation overnight at 55°C. Nucleic acids are then purified from the solution using a phenol:chloroform extraction [34] followed by isopropanol purification, or an alternative silica method such as Qiagen QIAquick spin columns. Although not explicitly investigated for insects, according to Rohland and coauthors [53] it is even possible to re-extract DNA from single tooth specimens on several occasions, although the amount of extractable DNA decreases with the increasing number of successive extractions.

19.4 REAGENTS AND EQUIPMENT

19.4.1 RECOMMENDED CHARACTERISTICS OF aDNA EXTRACTION FACILITY

The aDNA facility should be physically isolated from the post-PCR laboratory to reduce contamination of "ancient" specimens with modern DNA or previously amplified PCR products (aerosol DNA). Positive air pressure and nightly exposure of surfaces to UV-irradiation along with weekly cleaning of work surfaces, reagents, and equipment with diluted bleach are also important to minimize contamination and remove any DNA present. Researchers should wear full body suits, facemasks, dedicated clean room shoes, and gloves when entering and working in the laboratory.

19.4.2 SPECIALIZED EXTRACTION EQUIPMENT

1. FastPrep machine (Bio101 FastPrep system for rapid isolation of DNA, RNA, and proteins/FastPrep FP120A Instrument, QBiogene, Irvine, California)—ice/sediments/coprolites
2. Mikrodismembrator (Sartorius Stedim Biotech S.A., Aubagne, France)—bone/tooth
3. QIAquick PCR purification kit (Qiagen)—bone/tooth and keratinous/chitinous tissues
4. DNeasy blood and tissue kit (Qiagen)—bone/tooth
5. Centrifugal spin columns/Centricon microconcentrators (Amicon Ultra-4 Centrifugal Filter Unit with Ultracel-30 membrane, Millipore, Billerica, Massachusetts)—bone/tooth
6. FastPrep tubes (QBiogene, Irvine, California)—ice/sediments/coprolites

19.4.3 NONSPECIALIZED EXTRACTION EQUIPMENT

1. Pipettes—P10, P20, P100, P1000 (filter tips)
2. Centrifuges—rotors: 2.0, 15, 20 mL (size dependent on extraction size)
3. Rotor
4. Oven

19.4.4. REAGENTS NEEDED

Molecular biology grade/DNA-free is recommended for the reagents to prevent contamination arising from the extraction process. The reagents are listed at commonly available concentration, but this naturally can be modified at the user's discretion.

1. 0.5 M EDTA solution (pH 8.0)
2. Proteinase K (powdered or solution at known concentration)
3. 1 M Tris–HCl solution (pH 8.0)
4. 5 M Sodium chloride (NaCl) solution
5. 10% SDS solution
6. 1 M Calcium chloride ($CaCl_2$) solution
7. DTT (Cleland's reagent) (powder)
8. 10% N-lauroyl sarcosine solution
9. β-Mercaptoethanol

10. 100 mM PTB solution
11. Chloroform
12. Octanol
13. Phenol (Tris-buffered, pH 8.0)
14. Isopropanol
15. Ethanol
16. 3 M Sodium acetate (pH 5.2) solution
17. Water (double distilled)
18. Salton wash buffers 1 and 2 (individual kit component from Bio101's "Geneclean for aDNA kit," available from QBiogene, Irvine, California)
19. AW1/AW2 buffers (individual kit component from DNeasy extraction kit, available from Qiagen)
20. Commercially available bleach solution
21. Glycoblue (Ambion, Inc, Austin, Texas)

19.4.5 Buffer Recipes

Extraction buffer A: 0.45 M EDTA pH 8.0 and 0.25 mg/mL proteinase K
Extraction buffer B: 10 mM Tris-HCl pH 8.0; 10 mM NaCl; 2% w/v SDS; 5 mM CaCl$_2$; 2.5 mM EDTA pH 8.0; 40 mM DTT (Cleland's reagent); 10% proteinase K solution (e.g., > 600 mAU/mL, Qiagen)
Extraction buffer C: 2% N-lauroyl sarcosine; 50 mM Tris–HCl ph 8.0; 20 mM EDTA pH 8.0; 150 mM NaCl; 2 mM β-mercaptoethanol; ~10 mg proteinase K/10 mL buffer; 50 mM DTT and 2 mM PTB.
Note: The use of PTB is controversial. Some elect to omit it from the preparation due to the inherent difficulty in commercially obtaining PTB.

19.5 METHODS

The following are protocols that we find to be useful on a range of aDNA tissue sources, and are principally modified from many of the protocols discussed above. It is worth remembering that often multiple alternate methods are equally effective, so we caution the reader that the following are our preferential methods, but others may work equally well. Also the reader is reminded that the extractions often are only as good as the DNA quality in the tissue. Some samples simply have no surviving DNA, hence no DNA can be extracted regardless of the method!

19.5.1 General Procedures

19.5.1.1 QIAquick DNA Purification

Note: We (and many others in the aDNA community) find that Qiagen's "QIAquick" PCR clean up kits are an excellent and quick tool for purifying DNA, regardless of the source. Although the kit's manual can be directly followed, under the simple rule that one replaces the word "PCR product" with "DNA extract," we use a slightly modified version. As there are a large number of similar kits on the market, all working on the principal of DNA–silica binding coupled to centrifugation, it is likely that the Qiagen kit can be substituted to that of the reader's choice. This, of course, would require modifying the protocol suitably.

1. Prior to commencing, centrifuge the digestion mixture for 3–5 min at high speed to pellet any solids remaining. Carefully pipette the liquid fraction of the digestion into a new tube. Any solids

carried over into the next steps can block the spin filter, so care is advocated at this step. In our experience it is better to leave a small amount of liquid behind, than carry across solid.

2. Add 5 volumes QIAquick buffer PB (sometimes labeled by Qiagen as PBI) to the liquid.
3. Mix thoroughly.
4. Add 700 μL of this mixture to the QIAquick spin column.
5. Centrifuge for 1 min at 6000 g. We use this lower than recommended speed to try and limit how much target DNA passes through the filter without binding. However, if it is seen that the liquid does not pass through the filter in this time, the speed can be increased.
6. Empty the liquid waste from the spin column. Repeat Steps 5 and 6 with the remaining PB buffer–digestion mix, until all the liquid has been passed through the spin column.
7. Add 500 μL Qiagen wash buffer PE to the filter.
8. Centrifuge for 1 min at 10,000 g. Empty the waste and repeat if extra purity is required.
9. Centrifuge for 3 min at maximum speed to dry the filter. Any residual ethanol from the PE buffer will inhibit downstream applications.

19.5.1.2   Phenol: Chloroform Purification

1. Add a volume of phenol to the digestion mix at a ratio of 1:1 with the total digestion volume.
2. Agitate gently at room temperature for 5 min.
3. Centrifuge for 5 min to separate the layers. The speed will depend on the volumes and the centrifuge capacity. If after 5 min the layers have not fully separated, extend the centrifugation time.
4. Carefully remove the upper aqueous layer. Be careful not to remove the protein-containing interface.
5. Add to 1 volume of new phenol. Repeat Steps 2–4. After the second centrifugation, add the aqueous layer to 1 volume chloroform.
6. Agitate gently at room temperature for 5 min.
7. Centrifuge for 5 min to separate the layers. Remove the upper aqueous layer.

19.5.1.3   Isopropanol Precipitation

Note: Isopropanol precipitation in this context is most effective at relatively high centrifugal forces, and in small tubes (the area covered by the precipitated DNA that forms the observed DNA pellet is most concentrated and thus easiest to spot and resuspend if 1.5 mL tubes or smaller are used). Thus if large volumes are to be precipitated, we recommend first concentrating the liquid with a centrifugal concentrator.

1. Add 0.6–1 volume isopropanol and 0.1 volume 3 M sodium acetate (approximately pH 5). A small amount of commercial carrier solutions can also be added, if required, to facilitate pellet visualization, such as Glycoblue (Ambion, Inc, Austin, Texas), following the manufacturers' guidelines. Mix well.
2. Immediately centrifuge at high speed for 30 min. This can be done at room temperature, although some protocols recommend refrigeration. We have no evidence to support the observation that one is better than the other.
3. Immediately following centrifugation, decant the liquid from the tube carefully. The DNA will have precipitated into a pellet at the bottom of the tube, and may not be visible.
4. To rinse the pellet, gently add 500–1000 μL 85% ethanol, gently invert once, then centrifuge for 5 min at high speed.

5. Gently decant the ethanol. Repeat if necessary.
6. All ethanol must be removed from the pellet as any residual ethanol will inhibit downstream applications. This can be easily achieved with a small bore pipette, followed by a brief incubation at a relatively high temperature (e.g., 55°C–75°C)
7. Resuspend the pellet in TE buffer or ddH$_2$0. If the pellet has become very dry, this may require leaving the pellet at room temperature in the liquid for 5–10 min, followed by gentle pipetting.

19.5.2 Specific Procedures

19.5.2.1 Ethanol/Dried Tissues

1. Prior to digestion, ethanol preserved tissue must be dried of any residual ethanol, as any ethanol carryover will inhibit the tissue digestion. This can easily be achieved through incubation at 55°C until the tissue is obviously dry.
2. Cut up/macerate the tissue into small pieces to aid digestion.
3. We find that digestion and subsequent DNA purification is often sufficient using commercial kits, our favorite being Qiagen's DNeasy tissue extraction kit. However, we also find that DNA can equally easily be extracted following protocol given in Section 19.5.2.3 (keratinous/chitinous tissues) commencing at Step 4. An advantage of the latter is that larger digestion volumes can be used relatively easily to digest larger volumes of starting material. However, ultimately the method adopted ultimately comes down to user preference.

19.5.2.2 Bone/Tooth

In order to extract DNA from bone and teeth, the first step is to obtain a homogenized powder from the sample.
Day 1

1. Add 1.3 mL of digestion buffer A to the bone powder, and then incubate with rotation overnight at room temperature.

Day 2

1. Centrifuge the solution at 12,000 g for 5 min to pellet the nondigested powder. Carefully pipette the liquid into a Centricon microconcentrator (30 kD cutoff). Freeze the remaining pellet for later use if needed.
2. Spin the Centricons at 4000 g for 10 min. The liquid should concentrate down to about 200–250 μL. If after 10 min there is more left, spin it longer. If you find that all the liquid has gone, add 200 μL of ddH$_2$0 to the filter, let it sit for 10 min, then pipette up and down a bit to make sure all the DNA is redissolved.
3. The concentrated DNA can now be purified using the QIAquick purification kit. For QIAquick protocol refer to Section 19.5.1.1.

If after this procedure there is no DNA, the remaining powder can be digested and purified using the DNeasy blood and tissue kit (Qiagen). For protocol refer to the manufacturer's guidelines.

19.5.2.3 Keratinous/Chitinous Tissues

This protocol assumes the user is using pure keratinous tissue or chitin, e.g., hair, horn, nail, feather, or arthropod exoskeleton. For whole arthropods, specimens should be macerated and not bleached, to ensure that the soft internal tissues may contribute DNA. For nondestructive extraction from arthropods, refer to Section 19.5.2.5.

Day 1

1. For most materials proceed directly to Step 2. For large pieces of nail or horn, a suitable amount of powder (e.g., 100 mg) can be drilled directly from the specimen for use in the extraction.
2. For nonpowdered material, prior to DNA extraction the tissue should be cleaned by a brief wash in a dilute bleach solution (1:10 dilution commercial strength). Care should be taken to remove all obvious external sources of contaminant matter. For powdered material, clean by immersing the powder in the bleach solution for 10–20 s, then pellet the powder by brief centrifugation. Pour off the bleach.
3. Rinse material several times in ddH$_2$0 to remove all traces of bleach. Any bleach carryover will likely hinder the subsequent DNA extraction. For powdered material, use a vortex to ensure the pellet from Step 2 is homogenized after adding the water. After 10–20 s incubation, repellet the powder. Pour off the ddH$_2$0 then repeat.
4. Add sufficient extraction buffer B to the material, and then incubate with rotation overnight at 55°C. The volume added depends on the volume of starting material and in most cases is determined by previous experience.

Day 2

The extracted DNA can be purified in a number of different ways. We find that the method we use ultimately depends on convenience and the user's preference. For larger volumes of digestion mix (e.g., >1 mL) we find phenol:chloroform extractions preferable. For small volumes we use QIAquick silica spin-columns (Qiagen). As should be apparent from reading the QIAquick protocol, although rapid for small volumes, it can become extremely laborious for larger volumes due to the fivefold increase in volume at the buffer PB step. For QIAquick protocol refer to Section 19.5.1.1. For phenol:chloroform refer to Section 19.5.1.2.

Following the phenol:chloroform purification, the aqueous layer can be concentrated (if desired) or further purified using a number of alternative procedures including centrifugal spin columns (30 kD cutoff), Qiagen QIAquick silica columns, or isopropanol precipitation. For centrifugal filters follow the user manual of the columns. For QIAquick procedure refer to Section 19.5.1.1. For isopropanol precipitation refer to Section 19.5.1.3.

Note: Many keratinous tissues contain melanin, and this often co–purifies with the DNA during the DNA extraction. During centrifugal concentration, the concentrated solution often become dark colored as the melanin is retained with the DNA. Melanin also commonly precipitates with the DNA during the isopropanol precipitation method, leaving a brown pellet and resulting in a brown extract. As melanin can often inhibit PCR and other enzymatic reactions, it is often desirable to remove it. One way to achieve this is to undergo an additional purification at the end of the process, using a QIAquick procedure following Section 19.5.1.1.

19.5.2.4   Ice/Sediments/Coprolites

Day 1

1. Add 0.25 g ice/soil/coprolite to a FastPrep soil tube. Do not overload the tubes as this can lead to cross-contamination.
2. Suspend pellet in 600 μL extraction buffer C.
3. Agitate FastPrep tubes using a FastPrep (level 6 for 45 s × 4 sessions) and put the tubes on ice for 1–2 min between each session to allow to cool. Alternatively, a vortex can be used if no FastPrep is available, although this will not homogenize the particles as effectively.
4. Incubate at 55°C with gentle agitation overnight.

Day 2

  5. Centrifuge briefly (5000 rpm) to get rid of bubbles.
  6. Add 150 μL 5 M NaCl to enzymatic solution and agitate (level 6, 15 s).
  7. Add 375 μL chloroform/octanol solution (24:1).
  8. FastPrep/vortex tubes briefly to mix.
  9. Rotate tubes at room temperature and leave for 30 min (can be left overnight).
 10. Centrifuge the solution at 12,000 g for 2 min. Taking care not to disturb the interface, transfer the aqueous (top) phase from the tubes to a new 1.5 mL tube and incubate at 2°C–3°C for at least 1 h, preferably overnight.

Day 3

 11. Centrifuge the solution at 12,000 g for 2 min, and move the supernatant to a 15 mL tube.
 12. Add Qiagen PB buffer (five times the volume of the supernatant) and mix gently by several inversions of the tube.
 13. Filter this solution (700 μL a time) using a Qiagen QIAquick spin column. Take care not to overload the spin columns, or they may leak and cause cross-contamination. Each time, centrifuge for 1 min at 10,000 g. Discard filtrate in between spins. If required, samples can be split into multiple spin columns at this stage, although this ultimately leads to dilution of the final DNA extract.
 14. Add 500 μL Salton wash 1 buffer and centrifuge for 1 min at 10,000 g. Repeat it twice.
 15. Add 500 μL Salton wash 2 buffer and centrifuge for 1 min at 10,000 g. Repeat it twice.
 16. Add 500 μL AW 1 buffer and centrifuge for 1 min at 10,000 g. Repeat it twice.
 17. Add 500 μL AW 2 buffer and centrifuge for 3 min at 15,000 g. Repeat it twice. Discard filtrate and spin again to remove residual ethanol from base of spin column. Ethanol in the final solution will inhibit PCR.
 18. Place QIAquick column in a clean 1.5 mL tube with the lids removed.
 19. To elute the DNA, add 50 μL buffer EB to the centre of the QIAquick membrane and leave at room temperature for 10 min. Then centrifuge the column for 1 min at 10,000 g.
 20. Transfer the eluent to a new 1.5 mL tube.

Note: In Steps 14–17, number of repeats is dependent on anticipated/observed level of inhibitors in the extract and can be optimized.

19.5.2.5 Nondestructive Arthropod DNA Extraction

This protocol works best on small arthropods that have relatively thick exoskeletons, simply can be placed whole into common laboratory tubes. For larger samples, larger tubes and increased buffer volumes are required.

Day 1

  1. Place whole sample in a 2 mL tube. If pinned, the pin can often be left in the sample to prevent damage while removing the pin.
  2. Add a suitable volume of extraction buffer B. The volume required depends on the specimen size, and should minimally cover the specimen completely. Incubate at 55°C overnight (without agitation)

Day 2

  3. Carefully separate the sample from the digestion buffer by either pipetting off digestion mixture into a fresh tube, or carefully removing the sample from the digestion mix.

4. Place the sample into 100% ethanol. Leave for several hours. This halts further digestion of the sample. Following this step the sample can be removed and dried.
5. Purify DNA from the extraction buffer using the QIAquick protocol (Section 19.5.1.1).

19.6 FUTURE DEVELOPMENT TRENDS

Ancient DNA research is undergoing rapid development. However, many technical problems still remain that hinder the application of these techniques. Solving these problems is therefore currently a priority for researchers in the field, and improving the efficiency of the DNA extraction will be a major breakthrough for ancient DNA studies. Key problems that need to be addressed include the coextraction of PCR inhibitors, as well as the development of techniques to remove contamination from ancient (in particular human) samples.

Since most ancient DNA research is performed on extremely rare specimens, preservation of extracted DNA for extensive periods of time is also extremely important, as, once extracted, DNA continues to degrade. Preserving the extracted DNA involves stopping or delaying any adverse reactions, while still enabling the usability of the material for PCR and subsequent reactions.

Given the small amounts usually found in aDNA samples, increasing the concentration of DNA in the extracts will be essential to increase the amount of data that can be obtained from a single specimen. Techniques such as whole genome amplification, currently not widely used, will theoretically perpetuate the "laboratory-life" of rare specimens. This technique is currently available in commercial kits, like GenomiPhi (GE Healthcare). The GenomiPhi kit uses a bacteriophage Phi29 DNA polymerase to exponentially amplify single- or double-stranded DNA templates via a strand displacement reaction without the use of thermal cycling. The genomic DNA in the template is combined with a buffer containing random hexamer primers. This mixture is then denatured and cooled, so the hexamers can anneal to the DNA molecule, and all the other remaining components are added to the mixture. After overnight incubation at 30°C, high molecular weight fragment copies of the template DNA are obtained. This method is also very accurate, thanks to the proofreading activity of the used enzyme. Thus once the genome amplification is completed, various genotyping assays can be undertaken from a large base of synthetic DNA copies.

Another area that has been intensively studied is the possibility to repair damaged DNA using enzymes. To minimize sequence variation caused by miscoding lesions, DNA can be treated, prior to sequencing, with enzymes such as Uracil-*N*-Glycosylase (UNG). UNG treatment is often done to excise uracil bases caused by the hydrolytic deamination of cytosines. UNG reduces sequence artifacts caused by this common form of postmortem damage, which results in apparent G/C–A/T mutations and subsequent errors in sequence results [2].

To conclude, although much has been developed to help optimize the recovery of nucleic acids from ancient samples, over the next few years, much remains to be developed to help improve the way we work with ancient DNA.

Paula F. Campos received her BSc in animal applied biology and MSc in biodiversity and genetic resources from the Faculty of Sciences, University of Porto, Porto, Portugal. She is currently undertaking a PhD at the Ancient DNA and Evolution Group, University of Copenhagen, Copenhagen, Denmark. Her research interests center on the demographic history and migration paths of the Pleistocene musk ox.

Eske Willerslev received his MSc and DSc in biological sciences at the University of Copenhagen, Copenhagen, Denmark. During his thesis research, he focused predominantly on the recovery of ancient DNA directly from ice cores and sediments, including samples where no macrofossils were present. Following his doctoral work, he undertook a Wellcome Trust fellowship at the Department of Zoology, University of Oxford, Oxford, United Kingdom. During this time, he broadened his research areas into a range of different aspects of ancient DNA research. After 1.5 years in Oxford, he returned to Copenhagen to commence a position as full professor at the University of Copenhagen, first at the Niels Bohr Institute and later at the Department of Biology. His research interests focus on evolutionary biology, past environmental reconstructions, population genetics, and phylogenetics using both ancient and contemporary DNA sequence data.

M. Thomas P. Gilbert received his BA in biological sciences from Oriel College, Oxford University, Oxford, United Kingdom; he undertook a DPhil at the university's Department of Zoology, New College. During his thesis research, he focused predominantly on the challenges facing the field of ancient DNA, in particular with regard to the recovery of authentic human DNA from archaeological human specimens. After his DPhil, he undertook a 2 year postdoctoral fellowship at the Department of Ecology and Evolutionary Biology, University of Arizona, Tucson, Arizona. During this time, his research focused on developing methods to enhance the recovery of nucleic acids from formalin-fixed materials, with a specific aim toward the recovery of archival viral sequences for use in evolutionary analyses. He subsequently relocated to the ancient DNA and evolution group at the departments of biology and Niels Bohr institutes, University of Copenhagen, Copenhagen, Denmark. Initially, on a 2 year Marie Curie fellowship, he extended his stay in January 2008 upon gaining a Danish government Skou associate professor position. His research interests focus on the challenges offered by the study of nucleic acids in degraded materials, solutions to these problems, and the evolutionary biology, archaeological, and anthropological questions that can be answered through using such samples.

REFERENCES

1. Lindahl, T., Instability and decay of the primary structure of DNA, *Nature*, 362: 709, 1993.
2. Pääbo, S., Ancient DNA: Extraction, characterization, molecular cloning, and enzymatic amplification, *Proc. Natl. Acad. Sci. U. S. A.*, 86: 1939, 1989.
3. Hoss, M, et al., Molecular phylogeny of the extinct ground sloth *Mylodon darwinii*, *Proc. Natl. Acad. Sci. U. S. A.*, 93: 181, 1996.
4. Handt, O. et al., Molecular genetic analyses of the Tyrolean Ice Man, *Science,* 264: 1775, 1994.
5. Smith, C.I. et al. The thermal history of human fossils and the likelihood of successful DNA amplification, *J. Human Evol.*, 45: 203, 2003.
6. Paabo, S. et al. Ancient DNA and the polymerase chain reaction. The emerging field of molecular archaeology, *J. Biol. Chem.*, 264: 9709, 1989.
7. Stiller, M. et al., Inaugural article: Patterns of nucleotide misincorporations during enzymatic amplification and direct large-scale sequencing of ancient DNA, *Proc. Natl. Acad. Sci. U. S. A.*, 103: 13578, 2006.
8. Shanks, O.C. et al., DNA from ancient stone tools and bones excavated at Bugas-Holding, Wyoming, *J. Archaeol. Sci.*, 32: 27, 2005.
9. Gilbert, M.T.P. et al., Biochemical and physical correlates of DNA contamination in archaeological human bones and teeth excavated at Matera, Italy, *J. Archaeol. Sci.*, 32: 785, 2005.
10. Richards, M.B. et al., Authenticating DNA extracted from ancient skeletal remains, *J. Archaeol. Sci.*, 22: 291, 1995.
11. Nicholson, G.J. et al., Detection of bone glue treatment as a major source of contamination in ancient DNA analyses, *Am. J. Physic. Anthropol.*, 118: 117, 2002.
12. Leonard, J.A. et al., Animal DNA in PCR reagents plagues ancient DNA research, *J. Archaeol. Sci.*, 34: 1361, 2007.
13. Poinar, H.N. et al., Metagenomics to paleogenomics: Large-scale sequencing of mammoth DNA, *Science*, 311: 392, 2006.
14. Green, R.E. et al., Analysis of one million base pairs of Neanderthal DNA, *Nature*, 444: 330, 2006.
15. Wall, J.D. and Kim, S.K., Inconsistencies in Neanderthal genomic DNA sequences, *PLoS Genet.*, 3: e175, 2007.
16. Turner-Walker, G. et al., Sub-micron spongiform porosity is the major ultra-structural alteration occurring in archaeological bone, *Int. J. Osteoarchaeol.*, 12: 407, 2002.
17. Gilbert, M.T.P. et al., Insights into the processes behind the contamination of degraded human teeth and bone samples with exogenous sources of DNA, *Int. J. Osteoarchaeol.*, 16: 156, 2006.
18. Salamon, M. et al., Relatively well preserved DNA is present in the crystal aggregates of fossil bones, *Proc. Natl. Acad. Sci. U. S. A.*, 102: 13783, 2005.

19. Malmstrom, H., et al., Extensive human DNA contamination in extracts from ancient dog bones and teeth, *Mol. Biol. Evol.*, 22: 2040, 2005.
20. Gilbert, M.T.P. et al., Whole-genome shotgun sequencing of mitochondria from ancient hair shafts, *Science*, 317: 1927, 2007.
21. Gilbert, M.T.P. et al., Ancient mitochondrial DNA from hair, *Curr. Biol.*, 14: 463, 2004.
22. Rohland, N. and Hofreiter, M., Comparison and optimization of ancient DNA extraction, *Biotechniques*, 42: 343, 2007.
23. Schrooyen, P.M.M. et al., Stabilization of solutions of feather keratins by sodium dodecyl sulfate, *J. Colloid Interface Sci.*, 240: 30, 2001.
24. Clark, M., In: *Plant Molecular Biology—A Laboratory Manual.* pp. 305–325 Springer-Verlag, Berlin, Heidelberg, 1997.
25. Poinar, H.N. et al., Molecular coproscopy: Dung and diet of the extinct ground sloth *Nothrotheriops shastensis, Science*, 281: 402, 1998.
26. Boom, R. et al., Rapid and simple method for purification of nucleic acids, *J. Clin. Microbiol.*, 28: 495, 1990.
27. Fujiwara, M. et al., Adsorption of duplex DNA on mesoporous silicas: Possibility of inclusion of DNA into their mesopores, *Anal. Chem.*, 77: 8138, 2005.
28. Hummel, S. and Herrmann, B., Y-Chromosomal DNA from ancient bones. In: *Ancient DNA*, pp. 205–210, Springer-Verlag, 1994.
29. Bell, L.S. et al., Determining isotopic life history trajectories using bone density fractionation and stable isotope measurements: A new approach, *Am. J. Phys. Anthropol.*, 116: 66, 2001.
30. Oota, H., A genetic study of 2,000-year-old human remains from Japan using mitochondrial DNA sequences, *Am. J. Phys. Anthropol.*, 98: 133, 1995.
31. Krings, M. et al., Neandertal DNA sequences and the origin of modern humans, *Cell*, 90: 19, 1997.
32. Yang, D.Y. et al., Technical note: Improved DNA extraction from ancient bones using silica-based spin columns, *Am. J. Phys. Anthropol.*, 105: 539, 1998.
33. Bonnichsen, R. et al., Methods for the study of ancient hair: Radiocarbon dates and gene sequences from individual hairs, *J. Archaeol. Sci.*, 28: 775, 2001.
34. Sambrook, J. et al., *Molecular Cloning: A Laboratory Manual,* 3rd edn. (ed.). Cold Spring Harbor Laboratory Press, Cold Spring Harbor, NY, 2001.
35. Gugerli, F. et al., Ancient plant DNA: Review and prospects. *New Phytol.*, 166: 409, 2005.
36. Savolainen, V. et al., The use of herbarium specimens in DNA phylogenetics: Evaluation and improvement, *Plant Syst. Evol.*, 197: 87, 1995.
37. Pollmann, B. et al., Morphological and genetic studies of waterlogged *Prunus* species from the Roman vicus Tasgetium (Eschenz, Switzerland), *J. Archaeol. Sci.*, 32: 1471, 2005.
38. Parducci, L. et al., Ancient DNA from pollen: A genetic record of population history in Scots pine, *Mol. Ecol.*, 14: 2873, 2005.
39. Liepelt, S. et al., Authenticated DNA from ancient wood remains, *Ann. Botany*, 98: 1107, 2006.
40. Willerslev, E. et al., Ancient biomolecules from deep ice cores reveal a forested southern Greenland, *Science*, 317: 111, 2007.
41. Brutlag, D. et al., Properties of formaldehyde-treated nucleohistone, *Biochemistry*, 8: 3214, 1969.
42. Gilbert, M.T.P. et al., The isolation of nucleic acids from fixed, paraffin-embedded tissues, which methods are useful when? *PLoS ONE* 2, e537, 2007.
43. Shi, S.R. et al., DNA extraction from archival formalin-fixed, paraffin-embedded tissue sections based on the antigen retrieval principle: Heating under the influence of pH, *J. Histochem. Cytochem.*, 50: 1005, 2002.
44. Shi, S.R. et al., DNA extraction from archival formalin-fixed, paraffin-embedded tissues: Heat-induced retrieval in alkaline solution, *Histochem. Cell Biol.*, 122: 211, 2004.
45. Willerslev, E. et al., Diverse plant and animal genetic records from Holocene and Pleistocene sediments, *Science*, 300: 791, 2003.
46. Lydolph, M.C. et al., Beringian paleoecology inferred from permafrost-preserved fungal DNA, *Appl. Environ. Microbiol.*, 71: 1012, 2005.
47. Kemp, B.M. et al., Repeat silica extraction: A simple technique for the removal of PCR inhibitors from DNA extracts, *J. Archaeol. Sci.*, 33: 1680, 2006.
48. Poinar, H. et al., Nuclear gene sequences from a late Pleistocene sloth coprolite, *Curr. Biol.*, 13: 1150, 2003.
49. Gilbert, M.T.P. et al., DNA from pre-Clovis coprolites in Oregon, North America. *Science*, 320: 786, 2008.

50. Hofreiter, M. et al., A molecular analysis of ground sloth diet through the last glaciation, *Mol. Ecol.*, 9: 1975, 2000.
51. Poinar, H.N. et al., A molecular analysis of dietary diversity for three archaic Native Americans, *Proc. Natl. Acad. Sci. U. S. A.*, 98: 4317, 2001.
52. Shapiro, B., et al., Rise and fall of the Beringian Steppe bison, *Science*, 306: 1561, 2004.
53. Rohland, N., Siedel, H., and Hofreiter, M., Nondestructive DNA extraction method for mitochondrial DNA analyses of museum specimens, *Biotechniques*, 36: 814, 2004.
54. Gilbert, M.T.P. et al., DNA extraction from dry museum beetles without conferring external morphological damage, *PLoS ONE* 2, e272, 2007.
55. Rowley, D.L. et al., Vouchering DNA-barcoded specimens: Test of a nondestructive extraction protocol for terrestrial arthropods, *Mol. Ecol. Notes*, 7: 915, 2007.

20 Preparation of Forensic Samples for Direct Molecular Applications

Erin K. Hanson and Jack Ballantyne

CONTENTS

20.1   INTRODUCTION

20.1.1   FORENSIC BIOLOGICAL EVIDENCE

Forensic biology is the application of biology (genetics, biochemistry, and molecular biology) to the solution of certain problems that arise in connection with the administration of justice. It is science exercised in the service of the law. More practically, it is the study of blood and other physiological material as it relates to establishing a fact that may be at issue in a medical legal investigation. Forensic biological evidence (bodily fluids, tissues, hair, skin, etc.) may be useful in various situations, including crimes against a person (homicide, assault, rape or other sexual assault, criminal paternity, and terrorism), crimes against property (burglary), mass fatality incidents, motor vehicle incidents, and paternity/kinship analysis.

Every crime takes place at a certain time at a certain place or places (the scenes) and involves a victim and the person or persons committing the crime (perpetrators). The perpetrator may or may not use a weapon. Depending upon the case circumstances, biological material may be transferred between the scene, the victim, the perpetrator, and the weapon. Successful recovery of biological evidence from crime scenes, weapons, or people is often crucial in the identification and conviction of perpetrators and in the exoneration of falsely accused individuals. Typically, a DNA match between a crime scene sample and an individual would be an exceedingly rare event if the individual was not the true source of the crime scene sample.

In order to potentially identify the source or donor of any collected biological evidence, a variety of different DNA isolation methods have been developed. Standard forensic DNA isolation methods have been developed to ensure the recovery of nanogram quantities of DNA (i.e., 10^{-9} g) from relatively pristine samples. However, often times in forensic casework, the biological evidence recovered from crime scenes can have an extremely low number of starting templates (~1–15 cells) and, depending upon the circumstances, can often be environmentally compromised (degraded). While classic extraction methods are still the mainstay of many forensic laboratories, new analytical techniques and methods are continuously being developed that are more suitable for use with the nonpristine samples frequently encountered in forensic casework.

With the ability to obtain DNA of sufficient quantity and quality from forensic biological evidence, it is now a matter of routine for the forensic scientist to obtain the genetic profile of an individual from DNA recovered from a biological stain deposited at the crime scene. This is typically accomplished by the amplification and fluorescent labeling of autosomal short tandem repeat (STR) sequences followed by the laser-induced fluorescence detection of the electrophoretically separated alleles [1]. However, there is an additional nucleic acid type present in biological evidence that promises to gain widespread use in forensic casework. Recently, messenger ribonucleic acid (mRNA) has been utilized to identify the body fluid of origin, the age of an individual, and to provide a possible time of deposition for forensic stains [2–11]. While not routinely used in forensic casework yet, isolation methodologies have been developed to isolate RNA from forensic samples [4,11,12].

This chapter provides a brief overview of existing extraction methods, utilizing both in-house methods and commercially available extraction kits, for the recovery of DNA and RNA from forensic biological evidence.

20.1.2 STANDARD METHODS FOR PREPARING FORENSIC SAMPLES

For an isolation method to be suitable for use with forensic samples, it must be able to recover DNA of sufficient quality and quantity for analysis. Many factors can affect the quantity and quality of DNA in forensic samples including the initial size of the stain, environmental influences (heat, light, and humidity), and the nature of the laboratory extraction protocol itself. Shearing forces, nucleases, and normal extraction inefficiencies can all affect the amount and quality of DNA that is recovered. Extraction strategies need to effect the solubilization of cellular components, the denaturation or hydrolysis of proteins, and, ideally, to remove denatured proteins and other cell debris while precluding unnecessary sample loss during the physical manipulations required of the particular protocol.

Both organic solvent-based and nonorganic methods have been developed for use with forensic samples. Organic extraction methods involve the use of phenol/chloroform [13–16]. The extraction solution used is typically comprised of buffer, ethylenediaminetetraacetic acid (EDTA) (chelating Mg^{2+} ions required for nuclease activity), sodium dodecyl sulfate (SDS) (a detergent that solubilizes cell and organelle membranes), and proteinase K (acting on denatured proteins to produce oligopeptides and amino acids). Phenol/chloroform is then used to partition the DNA into an aqueous polar phase, while trapping proteins and other cell debris in the polar–nonpolar interphase. The DNA is then further purified of environmental contaminant and extraction solution reagents using an alcohol precipitation (ethanol or isopropanol) or filtration. The DNA is then dried and resolubilized in Tris-EDTA (TE) buffer, with the EDTA again protecting the isolate from environmental nucleases. Such classic organic extractions typically produce high-quality DNA due to the efficient manner in which proteins, lipids, and inhibitors are removed. However, organic extraction methodologies are time consuming, cannot be automated, and require the use of hazardous organic solvents.

Nonorganic isolation methods that are used in forensic analysis include ion-exchange resins (e.g., Chelex) [17–23] and silica-based methods [24–28]. Chelex-100 is a resin comprised of styrene divinylbenzene copolymers containing iminodiacetate ions that can chelate polyvalent metal cations. The Chelex DNA extraction is a quick and simple method involving the addition of a 5% Chelex solution, followed by incubation at 56°C for 30 min and boiling at 100°C for 8 min to lyse the cells and denature the proteins. The presence of Chelex during boiling prevents degradation of DNA by chelating metal ions that may act as catalysts in the breakdown of DNA at high temperatures. The Chelex extraction, unlike the organic extraction described above, does not involve additional purification steps. While simple to use, the DNA may not always be of sufficient purity for subsequent forensic DNA analysis. Cells are lysed directly into the Chelex solution and therefore extracts would still contain denatured proteins and any heat-stable environmental contaminants that may act as polymerase chain reaction (PCR) inhibitors. Additionally, the presence of any residual Chelex resin in the extracts can adversely affect Taq DNA polymerase activity in subsequent PCR-based assays due to Chelex-mediated chelation of Mg^{2+} ions required for DNA polymerase activity.

Silica-based methods take advantage of the ability of DNA to bind to silica particles in the presence of high salt concentrations. While DNA is bound to silica particles, proteins and other cellular debris can then be removed by washing. Once purified, the DNA can be eluted from the silica particles with a lower salt concentration solution and can be used directly in subsequent PCR-based assays without further purification. Silica particle formulations include slurries (e.g., glass milk or beads) or solid matrices such as spin columns. When glass milk or beads are used, DNA is readily eluted from the silica particles with a low salt concentration buffer. With silica-based spin columns, DNA is bound to a silica membrane or packed silica particles, washed several times to remove any remaining lysis buffer, and is then eluted directly into a collection tube by centrifugation in low salt buffer. More recently, silica-based paramagnetic bead resins have been used for forensic samples [29–38]. With this technology, samples are extracted with a lysis buffer, allowing DNA to bind to the paramagnetic silica beads. Once placed in a magnetic field, the beads are attracted to the magnet isolating the DNA from proteins and other cellular materials present in the extract, which are removed during subsequent wash steps. The DNA can then be eluted from the magnetic beads.

A significant number of samples processed in a forensic casework DNA laboratory will involve an analysis of male DNA from semen containing evidence due to the fact that males commit most of the reported sexual assaults. For example, in the United States, according to the 2005 U.S. Bureau of Justice Statistics, males committed 98% of all sexual assaults (http://www.ojp.usdoj.gov/bjs/). Failure to obtain the male donor profile can occur using standard DNA profiling techniques (i.e., STR analysis) if large quantities of female DNA are present. This is due to the kinetics of the PCR process itself that does not permit minor components to be detected at low levels (i.e., ≤1/20) because of titration of critical reagents by the major DNA component [39]. If a male/female mixed profile is obtained from these samples, interpretation of individual profiles without first separating out the component cell types prior to analysis can be complex and time consuming. In these situations, a differential extraction can be employed, which allows for a physical separation of sperm and non-sperm cells prior to analysis [40]. Sperm cell membranes contain a significant number of disulfide bonds that are not easily broken with standard extraction buffers. Disruption of these cellular membranes typically requires an additional reducing agent, such as dithiothreitol (DTT). An initial incubation with standard extraction buffer at a lower temperature results in lysis of female epithelial cells while sperm cells remain intact. After centrifugation, sperm cells are pelleted and the nonsperm (mainly female) DNA present in the supernatant can be removed. Subsequent extraction steps can be performed on both fractions in order to obtain single source DNA profiles (one from the female victim and one from the male perpetrator). A differential extraction can be incorporated into most of the aforementioned methods by performing an initial lysis and incubation in the absence of DTT in order to separate the sperm and nonsperm fractions.

Although DNA isolated using the above-mentioned methods can be employed to potentially identify a suspect, the tissue source of the DNA (blood, semen, etc.) is not revealed by the DNA profiling itself. In some cases, it is important to know the tissue origin of the questioned forensic sample (e.g., is it saliva or semen). Therefore, there is a need for the development of tissue or body fluid type identification protocols that are compatible with current molecular genetic DNA profiling techniques. Each cell type has a distinctive pattern of mRNA expression. The use of mRNA profiling technology could supersede current protein-based protocols for body fluid identification [3,5,8–10]. Extraction of RNA can be accomplished through cell lysis using guanidinium thiocyanate (which denatures proteins) and β-mercaptoethanol (reducing agent) followed by an isopropanol precipitation [12].

The extraction methods described above have all been successfully applied to forensic samples and can accommodate a wide range of tissues and body fluids that would be frequently encountered in forensic casework. The selection of the best extraction method depends largely on the needs of individual laboratories. Ideally, extraction methods for use with forensic samples should be simple, nonlabor-intensive methods that require the least amount of sample manipulation while still resulting in the recovery of highly purified DNA. Forensic molecular genetics analysis methods typically require ~1 ng of genomic DNA. Thus, the protocols described below have been developed to isolate nanogram (as opposed to microgram) quantities of nucleic acid from forensic samples.

20.1.3 In-House and Commercial Techniques for Preparing Forensic Samples

Extraction of DNA is a critical step in the forensic analysis of biological material collected during criminal investigations. If DNA of sufficient quantity and quality is not recovered, it may not be possible to establish a DNA profile of the donor using standard PCR-based assays. A majority of the methods described above require preparation of reagents and buffers and do not offer a completely standardized extraction protocol. In-house preparation of reagents and other materials can lead to batch to batch inconsistencies and can affect the efficiency of the extraction method. Standardized protocols and extensive validations are crucial to the development of optimized and reliable analysis methods. Therefore, numerous commercially available DNA isolation kits are utilized in forensic casework laboratories. Prior to being used in forensic casework, all kits must be fully validated as required by U.S. national DNA standards and validation guidelines [41]. Manufactured reagents included in these kits contain unique lot numbers and expiration dates allowing for improved quality control of reagents. Additionally, standardized protocols ensure that more consistent results can be obtained amongst various laboratories.

While commercially available extraction kits provide crime laboratories with standardized protocols and optimized reagents, they still require the physical manipulation of samples by the analyst. Therefore, a significant amount of time is still required for manual DNA extractions. In an attempt to increase efficiency of sample processing for high-throughput forensic casework and database laboratories, automated systems for DNA extraction have therefore been developed to allow for high-throughput sample preparation and processing and to reduce potential sources of error or contamination (e.g., GenoM M-48 robotic workstation and QIAsymphony SP [Qiagen, Valencia, CA], TECAN robotic workstations [Tecan, Durham, MC], and BioMek 2000 and BioMek FX laboratory automation workstations [Beckman Coulter, Brea, CA]). A majority of these systems relies on the use of silica-coated magnetic bead technology that involves cell lysis, binding of DNA to the magnetic beads to remove proteins and other inhibitors, and elution of DNA from the beads. Numerous studies have been conducted to validate these automated systems for use with forensic samples [42–50]. The ability to automate DNA extraction methods eliminates the need for manual steps subsequent to loading samples into the instrument thus reducing potential sources of human error during the extraction process. Some automated systems have enclosed sample chambers and UV sterilization systems in order to further reduce the potential for contamination.

The automated systems described above were designed for high-throughput sample processing. However, smaller laboratories may only require low or moderate throughput that may not warrant the purchase and implementation of the more sophisticated and costly high-throughput systems described above. A few smaller automated instruments, including the Qiagen BioRobot EZ1, the Qiagen QIAcube, and the Promega Maxwell 16 personal automation system, Madison, WI, have recently been designed that process up to 6, 10, or 16 samples at a time, respectively [51–53]. While the Qiagen BioRobot EZ1 and the Promega Maxwell 16 system depend on the silica-based magnetic bead method for DNA isolation, the Qiagen QIAcube utilizes silica-based spin columns to permit that company's extensive range of DNA and RNA isolation and purification kits to be automated.

The implementation of an automated system requires extensive validation and training measures to be undertaken as well as a significant cost to be incurred. This may not be feasible for some operational crime laboratories. As a result, automated systems are not yet used routinely for processing forensic casework samples in most laboratories. The significant advantages of these automated systems combined with continual research efforts and advancements will ensure the gradual replacement of manual isolation methods by automated systems.

20.2 METHODS

The following is a list of general equipment and supplies that are required for most extractions. Specific requirements for each particular method are provided in subsequent sections.

- Pipettes (0.5–10, 2–20, 20–200, and 100–1000 μL)
- Microcentrifuge tubes (0.5 and 1.5 mL)
- Microcentrifuge tube racks
- Extraction tubes (1.5 mL)
- Spin ease baskets
- Disposable transfer pipettes
- Water bath
- Heat blocks
- Microcentrifuge
- Vortex
- Vacuum source
- Refrigerators and freezers
- Gloves
- 15 and 50 mL conical tubes
- Tweezers and scissors
- Stir plates
- Stir bars
- pH meter
- Autoclave

20.2.1   DNA Extraction—Standard Organic

Preparation of reagents:

1. Prepare DNA extraction buffer (500 mL):
   a.   Combine the following reagents: 10 mL 5 M NaCl, 5 mL 1 M Tris–HCl pH 8.0, 25 mL 0.5 M EDTA pH 8.0, and 2.5 g SDS
   b.   Bring volume to 500 mL with deionized water
   c.   Stir until all reagents are dissolved
2. Prepare TE^{-4} (500 mL):
   a.   Dissolve 0.605 g Tris base in 400 mL of deionized water
   b.   Adjust pH to 7.5
   c.   Add 0.0185 g EDTA disodium salt
   d.   Recheck pH and adjust to 7.5 if needed
   e.   Bring final volume to 500 mL with deionized water
   f.   Autoclave

20.2.1.1   Ethanol Precipitation

Reagents and materials:

- 1.5 mL extraction tubes
- Spin baskets
- 1.5 mL microcentrifuge tubes
- Disposable transfer pipettes
- Pipettes and tips
- DNA extraction buffer
- Proteinase K (20 mg/mL)
- 0.39 M DTT
- TE^{-4}
- Ethanol, 70%
- Ethanol, 100%
- Water bath
- Vortex
- Centrifuge
- Freezer (−20°C)
- Vacuum centrifuge

Procedure:

1. Remove the swab from wooden stick or cut a portion of a dried stain and place sample in a 1.5 mL extraction tube.
2. To each sample, add the following:
   a.   400 µL DNA extraction buffer
   b.   13 µL proteinase K (20 mg/mL)
   c.   For semen containing samples only, 40 µL DTT (10% of extraction volume)
3. Mix samples on a vortex for ~2 s.
4. Incubate samples in a 56°C water bath overnight.
5. Remove the swab/stain pieces and place in a spin basket. Place the basket back into the original tube.
6. Centrifuge samples at 14,000 rpm (16,000 g) for 5 min. Discard basket.

7. Add 400 μL of phenol/chloroform/isoamyl alcohol (amount equal to the extract volume). Mix by inversion. Do not vortex.
8. Centrifuge samples at 14,000 rpm (16,000 g) for 5 min to separate the phases. The organic material will be trapped in the lower nonpolar layer, and the polar aqueous phase (top layer) will contain the DNA.
9. Carefully remove the top aqueous layer and transfer to a new 1.5 m tube. Be careful not to disturb the interface.
10. Add 1 mL cold absolute ethanol (100%) to transferred aqueous layer. Mix by inversion.
11. Place samples at −20°C for at least 1 h to precipitate DNA (may proceed overnight).
12. Centrifuge samples at 14,000 rpm (16,000 g) for 15 min to pellet the DNA.
13. Remove the ethanol with a disposable transfer pipette without disturbing the pellet.
14. Wash the pellet with 1 mL of room temperature 70% ethanol.
15. Centrifuge samples at 14,000 rpm (16,000 g) for 5 min.
16. Remove the ethanol with a pipette.
17. Steps 14–16 can be repeated for a total of two washes if desired.
18. Dry the pellet using a vacuum centrifuge for 10–15 min.
19. Add 100 μL (or desired volume) of TE^{-4} to each sample.
20. Place samples in a 56°C water bath overnight to resolubilize the DNA.

20.2.1.2 Microcon Purification (An Alternative to Ethanol Precipitation)

Reagents and materials:

- 1.5 mL extraction tubes
- Spin baskets
- 1.5 mL microcentrifuge tubes
- Disposable transfer pipettes
- Pipettes and tips
- Microcon YM-100 (Millipore, Bedford, Massachusetts)
- DNA extraction buffer
- Proteinase K (20 mg/mL)
- 0.39 M DTT
- TE^{-4}
- Water bath
- Vortex
- Centrifuge

Procedure:

1. Remove the swab from wooden stick or cut a portion of a dried stain and place sample in a 1.5 mL extraction tube.
2. To each sample, add the following:
   a. 400 μL DNA extraction buffer
   b. 13 μL proteinase K (20 mg/mL)
   c. For semen containing samples only, 40 μL DTT (10% of extraction volume)
3. Mix samples on a vortex for ~2 s.
4. Incubate samples in a 56°C water bath overnight.
5. Remove the swab/stain pieces and place in a spin basket. Place the basket back into the original tube.
6. Centrifuge samples at 14,000 rpm (16,000 g) for 5 min. Discard basket.
7. Add 400 μL of phenol/chloroform/isoamyl alcohol (amount equal to the extract volume). Mix by inversion. Do not vortex.

8. Centrifuge samples at 14,000 rpm (16,000 g) for 5 min to separate the phases. The organic material will be trapped in the lower layer, and the aqueous phase (top layer) will contain the DNA.
9. Carefully remove the top (aqueous layer) of the extract and transfer to a Microcon (YM-100) sample reservoir. Be careful not to disturb the interface as it may contain proteins.
10. Centrifuge the samples at 2500 rpm (550 g) until the fluid is filtered through the membrane, leaving a small volume. Be careful not to overspin or the membrane will break.
11. Add an equal volume of sterile water or TE^{-4} pH 7.5 to wash the extract.
12. Centrifuge samples at 2500 rpm (550 g) until the fluid is filtered through the membrane, leaving a small volume.
13. Repeat Steps 11 and 12 for a total of two washes.
14. Invert the sample reservoir into a new collection tube.
15. Centrifuge samples at 2500 rpm (550 g) for 3 min. Discard sample reservoir.

20.2.2   DNA Extraction—Organic Differential (Separation of Sperm/Nonsperm)

20.2.2.1   Ethanol Precipitation

Reagents and materials:

- 1.5 mL extraction tubes
- Spin baskets
- 1.5 mL microcentrifuge tubes
- Disposable transfer pipettes
- Pipettes and tips
- DNA extraction buffer (Section 20.2.1)
- Proteinase K (20 mg/mL)
- 0.39 M DTT
- TE^{-4} (Section 20.3.1)
- Ethanol, 70%
- Ethanol, 100%
- Water bath
- Vortex
- Centrifuge
- Freezer (−20°C)
- Vacuum centrifuge

Procedure:

1. Remove the swab from wooden stick or cut a portion of a dried stain and place sample in a 1.5 mL extraction tube.
2. To each sample, add the following:
   a.   400 μL DNA extraction buffer
   b.   13 μL proteinase K (20 mg/mL)
3. Mix samples on a vortex for ~2 s.
4. Incubate samples in a 37°C water bath overnight to lyse the nonsperm cells.
5. Remove swab/stain pieces and place into spin basket. Place basket back into original tube.
6. Centrifuge at 14,000 rpm (16,000 g) for 5 min. This step will remove any remaining liquid from the substrate and will also serve to pellet the nonlysed sperm cells contained in the sample.

7. Carefully remove the supernatant, which contains the nonsperm fraction, and transfer to a new 1.5 mL microcentrifuge tube. Save this portion for Step 13.
8. Wash the sperm pellet with 1 mL DNA extraction buffer. Centrifuge samples at 14,000 rpm (16,000 g) for 5 min. Remove supernatant and discard.
9. Repeat Step 8 for a total of two washes.
10. To the sperm pellet, add the following:
    a. 400 µL DNA extraction buffer
    b. 13 µL proteinase K
    c. 40 µL DTT (10% of the extraction volume)
11. Incubate samples in a 56°C water bath for at least 1 h to lyse sperm cells (may proceed overnight).
12. After incubation, continue through the protocol with both the sperm and nonsperm fractions.
13. To the crude extracts, add 400 µL of phenol/chloroform/isoamyl alcohol (amount equal to the extract volume). Mix thoroughly by inversion. Do not vortex.
14. Centrifuge the samples at 14,000 rpm (16,000 g) for 5 min to separate the phases. The organic material will be trapped in the lower layer, and the aqueous phase (top layer) will contain the DNA.
15. Carefully remove the top aqueous layer and transfer to a new 1.5 mL tube. Be careful not to disturb the interface.
16. Add 1 mL cold absolute ethanol (100%) to transferred aqueous layer. Mix by inversion.
17. Place samples at −20°C for at least 1 h to precipitate DNA (may proceed overnight).
18. Centrifuge samples at 14,000 rpm (16,000 g) for 15 min to pellet the DNA.
19. Remove the ethanol with a disposable transfer pipette without disturbing the pellet.
20. Wash the pellet with 1 mL of room temperature 70% ethanol.
21. Centrifuge samples at 14,000 rpm (16,000 g) for 5 min.
22. Remove the ethanol with a pipette.
23. Steps 14–16 can be repeated for a total of two washes if desired.
24. Dry the pellet using a vacuum centrifuge for 10–15 min.
25. Add 100 µL (or desired volume) of TE^{-4} to each sample.
26. Place samples in a 56°C water bath overnight to resolubilize the DNA.

20.2.2.2 Microcon Purification (An Alternative to Ethanol Precipitation)

Reagents and materials:

- 1.5 mL extraction tubes
- Spin baskets
- 1.5 mL microcentrifuge tubes
- Pipettes and tips
- DNA extraction buffer (Section 20.2.1)
- Proteinase K (20 mg/mL)
- 0.39 M DTT
- TE^{-4} (Section 20.2.1)
- Sterile water
- Microcon YM-100
- Ethanol, 70%
- Ethanol, 100%
- Water bath
- Vortex
- Centrifuge

Procedure:

1. Remove the swab from wooden stick or cut a portion of a dried stain and place sample in a 1.5 mL extraction tube.
2. To each sample, add the following:
   a. 400 μL DNA extraction buffer
   b. 13 μL proteinase K (20 mg/mL)
3. Mix samples on a vortex for ~2 s.
4. Incubate samples in a 37°C water bath overnight to lyse the nonsperm cells.
5. Remove swab/stain pieces and place into spin basket. Place basket back into original tube.
6. Centrifuge at 14,000 rpm (16,000 *g*) for 5 min. This step will remove any remaining liquid from the substrate and will also serve to pellet the nonlysed sperm cells contained in the sample.
7. Carefully remove the supernatant, which contains the nonsperm fraction, and transfer to a new 1.5 mL microcentrifuge tube. Save this portion for Step 13.
8. Wash the sperm pellet with 1 mL DNA extraction buffer. Centrifuge samples at 14,000 rpm (16,000 *g*) for 5 min. Remove supernatant and discard.
9. Repeat Step 8 for a total of two washes.
10. To the sperm pellet, add the following:
    a. 400 μL DNA extraction buffer
    b. 13 μL proteinase K
    c. 40 μL DTT (10% of the extraction volume)
11. Incubate samples in a 56°C water bath for at least 1 h to lyse sperm cells (may proceed overnight).
12. After incubation, continue through the protocol with both the sperm and nonsperm fractions.
13. To the crude extracts, add 400 μL of phenol/chloroform/isoamyl alcohol (amount equal to the extract volume). Mix thoroughly by inversion. Do not vortex.
14. Centrifuge the samples at 14,000 rpm (16,000 *g*) for 5 min to separate the phases. The organic material will be trapped in the lower layer, and the aqueous phase (top layer) will contain the DNA.
15. Carefully remove the top (aqueous layer) of the extract and transfer to a Microcon (YM-100) sample reservoir. Be careful not to disturb the interface as it may contain proteins.
16. Centrifuge the samples at 2500 rpm (550 *g*) until the fluid is filtered through the membrane, leaving a small volume. Be careful not to overspin or the membrane will break.
17. Add an equal volume of sterile water or TE^{-4} pH 7.5 to wash the extract.
18. Centrifuge samples at 2500 rpm (550 *g*) until the fluid is filtered through the membrane, leaving a small volume.
19. Repeat Steps 11 and 12 for a total of two washes.
20. Invert the sample reservoir into a new collection tube.
21. Centrifuge samples at 2500 rpm (550 *g*) for 3 min. Discard sample reservoir.

20.2.3 DNA Extraction—Organic Special Sample Types

20.2.3.1 Hair

Reagents and materials:

- 1.5 mL extraction tubes
- Spin baskets
- DNA extraction buffer (Section 20.2.1)

- TE^{-4} pH 7.5 (Section 20.2.1)
- Sterile water
- Phenol/chloroform/isoamyl alcohol (25:24:1)
- Proteinase K (20 mg/mL)
- Microcon YM-100
- Vortex
- Centrifuge
- Water bath
- Pipettes and tips

Procedure:

1. Take 3–5 body hairs (with follicle and root sheath tissue present) and place into a 1.5 mL extraction tube.
2. To each sample, add the following:
   a. 200 μL DNA extraction buffer
   b. 7.5 μL proteinase K
3. Mix the samples on a vortex for ~2 s. Make sure hairs are still in solution.
4. Incubate samples in a 56°C water bath overnight.
5. Remove the hairs and place them in a spin basket. Place the basket into the original tube.
6. Centrifuge samples at 14,000 rpm (16,000 *g*) for 3 min to remove any liquid remaining in the substrate. Discard the basket.
7. To the extract, add 200 μL of phenol/chloroform/isoamyl alcohol (amount equal to the extract volume). Mix thoroughly by inversion. Do not vortex.
8. Centrifuge the samples at 14,000 rpm (16,000 *g*) for 5 min to separate the phases. The organic material will be trapped in the lower layer, and the aqueous phase containing the DNA will be in the upper layer.
9. Carefully remove the top (aqueous) layer of the extract and add it to a Microcon (YM-100) sample reservoir. Do not disturb the interface as it may contain proteins.
10. Centrifuge the Microcon device at 2500 rpm (550 *g*) until the fluid is filtered through the membrane, leaving a small volume. Be careful not to overspin or the membrane will break.
11. Add an equal volume of sterile water or TE^{-4} pH 7.5 to wash the extract.
12. Centrifuge samples at 2500 rpm (550 *g*) until most of the fluid is filtered.
13. Repeat Steps 11 and 12 for a second wash.
14. Invert the sample reservoir into a new collection tube.
15. Centrifuge the samples at 2500 rpm (550 *g*) for 3 min to collect sample.
16. Discard sample reservoir.

20.2.3.2 Teeth

Reagents and materials:

- Dremel tool (with separating disk, emery disk)
- Tissue pulverizer
- 1.5 mL extraction tube
- Alcohol swabs
- Endodontic files
- Spoon excavators
- DNA extraction buffer (Section 20.2.1)
- TE^{-4} pH 7.5 (Section 20.2.1)

- Phenol/chloroform/isoamyl alcohol
- Proteinase K (20 mg/mL)
- Microcon YM-100 concentrators
- Pipettes and tips
- Centrifuge
- Vortex

Procedure:

1. Clean the outer surface of the tooth with an alcohol swab.
2. Using a Dremel tool with a new separating disk, cut the tooth horizontally at the cementoe-namel junction.
3. If soft pulp tissue is available:
   a. Remove the pulp tissue with endodontic files and spoon excavators (or, if not available, fine forceps).
   b. Place the tissue into a 1.5 mL extraction tube.
4. If no soft pulp tissue is available:
   a. Use the Dremel tool with the separating disk attached to cut a thin, horizontal section of tooth slightly apical to the original cut.
   b. Use the Dremel tool with an emery disk attached to sand the outer surface of the tooth section so that it appears free of dirt and debris.
   c. Place the sanded tooth section in a tissue pulverizer and insert pestle. Using a hammer, strike the pestle repetitively until the tooth is pulverized.
   d. Place 0.02–0.05 g of pulverized tooth into a 1.5 mL extraction tube.
5. Add 400 μL DNA extraction buffer and 13 μL proteinase K.
6. Incubate the sample in a 56°C water bath overnight.
7. In a fume hood, add 400 μL of phenol/chloroform/isoamyl alcohol to each sample. Shake until a milky appearance is observed.
8. Centrifuge samples at 14,000 rpm (16,000 g) for 5 min to separate the phases.
9. Carefully remove the top aqueous layer to the Microcon YM-100 sample reservoir.
10. Centrifuge the Microcon device at 2500 rpm (550 g) until the fluid is filtered through the membrane, leaving a small volume. Be careful not to overspin as the membrane will break.
11. Add an equal volume of TE^{-4} to wash the extract.
12. Centrifuge the Microcon device at 2500 rpm (550 g) until most of the fluid is filtered.
13. Repeat Steps 11 and 12 for a total of two washes.
14. Invert the sample reservoir into a collection tube.
15. Centrifuge the samples at 2500 rpm (550 g) for 3 min.

20.2.3.3   Bone

Reagents and materials:

- Scalpel
- Water bath
- Deionized water
- MilliQ water
- CertiPrep freezer/mill
- Craftsman high-speed rotary tool (small drum sander attachment, mandrel attachment)
- Vacuum centrifuge
- 1% sodium hypochlorite solution

- Ethanol, 70%
- Ethanol, 100%
- 0.5 M EDTA pH 8.0
- 2 mL microcentrifuge tubes
- 1.5 mL microcentrifuge tubes
- DNA extraction buffer (Section 20.2.1)
- TE^{-4} pH 7.5 (Section 20.2.1)
- Proteinase K (20 mg/mL)
- Phenol/chloroform/isoamyl alcohol (25:24:1)

Bone preparation:

1. Record general observations pertaining to the sample such as condition, color, size, weight, or distinguishing features since the morphological features of the bone may be evidentiary probative. Photographs should be taken if entire sample will be altered during sampling process.
2. Place absorbent mat on work surface of clean fume hood. Use a scalpel to scrape off adhering tissue and debris and then rinse the bone in deionized water. If tissue is not easily removed, soak sample in deionized water for a few hours to soften tissue and then continue mechanically removing tissue with a scalpel. Allow bones to air-dry. This may take up to a week for samples with high amounts of fat. Samples must be dry to prevent caking of bone powder after grinding.
3. Sand ~1–2 mm of the outer surfaces of each bone sample using a small drum sander attachment on a high-speed rotary tool. Pliers can be used to hold the sample while sanding (sterilize by rinsing in bleach and then ethanol and by flaming). A new sanding band should be used for each sample.
4. Wipe bone surfaces with 1% sodium hypochlorite solution.
5. Cut into small samples using a high-speed rotary tool with a mandrel attachment and a cut-off wheel. Pliers can be used to hold sample while cutting. Care should be taken when cutting small fragments as these are harder to hold and manipulate.
6. Weigh the cut sample pieces.

Bone cleaning:

1. Submerge samples in 1% sodium hypochlorite solution. Mix for 30 s and then decant.
2. Submerge samples in sterile milliQ. Mix for 30 s and then decant.
3. Repeat Step 2 two additional times for a total of three washes.
4. Transfer sample to a clean weight boat and allow to air-dry.
5. A control swab of the surface can be taken to make sure contaminating surface DNA was removed.

Bone grinding:

1. Place samples into a new or thoroughly cleaned polycarbonate grinding tube (one end sealed with steel end plug). Clean the impactor with bleach and ethanol and then add it to the tube. Swab the surface of the tube, impactor, and end plugs as controls.
2. Close the vial by adding the other end plug.
3. Fill tub of the freezer/mill with liquid nitrogen (use cryogenic gloves and face shield).
4. Program the freezer/mill: 3 cycles, a 10 min precool time, 2 min run time, 2 min cool time, and a rate of 10 cps.
5. Load the sample vial into the freezer/mill. Run the program.

6. Remove sample vial (use cryogenic gloves) and bring to near room temperature before opening.
7. Shake contents into a weigh boat and then divide the powder into 2 mL microcentrifuge tubes according to sample sizes desired for extraction (~0.1–0.25 g).
8. Store samples at 4°C if used within 3 weeks and −20°C for longer storage times.

Extraction:

1. Add 1.6 mL of 0.5 M EDTA pH 8.0 to tubes containing ~0.1 g of pulverized bone sample and 1.8 mL to tubes containing ~0.25 g of pulverized bone sample.
2. Vortex to suspend bone powder.
3. Incubate tubes on a platform shaker (90 rpm) at room temperature for 8–24 h.
4. Centrifuge tubes for 1 min at 7000 rpm (5200 g) to pellet bone powder. Discard supernatant.
5. Wash pellet with 1 mL deionized water. Vortex briefly. Centrifuge for 1 min at 7000 rpm (5200 g). Discard supernatant.
6. Repeat wash step two additional times for a total of three washes.
7. For samples containing ~0.1 g of pulverized bone, add 400 µL DNA extraction buffer and 13 µL proteinase K. For samples containing ~0.25 g of pulverized bone, add 600 µL of DNA extraction buffer and 19.5 µL of proteinase K.
8. Vortex and pulse spin tubes. Incubate samples overnight in a 56°C water bath.
9. Add a volume equal to the crude extract of phenol/chloroform/isoamyl alcohol (25:24:1) and mix by inversion.
10. Centrifuge at 14,000 rpm (16,000 g) for 5 min to separate the phases.
11. Transfer the top aqueous layer containing the DNA to a new 1.5 mL microcentrifuge tube.
12. Add 1 mL cold absolute ethanol and mix by inversion.
13. Incubate at −20°C for at least 1 h (can also proceed overnight).
14. Centrifuge samples at 14,000 rpm (16,000 g) for 15 min to pellet the DNA.
15. Remove ethanol with a pipette.
16. Wash the pellet with 1 mL room temperature 70% ethanol. Centrifuge samples at 14,000 rpm (16,000 g) for 5 min.
17. Remove ethanol with a pipette.
18. Dry the pellet using vacuum centrifugation for 10–15 min.
19. Add 100 µL of TE^{-4} pH 7.5 to each sample.
20. Place samples in a 56°C water bath overnight to resolubilize the DNA.

20.2.4 DNA Extraction—Ion Exchange (Chelex)

20.2.4.1 Nonsemen-Containing Samples

Reagents:

- 5% Chelex (heat-sterilized)
- 1.5 mL extraction tubes
- Sterile deionized water
- Water bath or heat blocks

Procedure:

1. Label 1.5 mL extraction tubes with appropriate information.
2. Cut swab or stain piece and place in extraction tube.

3. To each sample add 1 mL of sterile deionized water. Mix tubes by inversion or by vortexing briefly.
4. Incubate at room temperature for 15–30 min. Mix occasionally by inversion or vortexing.
5. Centrifuge for 2–3 min at 14,000 rpm (16,000 g).
6. Carefully remove supernatant and discard.
7. Add 200 μL of 5% Chelex to each sample. Vortex samples for 10 s.
8. Incubate samples at 56°C for 30 min. Vortex samples for 10 s.
9. Incubate samples at 100°C for 8 min.
10. Centrifuge sample at 14,000 rpm (16,000 g) for 3 min.
11. Store at 4°C.

20.2.4.2   Differential (Semen/Epithelial Cell Mixtures)

Reagents:

- 5% Chelex (heat-sterilized)
- 20% Chelex (heat-sterilized)
- Phosphate buffered saline (PBS)
- 1 M DTT
- 1.5 mL extraction tubes
- Spin basket
- Water bath or heat blocks
- Orbital shaker
- Centrifuge
- Vortex
- Proteinase K (20 mg/mL)
- Sperm wash buffer
- Sterile water

Preparation of reagents:

1. To prepare sperm wash buffer (2 L), combine the following reagents:
   a. 2.4 g Tris
   b. 40 mL 0.5 M EDTA pH 8.0
   c. 5.8 g sodium chloride
   d. 200 mL SDS (20%)
   e. 6 mL HCl (50%)
   f. 1.75 L deionized water

Procedure:

1. Remove cotton from swab or cut portion of stain and place in 1.5 mL extraction tube.
2. Add 500 μL of PBS to each tube.
3. Vortex on low speed and spin in a microcentrifuge tube for 2 s to force sample into extraction fluid.
4. Place on an orbital shaker for 30–60 min at room temperature. Vortex intermittently during the incubation time.
5. Remove the sample and place in a spin basket. Insert basket back into original tube.
6. Centrifuge at 14,000 rpm (16,000 g) for 5 min.
7. Contents of spin basket can be transferred to a new 1.5 mL tube and saved for future analysis if necessary.

8. Remove all but 20 μL of the supernatant fluid from the tube and discard. Do not disturb the pellet.
9. Resuspend the pellet by stirring with a pipette.
10. Add 150 μL of sterile water to each tube.
11. Add 1 μL proteinase K (20 mg/mL). Vortex briefly to resuspend pellet.
12. Incubate samples in 56°C water bath for 45 min to lyse epithelial cells.
13. Centrifuge samples for 5 min at 14,000 rpm (16,000 g).
14. Transfer 150 μL of supernatant to a new 1.5 mL tube (epithelial fraction). Add 50 μL of 20% Chelex to each "epithelial" fraction. Store at 4°C until Step 25.
15. Wash the sperm pellet with 500 μL sperm wash buffer. Vortex briefly.
16. Centrifuge samples at 14,000 rpm (16,000 g) for 5 min.
17. Remove and discard the supernatant, being careful not to disturb the pellet.
18. Repeat Steps 15–17 two additional times for a total of three washes.
19. Wash the cell pellet with 500 μL of sterile water. Vortex briefly.
20. Centrifuge samples at 14,000 rpm (16,000 g) for 5 min.
21. Remove and discard all but 20 μL of the supernatant, being careful not to disturb the pellet.
22. Resuspend the pellet by stirring with a pipette tip (sperm fraction).
23. Add 150 μL of 5% Chelex to each "sperm" fraction tube. Mix with a pipette tip.
24. Add 1 μL proteinase K (20 mg/mL) and 7 μL of 1 M DTT to each "sperm" fraction tube. Mix gently.
25. Mix epithelial and sperm fraction tubes by inversion or mixing briefly.
26. Incubate at 56°C for 60 min.
27. Vortex for 5–10 s.
28. Incubate at 100°C for 8 min. Place a weight on top of the tubes to prevent them from popping open.
29. Vortex for 5–10 s.
30. Centrifuge for 2–3 min at 14,000 rpm (16,000 g).
31. Store samples at 4°C.

20.2.5 RNA Extraction

Reagents and materials:

- Acid-phenol/chloroform (5:1)
- β-Mercaptoethanol
- Decontaminant (RNase zap spray and RNase zap wipes, Ambion, Austin, TX)
- Denaturing solution (4 M guanidine isothiocyanate, 0.02 M sodium citrate, and 0.5% sarkosyl)
- DNase I and buffer (Ambion)
- Ethanol (200 proof)
- Glycogen carrier (GlycoBlue, Ambion)
- Isopropanol
- Sodium acetate pH 4.0
- RNA secure resuspension solution (Ambion)
- Water, diethylpyrocarbonate (DEPC)-treated
- Water, nuclease-free
- 1.5 mL microcentrifuge tubes
- 1.5 mL extraction tubes
- Spin baskets
- Tweezers and scissors

- Water bath and heat block
- 50 mL conical tubes
- Vortex
- Centrifuge

Preparation of reagents:

1. Spray RNase zap on workspace and wipe off with a dry KimWipe. Wipe pipettes and equipment exteriors with an RNase zap wipe. Wipe the same surfaces with a fresh KimWipe.
2. Prepare denaturing solution (100 mL):
   a. Weigh out indicated quantity of each reagent: 47.264 g guanidine isothiocyanate, 0.5882 g sodium citrate isothiocyanate, and 0.5 g sarkosyl
   b. Add all reagents to a glass bottle
   c. Add nuclease-free water to the 80 mL mark
   d. Dissolve reagents
   e. Bring final volume to 100 mL using nuclease-free water
   f. Aliquot 25 mL into sterile 50 mL conical tubes
   g. Store at room temperature
3. Prepare sodium acetate (100 mL):
   a. Weigh out indicated quantity of each reagent: 27.216 g sodium acetate trihydrate or 16.406 g sodium acetate anhydrous
   b. Add reagent to glass bottle
   c. Add nuclease-free water to the 75 mL mark
   d. Using a calibrated pH meter, adjust the pH of the solution to 4.0 with HCl or NaOH
   e. Bring final volume to 100 mL with nuclease-free water
   f. Autoclave
   g. Aliquot 25 mL into sterile 50 mL conical tubes
   h. Store at room temperature
4. Remove 500 µL of β-mercaptoethanol from the stock bottle. Dispense into a sterile 1.5 mL microcentrifuge tube. Store at 4°C.
5. Remove 40 mL of isopropanol from the stock bottle. Dispense into a sterile 50 mL conical tube. Store at room temperature.
6. Prepare ethanol wash (75% ethanol and 25% DEPC-treated water):
   a. Add 30 mL molecular biology grade ethanol to a 50 mL conical tube
   b. Add 10 mL DEPC-treated water
   c. Close cap of tube and invert to mix
   d. Store at room temperature

Procedure:

1. Unless otherwise indicated, each step of the extraction procedure should be completed at a minimum in a preamplification hood or ideally in a hood designated for RNA procedures in the preamplification area of the laboratory.
2. Remove the appropriate number of nuclease-free extraction tubes from their container and place them on a rack in the RNA hood. Label tubes appropriately.
3. Prepare a master mix of denaturing solution and β-mercaptoethanol:
   a. 500 µL denaturing solution per sample
   b. 3.6 µL β-mercaptoethanol per sample
4. Place denaturing solution mixture into a 56°C water bath for 10 min.
5. Prepare samples while denaturing solution is heating:

    a. Place scissors and tweezers in 70% ethanol for at least 1 min
    b. Wipe scissors and tweezers with KimWipes to dry
    c. Cut stain or swab piece and place in nuclease-free 1.5 mL extraction tube
    d. Clean scissors and tweezers in between each sample

6. Add ~504 µL preheated denaturing solution (with β-mercaptoethanol) into each tube.
7. Place tubes in a 56°C water bath for 30 min.
8. Remove all pieces of stain/swab using tweezers and place into a spin basket. Place basket into original extraction tube.
9. Centrifuge the tubes for 10 min at 10,000 rpm (8,100 g) at room temperature.
10. Remove the spin basket with the stain/swab material and discard.
11. Add 50 µL of 2 M sodium acetate to the extraction tube.
12. Add 600 µL acid-phenol/chloroform (5:1) pH 4.5 (from bottom phase) to each tube. Cap tightly and vortex vigorously for 1 min to disrupt protein–nucleic acid interactions.
13. Place tubes in a rack and place in a 4°C refrigerator for 30 min–1 h to ease separation of phases.
14. Centrifuge samples for 20 min at 14,000 rpm (16,000 g). This step may be completed at 4°C or room temperature.
15. Prepare the appropriate number of sterile 1.5 mL microcentrifuge tubes and label them with the appropriate information.
16. Remove the upper (aqueous) phase of the samples into the new 1.5 mL microcentrifuge tubes. Care must be taken not to remove the interface or lower organic layer of the sample.
17. Discard the tube with interface and lower phase in biohazard waste.
18. Add 2 µL (30 µg) GlycoBlue glycogen carrier to the aqueous layer.
19. Add 500 µL of isopropanol. Close the cap of the tube. Mix by inversion. Do not vortex.
20. Place tubes in freezer (−20°C) for at least 2 h (alternatively can proceed overnight).
21. Centrifuge tubes for 20 min at 14,000 rpm (16,000 g) orienting the tube in such a way that the hinge is pointed out and away from the rotor. This will allow the RNA pellet to form along the hinge side of the tube.
22. Remove the supernatant. The RNA pellet may or may not be visible. Care should be taken not to disturb the pellet.
23. Add 1000 µL of 75% ethanol and 25% DEPC-treated water to the RNA pellet. Close cap of tube. Centrifuge the sample for 10 min at 14,000 rpm (16,000 g).
24. Remove supernatant completely.
25. Wipe the inside chamber and lid of a vacuum centrifuge with an RNase wipe. Wipe the same surfaces with a KimWipe wetted with deionized water. Immediately dry all surfaces with a dry KimWipe.
26. Place tubes (cap open) in the vacuum centrifuge. Dry samples for 5–7 min or longer if liquid is still visible. Care should be taken to not overdry the samples.
27. Preheat RNA secure resuspension solution in a 60°C heat block for 10 min (20 µL per sample).
28. Remove samples from vacuum centrifuge. Close all tubes and move tubes to the RNA hood.
29. Add 20 µL of preheated RNA secure resuspension solution to each tube.
30. Place tubes in a 60°C heat block for 10 min to ease resuspension.
31. Remove samples from heat block. Vortex and centrifuge samples briefly.
32. DNase I digestion:
    a. Add 2.55 µL of 10x DNase I buffer to each sample
    b. Add 3 µL RNase-free DNase I to each sample
    c. Briefly vortex and centrifuge samples (~5 s)
    d. Incubate samples in a 37°C heat block or water bath for 1 h
    e. Incubate samples in a 75°C heat block for 10 min to inactivate the DNase I enzyme
33. Allow samples to cool to room temperature. Store at −20°C until use.

20.2.6 Simultaneous DNA/RNA Extraction Procedure

Reagents and Materials:

- Acid-phenol/Chloroform (5:1)
- Phenol/chloroform/isoamyl alcohol (25:24:1)
- β-Mercaptoethanol
- Decontaminant (RNase zap spray and RNase zap wipes, Ambion)
- Denaturing solution (Section 20.3.5)
- DNase I and buffer (Ambion)
- Ethanol (200 proof)
- Glycogen carrier (GlycoBlue, Ambion)
- Isopropanol
- Sodium acetate pH 4.0
- RNA secure resuspension solution (Ambion)
- 75% ethanol/25% DEPC-treated water (Section 20.2.5)
- Water, nuclease-free
- 1.5 mL microcentrifuge tubes
- 1.5 mL extraction tubes
- Spin baskets
- Tweezers and scissors
- Water bath and heat block
- 50 mL conical tubes
- Vortex
- Centrifuge

Procedure:

1. Remove the cotton from a swab or cut portions of a dried stain and place into a 1.5 mL extraction tube.
2. To each sample, add the following:
   a. 500 μL nuclease-free DNA extraction buffer
   b. 16.25 μL proteinase K (20 mg/mL)
   c. For semen containing samples only, 50 μL DTT (0.39 M)
3. Vortex the samples and pulse spin.
4. Incubate samples in a 56°C water bath for at least 1 h.
5. Remove the swab/stain pieces and place them in a spin basket. Place the basket into the original tube.
6. Centrifuge at 14,000 rpm (16,000 g) for 5 min to remove any remaining liquid from the sample. Remove and discard the basket.
7. To the crude extract, add 50 μL of 2 M sodium acetate. Invert to mix.
8. Add 600 μL of acid-phenol/chloroform (5:1) solution. Vortex vigorously and mix by inversion. Do not centrifuge.
9. Incubate at 4°C for 20 min–1 h, until two layers are visible.
10. Centrifuge samples at 14,000 rpm (16,000 g) for 20 min to separate the two phases.
11. In the good, pipet the top layer (~400 μL) into a sterile 1.5 mL microcentrifuge tube (labeled "DNA"). Be careful not to disrupt the interface as it may contain proteins. Proceed to "DNA extraction" procedure below.
12. Mix the sample by pipetting up and down. Transfer ~200 μL to a new 1.5 mL microcentrifuge tube (labeled "RNA"). Proceed to "RNA extraction" procedure below.

RNA extraction:

1. To each sample, add the following:
   a. 2 μL GlycoBlue glycogen carrier (vortex each sample after adding)
   b. 250 μL isopropanol
2. Mix by inversion.
3. Incubate samples at −20°C for at least 1 h (may proceed overnight).
4. Centrifuge samples at 14,000 rpm (16,000 *g*) for 20 min to pellet the RNA.
5. Remove supernatant and discard.
6. Wash with 500 μL 75% ethanol/25% DEPC-treated water.
7. Centrifuge samples at 14,000 rpm (16,000 *g*) for 10 min.
8. Remove supernatant and discard.
9. Dry samples in a vacuum centrifuge for ~3–5 min.
10. Preheat RNA secure resuspension solution in a 60°C heat block for 5 min.
11. Resuspend RNA in 17 μL of preheated RNA secure resuspension solution.
12. Incubate samples in a 60°C heat block for 10 min.
13. Vortex vigorously and pulse spin.
14. DNase I digestion:
    a. Add 2.55 μL of 10x DNase I buffer to each sample
    b. Add 3 μL RNase-free DNase I to each sample
    c. Briefly vortex and centrifuge samples (~5 s)
    d. Incubate samples in a 37°C heat block or water bath for 1 h
    e. Incubate samples in a 75°C heat block for 10 min to inactivate the DNase I enzyme
15. Store at −20°C until needed.

DNA extraction:

1. Add 500 μL of cold absolute ethanol (100%) to each sample.
2. Mix by inversion.
3. Incubate samples at −20°C for at least 1 h (may proceed overnight).
4. Centrifuge samples at 14,000 rpm (16,000 *g*) for 15 min.
5. Remove ethanol with a pipette. Do not disturb the pellet.
6. Wash the pellet with 700 μL of room temperature 70% ethanol.
7. Centrifuge samples at 14,000 rpm (16,000 *g*) for 5 min.
8. Remove the ethanol with a pipette.
9. Steps 6–8 can be repeated for an additional wash if needed.
10. Dry the pellet using a vacuum centrifuge for ~10–15 min.
11. Add 50 μL of TE^{-4} pH 7.5 to each sample.
12. Incubate samples in a 56°C water bath for at least 45 min.

20.2.7 COMMERCIAL KIT PROCEDURES

20.2.7.1 QIAamp DNA Micro Kit (Qiagen, Cat. No. 56304)

Description:

The QIAamp DNA mini kit utilizes silica-based columns to bind DNA, allowing purified DNA to be recovered and PCR inhibitors such as divalent cations and proteins to be removed. The kit can be used for most types of forensic samples including hair, blood swabs, saliva, and sperm.

Reagents included in kit:

• QIAamp MinElute columns
• Collection tubes (2 mL)

- Buffer ATL
- Buffer AL
- Buffer AW1 (concentrate)
- Buffer AW2 (concentrate)
- Buffer AE
- Carrier RNA
- Proteinase K

Reagents and materials to be supplied by the user:

- Ethanol (96%–100%)
- 1.5 mL microcentrifuge tubes
- Water bath or heat block
- Vortex
- DTT (for semen containing samples)

Preparation of reagents:

1. To prepare buffer AW1:
   a. Mix the bottle of buffer AW1 concentrate by shaking
   b. Add 25 mL ethanol (96%–100%) to the buffer AW1 bottle
2. To prepare buffer AW2:
   a. Mix the bottle of buffer AW2 concentrate by shaking
   b. Add 30 mL ethanol (96%–100%) to the buffer AW2 bottle

Procedure:

1. To lyse blood, saliva and semen stains, cigarette butts, envelopes, stamps and nail clippings:
   a. Remove a small portion of the sample and place into a 2 mL microcentrifuge tube
   b. Add 300 μL buffer ATL and 20 μL proteinase K
   c. For semen containing samples only, add 20 μL of 1M DTT
   d. Mix by pulse vortexing for 10 s
2. To lyse hair roots or shafts:
   a. Add 300 μL buffer ATL, 20 μL proteinase K, and 20 μL 1 M DTT to a 1.5 mL micro-centrifuge tube
   b. Remove a 0.5–1 cm piece of hair starting from the hair bulb or 0.5–1 cm pieces of the hair shaft and place in the tube
   c. Mix by pulse vortexing for 10 s
3. Incubate samples at 56°C (water bath or heat block) for at least 1 h. Vortex samples for 10 s every 10 min. (Note: Incubation time may need to be increased for difficult samples like hairs or nail clippings.)
4. Briefly centrifuge samples to remove any liquid from the lid.
5. Add 300 μL of buffer AL to each sample. Pulse-vortex for 10 s. (Note: Dissolved carrier RNA of 1 μg can be added at this step, if necessary, to enhance binding of DNA to the MinElute column.)
6. Incubate samples at 70°C (water bath or heat block) for 10 min. Vortex samples for 10 s every 3 min.
7. Centrifuge samples at 14,000 rpm (16,000 g) for 1 min.
8. Transfer the supernatant to the QIAamp MinElute column. Centrifuge samples at 8000 rpm (6000 g) for 1 min.

9. Place the column in a new 2 mL collection tube and discard the previous collection tube with the flow through.
10. Add 500 μL buffer AW1 to each sample. Centrifuge at 8000 rpm (6000 g) for 1 min.
11. Place the column in a new 2 mL collection tube and discard the previous collection tube with the flow through.
12. Add 500 μL buffer AW2 to each sample. Centrifuge at 8000 rpm (6000 g) for 1 min.
13. Place the column in a new 2 mL collection tube and discard the previous collection tube with the flow through.
14. Centrifuge samples at 14,000 rpm (16,000 g) for 3 min to dry the membrane completely.
15. Place the column into a new 1.5 mL microcentrifuge tube and discard the previous collection tube with the flow through.
16. Add 20–50 μL of buffer AE to the center of the membrane in each column.
17. Incubate samples at room temperature for 1–5 min.
18. Centrifuge samples at 14,000 rpm (16,000 g) for 1 min.
19. Discard column. Store samples at desired temperature.

20.2.7.2 DNA IQ System (Promega, Cat. No. DC6700-400 Samples, DC6701-100 Samples)

Description:

The Promega DNA IQ system utilizes a paramagnetic resin for DNA isolation. Samples are first extracted in a quick, efficient extraction step that can be applied to various types of samples (blood, buccal swabs, semen, urine, bone, hair, and tissues). The extracted DNA is then purified using the magnetic resin to remove any excess extraction reagents or cell debris. The DNA IQ system can also be automated.

Reagents included in kit:

- Resin
- Lysis buffer
- 2x wash buffer
- Elution buffer

Reagents and materials to be supplied by the user:

- 95%–100% ethanol
- Isopropyl alcohol
- 1 M DTT
- Heat blocks, water baths, or thermal cycler
- Vortex
- 1.5 mL microcentrifuge tubes
- Spin baskets
- Pipettes and tips
- MagneSphere technology magnetic separation stand (Cat. No. Z5342)
- Proteinase K (for tissue, hair, bone, and differential extractions)

Preparation of reagents:

1. To prepare the 1x wash buffer:
   a. DC6701-100 samples: Add 15 mL each of 95%–100% ethanol and isopropyl alcohol to the 2x wash buffer

   b. DC6700-400 samples: Add 35 mL each of 95%–100% ethanol and isopropyl alcohol to
       the 2x wash buffer
   c. Mix by inverting several times
  2. To prepare the lysis buffer:
   a. Determine the total volume of prepared lysis buffer to be used: liquid samples, 200 μL;
       cotton swab, 350 μL; FTA paper or cloth, 250 μL
   b. Add 1 μL of 1 M DTT for every 100 μL of lysis buffer
   c. For semen containing samples, add 6 μL of 1 M DTT for every 100 μL of lysis buffer
   d. Mix by inverting several times

Procedure—Samples on solid materials:

  1. Place the sample in a 1.5 mL extraction tube.
  2. Add 150 μL of lysis buffer to samples on FTA paper or cloth and 250 μL of lysis buffer to
     samples on cotton swabs.
   a. For small stains, a spin basket can be placed in a 1.5 mL extraction tube. The sample
       and lysis buffer can then be added to the basket.
  3. Incubate the samples at 70°C for 30 min.
  4. If the sample is not already in a spin basket at this step, transfer the lysis buffer and sample
     to a spin basket and place into original tube.
  5. Centrifuge the samples at 14,000 rpm (16,000 *g*) for 2 min. Discard the spin basket.
  6. Vortex the stock DNA IQ resin bottle for 10 s or until resin is well mixed.
  7. Add 7 μL of the resin to each sample. Vortex for 3 min.
  8. Incubate the samples at room temperature for 5 min, vortexing briefly once every min.
  9. Vortex each sample for 2 s and then place in the magnetic stand.
 10. Remove all solution from the tube without disturbing the resin pellet.
 11. Add 100 μL of lysis buffer.
 12. Remove the tube from the stand and vortex for 2 s. Return tube to the magnetic stand.
 13. Remove and discard all solution from the tube without disturbing the resin pellet.
 14. Add 100 μL of the 1x wash buffer to each sample.
 15. Remove the tube from the stand and vortex for 2 s. Return tube to the magnetic stand.
 16. Remove and discard all solution from the tube without disturbing the pellet.
 17. Repeat Steps 14–16 two additional times for a total of three washes.
 18. With the tubes still in the magnetic stain, open the lids for all samples and let the samples
     dry at room temperature for 5 min (do not exceed 20 min).
 19. Add 25–100 μL of elution buffer depending on the desired concentration.
 20. Close the lid to each tube and vortex for 2 s.
 21. Incubate samples at 65°C for 5 min.
 22. Vortex samples for 2 s and immediately place them back into the magnetic stand.
 23. Carefully transfer the solution (containing the DNA) into a new 1.5 mL microcentrifuge
     tube.

Procedure—Liquid samples:

  1. Prepare a solution of resin and lysis buffer using 7 μL of resin and 93 μL of lysis buffer per
     sample.
  2. Place up to 40 μL of the liquid sample into a 1.5 mL microcentrifuge tube.
  3. Vortex the resin/lysis buffer mixture for 3 s. Add 100 μL of the mixture to each sample.
     Vortex samples for 3 s.
  4. Incubate the samples at room temperature for 5 min, vortexing briefly once every min.
  5. Vortex each sample for 2 s and then place in the magnetic stand.

6. Remove all solution from the tube without disturbing the resin pellet.
7. Add 100 µL of lysis buffer.
8. Remove the tube from the stand and vortex for 2 s. Return tube to the magnetic stand.
9. Remove and discard all solution from the tube without disturbing the resin pellet.
10. Add 100 µL of the 1x wash buffer to each sample.
11. Remove the tube from the stand and vortex for 2 s. Return tube to the magnetic stand.
12. Remove and discard all solution from the tube without disturbing the pellet.
13. Repeat Steps 10–12 two additional times for a total of three washes.
14. With the tubes still in the magnetic stain, open the lids for all samples and let the samples dry at room temperature for 5 min (do not exceed 20 min).
15. Add 25–100 µL of elution buffer depending on the desired concentration.
16. Close the lid to each tube and vortex for 2 s.
17. Incubate samples at 65°C for 5 min.
18. Vortex samples for 2 s and immediately place them back into the magnetic stand.
19. Carefully transfer the solution (containing the DNA) into a new 1.5 mL microcentrifuge tube.

Procedure—Semen/epithelial cell mixtures:

Additional reagents and materials required:

- DNA extraction buffer (Section 20.2.1)
- Proteinase K (20 mg/mL)
- 0.39 M DTT

Procedure:

1. Remove the swab from wooden stick or cut a portion of a dried stain and place sample in a 1.5 mL extraction tube.
2. To each sample, add the following:
   a. 400 µL DNA extraction buffer
   b. 13 µL proteinase K (20 mg/mL)
3. Mix samples on a vortex for ~2 s.
4. Incubate samples in a 37°C water bath overnight to lyse the nonsperm cells.
5. Remove swab/stain pieces and place into spin basket. Place basket back into original tube.
6. Centrifuge at 14,000 rpm (16,000 g) for 5 min. This step will remove any remaining liquid from the substrate and will also serve to pellet the nonlysed sperm cells contained in the sample.
7. Carefully remove the supernatant, which contains the nonsperm fraction, and transfer to a new 1.5 mL microcentrifuge tube. Save this portion for Step 13.
8. Wash the sperm pellet with 1 mL DNA extraction buffer. Centrifuge samples at 14,000 rpm (16,000 g) for 5 min. Remove supernatant and discard.
9. Repeat Step 8 for a total of two washes.
10. Extraction of DNA from sperm pellet:
    a. Add 100 µL of prepared lysis buffer (with DTT) and 7 µL of resin to each sample.
    b. Vortex for 3 s. Incubate at room temperature for 5 min.
    c. Vortex each sample for 2 s and then place in the magnetic stand.
    d. Remove all solution from the tube without disturbing the resin pellet.
    e. Add 100 µL of lysis buffer.
    f. Remove the tube from the stand and vortex for 2 s. Return tube to the magnetic stand.
    g. Remove and discard all solution from the tube without disturbing the resin pellet.

    h. Add 100 μL of the 1x wash buffer to each sample.

    i. Remove the tube from the stand and vortex for 2 s. Return tube to the magnetic stand.

    j. Remove and discard all solution from the tube without disturbing the pellet.

    k. Repeat Steps 10–12 two additional times for a total of three washes.

    l. With the tubes still in the magnetic stain, open the lids for all samples and let the samples dry at room temperature for 5 min (do not exceed 20 min).

    m. Add 25–100 μL of elution buffer depending on the desired concentration.

    n. Close the lid to each tube and vortex for 2 s.

    o. Incubate samples at 65°C for 5 min.

    p. Vortex samples for 2 s and immediately place them back into the magnetic stand.

    q. Carefully transfer the solution (containing the DNA) into a new 1.5 mL microcentrifuge tube.

11. Extraction of nonsperm fraction:

    a. Add 2 volumes of prepared lysis buffer and 7 μL of resin to each sample.

    b. Vortex for 3 s. Incubate at room temperature for 5 min.

    c. Vortex each sample for 2 s and then place in the magnetic stand.

    d. Remove all solution from the tube without disturbing the resin pellet.

    e. Add 100 μL of lysis buffer.

    f. Remove the tube from the stand and vortex for 2 s. Return tube to the magnetic stand.

    g. Remove and discard all solution from the tube without disturbing the resin pellet.

    h. Add 100 μL of the 1x wash buffer to each sample.

    i. Remove the tube from the stand and vortex for 2 s. Return tube to the magnetic stand.

    j. Remove and discard all solution from the tube without disturbing the pellet.

    k. Repeat Steps 10–12 two additional times for a total of three washes.

    l. With the tubes still in the magnetic stain, open the lids for all samples and let the samples dry at room temperature for 5 min (do not exceed 20 min).

    m. Add 25–100 μL of elution buffer depending on the desired concentration.

    n. Close the lid to each tube and vortex for 2 s.

    o. Incubate samples at 65°C for 5 min.

    p. Vortex samples for 2 s and immediately place them back into the magnetic stand.

    q. Carefully transfer the solution (containing the DNA) into a new 1.5 mL microcentrifuge tube.

20.3 FUTURE DEVELOPMENT AND TRENDS

20.3.1 LASER CAPTURE MICRODISSECTION

Laser capture microdissection (LCM) is a technique that permits the isolation and collection of individual, or groups of, cells [54–57]. LCM can be accomplished using a variety of platforms, which can be classified as direct versus nondirect cell contact. In the direct contact method a thermoplastic membrane polymer is melted by the laser over the cells of interest and the captured cells are removed by adhesion. This approach is less desirable as the direct contact of the laser could damage the genetic material in the targeted cells. The nondirect contact LCM method requires the dispersal of the cell suspension onto a proprietary membrane and the targeted cells are then circumscribed by, and cut out by, the laser.

The ability to isolate and recover a small number of cells could be ideal for forensic cases involving low copy number (LCN) samples, containing less than 100 pg of template DNA, equivalent to 15 diploid or 30 haploid cells [58]. The presence of such LCN samples could be due to several factors including damaged or degraded DNA, oligospermic or aspermic perpetrators, or from extended interval postcoital samples, where sperm have been lost over time due to the effects of drainage or host cell metabolism [59–61]. Other trace biological evidence will also contain small quantities of cells, including fingerprints, particulate matter, and aerosols [62,63]. Rather than subjecting these samples to the numerous physical manipulations required by most extraction methods that could result in a potential loss of the small amount of genetic material present, a direct lysis strategy can be employed using a small volume of lysis buffer. Subsequent reactions can typically be performed in the same tube thus eliminating the need for further manipulation of the sample. However, typical direct lysis strategies do not contain additional purification steps. Therefore, the resulting lysate contains denatured proteins and cell debris that may interfere with subsequent PCR-based assays.

The potential use of this technology for forensic casework has largely been focused on developing the ability to separate sperm and vaginal epithelial cells from sexual assault type evidence [64–68]. Differences in sperm and nonsperm cell type morphologies allow for facile separation of the cell types, and thus can eliminate the need for the use of differential extractions. Further work is needed in order to be able to utilize this technology to physically separate the contributors of non-easily-distinguishable cell type mixtures such as those involving vaginal and buccal epithelial cells.

20.3.2 MICROCHIPS

DNA chips are tiny (only a few square centimeters in size) analytical devices constructed of a variety of materials including silicon, glass, or plastic using microfabrication or microfluidic technologies that use nanoliter volumes of reagents and samples rather than the microliter volumes typically used in bioanalytical separations. Ideally, the following steps in the analytical process would be transferable in an integrated micro-total analysis system (μTAS) (or lab-on-a chip) format: (1) fractionation of reaction components (e.g., isolation and separation of different cell types), (2) precise volume measurement of reagents, (3) mixing of solutions, (4) controlled thermal reaction of the mixture, including PCR analysis, (5) hybridization to allele-specific oligonucleotide probes (for sequence polymorphisms), (6) electrophoretic separation (for length polymorphisms), (7) analyte detection, and (8) data analysis. The use of chip technology is expected to make the DNA profiling process faster, cheaper, more efficient, susceptible to full automation and perhaps, akin to the clinical concept of point-of-care, and deliverable at the point-of-use (i.e., the crime scene). Several studies have demonstrated the ability to perform DNA extraction and subsequent analysis of biological samples on microchips technologies [69–78]. Thus, it is likely that fully integrated μTAS systems will eventually supplant existing macroscale instruments for routine biomolecular forensic analysis.

Erin K. Hanson is a senior research scientist at the National Center for Forensic Science (DNA Laboratory) in Orlando, Florida. She has a bachelor of science degree and a masters degree in forensic science from the University of Central Florida (UCF), Orlando, Florida and is currently pursuing a PhD in biomolecular science (UCF). During her graduate education, she was the instructor for undergraduate forensic biochemistry laboratories for 6 years. Her research activities include development and validation of novel Y-chromosome STR multiplex systems, single cell/low copy number analysis using laser capture microdissection and whole genome amplification, and the determination of the age (time of deposition) of biological stains.

Jack Ballantyne is an associate professor of chemistry at the University of Central Florida, Orlando, Florida, and the associate director for research at the National Center for Forensic Science in Orlando, Florida. He has a BSc (with honors) in biochemistry from the University of Glasgow, Glasgow, Scotland; an MSc in forensic science from the University of Strathclyde, Glasgow, Scotland; and a PhD in genetics from the State University of New York, Stony Brook, New York. His current duties include teaching and conducting research in forensic molecular genetics. Before entering academia, he was a casework forensic scientist in Scotland, Hong Kong, and New York, where he proffered expert testimony in the criminal courts of these jurisdictions. He was the full-time DNA technical leader in Suffolk County, New York, and then served as a part-time consultant DNA technical leader for the states of Mississippi and Delaware, the city of Dallas, and Sedgwick County, Kansas. He is the chair of the New York State DNA subcommittee, a regular visiting guest at the Scientific Working Group on DNA Analysis Methods, a member of the DoD Quality Assurance Oversight Committee, and was a member of the World Trade Center Kinship and Data Analysis Panel. His research interests include Y-chromosome markers, the assessment and *in vitro* repair of damaged DNA templates, mRNA profiling for body fluid identification, the determination of physical characteristics by molecular genetic analysis, and single cell/low copy number analysis.

REFERENCES

1. Butler, J.M., Genetics and genomics of core short tandem repeat loci used in human identity testing, *J. Forensic Sci.*, 51: 253, 2006.
2. Alvarez, M. and Ballantyne, J., The identification of newborns using messenger RNA profiling analysis, *Anal. Biochem.*, 357: 21, 2006.
3. Bauer, M. and Patzelt, D., Evaluation of mRNA markers for the identification of menstrual blood, *J. Forensic Sci.*, 47: 1278, 2002.
4. Bauer, M. and Patzelt, D., A method for simultaneous RNA and DNA isolation from dried blood and semen stains, *Forensic Sci. Int.*, 136: 76, 2003.
5. Bauer, M. and Patzelt, D., Protamine mRNA as molecular marker for spermatozoa in semen stains, *Int. J. Legal Med.*, 117: 175, 2003.
6. Bauer, M., Polzin, S., and Patzelt, D., Quantification of RNA degradation by semi-quantitative duplex and competitive RT-PCR: A possible indicator of the age of bloodstains?, *Forensic Sci. Int.*, 138: 94, 2003.
7. Bauer, M. et al., Quantification of mRNA degradation as possible indicator of postmortem interval—A pilot study, *Leg. Med. (Tokyo)*, 5: 220, 2003.
8. Juusola, J. and Ballantyne, J., Messenger RNA profiling: A prototype method to supplant conventional methods for body fluid identification, *Forensic Sci. Int.*, 135: 85, 2003.
9. Juusola, J. and Ballantyne, J., Multiplex mRNA profiling for the identification of body fluids, *Forensic Sci. Int.*, 152: 1, 2005.
10. Juusola, J. and Ballantyne, J., mRNA profiling for body fluid identification by multiplex quantitative RT-PCR, *J. Forensic Sci.*, 52: 1252, 2007.
11. Alvarez, M., Juusola, J., and Ballantyne, J., An mRNA and DNA co-isolation method for forensic casework samples, *Anal. Biochem.*, 335: 289, 2004.
12. Chomczynski, P. and Sacchi, N., Single-step method of RNA isolation by acid guanidinium thiocyanate-phenol-chloroform extraction, *Anal. Biochem.*, 162: 156, 1987.
13. Ahmad, N.N., Cu-Unjieng, A.B., and Donoso, L.A., Modification of standard proteinase K/phenol method for DNA isolation to improve yield and purity from frozen blood, *J. Med. Genet.*, 32: 129, 1995.
14. Albarino, C.G. and Romanowski, V., Phenol extraction revisited: A rapid method for the isolation and preservation of human genomic DNA from whole blood, *Mol. Cell Probes*, 8: 423, 1994.
15. Kochl, S., Niederstatter, H., and Parson, W., DNA extraction and quantitation of forensic samples using the phenol-chloroform method and real-time PCR, *Methods Mol. Biol.*, 297: 13, 2005.
16. Comey, C.K. et al., Extraction strategies for amplified fragment length polymorphism analysis, *J. Forensic Sci.*, 39: 1254, 1994.
17. De, L. et al., A one-step microbial DNA extraction method using "Chelex 100" suitable for gene amplification, *Res. Microbiol.*, 143: 785, 1992.
18. Gunther, S., Herold, J., and Patzelt, D., Extraction of high quality DNA from bloodstains using diatoms, *Int. J. Legal Med.*, 108: 154, 1995.
19. Iwasa, M. et al., Y-chromosomal short tandem repeats haplotyping from vaginal swabs using a chelating resin-based DNA extraction method and a dual-round polymerase chain reaction, *Am. J. Forensic Med. Pathol.*, 24: 303, 2003.
20. Polski, J.M. et al., Rapid and effective processing of blood specimens for diagnostic PCR using filter paper and Chelex-100, *Mol. Pathol.*, 51: 215, 1998.
21. Sweet, D. et al., Increasing DNA extraction yield from saliva stains with a modified Chelex method, *Forensic Sci. Int.*, 83: 167, 1996.
22. Walsh, P.S., Metzger, D.A., and Higuchi, R., Chelex 100 as a medium for simple extraction of DNA for PCR-based typing from forensic material, *Biotechniques*, 10: 506, 1991.

23. Willard, J.M., Lee, D.A., and Holland, M.M., Recovery of DNA for PCR amplification from blood and forensic samples using a chelating resin, *Methods Mol. Biol.*, 98: 9, 1998.

24. Castella, V. et al., Forensic evaluation of the QIAshredder/QIAamp DNA extraction procedure, *Forensic Sci. Int.*, 156: 70, 2006.

25. Greenspoon, S.A. et al., QIAamp spin columns as a method of DNA isolation for forensic casework, *J. Forensic Sci.*, 43: 1024, 1998.

26. Hanselle, T. et al., Isolation of genomic DNA from buccal swabs for forensic analysis, using fully automated silica-membrane purification technology, *Leg. Med. (Tokyo)*, 5(Suppl 1), S145, 2003.

27. Scherczinger, C.A. et al., DNA extraction from liquid blood using QIAamp, *J. Forensic Sci.*, 42: 893, 1997.

28. Sinclair, K. and McKechnie, V.M., DNA extraction from stamps and envelope flaps using QIAamp and QIAshredder, *J. Forensic Sci.*, 45: 229, 2000.

29. Cowan, C., The DNA IQ system on the Tecan Freedom EVO 100, *Profiles in DNA*, 9: 8, 2006.

30. Deggerdal, A. and Larsen, F., Rapid isolation of PCR-ready DNA from blood, bone marrow and cultured cells, based on paramagnetic beads, *Biotechniques*, 22: 554, 1997.

31. Eminovic, I. et al., A simple method of DNA extraction in solving difficult criminal cases, *Med. Arch.*, 59: 57, 2005.

32. Komonski, D.I. et al., Validation of the DNA IQ system for use in the DNA extraction of high volume forensic casework, *Can. Soc. Forensic Sci. J.*, 37: 103, 2004.

33. Mandrekar, P.V., Krenke, B.E., and Tereba, A., DNA IQ: The intelligent way to purify DNA, *Profiles in DNA*, 4: 16, 2001.

34. Nakagawa, T. et al., Capture and release of DNA using aminosilane-modified bacterial magnetic particles for automated detection system of single nucleotide polymorphisms, *Biotechnol. Bioeng.*, 94: 862, 2006.

35. Ng, L.-K. et al., Optimization of recovery of human DNA from envelope flaps using DNA IQ system for STR genotyping, *Forensic Sci. Int. Genet.*, 1: 283–286, 2007.

36. Prodelalova, J. et al., Isolation of genomic DNA using magnetic cobalt ferrite and silica particles, *J. Chromatogr. A*, 1056: 43, 2004.

37. Rittich, B. et al., Isolation of microbial DNA by newly designed magnetic particles, *Colloids Surf. B Biointerfaces*, 52: 143, 2006.

38. Stemmer, C. et al., Use of magnetic beads for plasma cell-free DNA extraction: Toward automation of plasma DNA analysis for molecular diagnostics, *Clin. Chem.*, 49: 1953, 2003.

39. Prinz, M. et al., Multiplexing of Y chromosome specific STRs and performance for mixed samples, *Forensic Sci. Int.*, 85: 209, 1997.

40. Gill, P., Jeffreys, A.J., and Werrrett, D.J., Forensic application of DNA "fingerprints", *Nature*, 318: 577, 1985.

41. SWGDAM, Revised validation guidelines, *Forensic Sci. Comm.*, 6: 1, 2004.

42. Belgrader, P. et al., Automated DNA purification and amplification from blood-stained cards using a robotic workstation, *Biotechniques*, 19: 426, 1995.

43. Crouse, C.A. et al., Improving efficiency of a small forensic DNA laboratory: Validation of robotic assays and evaluation of microcapillary array device, *Croat. Med. J.*, 46: 563, 2005.

44. Greenspoon, S. and Ban, J., Robotic extraction of mock sexual assault samples using the Biomek 2000 and the DNA IQ system, *Profiles in DNA*, 5: 3, 2002.

45. Greenspoon, S.A. et al., Application of the BioMek 2000 laboratory automation workstation and the DNA IQ system to the extraction of forensic casework samples, *J. Forensic Sci.*, 49: 29, 2004.

46. Kishore, R. et al., Optimization of DNA extraction from low-yield and degraded samples using the BioRobot EZ1 and BioRobot M48, *J. Forensic Sci.*, 51: 1055, 2006.

47. Koller, S. et al., Automated 960Well purification of genomic DNA from blood, *Promega Notes*, 85: 7, 2003.

48. Montpetit, S.A., Fitch, I.T., and O'Donnell, P.T., A simple automated instrument for DNA extraction in forensic casework, *J. Forensic Sci.*, 50: 555, 2005.

49. Nagy, M. et al., Optimization and validation of a fully automated silica-coated magnetic beads purification technology in forensics, *Forensic Sci. Int.*, 152: 13, 2005.

50. Schiffner, L.A. et al., Optimization of a simple, automatable extraction method to recover sufficient DNA from low copy number DNA samples for generation of short tandem repeat profiles, *Croat. Med. J.*, 46: 578, 2005.

51. Anslinger, K. et al., Application of the BioRobot EZ1 in a forensic laboratory, *Leg. Med. (Tokyo)*, 7: 164, 2005.

52. Grunst, T., New Maxwell 16 instrument accessories increase the versatility of your Maxwell 16 system, *Promega Notes*, 97: 4–5, 2007.
53. Mandrekar, P. et al., Introduction to Maxwell 16 low elution volume configuration for forensic casework, *Profiles in DNA*, 10: 10, 2007.
54. Eltoum, I.A., Siegal, G.P., and Frost, A.R., Microdissection of histologic sections: Past, present, and future, *Adv. Anat. Pathol.*, 9: 316, 2002.
55. Emmert-Buck, M.R. et al., Laser capture microdissection, *Science*, 274: 998, 1996.
56. Heath, P., Laser capture microdissection: Methods and protocols, *Histopathology*, 48: 462, 2006.
57. Schütze, K., Pösl, H., and Lahr, G., Laser micromanipulation systems as universal tools in cellular and molecular biology and in medicine, *Cell. Mol. Biol.*, 44: 735, 1998.
58. Kirby, L., *Specimens, in DNA Fingerprinting: An Introduction*, Stockton Press, New York, 1990, Chapter 4.
59. Hall, A. and Ballantyne, J., Novel Y-STR typing strategies reveal the genetic profile of the semen donor in extended interval post-coital cervicovaginal samples, *Forensic Sci. Int.*, 136: 58, 2003.
60. Shewale, J.G. et al., DNA profiling of azoospermic semen samples from vasectomized males by using Y-PLEX 6 amplification kit, *J. Forensic Sci.*, 48: 127, 2003.
61. Sibille, I. et al., Y-STR DNA amplification as biological evidence in sexually assaulted female victims with no cytological detection of spermatozoa, *Forensic Sci. Int.*, 125: 212, 2002.
62. Schulz, M.M. and Reichert, W., Archived or directly swabbed latent fingerprints as a DNA source for STR typing, *Forensic Sci. Int.*, 127: 128, 2002.
63. van Oorschot, R.A. and Jones, M.K., DNA fingerprints from fingerprints, *Nature*, 387: 767, 1997.
64. Di Martino, D. et al., Single sperm cell isolation by laser microdissection, *Forensic Sci. Int.*, 146(Suppl): S151, 2004.
65. Di Martino, D. et al., LMD as a forensic tool in a sexual assault casework: LCN DNA typing to identify the responsible, *Progr. Forensic Genet.*, 11: 571, 2006.
66. Elliott, K. et al., Use of laser microdissection greatly improves the recovery of DNA from sperm on microscope slides, *Forensic Sci. Int.*, 137: 28, 2003.
67. Elliott, K. et al., Use of laser microdissection greatly improves the recovery of DNA from sperm on microscope slides, *Progr. Forensic Genet.*, 10: 45, 2004.
68. Sanders, C.T. et al., Laser microdissection separation of pure spermatozoa from epithelial cells for short tandem repeat analysis, *J. Forensic Sci.*, 51: 748, 2006.
69. Bienvenue, J.M. et al., Microchip-based cell lysis and DNA extraction from sperm cells for application to forensic analysis, *J. Forensic Sci.*, 51: 266, 2006.
70. Legendre, L.A. et al., A simple, valveless microfluidic sample preparation device for extraction and amplification of DNA from nanoliter-volume samples, *Anal. Chem.*, 78: 1444, 2006.
71. Panaro, N.J. et al., Micropillar array chip for integrated white blood cell isolation and PCR, *Biomol. Eng.*, 21: 157, 2005.
72. Chung, Y.C. et al., Microfluidic chip for high efficiency DNA extraction, *Lab Chip*, 4: 141, 2004.
73. Easley, C.J. et al., A fully integrated microfluidic genetic analysis system with sample-in-answer-out capability, *Proc. Natl. Acad. Sci. U. S. A.*, 103: 19272, 2006.
74. Goedecke, N. et al., A high-performance multilane microdevice system designed for the DNA forensics laboratory, *Electrophoresis*, 25: 1678, 2004.
75. Horsman, K.M. et al., Forensic DNA analysis on microfluidic devices: A review, *J. Forensic Sci.*, 52: 784, 2007.
76. Verpoorte, E., Microfluidic chips for clinical and forensic analysis, *Electrophoresis*, 23: 677, 2002.
77. Horsman, K.M. et al., Separation of sperm and epithelial cells in a micro-fabricated device: Potential application to forensic analysis of sexual assault evidence, *Anal. Chem.*, 77: 742, 2005.
78. Liu, P. et al., Integrated portable polymerase chain reaction-capillary electrophoresis microsystem for rapid forensic short tandem repeat typing, *Anal. Chem.*, 79: 1881, 2007.

Part VI

---

*Purification of Nucleic Acids
from Plants*

21 Isolation of Nucleic Acids from Plants

Song Weining and Dongyou Liu

CONTENTS

21.1 INTRODUCTION

21.1.1 CLASSIFICATION AND BIOLOGY OF PLANTS

Plants are a diverse group of organisms in the kingdom Plantae that differ from those in the kingdoms Animalia (animals), Fungi, Protista, and Monera (prokaryotes) by possessing chlorophyll for photosynthesis (which converts energy from sunlight into carbohydrate), having a rigid cell wall composed of cellulose (a complex carbohydrate made from glucose, which provides structural support while remains flexible), and being fixed in one place (nonmotile). However, plants are similar to other eukaryotes in having a nucleus, plasma membrane, mitochondria, and other organelles.

Based on their variable growth habits, plants are divided into succulent plants (i.e., herbaceous or herbs with succulent seeds possessing self-supporting stems), vine (climbing or trailing herbaceous plant), trees (having a single central axis), and shrubs (having several upright stems). According to the types of their leaf drops, plants are called deciduous (no living leaves during dormant winter season) or evergreen (retaining functional leaves throughout the year). Considering their differences in life span, plants are known as annual (completing its life cycle during a single growing season), biennial (completing its life cycle during a period of two growing seasons), and perennial (growing year after year and taking many years to mature). With their contrasting temperature tolerance, plants are differentiated into tender plant (which is damaged by low temperature) and hardy plant (which withstands low temperature). Depending on their site preference, plants are separated into xerophyte (preferring dry sites), shade plants (preferring low light intensity), acid loving (preferring low pH soils), and halophyte (preferring salty soils). In view of their distinct tissue structures, plants are grouped into vascular (which is further separated into seedless vascular gymnosperms and angiosperms) and nonvascular categories (Table 21.1). While vascular plants have tube-like structures (i.e., xylem for transporting water and dissolved minerals upward from the roots to the shoot, and phloem for transporting the essential products of photosynthesis throughout the plant body), nonvascular plants lack such structures and rely on their surfaces to absorb water (thus they usually live in moist places). Taxonomically, living plants in the kingdom Plantae belong to at least 10 phyla or divisions, which cover about 300,000 known species (Table 21.1). Once classified within the kingdom Plantae, algae are now included among the kingdom Protists, although they also produce energy through photosynthesis. In addition, fungi were historically treated as close relatives to plants, but they now form a separate Fungi kingdom as they are saprotrophs that are capable of obtaining food by breaking down and absorbing surrounding materials, and not by photosynthesis.

Plants can reproduce either asexually or sexually. For the asexual reproduction, plants make use of rhizoids (a filament that anchors the plant to the ground) and undergo fragmentation and budding processes. For the sexual reproduction, plants employ specific female (i.e., the stigma, the style, and the ovary) and male (i.e., the filament and the anther) structures. The style is a long and slender female reproductive structure in plants supporting the stigma, which forms the sticky portion of the pistil that captures pollen, whereas the ovary is composed of one or more ovules that house the eggs.

TABLE 21.1
Classification of Living Plants

Plant Group	Phylum	Common Name	Key Features	No. of Known Species
Nonvascular	*Bryophyta*	Mosses	Small nonvascular plants with restrict height; living in moist habitats; lacking true leaves, stems, roots, and rhizoids; and reproducing by spores	24,000
	Hepatophyta	Liverworts		
	Anthocerophyta	Hornworts		
Seedless vascular	*Pterophyta*	Ferns, horse tails, and whisk ferns	Vascular plants with stems mostly creeping; reproducing by spores	12,000
	Lycophyta	Club mosses	Vascular plants lacking true roots; reproducing by spores	1,000
Gymnosperms	*Coniferophyta*	Conifers	Vascular shrubs or trees producing naked seeds or cone (no fruits); leaves mostly needlelike or scalelike	600
	Ginkgophyta	Ginkgo	Vascular cone-bearing plant with deciduous fan-shaped leaves	1
	Gnetophyta	Gnetophytes	Vascular cone-bearing desert plants	70
	Cycadophyta	Cycads	Vascular perennial shrubs or palmlike trees with unbranched stems and pinnate leaves; reproducing by seeds	100
Angiosperms	*Anthophyta*	Flowering plants	Vascular herbaceous and woody plants producing flowers and seeds (enclosed in a fruit)	235,000

Source: Adapted from McMahon, M.E., Kofranek, A.M., and Rubatzky, V.E. (eds), *Hartmann's Plant Science: Growth, Development, and Utilization of Cultivated Plants*, 4th ed., Prentice Hall, 2007.

The filament is a male reproductive structure in plants that supports the anther, which produces and stores pollen. During pollination, a pollen grain is transferred from an anther to a stigma with assistance from wind, water, insects, birds, and small mammals. After the pollination event, one nucleus of the pollen grain forms a tube down through the style to the ovary, and the second nucleus moves down the tube and splits into two sperm nuclei to fertilize the egg and combine with polar bodies to form the endosperm (stored fruit).

When a plant grows, the plant cells generate a new cell wall (in the form of a network of microtubules called xylem ad phloem) along the axis of cell division. The extension of microtubes permits transfer of nutrients, water, and some other materials without going across the cell wall barriers between cells. A mature plant cell is a membrane-bound vacuole filled with fluid (containing ions, stored nutrients, and waste materials) together with the nucleus, plastids (containing chlorophyll), and other organelles located close to the cell membrane. The fluid-filled vacuole also serves to maintain the cell turgor pressure and rigidity. Plants utilize an elaborate process to capture the energy of sunlight to produce sugars, for which carbon dioxide is used as one of the raw materials and oxygen is released as a by-product. The resulting sugars can be a fuel source for continued growth and stored as a complex carbohydrate (or starch). This photosynthetic process is carried out in an organelle called plastid, which harbors chlorophyll. Besides sunlight, plants are dependent on soil for support and water supply, as well as nitrogen, phosphorus, and other crucial elemental nutrients for growth.

Due to their genotypic diversity, plants demonstrate enormous variations in growth rate, with some mosses growing less than 0.001 mm/h, most trees growing 0.025–0.250 mm/h, and some climbing species growing up to 12.5 mm/h (as they have no need to produce thick supportive tissue). Other factors affecting plant growth include the environmental factors (e.g., temperature, available water, light, and nutrients in the soil), biotic factors (living organisms competing with other plants for space, water, light, and nutrients), pollination (by birds and insects), and soil fertility (relating to the activity of bacteria and fungi). While most nonvascular plants require constant moisture for survival, vascular plants are more resistant to desiccation. This is attributable to the fact that vascular plants can use the xylem to move water and minerals from the root to the rest of the plant, and the phloem to provide the roots with sugars and other nutrient produced by the leaves. Additionally, plants often secrete antifreeze proteins, heat-shock proteins, and sugars (e.g., sucrose) to protect themselves from frost and dehydration stresses.

21.1.2 IMPORTANCE OF PLANTS

Environmental benefits. Plants are distributed in a wide range of habitats on earth, in which they often function as the primary source of energy and organic materials for virtually all types of ecosystems. The photosynthesis of plants as well as algae removes carbon dioxide from earth's atmosphere (which is produced by animals and humans) and releases oxygen as a by-product (which is essential for animal and human existence). Land plants also play a pivotal role in water cycle and other biogeochemical processes. Plant roots contribute significantly to soil development and prevention of soil erosion.

Food and shelter. Plants represent a primary source of energy for many animal species (in particular herbivores) and humans. A number of plant products (e.g., wheat, rice, corn, potato, and legumes) are stable foods for human population. Many other parts of plants (e.g., fruits, vegetables, nuts, and cooking oils) constitute integral components of human diets. Plants are also important ingredients for production of beverages (e.g., tea, coffee, cocoa, beer, and wine). In addition to their use as vital raw materials for building, furniture and paper making, and clothing, plants provide renewable fuels, and plants in the fossil form (coal and petroleum) are an essential source of power and ingredient for manufacturing industry.

Medicine. Plants are important source of medicines (e.g., aspirin, morphine, and quinine), medicinal ingredients (e.g., ginkgo), and nutritional supplements. Indeed, hundreds of herbs and plants have been described in folk medicines for treatment of a variety of diseases and infections throughout the world.

While the underlying pharmacological mechanisms for many of these medicinal plants remain unclear, application of modern analytical techniques has uncovered valuable details on how some of these herbal medicines function, which in turn have underscored the development of novel drugs against infectious agents and other human ailments. For example, artemisinin, isolated from Chinese wormwood (qinghao, which has been traditionally used for the treatment of skin diseases and malaria), has proven highly effective for control and prevention of multidrug resistant strains of falciparum malaria. Additionally, it is also under exploration to be an anticancer agent. Some pesticides (e.g., nicotine and pyrethrins) are also obtained from plants. Pyrethrins are present in the seed cases of the perennial plant pyrethrum (*Chrysanthemum cinerariaefolium*), and function as neurotoxins for all insects. Being easily broken down by exposure to light or oxygen and thus biodegradable, and relatively harmless toward mammals, pyrethrins represent one of the safest insecticides for agriculture and food industries.

Detrimental aspects of plants. Some plants and their products (e.g., ricin) are poisonous, and pollens produced by some plants cause allergy in humans. Ricin is a toxic protein found in the castor bean (*Ricinus communis*). Acting as a toxin through inhibition of protein synthesis, ricin is poisonous if inhaled, injected, or ingested. Pollens from some grasses (e.g., timothy grass, Kentucky bluegrass, Johnson grass, Bermuda grass, redtop grass, orchard grass, and sweet vernal grass); trees (e.g., oak, ash, elm, hickory, pecan, box elder, and mountain cedar); and weeds (e.g., ragweed, sagebrush, redroot pigweed, and tumbleweed) are renowned for their ability to elicit allergic reactions in humans with symptoms ranging from sneezing, coughing, itching, runny noses, watering eyes, and general malaises. Weeds and obnoxious plants often grow rapidly and compete with agricultural plants for nutrition and reduce food production. They also contaminate food, feed, fiber, ballast, and packing and bagging material.

21.2 GENERAL PRINCIPLES AND CURRENT TECHNIQUES FOR NUCLEIC ACID EXTRACTION FROM PLANTS

Being a major component of the plant cell wall, polysaccharide increases the cell rigidity and makes it difficult to break up. The existence of a range of secondary metabolites (e.g., polyphenols) and pigments within the plant cells further complicates the isolation of high quality and purity nucleic acids from plants in relation to other biological organisms such as bacteria. Below we discuss several key issues that are vital to improve the yield and purity of nucleic acids from plant specimens.

21.2.1 SAMPLE SELECTION AND PREPREPARATION

Although all types of plant materials (e.g., leaves, stem, roots, seeds, and embryos) are useful for nucleic acid extraction, fresh leaves from young plants offer added advantage as they are comparatively easy to break up, and contain lower amounts of starch, polyphenols, tannins, polysaccharides, and other secondary metabolites [2,3], contributing to an increased recovery of nucleic acids. Therefore, to generate young leaves for easy and efficient isolation of nucleic acids, an often-utilized strategy is to grow the plant in dark for 2–4 days before harvesting [4,5]. Similarly, when the processing of fresh leaves and other plant tissues is not possible immediately after harvesting, such as under field situations, the plant specimens should be kept on ice or in dry place, and then frozen at −80°C or in liquid nitrogen, to reduce the damaging activity of endogenous nuclease, and limit growth of contaminating organisms [6]. Alternatively, plant tissues may be maintained in preserving solutions such as saturated CTAB (cetyltrimethyl ammonium bromide)-NaCl-NaN3 [7], or with silica gel to decrease available moisture [8]. Furthermore, lyophilization can be employed for long-term preservation of plant tissues intended for DNA extraction.

Prior to proceeding with the nucleic acid extraction, it is often helpful to properly clean and sometimes sterilize plant tissues. Peterson et al. [9] employed ethyl ether to treat the fresh leaves followed by thorough wash. Krasova-Wade et al. [10] described a pre-preparation procedure for legume nodules, in which the legume nodules are first rehydrated in sterile distilled water and surface sterilized by immersion in 3.3% w/v $Ca(OCl)_2$ for 3 min, rinsed in sterile water, then in absolute ethanol for 2–3 min, and then finally rinsed in sterile water. The cleaned nodules are then crushed in TES/sucrose (20 mM Tris–HCl, pH 8.0, 50 mM EDTA disodium, pH 8.0, 50 mM NaCl, 8% w/v sucrose) buffer by using plastic pestles.

21.2.2 DISRUPTION OF PLANT CELL WALL

Give its polysaccharide content, plant cell wall is notably resistant to breakup and will only succumb after the use of physical and chemical means. During physical disruption, plant tissues are crushed or homogenized with the assistance of mortar and pestle, mechanical steel grinder, stainless steel ball bearings, or glass beads. Often physical disruption of plant cells is undertaken at a low temperature such as in the presence of liquid nitrogen or in cold room. This not only helps to achieve a rapid and efficient disruption of soft and hard plant tissues, which is critically important for RNA extraction, but also provides a condition (i.e., low temperature) that inhibits the activity of endogenous RNAse and other endonucleases [11,12]. When liquid nitrogen is not available, guanidine isothiocyanate (Tri reagent or Trizol) may be applied as a convenient alternative, since RNase and DNase are readily inactivated by guanidine isothiocyanate [13]. It is helpful to bear in mind that although homogenization serves to disrupt plant cell wall and reduce the viscosity of plant lysate resulting from the release of high molecular weight compounds such as complex carbohydrates, prolonged and excessive grinding (or disruption with glass beads) may contribute to the shearing of high molecular weight DNA, which may impact negatively on certain downstream applications and affect the authenticity of analysis using pulse-field gel electrophoresis and other molecular techniques.

Chemical disruption relies on the use of cellulases pectinases, cell wall macerases, and other hydrolyzing enzymes to digest plant cell wall of selective plant specimens [14,15]. While convenient, this approach can be costly. Other compounds that are useful for breaking up plant cell wall include sodium/potassium ethylxanthogenate, which form water-soluble polysaccharide xanthates with hydryl groups of plant cell wall polysaccharides, thus leading to the dissolution of plant cell wall.

21.2.3 REMOVAL OF POLYSACCHARIDES AND POLYPHENOLS

Following the disruption of plant cell wall and cell membrane, nucleic acids and other cellular components (e.g., polysaccharides, polyphenolics, proteins, and other secondary metabolites) are released in the lysate. As these cellular components tend to bind, sequester, and coprecipitate with nucleic acids, and interfere or inhibit the activity of enzymatic reactions (e.g., DNA polymerases, ligases, and restriction endonucleases) during the subsequent molecular applications, they need to be effectively removed during the nucleic acid isolation process [16].

Being a dominant component in plant cells, in particular of mucilaginous plants [12], polysaccharides increase the viscosity of the plant lysate, and show the tendency to coprecipitate with nucleic acids, resulting in sequestration of nucleic acids and interference with the activity of enzymes used in various molecular studies [17–24]. One common way to eliminate polysaccharides from the lysate is through the use of high concentration of NaCl (0.5–6 M), which increases the solubility of polysaccharides in ethanol [25–27]. In addition, besides assisting in the breakup of the cell membrane, detergents such as CTAB and sarkosyl have the capacity to form insoluble complexes with proteins and most acidic polysaccharides at high salt concentration, leaving nucleic acids in the solution, it can be employed to separate DNA from polysaccharides [2,3,24,28]. Furthermore, hydrolytic enzymes (e.g., pectinase) assist the digestion of pectin-like contaminants, and contribute to the reduction of polysaccharides in plant cell lysate [14,15]. Lastly, a few extra washing with extraction buffer have also proven to be effective to decrease polysaccharide residuals in the final preparation [24,29].

Polyphenols are a group of chemical compounds (including tannins, lignins, and flavonoids) that are commonly found in plants and act as important controllers of decomposition and nitrogen cycling processes. The condensed tannins are the most abundant polyphenols in plants, making up to 50% of the dry weight of leaves. After the disruption of plant cell wall, polyphenols are released from the vacuoles and then oxidized by cellular polyphenol oxidases. Subsequent interaction between the oxidized polyphenols and nucleic acids causes browning of the DNA pellets and degradation of DNA [9,19,21,23,30,33]. Therefore, it is essential to deal with the problem of polyphenolic contamination early in the nucleic acid isolation process. An effective way to counter the effects of polyphenols is to include adsorbents and antioxidants in the extraction buffer so that polyphenols can be removed and oxidation of polyphenols during cell lysis is controlled. Two commonly used adsorbents to eliminate polyphenols are polyvinyl pyrollidone (PVP, especially PVP 10) and polyvinyl polypyrrollidone (PVPP), both of which adsorb polyphenols at low pH. Both PVP and PVPP can form complexes with the polyphenols via hydrogen bonding, while leave the nucleic acids in the solution [21,30,33,34]. β-Mercaptoethanol (which prevents the polymerization of tannins), bovine serum albumin (BSA), sodium azide, and dithiothreitol (DTT, also known as Cleland's reagent) are examples of antioxidants that have also been applied along with PVP to limit the effects of polyphenols during nucleic acid purification [25,35,36]. Moreover, polyphenol oxidases are also employed as specific inhibitors to reduce DNA browning by polyphenols [37,38].

21.2.4 SEPARATION OF NUCLEIC ACIDS FROM PROTEINS AND OTHER SUBSTANCES

While removal of polysaccharides and polyphenols represents a vital initial step, proteins and other contaminating substances also need to be eliminated in order to generate high purity nucleic acids from plants. The most common approach for protein removal is through the use of proteinase K and pronase, which break up peptide bonds and inactivate endogenous DNases and RNases in the cell lysate. In addition, sodium dodecyl sulfate (SDS), DTT, and β-mercaptoethanol also destroy the structural organizations of proteins, and further enhance the efficiency of proteinase K treatment [39]. The degraded proteins and other contaminating substances including cell debris are then separated with organic solvents (e.g., phenol and chloroform) [23,40]. Given their caustic and toxic nature, phenol and chloroform have been increasingly superseded by organic solvent-free techniques such as silica- and diethylaminoethyl cellulose (DEAE)-based filter columns [8,41,42]. Indeed, an increasing number of commercial kits for DNA and RNA isolation for plants and other biological organisms are built on these types of columns for convenience and speed.

If only genomic/organellar DNA is intended, RNA can be digested with DNase-free RNase. Alternatively, lithium chloride can be utilized (at 0.5 M) for preferential precipitation of RNA [12]. On the other hand, if only nuclear, chloroplast, or mitochondrial DNA is preferred, it is possible to apply appropriate cell lysis and centrifugation conditions for step wise removal of nontarget organelles. Specifically, nuclei can be isolated from the plant cells using a homogenization buffer containing osmoprotectants (e.g., glucose and sucrose), nuclear membrane stabilizers (e.g., spermine, spermidine, and polylysine), and sometimes mono-ionic detergent Triton X-100 followed by centrifugation, as Triton X-100 lyses chloroplasts and mitochondria, leaving nuclei intact [9]. In addition, centrifugation of plant cell lysate at different speed and gravity force is also effective for step-wise removal of plant organelles (Section 21.3.1.2). The resulting organelles (nuclei, chloroplasts, and mitochondria) are then lysed and nucleic acids purified by phenol–chloroform and ethanol precipitation.

21.2.5 PRECIPITATION OF NUCLEIC ACIDS

Following the removal of degraded proteins and other substances (including the insoluble cell debris), the resulting nucleic acids in the lysate can be precipitated and concentrated with alcohol (e.g., 0.6 volume of isopropanol or 2 volumes of ethanol) or in the presence of salt (e.g., NaCl,

NaOAc, NH_4OAc, or LiCl). Alternatively, nucleic acids in the lysate can be purified via silica- and DEAE-based filter columns. The quantity and purity of the resulting nucleic acids can then be assessed by spectrophotometry at wavelengths of 260/280 nm (i.e., $OD_{260 nm}/OD_{280 nm}$), with an $OD_{260 nm}/OD_{280 nm}$ ratio of 1.7–1.8 indicating a pure nucleic acid preparation. Further, the integrity of the purified nucleic acids can be evaluated by agarose gel electrophoresis.

21.2.6 CURRENT TECHNIQUES FOR EXTRACTION OF PLANT NUCLEIC ACIDS

Since plants comprise a large number of diverse species whose tissues show considerable variations in structures and conformations, isolation of high purity and quality nucleic acids from plants requires specialized approaches and techniques that facilitate efficient disruption of plant cell wall, and removal of polysaccharides, polyphenols, proteins, and other substances. Elimination of these contaminating compounds is essential as they often impact negatively the quality and usability of resulting nucleic acid preparation. Not surprisingly, many in-house procedures have been developed over the years for the purification of DNA and RNA from plant tissues and organelles [43–56], in addition to the description of a large number of rapid procedures for mini-preparation of DNA for PCR and molecular analysis [57–87]. While the use of phenol–chloroform has provided a reliable method for nucleic acid extraction from plants, incorporation of CTAB and PVP in the protocols has further increased efficiency in the contaminant removal from the lysate [2,3]. In fact, since its first inclusion in the plant DNA isolation protocols in the 1980s [2,3], CTAB has become a vital reagent in many plant DNA extraction methods subsequently reported, due to its role in eliminating polysaccharides from plant lysate. Also, besides the use of mortar and pestle, blender, grinder, glass beads, and hydrolyzing enzymes for breakup of plant cell wall, other simpler approaches such disposable plastic pestles fitting into microcentrifuge tubes and pipette tips can be used to disrupt plant cell wall for mini-preparation of nucleic acids from plants.

In a recent study, Krasova-Wade and Neyra [10] examined four DNA isolation protocols for legume nodules: (1) modified DNAzol protocol [88]; (2) CTAB/PVPP with phenol–chloroform–isoamyl alcohol purification [89]; and (3) the GES protocol. The GES protocol involves crushing a legume nodule (3.3–4.0 mg) in 100 μL of TES/sucrose buffer with a plastic pestle sterilized in absolute ethanol in a 1.5 mL tube followed by addition of lysozyme and GES reagent (0.5 mM guanidine thiocyanate, 0.1 M EDTA disodium pH 8.0, 1% w/v N-lauroylsarcosine sodium salt), and subsequent ethanol precipitation of DNA. They showed that the GES protocol generated produced the highest DNA yield (of 10 kb in size) with 15 min incubation (4.8 mg per gram of dry matter) and with 0.5 mM guanidine thiocyanate, which compared favorably to the DNAzol method (2.2 mg per gram of dry matter) and the CTAB/PVPP method (1.8 mg per gram of dry matter) (Table 21.2). This procedure offers a reproductive method to isolate DNA from nodules of different legume species. As this method does not require phenol, chloroform, and CTAB, it is not only safer but also more economical.

Ribeiro and Lovato [90] compared five protocols for DNA extraction from fresh and herbarium leaves of the genus *Dalbergia*, which contain high amounts of secondary metabolites, and therefore present additional challenges for generating a clean DNA isolation. The five DNA extraction protocols assessed include (1) protocol A based on CTAB extraction protocol [3]; (2) protocol B utilizing PVP to bind the phenolic compounds, and a high molar concentration of sodium chloride to inhibit coprecipitation of the polysaccharides and DNA, and an improved method for removing RNA by selective precipitation with lithium chloride [19]; (3) protocol C utilizing SDS as detergent and potassium acetate to remove some proteins and polysaccharides as a complex with the potassium-SDS precipitate [91]; (4) protocol D, which is a modified CTAB method by isolating membrane-bound organelles in order to remove polysaccharides and secondary metabolites [92]; and (5) protocol E (Qiagen DNeasy mini plant kit) using silica-gel-membrane technology and simple spin procedures to isolate high-quality DNA. The authors found that the protocol B [19] gives the best DNA quality in both fresh and dried leaves of the *Dalbergia* species, although it is labor intensive

TABLE 21.2

Comparison of In-House Reagents and Commercial Kits for Extraction of Plant DNA

Method	Feature	Performance	Reference
GES protocol	Crush legume nodules with a plastic pestle followed by lysozyme and extraction with 0.5 mM guanidine thiocyanate and ethanol precipitation	4.8 mg DNA per gram of dry matter	[10]
Modified DNAzol protocol [88]	Guanidine thioisocyanate and silica particle column	2.2 mg DNA per gram of dry matter	[10]
CTAB/PVPP [89]	Hexadecyltrimethylammonium bromide/ polyvinylpolypyrrolidone with phenol:chloroform:isoamyl alcohol purification	1.8 mg DNA per gram of dry matter	[10]
Protocol A [3]	Lysis and purification with CTAB that selectively precipitates DNA while maintaining the solubility of many polysaccharides	Satisfactory with fresh and herbarium specimens, but DNA appears yellowish	[90]
Protocol B [19]	Utilizes PVP to bind the phenolic compounds, and sodium chloride to inhibit coprecipitation of the polysaccharides and DNA, and an improved method for removing RNA by selective precipitation with lithium chloride	High-quality DNA with fresh leaves and herbarium specimens, but is labor intensive	[90]
Protocol C [91]	Utilizes SDS as detergent and potassium acetate to remove some proteins and polysaccharides as a complex with the potassium-SDS precipitate followed by isopropanol precipitation	Satisfactory with fresh leaves and herbarium specimens, but DNA appears yellowish	[90]
Protocol D [92]	CTAB method by isolating membrane-bound organelles to remove polysaccharides and secondary metabolites followed by isopropanol precipitation	Satisfactory with most plant samples (except *Dalbergia* species)	[90]
Protocol E (Qiagen DNeasy mini plant kit)	Uses silica-gel-membrane technology and simple spin procedures to isolate high-quality DNA	High-quality DNA with fresh leaves, but no DNA with herbarium specimens	[90]
Sucrose prep	Grind leaves in 200 μL sucrose solution using pipette tip or pestle; heat at 100°C for 10 min and spin briefly; use supernatant for PCR	Plant DNA and RNA can be detected reliably by PCR and RT-PCR	[57]
Touch-and-go method	Leaves are punctured against a firm surface with pipette tip, transferred in PCR tube directly	Plant DNA (<1 bp only) can be detected reliably by PCR	[57]
CTAB method [93]	Hexadecyltrimethylammonium bromide with phenol–chloroform–isoamyl alcohol purification	Plant DNA and RNA can be detected reliably by PCR and RT-PCR, but it is relatively time-consuming	[57]

because several solutions needed to be prepared. In fact, in the presence of PVP, phenolics adhere to DNA in solution forming a colored extract around the DNA that becomes cleaner after the addition of the detergent SDS. The addition of high molar concentration of NaCl increases the solubility of polysaccharides in ethanol, effectively decreasing coprecipitation of the polysaccharides and DNA [25]. Finally, the addition of LiCl selectively precipitates large RNA molecules reducing the amount of RNA present in the final DNA solution. Selective precipitation has an advantage over RNase treatment in that the RNA is removed and not simply degraded into smaller units. Protocols A and C demonstrated reasonable results for *Dalbergia* species, as the DNA obtained was yellowish in most of the samples. Protocol D successfully extracted DNA from many genera including *Schefflera* (Araliaceae), *Macadamia* (Proteaceae), *Dysoxylum* (Meliaceae), *Flindersia* (Rutaceae), *Sarcopteryx* (Sapindaceae), *Acacia* (Mimosaceae), and *Melicope* (Rutaceae) [92]. However, this protocol did not exhibit satisfactory results for *Dalbergia* species, maybe due to the modification introduced in the protocol, as the samples were ground with liquid nitrogen instead of sand. Protocol E, Qiagen DNeasy mini plant kit, produced high-quality DNA and no degradation in fresh leaves; however, for herbarium specimens it did not yield DNA, and consequently there was no PCR amplification, probably due to high degradation of the samples.

With the aim to develop a simple and efficient mini-prep protocol for PCR and other phylogenetic analysis, Berendzen et al. [57] evaluated three rapid procedures (i.e., the sucrose prep, the "touch-and-go" prep, and the CTAB prep) for combined DNA/RNA extraction for PCR application. The sucrose prep involves crushing of plant leaves in 200 µL sucrose solution (50 mM Tris-HCl pH 7.5, 300 mM NaCl and 300 mM sucrose) using a pipette tip or pestle followed by heating at 100°C for 10 min and a spin at 2000–6000 g for 5 s; the touch-and-go method uses a fine pipette tip (TipOne from Starlab GMBH, catalog no. S1111-3000 or S1110-3000) to punch plant leaves against a firm surface and subsequently transfer the sample directly to the pre-prepared PCR solution (or 10 µL water) in the PCR tubes; and the CTAB method is a modified procedure of Murray and Thompson [2,55,93]. The authors showed that the sucrose prep provides a reliable way to simultaneously isolate plant DNA and RNA templates for direct PCR application while the touch-and-go method gives consistent PCR results with fragments of <1 kb. The CTAB method also works well, but is more labor intensive and costly (Table 21.2).

21.3   METHODS

21.3.1   EXTRACTION OF PLANT DNA

21.3.1.1   Total DNA

The following protocol is based on the method of Mishra et al. [53] that uses CTAB and PVP to remove polysaccharides and other contaminating substances from plant cell lysate, and phenol–chloroform to eliminate proteins. Under the specified conditions, proteinase K treatment is unnecessary. The method enables isolation of high-quality total DNA (including nuclear, mitochondrial, and chloroplast DNA) with yield ranging from 200 to 400 µg DNA per gram of plant tissue, which is essentially free of contamination from polysaccharides and polyphenols as well as yellowish coloration.

Reagents

Extraction buffer: 200 mM Tris–Cl pH 8.0, 50 mM EDTA pH 8.0, 2.2 mM NaCl, 2% CTAB, 0.06% sarkosyl, 50 mg PVP MW 40,000. Add 100 µL β-mercaptoethanol per 10 mL before use.
Chloroform/isoamyl alcohol (24:1)
Phenol:chloroform/isoamyl alcohol (25:24:1)
Isopropanol
80% Ethanol
5 M NaCl
RNase A (10 mg/mL)
TE: 10 mM Tris–Cl pH 8.0, 1 mM EDTA pH 8.0

Supplies and equipment

Blender (or mortar and pestle), 30 mL and 1.5 mL tubes, centrifuge, microfuge, incubator, pipette tips, and gloves

Procedure

1. Homogenize 1 g leaves (seeds or other plant tissues) in 10 mL freshly prepared extraction buffer using a blender (or mortar and pestle), and transfer the sample to a 30 mL centrifuge tube.
2. Incubate the tube at 65°C for 1 h with swirling every 15 min.

3. Add an equal volume of chloroform/isoamyl alcohol (24:1), mix by gentle inversion for 10 min, and centrifuge at 6000 rpm for 15 min at room temperature.

4. Transfer the supernatant to a new 30 mL tube, add 2/3 volume of ice-cold isopropanol and 1 mL of 5 M NaCl, and leave at −20°C for 1 h or longer.

5. Centrifuge at 5000 rpm for 5 min at 4°C and 8000 rpm for 15 min.

6. Discard the supernatant, add 5 mL of 80% ethanol, centrifuge at 5000 rpm for 5 min, discard supernatant, and air-dry the pellet for 10 min.

7. Resuspend in 500 μL of TE, transfer the DNA to a 1.5 mL tube, and remove contaminating RNA with 5 μL of RNase A (10 mg/mL) at 37°C for 60 min.

8. Add an equal volume of phenol:chlororform:isoamyl alcohol (25:24:1), and centrifuge at 10,000 rpm for 10 min at 4°C.

9. Transfer the supernatant to a new 1.5 mL tube, add an equal volume of chlororform:isoamyl alcohol (24:1), and centrifuge at 10,000 rpm for 10 min at 4°C.

10. Transfer the supernatant to a new 1.5 mL tube, add an equal volume of isopropanol and 50 μL of 5 M NaCl, leave at −20°C for 1 h, and centrifuge at 10,000 rpm for 10 min at 4°C.

11. Discard the supernatant, add 500 μL of 80% ethanol, and centrifuge at 10,000 rpm for 5 min at 4°C.

12. Discard the supernatant, air-dry for 10 min, dissolve the DNA in 200 μL of TE, and store at −20°C.

Note: After use, wash pestle and mortar in 0.25 M HCl for 30 min, and rinse in water and air-dry.

21.3.1.2   Total DNA (and RNA)

The following protocol is based on the method of Weining and Langridge [94], modified by Weining and Henry [95]. Compared with normal CTAB methods, sarkosyl is employed instead in the extraction buffer and no high temperature water-bath is necessary. Because of the milder conditions adopted with this approach, high-quality DNA and RNA can be isolated at the same time, especially with mini-prep scale (S. Weining, unpublished data). This protocol can also be easily upgraded for larger amounts of starting material.

Reagents

Extraction buffer: 2% sarkosyl, 0.1 M Tris–HCl, 10 mM EDTA pH 8.0
Chloroform
Phenol:chloroform/isoamyl alcohol (25:24:1)
Isopropanol
75% Ethanol
3M Sodium acetate (pH 4.8)
TE: 10 mM Tris–Cl pH 8.0, 1 mM EDTA pH 8.0

Supplies and equipment

Plastic pestles or knitting needles, 2.0 mL tubes, microfuge, pipette tips, and gloves

Procedure

1. Grind 50–200 mg of leaves to a powder in 2 mL microfuge tubes in liquid nitrogen.
2. Add 0.6 mL of extraction buffer and subsequently, 0.6 mL of phenol/chloroform/isoamyl alcohol
3. Shake the mixture well for 30 s and leave it on ice for 10–20 min.
4. Recover the aqueous phase by centrifugation for 1 min. (This phenol/chloroform extraction can be repeated if necessary).

5. Add 0.6mL of chloroform and shake well before centrifugation for 1 min.
6. Collect the upper phase, add 0.5 mL of isoproponol and 50 μL of 3 M NaAc
7. Invert the tube gently for a few times and centrifuge for 30 s to 1 min
8. Wash the pellet twice with 75% ethanol.
9. Air-dry the pellet before adding 50–100 μL of TE or H_2O.

21.3.1.3 Organellar DNA

Plant cells harbor several distinct types of organelles (e.g., nuclei, mitochondria, and chloroplasts or plastids) that possess nucleic acids of their own. Analysis of these organellar nucleic acids often helps reveal valuable biological details on plant cells. For example, mitochondrial DNA (mtDNA) and RNA (mtRNA) are useful for molecular investigation of plant cytoplasmic variability. A number of protocols have been reported for the isolation of nucleic acids from nuclei (Chapter 23) [8,9,96], chloroplasts [97,98], and mitochondria [97,99,100]. Plant organelles can be separated by differential centrifugation or by selective lysis. For example, Peterson et al. [9] employed Triton X-100 to preferentially eliminate chloroplasts and mitochondria and leave nuclei intact. The protocol shown below utilizes differential centrifugation for selective removal of organelles, permitting subsequent extraction of mitochondrial, chloroplast, and nuclear DNA [101].

Reagents

Homogenization buffer A: 50 mM Tris–Cl pH 8.0, 1.3 M NaCl, 25 mM EDTA pH 8.0, 0.2% BSA. Add 0.05% cysteine and 56 mM β–mecaptoethanol before use.
Homogenization buffer B: 100 mM Tris–Cl pH 8.0, 2.6 M NaCl, 50 mM EDTA pH 8.0, 0.4% BSA. Add 0.1% cysteine and 56 mM β–mecaptoethanol before use.
Lysis buffer: 25 mM Tris–Cl pH 8.0, 20 mM EDTA pH 8.0, 0.5% SDS
Proteinase K: 10 mg/mL
Ammonium acetate: 2 M
TE: 10 mM Tris–Cl pH 8.0, 1 mM EDTA pH 8.0
TE-saturated phenol:chloroform (50:50)
Water-saturated phenol:chloroform:isoamyl alcohol (50:49:1)
Isopropanol
75% Ethanol
RNase A: 10 mg/mL

Supplies and equipment

Waring blender (for leaves), nylon filter (100 μm mesh), centrifuge, tubes

Procedure

1. Homogenize 10 g plant material (leaves, flowers, seeds, or other plant tissues) in 50 mL of cold homogenization buffer A in a cold room.
2. Filter through a nylon filter (100 μm), and centrifuge at 500 g for 10 min to remove large cell fragments.
3. Centrifuge the supernatant twice at 2600 g for 15 and 10 min, respectively, to remove plastids (which contain chlorophyll in the pellet and can be used in the step 6 for extraction of chloroplast DNA).
4. Centrifuge the supernatant at 14,500 g for 15 min to pellet mitochondria (supernatant contains nuclei, which can be used in Step 6 for extraction of nuclear DNA).
5. Resuspend the pellet in 20 mL of homogenization buffer A, centrifuge at 14,500 g for 15 min, and use the mitochondrial pellet directly for mtDNA or mtRNA extraction (or store at −80°C after freezing in liquid nitrogen for later use).

6. Resuspend the mitochondrial pellet in 1 mL of lysis buffer, add proteinase K to a final concentration of 50 µg/mL, and incubate at 37°C for 1 h.
7. Add 0.1 volume of ammonium acetate, and an equal volume of TE-saturated phenol: chloroform (50:50), mix, and centrifuge at 13,000 g for 10 min.
8. Transfer supernatant to a new tube, add 1 volume of isopropanol, and centrifuge at 13,000 g, 4°C for 10 min.
9. Discard supernatant, add 500 µL of 70% ethanol, and centrifuge at 13,000 g, 4°C for 2 min.
10. Dissolve the pellet in 50 µL of water, add 2 µL of RNase A (10 mg/mL), and incubate at 37°C for 30 min. Re-extract the DNA with water-saturated phenol:chloroform:isoamyl alcohol and precipitate with isopropanol.

21.3.2 Extraction of Plant RNA

In comparison with DNA, RNA is intrinsically unstable at nonfreezing temperature and liable to destruction by endogenous and exogenous RNases. Therefore, a number of precautions should be heeded during the RNA isolation process. These include the use of RNase-free reagents and consumables, and chemicals that inactivate RNases upon cell lysis. Since its first documented use, guanidine thiocyanate has been widely applied for isolation of RNA from various biological samples including plants due to its ability to inactivate RNases upon contact [13]. Several commercial kits (e.g., Qiagen RNeasy plant kit, see Chapter 10) based on this principle have also become available. Besides giving batch-to-batch consistency, these kits do not require hazardous chemicals such as phenol–chloroform. However, the relative cost of these kits is high (especially when a large number of samples are involved), and there is also inflexibility (i.e., inscalability) relating to the amount of starting material that can be processed for each preparation. We present below the methods for isolation of total RNA and organellar RNA using guanidine thiocyanate together with phenol–chloroform extraction. In case that messenger RNA (mRNA) is required, use of a commercial mRNA isolation kit is recommended.

21.3.2.1 Total RNA

Reagents

Trizol reagent (Molecular Research Center, Inc.; Sigma; or Invitrogen) [100]
Chloroform
Isopropyl alcohol
75% Ethanol in RNase-free water
RNase-free water [treated with 0.01% DEPC (v/v)]
Sterile plasticware and glassware

Equipment

Liquid nitrogen and mortar and pestle

Procedure

1. Grind approximately 100 mg leaves (seeds or other plant tissues) in liquid nitrogen.
2. Add 1 mL of Trizol reagent to the ground powder, transfer into 1.5 mL tube, centrifuge at 12,000 g for 5 min at 4°C, transfer supernatant to new 1.5 mL tube.
3. Add 200 µL of chloroform, mix by inversion, centrifuge at 12,000 g for 15 min at 4°C, and transfer the upper aqueous phase to a new 1.5 mL tube.
4. Add 0.5 mL of isopropyl alcohol, mix, centrifuge at 12,000 g for 10 min at 4°C, and carefully discard supernatant (Note: The pellet may be transparent or not clearly visible).

5. Add 1 mL of 75% ethanol, vortex briefly, centrifuge at 12,000 g for 5 min at 4°C.
6. Discard the supernatant, air-dry for 10 min, dissolve the pellet in 20 μL of RNase-free water by very gently sucking with a pipette, and incubate for 10 min at 60°C.
7. Store at −20 or −80°C until use, or extract the mRNA using commercial kits.

Note: TRI Reagent isolates a whole spectrum of RNA molecules. Ethidium bromide staining of RNA separated in an agarose gel visualizes two predominant bands of small (~2 kb) and large (~5 kb) ribosomal RNA, low molecular weight (0.1–0.3 kb) RNA, and discrete bands of high molecular weight (7–15 kb) RNA. Although the final preparation of total RNA is essentially free of DNA and proteins and has an A260/280 ratio of 1.6–1.9, for RT-PCR analysis, DNase treatment may be necessary for optimal results. For optimal spectrophotometric measurements, RNA aliquots should be diluted with water or buffer with a pH >7.5 such as phosphate buffer. Distilled water with a pH <7.0 falsely decreases the A260/280 ratio and impedes the detection of protein contamination in RNA samples [101].

21.3.2.2 Total RNA from Small Amounts of Plant Tissue

This simple method helps to extract total RNA from small amounts of tissue, such as a single mature wheat seed embryo, employing guanidine thiocyanate, and chloroform [102]. A "jacket" of liquid nitrogen and simplified procedures are applied to ensure the thorough grinding of the minute amounts of tissue and to minimize the loss of samples. These measures substantially increase the recovery of total RNA in the extraction process. Reliable downstream molecular analysis can be successfully achieved with the total RNA obtained by this approach. This method makes it valuable to study gene expression and gene regulation in a single wheat seed embryo. It may also give researchers the ability to analyze mRNAs in tissues or organs, not only just confined to plants, which are previously too small for RNA isolation using conventional procedures.

Reagents

Total RNA isolation buffer (freshly prepared): 12 μL 2-mercaptoethanol, 488 μL 4 M guanidine thiocyanate, and 100 μL chloroform (total volume 600 μL)
Absolute ethanol
80% Ethanol in RNase-free water
RNase-free water [treated with 0.01% DEPC (v/v)]

Equipment

Liquid nitrogen, 1.5 mL centrifuge tubes, 5 mL polypropylene tube (Sarstedt), disposable pellet pestle (Kontes), foam plastic blocks (can be made from packaging material), and a centrifuge

Procedure

1. Fix a 5 mL polypropylene tube in a foam plastic block and fill it with liquid nitrogen.
2. Place a 1.5 mL centrifuge tube containing the plant tissue, such as an embryo, into this 5 mL tube.
3. Liquid nitrogen in both tubes is replenished, until it no longer boils.
4. Grind the sample into a fine powder using a prechilled pellet pestle.
5. Allow the sample to warm up to room temperature after it is fully ground.
6. Add 100 μL RNA isolation buffer to the ground material and use the plastic pestle to ensure that a homogenized suspension is formed.
7. Add 500 μL RNA isolation buffer and also use it to rinse the material remaining on the pestle into the tube.

8. Mix the solution in the tube and leave it at room temperature for about 30 min.
9. Centrifuge the tube at 16,100 g for 10 min and transfer the supernatant into a new tube.
10. Add 1 mL absolute ethanol to the supernatant and place the tube at −20°C for 2 h following mixing.
11. Centrifuge the tube at 16,100 g for 10 min and discard the supernatant. Wash the pellet twice with 200 μL 80% ethanol.
12. Add 40–50 μL DEPC-treated MilliQ water and stir the solution with a pipette tip to assist in redissolving the pellet.
13. Centrifuge the tube at 16,100 g for 5 min and transfer the supernatant to a new tube for DNase treatment.

21.3.2.3 Organellar RNA

Plant organelles (mitochondria, chloroplasts, and nuclei) are prepared as in Section 21.3.1.3, and RNA can be extracted using guanidine thiocyanate [101,102]

Reagents

RNA extraction buffer: 4 M guanidine thiocyanate, 25 mM sodium citrate pH 7.0, 0.5% sarkosyl. Add 0.1% β–mecaptoethanol before use.
LiCl: 4 M
TE-saturated phenol:chloroform (50:50)
Water-saturated phenol:chloroform:isoamyl alcohol (50:49:1)
Isopropanol
75% Ethanol

Procedure

1. Resuspend organellar pellet (mitochondrial, chloroplast, or nuclear) in 0.6 mL of RNA extraction buffer, and incubate at 65°C for 2 min.
2. Add 0.1 volume of 2 M sodium acetate pH 4.0, add 1 volume of water-saturated phenol:chlororform:isoamylalcohol (50:49:1), mix, and centrifuge at 10,000 g for 10 min.
3. Transfer the supernatant to a new 1.5 mL tube, add 1 volume of isopropanol, and leave at −20°C for 1 h.
4. Centrifuge at 13,000 g for 20 min, discard supernatant, resuspend the pellet in 0.4 mL of 4 M LiCl, and centrifuge again at 13,000 g for 20 min.
5. Discard supernatant, dissolve the pellet in 0.3 mL of RNA extraction buffer at 65°C for 2 min, precipitate the RNA with 1 volume of isopropanol at −20°C for 1 h, centrifuge at 13,000 g for 20 min and wash with 75% ethanol twice, air-dry, and dissolve in 10–20 μL of water.

21.4 FUTURE DEVELOPMENT TRENDS

Although it is possible to isolate high quality of nucleic acids from plants through the use of various in-house reagents and commercial kits, the current protocols are generally labor intensive and produce variable outcome at times. Therefore, further optimization is needed to streamline the extraction process, so that nucleic acids of consistent yield and purity can be obtained from a variety of plant species and sample types. Clearly, one useful approach is to comparatively evaluate the existing procedures, with the aim to identify and pinpoint methods that are more efficient than others for recovery of nucleic acids from a particular sample type as well as among different sample types (either of the same plant species or of different species) [10,57,90,103]. Another approach is to look for ways that provide more efficient disruption of plant cell wall without compromising the nucleic acid quality. This will help speed up the nucleic acid process and make subsequent automation possible. Finally, innovation in the separation of nucleic acids from polysaccharides, polyphenols, proteins, and other metabolites in the plant cell lysate will undoubtedly contribute to the improved nucleic acid purity and yield from plants.

Song Weining received his BAgSc in plant pathology at Hunan Agricultural University in Changsha, China; he obtained his MAgSc and PhD in microbial physiology and plant genomics, respectively, from the University of Adelaide, Adelaide, South Australia, Australia. He worked in government and university laboratories in Australia for several years before taking up the position of head of the molecular biology laboratory at the Queensland Wheat Research Institute (Leslie Research Centre after 2000), Queensland State Government, Australia. Later he moved to the Institute of Evolution, University of Haifa, Haifa, Israel, as a senior scientist. Currently, he is a professor of plant genomics at the College of Agronomy, Northwest A&F University near Xian, China. His research interests include genome evolution, gene expression and function, with particular focus on stress genomics in crops, wild cereals, and fungi, encompassing functional genomics, molecular biology, plant physiology, and bioinformatics. He also has a strong interest in the domestication process of wild cereals.

REFERENCES

1. McMahon, M.E., Kofranek, A.M., and Rubatzky, V.E. (eds), *Hartmann's Plant Science: Growth, Development, and Utilization of Cultivated Plants,* 4th ed., Prentice Hall, Englewood Cliffs, NJ, 2007.
2. Murray, M.G. and Thompson, W.F., Rapid isolation of high molecular weight plant DNA, *Nucleic Acids Res.,* 8: 4321, 1980.
3. Doyle, J.J. and Doyle, J.L., Isolation of plant DNA from fresh tissue, *Focus,* 12: 13, 1990.
4. Michiels, A. et al., Extraction of high-quality genomic DNA from latex-containing plants, *Anal. Biochem.,* 315: 85, 2002.
5. Puchooa, D. and Khoyratty, S.S., Genomic DNA extraction from *Victoria amazonica, Plant Mol. Biol. Rep.,* 22: 195a, 2004.
6. Jofuku, K.D. et al., Analysis of plant gene structure, in: Shaw, C.H. (ed.), *Plant Molecular Biology, A Practical Approach,* IRL Press, Oxford 1988, pp. 37–44.
7. Bhattacharjee, R. et al., An improved semiautomated rapid method of extracting genomic DNA for molecular marker analysis in cocoa, *Theobroma cacao* L, *Plant Mol. Biol. Rep.,* 22: 435a, 2004.
8. Wang, X.D., Wang, Z.P., and Zou, Y.P., An improved procedure for the isolation of nuclear DNA from the leaves of wild grapevine dried with silica gel, *Plant Mol. Biol. Rep.,* 14: 369, 1996.
9. Peterson, D.G., Boehm, K.S., and Stack, S.M, Isolation of milligram quantities of nuclear DNA from tomato *(Lycopersicon esculentum),* a plant containing high levels of polyphenolic compounds, *Plant Mol. Biol. Rep.,* 15: 148, 1997.
10. Krasova-Wade, T. and Neyra, M., Optimization of DNA isolation from legume nodules, *Lett. Appl. Microbiol.,* 45: 95, 2007.
11. Lodhi, M.A. et al., A simple and efficient method for DNA extraction from grapevine cultivars, *Vitis* species and *Ampelopsis, Plant Mol. Biol. Rep.,* 12: 6, 1994.
12. Barnwell, P. et al., Isolation of DNA from the highly mucilagenous succulent plant *Sedum telephium, Plant Mol. Biol. Rep.,* 16: 133, 1998.
13. Chomczynski, P. and Sacchi N., Single-step method of RNA isolation by acid guanidinium thiocyanate-phenol-chloroform extraction, *Anal. Biochem.,* 162: 156, 1987.
14. Rogstad, S.H. et al., DNA extraction from plants: The use of pectinase, *Plant Mol. Biol. Rep.,* 19: 353, 2001.
15. Manen, J.F. et al., A fully automatable enzymatic method for DNA extraction from plant tissues, *BMC Plant Biol.,* 5: 23, 2005.
16. Pirttila, A.M. et al., DNA isolation methods for medicinal and aromatic plants, *Plant Mol. Biol. Rep.,* 19: 273a, 2001.
17. Demeke, T. and Adams, R.P., The effects of plant polysaccharides and buffer additives on PCR, *Biotechniques,* 12: 332, 1992.
18. Do, N. and Adams, R.P., A simple technique for removing plant polysaccharide contaminants from plant DNA, *Biotechniques,* 10: 162, 1991.
19. Jobes, D.V., Hurley, D.L. and Thien, L.B., Plant DNA isolation: A method to efficiently remove polyphenolics, polysaccharides, and RNA, *Taxon,* 44: 349, 1995.

20. Marechal-Drouard, L. and Guillemaut, P., A powerful but simple technique to prepare polysaccharide-free DNA quickly and without phenol extraction, *Plant Mol. Biol. Rep.*, 13: 26, 1995.
21. Porebski, S., Bailey, L.G., and Baum, B.R., Modification of a CTAB DNA extraction protocol for plants containing high polysaccharide and polyphenol components, *Plant Mol. Biol. Rep.,* 15: 8, 1997.
22. Sharma, A.D., Gill, P.K., and Singh, P., DNA isolation from dry and fresh samples of polysaccharide-rich plants, *Plant Mol. Biol. Rep.*, 20: 415a, 2002.
23. Sangwan, R.S., Yadav, U., and Sangwan, N.S., Isolation of genomic DNA from defatted oil seed residue of opium poppy *(Papaver sominiferum), Plant Mol. Biol. Rep.*, 18: 265, 2000.
24. Tel-Zur, N. et al., Modified CTAB procedure for DNA isolation from *Epiphytic cacti* of the genera *Hylocereus* and *Selenicereus* (Cactaceae), *Plant Mol. Biol. Rep.*, 17: 249, 1999.
25. Fang, G., Hammer, S., and Grumet, R., A quick and inexpensive method for removing polysaccharide from plant genomic DNA, *Biotechniques,* 13: 52, 1992.
26. Zidani, S., Ferchichi, A., and Chaieb, M., Genomic DNA extraction method from pearl millet *(Pennisetum glaucum)* leaves, *Afr. J. Biotechnol.*, 4: 862, 2005.
27. Aljanabi, S.M. and Martinez, I., Universal and rapid salt-extraction of high quality genomic DNA for PCR-based techniques, *Nucleic Acids Res.*, 25: 4692, 1997.
28. Syamkumar, S., Lowarence, B., and Sasikumar, B., Isolation and amplification of DNA from rhizomes of turmeric and ginger, *Plant Mol. Biol. Rep.*, 21: 171a, 2003.
29. Sharma, K.K., Lavanya, M., and Anjaiah, V., A method for isolation and purification of peanut DNA suitable for analytical applications, *Plant Mol. Biol. Rep.*, 18: 393a, 2000.
30. John, M.E., An efficient method for isolation of RNA and DNA from plants containing polyphenolics, *Nucleic Acids Res.* 20: 2381, 1992.
31. Couch, J.A. and Fritz, P.J., Isolation of DNA from plants high in phenolics, *Plant Mol. Biol. Rep.*, 8: 8, 1990.
32. Loomis, W.D., Overcoming problems of phenolics and quinones in the isolation of plant enzymes and organelles, *Methods Enzymol.*, 31: 528, 1974.
33. Pich, U. and Schubert, I., Midiprep method for isolation of DNA from plants with a high content of polyphenolics, *Nucleic Acids Res.*, 1993: 21, 3328.
34. Kim, C.S. et al., A simple and rapid method for isolation of high quality genomic DNA from fruit trees and conifers using PVP, *Nucleic Acids Res.*, 25: 1085, 1997.
35. Puchooa, D., A simple rapid and efficient method for the extraction of genomic DNA from lychee *(Litchi chinensis* Sonn.), *Afr. J. Biotechnol.*, 3: 253, 2004.
36. Suzuki, Y. et al., Extraction of total RNA from leaves of Eucalyptus and some other woody and herbaceous plants using sodium isoascorbate, *Biotechniques,* 34: 988, 2003.
37. Chen, D.H. and Ronald, P.C., A rapid DNA minipreparation method suitable for AFLP and other PCR applications, *Plant Mol. Biol. Rep.*, 17: 53, 1999.
38. Permingeat, H.R., Romagnoli, M.V., and Vallejos, R.H., A simple method for isolating high yield and quality DNA from cotton *(Gossypium hirsutum* L.) leaves, *Plant Mol. Biol. Rep.,* 16: 1, 1998.
39. Thangjam, R., Maibam, D., and Sharma, J.G., A simple and rapid method for isolation of DNA from imbibed embryos of *Parkia timoriana* (DC.) Merr. for PCR analysis, *Food Agric. Environ.*, 1: 36, 2003.
40. Thomas, S.M., Moreno, R.F., and Tilzer, L.L., DNA extraction with organic solvents in gel barrier tubes, *Nucleic Acids Res.*, 17: 5411, 1989.
41. Rogstad, S.H., Plant DNA extraction using silica, *Plant Mol. Biol. Rep.*, 21: 463a, 2003.
42. Tanaka, J. and Ikeda, S., Rapid and efficient DNA extraction method form various plant species using diatomaceous earth and a spin filter, *Breeding Sci.*, 52: 151, 2002.
43. Ahmad, S.M. et al., Rapid DNA isolation protocol for angiospermic plants, *Bulg. J. Plant Physiol.*, 30: 25, 2004.
44. Bake, S.S., Rugh, C.L., and Kamalay, J.C., DNA and RNA isolation from recalcitrant plant tissues, *Biotechniques,* 9: 268, 1990.
45. Cheng, Y.J. et al., An efficient protocol for genomic DNA extraction from citrus species, *Plant Mol. Biol. Rep.*, 21: 177a, 2003.
46. Deshmukh, V.P. et al., A simple method for isolation of genomic DNA from fresh and dry leaves of *Terminalia arjuna* (Roxb.) Wight and Arnot, *Electron. J. Biotechnol.*, 10: 3, 2007.
47. Guillemaut, P. and MarcchaL-Drouard, L., Isolation of plant DNA: A fast, inexpensive and reliable method, *Plant Mol. Biol. Rep.*, 10: 60, 1992.
48. Hanania, U., Velcheva, M., Sahar, N., and Perl, A., An improved method for isolating high-quality DNA from *Vitis vinifera* nuclei, *Plant Mol. Biol. Rep.*, 22: 173, 2004.

49. Hansen, N.J.V. et al., A fast, economical and efficient method for DNA purification by use of a home-made bead column, *Biochem. Mol. Biol. Int.*, 35: 461, 1995.

50. Haymes, K.M. et al., Rapid isolation of DNA from chocolate and date palm tree crops, *J. Agric. Food Chem.*, 52: 5456, 2004.

51. Kang, T.J. and Yang, M.S., Rapid and reliable extraction of genomic DNA from various wild-type and transgenic plants, *BMC Biotechnol.*, 4: 20, 2004.

52. Li H. et al., A rapid and high yielding DNA miniprep for cotton (*Gossypium* spp.), *Plant Mol. Biol. Rep.* 19: 183a, 2001.

53. Mishra, M.K. et al., A simple method of DNA extraction from coffee seeds suitable for PCR analysis, *Afr. J. Biotechnol.*, 7: 409, 2008.

54. Mogg, R.J. and Bond, J.M., A cheap, reliable and rapid method of extracting high-quality DNA from plants, *Mol. Ecol. Notes*, 3: 666, 2003.

55. Rogers S. and Bendich A: Extraction of DNA from milligrams of fresh, herbarium and mummified plant tissues, *Plant Mol. Biol.*, 5: 69, 1985.

56. Smith, J.F. et al., A qualitative comparison of total cellular DNA extraction protocols, *Phytochem. Bull.*, 23: 2, 1991.

57. Berendzen, K. et al., A rapid and versatile combined DNA/RNA extraction protocol and its application to the analysis of a novel DNA marker set polymorphic between Arabidopsis thaliana ecotypes Col-0 and *Landsberg erecta, Plant Methods*, 1: 4, 2005.

58. Burr K., Harper R., and Linacre A., One-step isolation of plant DNA suitable for PCR amplification, *Plant Mol. Biol. Rep.*, 19: 367, 2001.

59. Chaudhry, B. et al., Mini-scale genomic DNA extraction from cotton, *Plant Mol. Biol. Rep.*, 17: 1, 1999.

60. Chakraborty, D. et al., Small and large scale genomic DNA isolation protocol for chickpea (*Cicer arietinum* L.), suitable for molecular marker and transgenic analyses, *Afr. J. Biotechnol.*, 5: 585, 2006.

61. Csaikl, U.M. et al., Comparative analysis of different DNA extraction protocols: A fast, universal maxi-preparation of high quality plant DNA for genetic evaluation and phylogenetic studies, *Plant Mol. Biol. Rep.* 16: 69, 1998.

62. Dilworth, E. and Frey, J., A rapid method for high throughput DNA extraction from plant material for PCR amplification, *Plant Mol. Biol. Rep.*, 18: 61, 2000.

63. Dixit, A., A simple and rapid procedure for isolation of *Amaranthus* DNA suitable for fingerprint analysis, *Plant Mol. Biol. Rep.*, 16: 1, 1998.

64. Doyle, J.J. and Doyle, J.L., A rapid isolation procedure for small quantities of fresh leaf tissue, *Phytochem. Bull.*, 19: 11, 1987.

65. Edwards, K., Johnstone, C., and Thompson, C., A simple and rapid method for the preparation of plant genomic DNA for PCR analysis, *Nucleic Acids Res.*, 19: 1349, 1991.

66. Guidet, F., Rogowsky, P., and Langridge, P., A rapid method of preparing megabase plant DNA, *Nucleic Acids Res.*, 18: 4955, 1990.

67. Hill-Ambroz, K.L., Brown-Guedira, G.L., and Fellers, J.P., Modified rapid DNA extraction protocol for high throughput microsatellite analysis in wheat, *Crop Sci.*, 42: 2088, 2002.

68. Hosaka, K., An easy, rapid and inexpensive DNA extraction method, one minute DNA extraction for PCR in potato, *Am. J. Potato Res.*, 81: 17, 2004.

69. Karakousis, A. and Langridge, P., A high-throughput Plant DNA extraction method for marker analysis, *Plant Mol. Biol. Rep.*, 21: 95a, 2003.

70. Kasajima, I. et al., A protocol for rapid DNA extraction from *Arabidopsis thaliana* for PCR analysis, *Plant Mol. Biol. Rep.*, 22: 49, 2004.

71. Klimyuk, V. et al., Alkali treatment for rapid preparation of plant material for reliable PCR analysis, *Plant* J., 3: 493, 1993.

72. Kuta, D.D. et al., Optimization of protocols for DNA extraction and RAPD analysis in West African fonio (*Digitaria exilis* and *Digitaria iburua*) germplasm characterization, *Afr. J. Biotechnol.*, 4: 1368, 2005.

73. Lange, D.A. et al., A plant DNA extraction protocol suitable for polymerase chain reaction based marker-assisted selection, *Crop Sci.*, 38: 217, 1998.

74. Lin, R.-C. et al., A rapid and efficient DNA minipreparation suitable for screening transgenic plants, *Plant Mol. Biol. Rep.*, 19: 379a, 2001.

75. Mace, E.S., Buhariwalla, H.K., and Crouch, J.H., A high-throughput DNA extraction protocol for tropical molecular breeding programs, *Plant Mol. Biol. Rep.*, 21: 459a, 2003.

76. Paris, M. and Carter, M., Cereal DNA: A rapid high-throughput extraction method for marker assisted selection, *Plant Mol. Biol. Rep.*, 18: 357, 2000.

77. Sangwan, N.S., Sangwan, R.S., and Kumar, S., Isolation of genomic DNA from the antimalarial plant *Artemisia annua, Plant Mol. Biol. Rep.,* 16: 1, 1998.
78. Sarwat, M., et al., A standardized protocol for genomic DNA isolation from *Terminalia arjuna* for genetic diversity analysis, *Electron. J. Biotechnol.,* 9: 86, 2006.
79. Shepherd, M., et al., High-throughput DNA extraction from forest trees, *Plant Mol. Biol. Rep.,* 20: 425a, 2002.
80. Singh, M., Bandana, and Ahuja, P.S., Isolation and PCR amplification of genomic DNA from market samples of dry tea, *Plant Mol. Biol. Rep.,* 17: 171, 1999.
81. Syamkumar, S., Lowarence, B., and Sasikumar, B., Isolation and amplification of DNA from rhizomes of turmeric and ginger, *Plant Mol. Biol. Rep.,* 21: 171a, 2003.
82. Thomson, D. and Henry, R., Single-step protocol for preparation of plant tissue for analysis by PCR, *Biotechniques* 19: 394, 1995.
83. Wang, H, Qi, M, and Cutler, A., A simple method of preparing plant samples for PCR, *Nucleic Acids Res.,* 21: 4153, 1993.
84. Warude, D. et al., DNA isolation from fresh and dry samples with highly acidic tissue extracts, *Plant Mol. Biol. Rep.,* 21: 467a, 2003.
85. Williams, C.E. and Ronald, P.C., PCR template-DNA isolated quickly from monocot and dicot leaves without tissue homogenization, *Nucleic Acids Res.,* 22: 1917, 1994.
86. Xin, Z. et al., High-throughput DNA extraction method suitable for PCR, *Biotechniques,* 34: 820, 2003.
87. Zhang, J. and Stewart, J., Economical and rapid method for extracting cotton genomic DNA, *J. Cotton Sci.,* 4: 193, 2000.
88. Ligozzi, M. and Fontana, R., Isolation of total DNA from bacteria and yeast, *Afr. J. Biotechnol.,* 2: 251, 2003.
89. Krasova-Wade, T. et al., Diversity of indigenous bradyrhizobia associated with three cowpea cultivars (*Vigna unguiculata* (L.) Walp.) grown under limited and favorable water conditions in Senegal (West Africa), *Afr. J. Biotechnol.,* 2: 13, 2003.
90. Ribeiro, R.A. and Lovato, M.B., Comparative analysis of different DNA extraction protocols in fresh and herbarium specimens of the genus *Dalbergia, Genet. Mol. Res.,* 6: 173, 2007.
91. Dellaporta, S.L., Wood, J., and Hicks, J.B., A plant minipreparation: Version II, *Plant Mol. Biol. Rep.,* 1: 19, 1983.
92. Scott, K.D. and Playford, J., DNA extraction technique for PCR in rain forest plant species, *Biotechniques,* 20: 974, 1996.
93. Ríos, G. et al., Rapid identification of Arabidopsis insertion mutants by non-radioactive detection of T-DNA tagged genes, *Plant J.,* 32: 243, 2002.
94. Weining, S. and Langridge, P., Identification and mapping polymorphism in cereals based on polymerase chain reaction, *Theor. Appl. Genet.,* 82: 209, 1991.
95. Weining, S. and Henry, R., Polymorphisms in α-*amy1* gene of wild and cultivated barley revealed by the polymerase chain reaction, *Theor. Appl. Genet.,* 89: 509, 1994.
96. Zhang, H.-B. et al., Preparation of megabase-size DNA from plant nuclei, *Plant J.,* 7: 175, 1995.
97. Triboush, S.O., Danilenko, N.G., and Davydenko, O.G., A method for isolation of chloroplast DNA and mitochondrial DNA from sunflower, *Plant Mol. Biol. Rep.,* 16: 183, 1998.
98. Mariac, C. et al., Chloroplast DNA extraction from herbaceous and woody plants for direct restriction fragment length polymorphism analysis, *Biotechniques,* 28: 110, 2000.
99. Wilson, A.J. and Chourey, P.S., A rapid inexpensive method for the isolation of restrictable mitochondrial DNA from various plant sources, *Plant Cell Rep.,* 3: 237, 1984.
100. Scotti, N. et al., Mitochondrial DNA and RNA isolation from small amounts of potato tissues, *Plant Mol. Biol. Rep.,* 19: 67a, 2001.
101. TRI Reagent—RNA, DNA, protein isolation reagent, Manufacturer's protocol, 1995, Molecular Research Center, Inc., Cincinnati, OH (http://www.mrcgene.com/tri.htm).
102. Chen, Y.H., Weining, S., and Daggard, G., Preparation of total RNA from a very small wheat embryo suitable for differential display, *Ann. Appl. Biol.,* 143: 261, 2003.
103. Drabkova, L., Kirschner, and J., Vlcek, C., Comparison of seven DNA extraction and amplification protocols in historical herbarium specimens of *Juncaceae, Plant Mol. Biol. Rep.,* 20: 161, 2002.

22 Isolation of Megabase-Sized DNA Fragments from Plants

Meiping Zhang, Yaning Li, and Hong-Bin Zhang

CONTENTS

22.1 INTRODUCTION

22.1.1 MEGABASE-SIZED DNA AND THEIR APPLICATIONS

With the rediscovery of Mendel's laws, determination of DNA as the genetic material, resolution of the double-helix structure of DNA and its implications for genetic behavior during the past century, biology research has entered the era of genomics, for which studies are emphasized at the genome-wide or biological process-wide levels. One of the basic, but essential prerequisites for such studies is the availability of techniques for preparation of DNA that is megabase-sized from organisms of interest, considering the fact that conventional DNA extraction procedures (including in-house reagents and commercial kits) usually result in the isolation of DNA under 120 kb in size. Recent development of procedures for isolation of megabase-sized DNA fragments has opened new avenues for genomics research on a variety of biological organisms. The potential use of megabase DNA fragments includes the analysis of long-genome spanning genes (e.g., the mammalian *dystrophin* gene that has a transcript of >2000 kb), characterization of gene regulatory elements and clusters of genes (e.g., plant disease resistance genes), development of large-insert genomic DNA libraries [1–5], genome physical mapping [6–10], long-range genome analysis [11,12], map-based or positional cloning of genes and quantitative trait loci (QTLs) [13], large-scale genome sequencing [2,14], and microbe genome karotyping [15–17]. Furthermore, a recent study [3] showed that DNA molecules contained in a genome are structured as linear "jigsaw puzzle" and that the content, array, and interaction of the fundamental functional elements constituting the DNA jigsaw puzzle structure are responsible for the abundance, diversity, and complexity of living organisms, indicating the importance of long-range genome analysis in biology research.

22.1.2 PRINCIPLES OF MEGABASE-SIZED DNA ISOLATION FROM PLANTS

Techniques have been developed to isolate DNA from a variety of organisms for different purposes of molecular research. However, the DNA isolated with the conventional procedures are usually about 120 kb or smaller [1] due to physical shearing during isolation, which is not suited for modern genomics research such as those indicated above. Therefore, it is essential to develop techniques that enable rapid and ready isolation of megabase-sized DNA from different species of plants, animals, insects, fishes, and microbes. For DNA isolation, three major steps are often included: (1) isolation of whole cells, protoplasts, or nuclei from organism's tissues; (2) lysis of the source cells to release the cell contents including DNA while simultaneously removing nucleases that potentially degrade DNA, histones that bind to DNA, and metabolic substances (such as polysaccharides and polyphenolics) that may inhibit the activity of commonly applied molecular biological reagents; and (3) harvesting and postpurification of DNA to make it suitable for different purposes of genetic research. In the procedures of conventional-sized DNA isolation, the whole cells, protoplasts, or nuclei that are used for DNA isolation can be isolated by directly harvesting culture or blood cells, digesting cell walls with hydrolases, or homogenizing the source tissues in liquid nitrogen or with a blender. The nucleases, histones, and metabolic substances are removed by extracting the lysate of the cells with chloroform and phenol. The DNA is harvested and further purified by precipitation with ethanol or isopropanol, followed by washing with 70% ethanol.

For the isolation of megabase-sized DNA, the DNA source cells, protoplasts, or nuclei can be generated as those used in the conventional-sized DNA isolation procedure. However, the procedures for preparation of megabase-sized DNA must include a method to protect DNA from physical shearing during DNA isolation. This can be facilitated by embedding the cells, protoplasts, or nuclei into a solid but porous supporting matrix such as low-melting-point (LMP) agarose before they are lysed to release their contents. The pores of the solid matrix would allow reaction reagents to access the DNA embedded in when it is used for different research purposes. For removal of the nucleases, histones, and metabolites in the cells, chloroform and phenol can no longer be used because they would destroy the supporting matrix as well. Instead, a protein degradation enzyme, proteinase K, is introduced into the procedures to degrade the nucleases and remove the histones from DNA, whereas the metabolites that may inhibit DNA manipulation in later use such as digestion could be minimized or eliminated by repeatedly washing the source nuclei prior to being embedded in LMP agarose since they exist in cytoplasm. The proteinase K that may also break up the enzymes to be used in later DNA manipulation and salts that are trapped in the agarose matrix are removed by treating with a proteinase inhibitor, phenylmethylsulfonyl fluoride (PMSF), followed by dialysis against Tris–EDTA (TE) buffer (for detail, see below).

There are differences in megabase-sized DNA isolation between plants and animals. This is because plant cells have walls, relative to animals, insects, fishes, and microbes whose cells have no or much less-specialized walls. Therefore, to isolate megabase-sized DNA from plants, the cell walls must be removed. However, the cells isolated from cultures, blood, or tissues of animals, insects, fishes, and microbes can be directly harvested and embedded in LMP agarose matrix. Moreover, unlike animals, insects, fishes, and microbes, plants often contain abundant metabolic substances due to their photosynthesis process. Although the products of photosynthesis such as polysaccharides, starches, and polyphenolics all could affect the subsequent manipulation of DNA for different purposes of genomics research, polyphenolics have been demonstrated to be much more problematic for megabse-sized DNA manipulation [18]. This is especially true for dicot, bush, and tree species, such as cotton, rose, and willows in which phenolics are abundantly present. The polyphenolics interact with DNA, making it no longer digestible with a restriction enzyme or no longer clonable. In addition, since the growing conditions of plants could significantly affect the status of plant metabolism, they could affect the quality of megabase-sized DNA isolated. Therefore, selective use of young tissues at proper developmental stage or pretreatment of the plants before sampling will be helpful for isolation of megabase-sized DNA of high quality.

22.1.3 DEVELOPMENT OF TECHNIQUES FOR MEGABASE-SIZED DNA ISOLATION FROM PLANTS

The isolation of megabase-sized DNA was first pursued in animals, especially human and mouse, mainly for construction of yeast artificial chromosome (YAC) libraries [19] and then yeast for purification of YAC DNAs. Since animals have no cell walls, the cells isolated from blood or other tissues are directly embedded in LMP agarose [20], whereas the cell walls of yeast are removed by lyticase or zymolase to form spheroplasts (similar to plant protoplasts) before being embedded into LMP agarose [21]. Therefore, plant scientists, following the concept of megabase-sized DNA isolation in animals and yeast, developed a procedure to isolate megabase-sized DNA from plant protoplasts by digesting the cell walls with hydrolases such as cellulase and pectinase, and then embedding into LMP agarose [22–29]. This so-called protoplast procedure worked well in some species such as *Arabidopsis* [26], tomato [22], and sorghum [29], but did not in others such as cotton [18]. This is because of several reasons. First of all, there must be a procedure available for isolation of large amount of protoplasts from a plant tissue. Second, the procedure of protoplast isolation, if available, is often species- or genotype-specific due to the use of cell wall hydrolases involved in the protoplast isolation. This implies that a protoplast isolation procedure may work for one species or genotype, but not for another species or genotype. Third, the most important is that the protoplast procedure does not work for isolation of megabase-sized DNA from species that are abundant in polyphenolics in cytoplasm because they interact with the DNA, making it no longer digestible, for instance, cotton and many tree species [18,30]. Fourth, the protoplast procedure is time consuming and costly. Finally, the DNA isolated with the protoplast procedure is often contaminated with cytoplast organelle DNA such as chloroplast and mitochondria DNA at a level of 10% or higher [30].

Therefore, a procedure, named here the nuclei procedure, was developed by Zhang et al. [30]. In this procedure, plant nuclei are isolated by simply grinding the source tissues in liquid nitrogen or homogenizing the tissues with a kitchen bleeder, followed by nuclei isolation, in which no cell wall hydrolases are involved as those of the protoplast procedure. Zhang et al. [30] showed that the nuclei isolated with the procedure were not only clean but also intact, thus being well suited for megabase-sized nuclear DNA isolation from plants. Furthermore, the organelles and metabolites contained in cytoplasm, such as polyphenolics, can be minimized in the procedure by repeatedly washing the nuclei with the nuclei isolation buffer before being embedded into LMP agarose. So, the nuclei procedure eliminates or minimizes all problems that are associated with megabase-sized DNA isolation from plant protoplasts [22,25,29]. Furthermore, it is simple, economical, and widely applicable for preparation of megabase-sized DNA from plants. Zhang et al. [30] also showed that the DNA isolated with the nuclei procedure was at least over 10 Mb in size, readily digestible, and low in contamination with organelle DNA. Therefore, this procedure has rapidly become the method of choice for preparation of megabase-sized DNA from plants and has been widely used in different areas of genomics research. Moreover, the concept and procedure of nuclei isolation and megabase-sized DNA preparation used in the plant nuclei procedure have been later adopted for megabase-sized DNA preparation from species of other kingdoms of organisms, including animals, insects, fishes, and microbes even though in some cases the nuclei isolation buffer may be modified. Therefore, the development of the nuclei procedure has made it possible to rapidly isolate megabase-sized DNA of high quality from a variety of organisms, including plants, animals, insects, fishes, and microbes, and thus rapidly and significantly promoted genomics research in these species. In this chapter, we present the nuclei procedure of Zhang et al. [30] with our further improvements and research experience in preparation of quality megabase-sized DNA from over 100 species of plants, animals, fishes, insects, and microbes.

22.2   METHOD FOR MEGABASE-SIZED DNA ISOLATION FROM PLANT NUCLEI

22.2.1   Reagents and Equipment

Reagents:

10x homogenization buffer (HB) stock: 0.1 M Trizma base, 0.8 M KCl, 0.1 M EDTA, 10 mM spermidine, 10 mM spermine, final pH 9.4–9.5 adjusted with NaOH. The stock is stored at 4°C.

Note: The pH value of the HB stock was found to be crucial to the success of quality megabase-sized DNA isolation. As DNA is acidic in chemistry, the pH value of the buffer, if lower than 8.0, would likely lead to partial degradation of the megebase-sized DNA. The DNA isolated using the buffer of pH < 8.0 is often only 100–200 kb in size.

1x HB: A suitable amount of sucrose is mixed with a suitable volume of 10x HB stock. The final concentration of sucrose is 0.5 M and HB stock is 1x. The resultant 1x HB is stored at 4°C.
1x HB plus 20% Triton X-l00: Triton X-100 is mixed with 1x HB to 20%. The solution is stored at 4°C.
Nuclei isolation buffer (1x HB plus 0.5% Triton X-100 and 0.15% β-mercaptoethanol): It is prepared by mixing 1x HB with 1x HB plus 20% Triton X-l00 and stored at 4°C. Before use, β-mercaptoethanol is added to 0.15%.
Lysis buffer: 0.5 M EDTA, pH 9.0–9.3, 1% sodium lauryl sarcosine, and 0.3 mg/mL proteinase K. The proteinase K powder is added just before use.
TE: 10 mM Tris–HCl, pH 8.0 and 1 mM EDTA pH 8.0.
100 mM PMSF: PMSF is a highly toxic acetylcholinesterase inhibitor. To avoid laboratory contamination, we purchase 250 mg aliquots of PMSF stored in a vial, then add 14.35 mL of isopropanol to the vial directly to make a final concentration of 100 mM and store at 4°C.

Note: The solution of 100 mM PMSF in isoproponal is stable at 4°C for at least 2 years. PMSF is recrystallized while it is stored at 4°C. The solution should be warmed up at 37°C for 15–20 min to dissolve the crystals before use.

Supplies and equipment:

Liquid nitrogen, large mortar and pestle (30 cm in diameter), 1000-mL beaker, −80°C freezer, magnetic stirrer and bars, cheesecloth, miracloth, refrigerated centrifuge, 250-mL centrifuge bottles, LMP agarose, small paintbrushes, 100-μL plug molds, and microwave.

22.2.2   Plant Material

Plant leaves or whole plants of taxa-diverse species, including grasses, legumes, vegetables, and trees, can be used as materials for preparation of megabase-sized DNA by using this procedure. However, following tissues are preferred: (1) whole plants or parts of plants that are growing at seedling stages; (2) young leaves or meristems that are less active in metabolism (for animals, sperms, and newly-hatched larvae are the desirable choice of tissues), or (3) tissues collected from plants that are pretreated in dark for two or more days. Whenever facilities are allowed, the plants of DNA sources should be planted in light-, temperature-, and humidity-controllable growth chamber or room as the first choice of plant growth, then in greenhouse, and in the fields as the last choice. The tissues can be either frozen and stored in a −80°C freezer or kept fresh on ice before DNA isolation.

22.2.3   Procedure

Preparation of intact nuclei

1. Grind about 50 g of the fresh or frozen tissue into fine powder in a large amount of liquid nitrogen with a large mortar and pestle (about 30 cm in diameter) and immediately

transfer into an ice-cold 1000 mL beaker containing about 500 mL of the nuclei isolation buffer.

Note: The mortar must be cooled in a −80°C freezer for at least 1 h before use; otherwise, it would be broken when liquid nitrogen is added.

2. Gently swirl the contents with a magnetic stir bar for 5 min on ice and filter into two or more ice-cold 250 mL centrifuge bottles through two layers of cheesecloth and one layer of miracloth by squeezing with gloved hands.

Note: In the case that there is no miracloth available, 3–4 layers of cheesecloth could be used to separate the cell debris out of the nuclei.

3. Pellet the homogenate contained in the 250 mL centrifuge bottles by centrifugation with fixed-angle rotor at 4500 rpm (3110 g) at 4°C for 20 min.

Notes: This step is to harvest and separate nuclei from organelles such as chloroplasts and mitochondria as well as the cytoplasm-containing metabolites such as polyphenolics. So, the speed of centrifugation is important to control the DNA quality. For most plant and animal species, 3500–5000 rpm are used for the centrifugation; however, an adjustment may be needed for different species, depending on their differences in nuclei or genome sizes [31]. In the cases of isolating nuclei from grass species, the nuclei pellet is loose; therefore, the pellet should be harvested as soon as the centrifugation is completed.

4. Discard the supernatant fluid and add approximately 1 mL of ice-cold nuclei isolation buffer to each bottle.

Note: Care should be taken to discard the supernatant fluid in the case of isolating nuclei from grass species because the pellet is loose. Some leftover of the supernatant would not affect the quality of the DNA.

5. Gently resuspend the pellet with assistance of a small paintbrush soaked in ice-cold nuclei isolation buffer and add additional 5 mL of the ice-cold nuclei isolation buffer to each centrifuge bottle.
6. Transfer the supernatant fluid in all centrifuge bottles into a fresh 40 mL centrifuge tube and pellet the nuclei by centrifugation at 4500 rpm (2790 g), 4°C for 15 min.
7. Wash the pellet 1–5 additional times by resuspending in ice-cold nuclei isolation buffer, followed by centrifugation at 4500 rpm (2790 g), 4°C for 15 min.

Note: Additional washes will allow maximal removal of the cytoplasm contents, such as chloroplasts, mitochondria and metabolic substances. This is especially important for isolation of high-quality megabase-sized DNA from plants that have abundant polyphenolic substances such as cotton and trees.

8. After the final wash, resuspend the pelleted nuclei in a small amount (about 1 mL) of 1x HB, count the nuclei, if possible, under the contrast phase of a microscope, bring the nuclei concentration to approximately 5×10^7 nuclei/mL (this is for a species having a genome size of 1000 Mb/1C) with addition of ice-cold 1x HB, and store on ice.

Note: The concentration of the nuclei could be estimated for making LMP plugs of megabase-sized DNA, and the concentration of nuclei should be proper if the nuclei suspension is just transparent under light. Nevertheless, the light transparence of the nuclei isolated from dicot plants is usually much worse than that of the nuclei isolated from monocot plants. Therefore, the desirable degree of the dicot plant nuclei suspension transparence for preparation of megabase-sized DNA LMP agarose plugs should be much lower than for that of the monocot plant nuclei suspension. This is because the dicot plant nuclei are often contaminated with larger amount of metabolites that may reduce the light transparence of the nuclei suspension.

The concentration of DNA embedded in LMP agarose plugs at 5–10 μg/100 μL plug has been proven to be desirable for most purposes of genome research. Therefore, how many nuclei should be embedded in a 100 μL plug is dependent on the genome size of species from which megabase-sized DNA is isolated. According to Arumuganathan and Earle [31], the genome size of 1000 Mb is approximately equivalent to about 1.04 pg of DNA. For instance, if the targeted species has a genome size of about 1000 Mb/1C or 1.04 pg/1C, the concentration of the nuclei suspension should be 5–10 μg/100 μL plug (containing 50 μL of the nuclei suspension as indicated below) × 10^6 pg/μg divided by 2 × 1.04 pg/cell = 2.4–4.8 × 10^6 nuclei per 50 μL, i.e., 2.4–4.8 × 10^6 × (1000/50) = 4.8–9.6 × 10^7 nuclei/mL.

Alternatively, for those who have no proper microscopes or no experience to estimate the concentration of nuclei suspension, a few dilutions of nuclei suspension could be made and used to prepare LMP agarose plugs. The plugs containing the most optimal concentration of DNA for targeted research are selected and used.

Embedding nuclei in LMP agarose plugs

10. Prewarm the nuclei to 45°C in a water bath (about 5 min) before being embedded in agarose.

Note: How long the nuclei suspension is incubated in the 45°C water bath before mixing with the 1% LMP agarose depends on the volume of the nuclei suspension. If it is too short, the agarose gel would be solidified in the tube, thus making it difficult to make uniform plugs using the plug molds; however, a little longer incubation of the nuclei in the 45°C water bath would not damage them significantly since the HB contains 0.5 M sucrose that stabilizes nuclei.

11. Mix the nuclei with an equal volume of 1% LMP agarose in 1x HB using a cutoff pepitte tip. The agarose should be melted in boiling water and kept at 45°C before use.

Note: At this step, the hardness of the LMP plugs could be adjusted. If harder plugs that are easier to manipulate in their use are preferred, more (55%–60%) LMP agarose could be added. The use of additional LMP agarose has been proven to not influence DNA digestion *in situ* significantly.

12. Aliquot the mixture into ice-cold plug molds (cat.: 170-3713, Bio-Rad, Hercules, California USA) on ice with the same pipette tip, 100 μL per plug.

Isolation and postpurification of DNA in LMP agarose plugs

13. When the agarose is completely solidified, transfer the plugs into 5–10 volumes of lysis buffer and incubate the agarose plugs in the lysis buffer for 24–48 h at 50°C with gentle shaking.
14. Wash the plugs once in 10–20 volumes of ice-cold TE and then three times in 10–20 volumes of ice-cold TE plus 0.1 mM PMSF (adding 1 μL 100 mM PMSF per 1 mL ice-cold TE) on ice, 1 h each wash.
15. Further wash the plugs three times in 10–20 volumes of ice-cold TE on ice, 1 h each wash.
16. Store the plugs in TE at 4°C before use. At this stage the plugs can be stored for several months without significant degradation. For a longer storage, the plugs should be stored in 50 mM EDTA; pH 9.0 at 4°C.

22.3 EXAMPLES OF MEGABASE-SIZED DNA ISOLATION FROM DIFFERENT PLANT SPECIES

Megabase-sized DNA that is suited for genomics research must meet at least the following criteria: Large in size (1000 kb or larger), high in yield with proper concentration (e.g., 5–10 μg DNA/100 μL LMP plug), readily digestible, and clonable. The nuclei procedure presented here has already been widely used for the preparation of megabase-sized DNA from a variety of plant species, including grasses, legumes, vegetables, and trees. These species include those reported by Zhang

FIGURE 22.1   Plant nuclei plugs prepared using the nuclei procedure presented in this chapter. The plugs are 100 μL in volume. The two plugs at left were prepared from cotton (dicot) nuclei whereas the two at right were from barley (monocot) nuclei.

et al. [30] and many other species from which megabase-sized DNA have been isolated in our laboratories in the past decade. Together, our laboratories have isolated megabase-sized DNA from at least 100 species of different taxa with the nuclei procedure. So far, megabase-sized DNA has been isolated successfully from all the species, all met the above criteria and used in different aspects of genomics research, including construction of large-insert bacterial artificial chromosome (BAC) and plant-transformation-competent binary BAC (BIBAC) libraries and long-range genome analysis [4,5,10,32].

To further demonstrate the process of the nuclei procedure and the quality of megabase-sized DNA of plants isolated, the 100 μL LMP agarose plugs of cotton and barley prepared with the nuclei procedure are illustrated in Figure 22.1; the megabase-sized DNA of cotton and common wheat, which are directly fractionated on a pulsed-field gel and digested with a restriction enzyme or no restriction enzyme, are displayed in Figure 22.2; and the sorghum BACs, which are constructed from

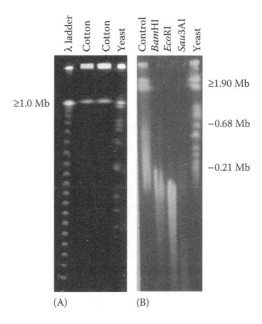

FIGURE 22.2   Megabase-sized DNA of cotton (dicot) (A) and common wheat (monocot) (B) isolated with the nuclei procedure presented in this chapter. The cotton DNA was directly fractionated on pulsed-field gel with no treatment. The common wheat DNA was digested with three different enzymes, respectively. For the control, the common wheat DNA contained in 1/3 of a 100 μL plug was incubated in the *Bam* HI reaction buffer as was for the digestions. For the digestions, the DNA contained in 1/3 of a 100 μL plug was digested with 10 units of each enzyme at 37°C for 2 h [32] and then subjected to pulsed-field gel electrophoresis. Note that the bands in the compressed zones contain unresolved DNA fragments that are equal to or larger than the size marked.

Insert size (kb)

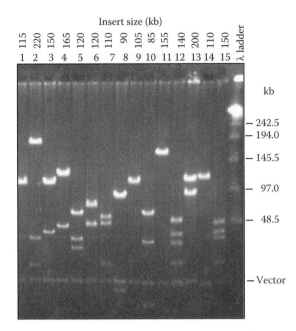

FIGURE 22.3 Sorghum BACs constructed from the megabase-sized DNA isolated with the nuclei proce-dure presented in this chapter and fractionated on a pulsed-field gel. BACs were constructed according to Wu et al. [32], isolated, digested with *Not* I to release the inserts from their cloning vector (pECBAC1) and fractionated on a pulsed-field gel. The BACs contain inserts ranging from 85 to 220 kb in size.

the megabase-sized DNA isolated with the nuclei procedure, are shown in Figure 22.3. These figures highlight that the nuclei procedure presented here is not only well suited for quality megabase-sized DNA isolation from plants but also applicable for the preparation of megabase-sized DNA from different plant species.

22.4 APPLICATIONS OF THE NUCLEI DNA ISOLATION PROCEDURE IN THE STUDIES OF NONPLANT SPECIES

The nuclei DNA isolation procedure has provided some concepts that are useful for preparation of megabase-sized DNA from other organisms, including animals, insects, fishes, scallops, and microbes. One key concept is that nuclei are not broken after DNA source tissues are frozen or while they are ground in liquid nitrogen. Triton X-100 is a nonionic surfactant and is used to solubilize the cytoplast membrane to release the nuclei while keeping them in intactness. The nuclei isolation buffer contains reagents that are widely used in DNA isolation from different organisms. Therefore, the method of isolating nuclei and the original nuclei procedure could be directly, or after modified in the nuclei isolation buffer to meet special needs of different species, used for preparation of megabase-sized DNA from different kingdoms of organisms. Based on this hypothesis and using the nuclei procedure presented here with and without modifications in the nuclei isolation buffer, we have isolated quality megabase-sized DNA from the frozen or fresh tissues of animals (e.g., bovine, swine, and chicken, unpublished), insects (e.g., mosquito [39], Drosophila, and coding moth, unpublished), shrimp (e.g., Pacific white shrimp and Hawaii shrimp, unpublished), scallop (e.g., Zhikong scallop) [33], fishes (e.g., catfish, unpublished), and microbes (e.g., *Penicillium chrysogenum* [17], *Phytophthora sojae* [34], and *Venturia inaequalis* [35]).

Notably, Hong et al. [39] constructed a BAC genomic DNA library of *Anopheles gambiae*, a mosquito transmitting human malaria pathogen in sub-Saharan Africa, using megabase DNA isolated, with estimated inserts of 133 kb. The subsequent characterization of this BAC library provides a valuable resource to the mosquito research community. In addition, Zhang et al. [33] described construction of two BAC libraries from nuclear DNA of Zhikong scallop using the widely applied vector pECBAC1. With resulting clones harboring scallop nuclear DNA inserts of 110–145 kb, the two scallop BAC libraries offer useful tools for gene cloning, genome physical mapping, and large-scale sequencing in the species. Also, Xu et al. [17] demonstrated the utility of large-insert BAC clones of physical map of *Penicillium chrysogenum*. This offers not only a platform for genomic studies of the penicillin-producing species but also strategies for genome physical mapping of other microbes as well as plants and animals. Similarly, Zhang et al. [34] applied the megabase DNA approach for preparation of two BAC libraries for integrated physical genome mapping of *Phytophthora sojae*, a serious microbial pathogen that threatens numerous cultivated crops, trees, and natural vegetation worldwide. The authors assembled 257 contigs, which collectively spanned ~132 Mb in physical length, and produced an integrated map consisting of 79 superscaffolds. This map represents the first genome-wide physical map of a *Phytophthora* species and paves the way for further genomics and molecular biology research in *P. sojae* and other *Phytophthora* spp. Moreover, Broggini et al. [35] prepared a BAC library from an Ascomycete *Venturia inaequalis AvrVg* isolate, the causal pathogen of apple scab. Comprising 7680 clones with an average insert size of 80 kb, this BAC library reveals new details of the size of the *V. inaequalis* genome (~100 Mb) and other important genetic features of this plant pathogen.

Although modified buffers were used in the preparation of megabase-sized DNA from some of these species, the nuclei isolation buffer was later approved to be suited for isolation of quality megabase-sized DNA from the species. Therefore, the nuclei procedure presented here, even though it was originally developed for preparation of megabase-sized DNA from plants [30], is applicable to isolation of megabase-sized DNA from other organisms, including animals, insects, fishes, scallops, and microbes.

22.5 CONCLUDING REMARKS

DNA fragments that are millions of base pairs in size (i.e., megabase) are essential for many aspects of modern genomics research. Here, we present a procedure of megabase-sized DNA isolation, named the nuclei procedure, with our latest improvements and research experience in the preparation of megabase-sized DNA from over 100 species. The nuclei procedure was originally developed for megabase-sized DNA isolation from plants and later has been employed for isolation of megabase-sized DNA from other organisms, including animals, insects, fishes, shrimps, scallops, and microbes. In this procedure, frozen or fresh tissues are homogenized in liquid nitrogen, nuclei are isolated and embedded in low-melting-point agarose plugs, and DNA is purified in the agarose plugs. Given that the nuclei procedure is not only capable of generating megabase-sized DNA of high quantity and high quality but also is simple, economical, and widely applicable to the isolation of megabase-sized DNA from a variety of organisms, it has emerged the method of choice for megabase-sized DNA isolation from different species. Furthermore, by including our laboratory notes on megabase-sized DNA isolation with the procedure obtained in our research of the past decade and provide technical discussion, the nuclei procedure offers a valuable approach for megabase-sized DNA isolation from different organisms, and it is readily adoptable by both experienced and new scientists.

Although the nuclei isolation procedure presented here represents an excellent method for the preparation of megabase-sized DNA that is well-suited for different aspects of genomics research, it will pay additional dividends by taking care of several issues below:

Polyphenolics and polysaccharides. Abundance of polysaccharides in the source tissues often leads to the stickiness of the tissue homogenate resulting from Step 2 of the nuclei procedure whereas that of polyphenolics leads to a brown color of the LMP agarose plugs. Both somehow interact with the DNA, thus significantly affecting digestion and cloning of the megabase-sized DNA. Modifications of the nuclei isolation buffer presented here with polyvinyl pyrrolidone (PVP 40) [18] and with ascorbic acid and diethyldithiocarbamic acid (DIECA) [36] have been reported in preparation of megabase-sized DNA from some species that are abundant in polyphenolics (e.g., cotton [18] and rose [36]) and polysaccharides (e.g., rose [36]) to reduce the problems that are caused by polyphenolics. Nevertheless, our studies showed that such modifications were not necessary to obtain quality megabase-sized DNA from these species using the procedure presented here (e.g., rose [37] and cotton [38]). In addition to sampling proper tissues as described above, repeatedly washing the nuclei at Step 7 of the nuclei procedure is crucial to minimize the problem that is associated with the cytoplasm metabolites such as polyphenolics and polysaccharides, and the contamination of cytoplasm organellar DNA.

Starches.   Abundance of starches that are often present in the plants growing in the fields or greenhouse with good light leads to white plugs in color if they are not removed from the nuclei. Although a thorough and repeated wash of the nuclei before being embedded in LMP agarose plugs would reduce the starches, source tissue sampling and pretreatment in dark as described above have been proven to be more effective for the purposes of removing starches. Nevertheless, starches do not seem significantly affecting digestion and cloning of megabase-sized DNA. In other words, the plugs that are white in color, indicating a significant contamination with starches, can be used for megabase-sized DNA digestion and cloning without significant problems.

Nuclei isolation buffers.   The nuclei isolation buffer presented here has been widely used to isolate megabase-sized DNA from a diverged taxa of species, including plants, animals, insects, fishes, scallops, and microbes as described above; however, different buffer systems could be employed to meet special needs of different organisms to isolate megabase-sized DNA using the nuclei procedure. For instance, we could isolate quality megabase-sized DNA using different buffers with the nuclei procedure in several species (mosquito [39], *P. chrysogenum* [17], *Ph. sojae* [34], and *V. inaequalis* [35] while the same quality megabase-sized DNA was isolated with the original nuclei isolation buffer. In our experience, use of the DNA isolation buffer for conventional-sized DNA isolation is also likely to result in the successful isolation of megabase-sized DNA from these species using the nuclei procedure described here.

LMP agarose plugs versus LMP agarose microbeads.   Both plugs and microbeads have been previously used to embed the nuclei in LMP agarose for preparation of megabase-sized DNA using the nuclei and protoplast procedure [18,27,29,30,32]. Although the microbeads containing megabase-sized DNA have much larger surface areas than plugs of the same volume, thus facilitating diffusion of reaction reagents to access the DNA and in turn, DNA digestion and cloning, embedding nuclei in LMP agarose plugs is much easier to perform and the resultant plugs are much easier to manipulate. Employment of BSA in the digestion of megabase-sized DNA embedded in LMP agarose has greatly facilitated the digestion of megabase-sized DNA embedded in LMP plugs [4,5,32]. This is because BSA functions as a restriction enzyme stabilizer that allows a reaction with an enzyme to be incubated on ice for a much longer time (2 h or more) and the enzyme could fully access the DNA, without significant loss of its activity. Consequently, megabase-sized DNA embedded in LMP plugs is readily digestible when BSA is included in the reaction. Therefore, LMP agarose plugs have emerged as the method of choice in embedding nuclei in LMP agarose.

ACKNOWLEDGMENTS

This work was supported by Research Grant No. US-3870-06C from BARD; The United States—Israel Binational Agricultural Research and Development Fund (HBZ); the USDA/CSCREES (No. 2008-35205-18720, HBZ), and the Changchun Bureau of Science and Technology, China (No. 007135, MPZ).

Meiping Zhang received her MS in botany from Harbin Normal University in Harbin, China; she became a lecturer at Jilin Agricultural University, Changchun, China in 1990 and taught cell biology, molecular biology, and plant tissue culture. While there she completed her PhD in plant biotechnology, and was promoted to professor in 2002. From 2007 to 2008, she conducted research in plant genomics and molecular genetics at Texas A&M University at College Station, Texas, as a visiting professor. Recently, she focuses her research on different aspects of medicinal plants, including regeneration via embryogenesis and organogenesis, application of hairy root culture to enhance secondary metabolites, application of biotechnology to infer secondary metabolic pathways, and applied functional genomics.

Yaning Li received her MS and PhD in plant pathology from Hebei Agricultural University in Hebei, China, with her PhD dissertation research conducted at Texas A&M University, College Station, Texas. The focus of her PhD research was on rice genomics, including the construction of large-insert plant-transformation-competent binary bacterial artificial chromosome (BIBAC) libraries of *Oryza barthii* and the development of a BIBAC-based integrated physical, genetic, and sequence map of japonica rice for the functional analysis and genetic engineering of the rice genome sequence. Her postdoctoral study on the identification and mapping of wheat leaf rust resistance genes was undertaken at Hebei Agricultural University, where she is currently an associate professor. Her main research interests include molecular plant pathology and biological control of plant fungal diseases, especially on plant genome library construction, molecular mapping and cloning of resistant genes, development of genetic transformation systems for biocontrol strains, and analysis of regulatory sequences of biocontrol strains.

Hong-Bin Zhang received his MS in genetics from the Chinese Academy of Agricultural Sciences, Beijing, China; he completed his PhD in genetics at the University of California at Davis, California in 1990. After undertaking his postdoctoral training for about 4 years in genomics and molecular genetics at the University of California at Davis and Texas A&M University at College Station, Texas, he joined the faculty of the Texas A&M University. He is currently a professor of genomics and molecular genetics, the director of the Laboratory for Plant Genomics and Molecular Genetics and the director of the GENEfinder Genomic Resources Center at the Texas A&M University, and the editor-in-chief of the *International Journal of Plant Genomics*. His research has focused on several aspects of genomics, including structural, evolutionary, and applied functional genomics in several plant and animal species. His laboratory helped pioneer the theory and technology of megabase-sized recombinant DNA, including megabase-sized DNA preparation and manipulation, and large-insert bacterial artificial chromosome (BAC) and plant-transformation-competent binary BAC (BIBAC) cloning and library construction; it also helped in pioneering the theory and technology of whole genome integrative physical mapping with BACs or BIBACs. His laboratory first proposed, discovered, and tested the DNA "jigsaw puzzle" structure theory and its roles in abundance, diversity, and complexity of living organisms; it was also the first to discover and establish the genetics and evolution of multigene families in plants and was among the groups that first proposed the research approaches and strategies of applied functional genomics.

REFERENCES

1. Zhang, H.-B., Woo, S.-S., and Wing, R.A., BAC, YAC and cosmid library construction. In: *Plant Gene Isolation: Principles and Practice*. Foster, G. and Twell, D. (eds.). John Wiley & Sons, Ltd. England, pp. 75–99, 1996.
2. Zhang, H.-B. and Wu, C.C., BACs as tools for genome sequencing, *Plant Physiol. Biochem.*, 39: 195, 2001.
3. Wu, C., Wang, S., and Zhang, H.-B., Interactions among genomic structure, function and evolution revealed by comprehensive analysis of the *Arabidopsis* genome, *Genomics*, 88: 394, 2006.
4. Ren, C., et al., Genomic DNA libraries and physical mapping. In: *The Handbook of Plant Genome Mapping: Genetic and Physical Mapping*. Meksem K. and Kahl G. (eds.). Wiley-VCH Verlag GmbH, Weinheim, Germany, pp. 173–213, 2005.

5. He, L. et al., Large-insert bacterial clone libraries and their applications. In: *Aquaculture Genome Technologies*. Liu, Z. (ed.). Blackwell Publishing, Ames, Iowa, pp. 215–244, 2007.

6. Gregory, S.G., Howell, G.R., and Bentley, D.R., Genome mapping by fluorescent fingerprinting, *Genome Res.*, 7: 1162, 1997.

7. Marra, M.A., et al., High throughput fingerprint analysis of large-insert clones, *Genome Res.*, 7: 1072, 1997.

8. Zhang, H.-B. and Wing, R.A., Physical mapping of the rice genome with BACs, *Plant Mol. Biol.*, 35: 115, 1997.

9. Zwick, S. et al., Physical mapping of the *liguleless* linkage group in *Sorghum bicolor* using rice RFLP-selected sorghum BACs, *Genetics*, 148: 1983, 1998.

10. Wu, C. et al., Whole genome physical mapping: An overview on methods for DNA fingerprinting. In: *The Handbook of Plant Genome Mapping: Genetic and Physical Mapping*. Meksem K. and Kahl G. (eds.). Wiley-VCH Verlag GmbH, Weinheim, Germany, pp. 257–284, 2005.

11. Chen, M. et al., Microcolinearity in *sh2*-homologous regions of the maize, rice and sorghum genomes, *Proc. Natl. Acad. Sci. U.S.A.*, 94: 3431, 1997.

12. Fu, H. and Dooner, H.K., A Gene-enriched BAC library for cloning large allele-specific fragments from maize: Isolation of a 240-kb contig of the *bronze* region, *Genome Res.*, 10: 866, 2000.

13. Zhang, H.-B., Map-based cloning of genes and quantitative trait loci. In: *Principles and Practices of Plant Genomics*, Vol. 1: *Genome Mapping*. Kole, C. and Abbott, A.G. (eds.). Science Publishers, New Hampshire, pp. 229–267, 2007.

14. Venter, J.C., Smith, H.O., and Hood, L., A new strategy for genome sequencing, *Nature*, 381: 364, 1996.

15. Fierro, F. et al., Resolution of four large chromosomes in penicillin-producing filamentous fungi: The penicillin gene cluster is located on chromosome II (9.6 Mb) in *Penicillin notatum* and chromosome I (10.4 Mb) in *Penicillin chrysogenum*, *Mol. Gen. Genet.*, 241: 573, 1993.

16. Hume, M.E. et al., Genotypic variation among *Arcobacter* isolates from a furrow-to-finish swine facility, *J. Food Prot.*, 64: 645, 2001.

17. Xu, Z. et al., Genome-wide physical mapping from large-insert clones by fingerprint analysis with capillary electrophoresis: A robust physical map of *Penicillium chrysogenum*, *Nucleic Acids Res.*, 33: e50, 2005.

18. Zhao, X.P. et al., A simple method for cotton megabase DNA isolation, *Plant Mol. Biol. Rep.*, 12: 126, 1994.

19. Burke, D.T., Carle, G.F., and Olson, M., Cloning of large segments of exogenous DNA into yeast by means of artificial chromosome vectors, *Science*, 236: 806, 1987

20. Overhauser, J. and Radic, M.Z., Encapsulation of cells in agarose beads for use with pulsed-field gel electrophoresis, *Focus*, 9.3: 8, 1987.

21. Schartz, D.C. and Cantor, C.R., Separation of yeast chromosome-sized DNAs by pulsed-field gel electrophoresis, *Cell*, 37: 67, 1984.

22. Ganal, M.W. and Tanksley, S.D., Analysis of tomato DNA by pulsed field gel electrophoresis, *Plant Mol. Biol. Rep.*, 7: 17, 1989.

23. Van Daelen, R.A.J., Jonkers, J.J., and Zabel, P., Preparation of megabase-sized tomato DNA and separation of large restriction fragments by field inversion gel electrophoresis (FIGE), *Plant Mol. Biol.*, 12: 341, 1989.

24. Cheung, W.Y. and Gale, M.D., The isolation of high molecular weight DNA from wheat, barley and rye for analysis of pulsed-field gel electrophoresis, *Plant Mol. Biol.*, 14: 881, 1990.

25. Wing, R.A. et al., An improved method of plant megabase DNA isolation in agarose microbeads suitable for physical mapping and YAC cloning, *Plant J.*, 4: 893, 1993.

26. Ecker, J.R., PFGE and YAC analysis of the *Arabidopsis* genome, *Methods*, 1: 186, 1990.

27. Honeycutt, R. et al., Analysis of large DNA from soybean (*Glycine max* L. Meer.) by pulsed-field gel electrophoresis, *Plant J.*, 2: 133, 1992.

28. Wing, A.R., Zhang, H.-B., and Tanksley, S.D., Map-based cloning in crop plants: Tomato as a model system I. Genetic and physical mapping of *jointless*, *Mol. Gen. Genet.*, 242: 681, 1994.

29. Woo, S.-S. et al., Isolation of megabase-size DNA from sorghum and applications for physical mapping and bacterial and yeast artificial chromosome library construction, *Plant Mol. Biol. Rep.*, 3: 82, 1995.

30. Zhang, H.-B. et al., Preparation of megabase-size DNA from plant nuclei, *Plant J.*, 7: 175, 1995.

31. Arumuganathan, K. and Earle, E.D., Nuclear DNA content of some important plant species, *Plant Mol. Biol. Rep.*, 9: 208, 1991.

32. Wu, C., Xu, Z., and Zhang, H.-B., DNA libraries. In: *Encyclopedia of Molecular Cell Biology and Molecular Medicine.* Meyers, R.A. (ed.). Vol. 3 (2nd edn.). Wiley-VCH Verlag GmbH, Weinheim, Germany, pp. 385–425, 2004.

33. Zhang, Y. et al., Construction and characterization of two bacterial artificial chromosome libraries of Zhikong Scallop, *Chlamys farreri* Jones et Preston, and identification of BAC clones containing the genes involved in its innate immune system, *Marine Biotechnol.,* 10: 358, 2008.

34. Zhang, X. et al., An integrated BAC and genome sequence physical map of *Phytophthora sojae, Mol. Plant Microbe Interact.,* 19: 1302, 2006.

35. Broggini, G.A.L. et al., Construction of a contig of BAC clones spanning the region of the apple scab avirulence gene *AvrVg, Fungal Genet. Biol.* 44: 44, 2007.

36. Kaufmann, H. et al., Construction of a BAC library of *Rosa rugosa* Thunb. and assembly of a contig spanning Rdr1, a gene that confers resistance to blackspot, *Mol. Gen. Genomics,* 268: 666, 2003.

37. Hess, G., Byrne, D.H., and Zhang, H-B., Toward positional cloning of everblooming gene (*evb*) in plants: A BAC library of *Rosa chinensis* cv. 'Old Blush', *Acta Hort.* (ISHS), 751: 169, 2007.

38. Zhang, H.-B. et al., Recent advances in cotton genomics, *Int. J. Plant Genomics,* Article ID 742304, 20 pages, 2008.

39. Hong, Y.S. et al., Construction and characterization of a BAC library and generation of BAC end sequence-tagged connectors for genome sequencing of the malaria mosquito, *Anopheles gambiae, Mol. Genet. Genomics,* 268: 720, 2003.

Part VII

Purification of Nucleic Acids from Miscellaneous Sources

23 Purification of Nucleic Acids from Gels

Chunfang Zhang and Theo Papakonstantinou

CONTENTS

23.1 INTRODUCTION

Deoxyribonucleic acid (DNA) and ribonucleic acid (RNA) fragments of high purity are often required in many molecular applications, including but not limited to the following:

Disclaimer: The commercial products mentioned in this chapter are the authors' personal preferences and should not be considered as endorsements in the place of other equivalent products.

1. In the case of constructing gene libraries following partial digestion of genomic DNA, a mixture of fragments covering a particular size range may be needed.
2. To be used for ligation, labeling, transformation into cultured mammalian cells, or injection to fertilized eggs, one or more DNA fragments may be purified from a restriction enzyme digestion reaction generating multiple fragments.
3. Following restriction enzyme digestion, alkaline phosphatase and nuclease treatments and so forth, DNA free of enzymes and other reagents can be prepared.
4. To remove the DNA polymerase, nucleotides, residual primers, primer dimers, and nonspecific amplification products that remain in a PCR mixture, a specific PCR product may be isolated.
5. An RNA fragment may need to be purified after being generated by *in vitro* transcription.

Although many procedures have been developed for the purification of nucleic acid fragments, the most widely applied technique involves their separation by electrophoresis on an agarose or polyacrylamide gel followed by recovery using a variety of methods. The nucleic acid fragments isolated with these procedures have different levels of purity and can be used in many subsequent molecular biological manipulations. This chapter examines the fundamental aspects concerning the separation of nucleic acids using agarose and polyacrylamide gel electrophoresis (PAGE), and reviews techniques for subsequent recovery of DNA and RNA from agarose and polyacrylamide gels. It then goes on to present several protocols that have proven reliable and highly efficient in our own and other authors' experience, and finally discusses areas where future refinements will lead to more speedy and efficient recovery of nucleic acid fragments from gels.

23.1.1 SEPARATION OF NUCLEIC ACIDS BY GEL ELECTROPHORESIS

Gel electrophoresis has become an indispensable tool in the field of molecular biology. It is a simple, yet powerful method to separate and analyze nucleic acids and proteins. Gel electrophoresis of nucleic acids started in the 1960s, using polyacrylamide for RNA [1–3] and agar for both RNA [4] and DNA [5] studies. Polyacrylamide gels were subsequently employed to fractionate DNA samples during the early 1970s [6]. However, it was not until the use of agarose to separate nucleic acids [7–9] and the fluorescent dye ethidium bromide to stain them [8,9] that gel electrophoresis became so popular. Nowadays, for the separation of nucleic acids both agarose and polyacrylamide are utilized in molecular biology laboratories all around the world, with the former much more commonly applied than the latter. Both types of gels can be run under native or denaturing conditions [10]. The gel's porosity is directly related to the concentration of the matrix in the medium, so various levels of effective separation can be achieved by selecting different concentrations of the agarose/polyacrylamide (Table 23.1).

Depending on the sizes of DNA and RNA fragments to be separated, agarose (generally for fragments greater than 100 bp) or polyacrylamide (less than 100 bp, or the size difference of fragments is less than 100 bp) matrix can be used (Table 23.1). When separation of very large DNA fragments (up to a few megabase pairs) is desired, a modified agarose gel electrophoresis technique called pulsed field gel electrophoresis (PFGE), which was initially developed by Schwartz and Cantor [11], can be applied (Section 23.1.1.1). The sizes of DNA and RNA fragments are often estimated in accordance with molecular weight markers (e.g., 1 kb ladder or 1 kb plus ladder; Invitrogen, Carlsbad, California) that are run in a separate lane at the same time.

23.1.1.1 Agarose Gel Electrophoresis

Agarose is one of the two polysaccharide classes (the other being agaropectin) that form a heterogeneous mixture called "agar," which is obtained from red algae or seaweed. Although both polysaccharide classes share the same galactose-based backbone, agaropectin is heavily modified with

TABLE 23.1

Relationship between Agarose or Polyacrylamide Gel Concentration (%) and Effectiveness of Nucleic Acid Separation

Agarose Gel		Nondenaturing Polyacrylamide Gel		Denaturing Polyacrylamide Gel	
Concentration (%, w/v)	Effective Separation (bp)[a]	Concentration (%, w/v)	Effective Separation (bp)	Concentration (%, w/v)	Effective Separation (nt)[b]
0.5	2,000–50,000	3.5	100–1,000	4.0	100–500
0.7	800–12,000	5.0	80–500	6.0	40–300
1.0	400–8,000	8.0	60–400	8.0	30–200
1.5	200–3,000	12.0	50–200	10.0	20–100
2.0	100–2,000	15.0	25–150	15.0	10–50
3.0	50–1,000	20.0	5–100	20.0	5–30

[a] bp, base pair.
[b] nt, nucleotide.

acidic side-groups such as sulfate and pyruvate, while agarose is essentially made up of long neutrally charged chains of cross-linked galactopyranose residues with a lower degree of chemical complexity [12,13]. This makes agarose less likely to interact with biomolecules such as proteins and nucleic acids.

Agarose is insoluble in cold water but readily dissolves (melts) in boiling water. A solid gel is formed by hydrogen bonding upon cooling. There are two main types of agaroses: standard (unmodified) agarose and low melting/gelling temperature agarose. Standard agarose generally melts at approximately 85°C–95°C and gels at about 35°C–45°C. Low melting/gelling temperature agarose has been modified by hydroxyethylation. This modification reduces the number of intrastrand hydrogen bonds and allows the agarose to melt and to solidify at lower temperatures than standard agarose. It normally melts at 45°C–65°C and gels at 25°C–35°C.

Agarose is often prepared in one of the two buffers, namely, Tris (Tris[hydroxymethyl] aminomethane)–acetate with EDTA (TAE) or Tris–borate with EDTA (TBE) (Table 23.2), although other buffers (e.g., Tris–phosphate–EDTA [TPE] and alkaline electrophoresis buffer) are used occasionally. These Tris–acid solutions provide slightly basic conditions to keep nucleic acids deprotonated, and EDTA (ethylenediamine tetraacetic acid) is a chelator of divalent cations such as Mg^{2+},

TABLE 23.2

Components of TAE and TBE Buffers

TAE			TBE		
Component	Stock (50x)	Working Concentration (1x)	Component	Stock (5x)	Working Concentration (0.5x)
Tris–base	242.2 g (2.0 M)	4,844 g (40 mM)	Tris–base	53.9 g (0.445 M)	5.39 g (44.5 mM)
EDTA disodium salt	18.6 g (0.05 M)[a]	0.372 g (1 mM)	EDTA disodium salt	3.72 g (0.01 M)	0.372 g (1 mM)
Acetic acid, glacial	57.1 mL (1.0 M)	1.142 mL (20 mM)	Boric acid	27.5 g (0.445 M)	2.75 g (44.5 mM)
Deionized water	To 1 L	To 1 L	Deionized water	To 1 L	To 1 L

[a] EDTA can be prepared ahead. For a 500 mL stock solution of 0.5 M EDTA, weigh out 93.05 g EDTA disodium salt. Dissolve in 400 mL deionized water and adjust the pH to 8.0 with NaOH. Top up the solution to a final volume of 500 mL and use 100 mL per 1 L 50x TAE. The pH of TAE (at around 8.5) and TBE (at 8.3) is not adjusted.

which are essential cofactors for enzymatic reactions. By sequestrating divalent cations, EDTA protects nucleic acids against degradation by enzymes such as nucleases. On the other hand, to minimize the effect of EDTA on DNA-modifying enzymes such as restriction enzymes and DNA polymerases that rely on divalent cations in subsequent manipulations, the EDTA concentration in TAE or TBE is generally kept low (typically at 1 mM). TAE buffer has a better resolving power for large DNA/RNA molecules than TBE, and thus it is recommended for resolution of genomic DNA, large supercoiled DNA, and DNA/RNA fragments >1500 bp/nt in size. As TAE has a lower buffering capacity than TBE, linear, double-stranded DNA runs 10% faster in TAE than in TBE. Because of its low buffering capacity, periodic replacement of TAE is recommended in some cases during prolonged electrophoresis such as PFGE. TBE is recommended for resolution of nucleic acids <1500 bp/nt in agarose (at 0.5x concentration) and for analysis of small DNA/RNA fragments on both native and denaturing polyacrylamide gels (at 1.0x concentration). TAE and TBE can both be stored at room temperature for a lengthy period.

A typical mistake a novice researcher can make is to use distilled water or TAE/TBE stock solution instead of 1x TAE or 0.5x TBE for gel preparation, which leads to failure in the electrophoretic process and unintended loss of sample; thus, it is important to clearly label all your solutions.

The samples are loaded into the wells of the gel together with a gel loading buffer (Table 23.3), of which sucrose, glycerol, or Ficoll increases the density of the sample for easy loading and the bromophenol blue and xylene cyanol FF add color to the sample for tracking during loading and running. In agarose gels prepared with 1x TAE buffer, bromophenol blue migrates as a 600 bp fragment and xylene cyanol FF as a 4500 bp fragment. In gels prepared with 0.5x TBE buffer, bromophenol blue migrates as a 300 bp fragment and xylene cyanol FF as a 4000 bp fragment. On the other hand, in polyacrylamide gels prepared in 1x TBE buffer the bromophenol blue migrates as a 65 bp fragment in 5% gel and a 15 bp fragment in 15% gel; and xylene cyanol FF migrates as a 260 bp fragment in 5% gel and a 60 bp fragment in 15% gel. The gel loading buffers can be prepared in distilled water, 1x TAE, or 0.5x TBE, and stored at 4°C (buffers 1 and 2) or room temperature (buffer 3).

Another common mistake a novice researcher tends to commit is to plug the cathode cable into the anode hole on the power pack and vice versa, which results in the sample running in the wrong direction within the gel and prompt loss of sample. Attention must be maintained at all times.

During electrophoresis, the negatively charged nucleic acid molecules migrate toward the anode, in the opposite direction to positive ions that move toward the cathode. With the migration rate of nucleic acid molecules being dependent on their size, the smaller molecules move toward the anode faster than the larger molecules do. As a result fractionation is achieved, with the smaller nucleic acid molecules toward the bottom of the gel, and the larger nucleic acid molecules closer to the top. The molecules can then be visualized with a fluorescent dye under UV light and their sizes estimated with reference to a known molecular weight marker. Ethidium bromide fluoresces under UV light

TABLE 23.3

Compositions of Commonly Used Gel Loading Buffers for Nucleic Acid Analysis

Component	Gel Loading Buffer (6x Concentration)		
	1	2	3
Bromophenol blue (0.25%)	+	+	+
Xylene cyanol FF (0.25%)	+	+	+
Sucrose (40%)	+		
Glycerol (30%)		+	
Ficoll (Type 400) (15%)			+

when intercalated into DNA or RNA, and is the traditional dye of choice. Investigators should follow all relevant safety precautions as ethidium bromide is a powerful carcinogen and moderately toxic. SYBR Green I (Invitrogen) is an alternative double-stranded DNA staining reagent, which is more expensive, but is 25 times more sensitive and possibly safer than ethidium bromide. Another nucleic acid stain is Gelstar (Lonza, Basel, Switzerland), which is also more sensitive and probably safer than ethidium bromide, but comes with a higher price. Ethidium bromide, SYBR Green I, and Gelstar can be either incorporated into the gel and buffer before electrophoresis or used after electrophoresis. There are still other commercially available dyes (e.g., SYBR Gold [Invitrogen]) to stain nucleic acids, which are used only after the completion of gel electrophoresis.

A novice researcher should beware of the possibility of dropping the gel during the process of transferring it from the gel apparatus to the UV light box for visualization, leading to partial or complete loss of valuable data.

To separate large DNA fragments, a technique called pulsed field gel electrophoresis is often used. PFGE employs selected restriction enzymes to yield between 8 and 25 large DNA bands of 40–5000 kb in size, and alternating currents to cause DNA fragments to move back and forth, resulting in a higher level of fragment resolution. A net forward direction of the DNA is achieved by setting the voltage in the forward direction for a certain duration, followed by a shorter pulse duration in the reverse direction. For example, one can run the gel in the forward direction for 0.5 s followed by a reverse pulse for 0.25 s. For higher resolutions of even larger bands, these times can be increased, such as 3 s forward and 1 s reverse. While in general small fragments can wind their way through the gel matrix more easily than large DNA fragments, a threshold length exists where all large fragments will run at the same rate. But with a continuous changing of directions every few second or fraction of a second, the various lengths of DNA react to the change at differing rates. That is, larger pieces of DNA will be slower to begin moving in the opposite direction while smaller pieces will be quicker to change direction. Over the course of time with the consistent changing of directions, each molecule will separate more and more from the others. Thus, separation of very large DNA pieces using PFGE is possible. For this method, bacteria (or other organisms) are first placed in agarose plugs, where they are lysed, and the DNA is then digested with selected restriction enzymes. The plugs containing the digested DNA are transferred into an agarose gel and electrophoresed for 30–50 h with alternating currents, and specific DNA fragments can then be purified [14].

23.1.1.2 Polyacrylamide Gel Electrophoresis

Polyacrylamide is a cross-linked polymer of acrylamide (or acrylic amide), which itself is a white odorless crystalline solid, soluble in water and ethanol, and shows incompatibility with some acids, bases, oxidizing agents, iron, and iron salts. The concentration (typically between 3.5% and 20%) of polyacrylamide dictates the resolving power of the gel matrix (Table 23.1). Polyacrylamide gels are formed by polymerization of the synthetic acrylamide monomer with a cross-linking agent N,N'-methylenebisacrylamide. This free radical-mediated polymerization is initiated by ammonium persulfate (APS) and accelerated by N,N,N,N'-tetramethylethylene diamine (TEMED). Because oxygen inhibits the polymerization process, polyacrylamide gels must be prepared between glass plates (or cylinders). Polyacrylamide gels have a rather small range of separation, but very high resolving power. In the case of double-stranded DNA, they are useful for separating fragments less than 500 bp in size. Under appropriate conditions, fragments of DNA differing in length by a single base pair can be resolved. Therefore, PAGE provides a versatile and gentle method of high resolution for fractionation and physical–chemical characterization of molecules on the basis of size, conformation, and net charge. For the separation of nucleic acids, TBE buffer is preferably used in PAGE (Table 23.2).

Unlike agarose gels, polyacrylamide gels cannot be cast in the presence of ethidium bromide because this intercalating dye is known to inhibit the polymerization of the acrylamide monomers. However, staining of the gel can take place after the electrophoresis process is complete. Polyacrylamide is known to quench the fluorescence of ethidium bromide, so the sensitivity with which nucleic acids can be detected is diminished compared to agarose gels. However, due to their higher resolving power these gels are the preferred option under certain conditions.

23.1.2 RECOVERY OF NUCLEIC ACIDS FROM GELS

Along with the development and improvement of gel electrophoretic procedures, various techniques have been developed to recover nucleic acids from gels. Such procedures have been and are still being modified, with new protocols constantly being generated. Here, we examine a few examples of different strategies to purify nucleic acids from agarose and polyacrylamide gels.

23.1.2.1 Electroelution of DNA

Following electrophoretic separation of the DNA fragments, a slice of gel containing the fragment of interest is excised and placed in a dialysis bag containing an appropriate buffer. The DNA can be run into the buffer inside the dialysis bag by electrophoresis (electroelution) and subsequently recovered from the electroeluent. This can be done for polyacrylamide gels [15], agarose gels [16], or polyacrylamide–agarose composite gels [17].

23.1.2.2 DNA Adsorption

After agarose gel electrophoresis, a slit or a trough can be made in the gel in front of the band of interest. DNA-adsorbing material is then placed into the slit or trough. Continuation of the electrophoresis will transfer the DNA onto the adsorbing material from which the DNA can be further purified. Examples of this technique include running the DNA into a dialysis membrane [18], diethylaminoethyl (DEAE)-cellulose membrane [19], DEAE-cellulose slurry [20], or hydroxyapatite [21].

23.1.2.3 Diffusion of DNA

After electrophoresis, the gel slice can be ground and soaked in a buffer. The gel residue is then removed by centrifugation and the DNA in the liquid medium is further purified. This can be applied to both polyacrylamide [22–24] and agarose gels [25]. Alternatively, the agarose gel slice containing the DNA of interest can be frozen in phenol and thawed, sometimes repeatedly. The DNA is then recovered from the supernatant after direct centrifugation [26] or from the filtrate after centrifugation through glass wool [27,28]. This freeze and squeeze method has been developed as a commercial product with a spin column format (BioRad, Hercules, California, Cat. No. 732-6165).

23.1.2.4 Gel Liquification to Recover DNA

A slice of agarose gel containing the DNA band of interest can be dissolved chemically by chaotropic agents and the DNA purified from the resultant liquid. The most common chaotropic agents for dissolving DNA-containing agarose include potassium iodide [29] and sodium iodide [30] as well as sodium perchlorate [31,32]. Passage through a hydroxyapatite column, along with organic solvent extraction and ethanol precipitation, yields DNA molecules free of chaotropic agent contamination [32]. Alternatively, the nucleic acids can be bound to a DNA-adsorbing silica material. This matrix can be in liquid form or in a spin column format (Sections 23.2.2.3 and 23.2.2.4). Contaminants such as salts and soluble macromolecular components are removed with a wash step using an ethanolic buffer. Pure DNA is finally eluted under low ionic strength conditions using either water or an appropriate elution buffer.

DNA fragments can also be separated in low melting/gelling temperature agarose. The gel slice containing the fragment of interest is melted at a relatively low temperature without denaturing the DNA. The DNA can then be purified from the liquid agarose solution [33]. Alternatively, β-agarase can be used to digest the low melting/gelling temperature agarose and the DNA can be subsequently purified [34–36].

23.1.2.5   Purification of RNA from Gels

Generally, RNA purification from gels is not as commonly undertaken as DNA purification, probably due to the difficulty in avoiding RNA degradation by contaminating RNases. Nevertheless, some procedures have been published for the isolation of RNA from polyacrylamide and agarose gels, mainly by diffusion [37], electroelution [38], or by using low melting/gelling agarose gel slices [33], similar to some of the techniques used for DNA purification from gels.

23.2   METHODS

General reagents, supplies, and equipment that are needed for isolation of nucleic acids from gels are listed in Table 23.4. Chemicals and reagents are of analytical grade unless otherwise stated. Water is distilled and de-ionized using a reverse-osmosis filtration system (Millipore, Bedford, Massachusetts).

23.2.1   PREPARATION AND RUNNING OF GEL ELECTROPHORESIS

23.2.1.1   Agarose Gel Electrophoresis

In our laboratory, standard molecular biology grade agarose is commonly employed for the DNA electrophoresis procedure, using appropriate agarose concentrations according to the sizes of the fragments (Table 23.1). Before pouring the molten agarose into the gel-forming tray, we add ethidium bromide (0.5 mg/mL) to the solution at a final concentration of 0.5 μg/mL. The gel is then submerged in buffer within the electrophoresis tank, the samples loaded onto the gel, and the electrophoresis performed. Note that treating DNA with restriction enzymes that release two similar-sized fragments is undesirable, as this can result in contamination when recovering the desired DNA species. If a restriction map is available, it is recommended to use another enzyme to digest the unwanted fragment into smaller pieces to achieve purification of the desired DNA.

RNA is generally run on denaturing gels to avoid nucleic acid secondary structures. To run RNA on agarose gels, there are two main procedures that are in use today. The first involves electrophoresis of RNA denatured with either glyoxal (also known as diformyl or ethanediol) or formamide, through an agarose gel [10,39,40]. The second involves pretreatment of RNA with formaldehyde and dimethyl sulfoxide, followed by electrophoresis through gels containing formaldehyde [41,42].

23.2.1.2   Polyacrylamide Gel Electrophoresis

Polyacrylamide gels are made by mixing appropriate volumes of 5x TBE buffer (Table 23.2), a commercial acrylamide solution (AppliChem, Darmstadt, Germany, Cat. No. A0951), and water to give a final concentration of 1x TBE and the desirable acrylamide monomer level. Investigators should follow all relevant safety precautions as acrylamide is a powerful neurotoxin. TEMED and a 10% (w/v) solution of APS are used to catalyze the polymerization process. The gel apparatus is assembled, the samples loaded onto the gel, and the electrophoresis performed. Afterward, the polyacrylamide gel is stained in 1x TBE buffer containing 0.5 μg/mL ethidium bromide at room temperature for 30 min, followed by destaining in water and blotting excess moisture away with Kimwipes.

For the separation of RNA molecules using polyacrylamide gels, urea or formamide or both are commonly used as the denaturants [10].

23.2.2   PURIFICATION OF DNA FROM AGAROSE GELS

23.2.2.1   Electroelution

There are a number of protocols available for electroelution of nucleic acids from agarose and polyacrylamide gels. For electroelution of DNA from agarose gels into dialysis tubing, we use an adaptation of the protocol present in Sambrook and Russell [14]. In our experience, the yield from

TABLE 23.4

General Reagents, Supplies, and Equipment Required for Nucleic Acid Isolation from Gels

Reagents	Supplies and Equipment
Acetic acid (glacial)	Agarose gel mini-subelectrophoresis system, BioRad,
Acrylamide and *N,N*'-methylenebisacrylamide	Hercules, California
monomer solution	Acrylamide gel electrophoresis system, Hoefer Inc.,
β-Agarase	San Francisco, California
Agarose I (standard agarose)	Centrifuge (microfuge)
Agarose II (low melting/gelling temperature agarose)	Centrifuge tubes (1.7 mL)
Ammonium acetate	Dialysis clips, Spectra/Por, Rancho Dominguez, California
APS	Dialysis tubing, Spectra/Por, Rancho Dominguez, California
Boric acid	Drop dialysis membrane, Millipore, Bedford, Massachusetts
2-Butanol	Glass wool (siliconized), Merck, Darmstadt, Germany
Chloroform	Gloves, disposable
DEAE-cellulose membrane, Sartorius,	Kimwipes
Goettingen, Germany	Pasteur pipettes
Diethylpyrocarbonate (DEPC)	Pipettors (20, 200, and 1000 µL) and pipette tips
Dimethyl sulfoxide	(20, 200, and 1000 µL)
Ethanol	Saran wrap
Ethidium bromide	UV light box
EDTA	
Formaldehyde	
Formamide	
GENECLEAN kit, Q-Biogene, Irvine, California	
Glyoxal	
HCl	
Isoamyl alcohol	
Magnesium acetate	
NucleoSpin extract II kit, Macherey-Nagel,	
Düren, Germany	
Perfectprep gel cleanup kit, Eppendorf AG,	
Hamburg, Germany	
Phenol (TE-saturated)	
QIAEX II kit, Qiagen, Düsseldorf, Germany	
Sodium acetate	
NaCl	
SDS	
NaOH	
Tris	
TEMED	
Urea	

the electroelution procedure is roughly 50%–75% of the starting material, so it is important to have at least 500 ng of the fragment of interest for downstream recovery.

1. Perform the agarose gel electrophoresis procedure as described in Section 23.2.1.1.
2. If possible, set the UV light source to a long-wavelength mode (some models call it "preparative") and place the agarose gel upon Saran wrap spread out on top of the light-box. If the investigator is using a handheld UV light source, place Saran wrap on a clean bench surface.

3. With the UV light on, cut around the fluorescent band of interest on the agarose gel using a sterile scalpel blade. Minimize the amount of time the DNA is exposed to the UV radiation by having all materials ready at hand. Switch off the UV light and carefully remove the desired gel fragment from the main body of the agarose gel. Trim off any excess agarose on the gel slice by brief exposure to UV again and then place the gel slice into the appropriate tube. The gel fragment can be frozen at −20°C for 30 min to make it easier to handle.

4. Dialysis tubing (Spectra/Por, Cat. No. 132645) that has been previously prepared according to the manufacturer's instructions is removed from its storage buffer and a piece approximately 3 cm longer than the gel slice is cut off from the rest. The tubing needs to be rinsed inside and out with distilled water, then rinsed the same way with TAE buffer. Secure one end with a dialysis clip and add enough TAE buffer before inserting the gel slice to cover it completely. Seal the other end of the tubing with another dialysis clip, making sure that there are no air bubbles present.

5. Immerse the sealed tubing in an electrophoresis tank containing fresh TAE buffer so that the gel slice is parallel to the electrodes. The rear of the gel slice should be close to the dialysis tubing, leaving room for the TAE buffer at the front. Make sure that the tubing is completely submerged and to prevent floating remove TAE buffer until the clip edges are resting on the bottom of the tank. Electroelute at 80–100 V and the DNA will migrate out of the gel slice toward the anode into the buffer within the tubing. Monitor the movement of the DNA with a long-wavelength UV radiation source and the electroelution should be complete after 45–60 min depending on the size of the DNA fragment.

6. Electroeluted DNA may stick to the dialysis tubing after the procedure is complete. To overcome this, we suggest reversing the electrodes and reapplying voltage for 15 s before opening the dialysis clips and removing the buffer containing the DNA. If low yields are suspected, add another 0.1 mL TAE buffer and rinse out the dialysis tubing. If the agarose slice is much larger than 4 cm in length, split the solution containing the DNA into various 1.7 mL centrifuge tubes so that no tube contains more than 0.6 mL in volume.

7. Extract the DNA by mixing an equal volume of aqueous liquid with an organic solution containing TE-saturated phenol, chloroform, and isoamyl alcohol in the ratio of 50:49:1, and centrifuging at 13,000 g for 5 min at room temperature. Investigators should follow all relevant safety precautions as phenol is a powerful corrosive. Transfer the aqueous phase containing the DNA into a fresh tube, leaving behind any debris at the interphase. Dispose of the organic solution waste according to local safety regulations.

7a. (Optional) If the DNA is too dilute to be efficiently precipitated by ethanol, it can be concentrated by several rounds of butanol extraction. Butanol sequesters some water molecules so that the aqueous volume is effectively decreased. Add an equal volume of 2-butanol to the sample and mix well. Centrifuge briefly or stand the tubes on the bench until the phases separate. Remove and discard the top organic layer. Repeat until the desired volume is achieved.

8. To the aqueous solution containing DNA, add 0.1 volumes of 3 M sodium acetate pH 5.2 and 2.5 volumes of absolute ethanol that has been prechilled at −20°C. Incubate the mix at −70°C for 15–20 min to precipitate nucleic acids, or else at −20°C overnight. Centrifuge at 13,000 g for 30 min at 4°C to pellet the DNA. Remove the supernatant; be careful, as the pellet is often so small that it is invisible. Add 0.5 mL of prechilled 70% (v/v) ethanol to the tube and centrifuge again for 5 min at room temperature. Remove the ethanol and invert the tubes to drain off as much residual liquid as possible. Leave the lid open for a few minutes at room temperature to allow the residual ethanol to evaporate but do not dry the pellet completely, as it is difficult to dissolve completely dry DNA. Add an appropriate volume of water or 1x TE buffer (10 mM Tris–HCl and 1 mM EDTA pH 8.0) to dissolve the DNA pellet.

To electroelute DNA fragments of appropriate size from polyacrylamide gels into dialysis tubing, the procedure is carried out as for agarose gels, substituting 0.5x TBE buffer for 1x TAE buffer where required.

23.2.2.2 Electrophoresis onto DEAE-Cellulose Membrane

Winberg and Hammarskjöld [20] initially transferred DNA fragments from an agarose gel to a DEAE-cellulose membrane, in a similar manner to Southern blotting. The DNA bands were located on the membrane by UV illumination and subsequently purified. Dretzen et al. [19] developed the procedure to run DNA into a small piece of a DEAE-cellulose membrane and the procedure was later modified as described in Sambrook and Russell [14]. The technique is based on the ability of the membrane to adsorb DNA at low ionic strength and to release DNA at high ionic strength.

1. Separate the DNA fragments by agarose gel electrophoresis (Section 23.2.1.1).
2. Under UV illumination, preferably long wavelength, locate the band of interest and make an incision in the gel in front of the band and slightly wider than the band with a sterile scalpel blade.
3. Cut a piece of DEAE-cellulose membrane (Sartorius, Goettingen, Germany, Cat. No. 94IEXD42-001) as wide as the incision and slightly higher than the gel. Soak the membrane in 10 mM EDTA (pH 8.0) for 10 min at room temperature and activate the membrane by soaking it in 0.5 M NaOH for a further 5 min. Wash the membrane several times with water.
4. Hold incision walls apart with blunt-ended forceps and insert the membrane into the incision. Close the incision by removing the forceps.
5. Resume electrophoresis and stop as soon as the band of DNA has been transferred to the membrane, as observed with a long-wavelength UV light source.
6. Remove the membrane and rinse it in either 1x TE or water at room temperature to remove residual agarose.
7. Move the membrane into a 1.7 mL centrifuge tube and add elution buffer (6.7 mM Tris–HCl, 0.07 mM EDTA, and 1.7 M NaCl pH 8.0) just to cover the membrane. Crush or fold the membrane loosely to reduce the volume of buffer required. Incubate at 65°C for 30 min with the tube lid closed.
8. Transfer the fluid to a fresh centrifuge tube. Extract once with phenol/chloroform/isoamyl alcohol and precipitate with ethanol as described in Section 23.2.2.1.

Note:

1. If DNA is to be recovered from bands of multiple lanes, it is recommended to leave a blank lane between samples.
2. If DNA is to be recovered from large area of a lane, such as all fragments covering a range of fragment sizes from a restriction digest of genomic DNA, make an incision parallel to the lane and place a piece of DEAE-cellulose membrane into the incision. Reorient the gel by 90° so that the DNA can be run into the membrane. If multiple size ranges are needed, the DNA from the whole lane can be transferred to a single piece of membrane, which can subsequently be cut to pieces corresponding to different size ranges.
3. Low amounts of DNA can be collected more efficiently by placing a small piece of membrane at one side of the band and reorientating the electrophoresis direction by 90°.
4. A polyacrylamide gel slice containing the DNA fragment of interest can be placed in the slit with the membrane.

23.2.2.3 Use of Liquid Silica Matrices

In 1986, the company then known as BIO101 introduced the commercial GENECLEAN kit (Q-Biogene) as a rapid and efficient method for purifying DNA from agarose gels. The basis of this technology was the dissolution of the gel by a chaotropic agent (sodium iodide) and the binding of nucleic acids to a silica matrix at high salt concentrations. The DNA could subsequently be eluted in a small volume of low salt buffer for downstream applications. The original GENECLEAN kit can only work for TAE-based gels, whereas the later GENECLEAN II and III kits are also compatible with TBE-based gels using a modifier solution. TBE-based gels are not as readily solubilized as TAE-based gels, due to the formation of complexes between the borate molecules and the *cis*-diol groups of sugar polymers. However, complete solubilization can eventually be achieved.

Alternative commercially available liquid matrices include the QIAEX and QIAEX II kits (Qiagen). These silica matrices are designed for the extraction of 0.4–50 kb DNA fragments using either TAE- or TBE-based gels. They work slightly differently from the GENECLEAN products in that DNA adsorption to this matrix stringently requires a pH ≤ 7.5, and the addition of 3 M sodium acetate pH 5.2 may be necessary. Elution efficiency is also temperature-dependent (Table 23.5). Other liquid matrix-based systems are available, such as NuClean silica matrix (MoBiTec, Göttingen, Germany) and matrix gel extraction system (Marligen Biosciences, Ijamsville, Maryland).

We have used the GENECLEAN and QIAEX II commercial kits and both work well in our experience. A general outline of the procedures is described in Table 23.5, for exact protocols and trouble-shooting see the manufacturers' instructions.

23.2.2.4 Use of Column Matrices

A number of commercial kits with spin-column matrices are currently available from many companies, including but not limited to Qiagen, Invitrogen, Sigma-Aldrich, (St. Louis, Missouri), Promega (Madison, Wisconsin), Stratagene, (Ladolla, California), Marligen, and Q-Biogene. In our laboratory, we use the Perfectprep gel cleanup kit (Eppendorf) and the NucleoSpin extract II kit (Macherey-Nagel), as both work well in our hands. A general description of the methodology is outlined in Table 23.6, for exact procedures and troubleshooting see the manufacturers' instructions.

23.2.2.5 Use of Low Melting/Gelling Temperature Agarose Gels

Low melting/gelling temperature agarose melts and solidifies at significantly lower temperatures than standard agarose. This ensures that when the gel melts the DNA contained within it is not denatured. DNA can thus be readily extracted from the liquefied agarose in its native form. Furthermore, the enzyme β-agarase can digest low melting/gelling temperature agarose, allowing the purification of DNA molecules from the resultant solution.

After melting, the DNA in the liquefied agarose can be used directly in subsequent applications, such as restriction enzyme digestion, ligation, and random priming to make radioactive probes. The efficiency of these reactions, however, is generally lower than with DNA purified from the agarose. In this section, we discuss two procedures used to purify DNA from low melting/gelling temperature agarose gels. The gel can be either melted or digested with β-agarase before organic solvent extraction and ethanol precipitation of the DNA.

23.2.2.5.1 Melting the Gel

1. Electrophorese the DNA sample in low melting/gelling agarose gel made up in TAE buffer containing 0.5 μg/mL ethidium bromide, using the procedure outlined in Section 23.2.1.1). Alternatively, DNA fragments can be first separated using a normal agarose gel. Locate the

TABLE 23.5
DNA Purification from Agarose Gels Using Liquid Matrices from Two Different Manufacturers

GENECLEAN	QIAEX II
Excise DNA fragment	Excise DNA fragment
Place in a 1.7 mL tube if it weighs less than 0.4 g. If the gel slice weighs more, use a larger tube.	Place in a 1.7 mL tube if it weighs less than 0.4 g. If the gel slice weighs more, use a larger tube.
Add 2.5–3 volumes of sodium iodide solubilization buffer. Incubate at 55°C until the gel slice is completely dissolved (5–10 min).	Solubilization buffer depends on the size of the DNA: For DNA fragments <0.1 kb, add 6 volumes of buffer QX1; for DNA fragments 0.1–4 kb, 3 volumes of buffer QX1 are required; and for DNA fragments >4 kb, 3 volumes of buffer QX1 and 2 volumes of water are added.
Vortex to resuspend the entire silica matrix. Add 5 μL matrix to the solubilized mix for solutions containing 5 μg of DNA or less, and add an additional 1 μL for each 0.5 μg of DNA above 5 μg.	Add QIAEX II matrix, 10 μL for up to 2 μg of DNA and 30 μL for 2–10 μg of DNA, before incubating at 50°C for 10 min to solubilize the agarose and bind the nucleic acids.
Pellet the silica matrix/DNA complex using a bench top centrifuge at 13,000 g for 30 s. Discard the supernatant solution.	Pellet the silica matrix/DNA complex using a bench top centrifuge at 13,000 g for 30 s. Discard the supernatant solution.
Wash the matrix 3 times with 10–50 volumes of ice-cold wash solution, resuspending the pellet each time. For DNA molecules larger than 15 kb, do not resuspend as shearing will occur; soak for 5 min instead. Be sure to remove all traces of new wash solution with a final centrifugation for 5 s.	Three washing steps are carried out with one wash of buffer QX1 (0.5 mL) followed by two washes with buffer PE. For molecules larger than 10 kb, resuspend the pellet by inverting and flicking the tube instead of vortexing, as shearing will otherwise occur. After the last wash step, air-dry the pellet for 10–15 min.
Elution of DNA is achieved by addition of a small volume of water or 1x TE buffer, and incubate at 55°C for 3 min. Centrifuge for 30 s and remove the aqueous liquid containing the DNA to another tube. Residual silica can often be transferred across, as it is difficult to remove the last bit of eluent from the top of the pellet without disturbing it. Centrifuge again and remove the supernatant to a new tube if this occurs.	Elute the DNA with 20 μL of 10 mM Tris–HCl pH 8.5 or water (pH between 7.0 and 8.5). For DNA fragments up to 4 kb in size, incubate at room temperature for 5 min; for DNA fragments 4–10 kb in size, incubate at 50°C for 5 min; for DNA fragments >10 kb, incubate at 50°C for 10 min. After centrifugation for 30 s the supernatant containing the DNA can be transferred to another tube. Repeating the elution process and combining the eluents increases the yield of DNA.

band of interest under long wavelength UV light and cut out a piece of the agarose gel immediately in front of the band to make a trough. Fill the trough with low melting/gelling temperature agarose and, after setting, resume electrophoresis until the DNA fragment enters the low melting/gelling temperature agarose. This is more cost-effective than electrophoresis using a gel made entirely with low melting/gelling temperature agarose.

2. Under long-wavelength UV illumination, cut out the slice of agarose containing the DNA of interest with a sterile scalpel blade and transfer it into a 1.7 mL centrifuge tube.
3. Add 5 gel volumes of 1x TE buffer and incubate at 65°C for 5 min, with the lid closed, in order to melt the gel.
4. Cool the solution to room temperature. Extract once with phenol, once with phenol/chloroform/isoamyl alcohol, and precipitate with ethanol (Section 23.2.2.1).

23.2.2.5.2   Digestion of the Gel with β-Agarase

1. Carry out Steps 1 and 2 described in Section 23.2.2.5.1.
2. Equilibrate the agarose slice by washing twice with 2 volumes of 1x β-agarase I reaction buffer (10 mM Bis Tris–HCl pH 6.5 and 1 mM EDTA) on ice for 30 min. Remove the buffer and melt the gel slice by incubation at 65°C for 10 min. Cool to 42°C.

TABLE 23.6

DNA Purification from Agarose Gels Using Spin Columns from Two Different Manufacturers

Perfectprep Gel Cleanup	NucleoSpin Extract II
Excise DNA fragment	Excise DNA fragment
Add 3 volumes of binding buffer for every 1 volume of agarose gel, with the maximal amount of gel being 0.4 g. Incubate at 50°C for 10 min, vortexing every 2–3 min.	Add 0.2 mL of buffer NT per 0.1 g agarose gel, with the maximal amount of gel being 0.4 g. Incubate at 50°C for 5–10 min. Invert to mix every 2–3 min.
Add 1 gel volume of isopropanol and mix thoroughly	
Transfer up to 0.8 mL into a spin column within a 2 mL collection tube.	Transfer up to 0.59 mL into a extract II column within a 2 mL collecting tube.
Centrifuge at 6,000–10,000 g for 1 min and discard the flow through. If the sample volume is larger than 0.8 mL, reload and spin.	Centrifuge at 11,000 g for 1 min and discard the flow through. If the sample volume is larger than 0.59 mL, reload and spin.
Add 0.75 mL of diluted wash buffer to the spin column/collection tube assembly, and centrifuge for 1 min.	Add 0.6 mL of buffer NT3 to the extract II column/collecting tube assembly, and centrifuge for 1 min.
After discarding the filtrate, centrifuge again for 1 min to remove any residual liquid.	After discarding the filtrate, centrifuge again for 2 min to remove any residual liquid.
Place the spin column into a new 2 mL collection tube. Add 30 μL of elution buffer or water.	Place the extract II column into a new 2 mL collecting tube. Add 15–50 μL of elution buffer NE, and incubate at room temperature for 1 min.
Centrifuge for 1 min and discard the spin column.	Centrifuge for 1 min and discard the extract II column.

3. Add 1–2 units of β-agarase (NEB, Ipswich, Massachusetts, Cat. No. M0392) to 200 μL of liquefied agarose solution containing DNA and incubate at 42°C for 1 h. For larger volumes, adjust enzyme levels accordingly.

   An alternative to Step 3 is to add 0.1 volumes of 10x β-agarase I reaction buffer and melt together with the agarose. Cool to 42°C and add twice the amount of enzyme in Step 3. Incubate at 42°C for 1 h. This alternative method is recommended when working with DNA fragments shorter than 500 bp because it avoids loss of DNA during the washing step.

4. After the agarose has been melted, the liquid can be used directly in subsequent reactions such as ligation or restriction enzyme digestion. But if desired, the DNA can be further purified as follows.

5. For DNA fragments smaller than 20 kb, extract twice with phenol. Add 2 volumes of 1x TE buffer and precipitate with ethanol (Section 23.2.2.1).

6. DNA fragments larger than 20 kb can be purified by dialysis with appropriate dialysis tubing or by drop dialysis on top of a membrane (Millipore, Cat. No. VSWP02500). This avoids the mechanical shearing of the DNA during organic solution extraction.

23.2.3 PURIFICATION OF DNA FROM POLYACRYLAMIDE GELS

23.2.3.1 Use of Liquid Silica Matrices

One advantage the QIAEX II kit has over the GENECLEAN technology is its ability to extract DNA from polyacrylamide gels. A general outline of the procedure is as follows:

1. Carry out the PAGE (Section 23.2.1.2). Excise the polyacrylamide gel slice containing the fluorescent DNA band of interest with a sterile scalpel blade.

2. Weigh the gel slice and add 1–2 volumes of solubilization buffer (0.5 M ammonium acetate, 10 mM magnesium acetate, 1 mM EDTA pH 8.0, and 0.1% [v/v] sodium dodecyl sulfate [SDS]) before incubating at 50°C for 30 min.

3. Once fully dissolved, centrifuge at 13,000 g for 1 min.
4. Carefully remove the supernatant and pass through a disposable plastic column or syringe barrel containing siliconized glass wool to remove residual polyacrylamide. Determine the volume of clarified supernatant.
5. For DNA fragments <0.1 kb in size, 6 volumes of buffer QX1 are added. For DNA fragments of 0.1–4 kb, only 3 volumes of buffer QX1 are required.
6. After resuspending the QIAEX II silica matrix by vortexing for 30 s, a total of 10 μL is added to the solution and incubated at room temperature for 10 min. It is necessary to vortex every 2 min or so to maintain the QIAEX II in suspension.
7. The sample is centrifuged again for 30 s and the liquid supernatant discarded.
8. The matrix pellet is washed twice with 0.5 mL of buffer PE.
9. The QIAEX II matrix/DNA is subsequently air-dried for 10–15 min.
10. The DNA is eluted from the resin by adding 20 μL of 10 mM Tris–HCl pH 8.5 or water (pH between 7.0 and 8.5) and resuspending the mix, followed by incubating at room temperature for 5 min.
11. After centrifugation for 30 s the supernatant containing the DNA can be transferred to another tube. Repeating the elution process and combining the eluates increase the yield of DNA.

23.2.3.2   Crush and Soak Method

The original method for recovering DNA from polyacrylamide gels was called the crush and soak method [22]. Faster methods for isolating double-stranded DNA include electroelution (Section 23.2.2.1) or electrophoresis onto a DEAE/cellulose membrane (Section 23.2.2.2). However, for single-stranded DNA molecules, crush and soak still remains the method of choice. The following procedure is a modification of that described in Sambrook and Russell [14]:

1. Carry out PAGE of the DNA sample as described in Section 23.2.1.2. Use a sterile scalpel blade to cut out a gel slice containing the fluorescent band of interest, observing all relevant safety precautions.
2. Transfer the gel slice to a preweighed 1.7 mL centrifuge tube, and use a disposable pipette tip to crush the polyacrylamide gel slice against the wall of the tube. Alternatively, it may be easier to slice the gel fragment into tiny pieces with the blade prior to transfer into the plastic tube.
3. Weigh the tube again, and calculate the approximate volume of the slice (assume 1 mg wet weight corresponds to 1 μL volume). Add 1–2 volumes of solubilization buffer (0.5 M ammonium acetate, 10 mM magnesium acetate, 1 mM EDTA pH 8.0, and 0.1 % [v/v] SDS) and incubate at 37°C on a rotating wheel/rotary platform. Small-sized fragments (<500 bp) are eluted within 3–4 h, but larger fragments take 12–16 h.
4. Centrifuge the sample at 13,000 g for 1 min at 4°C. Carefully transfer the supernatant to a fresh tube using a drawn-out Pasteur pipette.
5. Add 0.5 volumes of solubilization buffer to the polyacrylamide pellet, vortex, spin again, and combine the supernatants.
6. Pass the liquid through a disposable plastic column or syringe barrel containing siliconized glass wool, to remove residual polyacrylamide. Determine the volume of filtrate.
7. Extract with phenol/chloroform/isoamyl alcohol to remove the SDS (Section 23.2.2.1).
8. Add 2 volumes of chilled absolute ethanol and incubate at 4°C for 30 min. Collect precipitated nucleic acids by centrifugation at 13,000 g for 10 min at 4°C.
9. Discard the supernatant. Dissolve the DNA pellet in 0.2 mL of 1x TE buffer, then add 25 μL of 3 M sodium acetate pH 5.2 and 2 volumes of chilled absolute ethanol to precipitate the DNA once more. Centrifuge as described in Step 8.
10. Discard the supernatant. Wash the DNA once with 0.5 mL 70% (v/v) ethanol.
11. Air-dry for 10–15 min, and then dissolve the DNA in 10 μL of water or 1x TE buffer.

23.2.4 PURIFICATION OF RNA FROM GELS

In contrast to DNA, RNA comes in various forms: ribosomal RNA, transfer RNA, and poly(A)$^+$ messenger RNA. To obtain high-quality intact RNA molecules from agarose and polyacrylamide gels, solutions should be made using water treated with DEPC, which denatures RNases [14]. DEPC is highly toxic, so investigators should follow all relevant safety regulations.

We discuss below a number of commonly used procedures to purify RNA from agarose or polyacrylamide gels [33,37,43,44].

23.2.4.1 Electroelution of RNA

1. Fractionate RNA molecules in a polyacrylamide or agarose gel using the appropriate buffer and, under long-wavelength UV light, cut out the gel slice containing the band of interest as per DNA molecules (Section 23.2.2.1).
2. Place the gel slice in a piece of dialysis tubing and add an equal volume of fresh electrophoresis buffer. Seal both ends of the tubing with dialysis clips.
3. Place the dialysis tubing into the electrophoresis tank with the gel slice close to the anode, as discussed in Section 23.2.2.1.
4. Electrophorese at 90 V for 30 min. Monitor electroelution progress with a handheld long-wavelength UV light source.
5. Once the electroelution procedure is complete, reverse the polarity of the electrical field and electrophorese for 20 s to remove any RNA bound to the dialysis tubing.
6. Remove the electrophoresis buffer containing the RNA from the tubing to a 1.7 mL microcentrifuge tube (pretreated with DEPC-containing water). Extract twice by mixing with an equal volume of phenol and centrifuging at 10,000 g for 10 min.
7. Precipitate RNA by addition of 0.1 volumes of 3 M sodium acetate pH 4.7 and 2.5 volumes of prechilled 100% ethanol, and incubate for 20 min at −70°C or at least 2 h at −20°C. Centrifuge at 16,000 g for 30–45 min at 4°C. Wash the pellet with 0.5 mL of prechilled 70% (v/v) ethanol and centrifuge again for 10 min. Remove the ethanol and invert the tube to drain off as much residual liquid as possible. Leave the lid open for a few minutes at room temperature to allow the residual ethanol to evaporate. Add an appropriate volume of DEPC-treated water to dissolve the RNA pellet.

23.2.4.2 Diffusion of RNA

1. Fractionate RNA molecules using either polyacrylamide or agarose gel electrophoresis and an appropriate buffer. Under long-wavelength UV light, cut out the gel slice containing the band of interest. Transfer the gel slice to a preweighed and pretreated 1.7 mL centrifuge tube (Section 23.2.4.1).
1a. (Optional) Use a disposable pipette tip to crush the gel slice against the wall of the tube. Alternatively, it may be easier to slice the gel fragment into tiny pieces with a scalpel blade prior to transfer into the plastic tube.
2. Weigh the tube again, and calculate the approximate volume of the slice (assume 1 mg wet weight corresponds to 1 μL volume). Add 1–2 volumes of electrophoresis buffer and incubate at 30°C on a rotating wheel/rotary platform for a minimum of 5 h.
3. Centrifuge the sample at 10,000 g for 1 min at 4°C. Carefully transfer the supernatant to a fresh tube using a drawn-out Pasteur pipette.
4. Extract twice with an equal volume of phenol (Section 23.2.4.1).
5. Precipitate RNA by addition of 0.1 volumes of 3 M sodium acetate pH 4.7 and 2.5 volumes of ethanol (Section 23.2.4.1).

23.2.4.3   Isolation of RNA from Low Melting/Gelling Temperature Agarose Gels

1. Fractionate RNA molecules in a low melting/gelling temperature agarose gel using the appropriate buffer and, under long-wavelength UV light, cut out the gel slice containing the band of interest. Alternatively, instead of using a gel composed entirely of low melting/ gelling temperature agarose, the plug and trough method (Section 23.2.2.5.1) can be used in the interest of cost-effectiveness.
2. Add 5–10 volumes of electrophoresis buffer and melt the gel slice at 65°C for 5 min.
3. Extract twice with an equal volume of phenol (Section 23.2.4.1).
4. Precipitate RNA by addition of 0.1 volumes of 3M sodium acetate pH 4.7 and 2.5 volumes of ethanol (Section 23.2.4.1).

23.3   FUTURE DEVELOPMENT TRENDS

The various techniques described above to purify nucleic acids from agarose or polyacrylamide gels have evolved over several decades, and although some of them may be made redundant within a few years, others will no doubt remain irreplaceable in the foreseeable future.

In today's world, investigators are clamoring for safer, cheaper, and less time-consuming alternatives. One of the alternatives in the area of electrophoresis is to replace the toxic ethidium bromide reagent with other safer dyes such as GelStar nucleic acid stain (Lonza, Cat. No. 50535); this reagent can be used for sensitive fluorescent detection of both double-stranded and single-stranded DNA, oligonucleotides, and RNA in gels [45]. SYBR Gold, SYBR Safe, and SYBR Green I (Invitrogen) are also relatively new commercially available products that are generally considered to be less toxic than ethidium bromide [46]. Another recent development with the SYBR Safe technology is being able to visualize nucleic acids with a new prototype blue-light trans-illuminator, Safe Imager (Invitrogen). It is anticipated that other safer reagents and equipment will be developed in the future.

Some new nucleic acid purification procedures are also emerging. One such example is the E-Gel CloneWell SYBR Safe gel system (Invitrogen). In this system, the gel features two rows of wells. The samples are loaded into the top row and electrophoresed using a safe blue light to constantly monitor migration. When the band of interest moves into the second row of wells, the DNA can be removed by standard pipetting technique, bypassing any need to physically cut the gel and extract the nucleic acids with all the techniques mentioned above. This procedure is claimed by the manufacturer to be very efficient in terms of nucleic acid recovery; however, we have yet to personally assess it. It can be predicted that new and further refined purification protocols and commercial products will become available in the years ahead.

ACKNOWLEDGMENT

The authors wish to thank Professor Milton T.W. Hearn for his support in writing this chapter.

Chunfang Zhang obtained his BSc in microbiology from Beijing Agricultural University in Beijing, China. During his PhD study in bacterial genetics at Monash University, Clayton, Victoria, Australia, he developed novel transposon-based mutagenesis techniques, which led to the discovery and mapping of many new bacterial genes. In his earlier postdoctoral work, he optimized different techniques for the detection and quantification of mitochondrial DNA mutations in tissues of humans and animals of various ages and found the age-related accumulation of such mutations. He also used microarray and differential display techniques to study altered gene expression during human aging. He subsequently worked on yeast cells to enhance a biosynthetic pathway using molecular techniques in order to improve the yield of the end-product. Afterward, he moved to a newly established biotech company as the molecular biology team leader, where he designed alternative targeting vectors used in the generation of mouse gene knockouts and worked on the production of transgenic rats using adenovirus and lentivirus. He is currently in the Centre for Green Chemistry, Monash University, focusing on the development of novel peptide affinity tags for protein purification.

Theo Papakonstantinou received his BSc (first class honors) in biochemistry and immunology from Monash University, Clayton, Victoria, Australia and undertook his PhD training in biochemistry and molecular biology at the same institute. After conducting postdoctoral research on developing a vaccine vector based on attenuated *Salmonella typhimurium* and DNA vaccination, he worked on the production and purification of the key diabetes autoantigen glutamic acid decarboxylase using the yeast host *Pichia pastoris*, which later resulted in the first x-ray structure characterization of the autoantigen. His subsequent research at the Center for Molecular Biology and Medicine, involved manipulating a yeast biosynthetic pathway to improve the yield of a commercial end-product and later work at the Monash Medical Centre involved using mouse embryonic stem cells in the area of gene targeting. Since joining the Centre for Green Chemistry, Monash University, he has worked with a number of human activin genes for expression in *P. pastoris* and purified the expressed proteins using chromatographic techniques, with the goal to open up ways for making these proteins in a cost-effective manner for commercial use.

REFERENCES

1. Richards, E.G., Coll, J.A., and Gratzer, W.B., Disc electrophoresis of ribonucleic acid in polyacrylamide gels, *Anal. Biochem.*, 12: 452, 1965.
2. Gould, H., The specific cleavage of ribonucleic acid from reticulocyte ribosomal subunits, *Biochemistry*, 5: 1103, 1966.
3. Loening, U.E., The fractionation of high-molecular-weight ribonucleic acid by polyacrylamide-gel electrophoresis, *Biochem. J.*, 102: 251, 1967.
4. Tsanev, R., Fractionation of RNA in agar-gel electrophoresis studied by direct ultraviolet spectrophotometry, *Biokhimiia U. S. S. R.*, 30: 124, 1965.
5. Thorne, H.V., Electrophoretic separation of polyoma virus DNA from host cell DNA, *Virology*, 29: 234, 1966.
6. Danna, K. and Nathans, D., Specific cleavage of *Simian virus 40* DNA by restriction endonuclease of *Hemophilus influenzae, Proc. Natl. Acad. Sci. U.S.A.*, 68: 2913, 1971.
7. Takahashi, M., Ogino, T., and Baba, K., Estimation of relative molecular length of DNA by electrophoresis in agarose gel, *Biochim. Biophys. Acta*, 174: 183, 1969.
8. Aaij, C. and Borst, P., The gel electrophoresis of DNA, *Biochim. Biophys. Acta*, 269: 192, 1972.
9. Sharp, P.A., Sugden, B., and Sambrook, J., Detection of two restriction endonuclease activities in *Haemophilus* parainfluenzae using analytical agarose—Ethidium bromide electrophoresis, *Biochemistry*, 12: 3055, 1973.
10. McMaster, G.K. and Carmichael, G.G., Analysis of single- and double-stranded nucleic acids on polyacrylamide and agarose gels by using glyoxal and acridine orange, *Proc. Natl. Acad. Sci. U.S.A.*, 74: 4835, 1977.
11. Schwartz, D.C. and Cantor, C.R., Separation of yeast chromosome-sized DNAs by pulsed field gradient gel electrophoresis, *Cell*, 37: 67, 1984.
12. Hjertén, S., A new method for preparation of agarose for gel electrophoresis, *Biochim. Biophys. Acta*, 62: 445, 1962.
13. Fuse, T. and Suzuki, T., Preparation and properties of agar sulfates, *Agric. Biol. Chem.*, 39: 119, 1975.
14. Sambrook, J. and Russell, D.W., *Molecular Cloning: A Laboratory Manual*, 3rd ed., Cold Spring Harbor Laboratory Press, Cold Spring Harbor, New York, 2001.
15. Griffin, B.E., Friedt, M., and Cowie, A., Polyoma DNA: A physical map, *Proc. Natl. Acad. Sci. U.S.A.*, 71: 2077, 1974.
16. McDonell, M.W., Simon, M.N., and Studier, F.W., Analysis of restriction fragments of T7 DNA and determination of molecular weights by electrophoresis in neutral and alkaline gels, *J. Mol. Biol.*, 110: 119, 1977.
17. Pettersson, U. et al., Cleavage of adenovirus type 2 DNA into six unique fragments by endonuclease R·RI, *Proc. Natl. Acad. Sci. U.S.A.*, 70: 200, 1973.
18. Girvitz, S.C. et al., A rapid and efficient procedure for the purification of DNA from agarose gels, *Anal. Biochem.*, 106: 492, 1980.

19. Dretzen, G. et al., A reliable method for the recovery of DNA fragments from agarose and acrylamide gels, *Anal. Biochem.*, 112: 295, 1981.
20. Winberg, G. and Hammarskjöld, M.L., Isolation of DNA from agarose gels using DEAE-paper. Application to restriction site mapping of adenovirus type 16 DNA, *Nucleic Acids Res.*, 8: 253, 1980.
21. Tabak, H.F. and Flavell, R.A., A method for the recovery of DNA from agarose gels, *Nucleic Acids Res.*, 5: 2321, 1978.
22. Maxam, A.M. and Gilbert, W., A new method for sequencing DNA, *Proc. Natl. Acad. Sci. U.S.A.*, 74: 560, 1977.
23. Maxam, A.M. and Gilbert, W., Sequencing end-labeled DNA with base-specific chemical cleavages, *Methods Enzymol.*, 65: 499, 1980.
24. Subramanian, K.N. et al., The mapping and ordering of fragments of SV40 DNA produced by restriction endonucleases, *Nucleic Acids Res.*, 1: 727, 1974.
25. Chilton, M.D. et al., Stable incorporation of plasmid DNA into higher plant cells: The molecular basis of crown gall tumorigenesis, *Cell*, 11: 263, 1977.
26. Pramatarova, A., et al., Efficient recovery of cloned human cytomegalovirus DNA fragments from agarose gels, *J. Virol. Methods*, 46: 1, 1994.
27. Thuring, R.W. et al., A freeze-squeeze method for recovering long DNA from agarose gels, *Anal. Biochem.*, 66: 213, 1975.
28. Tautz, D. and Renz, M., An optimized freeze-squeeze method for the recovery of DNA fragments from agarose gels, *Anal. Biochem.*, 132: 14, 1983.
29. Blin, N., von Gabain, A., and Bujard, H., Isolation of large molecular weight DNA from agarose gels for further digestion by restriction enzymes, *FEBS Lett.*, 53: 84, 1975.
30. Vogelstein, B. and Gillespie, D., Preparative and analytical purification of DNA from agarose, *Proc. Natl. Acad. Sci. U.S.A.*, 76: 615, 1979.
31. Fuke, M. and Thomas, C.A., Isolation of open-circular DNA molecules by retention in agar gels, *J. Mol. Biol.*, 52: 395, 1970.
32. Lewis, J.B. et al., Mapping of late adenovirus genes by cell-free translation of RNA selected by hybridization to specific DNA fragments, *Proc. Natl. Acad. Sci. U.S.A.*, 72: 1344, 1975.
33. Wieslander, L., A simple method to recover intact high molecular weight RNA and DNA after electrophoretic separation in low gelling temperature agarose gels, *Anal. Biochem.*, 98: 305, 1979.
34. Bucan, M. et al., Genetic and cytogenetic localisation of the homeo box containing genes on mouse chromosome 6 and human chromosome 7, *EMBO J.*, 5: 2899, 1986.
35. Michiels, F., Burmeister, M., and Lehrach, H., Derivation of clones close to met by preparative field inversion gel electrophoresis, *Science*, 236: 1305, 1987.
36. Burmeister, M. and Lehrach, H., Isolation of large DNA fragments from agarose gels using agarase, *Trends Genet.*, 5: 41, 1989.
37. Jones, P., Qiu, J., and Rickwood, D., *RNA Isolation and Analysis*, BIOS Scientific Publishers Ltd., Oxford, United Kingdom, 1994.
38. Cattolico, R.A. and Jones, R.F., An improved technique for the preparative electrophoresis and electroelution of high molecular weight ribosomal RNA, *Anal. Biochem.*, 66: 35, 1975.
39. Thomas, P.S., Hybridization of denatured RNA and small DNA fragments transferred to nitrocellulose, *Proc. Natl. Acad. Sci. U.S.A.*, 77: 5201, 1980.
40. Thomas, P.S., Hybridization of denatured RNA transferred or dotted nitrocellulose paper, *Methods Enzymol.*, 100: 255, 1983.
41. Lehrach, H. et al., RNA molecular weight determinations by gel electrophoresis under denaturing conditions, a critical reexamination, *Biochemistry*, 16: 4743, 1977.
42. Rave, N., Crkvenjakov, R., and Boedtker, H., Identification of procollagen mRNAs transferred to diazobenzyloxymethyl paper from formaldehyde agarose gels, *Nucleic Acids Res.*, 6: 3559, 1979.
43. Gruegelsiepe, H. et al., Enzymatic RNA synthesis using bacteriophage T7 RNA polymerase, in *Handbook of RNA Biochemistry*, Hartmann, R.K. et al. (Eds.), Wiley-VCH Verlag GmbH & Co., KGaA, Weinheim, Germany, 2005, Chapter 1.
44. Frilander, M.J. and Turunen, J.J., RNA ligation using T4 DNA ligase, in *Handbook of RNA Biochemistry*, Hartmann, R.K. et al. (Eds.), Wiley-VCH Verlag GmbH & Co., KGaA, Weinheim, Germany, 2005, Chapter 3.
45. White, H.W. et al., GelStar nucleic acid gel stain: High sensitivity detection in gels, *Biotechniques*, 26: 984, 1999.
46. Huang, Q. and Fu, W.L., Comparative analysis of the DNA staining efficiencies of different fluorescent dyes in preparative agarose gel electrophoresis, *Clin. Chem. Lab. Med.*, 43: 841, 2005.

Index